Lecture Notes in Mathematics 2168

More information about this series at http://www.springer.com/series/304

Preface

After the exceptional 47th volume of the Séminaire de Probabilités dedicated to Marc Yor, we continue in this 48th volume with the usual formula: some of the contributions are related to talks given during the Journées de Probabilités held in Luminy (CIRM) in 2014 and in Toulouse in 2015, and the other ones come from spontaneous submissions. Apart from the traditional topics such as stochastic calculus, filtrations and random matrices, this volume continues to explore the subject of peacocks, recently introduced in previous volumes. Other particularly interesting papers involve harmonic measures, random fields and loop soups.

We hope that these contributions offer a good sample of the mainstreams of current research on probability and stochastic processes, in particular those active in France.

We would like to remind the reader that the website of the Séminaire is

http://portail.mathdoc.fr/SemProba/

and that all the articles of the Séminaire from Volume I in 1967 to Volume XXXVI in 2002 are freely accessible from the web site

http://www.numdam.org/numdam-bin/feuilleter?j=SPS.

We thank the Cellule MathDoc for hosting all these articles within the NUMDAM project.

Versailles, France — Catherine Donati-Martin
Vandoeuvre-lès-Nancy, France — Antoine Lejay
Versailles, France — Alain Rouault

Contents

Root to Kellerer

Mathias Beiglböck, Martin Huesmann, and Florian Stebegg

Abstract We revisit Kellerer's Theorem, that is, we show that for a family of real probability distributions $(\mu_t)_{t\in[0,1]}$ which increases in convex order there exists a Markov martingale $(S_t)_{t\in[0,1]}$ s.t. $S_t \sim \mu_t$.

To establish the result, we observe that the set of martingale measures with given marginals carries a natural compact Polish topology. Based on a particular property of the martingale coupling associated to Root's embedding this allows for a relatively concise proof of Kellerer's theorem.

We emphasize that many of our arguments are borrowed from Kellerer (Math Ann 198:99–122, 1972), Lowther (Limits of one dimensional diffusions. ArXiv e-prints, 2007), Hirsch-Roynette-Profeta-Yor (Peacocks and Associated Martingales, with Explicit Constructions. Bocconi & Springer Series, vol. 3, Springer, Milan; Bocconi University Press, Milan, 2011), and Hirsch et al. (Kellerer's Theorem Revisited, vol. 361, Prépublication Université dÉvry, Columbus, OH, 2012).

Mathematics Subject Classification (2010): Primary 60G42, 60G44; Secondary 91G20

M. Beiglböck (✉)
Institut für Stochastik und Wirtschaftsmathematik, Technische Universität Wien, Wiedner Hauptstraße 8, 1040 Wien, Austria
e-mail: mathias.beiglboeck@tuwien.ac.at

M. Huesmann
Institut für Angewandte Mathematik, Rheinische Friedrich-Wilhelms-Universität Bonn, Endenicher Allee 60, 53115 Bonn, Germany
e-mail: huesmann@iam.uni-bonn.de

F. Stebegg
Department of Statistics, Columbia University, 1255 Amsterdam Avenue, New York, 10025 NY, USA
e-mail: florian.stebegg@columbia.edu

C. Donati-Martin et al. (eds.), *Séminaire de Probabilités XLVIII*, Lecture Notes in Mathematics 2168, DOI 10.1007/978-3-319-44465-9_1

1 Introduction

1.1 Problem and Basic Concepts

We consider couplings between probabilities $(\mu_t)_{t\in T}$ on the real line, where t ranges over different choices of time sets T. Throughout we assume that all μ_t have a first moment. We represent these couplings as probabilities (usually denoted by π or $\mathbb{P}$) on the canonical space Ω corresponding to the set of times under consideration. More precisely Ω may be $\mathbb{R}^T$ or the space $\mathcal{D}$ of càdlàg functions if $T = [0,1]$. In each case we will write (S_t) for the canonical process and $\mathcal{F} = (\mathcal{F}_t)$ for the natural filtration. $\Pi((\mu_t))$ denotes the set of probabilities $\mathbb{P}$ for which $S_t \sim_{\mathbb{P}} \mu_t$. $\mathsf{M}((\mu_t))$ will denote the subset of probabilities ("martingale measures") for which S is a martingale wrt $\mathcal{F}$ resp. the right-continuous filtration $\mathcal{F}^+ = (\mathcal{F}_t^+)_{t\in[0,1]}$ in the case $\Omega = \mathcal{D}$. To have $\mathsf{M}((\mu_t)) \neq \emptyset$ it is *necessary* that (μ_t) increases in convex order, i.e. $\mu_s(\varphi) \leq \mu_t(\varphi)$ for all convex functions φ and $s \leq t$. This is an immediate consequence of Jensen's inequality. We denote the convex order by $\preceq$.

Our interest lies in the fact that this condition is also *sufficient*, and we shall from now on assume that $(\mu_t)_{t\in T}$ increases in convex order, i.e. that $(\mu_t)_{t\in T}$ is a *peacock* in the terminology of [5, 6]. The proof that $\mathsf{M}((\mu_t)_{t\in T}) \neq \emptyset$ gets increasingly difficult as we increase the cardinality of the set of times under consideration.

If $T = \{1, 2\}$, this follows from Strassen's Theorem [18] and we take this result for granted. The case $T = \{1, \dots, n\}$ immediately follows by composition of one-period martingale measures $\pi_k \in \mathsf{M}(\mu_k, \mu_{k+1})$.

If T is not finite, the fact that $\mathsf{M}((\mu_t)_{t\in T}) \neq \emptyset$ is less immediate and to establish that $\mathsf{M}((\mu_t)_{t\in T})$ contains a Markov martingale is harder still; these results were first proved by Kellerer in [11, 12] and now go under the name of Kellerer's theorem. We recover these classical results in a framework akin to that of martingale optimal transport.

1.2 Comparison with Kellerer's Approach

Kellerer [11, 12] works with peacocks indexed by a general totally ordered index set T and the corresponding natural filtration $\mathcal{F}$. He establishes compactness of martingale measures on $\mathbb{R}^T$ which correspond to the peacock $(\mu_t)_{t\in T}$. Then Strassen's theorem allows him to show the existence of a martingale with given marginals $(\mu_t)_{t\in T}$ for general T.

To show that $\mathsf{M}((\mu_t)_{t\in T})$ also contains a Markov martingale is more involved. On a technical level, an obstacle is that the property of being a Markovian martingale measure is not suitably closed. Kellerer circumvents this difficulty based on a stronger notion of Markov kernel, the concept of *Lipschitz* or *Lipschitz-Markov kernels* on which all known proofs of Kellerer's Theorem rely. The key step to showing that $\mathsf{M}((\mu_t)_{t\in T})$ contains a Markov martingale is to establish the existence of a two marginal Lipschitz kernel. Kellerer achieves this by showing that there are Lipschitz-Markov martingale kernels transporting a given distribution μ to the extremal points of the set $\mu \preceq \nu$ and subsequently obtaining an appealing Choquet-type representation for this set.

Our aim is to give a compact, self contained presentation of Kellerer's result in a framework that can be useful for questions arising in martingale optimal transport[1] for a continuum of marginals. While Kellerer is not interested in continuity properties of the paths of the corresponding martingales, it is favourable to work in the more traditional setup of martingales with càdlàg paths to make sense of typical path-functionals (based on e.g. running maximum, quadratic variation, etc.).

In Theorem 1 we make it a point to show that the space of *càdlàg* martingales corresponding to $(\mu_t)_{t\in[0,1]}$ carries a compact Polish topology. We then note that the Root solution of the Skorokhod problem yields an explicit Lipschitz-Markov kernel, establishing the existence of a Markovian martingale with prescribed marginals.

1.3 Further Literature

Lowther [14, 15] is particularly interested in martingales which have a property even stronger than being Lipschitz Markov: He shows that there exists a unique almost continuous diffusion martingale whose marginals fit the given peacock. Under additional conditions on the peacock he is able to show that this martingale has (a.s.) continuous paths.

Hirsch-Roynette-Profeta-Yor [5, 6] avoid constructing Lipschitz-Markov-kernels explicitly. Rather they establish the link to the works of Gyöngy [3] and Dupire [2] on mimicking process/local volatility models, showing that Lipschitz-Markov martingales exist for sufficiently regular peacocks. This is extended to general peacocks through approximation arguments. On a technical level, their arguments differ from Kellerer's approach in that ultrafilters rather than compactness arguments are used to pass to accumulation points. We also recommend [6] for a more detailed review of existing results.

2 The Compact Set of Martingales Associated to a Peacock

It is well known and in fact a simple consequence of Prohorov's Theorem that $\Pi(\mu_1, \mu_2)$ is compact wrt the weak topology induced by the bounded continuous functions (see e.g. [19, Sect. 4] for details). It is also straightforward that the continuous functions $f : \mathbb{R}^2 \to \mathbb{R}$ which are bounded in the sense that $|f(x, y)| \leq$

[1] An early article to study this continuum time version of the martingale optimal transport problem is the recent article [10] of Kallblad et al.

$\varphi(x) + \psi(y)$ for some $\varphi \in L^1(\mu_1), \psi \in L^1(\mu_2)$ induce the same topology on $\Pi(\mu_1, \mu_2)$.

A transport plan $\pi \in \Pi(\mu_1, \mu_2)$ is a martingale measure iff for all continuous, compact support functions h, $\int h(x)(y-x)\, d\pi = 0$. Hence, $\mathsf{M}(\mu_1, \mu_2)$ is a closed subset of $\Pi(\mu_1, \mu_2)$ and thus compact. Likewise, $\mathsf{M}(\mu_1, \dots, \mu_n)$ is compact.

2.1 The Countable Case

We fix a countable set $Q \ni 1$ which is dense in $[0, 1]$ and write M_Q for the set of all martingale measures on $\mathbb{R}^Q$. For $D \subseteq Q$ we set:

$$\mathsf{M}_Q((\mu_t)_{t\in D}) := \{\mathbb{P} \in \mathsf{M}_Q : S_t \sim_{\mathbb{P}} \mu_t \text{ for } t \in D\}.$$

We equip $\mathbb{R}^Q$ with the product topology and consider M_Q with the topology of weak convergence with respect to continuous bounded functions. Note that this topology is in fact induced by the functions $\omega \mapsto f(S_{t_1}(\omega), \dots, S_{t_n}(\omega))$, where $t_i \in Q$ and f is continuous and bounded.

Lemma 1 *For every finite $D \subseteq Q, D \ni 1$ the set $\mathsf{M}_Q((\mu_t)_{t\in D})$ is non-empty and compact. As a consequence, $\mathsf{M}((\mu_t)_{t\in Q}) = \mathsf{M}_Q((\mu_t)_{t\in Q})$ is non-empty and compact.*

Proof We first show that $\mathsf{M}_Q(\mu_1)$ is compact. To this end, we note that for every $\varepsilon > 0$ there exists n such that $\int (|x| - n)_+ \, d\mu_1 < \varepsilon$. We then also have

$$\mu(\mathbb{R} \setminus [-(n+1), (n+1)]) \le \int (|x| - n)_+ \, d\mu \le \int (|x| - n)_+ \, d\mu_1 < \varepsilon$$

for every $\mu \preceq \mu_1$.

For every $r : Q \to \mathbb{R}_+$ the set $K_r := \{g : Q \to \mathbb{R}, |g| \le r\}$ is compact by Tychonoff's theorem. Also, for given $\varepsilon > 0$ there exists r such that for all $\mathbb{P}$ on $\mathbb{R}^Q$ with $\mathrm{Law}_{\mathbb{P}}(S_t) \preceq \mu_1$ for all $t \in Q$ we have $\mathbb{P}(K_r) > 1 - \varepsilon$. Hence Prohoroff's Theorem implies that $\mathsf{M}_Q(\mu_1)$ is compact.

Next observe that for any finite set $D \subseteq Q, 1 \in D$ the set $\mathsf{M}_Q((\mu_t)_{t\in D})$ is non empty by Strassen's theorem. Clearly $\mathsf{M}_Q((\mu_t)_{t\in D})$ is also closed and hence compact. The family of all such sets $\mathsf{M}_Q((\mu_t)_{t\in D})$ has the finite intersection property, hence by compactness

$$\mathsf{M}_Q((\mu_t)_{t\in Q}) = \bigcap\nolimits_{D\subseteq Q, 1\in D, |D|<\infty} M_Q((\mu_t)_{t\in D}) \neq \emptyset.$$

2.2 The Right-Continuous Case

We will now extend this construction to right-continuous families of marginals on the whole interval $[0, 1]$.

We first note that it is not necessary to distinguish between the terms right-continuous and càdlàg in this context: fix a (not necessarily countable) set $Q \subseteq [0,1], Q \ni 1$, a peacock $(\mu_t)_{t\in Q}$ and a strictly convex function φ which grows at most linearly, e.g. $\varphi(x) = \sqrt{1+x^2}$. Then the following is straightforward: the mapping $\mu. : Q \to P(\mathbb{R}), q \mapsto \mu_q$ is càdlàg wrt the weak topology on $P(\mathbb{R})$ iff the increasing function $q \mapsto \int \varphi \, d\mu_q$ is right-continuous. In this case we say that $(\mu_t)_{t\in Q}$ is a right-continuous peacock.

As we have to deal with right limits we will recall the following:

Lemma 2 *Let $(X_n)_{n\in -\mathbb{N}\cup\{-\infty\}}$ be a martingale wrt $(\mathscr{G}_n)_{n\in -\mathbb{N}\cup\{-\infty\}}$ and write $\mu_n = \mathrm{Law}(X_n)$. If $\lim_{n\to-\infty} \mu_n = \mu_{-\infty}$, then $X_{-\infty} = \lim X_n$ a.s. and in L_1.*

Proof Set $Y := \lim_{n\to-\infty} X_n$ which exists (see for instance [16, Theorem II.2. 3]), has the same law as $X_{-\infty}$ and satisfies $\mathbb{E}[Y|X_{-\infty}] = X_{-\infty}$. This clearly implies that $X_{-\infty} = Y$.

As above, we fix a countable and dense set $Q \subseteq [0,1]$ with $1 \in Q$ and consider

$$\mathscr{D} = \{g : [0,1] \to \mathbb{R} : g \text{ is càdlàg }\},$$
$$\mathbb{D}_Q = \{f : Q \to \mathbb{R} : \exists g \in \mathscr{D} \text{ s.t. } g_{|Q} = f\}.$$

Note that $\mathbb{D}_Q$ is a Borel subset of $\mathbb{R}^Q$. Indeed a useful explicit description of $\mathbb{D}_Q$ can be given in terms of upcrossings. For $f : Q \to \mathbb{R}$ we write $UP(f,[a,b])$ for the number of upcrossings of f through the interval $[a,b]$. Then $f \in \mathbb{D}_Q$ iff f is càdlàg and bounded on Q and satisfies $UP(f,[a,b]) < \infty$ for arbitrary $a < b$ (clearly it is enough to take $a, b \in Q$). We also set

$$\bar{\mathscr{F}}_s := \bigcap\nolimits_{t\in Q, t>s} \mathscr{F}_t \tag{1}$$

for $s \in [0,1)$ and let $\bar{\mathscr{F}}_1 = \mathscr{F}_1$.

Proposition 1 *Assume that $(\mu_t)_{t\in Q}$ is a right-continuous peacock and let $\mathbb{P} \in \mathsf{M}((\mu_t)_{t\in Q})$. Then $\mathbb{P}(\mathbb{D}_Q) = 1$. For $q \in Q$, $\bar{S}_q := S_q = \lim_{t\downarrow q, t\in Q, t>q} S_t$ holds $\mathbb{P}$-a.s. For $s \in [0,1] \setminus Q$, $\lim_{t\downarrow s, t\in Q, t>s} S_t$ exists and we define it to be $\bar{S}_s$. The thus defined process $(\bar{S}_t)_{t\in[0,1]}$ is a càdlàg martingale wrt $(\bar{\mathscr{F}}_t)_{t\in[0,1]}$.*

Proof By Lemma 2, $S_q = \lim_{t\downarrow q, t>q, t\in Q} S_t$ for all $q \in Q$. Using standard martingale folklore (cf. [16, Theorem 2.8]), this implies that $(S_t)_{t\in Q}$ is a martingale under π wrt $(\bar{\mathscr{F}}_t)_{t\in Q}$ as well and the paths of $(S_t)_{t\in Q}$ are almost surely càdlàg. Moreover these are almost surely bounded by Doob's maximal inequality and have only finitely many upcrossings by Doob's upcrossing inequality. This proves $\mathbb{P}(\mathbb{D}_Q) = 1$. As the paths of $(S_t)_{t\in Q}$ are càdlàg the definition $\bar{S}_s := \lim_{t\downarrow s, t\in Q, t>s} S_t$ is well for $s \in [0,1] \setminus Q$ and $(\bar{S}_t)_{t\in[0,1]}$ is a càdlàg martingale under $\mathbb{P}$ wrt $(\bar{\mathscr{F}}_t)_{t\in[0,1]}$.

Identifying elements of $\mathscr{D}$ and $\mathbb{D}_Q$, the right-continuous filtration $\mathscr{F}^+$ on $\mathscr{D}$ equals the restriction of $\bar{\mathscr{F}}$ [cf. (1)] to $\mathbb{D}_Q$. Since any martingale measure $\mathbb{P}$

concentrated on $\mathbb{D}_Q$ corresponds to a martingale measure $\widetilde{\mathbb{P}}$ on $\mathscr{D}$ Proposition 1 yields:

Proposition 2 *Let $(\mu_t)_{t\in[0,1]}$ be a right-continuous peacock and $Q \ni 1, Q \subseteq [0,1]$ a countable dense set. Then the above correspondence*

$$\mathbb{P} \mapsto \widetilde{\mathbb{P}} \tag{2}$$

constitutes a bijection between $\mathsf{M}((\mu_t)_{t\in Q})$ and $\mathsf{M}((\mu_t)_{t\in[0,1]})$.

Through the identification $\mathbb{P} \mapsto \widetilde{\mathbb{P}}$, the set $\mathsf{M}((\mu_t)_{t\in[0,1]})$ carries a compact topology $\mathscr{T}_Q$. Superficially, this topology seems to depend on the particular choice of the set Q but this is not the case. To see this, consider another countable dense set $Q' \subseteq [0,1]$. The set $Q \cup Q'$ gives rise to a topology $\mathscr{T}_{Q\cup Q'}$ which is a priori finer than $\mathscr{T}_Q$ and $\mathscr{T}_{Q'}$ resp. Recall that whenever two compact Hausdorff topologies on a fixed space are comparable, they are equal. Since $\mathscr{T}_Q, \mathscr{T}_{Q'}, \mathscr{T}_{Q\cup Q'}$ are compact Hausdorff topologies, we conclude that $\mathscr{T}_Q = \mathscr{T}_{Q\cup Q'} = \mathscr{T}_{Q'}$. Hence we obtain:

Theorem 1 *Let $(\mu_t)_{t\in[0,1]}$ be a right-continuous peacock and consider the canonical process $(S_t)_{t\in[0,1]}$ on the Skorokhod space $\mathscr{D}$. The set $\mathsf{M}((\mu_t)_{t\in[0,1]})$ of martingale measures with marginals (μ_t) is non empty and compact wrt the topology induced by the functions*

$$\omega \mapsto f(S_{t_1}(\omega), \dots, S_{t_n}(\omega)),$$

where $t_1, \dots, t_n \in [0,1]$ and f is continuous and bounded.

2.3 General Peacocks

Kellerer [11] considers the more general case of a peacock $(\mu_t)_{t\in T}$ where $(T, <)$ is an abstract total order and $s < t$ implies $\mu_s \preceq \mu_t$, moreover no continuity assumptions on $t \mapsto \mu_t$ are imposed. Notably the existence of a martingale associated to such a general peacock already follows from the case treated in the previous section since every peacock can be embedded in a (right-) continuous peacock indexed by real numbers:

Lemma 3 *Let $(T, <)$ be a total order and $(\mu_t)_{t\in T}$ a peacock. Then there exist a peacock $(\nu_s)_{s\in\mathbb{R}^+}$ which is continuous (in the sense that $s \mapsto \nu_s$ is weakly continuous) and an increasing function $f : T \to \mathbb{R}_+$ such that*

$$\mu_t = \nu_{f(t)}.$$

If T has a maximal element we may assume that $f : T \to [0,1]$.

Proof Assume first that T contains a maximal element t^*. Consider again $\varphi(x) = \sqrt{1+x^2}$ and set $f(t) := \int \varphi\, d\mu_t$ for $t \in T$. On the image S of f we define (ν_s) through $\nu_{f(t)} := \mu_t$. Then $s \mapsto \nu_s$ is continuous on I and $s^* := f(t^*)$ is a maximal element of S.

Using tightness of $(\nu_s)_{s\in S}$ we obtain that $\nu_s := \lim_{r\in S, r\to s}$ exists for $s \in \overline{S}$. It remains to extend $(\nu_s)_{s\in\overline{S}}$ to $[0, s]$. The set $[0, s] \setminus S$ is the union of countably many intervals and on each of these we can define ν_s by linear interpolation. Finally it is of course possible to replace $[0, s]$ by $[0, 1]$ through rescaling.

If T does not have a maximal element, we first pick an increasing sequence $(t_n)_{n\geq 1}$ in T such that $\sup_n \int \varphi\, d\mu_{t_n} = \sup_{t\in T} \int \varphi\, d\mu_t$, then we apply the previous argument to the initial segments $\{s \in T : s \leq t_n\}$.

Above we have seen that $\mathsf{M}((\mu_t)_{t\in[0,1]}) \neq \emptyset$ for $(\mu_t)_{t\in[0,1]}$ right-continuous and pasting countably many martingales together this extends to the case of a right-continuous peacock $(\nu_s)_{s\in\mathbb{R}_+}$. By Lemma 3 this already implies $\mathsf{M}((\mu_t)_{t\in T}) \neq \emptyset$ for a peacock wrt to a general total order T.

3 Root to Markov

So far we have constructed martingales which are not necessarily Markov. To obtain the existence of a Markov-martingale with desired marginals, one might try to adapt the previous argument by restricting the sets $\mathsf{M}_Q((\mu_t)_{t\in D})$ to the set of Markov-martingales. As noted above, this strategy does not work in a completely straight forward way as being *Markovian* is not a closed property wrt weak convergence.

Example 1 The sequence $\mu_n = \frac{1}{2}(\delta_{(1,\frac{1}{n},1)} + \delta_{(-1,-\frac{1}{n},-1)})$ of Markov-measures weakly converge to the non-Markovian measure $\mu = \frac{1}{2}(\delta_{(1,0,1)} + \delta_{(-1,0,-1)})$.

3.1 *Lipschitz-Markov Kernels*

A solution τ to the two marginal Skorokhod problem $B_0 \sim \mu, B_\tau \sim \nu$ gives rise to the particular martingale transport plan (B_0, B_τ). Sometimes these martingale couplings induced by solutions to the Skorokhod embedding problem exhibit certain desirable properties. In particular we shall be interested in the Root solution to the Skorokhod problem.

Theorem 2 (Root [17]) *Let $\mu \preceq \nu$ be two probability measures on $\mathbb{R}$. There exists a closed set ("barrier") $\mathscr{R} \subseteq \mathbb{R}_+ \times \mathbb{R}$ (i.e. $(s,x) \in \mathscr{R}, s < t$ implies that $(t,x) \in \mathscr{R}$) such that for Brownian motion $(B_t)_{t\geq 0}$ started in $B_0 \sim \mu$ the hitting time τ_R of $\mathscr{R}$ embeds ν in the sense that $B_{\tau_R} \sim \nu$ and $(B_{t\wedge\tau_R})_t$ is uniformly integrable.*

Before we formally introduce the Lipschitz-Markov property we recall that the L^1- Wasserstein distance between two probabilities α, β on $\mathbb{R}$ is given by

$$W(\alpha,\beta) = \inf\left\{\int |x-y|\,d\gamma : \gamma \in \Pi(\alpha,\beta)\right\} = \sup\left\{\int f\,d\nu - \int f\,d\mu : f \in \mathrm{Lip}_1\right\},$$

where $\Pi(\alpha,\beta)$ denotes the set of all couplings between α and β and Lip_1 denotes the set of all 1-Lipschitz functions $\mathbb{R} \to \mathbb{R}$. The equality of the two terms is a consequence of the Monge-Kantorovich duality in optimal transport, see e.g. [19, Sect. 5].

A martingale coupling $\pi \in \mathsf{M}(\mu,\nu)$ is *Lipschitz-Markov* iff for some (and then any) disintegration $(\pi_x)_x$ of π wrt μ and some set $X \subseteq \mathbb{R}$, $\mu(X) = 1$ we have for $x, x' \in X$

$$W(\pi_x, \pi_{x'}) = |x - x'|. \tag{3}$$

We note that the inequality $W(\pi_x, \pi_{x'}) \geq |x - x'|$ is satisfied for arbitrary $\pi \in \mathsf{M}(\mu,\nu)$: for typical $x, x', x < x'$, the mean of π_x equals x and the mean of $\pi_{x'}$ equals x'. We thus find for arbitrary $\gamma \in \Pi(\pi_x, \pi_{x'})$

$$\begin{aligned} \int |y-y'|\,d\gamma(y,y') &\geq \left|\int y\,d\gamma(y,y') - \int y'\,d\gamma(y,y')\right| \\ &= \left|\int y\,d\pi_x(y) - \int y'\,d\pi_{x'}(y')\right| = |x-x'|, \end{aligned} \tag{4}$$

hence $W(\pi_x, \pi_{x'}) \geq |x - x'|$.

Note also that $W(\pi_x, \pi_{x'}) = |x - x'|$ holds iff the inequality in (4) is an equality for the minimizing coupling γ^*. This holds true iff there is a transport plan γ which is *isotone* in the sense that it transports π_x-almost all points y to some $y' \geq y$. This is of course equivalent to saying that π_x precedes $\pi_{x'}$ in first order stochastic dominance.

Lemma 4 *The Root coupling $\pi_R = Law(B_0, B_{\tau_R})$ is Lipschitz-Markov.*

Proof Write $(B_t)_t$ for the canonical process on $\Omega = C[0,\infty)$, $\mathbb{W}$ for Wiener measure started in μ and τ_R for the Root stopping time s.t. $(B_0, B_{\tau_R}) \sim_{\mathbb{W}} \pi_R \in \mathsf{M}(\mu,\nu)$.

It follows from the geometric properties of the barrier $\mathscr{R}$ that for all $x < x'$ and $\omega \in \Omega$ such that $\omega(0) = 0$

$$B_{\tau_R(x+\omega)}(x+\omega) \leq B_{\tau_R(x'+\omega)}(x'+\omega).$$

Write π_x for the distribution of B_{τ_R} given $B_0 = x$ and $\mathbb{W}_0$ for Wiener measure with start in 0. Then $(\pi_x)_x$ defines a disintegration (wrt the first coordinate) of π_R and for $x < x'$ an isotone coupling $\gamma \in \Pi(\pi_x, \pi_{x'})$ can be explicitly defined by

$$\gamma(A \times B) := \int 1_{A\times B}(B_{\tau_R(x+\omega)}(x+\omega), B_{\tau_R(x'+\omega)}(x'+\omega))\,\mathbb{W}_0(d\omega).$$

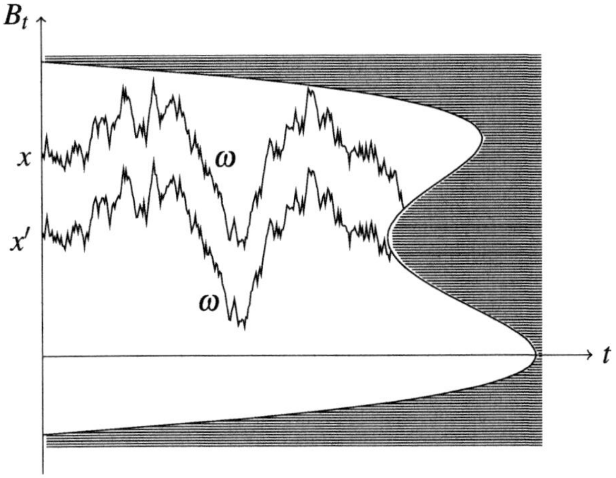

Remark 1 We thank David Hobson for pointing out that Lemma 4 remains true if we replace τ_R by Hobson's solution to the Skorokhod problem [7].[2]

We also note that this property is not common among martingale couplings. It is not present e.g. in the coupling corresponding to the Rost-embedding nor the various extremal martingale couplings recently introduced by Hobson–Neuberger [9], Hobson–Klimmek [8], Juillet (and one of the present authors) [1], and Henry-Labordere–Touzi [4].

3.2 Compactness of Lipschitz-Markov Martingales

To generalize the Lipschitz-Markov property to multiple time steps we first provide an equivalent formulation in the two step case. Using the Lipschitz-function characterization of the Wasserstein distance we find that (3) is tantamount to the following: for every $f \in \mathrm{Lip}_1(\mathbb{R})$ the mapping

$$x \mapsto \int f \, d\pi_x = \mathbb{E}[f(S_2)|S_1 = x] \tag{5}$$

is 1-Lipschitz (on a set of full μ-measure).

Let $Q \subseteq [0, 1]$ be a set which is at most countable. In accordance with (5) we call a measure/coupling $\mathbb{P}$ on $\mathbb{R}^Q$ *Lipschitz-Markov* if for any $s, t \in Q, s < t$ and $f \in \mathrm{Lip}_1(\mathbb{R})$ there exists $g \in \mathrm{Lip}_1(\mathbb{R})$ such that

$$\mathbb{E}_{\mathbb{P}}[f(S_t)|\mathcal{F}_s] = g(S_s). \tag{6}$$

[2]Hobson's solution [7] can be seen as an extension of the Azema-Yor embedding to the case of a general starting distribution.

The Lipschitz-Markov property is closed in the desired sense:

Lemma 5 *A martingale measure $\mathbb{P}$ on $\mathbb{R}^Q$ is Lipschitz-Markov iff*

$$\mathbb{E}_{\mathbb{P}}[Xf(S_t)]\,\mathbb{E}_{\mathbb{P}}[Y] - \mathbb{E}_{\mathbb{P}}[X]\,\mathbb{E}_{\mathbb{P}}[Yf(S_t)] \leq \int X(\omega)Y(\bar{\omega})|\omega_s - \bar{\omega}_s|\, d(\mathbb{P}\otimes\mathbb{P}) \tag{7}$$

for all $f \in Lip_1(\mathbb{R})$, $s < t \in Q$ and X, Y non-negative, bounded, and $\mathcal{F}_s$-measurable.

Proof If $\mathbb{P}$ is Lipschitz-Markov, then for a given 1-Lipschitz function f we can find by definition of a Lipschitz-Markov measure/coupling a 1-Lipschitz function g satisfying (6). Moreover, as $g \in \mathrm{Lip}_1$ we have for non-negative, bounded X, Y

$$(g(\omega_s) - g(\bar{\omega}_s))X(\omega)Y(\bar{\omega}) \leq |\omega_s - \bar{\omega}_s| X(\omega)Y(\bar{\omega}).$$

Integration with respect to $\mathbb{P}\otimes\mathbb{P}$ and an application of (6) yields (7).

For the reverse implication, by basic properties of conditional expectation there is a $\sigma((S_q)_{q\in Q\cap[0,s]})$-measurable function ψ such that $\mathbb{P}$-a.s.

$$\psi(\omega) = \mathbb{E}_{\mathbb{P}}[f(S_t)|\mathcal{F}_s](\omega).$$

Now from (7) we almost surely have $\psi(\omega) - \psi(\bar{\omega}) \leq |\omega_s - \bar{\omega}_s|$ which shows that ψ only depends on the s coordinate and is in fact 1-Lipschitz.

For $D \subseteq Q$ we set

$$\mathsf{L}_Q((\mu_t)_{t\in D}) := \{\mathbb{P} \in \mathsf{M}_Q : \mathbb{P} \text{ is Lipschitz-Markov}, S_t \sim_{\mathbb{P}} \mu_t \text{ for } t \in D\}.$$

Theorem 3 *Let $Q \subseteq [0,1]$, $Q \ni 1$ be countable. For every finite $1 \in D \subseteq Q$ the set $\mathsf{L}_Q((\mu_t)_{t\in D})$ is non-empty and compact. In particular, $\mathsf{L}((\mu_t)_{t\in Q}) := \mathsf{L}_Q((\mu_t)_{t\in Q})$ is non-empty and compact.*

Proof For finite $D \subseteq Q$ it is plain that $\mathsf{L}_Q((\mu_t)_{t\in D})$ is non-empty: this follows by composing of Lipschitz-Markov-kernels. Hence, $\mathsf{L}_Q((\mu_t)_{t\in Q}) = \bigcap_{D\subseteq Q,|D|<\infty} \mathsf{L}_Q((\mu_t)_{t\in D}) \neq \emptyset$ by compactness.

A martingale on $\mathcal{D}$ is *Lipschitz-Markov* if (6) holds for $s < t \in [0,1]$ wrt $\mathcal{F}^+$.

Theorem 4 *Assume that $(\mu_t)_{t\in[0,1]}$ is a right-continuous peacock and let $Q \ni 1$ be countable and dense in $[0,1]$. If $\mathbb{P} \in \mathsf{L}((\mu_t)_{t\in Q})$, then the corresponding [cf. (2)] martingale measure $\widetilde{\mathbb{P}} \in \mathsf{M}((\mu_t)_{t\in[0,1]})$ is Lipschitz-Markov.*

In particular, the set of all Lipschitz-Markov martingales with marginals $(\mu_t)_{t\in[0,1]}$ is compact and non-empty.

Proof The arguments in the proof of Lemma 5 work in exactly the same way to show that $\widetilde{\mathbb{P}}$ being Lipschitz-Markov is equivalent to conditions similar to (7) where X, Y are chosen to be measurable wrt $\mathcal{F}_s^+$ (or $\bar{\mathcal{F}}_s$, see the remark before Proposition 2).

For arbitrary $s, t \in [0, 1], s < t$ choose sequences $s_n \downarrow s, t_n \downarrow t$ in Q. Note that X, Y are in fact measurable wrt $\mathscr{F}_{s_n}$ and we thus have

$$\mathbb{E}_{\mathbb{P}}[Xf(S_{t_n})]\,\mathbb{E}_{\mathbb{P}}[Y] - \mathbb{E}_{\mathbb{P}}[X]\,\mathbb{E}_{\mathbb{P}}[Yf(S_{t_n})] \leq \int X(\omega)Y(\bar{\omega})|\omega_{s_n} - \bar{\omega}_{s_n}|\, d(\mathbb{P} \otimes \mathbb{P})(\omega, \bar{\omega})$$

by Lemma 5. Letting $n \to \infty$ concludes the proof.

3.3 Further Comments

It is plain that a Lipschitz-Markov kernel also has the Feller-property and in particular a Lipschitz-Markov martingales are strong Markov processes wrt $\mathscr{F}^+$ (see [13, Remark 1.70]). As in the previous section, the right-continuity of $(\mu_t)_{t\in[0,1]}$ is not necessary to establish the existence of a Lipschitz-Markov martingale, this follows from Lemma 3. We also remark that the arguments of Sect. 2 directly extend to the case of multidimensional peacocks, where the marginal distributions μ_t are probabilities on $\mathbb{R}^d$. However it remains open whether Theorem 4 extends to this multidimensional setup.

Acknowledgements Mathias Beiglböck and Florian Stebegg acknowledge support through FWF-projects P26736 and Y782-N25. Martin Huesmann acknowledges support through CRC 1060. We also thank the Hausdorff Research Institute for Mathematics (HIM) for its hospitality in spring 2015 and Nicolas Juillet and Christophe Profeta for many insightful comments.

References

1. M. Beiglböck, N. Juillet, On a problem of optimal transport under marginal martingale constraints. Ann. Probab. **44**(1), 42–106 (2016)
2. B. Dupire, Pricing with a smile. Risk **7**(1), 18–20 (1994)
3. I. Gyöngy, Mimicking the one-dimensional marginal distributions of processes having an Itô differential. Probab. Theory Relat. Fields **71**(4), 501–516 (1986)
4. P. Henry-Labordere, N. Touzi, An explicit Martingale version of Brenier's theorem. Finance Stochast. **20**(3), 635–668 (2016)
5. F. Hirsch, C. Profeta, B. Roynette, M. Yor, *Peacocks and Associated Martingales, with Explicit Constructions*. Bocconi & Springer Series, vol. 3 (Springer, Milan; Bocconi University Press, Milan, 2011)
6. F. Hirsch, B. Roynette, M. Yor, *Kellerer's Theorem Revisited*, vol. 361 (Prépublication Université dÉvry, Columbus, OH, 2012)
7. D. Hobson, The maximum maximum of a martingale, in *Séminaire de Probabilités, XXXII*. Lecture Notes in Mathematics, vol. 1686 (Springer, Berlin, 1998), pp. 250–263
8. D. Hobson, M. Klimmek, Model independent hedging strategies for variance swaps. Finance Stochast. **16**(4), 611–649 (2012)
9. D. Hobson, A. Neuberger, Robust bounds for forward start options. Math. Financ. **22**(1), 31–56 (2012)
10. S. Källblad, X. Tan, N. Touzi, Optimal Skorokhod embedding given full marginals and Azema-Yor peacocks. Ann. Appl. Probab. (2015, to appear)

11. H.G. Kellerer, Markov-Komposition und eine Anwendung auf Martingale. Math. Ann. **198**, 99–122 (1972)
12. H.G. Kellerer, Integraldarstellung von Dilationen, in *Transactions of the Sixth Prague Conference on Information Theory, Statistical Decision Functions, Random Processes (Tech. Univ., Prague, 1971; Dedicated to the Memory of Antonín Špaček)* (Academia, Prague, 1973), pp. 341–374
13. T. Liggett, *Continuous Time Markov Processes*. Graduate Studies in Mathematics, vol. 113 (American Mathematical Society, Providence, RI, 2010). An introduction.
14. G. Lowther, Limits of one dimensional diffusions. Ann. Probab. **37**(1), 78–106 (2009)
15. G. Lowther, Fitting martingales to given marginals (2008). arXiv:0808.2319
16. D. Revuz, M. Yor, *Continuous Martingales and Brownian Motion*. Grundlehren der Mathematischen Wissenschaften [Fundamental Principles of Mathematical Sciences], vol. 293, 3rd edn. (Springer, Berlin, 1999)
17. D.H. Root, The existence of certain stopping times on Brownian motion. Ann. Math. Stat. **40**, 715–718 (1969)
18. V. Strassen, The existence of probability measures with given marginals. Ann. Math. Stat. **36**, 423–439 (1965)
19. C. Villani, *Optimal Transport. Old and New*. Grundlehren der mathematischen Wissenschaften, vol. 338 (Springer, Berlin, 2009)

Peacocks Parametrised by a Partially Ordered Set

Nicolas Juillet

Abstract We indicate some counterexamples to the peacock problem for families of (a) real measures indexed by a partially ordered set or (b) vectorial measures indexed by a totally ordered set. This is a contribution to an open problem of the book (Peacocks and Associated Martingales, with Explicit Constructions, Bocconi & Springer Series, Springer, Milan, 2011) by Hirsch et al. and Yor (Problem 7a–7b: "Find other versions of Kellerer's Theorem").

Case (b) has been answered positively by Hirsch and Roynette (ESAIM Probab Stat 17:444–454, 2013) but the question whether a "Markovian" Kellerer Theorem hold remains open. We provide a negative answer for a stronger version: A "Lipschitz–Markovian" Kellerer Theorem will not exist.

In case (a) a partial conclusion is that no Kellerer Theorem in the sense of the original paper (Kellerer, Math Ann 198:99–122, 1972) can be obtained with the mere assumption on the convex order. Nevertheless we provide a sufficient condition for having a Markovian associate martingale. The resulting process is inspired by the quantile process obtained by using the inverse cumulative distribution function of measures $(\mu_t)_{t\in T}$ non-decreasing in the stochastic order.

We conclude the paper with open problems.

1 Introduction

The rich topic investigated by Strassen [16] in his fundamental paper of 1965 was to determine whether two probability measures μ and ν can be the marginals of a joint law satisfying some constraints. The most popular constraint on Law(X, Y) is probably $\mathbb{P}(X \leq Y) = 1$. In this case if $\leq$ is the usual order on $\mathbb{R}$, a necessary and sufficient condition on μ and ν to be the marginals of (X, Y) is $F_\mu \geq F_\nu$, where F_η denotes the cumulative distribution function of η. Actually if we note G_η the quantile function of η, the random variable (G_μ, G_ν) answers the question. Recall that the quantile function G_η is the generalised inverse of F_η, that

N. Juillet (✉)
Institut de Recherche Mathématique Avancée, UMR 7501, Université de Strasbourg et CNRS, 7 rue René Descartes, 67000 Strasbourg, France
e-mail: nicolas.juillet@math.unistra.fr

C. Donati-Martin et al. (eds.), *Séminaire de Probabilités XLVIII*, Lecture Notes in Mathematics 2168, DOI 10.1007/978-3-319-44465-9_2

is the unique nondecreasing functions on $]0,1[$ that is left-continuous and satisfies $(G_\eta)_\#\lambda|_{[0,1]} = \eta$. In the case of a general family $(\mu_t)_{t\in T}$, the family consisting of the quantile functions G_{μ_t} on $(]0,1[,\lambda|_{]0,1[})$ is also a process. It proves that measures are in stochastic order if and only if there exists a process $(X_t)_{t\in T}$ with $\mathbb{P}(t \mapsto X_t \text{ is non-decreasing}) = 1$ and $\text{Law}(X_t) = \mu_t$ for every $t \in T$. This result is part of the mathematical folklore on couplings. We name it *quantile process* or *Kamae–Krengel process* after the authors of Kamae and Krengel [12] because in this paper a generalisation for random variables valued in a partially ordered set E is proven. See also [15] where it appears.

Another type of constraint on $\text{Law}(X, Y)$ that is considered in Strassen article are the martingale and submartingale constraints, $\mathbb{E}(Y|\sigma(X)) = X$ and $\mathbb{E}(Y|\sigma(X)) \geq X$ respectively. Strassen proved that measures $(\mu_t)_{t\in\mathbb{N}}$ are the marginals of a martingale $(X_t)_{t\in\mathbb{N}}$ if and only if the measures μ_t are in the so-called convex order (see Definition 2). Kellerer extended this result to processes indexed by $\mathbb{R}$ and proved that the (sub)martingales can be assumed to be Markovian. Strangely enough, but for good reasons this famous result only concerns $\mathbb{R}$-valued processes indexed by $\mathbb{R}$ or another totally ordered set, which is essentially the same in this problem. Nevertheless, Strassen-type results have from the start been investigated with partially ordered set, both for the values of the processes or for the set of indices (see [5, 12, 13]). Hence the attempt of generalising Kellerer's theorem by replacing $\mathbb{R}$ by $\mathbb{R}^2$ for one of the two sets is a natural open problem that has been recorded as Problem 7 by Hirsch et al. in their book devoted to peacocks [9].

In Sects. 2 and 3 we define the different necessary concepts, state Kellerer Theorem and exam the possible generalised statement suggested in [9, Problem 7]. About Problem 7b we explain in Sect. 3.2 why Kellerer could not directly apply his techniques to the case of $\mathbb{R}^2$-valued martingales. Problem 7a is the topic of the last two parts. In Sect. 4 we exhibit counterexamples showing with several degrees of precision that one can not obtain a Kellerer theorem on the marginals of martingales indexed by $\mathbb{R}^2$, even if the martingales are not assumed to be Markovian. However, in Sect. 5 we provide a sufficient condition on $(\mu_t)_{t\in T}$ that is inspired by the quantile process. We conclude the paper with open problems.

2 Definitions

Let $(T, \leq)$ be a partially ordered set. In this note, the most important example may be $\mathbb{R}^2$ with the partial order: $(s,t) \leq (s',t')$ if and only if $s \leq s'$ and $t \leq t'$. We consider probability measures with finite first moment and we simply denote this set by $\mathscr{P}(\mathbb{R}^d)$.

We introduce the concepts that are necessary for our paper. Martingales indexed by a partially ordered set were introduced in the 1970. Two major contributions were [3, 17]. The theory was known under the name "two indices".

Definition 1 (Martingale Indexed by a Partially Ordered Set) Let $(X_t)_{t\in T}$ be the canonical process associated to some $\mathbb{P}$ on $(\mathbb{R}^d)^T$. For every $s \in T$ we introduce $\mathscr{F}_s = \sigma(X_r | r \le s)$.

A probability measure $\mathbb{P}$ on $(\mathbb{R}^d)^T$ is a *martingale* if and only if for every $(s,t) \in T^2$ satisfying $s \le t$ it holds $\mathbb{E}(X_t \mid \mathscr{F}_s) = X_s$. In other words it is a martingale if and only if for every $s \le t$, $n \in \mathbb{N}$ and $s_k \le s$ for $k \in \{0,1,\dots,n\}$ we have $\mathbb{E}_{\mathbb{P}}(X_t \mid X_s, X_{s_1}, \dots, X_{s_n}) = X_s$.

The convex order that we introduce now is also known under the names *second stochastic order* or *Choquet order*.

Definition 2 (Convex Order) The measures $\mu, \nu \in \mathscr{P}(\mathbb{R}^d)$ are said to be in *convex order* if for every convex function $\varphi : \mathbb{R}^d \to \mathbb{R}$, $\int \varphi \, d\mu \le \int \varphi \, d\nu$. This partial order is obviously transitive and we denote it by $\mu \preceq_C \nu$.

Note that in Definition 2, φ may not be integrable but the negative part is integrable because φ is convex.

The next concept of peacock is more recent. To our best knowledge it appeared the first time in [8] as the acronym PCOC, that is Processus Croissant pour l'Ordre Convexe. Both the writing peacock and the problem have been popularised in the book by Hirsch et al.: *Peacocks and Associated Martingales, with Explicit Constructions* [9].

Definition 3 (Peacock) The family $(\mu_t)_{t\in T}$ is said to be a *peacock* if for every $s \le t$ we have $\mu_s \preceq_C \mu_t$.

Because of the conditional Jensen inequality, if $(X_t)_{t\in T}$ is a martingale, the family $\mu_t = \mathrm{Law}(X_t)$ of marginals is a peacock. More generally if for some peacock $(\mu_t)_{t\in T}$ a martingale $(Y_t)_{t\in T}$ satisfies for every t, $\mathrm{Law}(Y_t) = \mu_t$, the martingale is said to be *associated* to the peacock $(\mu_t)_{t\in T}$.

Definition 4 (Kantorovich Distance) The Kantorovich distance between θ and $\theta' \in \mathscr{P}(\mathbb{R}^d)$ is

$$W(\theta, \theta') = \sup_f \left\| \int f \, \mathrm{d}\theta - \int f \, \mathrm{d}\theta' \right\|_{\mathbb{R}^d}$$

where f describes the set of 1-Lipschitz functions from $\mathbb{R}^d$ to $\mathbb{R}$.

Definition 5 (Lipschitz Kernel) A kernel $k : x \mapsto \theta_x$ transporting μ to $\nu = \mu k$ is called *Lipschitz* if there exist a set $A \subseteq \mathbb{R}^d$ satisfying $\mu(A) = 1$ such that $k|_A$ is Lipschitz of constant 1 from $(A, \|.\|_{\mathbb{R}^d})$ to $(\mathscr{P}(\mathbb{R}^d), W)$.

As $(\mathscr{P}(\mathbb{R}^d), W)$ is a complete geodesic metric space a simple extension procedure that we describe now permits us to extend k to a 1-Lipschitz function on $\mathbb{R}$. First the kernel k seen as a map is uniformly continuous so that one can extend it in a unique way on $\bar{A}$. The connected components of $\mathbb{R} \setminus \bar{A}$ are open intervals $]a, b[$ and the linear interpolation $t \mapsto (b-a)^{-1}((t-a)k(b) + (b-t)k(a))$ is also a geodesic curve for the Kantorovich distance. Therefore it gives a solution for extending k and making it a 1-Lipschitzian curve on $\mathbb{R}$.

To our best knowledge, the next concept is the key of all known proofs of Kellerer Theorem. Unlike Markov martingales, converging sequences of Lipschitz–Markov martingale have Markovian limits (in fact Lipschitz–Markov). In his original proof Kellerer uses a similar concept where the Kantorovich distance is replaced by the Kantorovich distance build on $d(x, y) = \min(1, |y - x|)$.

Definition 6 (Lipschitz–Markov Martingale) A process $(X_t)_{t\in T}$ is a *Lipschitz–Markov martingale* if it is a Markovian martingale and the Markovian transitions are Lipschitz kernels.

For surveys with examples of Lipschitz kernels and Lipschitz–Markov martingales, one can refer to [10] or [2].

3 The Kellerer Theorem and Trying to Generalise It

3.1 Problem 7a

Theorem 1 is a reformulation of Theorem 3 by Kellerer [14] in terms of the peacock terminology.

Theorem 1 ([14]) *Let $(\mu_t)_{t\in T}$ be a family of integrable probability measure on $\mathscr{P}(\mathbb{R})$ indexed by the totally ordered set T (for simplicity thing of $T = [0, +\infty[$). The following statements are equivalent*

1. *μ_t is a peacock,*
2. *μ_t is associated to a martingale process $(X_t)_{t\in T}$,*
3. *μ_t is associated to a Markovian martingale process $(X_t)_{t\in T}$,*
4. *μ_t is associated to a Lipschitz–Markovian martingale process $(X_t)_{t\in T}$.*

Note that the implications $4 \Rightarrow 3 \Rightarrow 2 \Rightarrow 1$ are obvious. Theorem 2 that we prove in Sect. 4 contradicts the converse implications if T is merely a partially ordered set. This is a negative answer to Problem 7a that we quote: *"Let $(X_{t,\lambda}; t, \lambda \geq 0)$ be a two-parameter peacock. Does there exist an associated two-parameter martingale $(M_{t,\lambda}; t, \lambda \geq 0)$?"*. Note that with our definition of peacock, one should read $\mathrm{Law}(X_{t,\lambda})$ in place of $X_{t,\lambda}$.

Theorem 2 *Let $(T, \leq)$ be $\{0, 1\}^2$, $\mathbb{R}^2_+$ or $\mathbb{R}^2$ with the partial order. For every choice of T, we have the following:*

- *There exists a peacock indexed by T that is not associated to a martingale,*
- *there exists a peacock indexed by T that is associated to a martingale process but not to a Markovian martingale process,*
- *there exists a peacock indexed by T that is associated to a Markovian martingale process but not to a Lipschitz-Markovian martingale process.*

3.2 Problem 7b

For completeness we explain what is known on Problem 7b: "Is a $\mathbb{R}^n$-valued peacock a $\mathbb{R}^n$-valued 1-martingale?", which with our notations means nothing but: Can any peacock on $\mathbb{R}^d$ be associated to an $\mathbb{R}^d$-valued martingale? Hirsch and Roynette provided a positive answer in [7].

Theorem 3 *Let $(\mu_t)_{t\in T}$ be a family of integrable probability measures on $\mathscr{P}(\mathbb{R}^d)$ indexed by the totally ordered set T. The following statements are equivalent*

1. *μ_t is a peacock,*
2. *μ_t is associated to a martingale process $(X_t)_{t\in E}$.*

Nevertheless it is to our knowledge still an open problem whether the full Kellerer theorem may hold in the vectorial case: Can every peacock be associated to a *Markovian* martingale? (equivalence of (1) and (3) in Theorem 1). We prove in Proposition 1 that (1) and (4) are not equivalent. Actually, the existence of a Lipschitz kernel for $\mu \preceq_C \nu$ is an essential step of each known proof of Kellerer Theorem, but for dimension $d > 1$ it does not exist for any pairs. This fact was very likely known by Kellerer (see the last paragraph of the introduction of Kellerer [14][1]). We provide a short proof of it.

Proposition 1 *There exists a peacock $(\mu_t)_{t\in T}$ indexed by $T = \{0, 1\}$ and with $\mu_t \in \mathscr{P}(\mathbb{R}^2)$ that is not associated to any Lipschitz-Markov martingale.*

As a trivial corollary, the same also holds for $T = [0, +\infty[$ defining $\mu_t = \mu_0$ on $[0, 1[$ and $\mu_t = \mu_1$ for $t \in [1, +\infty[$.

Proof Let $\mu_0 = \lambda|_{[0,1]} \times \delta_0 \in \mathscr{P}(\mathbb{R}^2)$ and k the dilation $(x, 0) \mapsto \frac{1}{2}(\delta_{(x,f(x))} + \delta_{(x,-f(x))})$. Let μ_1 be $\mu_0 k$. If $\mu_1 = \mu_0 k'$ for another dilation k', the projection of k' on the Ox-axis must be identity so that $k' = k$. We choose a non continuous function f as for instance $f = \chi_{[1/2,1]}$, and the proof is complete because k is not a Lipschitz kernel.

4 Proof of Theorem 2

In the three examples, we define a peacock on $T = \{0, 1, 1', 2\} \equiv \{0, 1\}^2$ where the indices 1, $1'$ stand for the intermediate elements, $0 \equiv (0, 0)$ is the minimal and $2 \equiv (1, 1)$ the maximal element. One will easily check that $(\mu_i)_{i\in T}$ is really a peacock from the fact that we indicate during the proof martingale transitions between μ_0 and μ_1, $\mu_{1'}$ as well as between μ_1, $\mu_{1'}$ and μ_2.

To complete the statement of Theorem 2 we need to explain what are the peacocks for $T = \mathbb{R}^2_+$ or $T = \mathbb{R}^2$. In fact for $(s, t) \in \{0, 1\}^2$, the measures $\mu_{s,t}$ are

[1]Kellerer: *"[...], während die Übertragung der im zweiten Teil enthaltenen Ergebnisse etwa auf den mehrdimensionalen Fall ein offenes Problem darstellt".*

defined exactly as in the three following constructions, and the peacock is extended in the following way

$$\mu_{s,t} = \begin{cases} \mu_{1,1} & \text{if } (s,t) \geq (1,1) \\ \mu_{1,0} & \text{if } s \geq 1 \text{ and } t < 1 \\ \mu_{0,1} & \text{if } t \geq 1 \text{ and } s < 1 \\ \mu_{0,0} & \text{otherwise: } \max(s,t) < 1. \end{cases}$$

With this bijection it is a direct check that the results for $\{0, 1, 1', 2\}$ will be transposed to the other sets of indices. Note that if a martingale $(X_{s,t})$ is defined for $(s,t) \in \{0,1\}^2$ it is extended in the same way as the peacock. For instance $X_{s,t} = X_{1,1}$ if $(s,t) \geq (1,1)$.

The three constructions are illustrated by figures where the amount of transported mass from x to y is the label of the arrow from x at time i to y at time j where $i \leq j$ (and (i,j) is not $(0,2)$). In order to write an integer we prefer to label with a multiple of the mass (factor 6 in Figs. 1 and 2, and 12 in Fig. 3).

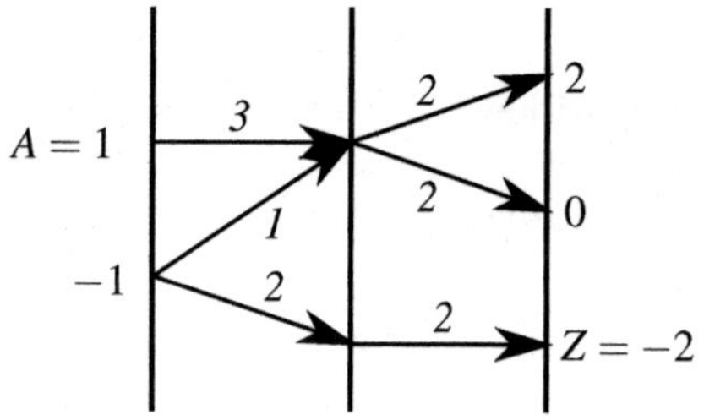

Fig. 1 The martingale associated to $(\mu_t)_{t\in\{0,1,2\}}$ in Sect. 4.1

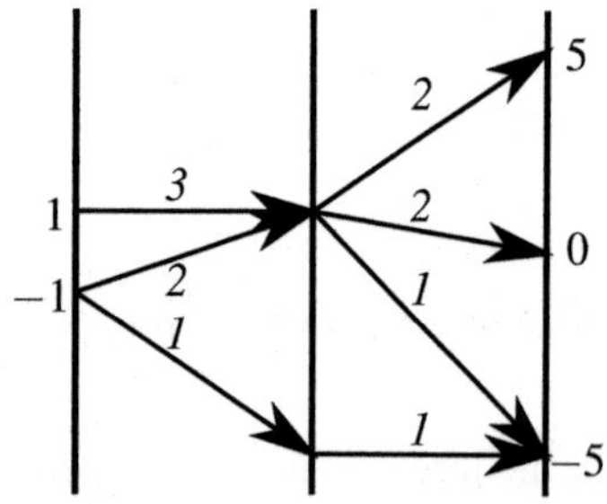

Fig. 2 The transition kernels of $(X_t)_{t\in\{0,1,2\}}$ in Sect. 4.2

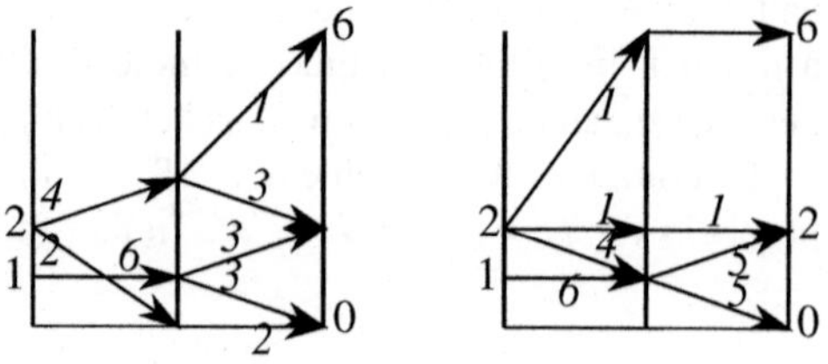

Fig. 3 The peacocks $(\mu_t)_{t\in\{0,1,2\}}$ and $(\mu_t)_{t\in\{0,'1,2\}}$

4.1 A Peacock Not Associated to a Martingale

We introduce the following peacock $(\mu_t)_{t\in T}$:

$$\begin{cases} 6\mu_0 = 3\delta_{-1} + 3\delta_1 \\ 6\mu_1 = 2\delta_{-2} + 4\delta_1 \\ 6\mu_2 = 2\delta_0 + 2\delta_{-2} + 2\delta_2 \\ 6\mu_{1'} = 4\delta_{-1} + 2\delta_2. \end{cases}$$

Note that the measures μ_0 and μ_2 are symmetric and μ_1 and $\mu_{1'}$ are obtained from the other by symmetry. On Fig. 1 we represent the (sub)peacock $(\mu_t)_{t\in\{0,1,2\}}$. It is easily seen that every martingale transition is uniquely determined. There exists an associated martingale that is forced to have the law

$$(1/3)\delta_{-1,-2,-2} + 1/12(\delta_{-1,1,0} + \delta_{-1,1,2}) + 1/4(\delta_{1,1,0} + \delta_{1,1,2}).$$

Hence, the law of the coupling between μ_0 and μ_2 is

$$\pi = (1/3)\delta_{-1-2} + 1/12(\delta_{-1,0} + \delta_{-1,2}) + 1/4(\delta_{1,0} + \delta_{1,2}).$$

Observe that the coefficient of $\delta_{1,-2}$ is zero. In other words, no mass is transported from 1 at time 0 (point A on Fig. 1) to -2 at time 2 (point Z). For the peacock $(\mu_t)_{t\in\{0,1',2\}}$ the coupling between μ_0 and μ_2 is obtained by symmetry from π. Thus some mass is transported from A to Z. Hence, there does not exist a martingale associated to both (sub)peacocks. Therefore, one can not associate a martingale to $(\mu_t)_{t\in T}$.

4.2 A Martingale Not Associated to a Markovian Martingale

For the second item of Theorem 2 we introduce a slight modification of the previous peacock where μ_0 does not change but the final peacock is concentrated on $\{-5, 0, 5\}$ instead of $\{-2, 0, 2\}$.

$$\begin{cases} 6\mu_0 = 3\delta_{-1} + 3\delta_1 \\ 6\mu_1 = \delta_{-5} + 5\delta_1 \\ 6\mu_2 = 2\delta_0 + 2\delta_{-5} + 2\delta_5 \\ 6\mu_{1'} = 5\delta_{-1} + \delta_5. \end{cases}$$

As in Sect. 4.1 the peacocks $(\mu_t)_{t\in\{0,1,2\}}$ and $(\mu_t)_{t\in\{0,1',2\}}$ are symmetric and the proof is similar. But the two symmetric martingales associated to the peacocks indexed by $\{0,1,2\}$ and $\{0,1',2\}$ are now unique only because one asks them to be Markovian. Let us see what is the first one that we call $(X_t)_{t\in\{0,1,2\}}$. It is obtained as the Markov composition of $\mathrm{Law}(X_0,X_1)$ and $\mathrm{Law}(X_1,X_2)$ that are uniquely determined as follows:

$$\mathrm{Law}(X_0,X_1) = (1/2)\delta_{1,1} + (1/6)(\delta_{-1,-5} + 2\delta_{-1,1})$$

and

$$\mathrm{Law}(X_1,X_2) = (1/6)\delta_{-5,-5} + (1/6)(\delta_{1,-5} + 2\delta_{1,0} + 2\delta_{1,5}).$$

This can be read on Fig. 2 where for the law of the Markovian martingale it remains to explain that at time 1 the mass is distributed independently from the past. For instance the coefficient of $\delta_{-1,1,5}$ is computed in the following way

$$\mathbb{P}(X_0 = -1)\mathbb{P}_{X_0=-1}(X_1 = 1)\mathbb{P}_{X_1=1}(X_2 = 5) = \frac{1}{2}\frac{2}{3}\frac{2}{5} = \frac{2}{15}.$$

Finally,

$$\begin{aligned}\mathrm{Law}(X_0,X_1,X_2) =&(1/6)\delta_{-1,-5,-5} + (1/15)\delta_{-1,1,-5} + (2/15)\delta_{-1,1,0}\\ &+ (2/15)\delta_{-1,1,5} + (1/10)\delta_{1,1,-5} + (1/5)\delta_{1,1,0} + (1/5)\delta_{1,1,5}.\end{aligned}$$

Observe that in $\mathrm{Law}(X_0,X_2)$ the coefficient of $\delta_{1,-5}$ and $\delta_{-1,5}$ are $1/10$ and $2/15$ respectively. Hence the measure is not symmetric, which completes the first part of the proof.

For the second part of the proof, it is enough to twist the composition of $\mathrm{Law}(X_0,X_1)$ and $\mathrm{Law}(X_1,X_2)$ at time 1 in a way that $\mathrm{Law}(X_0,X_2)$ becomes symmetric. This occurs exactly if

$$\mathbb{P}((X_0,X_2) = (1,0)) = \mathbb{P}((X_0,X_2) = (-1,0)) = 1/6 \tag{1}$$

because the space of martingales associated to μ_0, μ_2 depends only on one real parameter. The whole martingale $(X_t)_{t\in\{0,1,2\}}$ can be parametrised by the conditional law $\mathrm{Law}_{(X_0,X_1)=(1,1)}(X_2)$. We set $\mathrm{Law}_{(X_0,X_1)=(1,1)}(X_2) = \alpha^\theta$ defined as $\theta\alpha^1 + (1-\theta)\alpha^0$, where $\alpha^0 = (3/5)\delta_5 + (2/5)\delta_{-5}$ and $\alpha^1 = (1/5)\delta_5 + (4/5)\delta_0$ are the extreme admissible points. Notice that the Markovian composition would corresponds to the choice $\theta = 1/2$ because $\mathrm{Law}_{X_1=1}(X_2) = \alpha^{1/2}$. We have

$$\begin{aligned}\mathbb{P}((X_0, X_2) = (1,0)) = \mathbb{P}((X_0, X_1, X_2) = (1,1,0)) &= \frac{1}{2}\mathbb{P}_{(X_0,X_1)=(1,1)}(X_2 = 0) \\ &= \frac{1}{2}(\frac{4}{5}\theta).\end{aligned}$$

Thus we choose $\theta = 5/12$, which permits us to complete the proof.

4.3 A Markovian Martingale Not Associated to a Lipschitz–Markov Martingale

For the last item of Theorem 2 the peacocks $(\mu_t)_{t\in\{0,1,2\}}$ and $(\mu_t)_{t\in\{0,'1,2\}}$ are not symmetric in any way. That is why we represent both peacocks in Fig. 3.

$$\begin{cases} 12\mu_0 = 6\delta_1 + 6\delta_2 \\ 12\mu_1 = 2\delta_0 + 6\delta_1 + 4\delta_3 \\ 12\mu_2 = 5\delta_0 + 6\delta_2 + \delta_6 \\ 12\mu_{1'} = 10\delta_1 + \delta_2 + \delta_6 \end{cases}$$

Let see that there is a Markovian martingale associated to this peacock. Let us define the following joint laws:

$$\begin{cases} \mathrm{Law}(X_0, X_1) = (12)^{-1}(6\delta_{1,1} + 2\delta_{2,0} + 4\delta_{2,3}) \\ \mathrm{Law}(X_0, X_{1'}) = (12)^{-1}(6\delta_{1,1} + 4\delta_{2,1} + \delta_{2,2} + \delta_{2,6}) \\ \mathrm{Law}(X_1, X_2) = (12)^{-1}(2\delta_{0,0} + 3\delta_{1,0} + 3\delta_{1,2} + 3\delta_{3,2} + \delta_{3,6}) \\ \mathrm{Law}(X_{1'}, X_2) = (12)^{-1}(5\delta_{1,0} + 5\delta_{1,2} + \delta_{2,2} + \delta_{6,6}). \end{cases} \tag{2}$$

Assuming that the composition at times 1 and 1′ are Markovian, we obtain the same joint law $\mathrm{Law}(X_0, X_2)$. For this, as in Sect. 4.2 it suffices to compute one parameter of it in two manners. Let us do it for $\mathbb{P}((X_0, X_2) = (2,2))$:

$$\begin{cases} \mathbb{P}((X_0, X_2) = (2,2)) = \mathbb{P}((X_0, X_1, X_2) = (2,3,2)) = \frac{1}{2}\frac{2}{3}\frac{3}{4} \\ \mathbb{P}((X_0, X_2) = (2,2)) = \mathbb{P}((X_0, X_{1'}, X_2) = (2,2,2)) = \frac{1}{2}\frac{1}{6}1 + \frac{1}{2}\frac{2}{3}\frac{1}{2}. \end{cases}$$

Finally, we have proved that X_1 and $X_{1'}$ can be defined on the same probability space together with X_0 and X_2.

Note that the previous martingale is not Lipschitz–Markov because the Kantorovich distance between $\mathrm{Law}_{X_0=1}(X_1) = \delta_1$ and $\mathrm{Law}_{X_0=2}(X_1) = 1/3\delta_0 + 2/3\delta_2$ is $5/3$, which is strictly greater that $|1-0| = 1$. In (2), the marginal and martingale constraints uniquely determine all of the law apart from $\mathrm{Law}(X_0, X_1)$. It can be parametrised by

$$\mathrm{Law}(X_0, X_1) = \theta\pi^1 + (1-\theta)\pi^0$$

where $\pi^1 = (12)^{-1}(6\delta_{1,1} + 2\delta_{2,0} + 4\delta_{2,3})$ corresponds to the joint law in (2) and

$$\pi^0 = (12)^{-1}(2\delta_{1,0} + 3\delta_{1,1} + 1\delta_{1,3} + 0\delta_{2,0} + 3\delta_{2,1} + 3\delta_{2,3}).$$

The kernel is Lipschitz if and only if $\theta \in [0, 1/2]$. However we will not need to prove it because if $\theta \neq 1$, some mass is transported from 1 to 3 and part of this mass finishes in 6 at time $t = 2$. This leads to a joint law $\mathrm{Law}(X_0, X_2)$ that can not be associated to the peacock $(\mu_t)_{t\in\{0,'1,2\}}$ on the right part of Fig. 3. For the unique martingale law associated to this peacock, no mass is transported from 1 at time $t = 0$ to 6 at time $t = 2$.

5 A Positive Result

The aim of this section is to furnish sufficient conditions for Problem 7a. Under the hypothesis of Theorem 4, any peacock is associated to a martingale. Under the hypothesis of Theorem 5, this martingale is Markovian. Other examples are given by Hirsch et al. in Exercise 2.3 [9].

5.1 Disintegration of a Measure in $\{\mu \in \mathscr{P} : \mathbb{E}(\mu) = 0\}$

As in Choquet theorem, even if $\{\mu \in \mathscr{P} : \mathbb{E}(\mu) = 0\}$ is a noncompact set, any element can be decomposed as a mean of the extreme points. According to Douglas theorem [4] the extreme points are exactly the positive measures μ such that the affine functions are dense in $L^1(\mu)$. Hence, the extreme points are the diatomic measures $\theta_{a,b}$ with $a \le 0 \le b$ and $\theta_{a,b} = \frac{b}{b-a}\delta_a + \frac{-a}{b-a}\delta_b$. The decomposition is not unique as illustrated by $1/6(4\theta_{-1,1} + 2\theta_{-2,2}) = 1/6(3\theta_{-1,2} + 3\theta_{-2,1})$.

However one can give a canonical decomposition of μ. It relies on the order of its quantiles. It seems classical but we could not cite it from the literature. Hence we present some intuitive facts as consequences of the theory developed in [1] about the minimal shadow of positive measures into other measures. For every $q \in [0, 1]$, the set

$$F(\mu, q) = \{\eta |\ \eta \leq \mu \text{ and } \eta(\mathbb{R}) = q \text{ and } \int x \, \mathrm{d}\eta(x) = 0\}$$

has a minimal element for the convex order. We call it $S^{\mu}(q)$. It is the shadow of $q\delta_0$ in μ as defined in Lemma 4.6 of Beiglböck and Juillet [1]. This measure can be described as in Example 4.7 of Beiglböck and Juillet [1]: It is the unique measure $\eta \leq \mu$ of mass q and expectation 0 that can be written

$$\eta = \mu|_{]f(q),g(q)[} + a\delta_{f(q)} + b\delta_{g(q)}. \tag{3}$$

In other words it is the restriction of μ on a quantile interval

$$S^{\mu}(q) = (G_{\mu})_{\#}\lambda|_{[q_0,q_1]}$$

where we recall that G_{μ} is the quantile function defined before Sect. 2. In particular $q_1 - q_0 = q$.

Let f^{μ} and g^{μ} denote the functions

$$f^{\mu} : q \mapsto \max(\operatorname{spt}(S^{\mu}(q))) \text{ and } g^{\mu} : q \mapsto \min(\operatorname{spt}(S^{\mu}(q))).$$

We have the following properties

- If $q \leq q'$ it holds $S^{\mu}(q) \leq S^{\mu}(q')$,
- The function g^{μ} is left-continuous and nondecreasing,
- The function f^{μ} is left-continuous and nonincreasing.

Note that $[g^{\mu}(q), f^{\mu}(q)]$ is the smallest closed interval $[f(q), g(q)]$ of full mass for $S^{\mu}(q)$ and it is the unique choice if one demands that $q \mapsto g(q) - f(q)$ is left-continuous in (3). We will call it *the interval of quantile q* or the *q-interval*.

Let us now introduce a measure π on $\mathbb{R}^2$ such that for every $q \in [0, 1]$ the first marginal of $\pi|_{]-\infty,q]\times\mathbb{R}}$ is $\lambda|_{[0,q]}$ and the second marginal is $S^{\mu}(q)$. Such a measure exists because the family $(S^{\mu}(q))_{q\in[0,1]}$ is increasing and the mass of $S^{\mu}(q)$ is q. It is easy to check that $(\theta_{f^{\mu}(q),g^{\mu}(q)})_{q\in[0,1]}$ is an admissible disintegration of π with respect to $\lambda|_{[0,1]}$ and it is the only disintegration $(\theta_q)_{q\in[0,1]}$ such that $q \mapsto \theta_q$ is left-continuous for the weak topology. Finally we have obtained a canonical representation of μ. It writes

$$\mu = \int_0^1 \theta_{f^{\mu}(q),g^{\mu}(q)} \, \mathrm{d}q, \tag{4}$$

where $-f^{\mu}$ and g^{μ} are the unique nondecreasing and left-continuous functions with $S^{\mu}(q)([f^{\mu}(q), g^{\mu}(q)]) = q$. Note that as usual for a Choquet decomposition, Eq. (4)

has to be understood in the weak sense. For instance

$$\mu(A) = \int_0^1 \theta_{f^\mu(q),g^\mu(q)}(A)\,\mathrm{d}q$$

for every measurable A.

Remark 1 In this article we will sometime simply write F_t, G_t and $[f^t(q), g^t(q)]$ in place of F_{μ_t}, G_{μ_t} and $[f^{\mu_t}(q), g^{\mu_t}(q)]$ respectively.

5.1.1 Diatomic Convex Order

Let μ and ν be probability measures on $\mathbb{R}$ with expectation zero. We introduce the order $\preceq_{DC}$ with $\mu \preceq_{DC} \nu$ if and only if

$$\forall q \in [0,1],\ [f^\mu(q), g^\mu(q)] \subseteq [f^\nu(q), g^\nu(q)].$$

and call it the diatomic convex order. There exists a unique martingale law π_q between $\theta_{f^\mu(q),g^\mu(q)}$ and $\theta_{f^\nu(q),g^\nu(q)}$. Its formula is made explicit later in (6). In Sect. 5.3 we will consider the joint law

$$\pi = \int_0^1 \pi_q\,\mathrm{d}q \tag{5}$$

with marginals μ and ν. It is a martingale so that $\mu \preceq_{DC} \nu$ implies $\mu \preceq_C \nu$.

In the special case of symmetric measure, the order can be defined similarly as the stochastic order using the positive cone of even functions that are non-decreasing on $\mathbb{R}_+$ in place of the cone of convex functions. However, I could not find an appropriate cone for defining $\preceq_{DC}$ in the general case.

5.2 *Peacocks Consisting of Extreme Elements*

In this subsection we consider Problem 7a for peacocks $(\mu_t)_{t\in T}$ where every μ_t is an extreme element $\theta_{a,b}$. Observe that in this case, $\theta_{a,b} \preceq_C \theta_{a',b'}$ is equivalent to $[a,b] \subseteq [a',b']$, and as these two intervals are the q-intervals for every q, the relation $\theta_{a,b} \preceq_C \theta_{a',b'}$ is equivalent to $\theta_{a,b} \preceq_{DC} \theta_{a',b'}$. The set $\Pi_M(\theta_{a,b}, \theta_{a',b'})$ of martingales associated to a peacock of cardinal two is restricted to one element:

$$\begin{cases} \delta_0 \times \theta_{a',b'} & \text{if } a = b, \\ \frac{b}{b-a}\left(\frac{b'-a}{b'-a'}\delta_{a,a'} + \frac{a-a'}{b'-a'}\delta_{a,b'}\right) + \frac{-a}{b-a}\left(\frac{b'-b}{b'-a'}\delta_{b,a'} + \frac{b-a'}{b'-a'}\delta_{b,b'}\right) & \text{otherwise.} \end{cases} \tag{6}$$

We consider the totally ordered case before the general case.

5.2.1 Totally Ordered $(T, \leq)$

For $s \leq t$, the two-marginals joint law π_{st} between μ_s and μ_t is unique and has formula (6). Hence, for $s \leq t \leq u$, the Markovian and in fact any composition of π_{st} and π_{tu} is a martingale law with marginals μ_s and μ_u. Thus, it is π_{su} and the two-marginals joint laws $(\pi_{st})_{(s\leq t)}$ constitute a coherent family for the Markovian composition. Thus, there exists a Markovian martingale $(X_t)_{t\in T}$ with the wanted marginals and its law is the unique one among the associated martingales.

5.2.2 Partially Ordered $(T, \leq)$

It is less direct to associate a martingale when T is not totally ordered. It is no longer enough to check that the two-marginals laws constitute a coherent family. All finite families of marginals would have to be considered, also with elements noncomparable for $\preceq_C$ and their joint law can not uniquely be determined by the constraints of the problem.

Let us first reduce the problem to $(\mathbb{R}^2_+, \leq)$. We can map $t \in T$ to the element $\Phi(t) = (a, b) \in \mathbb{R}^2_+$ defined by $\mu_t = \theta_{-a,b}$. If we associate a Markovian martingale $(M_{x,y})_{(x,y)\in\mathbb{R}^2_+}$ to $(\mu_{x,y})_{(x,y)\in\mathbb{R}^2_+}$ with $\mu_{x,y} = \theta_{-x,y}$, it is easy to check that $(M_{\Phi(t)})_{t\in T}$ is a Markovian martingale associated to $(\mu_t)_{t\in T}$.

A martingale associated to $(\mu_{x,y})_{(x,y)\in\mathbb{R}^2_+}$ is the following: Consider the Wiener measure on $\mathscr{C}(\mathbb{R}_+)$ and let $M_{x,y}$ be the random variable

$$M_{x,y} = y.\mathbb{1}_{\{\tau_y<\tau_{-x}\}} - x.\mathbb{1}_{\{\tau_{-x}<\tau_y\}},$$

where τ_z is the hitting time of $z \in \mathbb{R}$. It is easy to check that $(M_{x,y})_{(x,y)\in\mathbb{R}^2_+}$ is a Markovian martingale associated to $(\mu_{x,y})_{(x,y)\in\mathbb{R}^2_+}$. For every restriction of the peacock to indices in a totally ordered set, the restriction of this martingale have the law described in Sect. 5.2.1.

Remark 2 The referee of this paper suggested to look at a peacock constructed from a reference measure ζ of barycenter 0 and defined by $\omega_{-a,b} = (\zeta|_{]-\infty,-a]\cup[b,+\infty[}) + \xi_{a,b}$ where ξ is the measure concentrated on $\{a, b\}$ with the same mass and barycenter as $\zeta|_{]a,b[}$. Note that in the case $\zeta = \delta_0$ it holds $\omega_{-a,b} = \theta_{-a,b}$.

The construction of this section generalises as follows. Let B_t be a Brownian motion with $\mathrm{Law}(B_0) = \zeta$ and for every $(a, b) \in \mathbb{R}^2_+$ let $\tau_{a,b}$ be the hitting time of $]-\infty, -a] \cup [b, +\infty[$. As $\mathrm{Law}(B_{\tau_{a,b}}) = \omega_{-a,b}$, we can simply associate the martingale $(B_{\tau_{a,b}})_{(a,b)\in\mathbb{R}^2_+}$ to the peacock $(\omega_{-a,b})_{(a,b)\in\mathbb{R}^2_+}$.

Notice finally that the measures $(\omega_{-a,b})_{(a,b)\in\mathbb{R}^2_+}$ are not non decreasing for $\preceq_{DC}$, as can be easily seen if ζ is uniform on $[-1, 1]$. Hence Theorem 4 does not apply.

5.3 A Positive Result for Peacocks Indexed by a Partially Ordered Set

In Sect. 5.3.2, for families of measures in the diatomic convex order we introduce the process similar to the quantile process in the martingale setting. We call it the *quantile martingale*. Recall that the quantile process relies on the quantile coupling, that is actually the model of (5) in the nonmartingale setting. In place of measures $\theta_{a,b}$ the extreme elements are Dirac masses that are parametrised by a quantile $q \in]0,1[$ for the two marginals. The quantile coupling couples them using the same parameter q in both disintegrations, as (5) does in the martingale case.

Theorem 5 states under which condition the quantile coupling has the Markov property. For completeness we start in Sect. 5.3.1 with the same question for the quantile process.

5.3.1 Characterisation of Markovian Kamae–Krengel Processes

We state the result on the quantile process and its relation to the stochastic order explained in the introduction of the present paper.

Proposition 2 *Let $(\mu_t)_{t\in T}$ be a family of real probability measures indexed by a partially order set $(T,\leq)$. The following statements are equivalent.*

- *The map $t \mapsto \mu_t$ is nondecreasing in stochastic order,*
- *the associated quantile process $t \mapsto X_t$ is almost surely nondecreasing.*

The following proposition characterises the Markovian quantile processes.

Proposition 3 *The quantile process is Markovian if and only if the following criterion is satisfied: for every $s \leq t \leq u$ and $q < q' \in]0,1[$, the conjunction of conditions*

$$\begin{cases} G_{\mu_s}(q) < G_{\mu_s}(q') \\ G_{\mu_t}(q) = G_{\mu_t}(q') \end{cases}$$

implies $G_{\mu_u}(q) = G_{\mu_u}(q')$. In other words $G_{\mu_t}(q) = G_{\mu_t}(q')$ implies $\{G_{\mu_s}(q) = G_{\mu_s}(q')$ or $G_{\mu_u}(q) = G_{\mu_u}(q')\}$.

Proof Assume that the property on the quantile function holds. Recall that $\mathscr{F}_t$ is the σ-algebra generated by all X_s where $s \leq t$. The Markov property holds if $\mathbb{E}(f(X_u)|\mathscr{F}_t) = \mathbb{E}(f(X_u)|\sigma(X_t))$ for every bounded measurable function f and $t \leq u$ elements of T. Let now t and u be fixed. It is enough to prove that for any $k \in \mathbb{N}$ and $s_1,\ldots,s_k \leq t$ the random vectors $(X_{s_1},\ldots,X_{s_k})$ and X_u are conditionally independent given X_t. For a family of conditional probabilities $(\mathbb{P}_{X_t=y})_{y\in\mathbb{R}}$ it is sufficient to prove that given real numbers $(x_1,\ldots,x_k)$ and z the events

$$A = \{X_{s_1} \leq x_1,\ldots,X_{s_k} \leq x_k\} \quad \text{and} \quad B = \{X_u \leq z\}$$

are independent under $\mathbb{P}_{X_t=y}$ for all y. We will define such a family. Recall that $\mathbb{P}$ is defined as $\int_0^1 \mathbb{P}_q \mathrm{d}q$ where the law of $(X_{s_1}, \ldots, X_{s_k}, X_t, X_u)$ under $\mathbb{P}_q$ is simply the Dirac mass in $(G_{s_1}(q), \ldots, G_{s_k}(q), G_t(q), G_u(q))$.

The events $\{X_t = y\}$ is of type $\{q \in]0,1[,\ G_t(q) = y\}$, that is $]q^-, q^+]$ or $]q^-, 1[$ where $q^+ = F_t(y)$ and $q^- = \lim_{\varepsilon \to 0+} F_t(y - \varepsilon)$. Recall that $\mu(y) = q^+ - q^-$. Thus $(\mathbb{P}_{X_t=y})_{y \in \mathbb{R}}$ defined by

$$\mathbb{P}_{X_t=y} = \begin{cases} \mathbb{P}_{q-} & \text{if } \mu_t(y) = 0, \\ \frac{1}{\mu(y)} \int_{q^-}^{q^+} \mathbb{P}_q \, \mathrm{d}q & \text{otherwise} \end{cases}$$

is a disintegration of $\mathbb{P}$ according to X_t.

For $\mu(y) = 0$ the measure of both A and B for $\mathbb{P}_{X_t=y}$ is zero or one so that A and B are independent. In the other case, let us prove that at least one of the two events has measure zero or one. In fact, the quantiles of $]q^-, q^+]$ are mapped on y by G_t. Hence, according to the criterion for every $i \leq k$ one of the two maps G_{s_i} or G_u is constant on $]q^-, q^+]$. Thus G_u is constant or $(G_{s_1}, \ldots, G_{s_k})$ is constant. Therefore $\mathbb{P}_q(A)$ or $\mathbb{P}_q(B)$ is constantly zero or one on $]q^-, q^+]$. We have proved that A and B are independent with respect to $P_{X_t=y}$. This completes the proof of the first implication.

For the second implication suppose that the criterion is not satisfied so that there exist $s \leq t \leq u$ and $q < q' \in]0,1[$ with $G_t(q') = G_t(q) := y$, $G_s(q') > G_s(q') := x$ and $G_u(q') > G_u(q) := z$. In this case $\mathbb{P}(X_t = y) > 0$. Let $q_s = F_s(x)$ and $q_u = F_u(z)$. Let also $q^+ = F_t(y)$ and $q^- = \lim_{\varepsilon \to 0+} F_t(y - \varepsilon)$ so that

$$q^- < \min(q_s, q_u) \leq \max(q_s, q_u) < q' \leq q^+.$$

We have on the one hand

$$\mathbb{P}_{X_t=y}(X_s \leq x,\ X_u \leq z) = \frac{\min(q_s, q_u) - q^-}{q^+ - q^-}$$

and on the other hand

$$\mathbb{P}_{X_t=y}(X_s \leq x,\ X_u \leq z) = \frac{q_s - q^-}{q^+ - q^-} \quad \text{and} \quad \mathbb{P}_{X_t=y}(X_s \leq x,\ X_u \leq z) = \frac{q_u - q^-}{q^+ - q^-}.$$

Hence $\{X_s \leq x\}$ and $\{X_u \leq z\}$ are not conditionally independent given $\{X_t = y\}$. This finishes the proof of the second implication.

5.3.2 Quantile Martingales and Characterisation of the Markov Property

Theorems 4 and 5 are the counterparts in the martingale setting of Propositions 2 and 3.

Theorem 4 *Let $(\mu_t)_{t\in T}$ be a peacock indexed by a partially ordered set $(T, \leq)$. Assume moreover that the measures have expectation zero and $t \mapsto \mu_t$ is nondecreasing for the diatomic convex order. Then there exists $(X_t)_{t\in T}$ a martingale associated to $(\mu_t)_{t\in T}$.*

Proof According to Sect. 5.1.1, the elements of the canonical decomposition of the measures are in convex order. Hence we can replace the peacock by a one-dimensional family of peacocks $(\theta^q_{-f^t(q),g^t(q)})_{t\in T}$. Each of them can be associated with a martingale (X^q_t) defined on the Wiener space as in Sect. 5.2.2. We consider the process on the probability space $[0,1]\times C([0,+\infty[)$ obtained using the conditioning in $q \in [0,1]$. It is a martingale with the correct marginal for every $t \in T$. We used the fact that convex combinations of martingale laws are martingale laws.

We call *quantile martingale* the martingale introduced during the proof of Theorem 4. In what follows we write $I^t(q)$ for the interval $[f^t(q), g^t(q)]$.

Theorem 5 *With the notation of Theorem 4 the quantile martingale is Markovian if and only if the following criterion is satisfied for every $s \leq t \leq u$ and $q < q' \in [0,1]$.*

1. *If $I^t(q) = I^t(q')$ it holds $I^s(q) = I^s(q')$ or $I^u(q) = I^u(q')$,*
2. *if $\{f^t(q) = f^t(q')$ and $g^t(q) \neq g^t(q')\}$ it holds $I^s(q') = [0,0]$ or $\{f^s(q) = f^s(q')$ and $g^s(q) = g^t(q)$ and $g^s(q') = g^t(q')\}$ or $f^t(q') = f^u(q')$ or $I^u(q) = I^u(q')$,*
3. *if $\{f^t(q) \neq f^t(q')$ and $g^t(q) = g^t(q')\}$ it holds $I^s(q') = [0,0]$ or $\{g^s(q) = g^s(q')$ and $f^s(q) = f^t(q)$ and $f^s(q') = f^t(q')\}$ or $g^t(q') = g^u(q')$ or $I^u(q) = I^u(q')$.*
4. *Nothing has to be satisfied in the case $\{f^t(q) \neq f^t(q')$ and $g^t(q) \neq g^t(q')\}$.*

Example 1 (Sufficient Conditions) The criterion for the Markov property in Theorem 5 applies for instance in the following situations.

- The measures are continuous (without atom). This is settled in (4).
- The measures μ_t are continuous or δ_0. If $\mu_t = \delta_0$ we check that the criterion is satisfied in (1) with $I^s(q) = I^s(q') = [0,0]$. The other case is (4).
- The measures are diatomic like in Sect. 5.2.2. For every $q < q'$ and $t \in T$ it holds $I^t(q) = I^t(q')$ so that the criterion is satisfied in (1).
- The measures are $\mu_t = 1/2(\theta_{-1,1+t} + \theta_{-1,2+t})$ for $T = [0,1]$. The peacock satisfies the criterion in (2) where $f^t(q') = f^u(q') = -1$.
- The measures are $\mu_t = 1/2(\theta_{-1-t,1} + \theta_{-1-t,2})$ for $T = [0,1]$. For $q \leq 1/2 < q'$ the criterion is satisfied in (2) because it holds $\{f^s(q) = f^s(q') = -1 - s$ and $g^s(q) = g^t(q) = 1$ and $g^s(q') = g^t(q') = 2\}$.

Proof (Proof of Theorem 5) The proof is similar to the one of Proposition 3 even if more technical. In particular even if μ_t is continuous, the value of X_t does not uniquely determine a trajectory. Nevertheless the law of the random trajectory is uniquely determined because it only depends on q and X_t. In fact, the quantile q is a function of X_t so that as described in Sect. 5.2.2 the law of the future is contained in the present position X_t. In the general case when μ_t has atoms the process that is Markovian when conditioned on q can loose the Markov property because X_t does not uniquely determine q.

Suppose that $(X_t)_{t\in T}$ described above is not Markovian. We will show that the criterion is not satisfied. Let us consider a time $t \in T$ and $y \in \mathbb{R}$ so that $\{X_t = y\}$ denies the Markov property: the future is not independent from the past. Observe that the previous remarks on continuous measures show that y must be an atom of μ_t. Moreover $y = 0$ is not possible because it would mean $X_s = 0$ for $s \le t$. Hence the past would be determined by the present so that no information on the past can change the law of the future.

Without loss of generality, we assume $X_t = y < 0$. Let $Q = \{q \in [0,1], f^t(q) = y\}$ be the interval of quantiles mapped in y. On Q the density of probability for the value of q conditioned on $X_t = y$ is proportional to $\frac{g^t}{g^t - f^t}$. As we supposed that the Markov property does not hold there exists an integer k and indices $s_1, \ldots, s_k \le t$ such that $(X_{s_1}, \ldots, X_{s_k})$ and X_u are not conditionally independent given $\{q \in Q\} \cap \{X_t = y\} = \{X_t = y\}$. As these random variables are independent with respect to the conditional probabilities $\mathbb{P}_{\{q\}\cap\{X_t=y\}}$ for any $q \in Q$, there exist two quantiles for which both the laws of the past and of the future are different respectively.[2] Let $q_1, q_2 \in Q$ with $q_1 < q_2$ be such quantiles.

Concerning the future first, the law β_i of X_u is the one of a Brownian motion starting in $y = f^t(q_i) < 0$ and stopped when hitting $f^u(q_i) \le y$ or $g^u(q_i) > 0$. Different future laws β_i are obtained for $i \in \{1,2\}$. Therefore $f^u(q_2) < y$ and $[f^u(q_1), g^u(q_1)] \neq [f^u(q_2), g^u(q_2)]$.

We consider now the past. For some $s \in \{s_1, \ldots, s_k\}$ we write α_i the law of X_s given $\{q_i\} \cap \{X_t = y\}$. It is the law of a Brownian motion stopped when it hits $\{f^s(q_i), g^s(q_i)\}$ conditioned on the fact that it hits $y = f^t(q_i)$ before $g^t(q_i)$. Recall also $f^t(q_i) \le f^s(q_i) \le 0 \le g^s(q_i) \le g^s(q_i)$. The support of α_i has cardinal one or two. It is one if and only if $I^s(q_i) = [0,0]$ or $g^s(q_i) = g^t(q_i)$ and then α_i is the Dirac mass in 0 or $f^s(q_i)$ respectively. If the support of α_i has two elements these are $\{f^s(q_i), g^s(q_i)\}$ and $\alpha_1 = \alpha_2$ if and only if $g^t(q_1) = g^t(q_2)$ and $I^s(q_1) = I^s(q_2)$. If $g^t(q_1) \neq g^t(q_2)$ the only possibility for $\alpha_1 = \alpha_2$ is that the supports are reduced to one point. Note now that if $\alpha_1 = \alpha_2$ the support of those measures uniquely determine $I^s(q_1)$ and $I^s(q_2)$ in both cases $g^t(q_1) = g^t(q_2)$ or $g^t(q_1) \neq g^t(q_2)$. But for $i \in \{1,2\}$ the law of $(X_{s_1}, \ldots, X_{s_k})$ given $\{q_i\} \cap \{X_t = y\}$ is uniquely determined by $\{I^s(q_i)\}_{s\in\{s_1,\ldots,s_k\}}$. As we supposed that these laws are different for $i = 1$ or $i = 2$, there exists $s \in \{s_1, \ldots, s_k\}$ such that $I^s(q_2) \neq [0,0]$ and $\{g^s(q_1) \neq g^t(q_1)$ or $g^s(q_2) \neq g^t(q_2)$ or $f^s(q_1) \neq f^t(q_1)\}$ in the case $g^t(q_1) \neq g^t(q_2)$ and $I^t(q_1) \neq I^t(q_2)$ in the case $g^t(q_1) = g^t(q_2)$.

In summary, let $y < 0$ be an atom of $\mathrm{Law}(X_t)$ such that the condition $\{X_t = y\}$ denies the Markov property of the quantile martingale. Concerning the future we have proved $f^u(q_2) < y$ and $[f^u(q_1), g^u(q_1)] \neq [f^u(q_2), g^u(q_2)]$. Concerning

[2] It is a general fact that if $\mathfrak{p}$ and $\mathfrak{f}$, a past and a future map defined on Q are both nonconstant, there exist $q_1, q_2 \in Q$ such that $\mathfrak{p}(q_1) \neq \mathfrak{p}(q_2)$ and $\mathfrak{f}(q_1) \neq \mathfrak{f}(q_2)$.

the past, we have proved $I^s(q_2) \neq [0,0]$ and $\{g^s(q_1) \neq g^t(q_1)$ or $g^s(q_2) \neq g^t(q_2)$ or $f^s(q_1) \neq f^t(q_1)\}$ in the case $g^t(q_1) \neq g^t(q_2)$ and $I^t(q_1) \neq I^t(q_2)$ in the case $g^t(q_1) = g^t(q_2)$. Symmetric conclusions happen in the symmetric situation $y > 0$. Hence one can carefully check that at least (1), (2) or (3) is not correct for the choice $(q, q') = (q_1, q_2)$. In fact if $g^t(q_1) = g^t(q_2)$ the criterion is not satisfied in (1). If $g^t(q_1) = g^t(q_2)$, it is not satisfied in (2). Finally we have proved that if the peacock satisfies the criterion the quantile martingale is Markovian.

Conversely, we assume that the criterion is not satisfied and will prove that the process is not Markovian. It is enough to assume that the criterion is not satisfied in (1) or (2). For $s \leq t \leq u$ and $q \in Q = (f^t)^{-1}\{y\}$ we denote as before $\alpha(q)$ the law of X_s given $\{q\} \cap \{X_t = y\}$ and $\beta(q)$ the law of X_u given the same condition. Let $q, q' \in Q$ such that the criterion is not satisfied. If the criterion is not satisfied in (1) we have $I^u(q) \neq I^u(q')$ and we can assume $f^u(q') < f^u(q) \leq y$ (if not $g^u(q') > g^u(q) \geq g^t(q) = y'$ and we can consider y' in place of y). Therefore $\beta(q) \neq \beta(q')$ and $\alpha(q) \neq \alpha(q')$ in the two cases. The joint law of (X_s, X_u) given $\{X_t = y\}$ is

$$\pi := Z^{-1} \int_Q \alpha(q) \times \beta(q) \frac{g^t(q)}{g^t(q) - f^t(q)} \mathrm{d}q$$

where $Z = \int_Q \frac{g^t(q)}{g^t(q)-f^t(q)} \mathrm{d}q$. We will prove that it is not the product of two probability measures, which will be enough for the implication. Recall that the support of $\alpha(q)$ and $\beta(q)$ are included in $\{f^s(q), g^s(q)\}$ and $\{f^u(q), g^u(q)\}$ respectively. The functions f are nonincreasing and left-continuous. The functions g are nondecreasing and left-continuous. In case (1) the measures $\alpha(q) \neq \alpha(q')$ and $\beta(q) \neq \beta(q')$ are not only different but their supports are also different. Hence one is easily convinced with a picture in $\mathbb{R}^2$ that π has not the support of a product measure. This argument does not work if (2) is denied because $\alpha(q)$ and $\alpha(q')$ may have the same support and be different. In fact they are different if and only if the support is made of the two points $f^s(q) = f^s(q')$ and $g^s(q) = g^s(q')$. With a simple Bayes formula the mass of $f^s(q)$ with respect to $\alpha(q)$ can be computed to be $(g^s(q)/g^s(q) - f^s(q))(g^t(q) - f^t(q)/g^t(q))$ and the same formula with primed letters holds for $\alpha(q')$. Note that the four quantities are the same except $g^t(q') > g^t(q)$. As $\beta(q) \neq \beta(q')$ and recalling the left continuity of f^u and g^u it follows that the conditional law of X_u with respect to $\{X_t = y$ and $X_s = f^s(q)\}$ is different to the conditional law of X_u with respect to $\{X_t = y$ and $X_s = g^s(q)\}$. Finally π is not a product measure and the martingale in not Markovian.

5.4 Questions

Even though Problem 7b is solved in [7], it is still an open question whether the full Kellerer theorem for measures on $\mathbb{R}^d$ hold, where "full" means with the Markov property. The following questions are rather related to our approach of Problem 7a in Sect. 5.3. To solve them may however bring some useful new ideas to Problem 7b.

- Let $t \mapsto \mu_t$ be nondecreasing for the stochastic order. Does it exist an associated process $(X_t)_{t\in T}$ that is Markovian? Recall that Proposition 3 is an exact account on the question whether the quantile process associated to $(\mu_t)_t$ is Markovian.
- Let $(\mu_t)_{t\in[0,1]}$ be a family of real measures. For any sequence of partitions of $[0, 1]$ we describe a procedure. We associate to the partition $0 = t_0 \leq \cdots \leq t_N = 1$ the Markovian process $(X_t)_{t\in[0,1]}$ constant on any $[t_k, t_{k+1}[$ such that $\mathrm{Law}(X_t) = \mathrm{Law}(X_{t_k})$, and $\mathrm{Law}(X_{t_k}, X_{t_{k+1}})$ is a quantile coupling. Under ad hoc general conditions on the peacock and the type of convergence, does it exist a sequence of partitions such that the sequence of processes converge to a Markovian limit process with marginal μ_t at any $t \in [0, 1]$? Is the Markovian limit unique? Is for instance the continuity of the peacock sufficient for these properties? This makes precise a question at the end of Juillet [11]. See this paper and also [6] for the same approach in the case of martingales.
- If $t \mapsto \mu_t$ is nondecreasing for the diatomic convex order $\preceq_{DC}$, does it exist an associated Markovian martingale? We proved in Theorem 5 that such a martingale can not systematically be the quantile martingale.

Of course the first and the third question have likely the same answer, yes or no. In the case $T = [0, 1]$ the second question suggests an approach for the first question. Recall that it is wrong that limit of Markovian processes are Markovian.

References

1. M. Beiglböck, N. Juillet, On a problem of optimal transport under marginal martingale constraints. Ann. Probab. **44**(1), 42–106 (2016)
2. M. Beiglböck, M. Huesmann, F. Stebegg, Root to Kellerer, in *Séminaire de Probabilités XLVIII*, ed. by C. Donati-Martin, A. Lejay, A. Rouault. Lecture Notes in Mathematics, vol. 2168 (Springer, Berlin, 2015)
3. R. Cairoli, J.B. Walsh, Stochastic integrals in the plane. Acta Math. **134**, 111–183 (1975)
4. R.G. Douglas, On extremal measures and subspace density. Mich. Math. J. **11**, 243–246 (1964)
5. J.A. Fill, M. Machida, Stochastic monotonicity and realizable monotonicity. Ann. Probab. **29**(2), 938–978 (2001)
6. P. Henry-Labordere, X. Tan, N. Touzi, An explicit martingale version of the one-dimensional Brenier's theorem with full marginals constraint (2014). Preprint
7. F. Hirsch, B. Roynette, On $\mathbb{R}^d$-valued peacocks. ESAIM Probab. Stat. **17**, 444–454 (2013)
8. F. Hirsch, M. Yor, Looking for martingales associated to a self-decomposable law. Electron. J. Probab. **15**(29), 932–961 (2010)
9. F. Hirsch, C. Profeta, B. Roynette, M. Yor, *Peacocks and Associated Martingales, with Explicit Constructions*. Bocconi & Springer Series (Springer, Milan, 2011)
10. F. Hirsch, B. Roynette, M. Yor, Kellerer's theorem revisited, in *Asymptotic Laws and Methods in Stochastics. Volume in Honour of Miklos Csorgo*, ed. by Springer. Fields Institute Communications Series (Springer, Berlin, 2014)
11. N. Juillet, Stability of the shadow projection and the left-curtain coupling. Ann. Inst. Henri Poincaré Probab. Stat. (2016, to appear). See http://imstat.org/aihp/accepted.html

12. T. Kamae, U. Krengel, Stochastic partial ordering. Ann. Probab. **6**(6), 1044–1049 (1978)
13. T. Kamae, U. Krengel, G.L. O'Brien, Stochastic inequalities on partially ordered spaces. Ann. Probab. **5**, 899–912 (1977)
14. H.G. Kellerer, Markov-Komposition und eine Anwendung auf Martingale. Math. Ann. **198**, 99–122 (1972)
15. B. Pass, On a class of optimal transportation problems with infinitely many marginals. SIAM J. Math. Anal. **45**(4), 2557–2575 (2013)
16. V. Strassen, The existence of probability measures with given marginals. Ann. Math. Statist. **36**, 423–439 (1965)
17. E. Wong, M. Zakai, Martingales and stochastic integrals for processes with a multi-dimensional parameter. Z. Wahrscheinlichkeitstheor. Verw. Geb. **29**, 109–122 (1974)

Convex Order for Path-Dependent Derivatives: A Dynamic Programming Approach

Gilles Pagès

Abstract We explore the functional convex order for various classes of martingales: Brownian or Lévy driven diffusions with respect to their diffusion coefficient, stochastic integrals with respect to their integrand. Each result is bordered by counterexamples. Our approach combines the propagation of convexity results through (simulable) discrete time recursive dynamics relying on a backward dynamic programming principle and powerful functional limit theorems to transfer the results to continuous time models. In a second part, we extend this approach to optimal stopping theory, namely to the *réduites* of adapted functionals of (jump) martingale diffusions. Applications to various types of bounds for the pricing of pathwise dependent European and American options in local volatility models are detailed. Doing so, earlier results are retrieved in a unified way and new ones are proved. This systematic paradigm provides tractable numerical methods preserving functional convex order which may be crucial for applications, especially in Finance.

1 Introduction and Motivation

The aim of this paper is to propose and develop a systematic and unified approach to establish *functional convex order* results for discrete and continuous time martingales based on the propagation of convexity through some backward dynamic programming principles in discrete time and, on the other hand, on weak functional limit theorems to make a transfer to continuous time setting. The term "functional" refers to the "parameter" we deal with: thus, for possibly jump diffusions processes, this parameter is the diffusion coefficient or, for stochastic integrals, their integrand. Doing so, we establish various results on functional convex order; some of them cover and extend existing results, others are new. As a second step, we will tackle the same question for Optimal Stopping problems, namely for the Snell envelopes and their means (*réduites*). Our main motivation is to propose *approximation schemes by simulable discrete time models which preserve this functional convex order.*

G. Pagès (✉)
Laboratoire de Probabilités et Modèles aléatoires, UMR 7599, UPMC, Case 188, 4 pl. Jussieu, F-75252 Paris Cedex 5, France
e-mail: gilles.pages@upmc.fr

C. Donati-Martin et al. (eds.), *Séminaire de Probabilités XLVIII*, Lecture Notes in Mathematics 2168, DOI 10.1007/978-3-319-44465-9_3

Let us first briefly recall that if X and Y are two integrable real-valued random variables, X is dominated by Y for the convex order—denoted $X \preceq_c Y$—if, for every convex function $f : \mathbb{R} \to \mathbb{R}$ such that $f(X), f(Y) \in L^1(\mathbb{P})$,

$$\mathbb{E} f(X) \leq \mathbb{E} f(Y).$$

Thus, if $(M_\lambda)_{\lambda>0}$ denotes a martingale indexed by a parameter λ, then $\lambda \mapsto M_\lambda$ is non-decreasing for the convex order as a straightforward consequence of Jensen's Inequality. The converse is clearly not true but, as first established by Kellerer in [21], whenever $\lambda \mapsto X_\lambda$ is non-decreasing for the convex order, there exists a martingale $(\widetilde{X}_\lambda)_{\lambda\geq 0}$ such that $(\widetilde{X}_\lambda)_{\lambda\geq 0}$ and $(X_\lambda)_{\lambda\geq 0}$ coincide in 1-marginal distributions ($X_\lambda \stackrel{d}{=} \widetilde{X}_\lambda$ for every $\lambda \geq 0$).

The connection with Quantitative Finance—or, to be more precise with the pricing and hedging of derivative products—is straightforward : let $(X_t^{(\theta)})_{t\in[0,T]}$ be a family of non-negative $\mathbb{P}$-martingales on a probability space $(\Omega, \mathscr{A}, \mathbb{P})$ indexed by θ. Such a family can be seen as possible parametrized models for the dynamics of the discounted price of a risky asset under its/a risk-neutral probability depending on θ. Temporarily assume θ is a real parameter, e.g. representative of the volatility. If $\theta \mapsto X_T^{(\theta)}$ is non-decreasing for the convex order, then for every convex *vanilla payoff* function $f : \mathbb{R}_+ \to \mathbb{R}_+$, the function $\theta \mapsto \mathbb{E} f(X_T^{(\theta)})$ is non-decreasing or equivalently, its "*θ-greek*" $\frac{\partial}{\partial\theta}\mathbb{E} f(X_T^{(\theta)})$ is non-negative, if it exists. Thus, in a discounted Black-Scholes model (coming back to the notation σ for the volatility),

$$X_t^{\sigma,x} = x e^{\sigma W_t - \frac{\sigma^2}{2}t}, \; x, \sigma > 0 \quad \text{(where } W \text{ is a standard Brownian motion)},$$

the function $\sigma \longmapsto \mathbb{E} f\big(x\, e^{\sigma W_T - \frac{\sigma^2 T}{2}}\big)$ is non-decreasing since

$$\forall\, \sigma > 0, \quad e^{\sigma W_T - \frac{\sigma^2 T}{2}} \stackrel{\mathscr{L}}{\sim} \Big[e^{W_u - \frac{u}{2}}\Big]_{|u=\sigma^2 T}$$

and $u \mapsto e^{W_u - \frac{u}{2}}$ is a martingale (as well as its composition with $\sigma \mapsto \sigma^2 T$). So $(X_T^{\sigma,x})_{\sigma\geq 0}$ coincides in 1-marginal distributions with a martingale. The same result holds true for the premium of convex *Asian payoff* functions of the form

$$\mathbb{E} f\left(\frac{1}{T}\int_0^T x e^{\sigma W_t - \frac{\sigma^2 t}{2}} dt\right).$$

By contrast, its proof is significantly more involved (see [6] or, more recently, the proof in [13] where an explicit marginals based on the Brownian sheet coinciding in 1-dimensional marginals is exhibited). Both results turn out to be examples of a general result dealing with convex pathwise dependent functionals (see e.g. [13] or [28] where a functional co-monotony argument is used).

A natural question at this stage is to try establishing a *functional version* of these results i.e. when θ is no longer a real number or an $\mathbb{R}^q$-valued vector but lives in a functional space or a space of stochastic processes. A first example of interest is to consider diffusion processes $X^{(\theta)}$, weak solution to a Stochastic Differential Equation (*SDE*)

$$dX_t^{(\theta)} = \theta(t, X_{t-}^{(\theta)})dZ_t,\ X_0^{(\theta)} = x, \quad t \in [0, T],$$

where $Z = (Z_t)_{t\in[0,T]}$ is a *martingale* Lévy process. Then the parameter θ is a continuous function. We will see it can also be a (predictable) stochastic process when considering

$$X_t^{(\theta)} = \int_0^t \theta_s dZ_s, \quad t \in [0, T].$$

As for optimal stopping problems, we deal with the *réduite* of a target process $Y_t = F(t, X^{(\theta),t})$, $t \in [0, T]$, where $X_s^{(\theta),t} = X_{s\wedge t}^{(\theta)}$ denotes the stopped process $X^{(\theta)}$ at t and all the functionals $F(t, .)$ are continuous convex functionals defined on the path space of the process X. In financial modeling, the *functional convex order* as defined above amounts to determine the sign of the *sensitivity* with respect to the parameter θ of a path-dependent American option with payoff $F(t, .)$ at time $t \in [0, T]$, "written" on $X^{(\theta)}$: if the holder of the American option contract exercises the option at time t, she receives the monetary flow $F(t, X^{(\theta),t})$.

More generally, various notions of convex order in Finance are closely related to risk modeling and come out in many other frameworks than the pricing and hedging of derivatives.

Many of these questions have already been investigated for a long time: thus, the first result known to us goes back to Hajek in [10] where convex order is established for Brownian martingale diffusions $X^{(\theta)}$ "parametrized" by their convex diffusion coefficient θ. He could extend the result to drifted diffusions with non-decreasing convex drifts, but only for *non-decreasing* convex functions f of X_T. The first application to the sensitivity of (vanilla) options of both European and American style, is due to [9]. It is shown that premium of an option with convex payoff in a local volatility model with volatility $\sigma(.) \in [\sigma_{\min}, \sigma_{\max}]$, can be lower- and upper-bounded by the premium of the payoffs in a Black-Scholes model with volatilities $\sigma_{\min}$ and $\sigma_{\max}$ respectively. Similar results can be obtained as a consequence of the maximal principle for parabolic *PDE*s and variational inequalities. See also [14] for a result on lookback options.

More recently, in a series of papers (see [1–3]) Bergenthum and Rüschendorf extensively investigated the above mentioned problems (for both fixed maturity and for optimal stopping problems) for various classes of continuous and jump processes, including general semi-martingales in [2]. The comparison is carried out in terms of their triplets of predictable local characteristics, once proved that one propagates convexity. In several of these papers, the convexity is often—but not always (see [1] for the use of an Euler scheme)—propagated directly in continuous time. This is clearly an elegant way to proceed but it also more heavily relies on specific features of the investigated class of processes (see [13]). In this paper, we propose as an alternative a generic and systematic twofold paradigm, which turns out to be efficient for various classes of stochastic dynamics and processes. It is based on a transfer from discrete times to continuous time using functional weak limit theorems "à la Jacod-Shiryaev" (see [17]). To be more precise, it can be described as follows:

- As first step, we establish the propagation of convexity "through" a discrete time dynamics—typically an "abstract" Euler scheme—in a very elementary way for path-dependent convex functionals relying on repeated elementary backward inductions and conditional Jensen's inequality. These inductions take advantage of the *linear* backward dynamical programming principle obtained by writing the step-by-step discrete time martingale property.
- As a second step, we consider time discretization schemes of the "target" continuous time dynamics (typically a standard Euler schemes for diffusion processes) to transfer to this target process the propagation of convexity property (in a functional form) by calling upon functional weak limit theorems (typically borrowed from [18] and/or [23]). A similar approach applies for stochastic integrals and Snell envelope in optimal stopping theory.

One important motivation for developing in a systematic manner this approach is related to Numerical Probability: the discrete time model in the first step often is a *simulable* discretization scheme of the continuous time dynamics of interest. It is important for applications, especially in Finance, to have at hand discretization schemes which *both* preserve convex order and can be simulated at a reasonable cost. So is the case of the Euler scheme of Lévy driven diffusions (provided the underling Lévy measure is itself simulable). But, for example, it is not true for the Milstein scheme for Brownian diffusions, in spite of its better performances in term of strong convergence rate.

Let us give a typical result that we obtain, here for jump diffusions (a more general statement is established in Theorems 1 and 2 in Sect. 2). Let $Z = (Z_t)_{t\in[0,T]}$ be a square integrable martingale Lévy process with Lévy measure ν. Let $\kappa_i : \mathbb{R} \to \mathbb{R}_+$, $i = 1, 2$, be continuous functions with linear growth and let $\kappa : \mathbb{R} \to \mathbb{R}_+$ be a convex function such that $0 \leq \kappa_1 \leq \kappa \leq \kappa_2$. Then the existing weak solutions $X^{(\kappa_i)}$, $i = 1, 2$, to the *SDE*s

$$X_t^{(\kappa_i)} = x + \int_{(0,t]} \kappa_i(X_{s-}^{(\kappa_i)})dZ_s$$

satisfy $X^{(\kappa_1)} \preceq_{fc} X^{(\kappa_2)}$ for the functional convex order defined on the Skorokhod space $\mathbb{D}([0,T],\mathbb{R})$ of right continuous left limited functions defined on $[0,T]$. Namely, for every convex functional $F : \mathbb{D}([0,T],\mathbb{R}) \to \mathbb{R}$, $\mathbb{P}_{X^{(\kappa_i)}}$-continuous, $i = 1, 2$, for the Skorokhod topology and with polynomial growth for the sup-norm,

$$\mathbb{E}\, F(X^{(\kappa_1)}) \leq \mathbb{E}\, F(X^{(\kappa_2)}).$$

Another way to formulate the result can be: if we assume this time that both κ_i, $i = 1, 2$, are convex, but possibly not κ, then $X^{(\kappa_1)} \preceq_{fc} X^{(\kappa)} \preceq_{fc} X^{(\kappa_2)}$.

Note that when $Z = W$ is a Brownian motion, on can consider functional F on the space $\mathscr{C}([0,T],\mathbb{R})$ of continuous functions and the continuity of the functional F is a consequence of its convexity under the polynomial growth assumption (see the remark in Sect. 2.2).

Results in the same spirit are obtained for stochastic integrals, Doléans exponentials (which unfortunately requires one of the two integrands H_1 and H_2 to be deterministic). Counter-examples to put the main results in perspective are exhibited to prove the consistency of these assumptions in both settings (see also [12]) for more counterexamples).

When we deal with optimal stopping problems, we use the same approach, taking advantage in discrete time of the classical Backward Dynamic Programming Principle (see also [11] in stochastic control) and using various convergence results for the Snell envelope (see [25]).

The paper is organized as follows. Section 2 is devoted to functional convex order for path-dependent functionals of Brownian and Lévy driven martingale diffusion processes. Section 3 is devoted to comparison results for Itô processes based on comparison of their integrands. Section 4 deals with *réduites*, Snell envelopes of path-dependent obstacle processes (American options) in both Brownian and Lévy driven martingale diffusions. In the twofold appendix, we provide short proofs of functional weak convergence of the Euler scheme toward a weak solution of *SDE*s in both Brownian and Lévy frameworks under natural continuity and linear growth assumptions on the diffusion coefficient.

Notations

- For every $T > 0$ and every integer $n \geq 1$, one denotes the uniform mesh of $[0,T]$ by $t_k^n = \frac{kT}{n}$, $k = 0, \ldots, n$. Then for every $t \in [\frac{kT}{n}, \frac{(k+1)T}{n})$, we set $\underline{t}_n = \frac{kT}{n}$ and $\bar{t}^n = \frac{(k+1)T}{n}$ with the convention $\underline{T}_n = T$. We also set $\underline{t}_{n-} = \lim_{s \to t} \underline{s}_n = \frac{kT}{n}$ if $t \in \left(\frac{kT}{n}, \frac{(k+1)T}{n}\right]$.
- For every $u = (u_1, \ldots, u_d)$, $v = (v_1, \ldots, v_d) \in \mathbb{R}^d$, $(u|v) = \sum_{i=1}^d u_i v_i$, $|u| = \sqrt{(u|u)}$.
- $x_{m:n} = (x_m, \ldots, x_n)$ (where $m \leq n$, $m, n \in \mathbb{N} \setminus \{0\}$).
- $\mathscr{F}([0,T],\mathbb{R})$ denotes the $\mathbb{R}$-vector space of $\mathbb{R}$-valued functions $f : [0,T] \to \mathbb{R}$ and $\mathscr{C}([0,T],\mathbb{R})$ denotes the subspace of $\mathbb{R}$-valued continuous functions defined over $[0,T]$.
- For every $\alpha \in \mathscr{F}([0,T],\mathbb{R})$, we define $\mathrm{Cont}(\alpha) = \big\{t \in [0,T] : \alpha \text{ is continuous at } t\big\}$ with the usual left- and right-continuity conventions at 0 and T respectively. We also define the *uniform continuity modulus* of α by $w(\alpha,\delta) = \sup\big\{|\alpha(u) - \alpha(v)|,\ u, v \in [0,T], |u - v| \leq \delta\big\}$ ($\delta \in [0,T]$).
- $L_T^p = L^p([0,T],dt)$, $1 \leq p \leq +\infty$, $|f|_{L_T^p} = \left(\int_0^T |f(t)|^p dt\right)^{\frac{1}{p}} \leq +\infty$, $1 \leq p < +\infty$ and $|f|_{L_T^\infty} = dt\text{-ess}\sup|f|$ where dt stands for the Lebesgue measure on $[0,T]$ equipped with its Borel σ-field.

- For a function $f : [0,T] \to \mathbb{R}$, we denote $\|f\|_{\sup} = \sup_{t\in[0,T]} |f(t)|$.
- Let $(\Omega, \mathcal{A}, \mathbb{P})$ be a probability space and let $p \in (0, +\infty)$. For every random vector $X : (\Omega, \mathcal{A}) \to \mathbb{R}^d$ we set $\|X\|_p = \big(\mathbb{E}|X|^p\big)^{\frac{1}{p}}$. $L^p_{\mathbb{R}^d}(\Omega, \mathcal{A}, \mathbb{P})$ denotes the vector space of (classes) of $\mathbb{R}^d$-valued random vectors X such that $\|X\|_p < +\infty$. $\|\,.\,\|_p$ is a norm on $L^p_{\mathbb{R}^d}(\Omega, \mathcal{A}, \mathbb{P})$ for $p \in [1, +\infty)$ (the mention of Ω, $\mathcal{A}$ and the subscript $_{\mathbb{R}^d}$ will be dropped when there is no ambiguity).
- If $\mathcal{F} = (\mathcal{F}_t)_{t\in[0,T]}$ denotes a filtration on $(\Omega, \mathcal{A}, \mathbb{P})$, let $\mathcal{T}^{\mathcal{F}}_{[0,T]} = \{\tau : \Omega \to [0,T], \mathcal{F}\text{-stopping time}\}$.
- $\mathcal{F}^Y = (\mathcal{F}^Y_t)_{t\in[0,T]}$ denotes the smallest right continuous filtration $(\mathcal{G}_t)_{t\in[0,T]}$ that makes the process $Y = (Y_t)_{t\in[0,T]}$ a $(\mathcal{G}_t)_{t\in[0,T]}$-adapted process.
- $\mathbb{D}([0,T], \mathbb{R}^d)$ denotes the set of $\mathbb{R}^d$-valued right continuous left limited (or càdlàg following the French acronym) functions defined on the interval $[0,T]$, $T > 0$. It is usually endowed with the Skorokhod topology denoted Sk (see [16, Chap. VI] or [4, Chap. 3], for an introduction to Skorokhod topology).
- If two random vectors U and V have the same distribution, we write $U \stackrel{d}{\sim} V$.
- The *weak convergence* (or *convergence in distribution* or *in law*) of a sequence $(Y_n)_{n\geq 1}$ of random variables having values in a Polish (S, d) equipped with its Borel σ-field $\mathcal{B}or(S)$ toward an (S,d)-valued random variable Y_∞ will be denoted by $Y_n \stackrel{\mathcal{L}(S,d_S)}{\longrightarrow} Y_\infty$ or, if no ambiguity, $Y_n \stackrel{\mathcal{L}(d_S)}{\longrightarrow} Y_\infty$.

We will make extensive use of the following classical result:

Let $(Y_n)_{n\geq 1}$ be a sequence of tight random variables taking values in a Polish space (S, d_s) (see [4, Chap. 1]). If Y_n weakly converges toward Y_∞ and $(\Phi(Y_n))_{n\geq 1}$ is uniformly integrable where $\Phi : S \to \mathbb{R}$ is a Borel function, then, for every $\mathbb{P}_{Y_\infty}$-*a.s.* continuous Borel functional $F : S \to \mathbb{R}_+$ such that $|F(u)| \leq C(1 + \Phi(u))$ for every $u \in S$, one has $\mathbb{E}\,F(Y_n) \to \mathbb{E}\,F(Y_\infty)$.

2 Functional Convex Order

2.1 Propagation of Convexity in Discrete Time Recursive Model

In this section, we give simple conditions to propagate convexity "through" a discrete time recursive "abstract" Euler scheme, simulable if Z is. The results are presented in the proposition below. It is the key result to be transferred to continuous time models, using weak approximation methods.

When the scheme (1) below is simulable, one may implement Monte Carlo simulations preserving convex order. This can be crucial when dealing with option pricing (see Sect. 2.3).

Proposition 1 *Let $(Z_k)_{1\leq k\leq n}$ be a sequence of independent, centered, $\mathbb{R}$-valued random vectors lying in $L^r(\Omega, \mathcal{A}, \mathbb{P})$, $r \geq 1$, and let $(\mathcal{F}^Z_k)_{k=0,\ldots,n}$ denote its natural* *filtration. Let* $(X_k)_{k=0,\ldots,n}$ *and* $(Y_k)_{k=0,\ldots,n}$ *be two sequences of random vectors recursively defined by*

$$X_{k+1} = X_k + \sigma_k(X_k)Z_{k+1},\ Y_{k+1} = Y_k + \theta_k(Y_k)Z_{k+1},\ 0 \le k \le n-1,\ X_0 = Y_0 = x \tag{1}$$

where σ_k, $\theta_k : \mathbb{R} \to \mathbb{R}$, $k = 0, \ldots, n-1$, *are Borel functions with linear growth i.e.* $|\sigma_k(x)| + |\theta_k(x)| \le C(1 + |x|)$, $x \in \mathbb{R}$, *for a real constant* $C \ge 0$.

(*a*) *Assume that, either* σ_k *is convex for every* $k = 0, \ldots, n-1$, *or* θ_k *is convex for every* $k = 0, \ldots, n-1$, *and that*

$$\forall\, k \in \{0, \ldots, n-1\},\ \ 0 \le \sigma_k \le \theta_k.$$

Then, for every convex function $\Phi : \mathbb{R}^{n+1} \to \mathbb{R}$ *with r-polynomial growth,* $r \ge 1$, *i.e. satisfying* $|\Phi(x)| \le C(1 + |x|^r)$, $x \in \mathbb{R}$, *for a real constant* $C \ge 0$,

$$\mathbb{E}\, \Phi(X_{0:n}) \le \mathbb{E}\, \Phi(Y_{0:n}).$$

(*b*) *If the random variable* Z_k *have symmetric distributions, if the functions* θ_k *are all convex and if*

$$\forall\, k \in \{0, \ldots, n-1\},\ \ |\sigma_k| \le \theta_k,$$

then the conclusion of claim (*a*) *remains valid.*

The proof relies on two ingredients: the first one is a simple revisited version of the celebrated Jensen Inequality, the second one is a "linear" Backward Dynamic Programming formula to the step-by-step dynamics of discrete time martingale (close in spirit of a binomial tree).

Lemma 1 *(Revisited Jensen's Lemma) Let* $Z : (\Omega, \mathcal{A}, \mathbb{P}) \to \mathbb{R}$ *be an integrable centered* $\mathbb{R}$*-valued random vector.*

(*a*) *Assume that* $Z \in L^r(\mathbb{P})$ *for an* $r \ge 1$. *For every Borel function* $\varphi : \mathbb{R} \to \mathbb{R}$ *such that* $|\varphi(x)| \le C(1 + |x|^r)$, $x \in \mathbb{R}$, *we define a function* $Q\varphi$ *by:*

$$\forall\, u \in \mathbb{R},\ Q\varphi(u) = \mathbb{E}\, \varphi\big(uZ\big). \tag{2}$$

If φ *is convex, then,* $Q\varphi$ *is convex and* $u \mapsto Q\varphi(u)$ *is non-decreasing on* $\mathbb{R}_+$, *non-increasing on* $\mathbb{R}_-$.

(*b*) *If* Z *has exponential moments in the sense that* $\mathbb{E}(e^{uZ}) < +\infty$ *for every* $u \in \mathbb{R}$, *or equivalently*

$$\forall\, a \ge 0,\ \mathbb{E}(e^{a|Z|}) < +\infty,$$

then claim (a) holds true for any convex function $\varphi : \mathbb{R} \to \mathbb{R}$ satisfying an exponential growth condition of the form $|\varphi(x)| \leq Ce^{C|x|}$, $x \in \mathbb{R}$, for a real constant $C \geq 0$.

(*c*) *If Z has a symmetric distribution (i.e. $Z \overset{d}{\sim} -Z$) and $\varphi : \mathbb{R} \to \mathbb{R}$ is convex, then $Q\varphi$ is an even function, hence satisfying the following* maximum principle*:*

$$\forall\, a \in \mathbb{R}_+, \quad \sup_{|u|\leq a} Q\varphi(u) = Q\varphi(a).$$

Proof (*a*)–(*b*) Existence and convexity of $Q\varphi$ are obvious. The function $Q\varphi$ is clearly finite on $\mathbb{R}$ and convex. Furthermore, Jensen's Inequality implies that

$$Q\varphi(u) = \mathbb{E}\,\varphi(uZ) \geq \varphi(\mathbb{E}\,uZ) = \varphi(0) = Q\varphi(0)$$

since Z is centered. Hence $Q\,\varphi$ is convex and minimum at $u = 0$ which implies that it is non-increasing on $\mathbb{R}_-$ and non-decreasing on $\mathbb{R}_+$.

(*c*) is obvious given the proof of (*a*)–(*b*).

We will establish in this first step the result for two *ARCH*(1) models living on the same probability space.

Proof (Proof of Proposition 1.)

(*a*) First we show by an easy induction that the random variables X_k and Y_k all lie in L^r. Let Q_k, $k = 1, \ldots, n$, denote the operator attached to Z_k by (2) in Lemma 1.

Then, one defines the following martingales

$$M_k = \mathbb{E}\big(\Phi(X_{0:n}) \,|\, \mathcal{F}_k^Z\big) \quad \text{and} \quad N_k = \mathbb{E}\big(\Phi(Y_{0:n}) \,|\, \mathcal{F}_k^Z\big),\ 0 \leq k \leq n.$$

Their existence follows from the growth assumptions on Φ, σ_k and θ_k, $k = 1, \ldots, n$. Now we define recursively in a backward way two sequences of functions Φ_k and $\Psi_k : \mathbb{R}^{k+1} \to \mathbb{R}$, $k = 0, \ldots, n$, by

$$\Phi_n = \Phi \text{ and } \Phi_k(x_{0:k}) = \big(Q_{k+1}\Phi_{k+1}(x_{0:k}, x_k+.)\big)(\sigma_k(x_k)),\ x_{0:k} \in \mathbb{R}^{k+1},\ k = 0, \ldots, n-1,$$

on the one hand and, on the other hand,

$$\Psi_n = \Phi \text{ and } \Psi_k(x_{0:k}) = \big(Q_{k+1}\Psi_{k+1}(x_{0:k}, x_k+.)\big)(\theta_k(x_k)),\ x_{0:k} \in \mathbb{R}^{k+1},\ k = 0, \ldots, n-1.$$

This appears as a *linear* Backward Dynamical Programming Principle. It is clear by a (first) backward induction and the definition of the operators Q_k that, for every $k \in \{0, \ldots, n\}$,

$$M_k = \Phi_k(X_{0:k}) \quad \text{and} \quad N_k = \Psi(Y_{0:k}).$$

Let $k \in \{0, \ldots, n-1\}$. One derives from the properties of the operator Q_{k+1} (and the representation below as an expectation) that, for any convex function $G : \mathbb{R}^{k+2} \to \mathbb{R}$ with r-polynomial growth, $r \geq 0$, the function

$$\widetilde{G} : (x_{0:k}, u) \longmapsto \big(Q_{k+1} G(x_{0:k}, x_k + \,.\,)\big)(u) = \mathbb{E}\, G(x_{0:k}, x_k + uZ_{k+1}) \tag{3}$$

is convex. Moreover, owing to Lemma 1(a), for fixed $x_{0:k}$, $\widetilde{G}$ is non-increasing on $(-\infty, 0)$, non-decreasing on $(0, +\infty)$ as a function of u. In turn, this implies that, if $\gamma : \mathbb{R} \to \mathbb{R}_+$ is convex (and non-negative), then $\xi \mapsto \widetilde{G} \circ \gamma(\xi) = Q_{k+1} G(x_{0:k}, x_k + .)\big(\gamma(\xi)\big)$ is convex in ξ.

▷ Assume all the functions σ_k, $k=0, \ldots, n-1$, are non-negative and convex. One shows by a (second) backward induction that the functions Φ_k are all convex.

Finally, we prove by a (third) backward induction on k that $\Phi_k \leq \Psi_k$ for every $k = 0, \ldots, n-1$. First note that $\Phi_n = \Psi_n = \Phi$. Now assume that $\Phi_{k+1} \leq \Psi_{k+1}$. Then

$$\begin{aligned}
\Phi_k(x_{0:k}) &= \big(Q_{k+1} \Phi_{k+1}(x_{0:k}, x_k + .)\big)(\sigma_k(x_k)) \\
&\leq \big(Q_{k+1} \Phi_{k+1}(x_{0:k}, x_k + .)\big)(\theta_k(x_k)) \\
&\leq \big(Q_{k+1} \Psi_{k+1}(x_{0:k}, x_k + .)\big)(\theta_k(x_k)) \\
&= \Psi_k(x_{0:k})
\end{aligned}$$

which completes the induction. In particular, when $k = 0$, we get $\Phi_0(x) \leq \Psi_0(x)$ or, equivalently, taking advantage of the martingale property, $\mathbb{E}\, \Phi(X_{0:n}) \leq \mathbb{E}\, \Phi(Y_{0:n})$.

▷ If all the functions θ_k, $k = 0, \ldots, n-1$ are convex, then all the functions Ψ_k, $k = 0, \ldots, n$, are convex and one shows likewise by induction that $\Phi_k \leq \Psi_k$ for every $k = 0, \ldots, n-1$.

(b) The proof follows the same lines as (a) calling upon Claim (c) of Lemma 1. In particular, the functions $u \mapsto \widetilde{G}(x_{0:k}, u)$ is also even so that $\sup_{u \in [-a,a]} \widetilde{G}(x_{0:k}, u) = \widetilde{G}(x_{0:k}, a)$ for any $a \geq 0$.

2.2 *Brownian Martingale Diffusion*

The main result of this section is the theorem below which show that martingale Brownian diffusions satisfy a functional convex order principle.

Theorem 1 *Let σ, θ : $[0, T] \times \mathbb{R} \to \mathbb{R}$ be two continuous functions with linear growth in x uniformly in $t \in [0, T]$. Let $X^{(\sigma)}$ and $X^{(\theta)}$ be two Brownian martingale diffusions, supposed to be the unique* weak *solutions starting from x at time 0, to the* *stochastic differential equations with zero drift*

$$dX_t^{(\sigma)} = \sigma(t, X_t^{(\sigma)})dW_t^{(\sigma)},\ X_0^{(\sigma)} = x \ \text{ and } \ dX_t^{(\theta)} = \theta(t, X_t^{(\theta)})dW_t^{(\theta)},\ X_0^{(\theta)} = x, \tag{4}$$

$t \in [0, T]$, respectively, where $W^{(\sigma)}$ and $W^{(\theta)}$ are standard one dimensional Brownian motions (possibly defined on different probability spaces).

(a) Convex Partitioning assumption*: Let $\kappa : [0, T] \times \mathbb{R} \to \mathbb{R}_+$ be a continuous function with (at most) linear growth in x uniformly in $t \in [0, T]$, satisfying*

$$\kappa(t, .) \text{ is convex for every } t \in [0, T] \text{ and } 0 \le \sigma \le \kappa \le \theta.$$

Then, for every convex functional $F : \mathscr{C}([0, T], \mathbb{R}) \to \mathbb{R}$ with $(r, \|\,.\,\|_{\sup})$-polynomial growth, $r \ge 1$, in the following sense

$$\forall\, \alpha \in \mathscr{C}([0, T], \mathbb{R}), \quad |F(\alpha)| \le C(1 + \|\alpha\|^r_{\sup}),$$

one has

$$\mathbb{E}\, F(X^{(\sigma)}) \le \mathbb{E}\, F(X^{(\theta)}).$$

From now on, the function κ will be called a convex partitioning function.

(a') *Claim (a) can be reformulated equivalently as follows: if either $\sigma(t, .)$ is convex for every $t \in [0, T]$ or $\theta(t, .)$ is convex for every $t \in [0, T]$ and $0 \le \sigma \le \theta$, then the conclusion of (a) still holds true.*

(b) Convex Domination assumption*: If $|\sigma| \le \theta$ and $\theta(t, .)$ is convex for every $t \in [0, T]$, then*

$$\mathbb{E}\, F(X^{(\sigma)}) \le \mathbb{E}\, F(X^{(\theta)}).$$

Remark 1

- The linear growth assumption made on the convex functional F implies its everywhere local boundedness on the Banach space $\big(\mathscr{C}([0, T], \mathbb{R}), \|\,.\,\|_{\sup}\big)$, hence its $\|\,.\,\|_{\sup}$-continuity (see e.g. Lemma 2.1.1 in [26, p. 22]).
- The introduction of two standard Brownian motions $W^{(\sigma)}$ and $W^{(\theta)}$ in claim (a) is just a way to recall that the two diffusions processes can be defined on different probability spaces, though it may be considered as an abuse of notation. By "unique weak solutions", we mean classically that any such two solutions, possibly defined on different probability spaces with respect to different Brownian motions, share the same distribution on the Wiener space.
- When strong uniqueness holds, typically because σ and θ are also Lipschitz continuous in x, uniformly in $t \in [0, T]$ (see e.g. Theorem A.3.3, p. 273, in [5]), then weak uniqueness holds as well.

- The extension of this purely one-dimensional result to a diffusion where σ, θ and the Brownian motion(s) are q-dimensional and the regular product is replaced by the canonical inner product in $\mathbb{R}^q$ is straightforward: the assumptions on σ and θ should simply be understood *component-wise*.

The proof of this theorem can be decomposed in two main steps: the first one consists in applying Proposition 1 to the Euler schemes of both diffusions, the second one relies on the functional weak convergence of the Euler schemes toward the diffusions in order to transport the functional convex order property. To this end, we introduce the notion of *piecewise affine interpolator* and recall an elementary weak convergence lemma. This remark applies throughout the paper.

Definition 1

(a) For every integer $n \geq 1$, let $i_n : \mathbb{R}^{n+1} \to \mathscr{C}([0,T],\mathbb{R})$ denote the *piecewise affine interpolator* defined by

$$\forall\, x_{0:n} \in \mathbb{R}^{n+1}, \forall\, k=0,\dots,n-1, \forall t \in [t_k^n, t_{k+1}^n],\; i_n(x_{0:n})(t) = \frac{n}{T}\big((t_{k+1}^n - t)x_k + (t - t_k^n)x_{k+1}\big).$$

(b) For every $n \geq 1$, let $I_n : \mathscr{F}([0,T],\mathbb{R}) \to \mathscr{C}([0,T],\mathbb{R})$ denote the *functional interpolator* defined by

$$\forall\, \alpha \in \mathscr{F}([0,T],\mathbb{R}), \quad I_n(\alpha) = i_n\big(\alpha(t_0^n),\dots,\alpha(t_n^n)\big).$$

For uniform integrability purpose, we will use extensively the following obvious fact

$$\sup_{t\in[0,T]} |I_n(\alpha)_t| \leq \sup_{t\in[0,T]} |\alpha(t)|.$$

Lemma 2 *Let X^n, $n \geq 1$, be a sequence of continuous processes weakly converging towards X for the $\|\,.\,\|_{\sup}$-norm. Then the sequence of interpolating processes $\widetilde{X}^n = I_n(X^n)$, $n \geq 1$, is weakly converging toward X for the $\|\,.\,\|_{\sup}$-norm topology.*

Proof For every integer $n \geq 1$ and every $\alpha \in \mathscr{F}([0,T],\mathbb{R}^d)$, the interpolation operators $I_n(\alpha)$ reads

$$I_n(\alpha) = \frac{n}{T}\big((t_{k+1}^n - t)\alpha(t_k^n) + (t - t_k^n)\alpha(t_{k+1}^n)\big),\; t \in [t_k^n, t_{k+1}^n],\; k = 0,\dots,n-1.$$

Note that I_n maps $\mathscr{C}([0,T],\mathbb{R}^d)$ into itself. One easily checks that $\|I_n(\alpha) - \alpha\|_{\sup} \leq w(\alpha, T/n)$, where w denotes the uniform continuity modulus of α, and $\|I_n(\alpha) - I_n(\beta)\|_{\sup} \leq \|\alpha - \beta\|_{\sup}$. We use the standard distance d_{wk} for weak convergence on Polish metric spaces defined by

$$d_{wk}\big(\mathscr{L}(X),\mathscr{L}(Y)\big) = \sup\big\{|\mathbb{E}\,F(X) - \mathbb{E}\,F(Y)|,\; [F]_{\text{Lip}} \leq 1,\; \|F\|_{\sup} \leq 1\big\}.$$

Then, X having continuous paths,

$$d_{wk}\big(\mathcal{L}(I_n(X^n)),\mathcal{L}(X)\big) \le d_{wk}\big(\mathcal{L}(I_n(X^n)),\mathcal{L}(I_n(X))\big) + d_{wk}\big(\mathcal{L}(I_n(X)),\mathcal{L}(X)\big)$$
$$\le d_{wk}\big(\mathcal{L}(X^n),\mathcal{L}(X)\big) + \mathbb{E}\left(w(X,T/n)\wedge 2\right) \to 0$$

as $n \to +\infty$

Proof (Proof of Theorem 1)

(a') We consider the continuous, or *genuine*, Euler schemes $\bar{X}^{(\sigma)}$ and $\bar{X}^{(\theta)}$ with step $\frac{T}{n}$, starting at x associated to the diffusion coefficients σ and θ and a given standard Brownian motion W defined on an appropriate probability space. Thus, the Euler scheme related to $X^{(\sigma)}$ reads:

$$\bar{X}^{(\sigma),n}_{t^n_{k+1}} = \bar{X}^{(\sigma),n}_{t^n_k} + \sigma(t^n_k,\bar{X}^{(\sigma),n}_{t^n_k})\big(W_{t^n_{k+1}} - W_{t^n_k}\big),\ k = 0,\dots,n-1,\ \bar{X}^{(\sigma),n}_0 = x$$
$$\bar{X}^{(\sigma),n}_t = \bar{X}^{(\sigma),n}_{t^n_k} + \sigma(t^n_k,\bar{X}^{(\sigma),n}_{t^n_k})\big(W_t - W_{t^n_k}\big),\quad t\in [t^n_k,t^n_{k+1}).$$

It is clear that both sequences $(\bar{X}^{(\sigma),n}_{t^n_k})_{k=0:n}$ and $(\bar{X}^{(\theta),n}_{t^n_k})_{k=0:n}$ are of the form (1) with $Z_k = W_{t^n_k} - W_{t^n_{k-1}}$, $k = 1,\dots,n$. Furthermore, owing to the linear growth assumption made on σ and θ, the sup-norm of these Euler schemes of Brownian diffusions lie in $L^p(\mathbb{P})$ for any $p\in(0,+\infty)$, uniformly in n, (see e.g. Lemma B.1.2, p. 275 in [5] or Proposition 10 in Appendix 1)

$$\sup_{n\ge1}\Big\|\sup_{t\in[0,T]}|\bar{X}^{(\sigma),n}_t|\Big\|_p + \sup_{n\ge1}\Big\|\sup_{t\in[0,T]}|\bar{X}^{(\theta),n}_t|\Big\|_p < +\infty.$$

Furthermore, $I_n(\bar{X}^{(\sigma),n}) = i_n\big((\bar{X}^{(\sigma),n}_{t^n_k})_{k=0:n}\big)$ is but the piecewise affine interpolated Euler scheme (which coincides with $\bar{X}^{(\sigma),n}$ at times t^n_k). Note that the sup-norm of $I_n(\bar{X}^{(\sigma),n})$ also has finite polynomial moments uniformly in n like the genuine Euler scheme.

Let $F : \mathscr{C}([0,T],\mathbb{R}) \to \mathbb{R}$ be a convex functional with $(r,\|\,.\,\|_{\sup})$-polynomial growth. For every integer $n\ge1$, we define on $\mathbb{R}^{n+1}$ the function F_n by

$$F_n(x_{0:n}) = F\big(i_n(x_{0:n})\big),\ x_{0:n}\in\mathbb{R}^{n+1}. \tag{5}$$

It is clear that the convexity of F on $\mathscr{C}([0,T],\mathbb{R})$ is transferred to the functions F_n, $n\ge1$ as well as the polynomial growth property. Moreover, F is $\|\,.\,\|_{\sup}$-continuous since it is convex with $\|\,.\,\|_{\sup}$-polynomial growth (see Lemma 2.1.1 in [26]). It follows from Proposition 1 applied with $\Phi = F_n$, $(Z_k)_{k=1:n} = (W_{t^n_k} - W_{t^n_{k-1}})_{k=1:n}$, $\sigma_k = \sigma(t^n_k,.)$ and $\theta_k = \theta(t^n_k,.)$, $k = 0,\dots,n-1$ which obviously satisfy the required linear growth and integrability assumptions, that, for every $n\ge1$,

$$\mathbb{E}\,F\big(I_n(\bar{X}^{(\sigma),n})\big) = \mathbb{E}\,F_n\big((\bar{X}^{(\sigma),n}_{t^n_k})_{k=0:n}\big) \le \mathbb{E}\,F_n\big((\bar{X}^{(\theta),n}_{t^n_k})_{k=0:n}\big) = \mathbb{E}\,F\big(I_n(\bar{X}^{(\theta),n})\big). \tag{6}$$

On the other hand, it is classical background that the *genuine (continuous) Euler schemes* $\bar{X}^{(\sigma),n}$ weakly converges for the $\|\,.\,\|_{\sup}$-norm topology toward $X^{(\sigma)}$ as $n \to +\infty$, unique weak solution to the *SDE* $\equiv dX_t = \sigma(X_t)dW_t$, $X_0 = x$. For a proof we refer e.g. to [30] (Exercise 23, p. 359) when σ is homogeneous in t, see also [18, 23]; a self-contained proof is provided in Proposition 10 in Appendix 1. All proofs rely on the weak convergence theorem for stochastic integrals first established in [18].

One derives from Lemma 2 and the $L^p(\mathbb{P})$-boundedness of the sup-norm of the sequence $(I_n(\bar{X}^{(\sigma),n}))_{n\geq 1}$ for $p > r$ that

$$\mathbb{E}\,F(X^{(\sigma)}) = \lim_n \mathbb{E}\,F\big(I_n(\bar{X}^{(\sigma),n})\big) = \lim_n \mathbb{E}\,F_n\big((\bar{X}^{(\sigma),n}_{t^n_k})_{k=0:n}\big).$$

The same holds true for the diffusion $X^{(\theta)}$ and its Euler scheme. The conclusion follows.

(a) Applying what precedes to both couples (σ, κ) and (κ, θ) until Eq. (6), we derive that

$$\mathbb{E}\,F\big(I_n(\bar{X}^{(\sigma),n})\big) \leq \mathbb{E}\,F\big(I_n(\bar{X}^{(\kappa),n})\big) \leq \mathbb{E}\,F\big(I_n(\bar{X}^{(\theta),n})\big).$$

One concludes likewise by letting n go to infinity in the resulting inequality

$$\mathbb{E}\,F\big(I_n(\bar{X}^{(\sigma),n})\big) \leq \mathbb{E}\,F\big(I_n(\bar{X}^{(\theta),n})\big).$$

(b) The proof follows the same lines by calling upon item (c) of the above Lemma 1, having in mind that the distribution of a standard Brownian increment is symmetric with polynomial moments at any order as a Gaussian random vector.

Remark 2

- Note that the *SDE* related to κ do not appear in this theorem.
- The Euler scheme has already been successfully used to establish convex order in [1].

The following corollaries hold when considering the *SDE* associated to κ, with an obvious proof.

Corollary 1 *Under the assumption of Claim* (a) *in Theorem 1 and if, furthermore, the SDE*

$$dX^{(\kappa)}_t = \kappa(t, X^{(\kappa)}_t)dW_t,\ X^{(\kappa)}_0 = x,\ t \in [0, T],$$

has a unique weak solution, then, for every convex functional $F : \mathscr{C}([0,T], \mathbb{R}) \to \mathbb{R}$ *with* $(r, \|\,.\,\|_{\sup})$*-polynomial growth,*

$$\mathbb{E}\,F(X^{(\sigma)}) \leq \mathbb{E}\,F(X^{(\kappa)}) \leq \mathbb{E}\,F(X^{(\theta)}).$$

Corollary 2 *If $\sigma, \theta : [0,T] \times I \to \mathbb{R}$, where I is a nontrivial interval of $\mathbb{R}$, are continuous with polynomial growth and if the related Brownian SDEs satisfy a weak uniqueness assumption for every I-valued weak solution starting from $x \in I$, at time $t=0$, the above theorem and results remain true.*

This approach based on the combination of a (linear) dynamic programming principle and a functional weak approximation argument also allows us to retrieve Hajek's result for drifted diffusions.

Proposition 2 (Extension to Drifted Diffusions, See [10], Theorem 4.1) *Let σ and θ be two functions on $[0,T]\times\mathbb{R}$ satisfying the convex partitioning or dominating assumptions (a) or (b) from Theorem 1 respectively. Let $b : [0,T] \times \mathbb{R} \to \mathbb{R}$ be a continuous function with linear growth in x uniformly in t and such that $b(t,.)$ is convex for every $t \in [0,T]$. Let $Y^{(\sigma)}$ and $Y^{(\theta)}$ be the* weak *solutions, supposed to be unique, starting from x at time 0 to the SDEs $dY_t^{(\sigma)} = b(t, Y_t^{(\sigma)})dt + \sigma(t, Y_t^{(\sigma)})dW_t^{(\sigma)}$ and $dY_t^{(\theta)} = b(t, Y_t^{(\theta)})dt + \theta(t, Y_t^{(\theta)})dW_t^{(\theta)}$. Then, for every* non-decreasing *convex function $f : \mathbb{R} \to \mathbb{R}$ with polynomial growth,*

$$\mathbb{E} f(X_T^{(\sigma)}) \le \mathbb{E} f(X_T^{(\theta)}).$$

Proof We have to introduce the operators $Q_{b,\gamma,t}$, $\gamma > 0$, $t \in [0,T]$, defined for every Borel function $f : \mathbb{R} \to \mathbb{R}$, satisfying the appropriate polynomial growth assumption in accordance with the existing moments of Z, by

$$Q_{b,\gamma,t}(f)(x,u) = \mathbb{E} f\big(x + \gamma b(t,x) + uZ\big).$$

One shows like in Lemma 1 that, if the function f is convex and non-decreasing, $Q_{b,\gamma,t}f$ is convex in (x,u), non-decreasing in u on $\mathbb{R}_+$, non-increasing in $u \in \mathbb{R}_-$. The extension of the functional weak convergence of the Euler scheme established in Proposition 10 of Appendix 1 under the above assumptions made on the drift b provides the "transfer". Details are left to the reader.

2.3 Applications to (Brownian) Functional Peacocks and Option Pricing

We consider a so-called local volatility model on the dynamics of a discounted risky asset given by

$$dS_t^{(\sigma)} = S_t^{(\sigma)} \sigma\big(t, S_t^{(\sigma)}\big) dW_t^{(\sigma)}, \; S_0^{(\sigma)} = s_0 > 0, \tag{7}$$

where $\sigma : [0,T] \times \mathbb{R} \to \mathbb{R}$ is a bounded continuous function. The above equation has at least a weak solution $(S_t^{(\sigma)})_{t\in[0,T]}$ defined on a probability space $(\Omega, \mathcal{A}, \mathbb{P})$ on which lives a Brownian motion $(W_t^{(\sigma)})_{\in[0,T]}$ (see Proposition 10 in Appendix 1, see also [31]). Then, $(S_t^{(\sigma)})_{t\in[0,T]}$ is a true $(\mathcal{F}_t^{W^{(\sigma)}})_{t\in[0,T]}$-martingale satisfying

$$S_t^{(\sigma)} = s_0 \exp\Big(\int_0^t \sigma(s, S_s^{(\sigma)})dW_s^{(\sigma)} - \frac{1}{2}\int_0^t \sigma^2(s, S_s^{(\sigma)})ds\Big)$$

so that $S_t^{(\sigma)} > 0$ for every $t \in [0,T]$. One introduces likewise the local volatility model $(S_t^{(\theta)})_{t\in[0,T]}$ related to the bounded volatility function $\theta : [0,T] \times \mathbb{R}_+ \to \mathbb{R}$, still starting from $s_0 > 0$. Then, the following proposition holds which appears as a functional non-parametric extension of the peacock property shared by $\Big(\int_0^T e^{\sigma B_t - \frac{\sigma^2 t}{2}} dt\Big)_{\sigma\geq 0}$ (see e.g. [6, 13]).

Proposition 3 (Functional Peacocks) *Let σ and θ be two real valued bounded continuous functions defined on $[0,T] \times \mathbb{R}$. Assume that $S^{(\sigma)}$ is the unique weak solution to (7) as well as $S^{(\theta)}$ for its counterpart involving θ. If one of the following additional conditions holds:*

(*i*) Convex Partitioning function*: there exists a function $\kappa : [0,T] \times \mathbb{R}_+ \to \mathbb{R}_+$ such that, for every $t \in [0,T]$,*

$$x \mapsto x\kappa(t,x) \text{ is convex on } \mathbb{R}_+ \text{ and } 0 \leq \sigma(t,.) \leq \kappa(t,.) \leq \theta(t,.) \text{ on } \mathbb{R}_+,$$

or

(*ii*) Convex Domination property*: for every $t \in [0,T]$ the function $x \mapsto x\theta(t,x)$ is convex on $\mathbb{R}_+$ and*

$$|\sigma(t,.)| \leq \theta(t,.),$$

then, for every convex (hence continuous) function $f : \mathbb{R} \to \mathbb{R}$ with polynomial growth

$$\mathbb{E} f\Big(\int_0^T S_s^{(\sigma)}\mu(ds)\Big) \leq \mathbb{E} f\Big(\int_0^T S_s^{(\theta)}\mu(ds)\Big)$$

where μ is a signed (finite) measure on $([0,T], \mathcal{B}or([0,T]))$. More generally, for every convex functional $F : \mathcal{C}([0,T], \mathbb{R}_+) \to \mathbb{R}$ with $(r, \|\,.\,\|_{\sup})$-polynomial growth,

$$\mathbb{E}\, F(S^{(\sigma)}) \leq \mathbb{E}\, F(S^{(\theta)}). \tag{8}$$

Proof We focus on the setting (*i*). The second one can be treated likewise. First note that κ is bounded since θ is. As a consequence, the function $x \mapsto x\kappa(t,x)$ is zero at $x = 0$ and can be extended into a convex function on the whole real line by setting $x\kappa(t,x) = 0$ if $x \leq 0$. One extends $x\sigma(t,x)$ and $x\theta(t,x)$ by zero on $\mathbb{R}_-$ likewise. Then, this claim appears as a straightforward consequence of Theorem 1 applied to the diffusion whose coefficients are given by the extension of $x\sigma(t,x)$ and $x\theta(t,x)$ on the whole real line. As above, the sup-norm continuity follows from the convexity and polynomial growth. In the end, we take advantage of the *a posteriori* positivity of $S^{(\theta)}$ and $S^{(\sigma)}$ when starting from $s_0 > 0$ to conclude.

Applications to Volatility Comparison Results The corollary below shows that comparison results for vanilla European options established in [9] appear as consequences of Proposition 3.

Corollary 3 *Assume* $\sigma \in \mathscr{C}([0,T]\times\mathbb{R},\mathbb{R}_+)$, $\sigma_{\min}, \sigma_{\max} \in \mathscr{C}([0,T],\mathbb{R})$ *satisfy*

$$0 \le \sigma_{\min}(t) \le \sigma(t,.) \le \sigma_{\max}(t),\; t \in [0,T],$$

then, for every convex functional $F : \mathscr{C}([0,T],\mathbb{R}_+) \to \mathbb{R}$ *with* $(r, \|\,.\,\|_{\sup})$*-polynomial growth* $(r \ge 1)$,

$$\mathbb{E}\,F\left(S_s^{(\sigma_{min})}\right) \le \mathbb{E}\,F\left(S_s^{(\sigma)}\right) \le \mathbb{E}\,F\left(S_s^{(\sigma_{\max})}\right). \tag{9}$$

Proof We successively apply the former Proposition 3 to the couple $(\sigma_{\min},\sigma)$ and the partitioning function $\kappa(t,x) = \sigma_{\min}(t)$ to get the left inequality and to the couple $(\sigma,\sigma_{\max})$ with $\kappa = \sigma_{\max}$ to get the right inequality.

Note that the left and right hand side of the above inequality are usually considered as quasi-closed forms since they correspond to a Hull-White model (or even to the regular Black-Scholes model if $\sigma_{\min}$, $\sigma_{\max}$ are constant). Moreover, let us emphasize that no convexity assumption on σ is requested.

2.4 *Counter-Example (Discrete Time Setting)*

The above comparison results for the convex order may fail when the assumptions of Theorem 1 are not satisfied by the diffusion coefficient. In fact, for simplicity, the counter-example below is developed in a discrete time framework corresponding to Proposition 1. We consider the 2-period dynamics $X = X^{\sigma,x} = (X^{\sigma,x}_{0:2})$ satisfying

$$X_1 = x + \sigma Z_1 \quad \text{and} \quad X_2 = X_1 + \sqrt{2v(X_1)}Z_2$$

where $Z_{1:2} \stackrel{\mathscr{L}}{\sim} \mathscr{N}(0;I_2)$, $\sigma \ge 0$, and $v : \mathbb{R} \to \mathbb{R}_+$ is a bounded $\mathscr{C}^2$-function such that v has a strict local maximum at x_0 satisfying $v'(x_0) = 0$ and $v''(x_0) < -1$. So is the case if $v(x) = v(x_0) - \rho(x-x_0)^2 + o((x-x_0)^2)$, $0 < \rho < \frac{1}{2}$, in the neighbourhood of x_0. Of course, this implies that $\sqrt{v}$ cannot be convex.

Let $f(x) = e^x$. It is clear that

$$\varphi(x,\sigma) := \mathbb{E}f(X_2) = e^x\mathbb{E}\big(e^{\sigma Z_1 + v(x+\sigma Z_1)}\big).$$

Elementary computations show that

$$\varphi'_{\sigma}(x,\sigma) = e^{x}\mathbb{E}\Big(e^{\sigma Z_1+v(x+\sigma Z_1)}\big(1+v'(x+\sigma Z_1)\big)Z_1\Big)$$
$$\varphi''_{\sigma^2}(x,\sigma) = e^{x}\Big[\mathbb{E}\Big(e^{\sigma Z_1+v(x+\sigma Z_1)}\big(1+v'(x+\sigma Z_1)\big)^2 Z_1^2\Big) + \mathbb{E}\Big(e^{\sigma Z_1+v(x+\sigma Z_1)}v''(x+\sigma Z_1)Z_1^2\Big)\Big].$$

In particular

$$\varphi'_{\sigma}(x,0) = e^{x+v(x)}(1+v'(x))\mathbb{E}\,Z_1 = 0 \;\text{ and }\; \varphi''_{\sigma^2}(x,0) = e^{x+v(x)}\Big((1+v'(x))^2+v''(x)\Big)$$

so that $\varphi''_{\sigma^2}(x_0,0) < 0$ which implies that there exists a small enough $\sigma_0 > 0$ such that $\varphi'_{\sigma}(x_0,\sigma) < 0$ on $(0,\sigma_0]$ so that

$$\sigma \longmapsto \varphi(x_0,\sigma) \text{ is decreasing on } (0,\sigma_0].$$

This clearly exhibits a counter-example to Proposition 1 when the convexity assumption is fulfilled neither by the functions $(\sigma_k)_{k=0:n}$ nor the functions $(\kappa_k)_{k=0:n}$ (here with $n=1$).

2.5 *Lévy Driven Diffusions*

Let $Z = (Z_t)_{t\in[0,T]}$ be a Lévy process with Lévy measure ν satisfying $\int_{|z|\geq 1}|z|^p\nu(dz) < +\infty$, $p\in[1,+\infty)$. Then $Z_t\in L^1(\mathbb{P})$ for every $t\in[0,T]$. Assume furthermore that $\mathbb{E}\,Z_1 = 0$ so that $(Z_t)_{t\in[0,T]}$ is an $\mathcal{F}^Z$-martingale.

Theorem 2 *Let $Z=(Z_t)_{t\in[0,T]}$ be a martingale Lévy process with Lévy measure ν satisfying $\nu(|z|^p) < +\infty$ for a $p\in(1,+\infty)$ if Z has no Brownian component and $\nu(z^2) < +\infty$ if Z has a Brownian component. Let $\kappa_i:[0,T]\times\mathbb{R}\to\mathbb{R}$, $i=1,2$, be continuous functions with linear growth in x uniformly in $t\in[0,T]$. For $i=1,2$, let $X^{(\kappa_i)}=(X_t^{(\kappa_i)})_{t\in[0,T]}$ be the weak solution, assumed to be unique, to*

$$dX_t^{(\kappa_i)} = \kappa_i(t,X_{t-}^{(\kappa_i)})dZ_t^{(\kappa_i)}, \quad X_0^{(\kappa_i)} = x\in\mathbb{R}, \tag{10}$$

where $Z^{(\kappa_i)}$, $i=1,2$ have the same distribution as Z. Let $F:\mathbb{D}([0,T],\mathbb{R})\to\mathbb{R}$ be a Borel convex functional, $\mathbb{P}_{X^{(\kappa_i)}}$-a.s. Sk-continuous, $i=1,2$, with $(r,\|.\|_{\sup})$-polynomial growth for some $r\in[1,p)$ i.e.

$$\forall\,\alpha\in\mathbb{D}([0,T],\mathbb{R}), \quad |F(\alpha)|\leq C(1+\|\alpha\|_{\sup}^r).$$

(*a*) Convex Partitioning function*: If there exists a function $\kappa : [0,T]\times\mathbb{R}\to\mathbb{R}_+$ such that $\kappa(t,.)$ is convex for every $t\in[0,T]$ and $0\le\kappa_1\le\kappa\le\kappa_2$, then*

$$\mathbb{E}\,F(X^{(\kappa_1)})\le\mathbb{E}\,F(X^{(\kappa_2)}).$$

(*a'*) *An equivalent form for claim (a) is: if $0\le\kappa_1\le\kappa_2$ and, either $\kappa_1(t,.)$ is convex for every $t\in[0,T]$, or $\kappa_2(t,.)$ is convex for every $t\in[0,T]$, then the conclusion of (a) still holds true.*

(*b*) Convex Domination property*: If Z has a symmetric distribution, $|\kappa_1|\le\kappa_2$ and κ_2 is convex, then*

$$\mathbb{E}\,F(X^{(\kappa_1)})\le\mathbb{E}\,F(X^{(\kappa_2)}).$$

Remark 3

- The $\mathbb{P}_{X^{(\kappa_i)}}$-*a.s. Sk*-continuity of the functional F, $i=1,2$, is now requested: *Sk*-continuity no longer follows from the convexity since $\big(\mathbb{D}([0,T],\mathbb{R}), Sk\big)$ is a Polish space but not a topological vector space. Thus, the *convex* function $\alpha\mapsto|\alpha(t_0)|$ for a fixed $t_0\in(0,T)$ is continuous at a given $\beta\in\mathbb{D}([0,T],\mathbb{R})$ if and only if β is continuous at t_0 (see [4, Chap. 3]).
- The result remains true under the less stringent moment assumption on the Lévy measure ν: $\nu(|z|^p\mathbf{1}_{\{|z|\ge 1\}})<+\infty$ but would require much more technicalities since one has to carry out the reasoning of the proof below between two large jumps of Z and "paste" these inter-jump results.

The following lemma is the key that solves the approximation part of the proof in this càdlàg setting.

Lemma 3 *Let $\alpha\in\mathbb{D}([0,T],\mathbb{R})$. The sequence of stepwise constant approximations defined by*

$$\alpha_n(t)=\alpha(\underline{t}_n),\ t\in[0,T],$$

converges toward α for the Skorokhod topology.

Proof See [17, Proposition VI.6.37, p. 387] (second edition).

Proof (Proof of Theorem 2)

Step 1. Let $(\bar{X}^n_t)_{t\in[0,T]}$ be the genuine Euler scheme defined by

$$\bar{X}^n_t=x+\int_{(0,t]}\kappa(\underline{s}_n,\bar{X}^n_{\underline{s}_n-})dZ_s$$

where $\kappa=\kappa_1$ or κ_2. Owing to the linear growth of κ, we derive (see e.g. Proposition 12 in Appendix 2) that

$$\Big\|\sup_{t\in[0,T]}|X_t|\Big\|_p+\sup_{n\ge1}\Big\|\sup_{t\in[0,T]}|\bar{X}^n_t|\Big\|_p<+\infty.$$

We know, e.g. from Proposition 11 in Appendix 2, that $(\bar{X}^n)_{n\geq 1}$ functionally weakly converges for the Skorokhod topology toward the unique weak solution X of the *SDE* $dX_k = \kappa(t, X_{t-})dZ_t$, $X_0 = x$. In turn, Lemma 3 implies that $(\bar{X}^n_{\underline{t}_n})_{t\in[0,T]}$ *Sk*-weakly converges toward X.

Step 2. Let $F : \mathbb{D}([0,T],\mathbb{R}) \to \mathbb{R}$ be a $\mathbb{P}_X$-*Sk*-continuous convex functional. For every integer $n \geq 1$, we still define the sequence of convex functionals $F_n : \mathbb{R}^{n+1} \to \mathbb{R}$ by $F_n(x_{0:n}) = F\Big(\sum_{k=0}^{n-1} x_k \mathbf{1}_{[t^n_k, t^n_{k+1})} + x_n \mathbf{1}_{\{T\}}\Big)$ so that $F_n\big((\bar{X}^n_{t^n_k})_{0:n}\big) = F\big((\bar{X}^n_{\underline{t}_n})_{t\in[0,T]}\big)$.

Now, for every $n \geq 1$, the discrete time Euler schemes $\bar{X}^{(\kappa_i),n}$, $i = 1, 2$, related to the jump diffusions with diffusion coefficients κ_1 and κ_2 are of the form (1) and $|F_n(x_{0:n})| \leq C(1 + \|x_{0:n}\|^r)$, $r \in [1, p)$.

(*a*) Assume $0 \leq \kappa_1 \leq \kappa_2$. Then, taking advantage of the partitioning function κ, it follows from Proposition 1(*a*) that, for every $n \geq 1$, $\mathbb{E}\, F_n\big((\bar{X}^{(\kappa_1),n}_{t^n_k})_{0:n}\big) \leq \mathbb{E}\, F_n\big((\bar{X}^{(\kappa_2),n}_{t^n_k})_{0:n}\big)$ i.e. $\mathbb{E}\, F\big((\bar{X}^{(\kappa_1),n}_{\underline{t}_n})_{t\in[0,T]}\big) \leq \mathbb{E}\, F\big((\bar{X}^{(\kappa_2),n}_{\underline{t}_n})_{t\in[0,T]}\big)$. Letting $n \to +\infty$ completes the proof like for Theorem 1 since F is $\mathbb{P}_X$-*a.s. Sk*-continuous.

(*b*) is an easy consequence of Proposition 1(*b*).

3 Convex Order for Non-Markovian Itô and Doléans Martingales

The results of this section illustrate another aspects of our paradigm in order to establish functional convex order for various classes of continuous time stochastic processes. Here we deal with (couples of) Itô integrals with the restriction that one of the two integrands needs to be deterministic.

3.1 Itô Martingales

Proposition 4 *Let $(W_t)_{t\in[0,T]}$ be a standard Brownian motion on a filtered probability space $(\Omega, \mathscr{A}, (\mathscr{F}_t)_{t\in[0,T]}, \mathbb{P})$ where $(\mathscr{F}_t)_{t\in[0,T]}$ satisfies the usual conditions and let $(H_t)_{t\in[0,T]}$ be an $(\mathscr{F}_t)$-progressively measurable process defined on the same probability space. Let $h = (h_t)_{t\in[0,T]} \in L^2_T$. Let $F : \mathscr{C}([0,T],\mathbb{R}) \to \mathbb{R}$ be a convex functional with $(r, \|.\|_{\sup})$-polynomial growth, $r \geq 1$.*

(*a*) *If $|H_t| \leq h_t$ $\mathbb{P}$-a.s. for every $t \in [0,T]$, then*

$$\mathbb{E}\, F\Big(\int_0^{\cdot} H_s dW_s\Big) \leq \mathbb{E}\, F\Big(\int_0^{\cdot} h_s dW_s\Big).$$

(b) If $H_t \geq h_t \geq 0$ $\mathbb{P}$-a.s. for every $t \in [0,T]$ and $|H|_{L^2_T} \in L^r(\mathbb{P})$, then

$$\mathbb{E}\,F\left(\int_0^{\cdot} H_s dW_s\right) \geq \mathbb{E}\,F\left(\int_0^{\cdot} h_s dW_s\right).$$

Remark 4

- In the "marginal" case where F is of the form $F(\alpha) = f(\alpha(T))$, it has been shown in [12] that the above assumptions on H and h in (*a*) and (*b*) are too stringent and can be relaxed into

$$\int_0^T \mathbb{E}\,H_t^2 dt \leq \int_0^T h_t^2 dt \quad \text{and} \quad \int_0^T \mathbb{E}\,H_t^2 dt \geq \int_0^T h_t^2 dt$$

  respectively. The main ingredient of the proof is the Dambis-Dubins-Schwartz representation theorem for one-dimensional Brownian martingales (see e.g. Theorem 1.6 in [31, p. 181]).
- The first step of the proof below is a variant of Proposition 1 in a non-Markov framework. It can be considered as an autonomous proposition devoted to discrete time dynamics.

Proof Step 1 (Discrete Time). Let $(Z_k)_{1\leq k\leq n}$ be an n-tuple of independent symmetric (hence centered) $\mathbb{R}$-valued random variables satisfying $Z_k \in L^r(\Omega, \mathcal{A}, \mathbb{P})$, $r \geq 1$, and let $\mathcal{F}_0^Z = \{\emptyset, \Omega\}$, $\mathcal{F}_k^Z = \sigma(Z_1, \ldots, Z_k)$, $k = 1, \ldots, n$ be its natural filtration. Let $(H_k)_{k=0,\ldots,n}$ be an $(\mathcal{F}_k^Z)_{k=0,\ldots,n}$-adapted sequence such that $H_k \in L^r(\mathbb{P})$, $k = 1, \ldots, n$.

Let $X = (X_k)_{k=0:n}$ and $Y = (Y_k)_{k=0:n}$ be two sequences of random variables recursively defined by

$$X_{k+1} = X_k + H_k Z_{k+1}, \quad Y_{k+1} = Y_k + h_k Z_{k+1}, \quad 0 \leq k \leq n-1, \quad X_0 = Y_0 = x_0.$$

These are the discrete time stochastic integrals of (H_k) and (h_k) with respect to the sequence of increments $(Z_k)_{k=1:n}$. It is clear by induction that X_k, $Y_k \in L^r(\mathbb{P})$ for every $k = 0, \ldots, n$ since H_k is $\mathcal{F}_k^Z$-measurable and Z_{k+1} is independent of $\mathcal{F}_k^Z$.

Let $\Phi : \mathbb{R}^{n+1} \to \mathbb{R}$ be a convex function with r-polynomial growth. Let us focus on the first inequality, discrete time counterpart of claim (*a*). We proceed like in the proof Proposition 1 to prove by three backward inductions that if $|H_k| \leq h_k$, for every $k = 0, \ldots, n$, then

$$\mathbb{E}\,\Phi(X) \leq \mathbb{E}\,\Phi(Y).$$

To be more precise, let us introduce by analogy with this proposition the sequence $(\Psi_k)_{k=0,\ldots,n}$ of functions recursively defined by

$$\Psi_n = \Phi, \; \Psi_k(x_{0:k}) = (Q_{k+1}\Psi_{k+1}(x_{0:k}, x_k + .))(h_k), \; x_{0:k} \in \mathbb{R}^{k+1}, \; k = 0, \ldots, n-1.$$

First note that the functions Ψ_k satisfy the following linear dynamic programming principle:

$$\Psi_k(Y_{0:k}) = \mathbb{E}\big(\Psi_{k+1}(Y_{0:k+1}) \,|\, \mathscr{F}_k^Z\big),\; k = 0, \ldots, n-1,$$

so that, by the chaining rule for conditional expectations, we have

$$\Phi_k(Y_{0:k}) = \mathbb{E}\big(\Phi(Y_{0:n}) \,|\, \mathscr{F}_k^Z\big),\; k = 0, \ldots, n.$$

Furthermore, owing to the properties of the operator Q_{k+1}, we already proved that for any convex function $G : \mathbb{R}^{k+2} \to \mathbb{R}$ with r-polynomial growth, the function

$$(x_{0:k}, u) \mapsto (Q_{k+1}G(x_{0:k}, x_k + .))(u) = \mathbb{E}\, G(x_{0:k}, x_k + uZ_{k+1})$$

is convex and even as a function of u for every fixed $x_{0:k}$. As a consequence, it also satisfies the maximum principle established in Lemma 1 (c) since the random variable Z_k have symmetric distributions.
Now, let us introduce the martingale induced by $\Phi(X_{0:n})$, namely

$$M_k = \mathbb{E}\big(\Phi(X_{0:n}) \,|\, \mathscr{F}_k^Z\big),\; k = 0, \ldots, n.$$

We show now by a backward induction that $M_k \le \Psi_k(X_{0:k})$ for every $k = 0, \ldots, n$. If $k = n$, this is trivial. Assume now that $M_{k+1} \le \Psi_{k+1}(X_{0:k+1})$ for a $k \in \{0, \ldots, n-1\}$. Then we get the following string of inequalities

$$\begin{aligned}
M_k = \mathbb{E}(M_{k+1} \,|\, \mathscr{F}_k^Z) &\le \mathbb{E}(\Psi_{k+1}(X_{0:k+1}) \,|\, \mathscr{F}_k^Z)\\
&= \mathbb{E}(\Psi_{k+1}(X_{0:k}, X_k + H_k Z_{k+1}) \,|\, \mathscr{F}_k^Z)\\
&= \Big(\mathbb{E}(\Psi_{k+1}(x_{0:k}, x_k + uZ_{k+1}) \,|\, \mathscr{F}_k^Z\Big)_{|x_{0;k}=X_{0:k}, u=H_k}\\
&= \Big(Q_{k+1}\Psi_{k+1}(x_{0:k}, x_k + .)(H_k)\Big)_{|x_{0;k}=X_{0:k}}\\
&\le \Big(Q_{k+1}\Psi_{k+1}(x_{0:k}, x_k + .)(h_k)\Big)_{|x_{0;k}=X_{0:k}} = \Psi_k(X_{0:k})
\end{aligned} \tag{11}$$

where we used in the fourth line that Z_{k+1} is independent of $\mathscr{F}_k^Z$ and, in the penultimate line, the assumption $|H_k| \le h_k$ and the maximum principle. Finally, at $k = 0$, we get $\mathbb{E}\, \Phi(X_{0:n}) = M_0 \le \Phi_0(x_0) = \mathbb{E}\Phi(Y_{0:n})$ which is the announced conclusion.

Step 2 (Approximation-Regularization). We temporarily assume that the function h has a modification which is bounded by a real constant so that $\mathbb{P}(d\omega)$-*a.s.* $\|H(\omega)\|_{\sup} \vee \|h\|_{\sup} \le K$. We first need a technical lemma adapted from

Lemma 2.4 in [20, p. 132] about approximation of progressively measurable processes by *simple* processes, with in mind the preservation of the domination property requested in our framework. The details of the proof of this lemma are left to the reader.

Lemma 4

(*a*) *For every* $\varepsilon \in (0, T)$ *and every* $g \in L^2([0,T], dt)$ *we define*

$$\Delta_\varepsilon g(t) \equiv t \longmapsto \frac{1}{\varepsilon}\int_{(t-\varepsilon)_+}^{t} g(s)ds \in \mathscr{C}([0,T],\mathbb{R}).$$

The operator $\Delta_\varepsilon : L^2_T \to \mathscr{C}([0,T],\mathbb{R})$ *is non-negative. In particular, if* $g,\ \gamma \in L^2_T$ *with* $|g| \le \gamma\ \lambda_1$*-a.e., then* $|\Delta_\varepsilon g| \le \Delta_\varepsilon \gamma$ *and* $\|\Delta_\varepsilon g\|_{\sup} \le |g|_{L^\infty_T}$.

(*b*) *If* $g \in \mathscr{C}([0,T],\mathbb{R})$*, define for every integer* $m \ge 1$*, the stepwise constant càglàd (for left continuous right limited) approximation* $\tilde{g}^m$ *of* g *by*

$$\tilde{g}^m(t) = g(0)\mathbf{1}_{\{0\}}(t) + \sum_{k=1}^{m} g\big(t^m_{k-1}\big)\mathbf{1}_{(t^m_{k-1},t^m_k]}.$$

Then $\tilde{g}^m \overset{\|\cdot\|_{\sup}}{\longrightarrow} g$ *as* $m \to +\infty$*. Furthermore, if* $g,\ \gamma \in \mathscr{C}([0,T],\mathbb{R})$ *and* $|g| \le \gamma$*, then* $|\tilde{g}^m| \le \tilde{\gamma}^m$ *for every* $m \ge 1$.

By the Lebesgue fundamental Theorem of Calculus we know that

$$\big|\Delta_{\frac{1}{n}} H - H\big|_{L^2_T} \longrightarrow 0 \quad \mathbb{P}\text{-}a.s.$$

Since $|\Delta_{\frac{1}{n}} H - H|_{L^2_T} \le 2K$, the Lebesgue dominated convergence Theorem implies that

$$\mathbb{E}\int_0^T |\Delta_{\frac{1}{n}} H_t - H_t|^2 dt \longrightarrow 0 \ \text{ as } \ n \to +\infty. \tag{12}$$

By construction, $\Delta_{\frac{1}{n}} H$ is an $(\mathscr{F}_t)$-adapted pathwise continuous process satisfying the domination property $|\Delta_{\frac{1}{n}} H| \le \Delta_{\frac{1}{n}} h$ so that, in turn, using this time claim (*b*) of the above lemma, for every $n,\ m \ge 1$,

$$|\widetilde{\Delta_{\frac{1}{n}} H_t}^m| \le \widetilde{\Delta_{\frac{1}{n}} h_t}^m.$$

On the other hand, for every $n \ge 1$, the *a.s.* uniform continuity of $\Delta_{\frac{1}{n}} H$ over $[0,T]$ implies

$$\int_0^T \big|\widetilde{\Delta_{\frac{1}{n}} H_t}^m - \Delta_{\frac{1}{n}} H_t\big|^2 dt \le T \sup_{t\in[0,T]} \big|\widetilde{\Delta_{\frac{1}{n}} H_t}^m - \Delta_{\frac{1}{n}} H_t\big|^2 \to 0 \ \text{ as } \ m \to +\infty\ \mathbb{P}\text{-}a.s.$$

One concludes again by the Lebesgue dominated convergence Theorem that, for every $n \geq 1$,

$$\mathbb{E}\int_0^T \left|\widetilde{\Delta_{\frac{1}{n}} H_t}^m - \Delta_{\frac{1}{n}} H_t\right|^2 dt \longrightarrow 0 \text{ as } m \to +\infty.$$

One shows likewise for the function h itself that

$$\left|\Delta_{\frac{1}{n}} h - h\right|_{L^2_T} \to 0 \text{ as } n \to +\infty$$

and, for every $n \geq 1$,

$$\left|\widetilde{\Delta_{\frac{1}{n}} h}^m - \Delta_{\frac{1}{n}} h\right|_{L^2_T} \to 0 \text{ as } m \to +\infty.$$

Consequently, there exists an increasing subsequence $m(n) \uparrow +\infty$ such that

$$\mathbb{E}\int_0^T \left|\widetilde{\Delta_{\frac{1}{n}} H_t}^{m(n)} - \Delta_{\frac{1}{n}} H_t\right|^2 dt + \int_0^T \left|\widetilde{\Delta_{\frac{1}{n}} h_t}^{m(n)} - \Delta_{\frac{1}{n}} h_t\right|^2 dt \longrightarrow 0 \quad \text{as} \quad n \to +\infty$$

which in turn implies, combined with (12) and its deterministic counterpart for h,

$$\mathbb{E}\int_0^T \left|\widetilde{\Delta_{\frac{1}{n}} H_t}^{m(n)} - H_t\right|^2 dt + \int_0^T \left|\widetilde{\Delta_{\frac{1}{n}} h_t}^{m(n)} - h_t\right|^2 dt \longrightarrow 0 \quad \text{as} \quad n \to +\infty.$$

At this stage, we set for every integer $n \geq 1$,

$$H_t^{(n)} = \widetilde{\Delta_{\frac{1}{n}} H_t}^{m(n)} \quad \text{and} \quad h_t^{(n)} = \widetilde{\Delta_{\frac{1}{n}} h_t}^{m(n)} \tag{13}$$

which satisfy

$$\mathbb{E}|H - H^{(n)}|^2_{L^2_T} + |h - h^{(n)}|_{L^2_T} \longrightarrow 0 \text{ as } n \to +\infty. \tag{14}$$

It should be noted that these processes $H^{(n)}$, H and these functions $h^{(n)}$, h are all bounded by $2K$.

We consider now the continuous modifications of the four (square integrable) Brownian martingales associated to the integrands $H^{(n)}$, H, $h^{(n)}$ and h, the last two being of Wiener type. It is clear by Doob's Inequality that

$$\sup_{t\in[0,T]}\left|\int_0^t H_s^{(n)} dW_s - \int_0^t H_s dW_s\right| + \sup_{t\in[0,T]}\left|\int_0^t h_s^{(n)} dW_s - \int_0^t h_s dW_s\right| \stackrel{L^2(\mathbb{P})}{\longrightarrow} 0 \text{ as } n \to +\infty.$$

In particular $\left(\int_0^t H_s^{(n)} dW_s\right)_{t\in[0,T]}$ functionally weakly converges to $\left(\int_0^t H_s dW_s\right)_{t\in[0,T]}$ for the $\|\,.\,\|_{\sup}$-norm topology. We also have, owing to the B.D.G. Inequality, that for every $p \in (0, +\infty)$,

$$\mathbb{E} \sup_{t\in[0,T]} \left| \int_0^t H_s^{(n)} dW_s \right|^p \leq c_p^p \, \mathbb{E}|H^{(n)}|^p_{L^2_T} \leq c_p^p K^p \tag{15}$$

where c_p is the universal constant involved in the B.D.G. Inequality. The same holds true for the three other integrals related to $h^{(n)}$, H, and h.

Let $n \geq 1$. Set $H_k^n = H^{(n)}_{t_k^{m(n)}}$, $h_k^n = h^{(n)}_{t_k^{m(n)}}$, $k = 0, \ldots, m(n) - 1$ and $Z_k^n = W_{t_k^{m(n)}} - W_{t_{k-1}^{m(n)}}$, $k = 1, \ldots, m(n)$. One easily checks that $\int_0^{t_k^{m(n)}} H_s^{(n)} dW_s = \sum_{\ell=1}^{k} H_{\ell-1}^n Z_\ell^n$, $k = 0, \ldots, m(n)$, so that

$$I_{m(n)} \Big(\int_0^{\cdot} H_s^{(n)} dW_s \Big) = i_{m(n)} \left(\Big(\sum_{\ell=1}^{k} H_{\ell-1}^n Z_\ell^n \Big)_{k=0:m(n)} \right).$$

Let $F_{m(n)}$ be defined by (5) from the convex functional F (with $(r, \|\,.\,\|_{\sup})$-polynomial growth). It is clearly convex. One derives from Step 1 applied with horizon $m(n)$ and discrete time random sequences $(Z_k^n)_{k=1:m(n)}$, $(H_k^n)_{k=0:m(n)-1}$, $(h_k)_{k=0:m(n)-1}$ that

$$\begin{aligned}
\mathbb{E}\, F \circ I_{m(n)} \Big(\int_0^{\cdot} H_s^{(n)} dW_s \Big) &= \mathbb{E}\, F_{m(n)} \left(\Big(\sum_{\ell=1}^{k} H_{\ell-1}^n Z_\ell^n \Big)_{k=0:m(n)} \right) \\
&\leq \mathbb{E}\, F_{m(n)} \left(\Big(\sum_{\ell=1}^{k} h_{\ell-1}^n Z_\ell^n \Big)_{k=0:m(n)} \right) \\
&= \mathbb{E}\, F \circ I_{m(n)} \Big(\int_0^{\cdot} h_s^{(n)} dW_s \Big).
\end{aligned}$$

Combining the above functional weak convergence, Lemma 2 and the uniform integrability derived form (15) (with any $p > r$) yields the expected inequality by letting n go to infinity.

Step 3 (Second Approximation). Let $K \in \mathbb{N}$ and $\chi_K : \mathbb{R} \to \mathbb{R}$ be the thresholding function defined by $\chi_K(u) = (u \wedge K) \vee (-K)$. It follows from the B.D.G. Inequality that, for every $p \in (0, +\infty)$,

$$\begin{aligned}
\mathbb{E} \sup_{t\in[0,T]} \left| \int_0^t H_s dW_s - \int_0^t \chi_K(H_s) dW_s \right|^p &\leq c_p^p \, \mathbb{E}|H - \chi_K(H)|^p_{L^2_T} \\
&= c_p^p \, \mathbb{E}|(|H| - K)_+|^p_{L^2_T} \qquad (16) \\
&\leq c_p^p \, |(|h| - K)_+|^p_{L^2_T} \qquad (17)
\end{aligned}$$

where $u_+ = \max(u, 0)$, $u \in \mathbb{R}$. The same bound obviously holds when replacing H by h. This shows that the convergence holds in every $L^p(\mathbb{P})$ space, $p \in (0, +\infty)$, as $K \to +\infty$. Hence, one gets the expected inequality by letting K go to infinity in the inequality

$$\mathbb{E}F\left(\int_0^{\cdot} \chi_K(H_s)dW_s\right) \leq \mathbb{E}F\left(\int_0^{\cdot} \chi_K(h_s)dW_s\right) = \mathbb{E}F\left(\int_0^{\cdot} h_s \wedge K dW_s\right). \tag{18}$$

(b) We consider the same steps as for the upper-bound established in (a) with the same notations.

Step 1. First, in a discrete time setting, we assume that $0 \leq h_k \leq H_k \in L^r(\mathbb{P})$ and we aim at showing by a backward induction that $M_k \geq \Psi_k(X_{0:k})$ where $M_k = \mathbb{E}\big(\Phi(X_{0:n}) \mid \mathcal{F}_k^Z\big)$.

If $k = n$, the inequality holds as an equality since $\Psi_n = \Phi$. Now assume $M_{k+1} \geq \Psi_{k+1}(X_{0:k+1})$. Then, like in (a), we have

$$\begin{aligned}
M_k &= \mathbb{E}\big(M_{k+1} \mid \mathcal{F}_k^Z\big)\\
&\geq \mathbb{E}\big(\Psi(X_{0:k+1}) \mid \mathcal{F}_k^Z\big)\\
&= \mathbb{E}\big(\Psi(X_{0:k}, X_k + H_k Z_{k+1}) \mid \mathcal{F}_k^Z\big) = \Big(Q_k \Psi_{k+1}(x_{0:k}, x_k + .)(H_k)\Big)_{|x_{0:k} = X_{0:k}}\\
&\geq \Big(Q_k \Psi_{k+1}(x_{0:k}, x_k + .)(h_k)\Big)_{|x_{0:k} = X_{0:k}} = \Psi_k(X_{0:k}).
\end{aligned}$$

Step 2. This step is devoted to the approximation in a bounded setting where $0 \leq h_t \leq H_t \leq K$. It follows the lines of its counterpart in claim (a), taking advantage of the global boundedness by K.

Step 3. This last step is devoted to the approximation procedure in the general setting. It differs from (a) since there is no longer a deterministic upper-bound provided by the function $h \in L_T^2$. Then, the key is to show that the process $\int_0^{\cdot} \chi_K(H_s)dW_s$ converges for the sup norm over $[0, T]$ in $L^r(\mathbb{P})$ toward the process $\int_0^{\cdot} H_s dW_s$ as $K \to +\infty$. In fact, it follows from (16) applied with $p = r$ that

$$\mathbb{E}\left(\sup_{t \in [0,T]} \left| \int_0^t H_s dW_s - \int_0^t \chi_K(H_s)dW_s \right|^r\right) \leq c_r^r\, \mathbb{E}\Big(\big|(|H| - K)_+\big|_{L_T^2}^r\Big).$$

As $|H|_{L_T^2} \in L^r(\mathbb{P})$, one concludes by the Lebesgue convergence Theorem by letting $K \to +\infty$.

Remark 5

- Step 1 can be extended to non-symmetric, centered independent random variables $(Z_k)_{1 \leq k \leq n}$ if the sequences $(H_k)_{0 \leq k \leq n-1}$ and $(h_k)_{0 \leq k \leq n-1}$ under consideration satisfy $0 \leq H_k \leq h_k$, $k = 0, \ldots, n-1$.

- When H has left continuous paths, the proof can be significantly simplified by considering the simpler approximating sequence $H_t^{(n)} = \widetilde{H}_t^n$ which clearly converges toward H $d\mathbb{P} \otimes dt$-*a.e.* (and in the appropriate $L^p(d\mathbb{P} \otimes dt)$-spaces as well).

3.2 Lévy-Itô Martingales

Proposition 5 *Let $Z = (Z_t)_{t\in[0,T]}$ be an integrable centered Lévy process with Lévy measure ν satisfying $\nu(|x|^p \mathbf{1}_{\{|x|\geq 1\}}) < +\infty$ for a real exponent $p > 1$. Let $F : \mathbb{D}([0,T],\mathbb{R}) \to \mathbb{R}$ be a convex Skorokhod continuous functional with $(p, \|\,.\,\|_{\sup})$-polynomial growth. Let $(H_t)_{t\in[0,T]}$ be an $(\mathcal{F}_t)$-predictable process and let $h = (h_t)_{t\in[0,T]}$ such that $|h|_{L_T^{p\vee 2}} < +\infty$.*

(*a*) *If $0 \leq H_t \leq h_t$ dt-a.e., $\mathbb{P}$-a.s. then*

$$\mathbb{E}\, F\left(\int_0^{\cdot} H_s dZ_s\right) \leq \mathbb{E}\, F\left(\int_0^{\cdot} h_s dZ_s\right).$$

If furthermore Z is symmetric, the result still holds if $|H_t| \leq h_t$ dt-a.e., $\mathbb{P}$-a.s..

(*b*) *If $H_t \geq h_t \geq 0$ dt-a.e., $\mathbb{P}$-a.s. and $|H|_{L_T^{p\vee 2}} \in L^p(\mathbb{P})$, then*

$$\mathbb{E}\, F\left(\int_0^{\cdot} H_s dZ_s\right) \geq \mathbb{E}\, F\left(\int_0^{\cdot} h_s dZ_s\right).$$

(*c*) *If the Lévy process Z has no Brownian component, the above claims* (*a*) *and* (*b*) *remain true if we only assume $h \in L_T^p$ and $|H|_{L_T^p} \in L^p(\mathbb{P})$ respectively.*

Proof (*a*) This proof follows the approach introduced for the Brownian case but requires more technicalities due to Lévy processes.

Step 1 (Discrete Time). This step does not differ from that developed for Brownian-Itô martingales, except that in the Lévy setting we rely on claim (*a*) of Lemma 1 since the marginal distribution of the increment of a Lévy process has no reason to be symmetric.

Step 2 (Approximation-Regularization). Temporarily assume that h is bounded. We consider the approximation procedure of H by stepwise constant càglàd $(\mathcal{F}_t)$-adapted, hence predictable, processes $H^{(n)}$ already defined by (13). Then, we first consider the Lévy-Khintchine decomposition of the Lévy martingale Z

$$\forall\, t \in [0,T], \qquad Z_t = a\, W_t + \widetilde{Z}_t^{\eta} + Z_t^{\eta}, \quad a \geq 0,$$

where $\widetilde{Z}^{\eta}$ is a martingale with jumps of size at most η and Lévy measure $\nu(\,.\cap\{|z| \leq \eta\})$ and Z^{η} is a compensated Poisson process with (finite) Lévy measure $\nu(.\cap\{|z| > \eta\})$. Let n be a positive integer. We will perform a "cascade" procedure to make p decrease thanks to the B.D.G. Inequality. This—classical—method is more detailed in the proof of Proposition 4 in Appendix 2 (higher moments of Lévy driven diffusions).

We first assume that $p \in (1,2]$. Combining Minkowski's and B.D.G.'s Inequalities yields

$$\left\| \sup_{t\in[0,T]} \left| \int_0^t H_s dZ_s - \int_0^t H_s^{(n)} dZ_s \right| \right\|_p \leq c_p\, a \left\| |H - H^{(n)}|_{L_T^2} \right\|_p$$

$$+ c_p \left\| \sum_{0<s\leq T} (H_s - H_s^{(n)})^2 (\Delta Z_s)^2 \mathbf{1}_{\{|\Delta Z_s|>\eta\}} \right\|_{\frac{p}{2}}^{\frac{1}{2}}$$

$$+ c_p \left\| \sum_{0<s\leq T} (H_s - H_s^{(n)})^2 (\Delta Z_s)^2 \mathbf{1}_{\{|\Delta Z_s|\leq\eta\}} \right\|_1^{\frac{1}{2}}$$

where we used in the last line the monotony of the $L^p(\mathbb{P})$-norm and $\frac{p}{2} \leq 1$.
Using now the compensation formula and again that $\frac{p}{2} \in (0,1]$, it follows

$$\begin{aligned}
\mathbb{E}\left| \sum_{0<s\leq T} (H_s - H_s^{(n)})^2 (\Delta Z_s)^2 \mathbf{1}_{\{|\Delta Z_s|>\eta\}} \right|^{\frac{p}{2}} &\leq \mathbb{E}\left[\sum_{0<s\leq T} |H_s - H_s^{(n)}|^p |\Delta Z_s|^p \mathbf{1}_{\{|\Delta Z_s|>\eta\}} \right] \\
&= \mathbb{E}\, |H - H^{(n)}|_{L_T^p}^p \, \nu(|z|^p \mathbf{1}_{\{|z|>\eta\}}) \\
&\leq T^{1-\frac{p}{2}} \mathbb{E}\big(|H - H^{(n)}|_{L_T^2}^p \big)\, \nu(|z|^p \mathbf{1}_{\{|z|>\eta\}}) \\
&\leq T^{1-\frac{p}{2}} \left(\mathbb{E}|H - H^{(n)}|_{L_T^2}^2 \right)^{\frac{p}{2}} \nu(|z|^p \mathbf{1}_{\{|z|>\eta\}}).
\end{aligned}$$

On the other hand,

$$\mathbb{E}\left| \sum_{0<s\leq T} (H_s - H_s^{(n)})^2 (\Delta Z_s)^2 \mathbf{1}_{\{|\Delta Z_s|\leq\eta\}} \right| = \mathbb{E}\, |H - H^{(n)}|_{L_T^2}^2 \, \nu(z^2 \wedge \eta).$$

We derive from (14) that the above three terms go to 0 as n goes to infinity so that

$$\sup_{t\in[0,T]} \left| \int_0^t H_s^{(n)} dZ_s - \int_0^t H_s dZ_s \right| \xrightarrow{L^p(\mathbb{P})} 0.$$

Then, Lemma 3 applied to the subsequence $(m(n))_{n\geq 1}$ implies that the stepwise constant process $\left(\int_0^{\underline{t}_{m(n)}} H_s^{(n)} dZ_s \right)_{t\in[0,T]}$ satisfies

$$\mathrm{dist}_{Sk}\left(\int_0^{\underline{\cdot}_{m(n)}} H_s^{(n)} dZ_s, \int_0^{\cdot} H_s dZ_s \right) \xrightarrow{\mathbb{P}} 0$$

which in turn implies the functional Sk-weak convergence. Furthermore, the above L^p-convergence implies that the sequence $\left(\sup_{t\in[0,T]}\left|\int_0^t H_s^{(n)}dZ_s\right|\right)_{n\geq 1}$ is uniformly L^p-integrable which is also clearly true for $\left(\sup_{t\in[0,T]}\left|\int_0^{\underline{t}_{m(n)}} H_s^{(n)}dZ_s\right|\right)_{n\geq 1}$. Following the same lines and still using Lemma 3, we get

$$\text{dist}_{Sk}\left(\int_0^{\cdot_{m(n)}} h_s^{(n)}dZ_s, \int_0^{\cdot} h_s dZ_s\right) \xrightarrow{\mathbb{P}\text{-}a.s.} 0 \text{ and } \left(\sup_{t\in[0,T]}\left|\int_0^t h_s^{(n)}dZ_s\right|\right)_{n\geq 1}$$

is uniformly L^p-integrable.

As $0 \leq H_t \leq h(t)$ dt-$a.e.$ $\mathbb{P}$-$a.s.$ (or $0 \leq |H_t| \leq h_t$ if Z is symmetric), for every fixed integer $n \geq 1$, we have, owing to Step 1 and following the lines of Step 3 of the proof of Proposition 4,

$$\mathbb{E}\left(F\Big(\int_0^{\underline{t}_{m(n)}} H_s^{(n)}dZ_s\Big)_{t\in[0,T]}\right) \leq \mathbb{E}\left(F\Big(\int_0^{\underline{t}_{m(n)}} h_s^{(n)}dZ_s\Big)_{t\in[0,T]}\right).$$

Letting $n \to +\infty$ yields the announced result since F is Sk-continuous with $(p, \|\,.\,\|_{\sup})$-polynomial growth (owing to the above uniform L^p-integrability results).

Assume now $p > 2$. First note that since h is bounded one can extend (14) as follows: there exists a sequence $m(n) \uparrow +\infty$ such that the processes $H^{(n)}$ and the functions $h^{(n)}$ defined by (13) satisfy

$$\mathbb{E}\,|H - H^{(n)}|^p_{L^p_T} + |h - h^{(n)}|_{L^p_T} \longrightarrow 0 \text{ as } n \to +\infty. \tag{19}$$

To this end, we introduce the dyadic logarithm of p i.e. the integer ℓ_p such that where $2^{\ell_p} < p \leq 2^{\ell_p+1}$. Thus, if $p\in (2,4]$ i.e. $\ell_p = 1$,

$$\left\|\sup_{t\in[0,T]}\left|\int_0^t H_s dZ_s - \int_0^t H_s^{(n)}dZ_s\right|\right\|_p \leq c_p\left(a\left\||H - H^{(n)}|_{L^2_T}\right\|_p\right.$$
$$\left.+\left\|\sum_{0<s\leq T}(H_s - H_s^{(n)})^2(\Delta Z_s)^2\right\|_{\frac{p}{2}}^{\frac{1}{2}}\right). \tag{20}$$

Now, Minkowski's Inequality applied with $\|.\|_{\frac{p}{2}}$ yields

$$\left\|\sum_{0<s\leq T}(H_s - H_s^{(n)})^2(\Delta Z_s)^2\right\|_{\frac{p}{2}} \leq \left\|\sum_{0<s\leq T}(H_s - H_s^{(n)})^2(\Delta Z_s)^2 - \nu(z^2)\int_0^T (H_s^{(n)} - H_s)^2 ds\right\|_{\frac{p}{2}}$$
$$+\nu(z^2)\left\||H^{(n)} - H|_{L^2_T}\right\|_p^2.$$

In turn, the B.D.G. Inequality applied to the martingale

$$M_t^{(1)} = \sum_{0<s\le t} (H_s - H_s^{(n)})^2(\Delta Z_s)^2 - \nu(z^2)\int_0^t (H_s^{(n)} - H_s)^2 ds,\ t \in [0,T],$$

yields

$$\begin{aligned}
&\Big\| \sum_{0<s\le T} (H_s - H_s^{(n)})^2(\Delta Z_s)^2 - nu(z^2)\int_0^T (H_s^{(n)} - H_s)^2 ds\Big)\Big\|_{\frac{p}{2}} \\
&\le c_{\frac{p}{2}} \Big\| \sum_{0<s\le T} (H_s - H_s^{(n)})^4(\Delta Z_s)^4 \Big\|_{\frac{p}{4}}^{\frac{1}{2}} \\
&\le c_{\frac{p}{2}} \Big(\mathbb{E} \sum_{0<s\le T} |H_s - H_s^{(n)}|^p |\Delta Z_s|^p\Big)^{\frac{2}{p}} \\
&= c_{\frac{p}{2}} \Big(\nu(|z|^p)\mathbb{E}\int_0^T |H_s - H_s^{(n)}|^p ds\Big)^{\frac{2}{p}} \\
&= c_{\frac{p}{2}} \Big(\nu(|z|^p)\Big)^{\frac{2}{p}} \big\| |H - H^{(n)}|_{L_T^p} \big\|_p^2
\end{aligned}$$

where we successively used that $\frac{p}{4} \le 1$ in the second line and the compensation formula in the third line. Finally, we note that, as $p \ge 2$, the convergence (19) implies

$$\big\| |H^{(n)} - H|_{L_T^2} \big\|_p \le T^{\frac{1}{2}-\frac{1}{p}} \big\| |H^{(n)} - H|_{L_T^p} \big\|_p \to 0 \text{ as } n \to +\infty.$$

This shows that both terms in the right hand side of (20) converge to 0 as $n \to +\infty$, so that

$$\Big\| \sup_{t\in[0,T]} \Big| \int_0^t H_s^{(n)} dZ_s - \int_0^t H_s dZ_s \Big| \Big\|_p \longrightarrow 0 \text{ as } n \to +\infty.$$

We show likewise

$$\Big\| \sup_{t\in[0,T]} \Big| \int_0^t h_s^{(n)} dZ_s - \int_0^t h_s dZ_s \Big| \Big\|_p \longrightarrow 0 \text{ as } n \to +\infty.$$

These two convergences imply the $L^p(\mathbb{P})$-uniform integrability of both sequences $\Big(\sup_{t\in[0,T]} \Big| \int_0^t H_s^{(n)} dZ_s \Big|\Big)_{n\ge1}$ and $\Big(\sup_{t\in[0,T]} \Big| \int_0^t h_s^{(n)} dZ_s \Big|\Big)_{n\ge1}$. At this stage, one concludes like in the case $p \in (1,2]$.

In the general case, one proceeds by a classical "cascade" argument based on repeated applications of the B.D.G. Inequality involving the martingales (see the proof of Proposition 12 in Appendix 2 for a more detailed implementation of this cascade procedure in a similar situation)

$$M_t^{(k)} = \sum_{0\le s\le t} (H_s^{(n)}-H_s)^{2^k}(\Delta Z_s)^{2^k} - \nu(|z|^{2^k})\int_0^t (H_s^{(n)}-H_s)^{2^k} ds,\ t\ge 0,\ k=1,\dots,\ell_p.$$

We show by switching from p to $p/2$, $p/2^2,\dots,p/2^k,\dots$ until we get $p/2^{\ell_p}\in(1,2]$ when $k=\ell_p$, that

$$\Big\|\sup_{t\in[0,T]}\Big|\int_0^t H_s dZ_s - \int_0^t H_s^{(n)} dZ_s\Big|\Big\|_p \le c_p\, a\Big\||H-H^{(n)}|_{L_T^2}\Big\|_p + \kappa_{p,\nu}\sum_{\ell=1}^{\ell_p}\Big\||H^{(n)}-H|_{L_T^{2^\ell}}\Big\|_p^2$$
$$+\kappa_{p,\nu}\Big\||H^{(n)}-H|_{L_T^p}\Big\|_p^2.$$

One shows likewise the counterpart related to h and $h^{(n)}$.

Step 3 (Second Approximation). Now we have to get rid of the boundedness of h. Like in the Brownian Itô case, we approximate h by $h\wedge K$ and H by $\chi_K(H)$ where the thresholding function χ_K have been introduced in Step 3 of the proof of Theorem 2 (to take into account at the same time the symmetric and the standard settings for the Lévy process Z). Let $p\in(1,+\infty)$.

$$\Big\|\sup_{t\in[0,T]}\Big|\int_0^t H_s dZ_s - \int_0^t \chi_K(H_s) dZ_s\Big|\Big\|_p$$
$$\le c_p\left(a\Big\||H-\chi_K(H)|_{L_T^2}\Big\|_p + \Big\|\sum_{0<s\le T}(H_s-\chi_K(H_s))^2(\Delta Z_s)^2\Big\|_{\frac p2}\right)$$
$$= c_p\left(a\Big\||(|H|-K)_+|_{L_T^2}\Big\|_p + \Big\|\sum_{0<s\le T}(|H_s|-K)_+^2(\Delta Z_s)^2\Big\|_{\frac p2}\right)$$
$$\le c_p\left(a|(h-K)_+|_{L_T^2} + \Big\|\sum_{0<s\le T}(h_s-K)_+^2(\Delta Z_s)^2\Big\|_{\frac p2}\right).$$

We derive again by this cascade argument that $\Big\|\sum_{0<s\le T}(h_s-K)_+^2(\Delta Z_s)^2\Big\|_{\frac p2}$ can be upper-bounded by linear combinations of quantities of the form

$$|(h-K)_+|_{L_T^{2^k}}\nu(z^{2^k}),\ 0\le k\le \ell_p,$$

and

$$\mathbb{E}\sum_{0<s\leq T}(h_s-K)_+^p|\Delta Z_s|^p = |(h-K)_+|^p_{L^p_T}\,\nu(|z|^p).$$

Consequently, if $h\in L^p_T$, all these quantities go to zero as $K\to+\infty$, owing to the Lebesgue dominated convergence Theorem. In turn this implies that

$$\left\|\sup_{t\in[0,T]}\left|\int_0^t H_s dZ_s-\int_0^t \chi_K(H_s)dZ_s\right|\right\|_p \longrightarrow 0 \ \text{ as } K\to+\infty.$$

The same holds with h and $h\wedge K$. So it is possible to let K go to infinity in the inequality

$$\mathbb{E}F\Big(\int_0^{\cdot}\chi_K(H_s)dZs\Big)\leq\mathbb{E}F\Big(\int_0^{\cdot}\chi_K(h_s)dZs\Big)$$

to get the expected result.

(b) is proved adapting the lines of the proof Proposition 4(b) as we did for (a). The main point is to get rid of the boundedness of h i.e. to obtain the conclusion of the above Step 3 without "domination property" of H by h. The additional assumption $|H|_{L^{p\vee 2}_T}\in L^p$ clearly yields the expected conclusion.

(c) This follows from a careful reading of the proof, having in mind that terms of the form $\left\||H-H^{(n)}|_{L^2_T}\right\|_p$ vanish when $a=0$.

3.2.1 Brownian Doléans Exponentials

Our paradigm applied to Doléans exponentials yields similar results with direct applications to the robustness of Black-Scholes formula for option pricing. First we recall that the Doléans exponential of a continuous local martingale $(M_t)_{t\in[0,T]}$ is defined by

$$\mathcal{E}(M)_t = e^{M_t-\frac{1}{2}\langle M\rangle_t},\ t\in[0,T].$$

It is a martingale on $[0,T]$ if and only if $\mathbb{E}\,e^{M_t-\frac{1}{2}\langle M\rangle_t}=1$. A practical criterion for "martingality", due to Novikov, is $\mathbb{E}\,e^{\frac{1}{2}\langle M\rangle_T}<+\infty$.

Proposition 6 *Let $(W_t)_{t\in[0,T]}$, $(H_t)_{t\in[0,T]}$ and $h=(h_t)_{t\in[0,T]}$ be like in Proposition 4. Let $F:\mathcal{C}([0,T],\mathbb{R}_+)\to\mathbb{R}$ be a convex functional with $(r,\|\,.\,\|_{\sup})$-polynomial growth, $r\geq 1$.*

(a) If $(\,|H_t|\leq h_t\ dt$-a.e.$)$ $\mathbb{P}$-a.s., then

$$\mathbb{E}\,F\left(\mathcal{E}\Big(\int_0^{\cdot}H_s dW_s\Big)\right)\leq\mathbb{E}\,F\left(\mathcal{E}\Big(\int_0^{\cdot}h_s dW_s\Big)\right).$$

(b) If $(H_t \geq h_t \geq 0$ dt-a.e.) $\mathbb{P}$-a.s. and there exists $\varepsilon > 0$ such that

$$\mathbb{E}\Big(e^{r^2(1-\frac{1}{2r}+\sqrt{\frac{r-1}{r}}+\varepsilon)|H|_{L^2_T}}\Big) < +\infty,$$

then

$$\mathbb{E}\,F\Big(\mathscr{E}\Big(\int_0^{\cdot} H_s dW_s\Big)\Big) \geq \mathbb{E}\,F\Big(\mathscr{E}\Big(\int_0^{\cdot} h_s dW_s\Big)\Big).$$

Proof

(a) Step 1. For a fixed integer $n \geq 1$, we consider the sequence of random variables $(\Xi_k^n)_{k=0:n}$ recursively defined in a forward way by

$$\Xi_0^n = 1 \quad \text{and} \quad \Xi_k^n = \Xi_{k-1}^n \exp\Big(H_{t_{k-1}^n}\Delta W_{t_k^n} - \frac{T}{2n}H^2_{t_{k-1}^n}\Big), \quad k = 1,\ldots,n,$$

where $\Delta W_{t_k^n} = W_{t_k^n} - W_{t_{k-1}^n}$ and the sequences $(\xi_\ell^{n,k})_{\ell=k:n}$ defined, still in a recursive forward way, by

$$\xi_k^{n,k} = 1,\ \xi_\ell^{n,k} = \xi_{\ell-1}^{n,k}\exp\Big(h_{t_{\ell-1}^n}\Delta W_{t_\ell^n} - \frac{T}{2n}h^2_{t_{\ell-1}^n}\Big),\ \ell = k+1,\ldots,n.$$

We denote by $\widetilde{Q}^{(n)}$ the operator defined on Borel functions $f : \mathbb{R}_+ \to \mathbb{R}$ with polynomial growth by

$$\forall\, x,\, h \in \mathbb{R}_+, \quad \widetilde{Q}^{(n)}(f)(x,h) = \mathbb{E}f\Big(x\exp\big(hW_{\frac{T}{n}} - \frac{T}{2n}h^2\big)\Big).$$

It is clear that $\Big(\exp\big(hW_{\frac{T}{n}} - \frac{T}{2n}h^2\big)\Big)_{h\geq 0}$ is increasing for the convex order (i.e. a peacock as already mentioned in the introduction) since $\exp\big(hW_{\frac{T}{n}} - \frac{T}{2n}h^2\big) \stackrel{d}{\sim} \exp\big(W_{h^2\frac{T}{n}} - \frac{1}{2}\frac{T}{n}h^2\big)$ and $(e^{W_u-\frac{u}{2}})_{u\geq 0}$ is a martingale. Hence, if f is convex,

$$h \mapsto \widetilde{Q}^{(n)}(f)(x,h) \text{ satisfies the } \textit{maximum principle} \tag{21}$$

i.e. is even and non-decreasing on $\mathbb{R}_+$. In turn, it implies that the function $(x,h) \mapsto \widetilde{Q}^{(n)}(f)(x,h)$ is convex on $\mathbb{R}\times\mathbb{R}_+$ since, for every $x, x' \in \mathbb{R}_+$, $h, h' \in \mathbb{R}$, $\lambda \in [0,1]$,

$$\begin{aligned}
&\mathbb{E}f\Big(\lambda x\exp\big(\lambda hW_{\frac{T}{n}} - \frac{T}{2n}(\lambda h)^2\big) + (1-\lambda)x'\exp\big((1-\lambda)h'W_{\frac{T}{n}} - \frac{T}{2n}((1-\lambda)h')^2\big)\Big)\\
&\leq \lambda\mathbb{E}f\Big(x\exp\big(\lambda hW_{\frac{T}{n}} - \frac{T}{2n}(\lambda h)^2\big)\Big)\\
&\quad +(1-\lambda)\mathbb{E}f\Big(x'\exp\big((1-\lambda)h'W_{\frac{T}{n}} - \frac{T}{2n}((1-\lambda)h')^2\big)\Big)
\end{aligned}$$

$$\leq \lambda \mathbb{E} f\Big(x \exp\big(|h| W_{\frac{T}{n}} - \frac{T}{2n} h^2\big)\Big)$$
$$+(1-\lambda)\mathbb{E} f\Big(x' \exp\big(|h'| W_{\frac{T}{n}} - \frac{T}{2n}(h')^2\big)\Big)$$
$$= \lambda \mathbb{E} f\Big(x \exp\big(h W_{\frac{T}{n}} - \frac{T}{2n} h^2\big)\Big) + (1-\lambda)\mathbb{E} f\Big(x' \exp\big(h' W_{\frac{T}{n}} - \frac{T}{2n}(h')^2\big)\Big)$$

where we used the convexity of f in the first inequality and (21) in the second one. From now on, we consider the discrete time filtration $\mathcal{G}_k^n = \mathcal{F}_{t_k^n}^W$ and set $\mathbb{E}_k = \mathbb{E}(\,.\,|\,\mathcal{G}_k^n)$, $k = 0, \ldots, n-1$.

We temporarily assume that for every $k = 0, \ldots, n$, $|H_{t_k^n}| \leq h_{t_k^n}$ $\mathbb{P}$-*a.s.*. Let $F : \mathcal{C}([0,T], \mathbb{R}) \to \mathbb{R}$ be a (Borel) functional with $(r, \|\,.\,\|_{\sup})$-polynomial growth and let $F_n = F \circ i_n$. Now let us show that, for every $k \in \{1, \ldots, n\}$,

$$\mathbb{E}_{k-1} F_n(\Xi_{0:k-1}^n, \Xi_k^n\, \xi_{k:n}^{n,k}) \leq \mathbb{E}_{k-1} F_n(\Xi_{0:k-2}^n, \Xi_{k-1}^n \xi_{k-1:n}^{n,k-1}) \tag{22}$$

with the convention $\Xi_{0:-1}^n = \emptyset$. Starting from the identity

$$F_n(\Xi_{0:k-1}^n, \Xi_k^n\, \xi_{k:n}^{n,k}) = F_n\Big(\Xi_{0:k-1}^n, \Xi_{k-1}^n \exp\big(H_{t_{k-1}^n} \Delta W_{t_k^n} - \frac{T}{2n} H_{t_{k-1}^n}^2\big)\xi_{k:n}^{n,k}\Big),$$

we derive

$$\mathbb{E}_{k-1} F_n(\Xi_{0:k-1}^n, \Xi_k^n\, \xi_{k:n}^{n,k})$$
$$= \Big[\mathbb{E}\Big(F(x_{0:k-1}, x_{k-1} \exp\big(\eta \Delta W_{t_k^n} - \frac{T}{2n}\eta^2\big)\xi_{k:n}^{n,k})\Big)\Big]_{\substack{|x_{0:k-1} = \Xi_{0:k-1}^n, \\ |\eta = H_{t_{k-1}^n}}}$$

since $(\Xi_{0:k-1}^n, H_{t_{k-1}^n})$ is $\mathcal{G}_{k-1}^n$-measurable and $(\Delta W_{t_k^n}, \xi_{k:n}^{n,k})$ is independent of $\mathcal{G}_{k-1}^n$. Now set, for every $x_{0:k-1} \in \mathbb{R}_+^k$, $\tilde{x}_k \in \mathbb{R}_+$,

$$G_{n,k}(x_{0:k-1}, \tilde{x}_k) = \mathbb{E} F_n\big(x_{0:k-1}, \tilde{x}_k\, \xi_{k:n}^{n,k}\big)$$

so that

$$\widetilde{Q}^{(n)}(G_{n,k}(x_{0:k-1}, .))(x_{k-1}, \eta) = \mathbb{E}\, F_n\Big(x_{0:k-1}, x_{k-1} \exp\big(\eta \Delta W_{t_k^n} - \frac{T}{2n}\eta^2\big)\xi_{k:n}^{n,k}\Big).$$

The function F_n being convex on $\mathbb{R}_+^{n+1}$, it is clear that $G_{n,k}$ is convex on $\mathbb{R}_+^{k+1}$ as well. It is in particular convex in the variable $\tilde{x}_k$ which in turn implies by (21) that $\eta \mapsto \widetilde{Q}^{(n)}(G_{n,k}(x_{0:k-1}, .))(x_{k-1}, \eta)$ satisfies the maximum principle i.e. is even and convex. As a consequence, $|H_{t_{k-1}^n}| \leq h_{t_{k-1}^n}$ implies

$$\mathbb{E}_{k-1} F_n(\Xi_{0:k-1}^n, \Xi_k^n\, \xi_{k:n}^{n,k}) = \big[\widetilde{Q}^{(n)}\big(G_{n,k}(x_{0:k-1},\ .)\big)(x_{k-1}, \eta)\big]_{|x_{0:k-1} = \Xi_{0:k-1}^n, \eta = H_{t_{k-1}^n}}$$
$$= \big[\widetilde{Q}^{(n)}\big(G_{n,k}(x_{0:k-1},\ .)\big)(x_{k-1}, \eta)\big]_{|x_{0:k-1} = \Xi_{0:k-1}^n, \eta = |H_{t_{k-1}^n}|}$$

$$\leq \left[\widetilde{Q}^{(n)}(G_{n,k}(x_{0:k-1},\,.))(x_{k-1},\eta)\right]_{|x_{0:k-1}=\Xi^n_{0:k-1},\eta=h_{t^n_{k-1}}}$$

$$= \mathbb{E}_{k-1}\Big(F_n\big(\Xi^n_{0:k-1},\Xi^n_{k-1}\exp\big(h_{t^n_{k-1}}\Delta W_{t^n_k}-\frac{T}{2n}h^2_{t^n_{k-1}}\big)\xi^{n,k}_{k:n}\big)\Big)$$

$$= \mathbb{E}_{k-1}\Big(F_n\big(\Xi^n_{0:k-2},\Xi^n_{k-1}\xi^{n,k-1}_{k-1:n}\big)\Big)$$

where we used once again that $\xi^{n,k}_{k:n}$ is independent of $\mathcal{G}^n_{k-1}$ in the penultimate line.

One derives by taking expectation of the resulting inequality that the sequence $\mathbb{E}F_n(\Xi^n_{0:k-1},\Xi^n_k\,\xi^{n,k}_{k:n})$, $k = 1 : n$, is non-increasing. Finally, by comparing the terms for $k = n$ and $k = 0$, we get

$$\mathbb{E}F\big(X^{n,n}\big) = \mathbb{E}\,F_n(\Xi^n_{0:n}) \leq \mathbb{E}\,F_n(\xi^{n,0}_{0:n}) = \mathbb{E}F\big(X^{n,0}\big).$$

Step 2 (Approximation-Regularization). We closely follow the approach developed in Steps 2 and 3 of Proposition 4. First, we temporarily assume that h is bounded by a real constant K and we introduce the stepwise constant càglàd processes $(H^{(n)})_{t\in[0,T]}$ and $(h^{(n)}_t)_{t\in[0,T]}$ defined by (13). Both satisfy (14), namely

$$\Big\|\,|H^{(n)}-H|_{L^2_T}\Big\|_2 + \big|h^{(n)}-h\big|_{L^2_T} \longrightarrow 0 \text{ as } n\to+\infty.$$

Now, as

$$\sup_{t\in[0,T]}\left|\int_0^t (H^{(n)}_s)^2 ds - \int_0^t H_s^2 ds\right| \leq 2K\big|H^{(n)}-H\big|_{L^1_T} \leq 2K\sqrt{T}\big|H^{(n)}-H\big|_{L^2_T},$$

we get

$$\sup_{t\in[0,T]}\left|\int_0^t H^{(n)}_s dW_s - \frac{1}{2}\int_0^t (H^{(n)}_s)^2 ds - \Big(\int_0^t H_s dW_s - \frac{1}{2}\int_0^t H_s^2 ds\Big)\right|$$
$$\leq \sup_{t\in[0,T]}\left|\int_0^t (H^{(n)}_s - H_s)dW_s\right| + K\sqrt{T}\big|H^{(n)}-H\big|_{L^2_T}.$$

For notational convenience, we temporarily set

$$X^{(n)}_t = \mathcal{E}\Big(\int_0^{\cdot} H^{(n)}_s dW_s\Big)_t \quad\text{and}\quad X_t = \mathcal{E}\Big(\int_0^{\cdot} H_s dW_s\Big)_t,\ t\in[0,T],$$

which are both true martingales owing to Novikov's criterion. The above inequality combined with Doob's Inequality implies that

$$\sup_{t\in[0,T]}\left|\log X^{(n)}_t - \log X_t\right| \xrightarrow{L^2} 0 \text{ as } n\to+\infty.$$

As a consequence, $X^{(n)} \overset{\mathcal{L}(\|\cdot\|_{\sup})}{\longrightarrow} X$ since the exponential function is continuous. Denoting by $x^{(n)}$ and x the counterpart of these processes for the functions $h^{(n)}$ and h, we get likewise $x^{(n)} \overset{\mathcal{L}(\|\cdot\|_{\sup})}{\longrightarrow} x$. Owing once again to Lemma 2, the continuity of the exponential, and the chain rule for weak convergence, we finally obtain

$$e^{I_{m(n)}(\log X^{(n)})} \overset{\mathcal{L}(\|\cdot\|_{\sup})}{\longrightarrow} e^{\log X} = X \text{ and } e^{I_{m(n)}(\log x^{(n)})} \overset{\mathcal{L}(\|\cdot\|_{\sup})}{\longrightarrow} e^{\log x} = x \text{ as } n \to +\infty.$$

Applying Step 1 with $X^{(n)}$ and $x^{(n)}$ yields

$$\forall\, n \in \mathbb{N}, \quad \mathbb{E}\, F(X^{(n)}) \leq \mathbb{E}\, F(x^{(n)}).$$

To let n go to infinity in this inequality, we again need a uniform integrability argument namely that $\|X^{(n)}\|_{\sup}$ and $\|x^{(n)}\|_{\sup}$ are both L^p-bounded for a $p > r$ since the functional F has at most a $(r, \|\,.\,\|_{\sup})$-polynomial growth. So, let $p > r \vee 1$. It follows from Doob's Inequality applied to the non-negative sub-martingale $(X^{(n)})^p$ that

$$\begin{aligned}
\mathbb{E}\Big(\sup_{t\in[0,T]} (X_t^{(n)})^p\Big) &\leq \Big(\frac{p}{p-1}\Big)^p \mathbb{E}(X_T^{(n)})^p \\
&\leq \Big(\frac{p}{p-1}\Big)^p \mathbb{E}\left(\mathscr{E}\Big(p\int_0^{\cdot} H_s^{(n)} dW_s\Big)_T e^{\frac{p(p-1)}{2}\int_0^T (H_s^{(n)})^2 ds}\right) \\
&\leq \Big(\frac{p}{p-1}\Big)^p e^{\frac{p(p-1)}{2}K^2 T}
\end{aligned}$$

where we used that $\Big(\mathscr{E}\big(p\int_0^{\cdot} H_s^{(n)} dW_s\big)_t\Big)_{t\geq 0}$ is a true martingale, owing to Novikov's criterion. The case of $F(x^{(n)})$ follows likewise.

Step 3. The extension to $h \in L_T^2$ is similar to that performed in the former propositions: first note that

$$\mathscr{E}\Big(\int_0^{\cdot} \chi_K(H_s)dW_s\Big) \overset{\mathcal{L}(\|\cdot\|_{\sup})}{\longrightarrow} \mathscr{E}\Big(\int_0^{\cdot} H_s dW_s\Big) \quad \text{as } K \to +\infty.$$

The uniform integrality of $\sup_{t\in[0,T]} \mathscr{E}\Big(\int_0^{\cdot} \chi_K(H_s)dW_s\Big)_t$ as K grows to infinity follows from its $L^p(\mathbb{P})$-boundedness for a $p \in (1, +\infty)$ which in turn is a consequence of Doob's inequality as above:

$$\sup_{K>0}\mathbb{E}\sup_{t\in[0,T]}\left(\mathcal{E}\left(\int_0^{\cdot}\chi_K(H_s)dW_s\right)_t\right)^p\le\left(\frac{p}{p-1}\right)^p e^{\frac{p(p-1)}{2}|h|^2_{L^2_T}}<+\infty$$

(*b*) The discrete time part is established by adapting item (*a*) in the spirit of Proposition 4(*b*). The approximation step follows like above as well, except for the final uniform integrability argument which needs specific care. It suffices to show that $\sup_{t\in[0,T]}\mathcal{E}\left(\int_0^{\cdot}\chi_K(H_s)dW_s\right)_t$ is L^p-bounded as $K\to+\infty$ for a $p>r$.

Let $p>r$. Combining successively Doob's Inequality and Hölder's Inequality for every Hölder conjugate exponents $\lambda,\mu=\frac{\lambda}{\lambda-1}>1$, leads to

$$\mathbb{E}\Big[\sup_{t\in[0,T]}\Big(\mathcal{E}\Big(\int_0^{\cdot}\chi_K(H_s)dW_s\Big)_t\Big)^p\Big]\le\Big(\frac{p}{p-1}\Big)^p\mathbb{E}\Big[\mathcal{E}\Big(\int_0^{\cdot}\chi_K(H_s)dW_s\Big)_T^p\Big]$$

$$\le\Big(\frac{p}{p-1}\Big)^p\Bigg[\underbrace{\mathbb{E}\Big[\mathcal{E}\Big(\lambda p\int_0^{\cdot}\chi_K(H_s)dW_s\Big)_T\Big]}_{=1}\Bigg]^{\frac{1}{\lambda}}\Big[\mathbb{E}\,e^{\frac{\lambda p(\lambda p-1)}{2(\lambda-1)}\int_0^T\chi_K(H_s)^2ds}\Big]^{\frac{\lambda-1}{\lambda}}$$

$$\le\Big(\frac{p}{p-1}\Big)^p\Big[\mathbb{E}\,e^{\frac{\lambda p(\lambda p-1)}{2(\lambda-1)}|H|^2_{L^2_T}}\Big]^{\frac{\lambda-1}{\lambda}}.$$

Now $\min_{\lambda>1}\frac{\lambda(\lambda p-1)p}{2(\lambda-1)}=p^2(1-\frac{1}{2p}+\sqrt{\frac{p}{p-1}})$ which can be made lower than $r^2(1-\frac{1}{2r}+\sqrt{\frac{r}{r-1}})+\varepsilon$ for p close enough to r.

3.2.2 A Counter-Example

The counter-example below shows that Theorem 4 is no longer true if we relax the assumption that the dominating process $(h_t)_{t\in[0,T]}$ is deterministic.

Let $X=X^\sigma=(X^\sigma_{0:2})$ be a two period process satisfying

$$X_0=0,\quad X_1=\sigma Z_1\ \text{ and }\ X_2=X_1+\sqrt{2v(Z_1)}Z_2$$

where $Z_{1:2}\overset{\mathcal{L}}{\sim}\mathcal{N}(0;I_2)$, $\sigma\ge 0$, and $v:\mathbb{R}\to\mathbb{R}_+$ is a bounded non-increasing function.

Let $f(x)=e^x$ and let $\varphi:\mathbb{R}_+\to\mathbb{R}$ be the function defined by

$$\varphi(\sigma):=\mathbb{E}f(X_2)=\mathbb{E}\big(e^{\sigma Z_1+v(Z_1)}\big).$$

Differentiating φ yields

$$\varphi'(\sigma) = \mathbb{E}\big(e^{\sigma Z_1 + v(Z_1)} Z_1\big)$$

so that

$$\varphi'(0) = \mathbb{E}\big(e^{v(Z_1)} Z_1\big) < \mathbb{E}\, e^{v(Z_1)} \mathbb{E}\, Z_1 = 0$$

by a standard one-dimensional co-monotony argument: the functions $z \mapsto e^{v(z)}$ and $z \mapsto z$ are non-increasing and non-decreasing respectively which implies $\varphi'(0) \le 0$ but none of them are $\mathbb{P}_{Z_1}$-$a.s.$ constant, hence equality cannot hold. As a consequence $\varphi'(0) < 0$ so that φ is (strictly) decreasing on a right neighbourhood $[0, \sigma_0]$, $\sigma_0 > 0$, of 0.

To include this into a Brownian stochastic integral framework, one proceeds as follows: let W be a standard Brownian motion and $\sigma, \tilde{\sigma} \in (0, \sigma_0]$, $\sigma < \tilde{\sigma}$.

$$H_t = \sigma \mathbf{1}_{[0,1]}(t) + \sqrt{2v(W_1)}\mathbf{1}_{(1,2]}(t), \; \widetilde{H}_t = \tilde{\sigma} \mathbf{1}_{[0,1]}(t) + \sqrt{2v(W_1)}\mathbf{1}_{(1,2]}(t).$$

It is clear that $0 \le H_t \le \widetilde{H}_t$, $t \in [0, 2]$, whereas

$$\mathbb{E}\Big(e^{\int_0^2 H_s dW_s}\Big) > \mathbb{E}\Big(e^{\int_0^2 \widetilde{H}_s dW_s}\Big)$$

which contradicts the conclusion of Proposition 4(a).

It has to be noted that if the function v is non-decreasing, then choosing $f(x) = e^{-x}$ leads to a similar result since $\psi(\sigma) := \mathbb{E} f(X_2) = \mathbb{E}\big(e^{-\sigma Z_1 + v(Z_1)}\big)$ satisfies $\Psi'(\sigma) = -\mathbb{E}\big(e^{-\sigma Z_1 + v(Z_1)}\big)$. In particular one still has by a co-monotony argument that $\psi'(0) < 0$ since v is not constant.

3.2.3 A Comparison Theorem for Laplace Transforms of Brownian Stochastic Integrals

Applying our paradigm, we start by a discrete time result with its own interest for applications.

Proposition 7 *Let $(Z_k)_{1 \le k \le n}$ be a sequence of $\mathcal{N}(0;1)$-distributed random variables. We set $S_0 = 0$ and $S_k = Z_1 \cdots + Z_k$, $k = 1, \ldots, n$ (partial sums). We consider the two discrete time stochastic integrals*

$$X_k = \sum_{\ell=1}^{k} f_\ell(S_{\ell-1}) Z_\ell \quad \text{and} \quad Y_k = \sum_{\ell=1}^{k} g_\ell(S_{\ell-1}) Z_\ell, \quad k = 1, \ldots, n, \quad X_0 = Y_0 = 0$$

where $f_k, g_k : \mathbb{R} \to \mathbb{R}_+$, $k = 1, \ldots, n$ are non-negative Borel functions satisfying: either all f_k, $k = 1, \ldots, n$, or all g_k, $k = 1, \ldots, n$, are non-decreasing.

If, furthermore, $0 \le f_k \le g_k$ *for all* $k = 1, \dots, n$*, then*

$$\forall \lambda \ge 0, \quad \mathbb{E}\, e^{\lambda X_n} \le \mathbb{E}\, e^{\lambda Y_n}.$$

Proof We start from the Cameron-Martin identity which reads on Borel functions $\varphi : \mathbb{R} \to \mathbb{R}$

$$\forall \sigma \in \mathbb{R}, \quad \mathbb{E}\, e^{\sigma Z + \varphi(Z)} = e^{\frac{\sigma^2}{2}} \mathbb{E}\, e^{\varphi(Z+\sigma)} \le +\infty.$$

First, we define in a backward way functions $\tilde{f}_k$ and $\tilde{g}_k$, $k = 1, \dots, n+1$ by $\tilde{f}_{n+1} = \tilde{g}_{n+1} \equiv 0$,

$$\tilde{f}_k(x) = \frac{\lambda^2}{2} \tilde{f}_k^2(x) + \log \mathbb{E}\Big(e^{\tilde{f}_{k+1}(x + \lambda f_k(x) + Z)}\Big), \quad k = 0, \dots, n, \tag{23}$$

where $Z \sim \mathcal{N}(0;1)$. The function $\tilde{g}_k$ is defined from the g_k the same way round. Then, relying on the chaining rule for conditional expectations, we check by a backward induction that

$$\mathbb{E}\, e^{\lambda X_n} = \mathbb{E}\, e^{\lambda X_k + \tilde{f}_{k+1}(S_k)}, \quad k = 1, \dots, n.$$

In particular, when $k = 0$, we get

$$\mathbb{E}\, e^{\lambda X_n} = e^{\tilde{f}_1(0)}.$$

It follows from (23) and a second backward induction that, if the functions f_k are non-decreasing for every $k = 1, \dots, n$, so are the functions $\tilde{f}_k$. The same holds for $\tilde{g}_k$ with respect to the functions g_k. Assume e.g. that all the functions $\tilde{f}_k$ are non-decreasing. Then, a third backward induction shows: that $\tilde{f}_k \le \tilde{g}_k$ for every $k = 0, \dots, n-1$. It is clear that $\tilde{f}_n \le \tilde{g}_n$. If $\tilde{f}_{k+1} \le \tilde{g}_{k+1}$, then for every $x \in \mathbb{R}$,

$$\tilde{f}_{k+1}(x + \lambda f_k(x) + Z) \le \tilde{f}_{k+1}(x + \lambda g_k(x) + Z) \le \tilde{g}_{k+1}(x + \lambda g_k(x) + Z).$$

Plugging this inequality in (23) combined with $f_k^2 \le g_k^2$, one concludes that $\tilde{f}_k \le \tilde{g}_k$. A similar reasoning can be carried out if the functions $\tilde{g}_k$ are non-decreasing.

By the standard weak approximation method detailed beforehand, we derive a following continuous time version involving (non-decreasing) *completely monotone* functions defined below.

Definition 2 A *non-decreasing* function $\varphi : \mathbb{R} \to \mathbb{R}$ is completely monotone if it is the Laplace transform of a non-negative Borel measure μ supported by the non-negative real line, namely

$$\forall x \in \mathbb{R}, \quad \varphi(x) = \int_{\mathbb{R}_+} e^{\lambda x} \mu(d\lambda).$$

Theorem 3 *Let $f, g : [0, T] \times \mathbb{R} \to \mathbb{R}_+$ two bounded Borel functions such that*

$$\begin{cases} (i) \;\; f,\, g \text{ are } dt \otimes dx\text{-a.e. continuous}, \\ (ii) \;\; 0 \le f \le g, \\ (iii) \;\; \Big(\forall\, t \in [0,T],\; f(t,.) \text{ is non-decreasing}\Big) \\ \qquad\quad \text{or } \Big(\forall\, t \in [0,T],\; g(t,.) \text{ is non-decreasing}\Big). \end{cases} \tag{24}$$

Then,

$$\forall\, \lambda \ge 0, \quad \mathbb{E}\, e^{\lambda \int_0^T f(t,W_t)dW_t} \le \mathbb{E}\, e^{\lambda \int_0^T g(t,W_t)dW_t}$$

so that, for every non-decreasing completely monotone *function $\varphi : \mathbb{R} \to \mathbb{R}_+$*

$$\mathbb{E}\,\varphi\left(\int_0^T f(t, W_t)dW_t\right) \le \mathbb{E}\,\varphi\left(\int_0^T g(t, W_t)dW_t\right).$$

Remark 6

- The finiteness of these integrals follows from Novikov's criterion.
- One derives from (24) the seemingly more general result

$$\begin{cases} (i) \;\; f, g \text{ are } dt \otimes dx\text{-}a.e. \text{ continuous}, \\ (ii) \;\; \exists\, h : [0,T] \times \mathbb{R} \overset{Borel}{\longrightarrow} \mathbb{R}_+ \text{ such that } \begin{cases} (a) \;\; 0 \le f \le h \le g \text{ and} \\ (b) \;\; \forall\, t \in [0,T],\; h(t,.) \text{ is non-decreasing}. \end{cases} \end{cases} \tag{25}$$

Proof Assume e.g. that $f(t, .)$ is non-decreasing for every $t \in [0, T]$. First note that by Fubini's Theorem and Itô's isometry

$$\left\| \int_0^T f(s, W_s)dW_s - \int_0^T f(\underline{s}_n, W_{\underline{s}_n})dW_s \right\|_2^2 = \int_0^T \mathbb{E}\big(f(s, W_s) - f(\underline{s}_n, W_{\underline{s}_n})\big)^2 ds.$$

Now, if we denote $C_s = \{x \in \mathbb{R} \,|\, f \text{ is continuous at } (s, x)\}$ for every $t \in [0, T]$, it follows from Assumption (24)(i) that $\lambda(^cC_s) = 0$ *ds-a.e.* still by Fubini's Theorem. As $\mathbb{P}_{X_s}$ is equivalent to the Lebesgue measure, one derives that $\mathbb{P}_s(C_s) = 1$ *ds-a.e.*. As a consequence, $\mathbb{E}\big(f(s, W_s) - f(\underline{s}_n, W_{\underline{s}_n})\big)^2 \to 0$ *ds-a.e.* as $n \to +\infty$ since $(\underline{s}_n, W_{\underline{s}_n}) \to (s, W_s)$. One concludes by the dominated Lebesgue theorem that $\left\| \int_0^T f(s, W_s)dW_s - \int_0^T f(\underline{s}_n, W_{\underline{s}_n})dW_s \right\|_2 \to 0$ as $n \to +\infty$ since f is bounded.

Now, define for every $k = 1, \ldots, n$,

$$X_k = \int_0^{t_k^n} f(\underline{s}_n, W_{\underline{s}_n})dW_s = \sqrt{\frac{T}{n}} \sum_{\ell=1}^{k} f(t_{\ell-1}^n, W_{t_{\ell-1}^n})U_\ell^n$$

where $U_\ell^n = \sqrt{\frac{n}{T}}(W_{t_\ell^n} - W_{t_{\ell-1}^n})$, $\ell = 1, \dots, n$. We define likewise $(Y_k)_{k=0:n}$ with respect to the function g. It is clear that both (X_k) and (Y_k) satisfy the assumptions of the above Proposition 7 so that

$$\forall\,\lambda \geq 0, \qquad \mathbb{E}\, e^{\lambda \int_0^T f(\underline{s}_n, W_{\underline{s}_n})dW_s} \leq \mathbb{E}\, e^{\lambda \int_0^T g(\underline{s}_n, W_{\underline{s}_n})dW_s}.$$

One concludes by combining the above quadratic (hence weak) convergence and the uniform integrability argument which follows from

$$\forall\,\lambda > 0, \qquad \sup_n \mathbb{E}\, e^{\lambda \int_0^T f(\underline{s}_n, W_{\underline{s}_n})dW_s} \leq e^{\frac{\lambda^2}{2}\|f\|_{\sup} T} < +\infty.$$

4 Convex Order for the *Réduites* and Applications to Path-Dependent American Options

In this section, we aim at applying the paradigm developed in the former sections to Optimal Stopping Theory, which corresponds in Quantitative finance to Bermuda and American style options. For background on Optimal stopping theory, we refer to [27, Chap. VI] and [7, Chap. 5.1] in discrete time and, among others, to [8, 19, 33] in continuous time.

4.1 Bermuda Options

We start from the discrete time dynamics introduced in Sect. 2. Let $(Z_k)_{k=1:n}$ be a sequence of centered independent random variables satisfying $Z_k \in L^r(\Omega, \mathscr{A}, \mathbb{P})$, $r \geq 1$ and $\mathbb{E}\, Z_k = 0$, $k = 1, \dots, n$. Let $(X_k^x)_{k=0:n}$ and $(Y_k^x)_{k=0:n}$ be the two sequences of random vectors defined by (1) i.e.

$$X_{k+1}^x = X_k^x + \sigma_k(X_k^x)Z_{k+1}, \quad Y_{k+1}^x = Y_k^x + \theta_k(Y_k^x)Z_{k+1},\ 0 \leq k \leq n-1,\ X_0^x = Y_0^x = x$$

where σ_k, θ_k, $k = 0, \dots, n$ are functions from $\mathbb{R}$ to $\mathbb{R}$, all with linear growth. This implies by a straightforward induction that the random variable X_k^x and Y_k^x all lie in L^r since, e.g., $\sigma_k(X_k^x)$ is $\mathscr{F}_k^Z$-measurable, hence independent of Z_{k+1}, $k = 0, \dots, n-1$.

Let $\mathscr{F} = (\mathscr{F}_k)_{k=0,\dots,n}$ and $\mathscr{G} = (\mathscr{G}_k)_{k=0,\dots,n}$ be two filtrations on $(\Omega, \mathscr{A}, \mathbb{P})$ such that X^x is $\mathscr{F}$-adapted and Y^x is $\mathscr{G}$-adapted. Let $F_k : \mathbb{R}^{k+1} \to \mathbb{R}_+$, $k = 0 : n$ be a sequence of non-negative functions with *r -polynomial growth* (i.e. $0 \leq F_k(x_{0:k}) \leq C(1 + |x_{0:k}|^r)$, $k = 0 : n$). Then, the processes $\big(F_k(X_{0:k}^x)\big)_{k=0:n}$ and $(F_k(Y_{0:k}^x))_{k=0:n}$, called *payoff* or *obstacle* processes, are $\mathscr{F}$-adapted and $\mathscr{G}$-adapted respectively.

We define the $\mathcal{F}$- and $\mathcal{G}$-"*réduites*" associated to these obstacle processes by

$$u_0(x) = \sup\left\{\mathbb{E}\,F_\tau(X^x_{0:\tau}),\ \tau\ \mathcal{F}\text{-stopping time}\right\}$$

and

$$v_0(x) = \sup\left\{\mathbb{E}\,F_\tau(Y^x_{0:\tau}),\ \tau\ \mathcal{G}\text{-stopping time}\right\}$$

respectively. Each quantity is closely related to the optimal stopping problem attached to its underlying dynamics X^x (Y^x respectively) since it represents the supremum of all possible gains among "honest" (i.e. non-anticipative with respect to the filtration) stopping strategies in a game where the player wins $F_k(X^x_{0:k})$ ($F_k(Y^x_{0:k})$ respectively) when leaving at time k. Owing to the dynamic programming formula and the Markov property shared by both dynamics X^x and Y^x (see the proof of Proposition 8 below), it is clear that we may assume without loss of generality that $\mathcal{F} = \mathcal{F}^X$ (natural filtration of X^x), $\mathcal{G} = \mathcal{F}^Y$ or even $\mathcal{F} = \mathcal{G} = \mathcal{F}^Z$ without changing the value of the *réduites*.

The proposition below is the counterpart of Proposition 1 from Sect. 2.

Proposition 8 *Let* $F_k : \mathbb{R}^{k+1} \to \mathbb{R}_+$, $k = 0,\ldots,n$, *be a sequence of non-negative* convex *functions with r-polynomial growth,* $r \geq 1$.

(a) Convex Partitioning function*: If, for every* $k \in \{0,\ldots,n-1\}$, *there exists a convex function* κ_k *such that* $0 \leq \sigma_k \leq \kappa_k \leq \theta_k$, *then, for every* $x \in \mathbb{R}$,

$$u_0(x) \leq v_0(x).$$

(b) Convex Dominating function*: If the random variables* Z_k *have symmetric distributions, the functions* θ_k, $k = 0,\ldots,n-1$, *are convex and* $|\sigma_k| \leq \theta_k$, $k = 0,\ldots,n-1$, *then the above inequality remains true.*

Remark 7 An equivalent formulation of claim (a) is: assume that both $(\sigma_k)_{k=0,\ldots,n-1}$ and $(\theta_k)_{k=0,\ldots,n-1}$ are non-negative convex functions with r-linear growth, then for every sequence $(\kappa_k)_{k=0,\ldots,n-1}$ of functions such that $\sigma_k \leq \kappa_k \leq \theta_k$, $k = 0,\ldots,n-1$,

$$u_0(x) \leq c_\kappa(x) \leq v_0(x)$$

where $c_\kappa(x)$ is the *réduite* of $(F_k(K^x_{0:k}))_{k=0,\ldots,n}$ where $(K^x_k)_{k=0,\ldots,n}$ satisfies the discrete time dynamics

$$K^x_{k+1} = K^x_k + \kappa_k(K^x_k)Z_{k+1},\ k = 0,\ldots,n-1,\ K^x_0 = x.$$

This follows from claim (a) applied successively to the pairs $(\sigma_k,\kappa_k)_{k=0,\ldots,n-1}$ and $(\kappa_k,\theta_k)_{k=0,\ldots,n-1}$.

Proof

(a) It is clear that this claim is equivalent to proving the expected inequality, either if all the functions $(\sigma_k)_{k=0,\ldots,n}$, or all the functions $(\theta_k)_{k=0,\ldots,n}$ are convex.

We introduce $U^x = (U_k^x)_{k=0,\dots,n}$ and $V^x = (V_k^x)_{k=0,\dots,n}$ the $(\mathbb{P},\mathcal{F})$- and $(\mathbb{P},\mathcal{G})$-Snell envelopes of $\big(F_k(X_{0:k}^x)\big)_{k=0,\dots,n}$ and $\big(F_k(Y_{0:k}^x)\big)_{k=0,\dots,n}$ respectively i.e.

$$U_k^x = \mathbb{P}\text{- ess sup}\Big\{\mathbb{E}\big(F_\tau(X_{0:\tau}^x) \mid \mathcal{F}_k\big),\ \tau\ \mathcal{F}\text{-stopping time}, \tau \geq k\Big\}$$

and

$$V_k^x = \mathbb{P}\text{- ess sup}\Big\{\mathbb{E}\big(F_\tau(Y_{0:\tau}^x) \mid \mathcal{G}_k\big),\ \tau\ \mathcal{G}\text{-stopping time}, \tau \geq k\Big\}.$$

The connection between *réduite* and Snell envelope is a classical fact from Optimal Stopping Theory for which we refer e.g. to [27, Chap. VI], namely

$$u_0(x) = \mathbb{E}\, U_0^x$$

(idem for v_0, V_0^x for Y^x). It is also classical background on Optimal stopping theory (see again e.g. [27, Chap. VI]) that the $(\mathbb{P},\mathcal{F})$-Snell envelope U^x satisfies the following Backward Dynamic Programming principle

$$U_n^x = F_n(X_{0:n}^x),\ U_k^x = \max\Big(F_k(X_{0:k}^x), \mathbb{E}(U_{k+1}^x \mid \mathcal{F}_k)\Big),\ k = 0,\dots,n-1.$$

Then, we derive by a backward induction from the Markov dynamics satisfied by the X_k^x that $U_k^x = u_k(X_{0:k}^x)$ *a.s.*, $k = 0,\dots,n$, where $u_k : \mathbb{R}^{k+1} \to \mathbb{R}_+$, $k = 0,\dots,n$ are Borel functions satisfying

$$u_n = F_n,\ u_k(x_{0:k}) = \max\Big(F_k(x_{0:k}), \big(Q_{k+1}u_{k+1}(x_{0:k}, x_k + .)\big)(\sigma_k(x_k))\Big),\ k = 0,\dots,n-1. \tag{26}$$

that is a backward dynamic programming principle in distribution. We define likewise the functions $v_k : \mathbb{R}^{k+1} \to \mathbb{R}$, $k = 0,\dots,n$, related to the $(\mathbb{P},\mathcal{G})$-Snell envelopes of $\big(F_k(Y_{0:k}^x)\big)_{k=0,\dots,n}$.

To emphasize the analogy with the proof of Proposition 1 we will detail the case where all the functions $\sigma_k = \kappa_k$ are convex and satisfy $0 \leq \sigma_k \leq \theta_k$, $k = 0,\dots,n-1$. Following the lines of the proof of this proposition, we show, still by induction, that the functions u_k are convex by combining Lemma 1 and (26). The additional argument to ensure the propagation of convexity is to note that the function $(u,v) \mapsto \max(u,v)$ is convex and increasing in each of its variable u and v.

On the other hand, as $0 \leq \sigma_k \leq \theta_k$, $k = 0,\dots,n-1$ and σ_k are all convex, we can show by a new backward induction that $u_k \leq v_k$, $k = 0,\dots,n$. If $k = n$, this is obvious. If it holds true for $k+1 \leq n$, then for every $x_{0:k} \in \mathbb{R}^{k+1}$,

$$\begin{aligned} u_k(x_{0:k}) &\leq \max\Big(F_k\big(x_{0:k}\big), \big(Q_{k+1}u_{k+1}(x_{0:k}, x_k + .)\big)(\theta_k(x_k))\Big) \\ &\leq \max\Big(F_k\big(x_{0:k}\big), \big(Q_{k+1}v_{k+1}(x_{0:k}, x_k + .)\big)(\theta_k(x_k))\Big) = v_k(x_{0:k}) \end{aligned}$$

where we used successively that $u \mapsto \big(Q_{k+1}u_{k+1}(x_{0:k}, x_k + .)\big)(u)$ is non-decreasing on $\mathbb{R}_+$ since u_{k+1} is convex and that $u_{k+1} \le v_{k+1}$. Finally, the inequality for $k = 0$ reads

$$u_0(x) = \mathbb{E}\, U_0^x \le \mathbb{E}\, V_0^x = v_0(x)$$

which yields the announced result. Other cases follow the same way round, following the lines of the proof of Proposition 1.

4.2 Continuous Time Optimal Stopping and American Options

4.2.1 Brownian Diffusions

In this section, we switch to continuous time. We will investigate the functional convex order properties of the *réduite* of obstacle/payoff processes of the form $(F(t, X^t))_{t\in[0,T]}$, where X^t is a Brownian martingale diffusion [like that defined in (4)] stopped process at time $t \in [0, T]$. The functional F defined on $[0, T] \times \mathscr{C}([0, T], \mathbb{R})$ satisfies: $F(t, .)$ is convex for every $t \in [0, T]$.

As far as pricing American options in local volatility models is concerned, the results of this section appear as an extension to path-dependent payoffs of El Karoui-Jeanblanc-Shreve's Theorem (see [9]) which mainly deals with convex functions of the marginal of the processes at time T (see also [14] devoted to path-dependent lookback options). The proposition below is also close to former results by Bergenthum and Rüschendorf by combining Theorems 3.2 and 3.6 from [1] with Theorem 4.1 from [3]. Here, we focus on the convex partitioning function.

Proposition 9 *Let σ, θ : $[0, T] \times \mathbb{R} \to \mathbb{R}_+$ be two Lipschitz continuous functions in (t, x) and let W be a standard $\mathscr{F} = (\mathscr{F}_t)_{t\ge0}$-Brownian motion defined on a probability space $(\Omega, \mathscr{A}, \mathbb{P})$ where $\mathscr{F}$ is a filtration satisfying the usual conditions. Let $(X_t^{(\sigma),x})_{t\in[0,T]}$ and $(X_t^{(\theta),x})_{t\in[0,T]}$ be the martingale unique strong solutions to* (4) *starting at $x \in \mathbb{R}$ (where $W^{(\sigma)} = W^{(\theta)} = W$).*

Assume that there exists a partitioning function κ : $[0, T] \times \mathbb{R} \to \mathbb{R}_+$ such that $\kappa(t, .)$ is convex for every $t \in [0, T]$ with linear growth in x uniformly in $t \in [0, T]$ and

$$0 \le \sigma(t, .) \le \kappa(t, .) \le \theta(t, .),\ t \in [0, T].$$

Let F : $[0, T] \times \mathscr{C}([0, T], \mathbb{R}) \to \mathbb{R}_+$ be a continuous functional for the product topology $|.|\otimes\|\,.\,\|_{\sup}$, with $(r, \|\,.\,\|_{\sup})$-polynomial growth, $r \ge 1$, in $\alpha \in \mathscr{C}([0, T], \mathbb{R})$, uniformly in $t \in [0, T]$. Moreover, assume that, for every $t \in [0, T]$, $F(t, .)$ is convex *on $\mathscr{C}([0, T], \mathbb{R})$. Let $u_0(x)$ and $v_0(x)$ denote the $\mathscr{F}$*-réduites *of $\big(F(t, (X^{(\sigma),x})^t)\big)_{t\in[0,T]}$*

and $\big(F(t,(X^{(\theta),x})^t)\big)_{t\in[0,T]}$ *respectively defined by*

$$u_0(x)=\sup\Big\{\mathbb{E}\,F\big(\tau,(X^{(\sigma),x})^{\tau}\big),\ \tau\in\mathcal{T}^{\mathcal{F}}_{0,T}\Big\}$$

and

$$v_0(x)=\sup\Big\{\mathbb{E}\,F\big(\tau,(X^{(\theta),x})^{\tau}\big),\ \tau\in\mathcal{T}^{\mathcal{F}}_{0,T}\Big\}$$

where $\mathcal{T}^{\mathcal{F}}_{[0,T]}=\{\tau:\Omega\to[0,T],\ \mathcal{F}\text{-stopping time}\}$. *Then, one can replace* $\mathcal{T}^{\mathcal{F}}_{0,T}$ *by* $\mathcal{T}^{\mathcal{F}^W}{}_{[0,T]}$ *and*

$$u_0(x)\le v_0(x).$$

Remark 8 All the quantities involved in the above theorem are well-defined since $\|X^{(\sigma),x}\|_{\sup}$ and $\|X^{(\theta),x}\|_{\sup}$ have polynomial moments at any order. Moreover, the Lipschitz continuity assumption is most likely too stringent: we adopt it to shorten the proof of the transfer from discrete to continuous time by reasoning on strong solutions.

Proof Step 1 (Euler Schemes). We consider the Euler schemes $\bar{X}^{(\sigma),n}$ and $\bar{X}^{(\theta),n}$ with step $\frac{T}{n}$ of both diffusions (we drop the dependence on the starting value x). Both schemes are adapted to the filtration $\mathcal{F}^{(n)}:=(\mathcal{F}_{t_k^n})_{k=0,\ldots,n}$.

It follows from Proposition 8 that the $(\mathbb{P},\mathcal{F}^{(n)})$-Snell envelopes $\bar{U}^{(n)}=(\bar{U}^{(n)}_{t_k^n})_{k=0,\ldots,n}$, $\bar{K}^{(n)}=(\bar{K}^{(n)}_{t_k^n})_{k=0,\ldots,n}$ and $\bar{V}^{(n)}=(\bar{V}^{(n)}_{t_k^n})_{k=0,\ldots,n}$ of the $\mathcal{F}^{(n)}$-adapted obstacle processes $\Big(F\big(t_k^n,[I_n(\bar{X}^{(\sigma),n})]^{t_k^n}\big)\Big)_{k=0,\ldots,n}$, $\Big(F\big(t_k^n,[I_n(\bar{X}^{(\kappa),n})]^{t_k^n}\big)\Big)_{k=0,\ldots,n}$ and $\Big(F\big(t_k^n,[I_n(\bar{X}^{(\theta),n})]^{t_k^n}\big)\Big)_{k=0,\ldots,n}$ satisfy

$$\mathbb{E}\,\bar{U}_0^n\le\mathbb{E}\,\bar{K}_0^n\le\mathbb{E}\,\bar{V}_0^n. \tag{27}$$

Note that it is always possible to define the Euler scheme associated to the function κ regardless of its convergence toward the related SDE.

Step 2 (Convergence). First, set for convenience $\bar{Y}^{(n)}_{t_k^n}=\Big(F\big(t_k^n,[I_n(\bar{X}^{(\sigma),n})]^{t_k^n}\big)\Big)_{k=0,\ldots,n}$, so that

$$\bar{U}^{(n)}_{t_k^n}=\mathbb{P}\text{-}\operatorname{ess\,sup}\Big\{\mathbb{E}(\bar{Y}^{(n)}_{\tau}\mid\mathcal{F}_{t_k^n}),\ \tau\in\mathcal{T}^{(n)}_{t_k^n,T}\Big\},\ k=0,\ldots,n,$$

where $\mathcal{T}^{(n)}_{t_k^n,T}=\Big\{\tau:\Omega\to\{t_k^n,\ldots,t_\ell^n,\ldots,t_n^n\},\ \mathcal{F}^{(n)}\text{-stopping time}\Big\}$; we also know that the $(\mathbb{P},\mathcal{F})$-Snell envelope of the process $Y_t=F(t,X^t)$, $t\in[0,T]$, is defined by

$$U_t=\mathbb{P}\text{-}\operatorname{ess\,sup}\Big\{\mathbb{E}\big(Y_\tau\mid\mathcal{F}_t\big),\ \tau\in\mathcal{T}^{\mathcal{F}}_{t,T}\Big\},\ t\in[0,T],$$

where $\mathcal{T}^{\mathcal{F}}_{t,T} = \left\{\tau : \Omega \to [t,T],\ \mathcal{F}\text{-stopping time}\right\}$. This Snell envelope is well-defined since $\|X\|_{\sup}$ lies in every $L^p(\mathbb{P})$, $p \in (0,+\infty)$, which implies in turn that $\|Y\|_{\sup}$ lies in every $L^p(\mathbb{P})$. Note that the obstacle process $(F(t,X^t))_{t\in[0,T]}$ has continuous paths since $\alpha \mapsto (t,\alpha^t)$ is continuous from $[0,T]$ to $([0,T]\times \mathcal{C}([0,T],\mathbb{R}), |.|\otimes\|.\|_{\sup})$ and F is continuous. Being uniformly integrable, it is regular for optimal stopping and $t\mapsto \mathbb{E}\,U_t$ is continuous (see [8, 24]). Hence, the super-martingale $(U_t)_{t\in[0,T]}$ has a non-negative càd modification whose non-decreasing compensator is continuous. More generally, if a sequence of stopping times $\tau_n \uparrow \tau < +\infty$ and $U_\tau \in L^1$, then $\mathbb{E}\,U_{\tau_n} \to \mathbb{E}\,U_\tau$. As a temporary intermediate quantity, we introduce an intermediate quantity defined by

$$\widetilde{U}^{(n)}_{t^n_k} = \mathbb{P}\text{-}\operatorname{ess\,sup}\left\{\mathbb{E}(Y_\tau \mid \mathcal{F}_{t^n_k}),\ \tau\in\mathcal{T}^{(n)}_{t^n_k,T}\right\} \le U_{t^n_k},\ k=0,\dots,n.$$

We will prove, after having extended $\bar{U}^{(n)}$ into a càdlàg stepwise constant process by setting $\bar{U}^{(n)}_t = \bar{U}^{(n)}_{t^n_k}$, $t\in[t^n_k,t^n_{k+1})$, that $\mathbb{E}\,\bar{U}^{(n)}_t$ converges to $\mathbb{E}\,U_t$ for every $t\in[0,T]$. We start from the fact that

$$|\mathbb{E}\,U_t - \mathbb{E}\,\bar{U}^{(n)}_{\underline{t}_n}| \le |\mathbb{E}\,U_t - \mathbb{E}\,U_{\underline{t}_n}| + \mathbb{E}\,U_{\underline{t}_n} - \mathbb{E}\,\widetilde{U}^{(n)}_{\underline{t}_n} + \mathbb{E}\,|\widetilde{U}^{(n)}_{\underline{t}_n} - \bar{U}^{(n)}_{\underline{t}_n}|. \tag{28}$$

The regularity of U for optimal stopping implies that $\mathbb{E}\,U_{t^n_k} \to \mathbb{E}\,U_t$ as $n\to+\infty$. As concerns the second term in the right hand side of (28), we proceed as follows

$$0 \le U_{t^n_k} - \widetilde{U}^{(n)}_{t^n_k} \le \mathbb{P}\text{-}\operatorname{ess\,sup}\left\{\mathbb{E}\big(Y_\tau - Y_{\tau^{(n)}} \mid \mathcal{F}^{(n)}_{t^n_k}\big),\ \tau\in\mathcal{T}_{t^n_k,T}\right\}$$

where $\tau^{(n)} = \sum_{\ell=k}^n \frac{\ell T}{n}\mathbf{1}_{\{\frac{(\ell-1)T}{n}<\tau\le\frac{\ell T}{n}\}} = \sum_{\ell=k}^n \bar{t}^n\mathbf{1}_{\{t^n_{\ell-1}<\tau\le t^n_\ell\}} \in \mathcal{T}^{(n)}_{t^n_k,T} \subset \mathcal{T}_{t^n_k,T}$ so that

$$0 \le U_{t^n_k} - \widetilde{U}^{(n)}_{t^n_k} \le \mathbb{E}\big(\sup_{t\ge t^n_k}|Y_t - Y_{\bar{t}^n}|\,|\,\mathcal{F}_{t^n_k}\big) \le \mathbb{E}\big(\sup_{t\in[0,T]}|Y_t - Y_{\bar{t}^n}|\,|\,\mathcal{F}_{t^n_k}\big).$$

Doob's Inequality applied to the martingale $M_n = \mathbb{E}\big(\sup_{t\in[0,T]}|Y_t - Y_{\bar{t}^n}|\,|\,\mathcal{F}_{t^n_k}\big)$, $n\ge1$, implies that for every $p\in(1,+\infty)$,

$$\Big\|\max_{k=0,\dots,n}(U_{t^n_k} - \widetilde{U}^{(n)}_{t^n_k})\Big\|_p \le \frac{p}{p-1}\|M_n\|_p = \frac{p}{p-1}\Big\|\sup_{t\in[0,T]}|Y_t - Y_{\bar{t}_n}|\Big\|_p \to 0 \text{ as } n\to+\infty$$

since $X^{\bar{t}_n}$ *a.s.* converges towards X^t for the sup-norm owing to the pathwise continuity of X. In turn, this implies that $F(\bar{t}^n, X^{\bar{t}^n})$ *a.s.* converges toward $F(t,X^t)$ since F is continuous. The L^p-convergence follows by uniform integrability, still since $\|Y\|_{\sup}$ has polynomial moments at any order.

Now we investigate the third term in the right hand side of (28).

$$|\widetilde{U}^{(n)}_{t^n_k} - \bar{U}^{(n)}_{t^n_k}| \le \mathbb{P}\text{-}\operatorname{ess\,sup}\left\{\mathbb{E}\big(|Y_\tau - \bar{Y}^{(n)}_\tau|\,|\,\mathcal{F}_{t^n_k}\big),\ \tau\in\mathcal{T}^{(n)}_{t^n_k,T}\right\} \le \mathbb{E}\Big[\max_{k=0,\dots,n}|\bar{Y}^{(n)}_{t^n_k} - Y_{t^n_k}|\,|\,\mathcal{F}_{t^n_k}\Big].$$

On the other hand,

$$\max_{k=0,\dots,n} |\bar{Y}^{(n)}_{t^n_k} - Y_{t^n_k}| \le \max_{k=0,\dots,n} \left|F(t^n_k, (I_n(\bar{X}^{(\sigma),n}))^{t^n_k}) - F(t^n_k, (X^{(\sigma)})^{t^n_k})\right|$$

$$\le \sup_{t\in[0,T]} \left|F\big(t, (I_n(\bar{X}^{(\sigma),n}))^t\big) - F\big(t, (X^{(\sigma)})^t\big)\right|. \tag{29}$$

Now, note that the functional $\alpha \mapsto \Big(t \mapsto F\big(t, \alpha^t\big)\Big)$ defined from $(\mathscr{C}([0,T],\mathbb{R}), \|\,.\,\|_{\sup})$ into itself is continuous: if $(t_n, \alpha_n) \to (t, \alpha)$ for the product topology on the product space $[0,T] \times (\mathscr{C}([0,T],\mathbb{R})$, then

$$\|\alpha_n^{t_n} - \alpha^t\|_{\sup} \le \|\alpha_n - \alpha\|_{\sup} + w(\alpha, |t - t_n|)$$

so that $(t_n, \alpha_n^{t_n}) \to (t, \alpha^t)$. As a consequence, the functional F being continuous on $[0,T] \times \mathscr{C}([0,T],\mathbb{R})$, $F(t_n, \alpha_n^{t_n}) \to F(t, \alpha^t)$ which in turn implies that $\|F(t, \alpha_n^t) - F(t, \alpha^t)\|_{\sup} \to 0$. As $\lim_n \|I_n(\alpha) - \alpha\|_{\sup} = 0$, we derive that, if $\alpha_n \to \alpha$ for the sup norm, then $\|F(t, I_n(\alpha_n)^t) - F(t, \alpha^t)\|_{\sup} \to 0$ as $n \to +\infty$.

Then, under the Lipschitz continuity assumption on σ, we know that the Euler scheme $\bar{X}^{(\sigma),x,n} \to X^{(\sigma),x}$ $\mathbb{P}$-*a.s.* as $n \to +\infty$ *a.s.* (see e.g. [5, Theorem B.14, p. 276]). The $(r, \|\,.\,\|_{\sup})$-polynomial growth assumption made on F and the fact that $\sup_{n\ge 1} \mathbb{E}\|\bar{X}^{(\sigma),x,n}\|^p_{\sup} < +\infty$ for any $p > r$ implies the L^1-convergence to 0 of the term in (29). Finally, this shows $\lim_n |\mathbb{E}\, U_t - \mathbb{E}\bar{U}_{\underline{t}_n}| = 0$ so that, for $t = 0$,

$$\mathbb{E}\, \bar{U}^n_0 \to u_0(x) \quad \text{as} \quad n \to +\infty.$$

The conclusion follows from (27) in Step 1 by letting $n \to +\infty$ in the inequality $\mathbb{E}\, \bar{U}^n_0 \le \mathbb{E}\, \bar{V}^n_0$.

Applications to Comparison Theorems for American Options in Local Volatility Models By specifying our diffusion dynamics as a local volatility model as defined by (7), we can extend the comparison result (8) to path-dependent American options provided the "payoff" functionals $F(t, .)$ are convex with polynomial growth as specified in the above theorem.

4.2.2 The Case of Jump Martingale Diffusions

In what follows the product space $[0,T] \times \mathbb{D}([0,T],\mathbb{R})$ is endowed with the product topology $|\,.\,| \otimes Sk$. The notation $X_t(\alpha) = \alpha(t)$, $\alpha \in \mathbb{D}([0,T],\mathbb{R})$ still denotes the canonical process on $\mathbb{D}([0,T],\mathbb{R})$ and θ denotes the canonical random variable on $[0,T]$ (i.e. $\theta(t) = t$, $t \in [0,T]$).

Let $(\mathscr{F}_t)_{t\in[0,T]}$ be a right continuous filtration on a probability space $(\Omega, \mathscr{A}, \mathbb{P})$ and let Y be an $(\mathscr{F}_t)_{t\in[0,T]}$-adapted càdlàg process defined on this probability space. We introduce the filtration enlargement assumption, the so-called $(\mathscr{H})$-assumption which reads as follows:

$$(\mathscr{H}) \equiv \forall H : \Omega \to \mathbb{R}, \text{ bounded and } \mathscr{F}_T^Y\text{-measurable}, \ \mathbb{E}\big(H \,|\, \mathscr{F}_t\big) = \mathbb{E}\big(H \,|\, \mathscr{F}_t^Y\big) \ \mathbb{P}\text{-}a.s.$$

This filtration enlargement assumption is equivalent to the following more tractable condition: there exists $D \subset [0, T]$, everywhere dense in $[0, T]$, with $T \in D$, such that

$$\forall n \geq 1, \ \forall t_1, \ldots, t_n \in D, \ \forall h \in \mathscr{C}_0(\mathbb{R}^n, \mathbb{R}), \ \mathbb{E}\big(h(Y_{t_1}, \ldots, Y_{t_n}) \,|\, \mathscr{F}_t\big) \overset{\mathbb{P}\text{-}a.s}{=} \mathbb{E}\big(h(Y_{t_1}, \ldots, Y_{t_n}) \,|\, \mathscr{F}_t^Y\big)$$

where $\mathscr{C}_0(\mathbb{R}^n, \mathbb{R}) = \{f \in \mathscr{C}(\mathbb{R}^n, \mathbb{R}) \text{ such that } \lim_{|x| \to +\infty} f(x) = 0\}$.

We still consider in this section jump diffusions of the form (10) i.e.

$$dX_t = \kappa(t, X_{t-}) dZ_t$$

where $\kappa : [0, T] \times \mathbb{R} \to \mathbb{R}$ is a continuous function, Lipschitz continuous in x uniformly in $t \in [0, T]$.

The aim of this section is to extend the result obtained for functional convex order for Brownian diffusions to such jump diffusions. We will rely on an abstract convergence result for *réduites* established in [25] (Theorem 3.7 and the remark that follows) that we recall below. To this end, we need to recall two classical definitions on stochastic processes.

Definition 3

(*a*) *Class* (*D*) *processes:* A càdlàg process $(Y_t)_{t \in [0,T]}$ is of class (D) if

$$\big\{Y_\tau, \ \tau \in \mathscr{T}_{[0,T]}\big\} \text{ is uniformly integrable.} \tag{30}$$

(*b*) *Aldous tightness criterion (see e.g. [17, Chap. VI, Theorem 4.5, p. 356]):* A sequence of $\mathscr{F}^n$-adapted càdlàg processes $Y^n = (Y_t^n)_{t \in [0,T]}$, $n \geq 1$, defined on filtered stochastic spaces $(\Omega^n, \mathscr{A}^n, \mathscr{F}^n, \mathbb{P}^n)$, $n \geq 1$, satisfies Aldous' tightness criterion with respect to the filtrations $\mathscr{F}^n$, $n \geq 1$, if

$$\forall \eta > 0, \ \lim_{\delta \to 0} \overline{\lim_{n}} \sup_{\tau_n \leq \tau_n' \leq (\tau_n + \delta) \wedge T} \mathbb{P}^n\big(|Y_{\tau_n}^n - Y_{\tau_n'}^n| \geq \eta\big) = 0 \tag{31}$$

where $\tau_n$and τ_n' run over $[0, T]$-valued $\mathscr{F}^{Y^n}$-stopping times.

Then, the sequence $(Y^n)_{n \geq 1}$ is tight for the Skorokhod topology.

Theorem 4

(*a*) *Let $(X^n)_{n\geq 1}$ be a sequence of adapted quasi-left càdlàg processes*[1] *defined on probability spaces $(\Omega^n, \mathscr{F}^n, \mathbb{P}^n)$ of class (D) and satisfying the above Aldous tightness criterion* (31). *For every $n \geq 1$, let*

$$u_0^n = \sup\{\mathbb{E}\, X_\tau^n,\ \tau\ [0,T]\text{-valued } \mathscr{F}^n\text{-stopping time}\}$$

*denote the $\mathscr{F}^n$-*réduite *of X^n. Let $(\tau_n^*)_{n\geq 1}$ be a sequence of $\big(\mathscr{F}^{X^n}, \mathbb{P}^n\big)$-optimal stopping times* (²). *Assume furthermore that $(X^n)_{n\geq 1}$ satisfies*

$$X^n \xrightarrow{\mathscr{L}} \mathbb{P},\ \mathbb{P} \text{ probability measure on } (\mathbb{D}([0,T],\mathbb{R}), \mathscr{D}_T) \text{ such that } \mathbb{E}_{\mathbb{P}} \sup_{t\in[0,T]} |X_t| < +\infty.$$

*If every limiting value $\mathbb{Q}$ of $\mathscr{L}(X^n, \tau_n^*)$ on $\mathbb{D}([0,T],\mathbb{R}) \times [0,T]$ satisfies the filtration enlargement $(\mathscr{H})$ property, then the $(\mathscr{F}^n, \mathbb{P}^n)$-*réduites *$u_0^n$ of X^n converge toward the $(\mathscr{D}, \mathbb{P})$-*réduite *$u_0$ of X i.e.*

$$\lim_n u_0^n = u_0.$$

Moreover, if the optimal stopping problem related to $(X, \mathbb{Q}, \mathscr{D}^\theta)$ has a unique solution in distribution, i.e. $\theta \stackrel{d}{=} \mu_{\tau^*}^*$, not depending on $\mathbb{Q}$, then $\tau_n^* \xrightarrow{\mathscr{L}([0,T])} \mu_{\tau^*}^*$.

(*b*) *The same result holds when considering a sequence of* companion processes *Y^n having values in a Polish metric space (E, d_E). To be more precise, we consider that the filtration of interest at finite range n is now $\big(\mathscr{F}_t^{(X^n,Y^n)}\big)_{t\in[0,T]}$. We assume that X^n is quasi-left continuous with respect to this enlarged filtration. We will only ask the couple (X^n, Y^n) to converge for the* product topology *i.e. on $(\mathbb{D}([0,T],\mathbb{R}), Sk_{\mathbb{R}}) \times (\mathbb{D}([0,T],E), Sk_E)$ since this product topology spans the same Borel σ-field as the regular Skorokhod topology on $\mathbb{D}([0,T], \mathbb{R}\times E)$.*

The main result of this section is the following:

Theorem 5 *Let $Z = (Z_t)_{t\in[0,T]}$ be a martingale Lévy process with Lévy measure ν satisfying $\nu(|z|^p) < +\infty$ for $p\in [2,+\infty)$, so that the process Z is an L^2-martingale null at* 0. *Let $\kappa_i : [0,T]\times\mathbb{R} \to \mathbb{R}_+$, $i = 1,2$, be two continuous functions, Lipschitz continuous in x uniformly in $t\in[0,T]$ and let $X^{(\kappa_i,x)}$ be the martingale jump diffusion solution to* (10) *(driven by Z with coefficient κ_i) starting at (the same) $x\in\mathbb{R}$, $i = 1,2$. Let $F : [0,T]\times\mathbb{D}([0,T],\mathbb{R}) \to \mathbb{R}_+$ be a functional satisfying the following local*

---

[1] A càdlàg $(\mathscr{F}_t)_{t\in[0,T]}$-adapted process $X = (X_t)_{t\in[0,T]}$ is *quasi-left continuous* with respect to the right continuous filtration $\mathscr{F} = (\mathscr{F}_t)_{t\in[0,T]}$ if, for every $\mathscr{F}$-stopping time τ having values in $[0,T]\cup\{+\infty\}$ and every increasing sequence of $\mathscr{F}$-stopping times $(\tau_k)_{k\geq 1}$ with limit τ, $\lim_k X_{\tau_k} = X_\tau$ on the event $\{\tau < +\infty\}$ (see e.g. [17, Chap. I.2.25, p. 22]).

[2] i.e. satisfying $\mathbb{E}\, X_{\tau_n^*}^n = u_0^n$.

Lipschitz continuity, convexity and Skorokhod continuity assumptions, namely

$$\begin{cases} (i) \quad \forall\, t \in [0,T],\ F(t,.) \text{ is a convex functional on } \mathbb{D}([0,T],\mathbb{R}), \\ (ii) \quad F : [0,T] \times \mathbb{D}([0,T],\mathbb{R}) \to \mathbb{R}_+ \text{ is Sk-continuous}, \\ (iii) \quad |F(t,\beta) - F(s,\alpha)| \le C\Big(|t-s|^{\rho'} + \|\alpha-\beta\|_{\sup}^{\rho}\big(1+\|\alpha\|_{\sup}^{r-\rho} + \|\beta\|_{\sup}^{r-\rho}\big)\Big), \\ \qquad \rho,\ \rho' \in (0,1],\ r \in [1,p). \end{cases} \tag{32}$$

Let $U^{(\kappa_i)}$ *denote the Snell envelopes of the processes* $\big(F(t,(X^{\kappa_i}))^t\big)_{t\in[0,T]}$, $i = 1,2$ *respectively.*

If there exists a partitioning function $\kappa : [0,T] \times \mathbb{R} \to \mathbb{R}$, *convex in* x *for every* $t \in [0,T]$, *such that*

$$0 \le \kappa_1 \le \kappa \le \kappa_2.$$

Then

$$U_0^{(\kappa_1)} \le U_0^{(\kappa_2)}.$$

Remark 9

- Note that, as $p \ge 2$,

$$\nu(|z|^p) < +\infty \iff \nu(|z|^p \mathbf{1}_{\{|z|\ge 1\}}) < +\infty \iff Z_t \in L^p \iff \sup_{t\in[0,T]} |Z_t| \in L^p.$$

- One proves likewise that, for every $t \in [0,T]$,

$$\mathbb{E}(U_t^{(\kappa_1)}) \le \mathbb{E}(U_t^{(\kappa_2)}).$$

- If the functions $\kappa(t,.)$, $t \in [0,T]$ are all convex (but possibly not the functions $\kappa_i(t,.)$) then the same proof shows by coupling (κ_1,κ) and (κ,κ_2) that

$$\mathbb{E}(U_0^{(\kappa_1)}) \le \mathbb{E}(U_0^{(\kappa)}) \le \mathbb{E}(U_0^{(\kappa_2)}).$$

Lemma 5 *Let* $X = (X_t)_{t\in[0,T]}$ *be an* $(\mathscr{F}_t)_{t\in[0,T]}$*-adapted càdlàg process defined on a probability space* $(\Omega,\mathscr{A},\mathbb{P})$ *where* $(\mathscr{F}_t)_{t\in[0,T]}$ *is a càd filtration. Let* $G : [0,T] \times \mathbb{D}([0,T],\mathbb{R}) \to \mathbb{R}_+$ *be a Skorokhod continuous functional such that* $|G(\alpha)| \le C(1+\|\alpha\|_{\sup}^r)$, $r \in (0,p)$. *If* X *is quasi-left continuous and if* $\|X\|_{\sup} \in L^p$, *then the obstacle process* $(G(t,X^t))_{t\in[0,T]}$ *is* regular for optimal stopping *i.e.* $\mathbb{E}\, G(\tau_n,X^{\tau_n}) \to \mathbb{E}\, G(\tau,X^\tau)$ *as soon as* $\tau < +\infty$ $\mathbb{P}$*-a.s.*

Proof First one easily proves by coming back to the very definition of Skorokhod topology that $\alpha_n \stackrel{Sk}{\longrightarrow} \alpha$ and $t_n \to t \in \mathrm{Cont}(\alpha)$ then $\alpha_n^{t_n} \stackrel{Sk}{\longrightarrow} \alpha^t$. Let $(\tau_n)_{n\ge1}$ be a sequence of $\mathscr{F}_t$-stopping times satisfying $\tau_n \uparrow \tau < +\infty$ $\mathbb{P}$*-a.s.*, then $X_\tau = X_{\tau_-}$ $\mathbb{P}$*-a.s.* i.e. $\tau(\omega) \in \mathrm{Cont}(X(\omega))$ $\mathbb{P}(d\omega)$*-a.s.*. It follows that $(\tau_n, X^{\tau_n}) \to (\tau, X^\tau)$ $\mathbb{P}$*-a.s.*.

The continuity assumption made on G implies that $G(\tau_n, X^{\tau_n}) \xrightarrow{Sk} G(\tau, X^{\tau})$. One concludes by a uniform integrability argument that $\mathbb{E}\, G(\tau_n, X^{\tau_n}) \to \mathbb{E}\, G(\tau, X^{\tau})$ since $\|X\|_{\sup} \in L^p$ implies that $\big(G(\tau_n, X^{\tau_n})\big)_{n\geq 1}$ is $L^{p/r}$-bounded.

Proof Step 1 (Aldous Tightness Criterion). We still consider the stepwise constant Euler scheme $\bar{X}^n = (\bar{X}^n_t)_{t\in[0,T]}$ with step $\frac{T}{n}$ defined for every $t \in [0, T]$ by $\bar{X}^n_t = \bar{X}^n_{\underline{t}_n}$ where, for every $k \in \mathbb{N}^*$,

$$\bar{X}^n_{t^n_k} = \bar{X}^n_{t^n_{k-1}} + \kappa(t^n_{k-1}, \bar{X}^n_{t^n_{k-1}})(Z_{t^n_k} - Z_{t^n_{k-1}}),\ \ \bar{X}^n_0 = X_0.$$

Let $\sigma_n, \tau_n \in \mathscr{T}^{\mathscr{F}^n}_{[0,T]}$, such that $\sigma_n \leq \tau_n \leq (\sigma_n + \delta) \wedge T$. In fact, following Lemma 3, we may assume without loss of generality that σ_n and τ_n take values in $\{t^n_k,\ k = 0, \ldots, n\}$. Then, owing to (32),

$$\mathbb{E}\big|F(\tau_n, (\bar{X}^n)^{\tau_n}) - F(\sigma_n, (\bar{X}^n)^{\sigma_n})\big| \leq C\delta^{\rho'} + C\,\mathbb{E}\big(\|(\bar{X}^n)^{\tau_n} - (\bar{X}^n)^{\sigma_n}\|^{\rho}_{\sup}(1 + 2\|\bar{X}^n\|^{r-\rho}_{\sup})\big).$$

Hölder's Inequality applied with the conjugate exponents $a = \frac{r}{\rho}$ and $b = \frac{r}{r-\rho}$ yields

$$\mathbb{E}\Big(\|(\bar{X}^n)^{\tau_n} - (\bar{X}^n)^{\sigma_n}\|^{\rho}_{\sup}\big(1 + 2\|\bar{X}^n\|^{r-\rho}_{\sup}\big)\Big)$$
$$\leq \Big\|\sup_{\sigma_n \leq s \leq (\sigma_n+\delta)\wedge T} |\bar{X}^n_s - \bar{X}^n_{\sigma_n}|\Big\|^{\rho}_r \Big(1 + 2\Big\|\sup_{t\in[0,T]} |\bar{X}^n_t|\Big\|^{r-\rho}_r\Big).$$

As $\nu(z^2) < +\infty$, we can decompose the Lévy process Z into $Z_t = a\,W_t + \widetilde{Z}_t$, $a \geq 0$ where W is a standard Brownian motion and $\widetilde{Z}$ is a pure jump square integrable martingale Lévy process.

- If $r \in [1, 2]$: the B.D.G. Inequality applied to the local martingale $(\bar{X}_{\sigma_n + \frac{iT}{n}} - \bar{X}^n_{\sigma_n})_{i\geq 0}$ implies.

$$\Big\|\sup_{\sigma_n \leq t^n_k \leq (\sigma_n+\delta)\wedge T} |\bar{X}^n_{t^n_k} - \bar{X}^n_{\sigma_n}|\Big\|^r_r$$
$$\leq c_r a^r \Big\|\sum_{\sigma_n < t^n_k \leq (\sigma_n+\delta)\wedge T} \kappa(t^n_{k-1}, \bar{X}^n_{t^n_{k-1}})^2 (W_{t^n_k} - W_{t^n_{k-1}})^2\Big\|^{\frac{r}{2}}_{\frac{r}{2}}$$
$$+ c_r \Big\|\sum_{\sigma_n < t^n_k \leq (\sigma_n+\delta)\wedge T} \kappa(t^n_{k-1}, \bar{X}^n_{t^n_{k-1}})^2 \big(\widetilde{Z}_{t^n_k} - \widetilde{Z}_{t^n_{k-1}}\big)^2\Big\|^{\frac{r}{2}}_{\frac{r}{2}}$$
$$\leq c_r a^r \Big\|\sum_{\sigma_n < t^n_k \leq (\sigma_n+\delta)\wedge T} \kappa(t^n_{k-1}, \bar{X}^n_{t^n_{k-1}})^2 (W_{t^n_k} - W_{t^n_{k-1}})^2\Big\|^{\frac{r}{2}}_{1}$$
$$+ c_r \mathbb{E}\Big[\sum_k \mathbf{1}_{\{\sigma_n < t^n_k \leq (\sigma_n+\delta)\wedge T\}} |\kappa(t^n_{k-1}, \bar{X}^n_{t^n_{k-1}})|^r |Z_{t^n_k} - Z_{t^n_{k-1}}|^r\Big].$$

Now

$$\begin{aligned}
&\mathbb{E}\Big[\sum_{\sigma_n<t_k^n\le(\sigma_n+\delta)\wedge T}\kappa(t_{k-1}^n,\bar X^n_{t_{k-1}^n})^2\big(W_{t_k^n}-!W_{t_{k-1}^n}\big)^2\Big]\\
&=\frac{T}{n}\mathbb{E}\Big[\sum_{\sigma_n<t_k^n\le(\sigma_n+\delta)\wedge T}\kappa(t_{k-1}^n,\bar X^n_{t_{k-1}^n})^2\Big]\\
&\le\frac{T}{n}\mathbb{E}\Big[\max_{1\le k\le n}|\kappa(t_{k-1}^n,\bar X^n_{t_{k-1}^n})|^2\times\mathrm{card}\{k:\sigma_n<t_k^n\le(\sigma_n+\delta)\wedge T\}\Big]\\
&\le\frac{T}{n}\mathbb{E}\Big[\max_{1\le k\le n}|\kappa(t_{k-1}^n,\bar X^n_{t_{k-1}^n})|^2\Big]\frac{\delta n}{T}\\
&=\delta\Big\|\max_{1\le k\le n}|\kappa(t_{k-1}^n,\bar X^n_{t_{k-1}^n})|\Big\|_2^2.
\end{aligned}$$

On the other hand,

$$\begin{aligned}
&\Big\|\sum_{\sigma_n<t_k^n\le(\sigma_n+\delta)\wedge T}\kappa(t_{k-1}^n,\bar X^n_{t_{k-1}^n})^2\big(\widetilde Z_{t_k^n}-\widetilde Z_{t_{k-1}^n}\big)^2\Big\|_{L^{\frac r2}}^{\frac r2}\\
&\le\mathbb{E}\sum_k\mathbf{1}_{\{\sigma_n<t_k^n\le(\sigma_n+\delta)\wedge T\}}|\kappa(t_{k-1}^n,\bar X^n_{t_{k-1}^n})|^r|\widetilde Z_{t_k^n}-\widetilde Z_{t_{k-1}^n}|^r\\
&=\mathbb{E}|\widetilde Z_{\frac Tn}|^r\mathbb{E}\Big[\sum_k\mathbf{1}_{\{\sigma_n<t_k^n\le(\sigma_n+\delta)\wedge T\}}|\kappa(t_{k-1}^n,\bar X^n_{t_{k-1}^n})|^r\Big]\\
&=\mathbb{E}|\widetilde Z_{\frac Tn}|^r\mathbb{E}\Big[\max_{1\le k\le n}|\kappa(t_{k-1}^n,\bar X^n_{t_{k-1}^n})|^r\times\mathrm{card}\{k:\sigma_n<t_k^n\le(\sigma_n+\delta)\wedge T\}\Big]\\
&\le\mathbb{E}|\widetilde Z_{\frac Tn}|^r\mathbb{E}\Big[\max_{1\le k\le n}|\kappa(t_{k-1}^n,\bar X^n_{t_{k-1}^n})|^r\Big]\frac{\delta n}{T}\\
&\le\delta\Big(\frac nT\mathbb{E}|\widetilde Z_{\frac Tn}|^r\Big)\mathbb{E}\Big[\max_{0\le k\le n-1}|\kappa(t_{k-1}^n,\bar X^n_{t_{k-1}^n})|^r\Big]\\
&\le C_{\kappa,\widetilde Z,T}\,\delta\Big\|\max_{1\le k\le n}|\kappa(t_{k-1}^n,\bar X^n_{t_{k-1}^n})|\Big\|_r^r
\end{aligned}$$

where we used that $t\mapsto\frac1t\mathbb{E}|\widetilde Z_t|^r$ remains bounded on the whole interval $(0,T]$. Under the assumptions $\nu(z^2)<+\infty$ and κ with linear growth (in x uniformly in $t\in[0,T]$), it follows from Proposition 12 in Appendix 2 that

$$\sup_{n\ge1}\Big\|\sup_{k=0,\dots,n}|\kappa(t_k^n,\bar X^n_{t_k^n})|\Big\|_r<+\infty$$

since $r\le2$ (see the first remark below the statement of the theorem), we get

$$\Big\|\sup_{\sigma_n\le t_k^n\le(\sigma_n+\delta)\wedge T}|\bar X^n_{t_k^n}-\bar X^n_{\sigma_n}|\Big\|_r\le C_{\rho,r,\kappa,Z,T}\big(\delta^{\frac14}+\delta^{\frac1r}\big)$$

where the real constant $C_{\rho,r,\kappa,Z,T}$ does not depend on n, σ_n, τ_n and δ. This implies in turn that

$$\lim_{\delta\to 0}\overline{\lim_{n}} \sup_{\sigma_n<\tau_n\le(\sigma_n+\delta)\wedge T} \mathbb{E}\left|F(\tau_n,\bar{X}^{n,\tau_n})-F(\sigma_n,\bar{X}^{n,\sigma_n})\right| = 0$$

and the conclusion follows.

- If $r\in[2,4]$: One writes

$$\sum_{\sigma_n<t_k^n\le(\sigma_n+\delta)\wedge T}\kappa(t_{k-1}^n,\bar{X}^n_{t_{k-1}^n})^2\big(Z_{t_k^n}-Z_{t_{k-1}^n}\big)^2$$

$$=\sum_{\sigma_n<t_k^n\le(\sigma_n+\delta)\wedge T}\kappa(t_{k-1}^n,\bar{X}^n_{t_{k-1}^n})^2\big((Z_{t_k^n}-Z_{t_{k-1}^n})^2-\mathbb{E}|Z_{\frac{T}{n}}|^2\big)$$

$$+\mathbb{E}|Z_{\frac{T}{n}}|^2\sum_{\sigma_n<t_k^n\le(\sigma_n+\delta)\wedge T}\kappa(t_{k-1}^n,\bar{X}^n_{t_{k-1}^n})^2\big(Z_{t_k^n}-Z_{t_{k-1}^n}\big)^2.$$

The second term of the sum in the right hand side of the above equality can be treated as above (it corresponds to $r=2$). As concerns the first one, note that the *i.i.d.* sequence $\big((Z_{t_k^n}-Z_{t_{k-1}^n})^2-\mathbb{E}|Z_{\frac{T}{n}}|^2\big)_{1\le k\le n}$ is centered and lies in $L^{\frac{r}{2}}(\mathbb{P})$ with $\frac{r}{2}\in[1,2]$. Hence, it can be controlled like in the former case. Carrying on the process by a cascade induction as detailed in the proof of Proposition 12 (Appendix 2), one can lower r to $r/2,\dots,r/2^{\ell_r}\in(1,2]$, owing to the B.D.G. inequality.

Step 2. It follows from Step 1 of Theorem 2, adapted to a 2-dimensional framework with $(\kappa,\mathbf{1})$ as a drift, that

$$\Big(\bar{X}^n,I_n(Z)\Big)\stackrel{\mathcal{L}(Sk)}{\longrightarrow}(X,Z)\quad\text{as}\quad n\to+\infty.$$

If we consider the discrete time Optimal Stopping problems related to the Euler schemes $\bar{X}^{(n,\kappa_i)}$, $i=1,2$, which turns out the be the same as in Step 1 of the proof of Proposition 9, the existence of optimal stopping times $\tau_n^{(i)}$, $i=1,2$, taking values in $\{t_k^n,\,k=0:n\}$ is straightforward since it is a discrete time optimal stopping problem with a finite horizon (see e.g. [27, Chap. VI]).

Step 3. Let $\Omega_c=\mathbb{D}([0,T],\mathbb{R})^2\times[0,T]$ be the canonical space of the distribution of the sequence $(\bar{X}^n,I_n(Z),\tau_n^*)_{n\ge1}$. For every $(\alpha,u)\in\mathbb{D}([0,T],\mathbb{R})^2\times[0,T]$, the canonical process is defined by $\Xi_t(\alpha,u)=\alpha(t)=(\alpha^1(t),\alpha^2(t))\in\mathbb{R}^2$ and the canonical random times is given by $\theta(\alpha,u)=u$. Furthermore, we will denote by $\Xi=(\Xi^1,\Xi^2)$ the two components of Ξ. Let

$$\mathcal{D}_t^\theta=\cap_{s>t}\sigma(\Xi_u,\{\theta\le u\},0\le u\le s\}\ \text{if}\ t\in[0,T),\ \mathcal{D}_T^\theta=\sigma\big(\Xi_s,\{\theta\le s\},0\le s\le T\big)$$

denote the canonical right-continuous filtration on Ω_c. This canonical space Ω_c is equipped with the product metric topology $Sk^{\otimes 2} \otimes |.|$ where $|.|$ denotes the standard topology on $[0, T]$ induced by the absolute value.

In order to conclude to the convergence of the *réduites*, we need to show, following Theorem 4 from [25], that any limiting distribution $\mathbb{Q} = \lim_n \mathbb{P}_{((\bar{X}^n, I_n(Z)), \tau_n^*)}$ on the canonical space $\big(\mathbb{D}([0, T], \mathbb{R}^2) \times [0, T], Sk^{\otimes 2} \otimes |.|\big)$ satisfies the $(\mathscr{H})$-assumption, namely

$$\mathbb{E}_{\mathbb{Q}}\big(H \mid \mathscr{D}_t^{\theta}\big) = \mathbb{E}_{\mathbb{Q}}\big(H \mid \mathscr{D}_t\big)\ \mathbb{Q}\text{-}a.s.$$

for every random variable H defined on Ω_c.

Let $\mathrm{Atom}_{\mathbb{Q}}(\theta) = \{s \in [0, T],\ \mathbb{Q}_{\theta}(\{s\}) > 0\}$ be the set, possibly empty, of $\mathbb{Q}$-atoms of θ. Let $\Phi : \mathbb{D}([0, T], \mathbb{R}^2) \to \mathbb{R}$ and $\Psi : \mathbb{D}([0, T], \mathbb{R}) \to \mathbb{R}$ be two bounded functionals, $Sk^{\otimes 2}$- and Sk-continuous respectively and let $u \notin \mathrm{Atom}_{\mathbb{Q}}(\theta)$, $u \le s \le T$. As $\Psi(I_n(Z)^s)\mathbf{1}_{\{\tau_n^* \le u\}}$ is $\mathscr{F}_s^n$-measurable, we get

$$\begin{aligned}\mathbb{E}_{\mathbb{Q}}\big[\Phi(\Xi)\Psi(\Xi^{2,s})\mathbf{1}_{\{\theta \le u\}}\big] &= \lim_n \mathbb{E}\Big[\Phi(\bar{X}^n, I_n(Z))\Psi(I_n(Z)^s)\mathbf{1}_{\{\tau_n^* \le u\}}\Big]\\ &= \lim_n \mathbb{E}\Big[\mathbb{E}\big(\Phi(\bar{X}^n, I_n(Z))|\mathscr{F}_s^Z\big)\Psi(I_n(Z)^s)\mathbf{1}_{\{\tau_n^* \le u\}}\Big].\end{aligned}$$

Up to an extraction (n'), we may assume that $\mathbb{E}\big[\Phi(\bar{X}^{n'}, I_{n'}(Z))|\mathscr{F}_s^Z\big]$ weakly converges to $\mathbb{E}\big[\Phi(X, Z)|\mathscr{F}_s^Z\big]$ since $\Phi(\bar{X}^{n'}, I_{n'}(Z))$ weakly converges toward $\Phi(X, Z)$. Up to a second extraction, still denoted (n'), we may assume that $\Psi(I_{n'}(Z)^s)$ *a.s.* converges toward $\Psi(Z^s)$ for the Skorokhod topology since $\mathbb{P}(\Delta Z_s \neq 0) = 0$ (the stopping operator at time s, $\alpha \mapsto \alpha^s$, is Sk-continuous at functions α which are continuous at s). Consequently, going back on the canonical space Ω_c, we obtain

$$\Big(\mathbb{E}\big(\Phi(\bar{X}^n, I_n(Z))|\mathscr{F}_s^Z\big), \Psi(I_n(Z)^s), \mathbf{1}_{\{\tau_n^* \le u\}}\Big) \xrightarrow{\mathscr{L}} \mathscr{L}_{\mathbb{Q}}\Big(\mathbb{E}\big(\Phi(\Xi)|\mathscr{F}_s^{\Xi^2}\big), \Psi(\Xi^{2,s}), \mathbf{1}_{\{\theta \le u\}}\Big)$$

which ensures that

$$\mathbb{E}_{\mathbb{Q}}\left[\Phi(\Xi)\psi(\Xi^{2,s})\mathbf{1}_{\{\theta \le u\}}\right] = \mathbb{E}_{\mathbb{Q}}\left[\mathbb{E}_{\mathbb{Q}}\big(\Phi(\Xi)|\mathscr{D}_{s-}\big)\Psi(\Xi^{2,s})\mathbf{1}_{\{\theta \le u\}}\right].$$

One concludes by standard functional monotone approximation arguments that the equality holds true for any bounded measurable functional Φ, Ψ and any $u \in [0, T]$. Then, by considering a sequence $s_n \downarrow s$, $s_n > s$, we derive by a standard reverse martingale argument that

$$\mathbb{E}_{\mathbb{Q}}\left(\Phi(\Xi) \mid \mathscr{D}_s^{\theta}\right) = \mathbb{E}_{\mathbb{Q}}\Big(\Phi(\Xi) \mid \mathscr{D}_s\Big).$$

The $(\mathscr{H})$-assumption being fulfilled, Theorem 4 applies i.e. $\mathbb{E}\, U_0^n$ converges toward $\mathbb{E}\, U_0$. □

Acknowledgements I am indebted to Marc Yor for helpful discussions and comments on a very preliminary version of this work. I also thank the referee for a careful reading of the manuscript.

Appendix 1: Euler Scheme for Brownian Martingale Diffusions

Proposition 10 *Let $(\bar{X}^n_t)_{t\in[0,T]}$ be the genuine Euler scheme of step $\frac{T}{n}$ of the SDE $\equiv dX_t = \sigma(t, X_t)dW_t$, $X_0 = x$ defined as the solution to*

$$d\bar{X}^n_t = \sigma(\underline{t}_n, \bar{X}^n_{\underline{t}_n})dW_t,\ \bar{X}^n_0 = x,\ t\in[0,T].$$

If $\sigma : [0,T]\times\mathbb{R}\to\mathbb{R}$ is continuous and satisfies the linear growth assumption

$$\forall\, t\in[0,T],\ \forall\, x\in\mathbb{R},\quad |\sigma(t,x)|\le C_\sigma(1+|x|),$$

then the sequence $(\bar{X}^n)_{n\ge1}$ is C-tight on $\mathscr{C}([0,T],\mathbb{R})$ and any of its limiting distributions is a weak solution to the above SDE. In particular if a weak uniqueness assumption holds, then $\bar{X}^n \overset{\mathscr{L}(\|\,.\,\|_{\sup})}{\longrightarrow} X$.

Following e.g. [5] (Lemma B.1.2, p. 275, see also [22, 29]), we first show that, owing to the linear growth assumption, the non-decreasing function $\varphi_{p,n}(t) = \mathbb{E}\big(\sup_{s\in[0,t]}|\bar{X}^n_s|^p\big)$, $p\in[1,+\infty)$, is finite for every $t\in[0,T]$. Using Doob's Inequality and Gronwall's Lemma, it follows that

$$\varphi_{p,n}(t)\le\varphi_p(t) := Ce^{Ct}(1+|x|^p)$$

for a real constant $C = C'_{p,\sigma} > 0$. Consequently, it follows from the L^p-B.D.G. and Hölder inequalities, applied successively that, for every for $p\in(2,+\infty)$ and every $s,t\in[0,T]$, $s\le t$,

$$\mathbb{E}|\bar{X}^n_t-\bar{X}^n_s|^p\le c_p^p\,\mathbb{E}\left(\int_s^t|\sigma(\underline{u}_n,\bar{X}^n_{\underline{u}_n})|^2du\right)^{\frac{p}{2}}\le c_p^p|t-s|^{\frac{p}{2}}\big(1+\varphi_p(T)\big).$$

Kolmogorov's criterion (see [4, Theorem 12.3, p. 95]) implies that the sequence $M_n = (W_t, \bar{X}^n_t)_{t\in[0,T]}$ is C-tight, i.e. tight on $(\mathscr{C}([0,T],\mathbb{R}^2), \|\,.\,\|_{\sup})$. From now on, we mainly rely on the results established in [18]. Let n' be a subsequence such that $(\bar{X}^{n'}, W)$ functionally weakly converges to a probability $\mathbb{Q}$ on $(\mathscr{C}([0,T],\mathbb{R}^2), \|\,.\,\|_{\sup})$; hence it satisfies the $U.T.$ (for *Uniform Tightness*) assumption (see Proposition 3.2 in [18]). The function σ being continuous on $[0,T]\times\mathbb{R}$, the sequence of càdlàg processes $(\sigma(\underline{t}_n,\bar{X}^n_{\underline{t}_n}))_{n\ge1}$ is C-tight on the Skorokhod space since $\big((\underline{t}_n,\bar{X}^n_{\underline{t}_n})_{t\in[0,T]}\big)_{n\ge1}$ clearly is. One derives that, up to a new extraction still denoted (n'), we may assume that $\big((\sigma(\underline{t}_{n'},\bar{X}^{n'}_{\underline{t}_{n'}}))_{t\in[0,T]},\bar{X}^{n'},W\big)_{n\ge1}$

functionally converges toward a probability $\mathbb{P}$ on $\mathbb{D}([0,T],\mathbb{R}^3)$. By Theorem 2.6 from [18]—the functional weak convergence of stochastic integrals theorem—we know that

$$\left(\sigma(\underline{t}_{n'},\bar{X}^{n'}_{\underline{t}_{n'}}),(\bar{X}^{n'}_t,W_t),\int_0^t \sigma(\underline{s}_{n'},\bar{X}^{n'}_{\underline{s}_{n'}})dW_s\right)_{t\in[0,T]} \overset{\mathcal{L}(Sk)}{\longrightarrow} \mathbb{Q} \quad \text{as} \quad n\to+\infty$$

where $\mathbb{Q}$ is a probability distribution on $\mathbb{D}([0,T],\mathbb{R}^4)$ such that the canonical process $Y=(Y^i)_{i=1:4}$ satisfies $Y \overset{\mathcal{L}}{\sim} \left(Y^1,(Y^2,B),\int_0^{\cdot} Y^2_s dB_s\right)$ where $B := Y^3$ is a standard $\mathbb{Q}$-Brownian motion with respect to the $\mathbb{Q}$-completed right continuous canonical filtration $(\mathscr{D}^4_t)_{t\in[0,T]}$ on $\mathbb{D}([0,T],\mathbb{R}^4)$. Furthermore, we know that $Y^1=\sigma(.,Y^2)$ $\mathbb{Q}$-*a.s.* since $\sup_{t\in[0,T]}|\sigma(\underline{t}_{n'},\bar{X}^{n'}_{\underline{t}_{n'}})-\sigma(t,\bar{X}^{n'}_t)|$ converges to 0 in probability. The former claim follows from the facts that $\sup_{t\in[0,T]}|\bar{X}^n_t|$ is tight and $\sigma(t,\xi)$ is uniformly continuous on every compact set of $[0,T]\times\mathbb{R}$, with linear growth in ξ uniformly in $t\in[0,T]$. On the other hand, we know that $\bar{X}^{n'}=x+\int_0^{\cdot}\sigma(\underline{s}_{n'},\bar{X}^{n'}_{\underline{s}_{n'}})dW_s$, which in turn implies that $Y^2_{\cdot}=x+\int_0^{\cdot}\sigma(s,Y^2_s)dW_s$. This shows the existence of a weak solution to the *SDE* $X_t=x+\int_0^t\sigma(s,X_s)\,dW_s$, $t\in[0,T]$.

Under the weak uniqueness assumption, this distribution is unique, hence is the only functional weak limiting distribution for the tight sequence $(\bar{X}^n)_{n\geq1}$. The convergence in distribution on $\mathscr{C}([0,T],\mathbb{R})$ follows.

Remark 10 If the original *SDE* has a unique strong solution, the same proof leads to establish the convergence in probability of the Euler scheme toward X. One just has to add the process X itself to the sequence $\left((\sigma(\underline{t}_n,\bar{X}^n_{\underline{t}_n}))_{t\in[0,T]},\bar{X}^n,W\right)_{n\geq1}$.

Appendix 2: Euler Scheme for a Lévy Driven Martingale Diffusion

We consider the following *SDE* driven by a martingale Lévy process Z with Lévy measure ν:

$$X_t = x+\int_{(0,t]}\kappa(s,X_{s-})dZ_s,\ t\in[0,T],\ X_0=x, \tag{33}$$

where κ is a Borel function on $[0,T]\times\mathbb{R}$. Its *genuine* Euler scheme is defined by

$$\bar{X}^n_{t_{k+1}} = \bar{X}^n_{t_k}+\kappa(t_k,\bar{X}_{t_k})(Z_{t_{k+1}}-Z_{t_k}),\ k=1,\dots,n,\ \bar{X}_0=X_0=x \tag{34}$$

at discrete times t^n_k and extended into a continuous time càdlàg process by setting

$$\bar{X}^n_t = x+\int_{(0,t]}\kappa(\underline{s}_{n}-,\bar{X}^n_{\underline{s}_n}-)dZ_s,\ t\in[0,T]. \tag{35}$$

Convergence of the Euler Scheme Toward a Solution to the Lévy Driven SDE

Proposition 11

(*a*) *Let $p \in (1,2]$. Assume that $\nu(|z|^p) < +\infty$ and that Z has no Brownian component and $\kappa(t,\xi)$ has linear growth in ξ, uniformly in $t \in [0,T]$. Then*

$$\sup_{n\geq 1}\Big\| \sup_{t\in[0,T]} |\bar{X}^n_t|\Big\|_p + \Big\| \sup_{t\in[0,T]} |X_t|\Big\|_p < +\infty.$$

If moreover κ is continuous, then SDE (33) has at least one weak solution. Finally, under a weak uniqueness assumption, one has

$$\bar{X}^n \stackrel{\mathcal{L}(Sk)}{\longrightarrow} X.$$

(*b*) *If $\nu(z^2) < +\infty$, the same result remains true* mutatis mutandis *if Z has a non-zero Brownian component.*

Remark 11 In fact, if (33) has a strong solution, one shows using arguments similar to those developed below, the stronger result

$$\sup_{t\in[0,T]} |\bar{X}^n_t - X_t| \stackrel{\mathbb{P}}{\longrightarrow} 0 \text{ as } n \to +\infty.$$

We refer to [15] (devoted to error bounds) for a simpler proof when κ is homogeneous and $\mathscr{C}$ on the real line.

Proof (*a*) We consider the Lévy-Khintchine decomposition of the Lévy process $Z = (Z_t)_{t\in[0,T]}$, namely

$$Z_t = \widetilde{Z}_t + Z^1,\ t \in [0,T],$$

where $\widetilde{Z}$ is a pure jump, *square integrable* martingale with jumps of size at most 1 and Lévy measure $\nu(\,.\cap\{|z| \leq 1\})$ and Z^1 is a compensated (hence martingale) Poisson process with (finite) Lévy measure $\nu(\,.\cap\{|z| > 1\})$.

It is clear from (34) that $\bar{X}^n_{t^n_k} \in L^p$ for every $k = 0,\ldots,n$. Then, as $\nu(|z|^p) < +\infty$, it follows classically that $\sup_{u\in[t^n_k,t^n_{k+1}]} |Z^1_u - Z^1_{t^n_k}| \stackrel{d}{\sim} \sup_{[0,\frac{T}{n}]} |Z^1_u| \in L^p$ (see e.g. [32]). Combining these two results implies that $\varphi_{p,n}(t) := \big\| \sup_{s\in[0,t]} |\bar{X}^n_s|\big\|_p$ is finite for every $t \in [0,T]$.

It follows from Eq. (35) satisfied by $\bar{X}$ that

$$\varphi_{p,n}(t) \leq |x| + \Big\| \sup_{s\in[0,t]} \Big| \int_{(0,s]} \kappa(\underline{u}_{n-}, \bar{X}^n_{\underline{u}_{n-}}) dZ_u \Big|\Big\|_p$$

The L^p-B.D.G. Inequality implies (since $p > 1$)

$$\Big\| \sup_{s\in[0,t]} \Big| \int_{(0,s]} \kappa(\underline{u}_n, \bar{X}_{\underline{u}_n-}) dZ_u \Big| \Big\|_p \le c_p \Big\| \sum_{0<s\le t} \kappa(\underline{s}_n, \bar{X}_{\underline{s}_n-})^2 (\Delta Z_s)^2 \Big\|_{\frac{p}{2}}^{\frac{1}{2}}.$$

Using that $\frac{p}{2} \le 1$, we derive

$$\begin{aligned}\Big\| \sum_{0<s\le t} \kappa(\underline{s}_n, \bar{X}_{\underline{s}_n-})^2 (\Delta Z_s)^2 \Big\|_{\frac{p}{2}}^{\frac{1}{2}} &\le \Big(\mathbb{E} \sum_{0<s\le t} |\kappa(\underline{s}_n, \bar{X}_{\underline{s}_n-})|^p |\Delta Z_s|^p\Big)^{\frac{1}{p}} \\ &= \Big(\nu(|z|^p)\, \mathbb{E} \int_0^t |\kappa(\underline{s}_n, \bar{X}_{\underline{s}_n-})|^p ds\Big)^{\frac{1}{p}} \\ &\le C^p_{\kappa,p} \nu(|z|^p)^{\frac{1}{p}} \Big(\int_0^t (1 + \varphi_{p,n}(s)^p) ds\Big)^{\frac{1}{p}}\end{aligned}$$

where $C_{\kappa,p}$ is a real constant satisfying $|\kappa(s,\xi)| \le C_{\kappa,p}(1+|\xi|^p)^{\frac{1}{p}}$, $(s,\xi) \in [0,T]\times\mathbb{R}$.

Finally, there exists a positive real constant $C' = C'_{\kappa,p,\nu}$ such that the function $\varphi_{p,n}$ satisfies

$$\varphi_{p,n}(t)^p \le C'\Big(|x|^p + t + \int_0^t \varphi_{p,n}(s)^p ds\Big).$$

One concludes by Gronwall's Lemma that

$$\forall\, t \in [0,T], \quad \varphi_{p,n}(t)^p \le e^{C't} C'(T + |x|^p)$$

or, equivalently, there exists a real constant $C'' = C''_{T,\kappa,p,\nu}$ such that

$$\forall\, t \in [0,T], \quad \varphi_{p,n}(t) \le \varphi_p(t) = e^{C''t} C''(1 + |x|).$$

To establish the Skorokhod tightness of the sequence $(\bar{X}^n)_{n\ge1}$, we rely on the Aldous tightness criterion (see Definition 3(b) or [17, Theorem 4.5, p. 356]). Let $\rho \in (0,1]$. Let σ and τ be two $[0,T]$-valued $\mathcal{F}^Z$-stopping times such that $\sigma \le \tau \le (\sigma + \delta) \wedge T$.

$$\begin{aligned}\mathbb{E}|\bar{X}^n_\tau - \bar{X}^n_\sigma|^\rho = \mathbb{E}\Big| \sum_{\sigma<u\le\tau} \kappa(\underline{u}_n, \bar{X}^n_{\underline{u}_n-}) \Delta Z_u \Big|^\rho &\le \mathbb{E}\Big(\sum_{\sigma<u\le\tau} |\kappa(\underline{u}_n, \bar{X}^n_{\underline{u}_n-})|^\rho |\Delta Z_u|^\rho \Big) \\ &= \nu(|z|^\rho) \mathbb{E} \int_\sigma^{(\sigma+\delta)\wedge T} |\kappa(\underline{u}_n, \bar{X}^n_{\underline{u}_n-})|^\rho du \\ &\le \delta\, \nu(|z|^\rho)\, \mathbb{E}\big[\sup_{t\in[0,T]} |\kappa(t, \bar{X}^n_t)|^\rho\big] \\ &\le \delta\, \nu(|z|^\rho)\, C_\kappa (1 + \varphi_p(T))^{\frac{\rho}{p}}\end{aligned}$$

where we used that $\rho \le 1 \le p$ and $\nu(|z|^p) \le \nu(|z|^2 \wedge 1) + \nu(|z|^p) < +\infty$. Then

$$\sup\left\{\mathbb{E}|\bar{X}^n_\tau - \bar{X}^n_\sigma|^\rho + \mathbb{E}|Z_\tau - Z_\sigma|^\rho,\ \sigma \le \tau \le (\sigma+\delta)\wedge T, \mathscr{F}^Z\text{-stopping times}\right\}$$
$$\le \nu(|z|^\rho)(1 + C_\kappa(1+\varphi_p(T))^{\frac{\rho}{p}})\delta$$

which goes to 0 as $\delta \to 0$. This implies that the sequence $M_n = (\bar{X}^n, Z)$, $n \ge 1$, is *Sk*-tight. Moreover, following Proposition 3.2 from [18], the sequence $(M_n)_{n\ge 1}$ satisfies the *U.T.* condition since it is *Sk*-tight and

$$\begin{aligned}\mathbb{E}\sup_{t\in[0,T]}\left(|\Delta\bar{X}^n_t| \vee |\Delta Z_t|\right) &\le \left[\mathbb{E}\Big(\sum_{0<t\le T}|\Delta\bar{X}^n_t|^p + |\Delta Z_t|^p\Big)\right]^{\frac{1}{p}}\\ &\le \left[\nu(|z|^p)\mathbb{E}\int_0^T\left(1+|\kappa(\underline{t}_n,\bar{X}^n_{\underline{t}_n})|^p\right)dt\right]^{\frac{1}{p}}\\ &\le \left(\nu(|z|^p)\right)^{\frac{1}{p}}\left(T + C^p_{\kappa,p}(1+\varphi_p(T))\right)^{\frac{1}{p}} < +\infty.\end{aligned}$$

On the other hand, the sequence $\left((\kappa(\underline{t}_n,\bar{X}^n_{\underline{t}_n}))_{t\in[0,T]}, M_n\right)_{n\ge 1}$ is *Sk*-tight, owing to the following lemma.

Lemma 6 *Let $\mathscr{V}^+_{[0,T]}$ be the set of functions $\mu : [0,T] \to [0,T]$ such that $\mu(0) = 0$ and $\mu(T) = T$ endowed with the sup norm. Assume $\kappa : [0,T]\times\mathbb{R} \to \mathbb{R}$ is continuous. Then the mapping $\Psi : \mathscr{V}^+_{[0,T]} \times \boldsymbol{D}([0,T],\mathbb{R}^d) \to \boldsymbol{D}([0,T],\mathbb{R}^{1+d})$ defined by $\Psi(\mu,\alpha) = \left(\kappa(\mu(.),\alpha^1(.)),\alpha\right)$ is continuous ($\alpha = (\alpha^1,\dots,\alpha^d)$) for the product topology.*

Proof (Proof of the Lemma) Let $(\lambda_n)_{n\ge1}$ be a sequence of increasing homeomorphisms of $[0,T]$ such that $\lambda_n \to Id_{[0,T]}$ and $\alpha_n \circ \lambda_n \to \alpha$ uniformly and let $\mu_n \to \mu$ in $\mathscr{V}^+_{[0,T]}$ where $Id_{[0,T]}$ denotes the identity on $[0,T]$. Then the closure of $(\alpha_n \circ \lambda_n(t))_{n\ge1,t\in[0,T]}$ is a compact set K of $\mathbb{R}^d$ so that the function κ is uniformly continuous on $[0,T]\times K$. On the other hand

$$\|\mu_n \circ \lambda_n - Id_{[0,T]}\|_{\sup} \le \|\mu_n - Id_{[0,T]}\|_{\sup} + \|\lambda_n - Id_{[0,T]}\|_{\sup} \quad \text{as} \quad n \to +\infty$$

and $\|\alpha_n \circ \lambda_n - \alpha\|_{\sup} \to 0$ as $n \to +\infty$. The conclusion follows.

Up to an extraction, we may assume that the triplet $\left((\kappa(\underline{t}_{n'},\bar{X}^{n'}_{\underline{t}_{n'}})_{t\in[0,T]}, M_{n'}\right)_{n\ge1}$ weakly converges for the Skorokhod topology toward a probability $\mathbb{P}$ on the canonical Skorokhod space $(\boldsymbol{D}([0,T],\mathbb{R}^3), (\mathscr{D}_t)_{t\in[0,T]})$.

By Theorem 2.6 from [18] for the functional convergence of stochastic integrals, we know that

$$\Big(\kappa(\underline{t}_{n'}, \bar{X}^{n'}_{\underline{t}_{n'}}), (\bar{X}^{n'}_t, Z_t), \int_0^t \kappa(\underline{s}_{n'-}, \bar{X}^{n'}_{\underline{s}_{n'}-})dZ_s\Big)_{t\in[0,T]} \xrightarrow{\mathscr{L}(Sk)} \mathbb{Q}$$

where $\mathbb{Q}$ is a probability on $\mathbb{D}([0,T],\mathbb{R}^4)$ such that the canonical process $Y = (Y^i)_{i=1:4}$ satisfies $Y \overset{d}{\sim} \big(Y^1, (Y^2, Y^3), \int_0^{\cdot} Y^2_s dY^3_s\big)$ where Y^3 is a Lévy process with respect to the $\mathbb{Q}$ and the $\mathbb{Q}$-completed right continuous canonical filtration $(\mathscr{D}^{\mathbb{Q}}_t)_{t\in[0,T]}$ on $\mathbb{D}([0,T],\mathbb{R}^4)$ having the distribution of Z (i.e. $\mathbb{Q}_{Y^3} = \mathscr{L}(Z)$). Furthermore, we know that $Y^1 = \kappa(.,Y^2.)$ $\mathbb{Q}$-*a.s.* since the mapping $(\mu,(\alpha^i)_{i=1:4}) \mapsto \alpha^1 - \kappa(\mu,\alpha^2)$ is continuous from $\mathscr{V}^+_{[0,T]} \times \mathbb{D}([0,T],\mathbb{R}^4)$ to $\mathbb{D}([0,T],\mathbb{R})$ (and $\underline{t}_n$ converges uniformly to $Id_{[0,T]}$).

On the other hand we know that $\bar{X}^{n'}_t = x + \int_0^t \kappa(\underline{s}_{n'-}, \bar{X}^{n'}_{\underline{s}_{n'}-})dZ_s$, $t\in[0,T]$ which in turn implies that $(Y^2_t = x + \int_0^t \kappa(s, Y^2_{s-})dZ_s$, $t\in[0,T])$ $\mathbb{Q}$-*a.s.*. This shows the existence of a weak solution to the *SDE* $X_t = x + \int_0^t \kappa(s, X_{s-})dZ_s$, $t\in[0,T]$.

Under the weak uniqueness assumption, the distribution $\mathbb{Q}_{Y^2}$ of Y^2 is unique equal, say, to $\mathbb{P}_X$.

(*b*) We assume that the Lévy measure has a finite second moment $\nu(z^2) < +\infty$ on the whole real line. Then one can decompose Z as

$$Z_t = a\,W_t + \widetilde{Z}_t,\ t\in[0,T], \quad (a \geq 0)$$

where $a \geq 0$ and $\widetilde{Z}$ is a pure jump martingale Lévy process with Lévy measure ν. Then one shows like in the Brownian case that $\varphi(t) = \mathbb{E}\big(\sup_{s\in[0,t]}|\bar{X}^n_s|^2\big)$ is finite over $[0,T]$ using that all $\bar{X}_{t_k}$ are square integrable and $\mathbb{E}\big(\sup_{s\in[t_k,t_{k+1})}|Z_s - Z_{t_k}|^2\big) = \mathbb{E}\big(\sup_{s\in[0,\frac{T}{n}]}|Z_s|^2\big) < +\infty$. Then, using Doob's Inequality, we show that

$$\varphi(t) \leq 4C_\kappa^2(a^2 + \nu(z^2))\Big(t + \int_0^t \varphi(s)ds\Big)$$

where C_κ is a real constant satisfying $\kappa(t,\xi) \leq C_\kappa(1+|\xi|^2)^{\frac{1}{2}}$, $\xi\in\mathbb{R}$.

To establish the Skorokhod tightness of the sequence, we rely again on Aldous' tightness criterion (see Definition 3(*b*) or[17, Theorem 4.5, p. 356]). Let σ, τ be two $[0,T]$-valued $\mathscr{F}^Z$-stopping times such that $\sigma \leq \tau \leq (\sigma+\delta)\wedge T$. Applying Doob's

Inequality, to the martingale $\left(\int_\sigma^{\sigma+s}\kappa(\underline{u}_n,\bar X_{\underline{u}_n-})dZ_u\right)_{s\ge 0}$ yields

$$\begin{aligned}
\mathbb{E}|\bar X^n_\tau-\bar X^n_\sigma|^2 &\le 4a^2\mathbb{E}\left(\int_\sigma^\tau|\kappa(\underline{u}_n,\bar X^n_{\underline{u}_n-})|^2du\right)+4\,\mathbb{E}\left(\sum_{\sigma<u\le\tau}|\kappa(\underline{u}_n,\bar X^n_{\underline{u}_n-})|^2|\Delta Z_u|^2\right)\\
&=4\big(a^2+\nu(z^2)\big)\mathbb{E}\left(\int_\sigma^\tau|\kappa(\underline{u}_n,\bar X^n_{\underline{u}_n-})|^2du\right)\\
&\le 4\big(a^2+\nu(z^2)\big)\mathbb{E}\left(\int_\sigma^{(\sigma+\delta)\wedge T}|\kappa(\underline{u}_n,\bar X^n_{\underline{u}_n-})|^2du\right)\\
&\le 4\big(a^2+\nu(z^2)\big)\delta\, C_\kappa^2(1+\varphi(T)).
\end{aligned}$$

It follows that $\mathbb{E}|\bar X_\tau-\bar X_\sigma|^2+\mathbb{E}|Z_\tau-Z_\sigma|^2\le 4\,C_\kappa^2(a^2+\nu(z^2))\nu(z^2)(1+\varphi(T))\delta$ which clearly implies the *Sk*-tightness of the sequence $M_n=(\bar X^n,Z)$, $n\ge 1$.

The sequence satisfies the $U.T.$ condition from [18] since $(M_n)_{n\ge1}$ is *Sk*-tight and (see Proposition 3.2 from [18])

$$\begin{aligned}
\mathbb{E}\Big[\sup_{t\in[0,T]}\big(|\Delta\bar X^n_t|\vee|\Delta Z_t|\big)\Big] &\le \Big(\mathbb{E}\Big[\sum_{0<t\le T}|\Delta\bar X^n_t|^2+|\Delta Z_t|^2\Big]\Big)^{\frac12}\\
&\le\Big(\nu(z^2)\mathbb{E}\int_0^T\big(1+|\kappa(\underline{t}_n,\bar X^n_{\underline{t}_n})|^2\big)dt\Big)^{\frac12}\\
&\le\Big(\nu(z^2)\big(T+C_\kappa\big(1+\varphi(T)\big)\big)\Big)^{\frac12}<+\infty.
\end{aligned}$$

From this point, the proof is similar to that of claim (a).

Higher Moments

Let $Z_t=aW_t+\widetilde Z_t$, $t\in[0,T]$, be the decomposition of the Lévy process Z where W is a standard B.M. and $\widetilde Z$ is a pure jump Lévy process independent of W.

Proposition 12 *Let $p\in[2,+\infty)$. If $\nu(|z|^p)<+\infty$ then*

$$\sup_{n\ge1}\Big\|\sup_{t\in[0,T]}|\bar X^n_t|\Big\|_p<+\infty.$$

Proof If $p\in(1,2]$, the claim follows from the above Proposition 11. Assume from now on $p\in[2,+\infty)$. Let $\varphi_{p,n}(t)=\mathbb{E}\big(\sup_{t\in[0,T]}|\bar X^n_t|^p\big)$. Let ℓ_p be the unique integer defined by the inequality $2^{\ell_p}<p\le 2^{\ell_p+1}$. It is straightforward, using the same arguments as above, that $\varphi_{p,n}(T)<+\infty$ since $\sup_{t\in[0,T]}|Z_t|^p\in L^1$ (see [32,

Theorem 25.18, p. 166]) and $X_{t_k} \in L^p$ by induction using (34). For convenience, we set $\kappa_{s_-} = \kappa(\underline{s}_n, \bar{X}^n_{\underline{s}_n}-)$.

Now, combining the integral and the regular Minkowski Inequalities with the B.D.G. Inequality implies

$$\begin{aligned}\varphi_{p,n}(t)^{\frac{1}{p}} &\leq |x| + c_p \Big\| a^2 \int_0^t \kappa_{s_-}^2 ds + \sum_{0<s\leq t} \kappa_{s_-}^2 (\Delta Z_s)^2 \Big\|_{\frac{p}{2}}^{\frac{1}{2}} \\ &\leq |x| + c_p \Big(a \Big\| \int_0^t \kappa_{s_-}^2 ds \Big\|_{\frac{p}{2}}^{\frac{1}{2}} + \Big\| \sum_{0<s\leq t} \kappa_{s_-}^2 (\Delta Z_s)^2 \Big\|_{\frac{p}{2}}^{\frac{1}{2}} \Big) \end{aligned} \tag{36}$$

where we used in the second inequality that $\sqrt{u+v} \leq \sqrt{u} + \sqrt{v}$, $u,\ v \geq 0$. First note that by two successive applications of Hölder Inequality to dt and $d\mathbb{P}$, we obtain

$$\Big\| \int_0^t \kappa_{s_-}^2 ds \Big\|_{\frac{p}{2}}^{\frac{1}{2}} \leq T^{\frac{1}{2}-\frac{1}{p}} \Big(\int_0^t \mathbb{E}\, |\kappa_{s_-}|^p ds \Big)^{\frac{1}{p}}. \tag{37}$$

Using that for every $\ell \in \{1, \ldots, \ell_p\}$,
$\Big(\sum_{0<s\leq t} |\kappa_{s_-}|^{2^\ell} |\Delta Z_s|^{2^\ell} - \int_0^t |\kappa_{s_-}|^{2^\ell} ds\, \nu(|z|^{2^\ell}) \Big)_{t\in[0,T]}$, is a true martingale, we have by combining this time the Minkowski inequality, the B.D.G. Inequality applied with $\frac{p}{2^\ell} > 1$ and the elementary inequality $(u+v)^r \leq u^r + v^r$, $u,\ v \geq 0$, $r \in (0, 1]$ that:

$$\begin{aligned}\Big\| \sum_{0<s\leq t} |\kappa_{s_-}|^{2^\ell} (\Delta Z_s)^{2^\ell} \Big\|_{\frac{p}{2^\ell}}^{\frac{1}{2^\ell}} &\leq \Big\| \sum_{0<s\leq t} |\kappa_{s_-}|^{2^\ell} (\Delta Z_s)^{2^\ell} - \int_0^t |\kappa_{s_-}|^{2^\ell} ds\, \nu(|z|^{2^\ell}) \Big\|_{\frac{p}{2^\ell}}^{\frac{1}{2^\ell}} \\ &\quad + \Big\| \int_0^t |\kappa_{s_-}|^{2^\ell} ds \Big\|_{\frac{p}{2^\ell}}^{\frac{1}{2^\ell}} \nu(|z|^{2^\ell})^{\frac{1}{2^\ell}} \\ &\leq c_{\frac{p}{2^\ell}}^{\frac{1}{2^\ell}} \Big\| \sum_{0<s\leq t} |\kappa_{s_-}|^{2^{\ell+1}} (\Delta Z_s)^{2^{\ell+1}} \Big\|_{\frac{p}{2^{\ell+1}}}^{\frac{1}{2^{\ell+1}}} \\ &\quad + \Big\| \int_0^t |\kappa_{s_-}|^{2^\ell} ds \Big\|_{\frac{p}{2^\ell}}^{\frac{1}{2^\ell}} \nu(|z|^{2^\ell})^{\frac{1}{2^\ell}}. \end{aligned}$$

Then two applications of Hölder Inequality applied to dt and $d\mathbb{P}$ successively imply

$$\Big\| \int_0^t |\kappa_{s_-}|^{2^\ell} ds \Big\|_{\frac{p}{2^\ell}}^{\frac{1}{2^\ell}} \leq T^{\frac{1}{2^\ell}-\frac{1}{p}} \Big(\int_0^t \mathbb{E}\, |\kappa_{s_-}|^p ds \Big)^{\frac{1}{p}}.$$

Summing up these inequalities in cascade finally yields a positive real constant $K^{(0)}_{p,\nu,a,T}$ such that

$$\Big\|\sum_{0<s\le t}|\kappa_{s-}|^2(\Delta Z_s)^2\Big\|^{\frac12}_{\frac p2}\le K^{(0)}_{p,\nu,a,T}\bigg(\Big(\int_0^t \mathbb{E}\,|\kappa_{s-}|^p ds\Big)^{\frac1p}$$

$$+\Big\|\sum_{0<s\le t}|\kappa_{s-}|^{2^{\ell_p+1}}(\Delta Z_s)^{2^{\ell_p+1}}\Big\|^{\frac{1}{2^{\ell_p+1}}}_{\frac{p}{2^{\ell_p+1}}}\bigg).$$

Now, as $\frac{p}{2^{\ell_p+1}}\le 1$, one gets by the compensation formula

$$\Big\|\sum_{0<s\le t}|\kappa_{s-}|^{2^{\ell_p+1}}|\Delta Z_s|^{2^{\ell_p+1}}\Big\|^{\frac{1}{2^{\ell_p+1}}}_{\frac{p}{2^{\ell_p+1}}}\le\Big(\mathbb{E}\sum_{0<s\le t}|\kappa_{s-}|^p(\Delta Z_s)^p\Big)^{\frac1p}$$

$$=\Big(\int_0^t \mathbb{E}|\kappa_{s-}|^p ds\Big)^{\frac1p}\nu(|z|^p)^{\frac1p}.$$

Hence, there exists a real constant $K^{(1)}_{p,\nu,a,T}>0$

$$\Big\|\sum_{0<s\le t}|\kappa_{s-}|^2(\Delta Z_s)^2\Big\|^{\frac12}_{\frac p2}\le K^{(1)}_{p,\nu,a,T}\Big(\int_0^t \mathbb{E}|\kappa_{s-}|^p ds\Big)^{\frac1p}. \tag{38}$$

Finally, plugging (37) and (38) in (36), there exist positive real constants $K^{(\ell)}_{p,\nu,a,T}$, $\ell=2,3$, such that

$$\varphi_{p,n}(t)^{\frac1p}\le K^{(2)}_{p,\nu,a,T}\Big(|x|+\Big(\int_0^t \mathbb{E}|\kappa_{s-}|^p ds\Big)^{\frac1p}\Big)\le K^{(3)}_{p,\nu,a,T}\Big(|x|+1+\Big(\int_0^t \varphi_{p,n}(s)ds\Big)^{\frac1p}\Big)$$

where we used in the second inequality that κ has linear growth. Hence

$$\varphi_{p,n}(t)\le 2^{p-1}(K'^{(3)}_{p,\nu,a,T})^p\Big((|x|+1)^p+\int_0^t\varphi_{p,n}(s)ds\Big).$$

Gronwall's lemma completes the proof since it implies that

$$\varphi_{p,n}(t)\le e^{2^{p-1}(K'^{(3)}_{p,\nu,a,T})^p t}2^{p-1}(K'^{(3)}_{p,\nu,a,T})^p(|x|+1)^p.$$

References

1. J. Bergenthum, L. Rüschendorf, Comparison of option prices in semi-martingales models. Finance Stoch. **10**(2), 222–249 (2006)
2. J. Bergenthum, L. Rüschendorf, Comparison of semi-martingales and Lévy processes. Ann. Probab. **35**, 228–254 (2007)
3. J. Bergenthum, L. Rüschendorf, Comparison results for path-dependent options. Stat. Decis. **26**, 53–72 (2008)
4. P. Billingsley, *Convergence of Probability Measures*. Wiley Series in Probability and Statistics: Probability and Statistics, 2nd edn. (Wiley, New York, 1999), 277pp.
5. N. Bouleau, D. Lépingle, *Numerical Methods for Stochastic Processes*. Wiley Series in Probability and Mathematical Statistics: Applied Probability and Statistics (Wiley, New York, 1994), 359pp.
6. P. Carr, C.-O. Ewald, Y. Xiao, On the qualitative effect of volatility and duration on prices of asian options. Finance Res. Lett. **5**, 162–171 (2008)
7. D. Dacunha-Castelle, M. Duflo, *Probabilités et Statistique II, problèmes à temps mobile* (Masson, Paris, 1982), xiv+286pp.
8. N. El Karoui, Les aspects probabilistes du contrôle stochastique (French) [The probabilistic aspects of stochastic control], in 9th *Saint Flour Probability Summer School, 1979 (Saint Flour, 1979)*. Lecture Notes in Mathematics, vol. 876 (Springer, Berlin/Heidelberg, 1981), pp. 73–238
9. N. El Karoui, M. Jeanblanc, S. Shreve, Robustness of the Black-Scholes formula. Math. Finance **8**(2), 93–126 (1998)
10. B. Hajek, Mean stochastic comparison of diffusions. Probab. Relat. Fields **68**(3), 315–329 (1985)
11. O. Hernández-Lerma, W.J. Runggaldier, Monotone approximations for convex stochastic control problems. J. Math. Syst. Estimation Control **4**(4), 99–140 (1994)
12. F. Hirsch, M. Yor, Comparing Brownian stochastic integrals for the convex order, in *Modern Stochastics and Applications*. Springer Optimization and Its Applications, vol. 90 (Springer, Cham, 2014), pp. 3–19
13. F. Hirsch, C. Profeta, B. Roynette, M. Yor, *Peacocks and Associated Martingales, with Explicit Constructions* (Bocconi, Milan; Springer, Milan, 2011), 430pp.
14. D.G. Hobson, Robust hedging of the loopback option. Finance Stoch. **2**(4), 329–347 (1998)
15. J. Jacod, The Euler scheme for Lévy driven stochastic differential equations: limit theorems. Ann. Probab. **32**(3), 1830–1872 (2004)
16. J. Jacod, A.N. Shiryaev, *Limit Theorems for Stochastic Processes*. Grundlehren der Mathematischen Wissenschaften [Fundamental Principles of Mathematical Sciences], vol. 288 (Springer, Berlin, 1987), 601pp.
17. J. Jacod, A.N. Shiryaev, *Limit Theorems for Stochastic Processes*. Grundlehren der Mathematischen Wissenschaften [Fundamental Principles of Mathematical Sciences], vol. 288, 2nd edn. (Springer, Berlin, 2003), 661pp.
18. A. Jakubowski, J. Mémin, G. Pagès, Convergence en loi des suites d'intégrales stochastiques sur l'espace $\mathbb{D}^1$ de Skorokhod (French) [Convergence in law of sequences of stochastic integrals on the Skorokhod space $\mathbb{D}^1$]. Probab. Theory Relat. Fields **81**(1), 111–137 (1989)
19. I. Karatzas, On the pricing of American options. Appl. Math. Optim. **17**(1), 37–60 (1988)
20. I. Karatzas, S.E. Shreve, *Brownian Motion and Stochastic Calculus*. Graduate Texts in Mathematics, vol. 113, 2nd edn. (Springer, New York, 1991), 470pp.
21. H.G. Kellerer, Markov-Komposition und eine Anwendung auf Martingale (German). Math. Ann. **198**, 99–122 (1972)
22. P.E. Kloeden, E. Platen, *Numerical Solution of Stochastic Differential Equations*. Applications of Mathematics, vol. 23 (Springer, Berlin, 1992), xxxvi+632pp.
23. T. Kurtz, P. Protter, Weak limit theorems for stochastic integrals and stochastic differential equations. Ann. Probab. **19**, 1035–1070 (1991)

24. D. Lamberton, Optimal stopping and American options (2009), Ljubljana Summer School on Financial Mathematics, http://www.fmf.uni-lj.si/finmath09/ShortCourseAmericanOptions.pdf
25. D. Lamberton, G. Pagès, Sur la convergence des *réduites*. Ann. de l'IHP, série B **26**(2), 331–35 (1990)
26. R. Lucchetti, *Convexity and Well-Posed Problems*. CMS Books in Mathematics (Springer, Berlin, 2006), 305pp.
27. J. Neveu, *Martingales à temps discret* (Masson, Paris, 1972), 218pp. English translation: *Discrete-Parameter Martingales* (North-Holland, New York, 1975), 236pp.
28. G. Pagès, Functional co-monotony of processes with an application to peacocks, in *Séminaire de Probabilités XLV*, ed. by C. Donati, A. Lejay, A. Rouault. Lecture Notes in Mathematics, vol. 2078 (Springer, Berlin, 2013), pp. 365–400
29. G. Pagès, *Introduction to Numerical Probability and Applications to Finance*, coll. Universitext, Springer (2014, to appear)
30. P. Protter, *Stochastic Integration and Differential Equation*. Stochastic Modeling and Applied Probability, vol. 21, 2nd edn. (3rd corrected printing) (Springer, New York, 2006), 419pp.
31. D. Revuz, M. Yor, *Continuous Martingales and Brownian Motion*. Grundlehren der Mathematischen Wissenschaften [Fundamental Principles of Mathematical Sciences], vol. 293, 3rd edn. (Springer, Berlin, 1999), 560pp.
32. K.I. Sato, *Lévy Distributions and Infinitely Divisible Distributions*. Cambridge Studies in Advanced Mathematics (Cambridge University Press, Cambridge, 1999), 486pp.
33. A.N. Shiryaev, *Optimal Stopping Rules* (Translated from the 1976 Russian *2nd* edition by A.B. Aries). Stochastic Modeling and Applied Probability, vol. 8 (Springer, Berlin, 2008), xii+216pp. Reprint of the 1978 translation

Stability Problem for One-Dimensional Stochastic Differential Equations with Discontinuous Drift

Dai Taguchi

Abstract We consider one-dimensional stochastic differential equations (SDEs) with irregular coefficients. The goal of this paper is to estimate the $L^p(\Omega)$-difference between two SDEs using a norm associated to the difference of coefficients. In our setting, the (possibly) discontinuous drift coefficient satisfies a one-sided Lipschitz condition and the diffusion coefficient is bounded, uniformly elliptic and Hölder continuous. As an application of this result, we consider the stability problem for this class of SDEs.

2010 Mathematics Subject Classification: 58K25, 41A25, 65C30

1 Introduction

Let $X = (X_t)_{0 \leq t \leq T}$ be a solution of the one-dimensional stochastic differential equation (SDE)

$$X_t = x_0 + \int_0^t b(X_s)ds + \int_0^t \sigma(X_s)dW_s,\ x_0 \in \mathbb{R},\ t \in [0,T], \tag{1}$$

where $W := (W_t)_{0 \leq t \leq T}$ is a standard one-dimensional Brownian motion on a probability space $(\Omega, \mathscr{F}, \mathbb{P})$ with a filtration $(\mathscr{F}_t)_{0 \leq t \leq T}$ satisfying the usual conditions. The drift coefficient b and the diffusion coefficient σ are Borel-measurable functions from $\mathbb{R}$ into $\mathbb{R}$. The diffusion process X is used in many fields of application, for example, mathematical finance, optimal control and filtering.

Let $X^{(n)}$ be a solution of the SDE (1) with drift coefficient b_n and diffusion coefficient σ_n. We consider the stability problem for $(X, X^{(n)})$ when the pair of coefficients (b_n, σ_n) converges to (b, σ). Stroock and Varadhan introduced the

D. Taguchi (✉)
Department of Mathematical Sciences, Ritsumeikan University, 1-1-1 Nojihigashi, Kusatsu, Shiga 525-8577, Japan
e-mail: dai.taguchi.dai@gmail.com

C. Donati-Martin et al. (eds.), *Séminaire de Probabilités XLVIII*, Lecture Notes in Mathematics 2168, DOI 10.1007/978-3-319-44465-9_4

stability problem in the weak sense in order to consider the martingale problem with continuous and locally bounded coefficients (see Chap. 11 of [17]). In [11], Kawabata and Yamada consider the strong convergence of the stability problem under the condition that the drift coefficients b and b_n are Lipschitz continuous functions, the diffusion coefficients σ and σ_n are Hölder continuous and (b_n, σ_n) locally uniformly converges to (b, σ) (see [11, Example 1]). Kaneko and Nakao [10] prove that if the coefficients b_n and σ_n are uniformly bounded, σ_n is uniformly elliptic and (b_n, σ_n) tends to (b, σ) in L^1-sense, then $(X^{(n)})_{n\in\mathbb{N}}$ converges to X in L^2-sense. Moreover they also prove that the solution of the SDE (1) can be constructed as the limit of the Euler-Maruyama approximation under the condition that the coefficients b and σ are continuous and of linear growth (see [10, Theorem D]). Recently, under the Nakao-Le Gall condition, Hashimoto and Tsuchiya [8] prove that $(X^{(n)})_{n\in\mathbb{N}}$ converges to X in L^p sense for any $p \geq 1$ and give the rate of convergence under the condition that $b_n \to b$ and $\sigma_n \to \sigma$ in L^1 and L^2 sense, respectively. Their proof is based on the Yamada-Watanabe approximation technique which was introduced in [19] and some estimates for the local time.

On a related study, the convergence for the Euler-Maruyama approximation with non-Lipschitz coefficients has been studied recently. Yan [18] has proven that if the sets of discontinuous points of b and σ are countable, then the Euler-Maruyama approximation converges weakly to the unique weak solution of the corresponding SDE. Kohatsu-Higa et al. [12] have studied the weak approximation error for the one-dimensional SDE with the drift $\mathbf{1}_{(-\infty,0]}(x) - \mathbf{1}_{(0,+\infty)}(x)$ and constant diffusion. Gyöngy and Rásonyi [7] give the order of the strong rate of convergence for a class of one-dimensional SDEs whose drift is the sum of a Lipschitz continuous function and a monotone decreasing Hölder continuous function and its diffusion coefficient is a Hölder continuous function. The Yamada-Watanabe approximation technique is a key idea to obtain their results. In [15], Ngo and Taguchi extend the results in [7] for SDEs with discontinuous drift. They prove that if the drift coefficient b is bounded and one-sided Lipschitz function, and the diffusion coefficient is bounded, uniformly elliptic and η-Hölder continuous, then there exists a positive constant C such that

$$\sup_{0\leq t\leq T} \mathbb{E}[|X_t - \overline{X}_t^{(n)}|] \leq \begin{cases} \dfrac{C}{n^{\eta-1/2}}, & \textit{if } \eta \in (1/2, 1], \\ \dfrac{C}{\log n}, & \textit{if } \eta = 1/2, \end{cases}$$

where $\overline{X}^{(n)}$ is the Euler-Maruyama approximation for SDE (1). This fact implies that the strong rate of convergence for the stability problem may also depend on the Hölder exponent of the diffusion coefficient.

The goal of this paper is to estimate the difference between two SDEs using the norm of the difference of coefficients. More precisely, let us consider another SDE given by

$$\hat{X}_t = x_0 + \int_0^t \hat{b}(\hat{X}_s)ds + \int_0^t \hat{\sigma}(\hat{X}_s)dW_s. \tag{2}$$

We will prove the following inequality:

$$\sup_{0\le t\le T} \mathbb{E}[|X_t - \hat{X}_t|] \le \begin{cases} C(||b-\hat{b}||_1 \vee ||\sigma - \hat{\sigma}||_2^2)^{(2\eta-1)/(2\eta)}, & \text{if } \eta \in (1/2, 1], \\ \dfrac{C}{\log(1/(||b-\hat{b}||_1 \vee ||\sigma-\hat{\sigma}||_2^2))}, & \text{if } \eta = 1/2, \end{cases} \tag{3}$$

where η is the Hölder exponent of the diffusion coefficients, C is a positive constant and $||\cdot||_p$ is a L^p-norm which will be defined by (4). We will also estimate $\mathbb{E}[\sup_{0\le t\le T}|X_t - \hat{X}_t|^p]$ for any $p \ge 1$. It is worth noting that in the papers [10] and [11], the authors only prove the strong convergence for the stability problem. On the other hand, applying our main results, we are able to establish the strong rate of convergence for the stability problem (see Sect. 4). In order to obtain (3), we use the Yamada-Watanabe approximation technique and a Gaussian upper bound for the density of SDE (2) (see [2, 16] and [14]).

Finally, we note that SDEs with discontinuous drift coefficient have many applications in mathematical finance [1] and [9], optimal control problems [4] and other domains (see also [5] and [13]).

This paper is organized as follows: Sect. 2 introduces our framework and main results. All the proofs are shown in Sect. 3. In Sect. 4, we apply the main results to the stability problem.

2 Main Results

2.1 Notations and Assumptions

We will assume that the drift coefficient b belongs to the class of one-sided Lipschitz functions which is defined as follows.

Definition 1 A function $f : \mathbb{R} \to \mathbb{R}$ is called a *one-sided Lipschitz function* if there exists a positive constant L such that for any $x, y \in \mathbb{R}$,

$$(x-y)(f(x) - f(y)) \le L|x-y|^2.$$

Let $\mathscr{L}$ be the class of all one-sided Lipschitz functions.

Remark 1 By the definition of the class $\mathscr{L}$, if $f, g \in \mathscr{L}$ and $\alpha \ge 0$, then $f+g$, $\alpha f \in \mathscr{L}$. The one-sided Lipschitz property is closely related to the monotonicity condition. Actually, any monotone decreasing function is one-sided Lipschitz. Moreover, any Lipschitz continuous function is also a one-sided Lipschitz.

Now we give assumptions for the coefficients $b, \hat{b}, \sigma$ and $\hat{\sigma}$.

Assumption 1 We assume that the coefficients $b, \hat{b}, \sigma$ and $\hat{\sigma}$ satisfy the following conditions:

A-(i): $b \in \mathscr{L}$.

A-(ii): b and $\hat{b}$ are measurable and there exists $K > 0$ such that

$$\sup_{x\in\mathbb{R}} \left(|b(x)| \vee |\hat{b}(x)|\right) \leq K.$$

A-(iii): σ and $\hat{\sigma}$ are $\eta := (1/2 + \alpha)$-Hölder continuous with some $\alpha \in [0, 1/2]$, i.e., there exists $K > 0$ such that

$$\sup_{x,y\in\mathbb{R},x\neq y} \left(\frac{|\sigma(x) - \sigma(y)|}{|x-y|^{\eta}} \vee \frac{|\hat{\sigma}(x) - \hat{\sigma}(y)|}{|x-y|^{\eta}}\right) \leq K.$$

A-(iv): $a = \sigma^2$ and $\hat{a} = \hat{\sigma}^2$ are bounded and uniformly elliptic, i.e., there exists $\lambda \geq 1$ such that for any $x \in \mathbb{R}$,

$$\lambda^{-1} \leq a(x) \leq \lambda \text{ and } \lambda^{-1} \leq \hat{a}(x) \leq \lambda.$$

Remark 2 Assume that A-(ii), A-(iii) and A-(iv) hold. Then the SDE (1) and the SDE (2) have unique strong solution (see [20]). Note that the one-sided Lipschitz property is used only in (11) for b, so we don't need to assume $\hat{b} \in \mathscr{L}$.

2.2 *Gaussian Upper Bound for the Density of SDE*

A Gaussian upper bounded for the density of X_t is well-known under suitable conditions for the coefficients. If coefficients b and σ are Hölder continuous and σ is bounded and uniformly elliptic, then a Gaussian type estimate holds for the fundamental solution of parabolic type partial differential equations (see [6, Theorem 11, Chap. 1]). Under A-(ii), (iii) and (iv), the density function $p_t(x_0, \cdot)$ of X_t exists for any $t \in (0, T]$ and there exist positive constants $\overline{C}$ and c_* such that for any $y \in \mathbb{R}$ and $t \in (0, T]$,

$$p_t(x_0, y) \leq \overline{C} p_{c_*}(t, x_0, y),$$

where $p_c(t, x, y) := \frac{e^{-\frac{(y-x)^2}{2ct}}}{\sqrt{2\pi ct}}$ (see [14, Remark 4.1]).

Using a Gaussian upper bound for the density of X_t, we can prove the following estimate.

Lemma 1 *Let $p \geq 1$. Assume that A-(ii), A-(iii) and A-(iv) hold. Then we have*

$$\int_0^T \mathbb{E}[|b(\hat{X}_s) - \hat{b}(\hat{X}_s)|^p]ds \leq C_T ||b - \hat{b}||_p^p$$

and

$$\int_0^T \mathbb{E}[|\sigma(\hat{X}_s) - \hat{\sigma}(\hat{X}_s)|^{2p}]ds \leq C_T ||\sigma - \hat{\sigma}||_{2p}^{2p},$$

where $C_T := \overline{C}\sqrt{\frac{2T}{\pi c_*}}$ and for any bounded measurable function f, $||\cdot||_p$ is defined by

$$||f||_p := \left(\int_{\mathbb{R}} |f(x)|^p e^{-\frac{|x-x_0|^2}{2c_* T}} dx \right)^{1/p}. \tag{4}$$

Proof We only prove the first estimate. The second one can be obtained by using a similar argument. From a Gaussian upper bound for the density of $\hat{X}_t$, for any $x \in \mathbb{R}$ and $s \in (0, T]$, we have

$$\hat{p}_s(x_0, x) \leq \overline{C} p_{c_*}(s, x_0, x) \leq \frac{\overline{C}}{\sqrt{2\pi c_* s}} e^{-\frac{|x-x_0|^2}{2c_* T}},$$

where $\hat{p}_s(x_0, \cdot)$ is a density function of $\hat{X}_s$. Hence we obtain

$$\begin{aligned} \int_0^T \mathbb{E}[|b(\hat{X}_s) - \hat{b}(\hat{X}_s)|^p]ds &= \int_0^T ds \int_{\mathbb{R}} dx |b(x) - \hat{b}(x)|^p \hat{p}_s(x_0, x) \\ &\leq \int_0^T ds \frac{\overline{C}}{\sqrt{2\pi c_* s}} \int_{\mathbb{R}} dx |b(x) - \hat{b}(x)|^p e^{-\frac{|x-x_0|^2}{2c_* T}} \\ &= C_T ||b - \hat{b}||_p^p. \end{aligned} \tag{5}$$

This concludes the proof.

Remark 3 Our proof of Lemma 1 is based on the fact that we are in the one-dimensional setting. In multi-dimensional case, the integrand of (5) is not integrable with respect to s in general. This is the main reason for restricting our discussion to the one-dimensional SDE case.

2.3 Rate of Convergence

For any $p \geq 1$, we define

$$\varepsilon_p := \|b - \hat{b}\|_p^p \vee \|\sigma - \hat{\sigma}\|_{2p}^{2p}.$$

Then we have the following estimate for the difference between two SDEs.

Theorem 1 *Suppose that Assumption 1 holds. We assume that* $\varepsilon_1 < 1$ *if* $\alpha \in (0, 1/2]$ *and* $1/\log(1/\varepsilon_1) < 1$ *if* $\alpha = 0$. *Then there exists a positive constant* C *which depends on* $\overline{C}, c_*, K, L, T, \alpha, \lambda$ *and* x_0 *such that*

$$\sup_{\tau \in \mathscr{T}} \mathbb{E}[|X_\tau - \hat{X}_\tau|] \leq \begin{cases} C\varepsilon_1^{2\alpha/(2\alpha+1)} & \textit{if } \alpha \in (0, 1/2], \\ \dfrac{C}{\log(1/\varepsilon_1)} & \textit{if } \alpha = 0, \end{cases}$$

where $\mathscr{T}$ *is the set of all stopping times* $\tau \leq T$.

Theorem 2 *Suppose that Assumption 1 holds. We assume that* $\varepsilon_1 < 1$ *if* $\alpha \in (0, 1/2]$ *and* $1/\log(1/\varepsilon_1) < 1$ *if* $\alpha = 0$. *Then there exists a positive constant* C *which depends on* $\overline{C}, c_*, K, L, T, \alpha, \lambda$ *and* x_0 *such that*

$$\mathbb{E}[\sup_{0 \leq t \leq T} |X_t - \hat{X}_t|] \leq \begin{cases} C\varepsilon_1^{4\alpha^2/(2\alpha+1)} & \textit{if } \alpha \in (0, 1/2], \\ \dfrac{C}{\sqrt{\log(1/\varepsilon_1)}} & \textit{if } \alpha = 0. \end{cases}$$

Theorem 3 *Suppose that Assumption 1 holds and* $p \geq 2$. *We assume that* $\varepsilon_p < 1$ *if* $\alpha \in (0, 1/2]$ *and* $1/\log(1/\varepsilon_p) < 1$ *if* $\alpha = 0$. *Then there exists a positive constant* C *which depends on* $\overline{C}, c_*, K, L, T, p, \alpha, \lambda$ *and* x_0 *such that*

$$\mathbb{E}[\sup_{0 \leq t \leq T} |X_t - \hat{X}_t|^p] \leq \begin{cases} C\varepsilon_p^{1/2} & \textit{if } \alpha = 1/2, \\ C\varepsilon_1^{2\alpha/(2\alpha+1)} & \textit{if } \alpha \in (0, 1/2), \\ \dfrac{C}{\log(1/\varepsilon_1)} & \textit{if } \alpha = 0. \end{cases}$$

Using Jensen's inequality, we can extend Theorem 3 as follows.

Corollary 1 *Suppose that Assumption 1 holds and* $p \in (1, 2)$. *We assume that* $\varepsilon_{2p} < 1$ *if* $\alpha \in (0, 1/2]$ *and* $1/\log(1/\varepsilon_{2p}) < 1$ *if* $\alpha = 0$. *Then there exists a positive constant* C *which depends on* $\overline{C}, c_*, K, L, T, p, \alpha, \lambda$ *and* x_0 *such that*

$$\mathbb{E}[\sup_{0 \leq t \leq T} |X_t - \hat{X}_t|^p] \leq \begin{cases} C\varepsilon_{2p}^{1/2} & \textit{if } \alpha = 1/2, \\ C\varepsilon_1^{\alpha/(2\alpha+1)} & \textit{if } \alpha \in (0, 1/2), \\ \dfrac{C}{\sqrt{\log(1/\varepsilon_1)}} & \textit{if } \alpha = 0. \end{cases}$$

Next, we will find a bound for $\mathbb{E}[|g(X_T) - g(\hat{X}_T)|^r]$ where g is a function of bounded variation and $r \geq 1$.

Definition 2 For a function $f : \mathbb{R} \to \mathbb{R}$, we define

$$T_f(x) := \sup \sum_{j=1}^{N} |f(x_j) - f(x_{j-1})|.$$

Here the supremum is taken over all positive integers N and all partitions $-\infty < x_0 < x_1 < \cdots < x_N = x < \infty$. We call f a function of bounded variation, if

$$V(f) := \lim_{x \to \infty} T_f(x) < \infty.$$

Denote by BV the class of all functions of bounded variation.

Corollary 2 *Suppose that Assumption 1 holds. Furthermore assume that $\varepsilon_1 < 1$ if $\alpha \in (0, 1/2]$ and $1/\log(1/\varepsilon_1) < 1$ if $\alpha = 0$. Then there exists a positive constant C which depends on $\overline{C}, c_*, K, L, T, \alpha, \lambda$ and x_0 such that for any $g \in BV$ and $r \geq 1$,*

$$\mathbb{E}[|g(X_T) - g(\hat{X}_T)|^r] \leq \begin{cases} 3^{r+1} V(g)^r C \varepsilon_1^{\alpha/(2\alpha+1)} & \text{if } \alpha \in (0, 1/2], \\ \dfrac{3^{r+1} V(g)^r C}{\sqrt{\log(1/\varepsilon_1)}} & \text{if } \alpha = 0. \end{cases}$$

Remark 4 In the proof of all results, we calculate the constant C explicitly. In Theorems 1–3 and Corollary 1, the constant C does not blow up when $T \to 0$. On the other hand, in Corollary 2, the constant C may tend to infinity as $T \to 0$ because we use a Gaussian upper bound for the density of X_T in (17).

3 Proofs

3.1 Yamada-Watanabe Approximation Technique

In this section, we introduce the approximation method of Yamada and Watanabe (see [19] and [7]) which is the key technique for our proof. We define an approximation for the function $\phi(x) = |x|$. For each $\delta \in (1, \infty)$ and $\kappa \in (0, 1)$, there exists a continuous function $\psi_{\delta,\kappa} : \mathbb{R} \to \mathbb{R}^+$ with *supp* $\psi_{\delta,\kappa} \subset [\kappa/\delta, \kappa]$ such that

$$\int_{\kappa/\delta}^{\kappa} \psi_{\delta,\kappa}(z)dz = 1 \text{ and } 0 \leq \psi_{\delta,\kappa}(z) \leq \frac{2}{z \log \delta}, \quad z > 0.$$

For example, we can take

$$\psi_{\delta,\kappa}(z) := \mu_{\delta,\kappa} \exp\left[-\frac{1}{(\kappa - z)(z - \kappa/\delta)}\right] \mathbf{1}_{(\kappa/\delta,\kappa)}(z),$$

where $\mu_{\delta,\kappa}^{-1} := \int_{\kappa/\delta}^{\kappa} \exp(-\frac{1}{(\kappa-z)(z-\kappa/\delta)})dz$. We define a function $\phi_{\delta,\kappa} \in C^2(\mathbb{R};\mathbb{R})$ by

$$\phi_{\delta,\kappa}(x) := \int_0^{|x|} \int_0^y \psi_{\delta,\kappa}(z) dz dy.$$

It is easy to verify that $\phi_{\delta,\kappa}$ has the following useful properties:

$$\frac{\phi'_{\delta,\kappa}(x)}{x} > 0, \text{ for any } x \in \mathbb{R} \setminus \{0\}. \tag{6}$$

$$0 \leq |\phi'_{\delta,\kappa}(x)| \leq 1, \text{ for any } x \in \mathbb{R}. \tag{7}$$

$$|x| \leq \kappa + \phi_{\delta,\kappa}(x), \text{ for any } x \in \mathbb{R}. \tag{8}$$

$$\phi''_{\delta,\kappa}(\pm|x|) = \psi_{\delta,\kappa}(|x|) \leq \frac{2}{|x| \log \delta} \mathbf{1}_{[\kappa/\delta,\kappa]}(|x|), \text{ for any } x \in \mathbb{R} \setminus \{0\}. \tag{9}$$

The property (8) implies that the function $\phi_{\delta,\kappa}$ approximates ϕ.

3.2 *Proof of Theorem 1*

To simplify the discussion, we set

$$Y_t := X_t - \hat{X}_t, \ t \in [0, T].$$

Proof (Proof of Theorem 1) Let $\delta \in (1, \infty)$ and $\kappa \in (0, 1)$. From Itô's formula, (7) and (8), we have

$$\begin{aligned}
|Y_t| &\leq \kappa + \phi_{\delta,\kappa}(Y_t) \\
&= \kappa + \int_0^t \phi'_{\delta,\kappa}(Y_s)(b(X_s) - \hat{b}(\hat{X}_s))ds \\
&\quad + \frac{1}{2} \int_0^t \phi''_{\delta,\kappa}(Y_s)|\sigma(X_s) - \hat{\sigma}(\hat{X}_s)|^2 ds + M_t^{\delta,\kappa} \\
&= \kappa + \int_0^t \phi'_{\delta,\kappa}(Y_s)(b(X_s) - b(\hat{X}_s))ds + \int_0^t \phi'_{\delta,\kappa}(Y_s)(b(\hat{X}_s) - \hat{b}(\hat{X}_s))ds \\
&\quad + \frac{1}{2} \int_0^t \phi''_{\delta,\kappa}(Y_s)|\sigma(X_s) - \hat{\sigma}(\hat{X}_s)|^2 ds + M_t^{\delta,\kappa}
\end{aligned}$$

$$
\leq \kappa + \int_0^t \phi'_{\delta,\kappa}(Y_s)(b(X_s) - b(\hat{X}_s))ds + \int_0^T |b(\hat{X}_s) - \hat{b}(\hat{X}_s)|ds
$$

$$
+ \frac{1}{2}\int_0^t \phi''_{\delta,\kappa}(Y_s)|\sigma(X_s) - \hat{\sigma}(\hat{X}_s)|^2 ds + M_t^{\delta,\kappa}, \tag{10}
$$

where

$$
M_t^{\delta,\kappa} := \int_0^t \phi'_{\delta,\kappa}(Y_s)(\sigma(X_s) - \hat{\sigma}(\hat{X}_s))dW_s.
$$

Note that since σ, $\hat{\sigma}$ and $\phi'_{\delta,\kappa}$ are bounded, $(M_t^{\delta,\kappa})_{0\leq t\leq T}$ is a martingale so $\mathbb{E}[M_t^{\delta,\kappa}] = 0$. Since $b \in \mathscr{L}$, for any $x, y \in \mathbb{R}$ with $x \neq y$, we have, from (6) and (7),

$$
\begin{aligned}
\phi'_{\delta,\kappa}(x-y)(b(x) - b(y)) &= \frac{\phi'_{\delta,\kappa}(x-y)}{x-y}(x-y)(b(x)-b(y)) \\
&\leq L\frac{\phi'_{\delta,\kappa}(x-y)}{x-y}|x-y|^2 \\
&\leq L|x-y|.
\end{aligned} \tag{11}
$$

Therefore we get

$$
\int_0^t \phi'_{\delta,\kappa}(Y_s)(b(X_s) - b(\hat{X}_s))ds \leq L\int_0^t |Y_s|ds. \tag{12}
$$

Using Lemma 1 with $p = 1$, we have

$$
\int_0^T \mathbb{E}[|b(\hat{X}_s) - \hat{b}(\hat{X}_s)|]ds \leq C_T||b - \hat{b}||_1. \tag{13}
$$

From (9) and $(x+y)^2 \leq 2x^2 + 2y^2$ for any $x, y \geq 0$, we have

$$
\begin{aligned}
&\frac{1}{2}\int_0^t \phi''_{\delta,\kappa}(Y_s)|\sigma(X_s) - \hat{\sigma}(\hat{X}_s)|^2 ds \leq \int_0^t \frac{\mathbf{1}_{[\kappa/\delta,\kappa]}(|Y_s|)}{|Y_s|\log\delta}|\sigma(X_s) - \hat{\sigma}(\hat{X}_s)|^2 ds \\
&\leq 2\int_0^t \frac{\mathbf{1}_{[\kappa/\delta,\kappa]}(|Y_s|)}{|Y_s|\log\delta}|\sigma(X_s) - \sigma(\hat{X}_s)|^2 ds + 2\int_0^t \frac{\mathbf{1}_{[\kappa/\delta,\kappa]}(|Y_s|)}{|Y_s|\log\delta}|\sigma(\hat{X}_s) - \hat{\sigma}(\hat{X}_s)|^2 ds \\
&\leq 2\int_0^t \frac{\mathbf{1}_{[\kappa/\delta,\kappa]}(|Y_s|)}{|Y_s|\log\delta}|\sigma(X_s) - \sigma(\hat{X}_s)|^2 ds + \frac{2\delta}{\kappa\log\delta}\int_0^T |\sigma(\hat{X}_s) - \hat{\sigma}(\hat{X}_s)|^2 ds.
\end{aligned} \tag{14}
$$

Again using Lemma 1 with $p = 1$, we have

$$\frac{2\delta}{\kappa \log \delta} \int_0^T \mathbb{E}[|\sigma(\hat{X}_s) - \hat{\sigma}(\hat{X}_s)|^2] ds \leq \frac{2C_T\delta}{\kappa \log \delta} ||\sigma - \hat{\sigma}||_2^2. \tag{15}$$

Since σ is $(1/2 + \alpha)$-Hölder continuous, we have

$$\begin{aligned} 2\int_0^T \frac{\mathbf{1}_{[\kappa/\delta,\kappa]}(|Y_s|)}{|Y_s| \log \delta} |\sigma(X_s) - \sigma(\hat{X}_s)|^2 ds &\leq 2K^2 \int_0^T \frac{\mathbf{1}_{[\kappa/\delta,\kappa]}(|Y_s|)}{|Y_s| \log \delta} |Y_s|^{1+2\alpha} ds \\ &\leq \frac{2TK^2\kappa^{2\alpha}}{\log \delta}. \end{aligned} \tag{16}$$

Let τ be a stopping time with $\tau \leq T$ and $Z_t := |Y_{t\wedge\tau}|$. From (10), (12), (13), (15) and (16), we obtain

$$\begin{aligned} \mathbb{E}[Z_t] &\leq \kappa + L \int_0^t \mathbb{E}[Z_s] ds + C_T ||b - \hat{b}||_1 + \frac{2C_T\delta}{\kappa \log \delta} ||\sigma - \hat{\sigma}||_2^2 + \frac{2TK^2\kappa^{2\alpha}}{\log \delta} \\ &\leq \kappa + L \int_0^t \mathbb{E}[Z_s] ds + C_T \varepsilon_1 + \frac{2C_T\delta}{\kappa \log \delta} \varepsilon_1 + \frac{2TK^2\kappa^{2\alpha}}{\log \delta}. \end{aligned}$$

If $\alpha \in (0, 1/2]$, then since $\varepsilon_1 < 1$, by choosing $\delta = 2$ and $\kappa = \varepsilon_1^{1/(2\alpha+1)}$, we have

$$\begin{aligned} \mathbb{E}[Z_t] &\leq L \int_0^t \mathbb{E}[Z_s] ds + \varepsilon_1^{1/(2\alpha+1)} + C_T \varepsilon_1 + \frac{4C_T \varepsilon_1^{1-1/(2\alpha+1)}}{\log 2} + \frac{2TK^2 \varepsilon_1^{2\alpha/(2\alpha+1)}}{\log 2} \\ &\leq L \int_0^t \mathbb{E}[Z_s] ds + C_1(\alpha, T) \varepsilon_1^{2\alpha/(2\alpha+1)}, \end{aligned}$$

where

$$C_1(\alpha, T) := 1 + C_T + \frac{4C_T}{\log 2} + \frac{2TK^2}{\log 2}.$$

By Gronwall's inequality, we get

$$\mathbb{E}[Z_t] \leq C_1(\alpha, T) e^{LT} \varepsilon_1^{2\alpha/(2\alpha+1)}.$$

Therefore by the dominated convergence theorem, we conclude the statement by taking $t \to T$.

If $\alpha = 0$, then since $1/\log(1/\varepsilon_1) < 1$, by choosing $\delta = \varepsilon_1^{-1/2}$ and $\kappa = 1/\log(1/\varepsilon_1)$, we have

$$\begin{aligned}\mathbb{E}[Z_t] &\leq L\int_0^t \mathbb{E}[Z_s]ds + \frac{1}{\log(1/\varepsilon_1)} + C_T\varepsilon_1 + 4C_T\varepsilon_1^{1/2} + \frac{4TK^2}{\log(1/\varepsilon_1)} \\ &\leq L\int_0^t \mathbb{E}[Z_s]ds + \frac{C_1(0,T)}{\log(1/\varepsilon_1)},\end{aligned}$$

where

$$C_1(0,T) := 1 + 5C_T + 4TK^2.$$

By Gronwall's inequality, we obtain

$$\mathbb{E}[Z_t] \leq \frac{C_1(0,T)e^{LT}}{\log(1/\varepsilon_1)}.$$

Therefore by the dominated convergence theorem, we conclude the statement by taking $t \to T$.

3.3 *Proof of Corollary 2*

To prove Corollary 2, we recall the upper bound for $\mathbb{E}[|g(X) - g(\hat{X})|^r]$ where g is a function of bounded variation, $r \geq 1$, X and $\hat{X}$ are random variables.

Lemma 2 ([3], Theorem 4.3) *Let X and $\hat{X}$ be random variables. Assume that X has a bounded density p_X. If $g \in BV$ and $r \geq 1$, then for every $q \geq 1$, we have*

$$\mathbb{E}[|g(X) - g(\hat{X})|^r] \leq 3^{r+1}V(g)^r\left(\sup_{x\in\mathbb{R}} p_X(x)\right)^{\frac{q}{q+1}} \mathbb{E}[|X - \hat{X}|^q]^{1/(q+1)}.$$

Using the above Lemma, we can prove Corollary 2.

Proof (Proof of Corollary 2) From the Gaussian upper bound for the density $p_T(x_0,\cdot)$ of X_T, we have for any $y \in \mathbb{R}$,

$$p_T(x_0,y) \leq \overline{C}p_{c_*}(T,x_0,y) \leq \frac{\overline{C}}{\sqrt{2\pi c_* T}}. \tag{17}$$

This means that the density $p_T(x_0, \cdot)$ of X_T is bounded. Hence from Lemma 2 with $q = 1$ and Theorem 1 with $\tau = T$, for any $g \in BV$ and $r \geq 1$, we have

$$\mathbb{E}[|g(X_T) - g(\hat{X}_T)|^r] \leq \frac{3^{r+1}V(g)^r\overline{C}^{1/2}}{(2\pi c_* T)^{1/4}}\mathbb{E}[|X_T - \hat{X}_T|]^{1/2}$$

$$\leq \begin{cases} 3^{r+1}V(g)^r C_2(\alpha, T)\varepsilon_1^{\alpha/(2\alpha+1)} & \text{if } \alpha \in (0, 1/2], \\ \dfrac{3^{r+1}V(g)^r C_2(0, T)}{\sqrt{\log(1/\varepsilon_1)}} & \text{if } \alpha = 0, \end{cases}$$

where

$$C_2(\alpha, T) := \frac{\overline{C}^{1/2} C_1(\alpha, T)^{1/2} e^{LT/2}}{(2\pi c_* T)^{1/4}}, \text{ for } \alpha \in [0, 1/2].$$

This concludes the proof of statement.

3.4 *Proof of Theorem 2*

Let $V_t := \sup_{0 \leq s \leq t} |Y_s|$. Recall that for each $\delta \in (1, \infty)$ and $\kappa \in (0, 1)$,

$$M_t^{\delta,\kappa} = \int_0^t \phi'_{\delta,\kappa}(Y_s)(\sigma(X_s) - \hat{\sigma}(\hat{X}_s))dW_s.$$

Hence the quadratic variation of $M_t^{\delta,\kappa}$ is given by

$$\langle M^{\delta,\kappa}\rangle_t = \int_0^t |\phi'_{\delta,\kappa}(Y_s)|^2 |\sigma(X_s) - \hat{\sigma}(\hat{X}_s)|^2 ds.$$

Before proving Theorem 2, we estimate the expectation of $\sup_{0 \leq s \leq t} |M_s^{\delta,\kappa}|$ for any $t \in [0, T]$, $\delta \in (1, \infty)$ and $\kappa \in (0, 1)$.

Lemma 3 *Suppose that the assumption of Theorem 2 hold. Then for any $t \in [0, T]$, $\delta \in (1, \infty)$ and $\kappa \in (0, 1)$, we have*

$$\mathbb{E}[\sup_{0 \leq s \leq t} |M_s^{\delta,\kappa}|] \leq \begin{cases} \dfrac{1}{2}\mathbb{E}[V_t] + C_3(\alpha, T)\varepsilon_1^{4\alpha^2/(2\alpha+1)} & \text{if } \alpha \in (0, 1/2], \\ \dfrac{C_3(0, T)}{\sqrt{\log(1/\varepsilon_1)}} & \text{if } \alpha = 0, \end{cases}$$

where

$$C_3(\alpha, T) := \begin{cases} \hat{C}_1^2 K^2 T C_1(\alpha, T)^{2\alpha} e^{2\alpha LT} + \sqrt{2}\hat{C}_1 C_T^{1/2}, & \text{if } \alpha \in (0, 1/2], \\ \sqrt{2}\hat{C}_1 K T^{1/2} C_1(0, T)^{1/2} e^{LT/2} + \sqrt{2}\hat{C}_1 C_T^{1/2}, & \text{if } \alpha = 0, \end{cases}$$

and $\hat{C}_p$ *is the constant of Burkholder-Davis-Gundy's inequality with* $p > 0$.

Proof From Burkholder-Davis-Gundy's inequality, we have

$$\begin{aligned} \mathbb{E}[\sup_{0\le s\le t} |M_s^{\delta,\kappa}|] &\le \hat{C}_1 \mathbb{E}[\langle M^{\delta,\kappa}\rangle_t^{1/2}] \le \hat{C}_1 \mathbb{E}\left[\left(\int_0^t |\sigma(X_s) - \hat{\sigma}(\hat{X}_s)|^2 ds\right)^{1/2}\right] \\ &\le \sqrt{2}\hat{C}_1 \mathbb{E}\left[\left(\int_0^t |\sigma(X_s) - \sigma(\hat{X}_s)|^2 ds\right)^{1/2}\right] \\ &\quad + \sqrt{2}\hat{C}_1 \mathbb{E}\left[\left(\int_0^T |\sigma(\hat{X}_s) - \hat{\sigma}(\hat{X}_s)|^2 ds\right)^{1/2}\right]. \end{aligned}$$

From Jensen's inequality and Lemma 1, we have

$$\begin{aligned} \mathbb{E}\left[\left(\int_0^T |\sigma(\hat{X}_s) - \hat{\sigma}(\hat{X}_s)|^2 ds\right)^{1/2}\right] &\le \left(\int_0^T \mathbb{E}\left[|\sigma(\hat{X}_s) - \hat{\sigma}(\hat{X}_s)|^2\right] ds\right)^{1/2} \\ &\le C_T^{1/2} ||\sigma - \hat{\sigma}||_2. \end{aligned}$$

Since σ is $(1/2 + \alpha)$-Hölder continuous, we obtain

$$\mathbb{E}[\sup_{0\le s\le t} |M_s^{\delta,\kappa}|] \le \sqrt{2}\hat{C}_1 K \mathbb{E}\left[\left(\int_0^t |Y_s|^{1+2\alpha} ds\right)^{1/2}\right] + \sqrt{2}\hat{C}_1 C_T^{1/2} ||\sigma - \hat{\sigma}||_2. \tag{18}$$

If $\alpha \in (0, 1/2]$, then we get

$$\sqrt{2}\hat{C}_1 K \mathbb{E}\left[\left(\int_0^t |Y_s|^{1+2\alpha} ds\right)^{1/2}\right] \le \sqrt{2}\hat{C}_1 K \mathbb{E}\left[V_t^{1/2}\left(\int_0^t |Y_s|^{2\alpha} ds\right)^{1/2}\right].$$

Using Young's inequality $xy \leq \frac{x^2}{2\sqrt{2}\hat{C}_1 K} + \frac{\sqrt{2}\hat{C}_1 K y^2}{2}$ for any $x, y \geq 0$ and Jensen's inequality, we obtain

$$\sqrt{2}\hat{C}_1 K \mathbb{E}\left[\left(\int_0^t |Y_s|^{1+2\alpha} ds\right)^{1/2}\right] \leq \frac{1}{2}\mathbb{E}[V_t] + \frac{2\hat{C}_1^2 K^2}{2} \int_0^T \mathbb{E}[|Y_s|^{2\alpha}] ds$$
$$\leq \frac{1}{2}\mathbb{E}[V_t] + \hat{C}_1^2 K^2 T^{1-2\alpha} \left(\int_0^T \mathbb{E}[|Y_s|] ds\right)^{2\alpha}.$$

From Theorem 1 with $\tau = s$, we have

$$\sqrt{2}\hat{C}_1 K \mathbb{E}\left[\left(\int_0^t |Y_s|^{1+2\alpha} ds\right)^{1/2}\right] \leq \frac{1}{2}\mathbb{E}[V_t] + \hat{C}_1^2 K^2 T C_1(\alpha, T)^{2\alpha} e^{2\alpha LT} \varepsilon_1^{4\alpha^2/(2\alpha+1)}. \tag{19}$$

Since $4\alpha^2/(2\alpha + 1) \leq \alpha \leq 1/2$, from (18) and (19), we get

$$\mathbb{E}[\sup_{0 \leq s \leq t} |M_s^{\delta,\kappa}|] \leq \frac{1}{2}\mathbb{E}[V_t] + C_3(\alpha, T)\varepsilon_1^{4\alpha^2/(2\alpha+1)}$$

which concludes the statement for $\alpha \in (0, 1/2]$.

If $\alpha = 0$, then from Jensen's inequality and Theorem 1 with $\tau = s$, we get

$$\sqrt{2}\hat{C}_1 K \mathbb{E}\left[\left(\int_0^t |Y_s| ds\right)^{1/2}\right] \leq \sqrt{2}\hat{C}_1 K \left(\int_0^T \mathbb{E}[|Y_s|] ds\right)^{1/2}$$
$$\leq \frac{\sqrt{2}\hat{C}_1 K T^{1/2} C_1(0, T)^{1/2} e^{LT/2}}{\sqrt{\log(1/\varepsilon_1)}}.$$

Therefore we have

$$\mathbb{E}[\sup_{0 \leq s \leq T} |M_s^{\delta,\kappa}|] \leq \frac{\sqrt{2}\hat{C}_1 K T^{1/2} C_1(0, T)^{1/2} e^{LT/2}}{\sqrt{\log(1/\varepsilon_1)}} + \sqrt{2}\hat{C}_1 C_T^{1/2} ||\sigma - \hat{\sigma}||_2$$
$$\leq \frac{C_3(0, T)}{\sqrt{\log(1/\varepsilon_1)}}.$$

This concludes the statement for $\alpha = 0$.

Using the above estimate, we can prove Theorem 2.

Proof (Proof of Theorem 2) From (10), (12), (14) and (16), we have

$$V_t \le \kappa + L\int_0^t V_s ds + \int_0^T |b(\hat{X}_s) - \hat{b}(\hat{X}_s)|ds + \frac{2\delta}{\kappa \log \delta}\int_0^T |\sigma(\hat{X}_s) - \hat{\sigma}(\hat{X}_s)|^2 ds + \frac{2TK^2\kappa^{2\alpha}}{\log \delta} + \sup_{0\le s\le t} |M_s^{\delta,\kappa}|. \tag{20}$$

If $\alpha \in (0, 1/2]$, then from (20), Lemmas 1 and 3, we have

$$\begin{aligned}\mathbb{E}[V_t] &\le \kappa + L\int_0^t \mathbb{E}[V_s]ds + C_T||b - \hat{b}||_1 + \frac{2C_T\delta}{\kappa \log \delta}||\sigma - \hat{\sigma}||_2^2 + \frac{2TK^2\kappa^{2\alpha}}{\log \delta} \\ &\quad + \frac{1}{2}\mathbb{E}[V_t] + C_3(\alpha, T)\varepsilon_1^{4\alpha^2/(2\alpha+1)} \\ &\le \kappa + L\int_0^t \mathbb{E}[V_s]ds + C_T\varepsilon_1 + \frac{2C_T\delta}{\kappa \log \delta}\varepsilon_1 + \frac{2TK^2\kappa^{2\alpha}}{\log \delta} \\ &\quad + \frac{1}{2}\mathbb{E}[V_t] + C_3(\alpha, T)\varepsilon_1^{4\alpha^2/(2\alpha+1)}.\end{aligned}$$

Hence we get

$$\begin{aligned}\mathbb{E}[V_t] &\le 2\kappa + 2L\int_0^t \mathbb{E}[V_s]ds + 2C_T\varepsilon_1 + \frac{4C_T\delta}{\kappa \log \delta}\varepsilon_1 \\ &\quad + \frac{4TK^2\kappa^{2\alpha}}{\log \delta} + 2C_3(\alpha, T)\varepsilon_1^{4\alpha^2/(2\alpha+1)}.\end{aligned}$$

Note that $0 < 4\alpha^2/(2\alpha + 1) \le \alpha \le 1/2$. Taking $\delta = 2$ and $\kappa = \varepsilon_1^{1/2}$, we have

$$\begin{aligned}\mathbb{E}[V_t] &\le 2L\int_0^t \mathbb{E}[V_s]ds + 2\left(1 + C_T + \frac{4C_T}{\log 2}\right)\varepsilon_1^{1/2} \\ &\quad + \frac{4TK^2}{\log 2}\varepsilon_1^{\alpha} + 2C_3(\alpha, T)\varepsilon_1^{4\alpha^2/(2\alpha+1)} \\ &\le 2L\int_0^t \mathbb{E}[V_s]ds + C_4(\alpha, T)\varepsilon_1^{4\alpha^2/(2\alpha+1)},\end{aligned}$$

where

$$C_4(\alpha, T) := 2\left(1 + C_T + \frac{4C_T + 2TK^2}{\log 2} + C_3(\alpha, T)\right).$$

By Gronwall's inequality, we obtain

$$\mathbb{E}[V_t] \leq C_4(\alpha, T) e^{2LT} \varepsilon_1^{4\alpha^2/(2\alpha+1)}.$$

If $\alpha = 0$, then from (20), Lemmas 1 and 3, we have

$$\mathbb{E}[V_t] \leq \kappa + L \int_0^t \mathbb{E}[V_s] ds + C_T \varepsilon_1 + \frac{2C_T \delta}{\kappa \log \delta} \varepsilon_1 + \frac{2TK^2}{\log \delta} + \frac{C_3(0,T)}{\sqrt{\log(1/\varepsilon_1)}}.$$

Taking $\delta = \varepsilon_1^{-1/2}$ and $\kappa = 1/\log(1/\varepsilon_1)$, we get

$$\mathbb{E}[V_t] \leq L \int_0^t \mathbb{E}[V_s] ds + \frac{C_4(0,T)}{\sqrt{\log(1/\varepsilon_1)}},$$

where

$$C_4(0,T) := 1 + 5C_T + 4TK^2 + C_3(0,T).$$

By Gronwall's inequality, we obtain

$$\mathbb{E}[V_t] \leq \frac{C_4(0,T) e^{LT}}{\sqrt{\log(1/\varepsilon_1)}}.$$

Hence we conclude the proof of Theorem 2.

3.5 *Proof of Theorem 3*

In this section, we also estimate the expectation of $\sup_{0 \leq s \leq t} |M_s^{\delta,\kappa}|^p$ for any $p \geq 2$, $t \in [0,T]$, $\delta \in (1,\infty)$ and $\kappa \in (0,1)$.

Lemma 4 *Let $p \geq 2$. Assume that A-(ii), A-(iii) and A-(iv) hold. Then for any $t \in [0,T]$, $\delta \in (1,\infty)$ and $\kappa \in (0,1)$, we have*

$$\mathbb{E}[\sup_{0 \leq s \leq t} |M_s^{\delta,\kappa}|^p] \leq C_5(p,T) \mathbb{E}\left[\left(\int_0^t |Y_s|^{1+2\alpha} ds\right)^{p/2}\right] + C_6(p,T) ||\sigma - \hat{\sigma}||_{2p}^p,$$

where $C_5(p,T) := 2^{p/2} C_p K^p$ and $C_6(p,T) := 2^{p/2} T^{\frac{p-1}{2}} C_p C_T^{1/2}$. In particular, if $\alpha = 1/2$, we have

$$\mathbb{E}[\sup_{0 \leq s \leq t} |M_s^{\delta,\kappa}|^p] \leq \frac{1}{2 \cdot 5^{p-1}} \mathbb{E}[V_t^p] + \frac{5^{p-1} C_5(p,T)^2 T^{p-1}}{2} \int_0^t \mathbb{E}[V_s^p] ds$$
$$+ C_6(p,T) ||\sigma - \hat{\sigma}||_{2p}^p.$$

Proof (Proof of Lemma 4) From Burkholder-Davis-Gundy's inequality, we have

$$\mathbb{E}[\sup_{0\le s\le t}|M_s^{\delta,\kappa}|^p] \le C_p\mathbb{E}[\langle M^{\delta,\kappa}\rangle_t^{p/2}] \le C_p\mathbb{E}\left[\left(\int_0^t|\sigma(X_s)-\hat{\sigma}(\hat{X}_s)|^2ds\right)^{p/2}\right]$$

$$\le 2^{p/2}C_p\left(\mathbb{E}\left[\left(\int_0^t|\sigma(X_s)-\sigma(\hat{X}_s)|^2ds\right)^{p/2}\right]+\mathbb{E}\left[\left(\int_0^T|\sigma(\hat{X}_s)-\hat{\sigma}(\hat{X}_s)|^2ds\right)^{p/2}\right]\right).$$

From Jensen's inequality and Lemma 1, we have

$$\mathbb{E}\left[\left(\int_0^T|\sigma(\hat{X}_s)-\hat{\sigma}(\hat{X}_s)|^2ds\right)^{p/2}\right] \le T^{\frac{p-1}{2}}\left(\int_0^T\mathbb{E}[|\sigma(\hat{X}_s)-\hat{\sigma}(\hat{X}_s)|^{2p}]ds\right)^{1/2}$$

$$\le T^{\frac{p-1}{2}}C_T^{1/2}||\sigma-\hat{\sigma}||_{2p}^p.$$

Since σ is $(1/2+\alpha)$-Hölder continuous, we get

$$\mathbb{E}[\sup_{0\le s\le t}|M_s^{\delta,\kappa}|^p] \le C_5(p,T)\mathbb{E}\left[\left(\int_0^t|Y_s|^{1+2\alpha}ds\right)^{p/2}\right]+C_6(p,T)||\sigma-\hat{\sigma}||_{2p}^p.$$

This concludes the first statement.

In particular, if $\alpha=1/2$, then we get from definition of V_t,

$$C_5(p,T)\mathbb{E}\left[\left(\int_0^t|Y_s|^2ds\right)^{p/2}\right] \le C_5(p,T)\mathbb{E}\left[(V_t)^{p/2}\left(\int_0^t|Y_s|ds\right)^{p/2}\right].$$

Using Young's inequality $xy \le \frac{x^2}{2\cdot 5^{p-1}C_5(p,T)}+\frac{5^{p-1}C_5(p,T)y^2}{2}$ for any $x,y\ge 0$ and Jensen's inequality, we obtain

$$C_5(p,T)\mathbb{E}\left[\left(\int_0^t|Y_s|^2ds\right)^{p/2}\right] \le \frac{1}{2\cdot 5^{p-1}}\mathbb{E}[V_t^p]+\frac{5^{p-1}C_5(p,T)^2}{2}\mathbb{E}\left[\left(\int_0^t|Y_s|ds\right)^p\right]$$

$$\le \frac{1}{2\cdot 5^{p-1}}\mathbb{E}[V_t^p]+\frac{5^{p-1}C_5(p,T)^2T^{p-1}}{2}\int_0^t\mathbb{E}[V_s^p]ds,$$

which concludes the second statement.

To prove Theorem 3, we recall the following Gronwall type inequality.

Lemma 5 ([7] Lemma 3.2.-(ii)) *Let* $(A_t)_{0\le t\le T}$ *be a nonnegative continuous stochastic process and set* $B_t:=\sup_{0\le s\le t}A_s$. *Assume that for some* $r>0$, $q\ge 1$,

$\rho \in [1, q]$ and $C_1, \xi \geq 0$,

$$\mathbb{E}[B_t^r] \leq C_1 \mathbb{E}\left[\left(\int_0^t B_s ds\right)^r\right] + C_1 \mathbb{E}\left[\left(\int_0^t A_s^\rho ds\right)^{r/q}\right] + \xi < \infty$$

for all $t \in [0, T]$. If $r \geq q$ or $q + 1 - \rho < r < q$ hold, then there exists constant C_2 depending on r, q, ρ, T and C_1 such that

$$\mathbb{E}[B_T^r] \leq C_2 \xi + C_2 \int_0^T \mathbb{E}[A_s] ds.$$

Now using Lemmas 4 and 5, we can prove Theorem 3.

Proof (Proof of Theorem 3) From (20) and the inequality $\left(\sum_{i=1}^m a_i\right)^p \leq m^{p-1} \sum_{i=1}^m a_i^p$ for any $p \geq 2$ $a_i > 0$ and $m \in \mathbb{N}$, and Jensen's inequality, we have

$$V_t^p \leq 5^{p-1}\left(\kappa^p + \left(L \int_0^t V_s ds\right)^p + T^{p-1} \int_0^T |b(\hat{X}_s) - \hat{b}(\hat{X}_s)|^p ds \right.$$
$$\left. + \frac{2T^{p-1}\delta^p}{\kappa^p (\log \delta)^p} \int_0^T |\sigma(\hat{X}_s) - \hat{\sigma}(\hat{X}_s)|^{2p} ds + \frac{(2TK^2)^p \kappa^{2p\alpha}}{(\log \delta)^p} + \sup_{0 \leq s \leq t} |M_s^{\delta,\kappa}|^p\right).$$

From Lemma 1 with $p \geq 2$, we have

$$\mathbb{E}[V_t^p] \leq 5^{p-1}\kappa^p + 5^{p-1} L^p \mathbb{E}\left[\left(\int_0^t V_s ds\right)^p\right] + (5T)^{p-1} C_T ||b - \hat{b}||_p^p$$
$$+ \frac{2(5T)^{p-1} C_T \delta^p}{\kappa^p (\log \delta)^p} ||\sigma - \hat{\sigma}||_{2p}^{2p} + \frac{5^{p-1}(2TK^2)^p \kappa^{2p\alpha}}{(\log \delta)^p} + 5^{p-1}\mathbb{E}[\sup_{0 \leq s \leq t} |M_s^{\delta,\kappa}|^p].$$

If $\alpha = 1/2$, using Lemma 4, we have

$$\mathbb{E}[V_t^p] \leq 5^{p-1}\kappa^p + (5T)^{p-1}\left(L^p + \frac{C_5(p,T)^2}{2}\right) \int_0^t \mathbb{E}[V_s^p] ds + (5T)^{p-1} C_T ||b - \hat{b}||_p^p$$
$$+ \frac{2(5T)^{p-1} C_T \delta^p}{\kappa^p (\log \delta)^p} ||\sigma - \hat{\sigma}||_{2p}^{2p} + \frac{5^{p-1}(2TK^2)^p \kappa^p}{(\log \delta)^p}$$
$$+ \frac{1}{2}\mathbb{E}[V_T^p] + 5^{p-1} C_6(p,T) ||\sigma - \hat{\sigma}||_{2p}^p.$$

Hence we get

$$
\begin{aligned}
\mathbb{E}[V_t^p] &\le 2\cdot 5^{p-1}\kappa^p + (5T)^{p-1}\left(2L^p + C_5(p,T)^2\right)\int_0^t \mathbb{E}[V_s^p]ds \\
&\quad + 2(5T)^{p-1}C_T||b-\hat{b}||_p^p + \frac{4(5T)^{p-1}C_T\delta^p}{\kappa^p(\log\delta)^p}||\sigma-\hat{\sigma}||_{2p}^{2p} \\
&\quad + \frac{2\cdot 5^{p-1}(2TK^2)^p\kappa^p}{(\log\delta)^p} + 2\cdot 5^{p-1}C_6(p,T)||\sigma-\hat{\sigma}||_{2p}^p \\
&\le 2\cdot 5^{p-1}\kappa^p + (5T)^{p-1}\left(2L^p + C_5(p,T)^2\right)\int_0^t \mathbb{E}[V_s^p]ds + 2(5T)^{p-1}C_T\varepsilon_p \\
&\quad + \frac{4(5T)^{p-1}C_T\delta^p}{\kappa^p(\log\delta)^p}\varepsilon_p + \frac{2\cdot 5^{p-1}(2TK^2)^p\kappa^p}{(\log\delta)^p} + 2\cdot 5^{p-1}C_6(p,T)\varepsilon_p^{1/2}.
\end{aligned}
$$

Taking $\delta = 2$ and $\kappa = \varepsilon_p^{1/(2p)}$, we have

$$
\mathbb{E}[V_t^p] \le (5T)^{p-1}\left(2L^p + C_5(p,T)^2\right)\int_0^t \mathbb{E}[V_s^p]ds + C_7(1/2,p,T)\varepsilon_p^{1/2},
$$

where

$$
\begin{aligned}
C_7(1/2,p,T) &:= 2\cdot 5^{p-1} + 2(5T)^{p-1}C_T + \frac{4\cdot 2^p(5T)^{p-1} + 2\cdot 5^{p-1}(2TK^2)^p}{(\log 2)^p} \\
&\quad + 2\cdot 5^{p-1}C_6(p,T).
\end{aligned}
$$

By Gronwall's inequality, we obtain

$$
\mathbb{E}[V_t^p] \le C_7(1/2,p,T)\exp(5^{p-1}T^p\left(2L^p + C_5(p,T)^2\right))\varepsilon_p^{1/2}.
$$

If $\alpha \in [0, 1/2)$, using Lemma 4, we have

$$
\begin{aligned}
\mathbb{E}[V_t^p] &\le 5^{p-1}\kappa^p + 5^{p-1}L^p\mathbb{E}\left[\left(\int_0^t V_s ds\right)^p\right] + (5T)^{p-1}C_T||b-\hat{b}||_p^p \\
&\quad + \frac{2(5T)^{p-1}C_T\delta^p}{\kappa^p(\log\delta)^p}||\sigma-\hat{\sigma}||_{2p}^{2p} + \frac{5^{p-1}(2TK^2)^p\kappa^{2p\alpha}}{(\log\delta)^p} \\
&\quad + 5^{p-1}C_5(p,T)\mathbb{E}\left[\left(\int_0^t |Y_s|^{1+2\alpha}ds\right)^{p/2}\right] + 5^{p-1}C_6(p,T)||\sigma-\hat{\sigma}||_{2p}^p \\
&\le 5^{p-1}L^p\mathbb{E}\left[\left(\int_0^t V_s ds\right)^p\right] + 5^{p-1}C_5(p,T)\mathbb{E}\left[\left(\int_0^t |Y_s|^{1+2\alpha}ds\right)^{p/2}\right]
\end{aligned}
$$

$$+ 5^{p-1}\kappa^p + ((5T)^{p-1}C_T + 5^{p-1}C_6(p,T))\varepsilon_p^{1/2}$$
$$+ \frac{2(5T)^{p-1}C_T\delta^p}{\kappa^p(\log\delta)^p}\varepsilon_p + \frac{5^{p-1}(2TK^2)^p\kappa^{2p\alpha}}{(\log\delta)^p}.$$

Now we apply Theorem 1 with $\tau = s$ and Lemma 5 with $r = p$, $q = 2$, $\rho = 1 + 2\alpha$ and

$$\xi = 5^{p-1}\kappa^p + ((5T)^{p-1}C_T + 5^{p-1}C_6(p,T))\varepsilon_p^{1/2}$$
$$+ \frac{2(5T)^{p-1}C_T\delta^p}{\kappa^p(\log\delta)^p}\varepsilon_p + \frac{5^{p-1}(2TK^2)^p\kappa^{2p\alpha}}{(\log\delta)^p}.$$

Then there exists $C_7(\alpha, p, T)$ which depends on p, α, T, L and $C_5(p,T)$ such that

$$\begin{aligned}
\mathbb{E}[V_T^p] &\leq C_7(\alpha,p,T)\left(\kappa^p + \varepsilon_p^{1/2} + \frac{\delta^p\varepsilon_p}{\kappa^p(\log\delta)^p} + \frac{\kappa^{2p\alpha}}{(\log\delta)^p}\right) \\
&\quad + C_7(\alpha,p,T)\int_0^T \mathbb{E}[|Y_s|]ds \\
&\leq C_7(\alpha,p,T)\left(\kappa^p + \varepsilon_p^{1/2} + \frac{\delta^p\varepsilon_p}{\kappa^p(\log\delta)^p} + \frac{\kappa^{2p\alpha}}{(\log\delta)^p}\right) \\
&\quad + \begin{cases} C_7(\alpha,p,T)C_1(\alpha,T)e^{LT}T\varepsilon_1^{2\alpha/(2\alpha+1)} & \textit{if } \alpha \in (0,1/2), \\ \dfrac{C_7(0,p,T)C_1(0,T)e^{LT}T}{\log(1/\varepsilon_1)} & \textit{if } \alpha = 0. \end{cases}
\end{aligned}$$

Taking $\delta = 2$ and $\kappa = \varepsilon_p^{1/(2p)}$ if $\alpha \in (0,1/2)$ and $\delta = \varepsilon_p^{-1/(2p)}$ and $\kappa = 1/\log(1/\varepsilon_p)$ if $\alpha = 0$, we get

$$\mathbb{E}[V_T^p] \leq \begin{cases} C_8(\alpha,p,T)\varepsilon_1^{2\alpha/(2\alpha+1)} & \textit{if } \alpha \in (0,1/2), \\ \dfrac{C_8(\alpha,p,T)}{\log(1/\varepsilon_1)} & \textit{if } \alpha = 0, \end{cases}$$

where

$$C_8(\alpha,p,T) := \begin{cases} C_7(\alpha,p,T)\left(2 + \dfrac{2^p+1}{(\log 2)^p} + C_1(\alpha,T)e^{LT}T\right) & \textit{if } \alpha \in (0,1/2), \\ C_7(\alpha,p,T)\left(2 + 2(2p)^p + C_1(\alpha,T)e^{LT}T\right) & \textit{if } \alpha = 0, \end{cases}$$

Hence we conclude the proof of Theorem 3.

4 Application to the Stability Problem

In this section, we apply our main results to the stability problem. For any $n \in \mathbb{N}$, we consider the one-dimensional stochastic differential equation

$$X_t^{(n)} = x_0 + \int_0^t b_n(X_s^{(n)})ds + \int_0^t \sigma_n(X_t^{(n)})dW_s.$$

Assumption 2 We assume that the coefficients b, σ and the sequence of coefficients $(b_n)_{n\in\mathbb{N}}$ and $(\sigma_n)_{n\in\mathbb{N}}$ satisfy the following conditions:

A'-(i): $b \in \mathscr{L}$.
A'-(ii): b and b_n are bounded measurable i.e., there exists $K > 0$ such that

$$\sup_{n\in\mathbb{N},x\in\mathbb{R}} (|b_n(x)| \vee |b(x)|) \leq K.$$

A'-(iii): σ and σ_n are $\eta = 1/2 + \alpha$-Hölder continuous with $\alpha \in [0, 1/2]$, i.e., there exists $K > 0$ such that

$$\sup_{n\in\mathbb{N},x,y\in\mathbb{R},x\neq y} \left(\frac{|\sigma(x) - \sigma(y)|}{|x-y|^\eta} \vee \frac{|\sigma_n(x) - \sigma_n(y)|}{|x-y|^\eta} \right) \leq K.$$

A'-(iv): $a = \sigma$ and $a_n := \sigma_n^2$ are bounded and uniformly elliptic, i.e., there exists $\lambda \geq 1$ such that for any $x \in \mathbb{R}$ and $n \in \mathbb{N}$,

$$\lambda^{-1} \leq a(x) \leq \lambda \textit{ and } \lambda^{-1} \leq a_n(x) \leq \lambda.$$

A'-(p): For given $p > 0$,

$$\varepsilon_{p,n} := ||b - b_n||_p^p \vee ||\sigma - \sigma_n||_{2p}^{2p} \to 0$$

as $n \to \infty$.

For $p \geq 1$ and $\alpha \in [0, 1/2]$, we define $N_{\alpha,p}$ by

$$N_{\alpha,p} := \begin{cases} \min\{n \in \mathbb{N} : \varepsilon_{p,m} < 1, \forall m \geq n\}, & \textit{if } \alpha \in (0, 1/2], \\ \min\{n \in \mathbb{N} : \varepsilon_{p,m} < 1/e, \forall m \geq n\}, & \textit{if } \alpha = 0. \end{cases}$$

Then using Theorem 1–3 and Corollary 1, 2, we have the following corollaries.

Corollary 3 *Suppose that Assumption 2 holds with $p = 1$. Then there exists a positive constant C which depends on $\overline{C}, c_*, K, L, T, \alpha, \lambda$ and x_0 such that for* *any* $n \geq N_{\alpha,1}$,

$$\sup_{\tau \in \mathscr{T}} \mathbb{E}[|X_\tau - X_\tau^{(n)}|] \leq \begin{cases} C\varepsilon_{1,n}^{2\alpha/(2\alpha+1)} & \text{if } \alpha \in (0,1/2], \\ \dfrac{C}{\log(1/\varepsilon_{1,n})} & \text{if } \alpha = 0 \end{cases}$$

and

$$\mathbb{E}[\sup_{0\leq t\leq T} |X_t - X_t^{(n)}|] \leq \begin{cases} C\varepsilon_{1,n}^{4\alpha^2/(2\alpha+1)} & \text{if } \alpha \in (0,1/2], \\ \dfrac{C}{\sqrt{\log(1/\varepsilon_{1,n})}} & \text{if } \alpha = 0 \end{cases}$$

and for any $g \in BV$ *and* $r \geq 1$, *we have*

$$\mathbb{E}[|g(X_T) - g(X_T^{(n)})|^r] \leq \begin{cases} 3^{r+1}V(g)^r C\varepsilon_{1,n}^{\alpha/(2\alpha+1)} & \text{if } \alpha \in (0,1/2], \\ \dfrac{3^{r+1}V(g)^r C}{\sqrt{\log(1/\varepsilon_{1,n})}} & \text{if } \alpha = 0. \end{cases}$$

Corollary 4 *Suppose that Assumption 2 holds with* $p \geq 2$. *Then there exists a positive constant* C *which depends on* $\overline{C}, c_*, K, L, T, p, \alpha, \lambda$ *and* x_0 *such that for any* $n \geq N_{\alpha,p}$,

$$\mathbb{E}[\sup_{0\leq t\leq T} |X_t - X_t^{(n)}|^p] \leq \begin{cases} C\varepsilon_{p,n}^{1/2} & \text{if } \alpha = 1/2, \\ C\varepsilon_{1,n}^{2\alpha/(2\alpha+1)} & \text{if } \alpha \in (0,1/2), \\ \dfrac{C}{\log(1/\varepsilon_{1,n})} & \text{if } \alpha = 0. \end{cases}$$

Corollary 5 *Suppose that Assumption 2 holds with* $2p$ *for* $p \in (1,2)$. *Then there exists a positive constant* C *which depends on* $\overline{C}, c_*, K, L, T, p, \alpha, \lambda$ *and* x_0 *such that for any* $n \geq N_{\alpha,2p}$,

$$\mathbb{E}[\sup_{0\leq t\leq T} |X_t - X_t^{(n)}|^p] \leq \begin{cases} C\varepsilon_{2p,n}^{1/2} & \text{if } \alpha = 1/2, \\ C\varepsilon_{1,n}^{\alpha/(2\alpha+1)} & \text{if } \alpha \in (0,1/2), \\ \dfrac{C}{\sqrt{\log(1/\varepsilon_{1,n})}} & \text{if } \alpha = 0. \end{cases}$$

The next proposition shows that there exist the sequences $(b_n)_{n\in\mathbb{N}}$ and $(\sigma_n)_{n\in\mathbb{N}}$ satisfying Assumption 2.

Proposition 1

(i) *Assume* $\sup_{x\in\mathbb{R}} |b(x)| \leq K$. *If the set of discontinuity points of* b *is a null set with respect to the Lebesgue measure, then there exists a differentiable and bounded sequence* $(b_n)_{n\in\mathbb{N}}$ *such that for any* $p \geq 1$,

$$\int_{\mathbb{R}} |b(x) - b_n(x)|^p e^{-\frac{|x-x_0|^2}{2c_*T}} dx \to 0 \tag{21}$$

as $n \to \infty$*. Moreover, if* b *is a one-sided Lipschitz function, we can construct an explicit sequence* $(b_n)_{n\in\mathbb{N}}$ *which satisfies a one-sided Lipschitz condition.*

(ii) If the diffusion coefficient σ *satisfies* A'*-(ii) and* A'*-(iii), then there exists a differentiable sequence* $(\sigma_n)_{n\in\mathbb{N}}$ *such that for any* $n \in \mathbb{N}$, σ_n *satisfies* A'*-(iii),* A'*-(iv) and for any* $p \geq 1$,

$$\int_{\mathbb{R}} |\sigma(x) - \sigma_n(x)|^{2p} e^{-\frac{|x-x_0|^2}{2c_*T}} dx \leq \frac{K^{2p}\sqrt{2\pi c_* T}}{n^{2p\eta}}.$$

Proof Let $\rho(x) := \mu e^{-1/(1-|x|^2)}\mathbf{1}(|x| < 1)$ with $\mu^{-1} = \int_{|x|<1} e^{-1/(1-|x|^2)}dx$ and a sequence $(\rho_n)_{n\in\mathbb{N}}$ be defined by $\rho_n(x) := n\rho(nx)$. We set $b_n(x) := \int_{\mathbb{R}} b(y)\rho_n(x-y)dy$ and $\sigma_n(x) := \int_{\mathbb{R}} \sigma(y)\rho_n(x-y)dy$. Then for any $n \in \mathbb{N}$ and $x \in \mathbb{R}$, we have $|b_n(x)| \leq K$ and $\lambda^{-1} \leq a_n(x) := \sigma_n^2(x) \leq \lambda$, b_n and σ_n are differentiable.

Proof of (i). From Jensen's inequality, we have

$$\begin{aligned}\int_{\mathbb{R}} |b(x) - b_n(x)|^p e^{-\frac{|x-x_0|^2}{2c_*T}} dx &\leq \int_{\mathbb{R}} dx \left(\int_{\mathbb{R}} dy |b(x) - b(y)|\rho_n(x-y)\right)^p e^{-\frac{|x-x_0|^2}{2c_*T}} \\ &= \int_{\mathbb{R}} dx \left(\int_{|z|<1} dz |b(x) - b(x-z/n)|\rho(z)\right)^p e^{-\frac{|x-x_0|^2}{2c_*T}} \\ &\leq \int_{|z|<1} dz \int_{\mathbb{R}} dx |b(x) - b(x-z/n)|^p e^{-\frac{|x-x_0|^2}{2c_*T}} \rho(z).\end{aligned}$$

Since b is bounded, we have

$$\int_{\mathbb{R}} |b(x) - b(x-z/n)|^p e^{-\frac{|x-x_0|^2}{2c_*T}} dx \leq (2K)^p \int_{\mathbb{R}} e^{-\frac{|x-x_0|^2}{2c_*T}} dx = (2K)^p \sqrt{2\pi c_* T}. \tag{22}$$

On the other hand, since the set of discontinuity points of b is a null set with respect to the Lebesgue measure, b is continuous almost everywhere. From (22), using the dominated convergence theorem, we have

$$\int_{\mathbb{R}} |b(x) - b(x-z/n)|^p e^{-\frac{|x-x_0|^2}{2c_*T}} dx \to 0$$

as $n \to \infty$. From this fact and the dominated convergence theorem, $(b_n)_{n\in\mathbb{N}}$ satisfies (21).

Let b be a one-sided Lipschitz function. Then, we have

$$(x-y)(b_n(x)-b_n(y)) = \int_{\mathbb{R}} (x-y)(b(x-z)-b(y-z))\rho_n(z)dz$$
$$= \int_{\mathbb{R}} \{(x-z)-(z-y)\}(b(x-z)-b(y-z))\rho_n(z)dz$$
$$\leq L|x-y|^2,$$

which implies that $(b_n)_{n\in\mathbb{N}}$ satisfies the one-sided Lipschitz condition.

Proof of (ii). In the same way as in the proof of (i), we have from Hölder continuity of σ

$$\int_{\mathbb{R}} |\sigma(x)-\sigma_n(x)|^{2p} e^{-\frac{|x-x_0|^2}{2c_* T}} dx \leq \int_{|z|<1} dz \int_{\mathbb{R}} dx |\sigma(x)-\sigma(x-z/n)|^{2p} e^{-\frac{|x-x_0|^2}{2c_* T}} \rho(z)$$
$$\leq \frac{K^{2p}}{n^{2p\eta}} \int_{|z|<1} dz \int_{\mathbb{R}} dx e^{-\frac{|x-x_0|^2}{2c_* T}} \rho(z) = \frac{K^{2p}\sqrt{2\pi c_* T}}{n^{2p\eta}}.$$

Finally, we show that σ_n is η-Hölder continuous. For any $x, y \in \mathbb{R}$,

$$|\sigma_n(x)-\sigma_n(y)| \leq \int_{\mathbb{R}} |\sigma(x-z)-\sigma(y-z)|\rho_n(z)dz \leq K|x-y|^{\eta},$$

which implies that σ_n is η-Hölder continuous. This concludes that $(\sigma_n)_{n\in\mathbb{N}}$ satisfies (ii).

Acknowledgements The author is very grateful to Professor Arturo Kohatsu-Higa for his supports and fruitful discussions. The author would also like to thank Hideyuki Tanaka and Takahiro Tsuchiya for their useful advices. The author would like to express my thanks to Professor Toshio Yamada for his encouragement and comments. The author also thanks the referees for their comments which helped to improve the paper.

References

1. J. Akahori, Y. Imamura, On a symmetrization of diffusion processes. Quant. Finan. **14**(7), 1211–1216 (2014). doi:10.1080/14697688.2013.825923
2. D.G. Aronson, Bounds for the fundamental solution of a parabolic equation. Bull. Am. Math. Soc. **73**, 890–896 (1967)
3. R. Avikainen, On irregular functionals of SDEs and the Euler scheme. Finance Stochast. **13**, 381–401 (2009)
4. V.E. Beneš, L.A. Shepp, H.S. Witsenhausen, Some Solvable stochastic control problems. Stochastics **4**, 39–83 (1980)
5. K.S. Chan, O. Stramer, Weak consistency of the Euler method for numerically solving stochastic differential equations with discontinuous coefficient. Stochast. Process. Appl. **76**, 33–44 (1998)

6. A. Friedman, *Partial Differential Equations of Parabolic Type* (Dover, New York, 1964)
7. I. Gyöngy, M. Rásonyi, A note on Euler approximations for SDEs with Hölder continuous diffusion coefficients. Stochast. Process. Appl. **121**, 2189–2200 (2011)
8. H. Hashimoto, T. Tsuchiya, Convergence rate of stability problems of SDEs with (Dis-) continuous coefficients. Preprint arXiv:1401.4542v1 (2014)
9. Y. Imamura, Y. Ishigaki, T. Kawagoe, T. Okumura, A numerical scheme based on semi-static hedging strategy. Monte Carlo Methods Appl. **20**(4), 223–235 (2014) doi:10.1515/mcma-2014-0002
10. H. Kaneko, S. Nakao, A note on approximation for stochastic differential equations. Séminaire de Probabiliteés de Strasbourg **22**, 155–162 (1988)
11. S. Kawabata, T. Yamada, On some limit theorems for solutions of stochastic differential equations, in *Seminaire de Probabilities XVI, University of Strasbourg 1980/81*. Lecture Notes in Mathematics, vol. 920 (Springer, New York, 1982), pp. 412–441
12. A. Kohatsu-Higa, A. Lejay, K. Yasuda, Weak approximation errors for stochastic differential equations with non-regular drift. Preprint (2013)
13. N.V. Krylov, M. Röckner, Strong solutions of stochastic equations with singular time dependent drift. Probab. Theory Relat. Fields **131**, 154–196 (2005)
14. V. Lemaire, S. Menozzi, On some non asymptotic bounds for the Euler scheme. Electron. J. Probab. **15**, 1645–1681 (2010)
15. H.-L. Ngo, D. Taguchi, Strong rate of convergence for the Euler-Maruyama approximation of stochastic differential equations with irregular coefficients. Math. Comput. **85**(300), 1793–1819 (2016)
16. D.W. Stroock, Diffusion semigroups corresponding to uniformly elliptic divergence form operators, in *Séminaire de Probabilités, XXII* (Springer, Berlin, 1988), pp. 316–347
17. D.W. Stroock, R.S. Varadhan, in *Multidimensional Diffusion Processes*. Die Grundlehren der Mathematischen Wissenschaften (Springer, Berlin/Heidelberg/New York, 1979)
18. B.L. Yan, The Euler scheme with irregular coefficients. Ann. Probab. **30**(3), 1172–1194 (2002)
19. T. Yamada, S. Watanabe, On the uniqueness of solutions of stochastic differential equations. J. Math. Kyoto Univ. **11**, 155–167 (1971)
20. A.K. Zvonkin, A transformation of the phase space of a diffusion process that removes the drift. Math. USSR Sb. **22**, 129–148 (1974)

The Maximum of the Local Time of a Diffusion Process in a Drifted Brownian Potential

Alexis Devulder

Abstract We consider a one-dimensional diffusion process X in a $(-\kappa/2)$-drifted Brownian potential for $\kappa \neq 0$. We are interested in the maximum of its local time, and study its almost sure asymptotic behaviour, which is proved to be different from the behaviour of the maximum local time of the transient random walk in random environment. We also obtain the convergence in law of the maximum local time of X under the annealed law after suitable renormalization when $\kappa \geq 1$. Moreover, we characterize all the upper and lower classes for the hitting times of X, in the sense of Paul Lévy, and provide laws of the iterated logarithm for the diffusion X itself. To this aim, we use annealed technics.

AMS Classification (2010): 60K37, 60J60, 60J55, 60F15

1 Introduction

In this section, we successively present the model, the maximum local time, and the main results of the paper.

1.1 Presentation of the Model

We consider a diffusion process in random environment, defined as follows. For $\kappa \in \mathbb{R}$, we introduce the random potential

$$W_\kappa(x) := W(x) - \frac{\kappa}{2}x, \qquad x \in \mathbb{R}, \tag{1}$$

A. Devulder (✉)
Laboratoire de Mathématiques de Versailles, UVSQ, CNRS, Université Paris-Saclay, 78035 Versailles, France
e-mail: devulder@math.uvsq.fr

C. Donati-Martin et al. (eds.), *Séminaire de Probabilités XLVIII*, Lecture Notes in Mathematics 2168, DOI 10.1007/978-3-319-44465-9_5

where $(W(x),\ x \in \mathbb{R})$ is a standard two-sided Brownian motion. Informally, a diffusion process $(X(t),\ t \geq 0)$ in the random potential W_κ is defined by

$$\begin{cases} \mathrm{d}X(t) = \mathrm{d}\beta(t) - \frac{1}{2}W_\kappa'(X(t))\mathrm{d}t, \\ X(0) = 0, \end{cases}$$

where $(\beta(t),\ t \geq 0)$ is a Brownian motion independent of W. More rigorously, $(X(t),\ t \geq 0)$ is a diffusion process such that $X(0) = 0$, and whose conditional generator given W_κ is

$$\frac{1}{2}e^{W_\kappa(x)}\frac{\mathrm{d}}{\mathrm{d}x}\left(e^{-W_\kappa(x)}\frac{\mathrm{d}}{\mathrm{d}x}\right).$$

Let P be the probability measure associated to W_κ. We denote by P_{W_κ} the law of X conditionally on the environment W_κ, and call it the *quenched law*. We also define the *annealed law* $\mathbb{P}$ as follows:

$$\mathbb{P}(\cdot) := \int P_{W_\kappa}(\cdot)P(W_\kappa \in \mathrm{d}\omega).$$

Notice in particular that X is a Markov process under P_{W_κ}, but not under $\mathbb{P}$. Such a diffusion can also be constructed from a Brownian motion through (random) changes of time and scale (see (90) below). This diffusion X, introduced by Schumacher [42] and Brox [11], is generally considered as the continuous time analogue of random walks in random environment (RWRE), which have many applications in physics and biology (see e.g. Le Doussal et al. [37]); for an account of general properties of RWRE, we refer to Révész [39] and Zeitouni [53]. This diffusion has been studied for example by Kawazu and Tanaka [36], see Theorem 1 below, later improved by Hu et al. [33]. Large deviations results are proved in Taleb [47] and Talet [48] (see also Devulder [20] for some properties of the rate function), and moderate deviations are given by Hu and Shi [32] in the recurrent case, and by Faraud [25] in the transient case. A localization result and an aging theorem are provided by Andreoletti and Devulder [3] in the case $0 < \kappa < 1$. For a relation between RWRE and the diffusion X, see e.g. Shi [44]. See also Carmona [12], Cheliotis [13], Mathieu [38], Singh [45, 46] and Tanaka [49] for diffusions in other potentials.

In this paper, we are interested in the transient case, that is, we suppose $\kappa \neq 0$. If X is a diffusion in the random potential W_κ, then $-X$ is a diffusion in the random potential $(W_\kappa(-x),\ x \in \mathbb{R})$ which has the same law as $(W_{-\kappa}(x),\ x \in \mathbb{R})$. Hence we may assume without loss of generality that $\kappa > 0$. In this case, $X(t) \to_{t\to+\infty} +\infty$ $\mathbb{P}$-almost surely.

Our goal is to study the asymptotics of the maximum of the local time of X. Corresponding problems for RWRE have attracted much attention, and have been studied, for example, in Révész [39, Chap. 29], Shi [43], Gantert et al. [26, 27], Hu et al. [30], Dembo et al. [18] and Andreoletti ([2], see also [1]). Moreover the local time of such processes in random environment plays an important role in estimation problems (see e.g. Comets et al. [15]), in persistence (see Devulder [21]) and in the study of processes in random scenery (see Zindy [54]).

1.2 Maximum Local Time

We denote by $(L_X(t,x),\ t \geq 0,\ x \in \mathbb{R})$ the local time of X, which is the jointly continuous process satisfying, for any positive measurable function f,

$$\int_0^t f(X(s))\mathrm{d}s = \int_{-\infty}^{+\infty} f(x)L_X(t,x)\mathrm{d}x, \qquad t \geq 0. \tag{2}$$

The existence of such a process was proved by Hu and Shi [30, Eq. (2.6)]; see (91) below for an expression of L_X. We are interested in the *maximum local time* of X at time t, defined as

$$L_X^*(t) := \sup_{x\in\mathbb{R}} L_X(t,x), \qquad t \geq 0.$$

In the recurrent case $\kappa = 0$, Hu and Shi [30] first proved that for any $x \in \mathbb{R}$,

$$\frac{\log L_X(t,x)}{\log t} \xrightarrow{\mathscr{L}} U \wedge \hat{U},$$

where U and $\hat{U}$ are two independent random variables uniformly distributed in $[0,1]$, and "$\xrightarrow{\mathscr{L}}$" denotes convergence in law under the annealed law $\mathbb{P}$. Moreover, throughout the paper, log denotes the natural logarithm. The limit law of $L_X^*(t)$, suitably renormalized, is determined by Andreoletti and Diel [4] when $\kappa = 0$:

$$\frac{L_X^*(t)}{t} \xrightarrow{\mathscr{L}} \left(\int_{-\infty}^{\infty} e^{-\widetilde{W}(x)}\mathrm{d}x\right)^{-1}, \tag{3}$$

where $\left(\widetilde{W}(x),\ x \in \mathbb{R}\right)$ is a two-sided Brownian motion conditioned to stay positive. Furthermore, Shi [43] proved the following surprising result: $\mathbb{P}$-almost surely when $\kappa = 0$,

$$\limsup_{t\to+\infty} L_X^*(t)/(t\log\log\log t) \geq 1/32. \tag{4}$$

The question whether this is the good renormalization remained open during 13 years, until Diel [22] gave a positive answer to this question. He proved indeed that in this recurrent case $\kappa = 0$,

$$\limsup_{t\to+\infty} L_X^*(t)/(t\log\log\log t) \leq e^2/2,$$

$$j_0^2/64 \leq \liminf_{t\to+\infty} L_X^*(t)/[t/(\log\log\log t)] \leq e^2\pi^2/4$$

$\mathbb{P}$-almost surely, where j_0 is the smallest strictly positive root of the Bessel function J_0. Moreover, the convergence in law (3) is extended to the case of stable Lévy environment by Diel and Voisin [23]. Finally, related questions about favorite sites, that is, locations in which the local time is maximum at time t, are considered by Hu and Shi [31], Cheliotis [14], and Andreoletti et al. [5].

1.3 Results

We define the first hitting time of r by X as follows:

$$H(r) := \inf\{t \geq 0, \quad X(t) > r\}, \qquad r \geq 0. \tag{5}$$

We recall that there are three different regimes for H in the transient case $\kappa > 0$:

Theorem 1 (Kawazu and Tanaka, [36]) *When r tends to infinity,*

$$\frac{H(r)}{r^{1/\kappa}} \xrightarrow{\mathcal{L}} c_0 S_\kappa^{ca}, \qquad 0 < \kappa < 1, \tag{6}$$

$$\frac{H(r)}{r \log r} \xrightarrow{P.} 4, \qquad \kappa = 1, \tag{7}$$

$$\frac{H(r)}{r} \xrightarrow{a.s.} \frac{4}{\kappa - 1}, \qquad \kappa > 1, \tag{8}$$

where $c_0 = c_0(\kappa) > 0$ is a finite constant, the symbols "$\xrightarrow{\mathcal{L}}$", "$\xrightarrow{P.}$" and "$\xrightarrow{a.s.}$" denote respectively convergence in law, in probability and almost sure convergence, with respect to the annealed probability $\mathbb{P}$. Moreover, for $0 < \kappa < 1$, S_κ^{ca} is a completely asymmetric stable variable of index κ, and is a positive variable (see (14) *for its characteristic function).*

The asymptotics of the maximum local time $L_X^*(t)$ heavily depend on the value of κ. We start with the upper asymptotics of $L_X^*(t)$:

Theorem 2 *If $0 < \kappa < 1$, then*

$$\limsup_{t \to +\infty} \frac{L_X^*(t)}{t} = +\infty \qquad \mathbb{P}\text{-}a.s.$$

Theorem 2 tells us that in the case $0 < \kappa < 1$, the maximum local time of X has a completely different behaviour from the maximum local time of RWRE (the latter is trivially bounded by $t/2$ for any positive integer t, for example). Such a peculiar phenomenon has already been observed (see (4)) by Shi [43] in the recurrent case, and is even more surprising here since X is transient.

Theorem 3 gives, in the case $\kappa > 1$, an integral test which completely characterizes the upper functions of $L_X^*(t)$, in the sense of Paul Lévy.

Theorem 3 *Let $a(\cdot)$ be a positive nondecreasing function. If $\kappa > 1$, then*

$$\sum_{n=1}^{\infty} \frac{1}{na(n)} \begin{cases} < +\infty \\ = +\infty \end{cases} \Longleftrightarrow \limsup_{t\to\infty} \frac{L_X^*(t)}{[ta(t)]^{1/\kappa}} = \begin{cases} 0 \\ +\infty \end{cases} \qquad \mathbb{P}\text{-}a.s.$$

This is in agreement with a result of Gantert and Shi [26] for RWRE. We notice in particular that $\limsup_{t\to+\infty} L_X^*(t)/t$ is almost surely $+\infty$ when $0 < \kappa < 1$ by Theorem 2, whereas it is 0 when $\kappa > 1$ by Theorem 3. We have not been able to prove whether $\limsup_{t\to+\infty} L_X^*(t)/t$ is infinite in the very delicate case $\kappa = 1$, since a proof similar to that of Theorem 2 just shows that it is greater than a positive deterministic constant (see Remark 1 page 158 for more details).

We now turn to the lower asymptotics of $L_X^*(t)$.

Theorem 4 *We have*

$$\liminf_{t\to\infty} \frac{L_X^*(t)}{t/\log\log t} \le \kappa^2 c_1(\kappa) \qquad \mathbb{P}\text{-}a.s. \quad \text{if } 0 < \kappa < 1,$$

$$\liminf_{t\to\infty} \frac{L_X^*(t)}{t/[(\log t)\log\log t]} \le \frac{1}{2} \qquad \mathbb{P}\text{-}a.s. \quad \text{if } \kappa = 1,$$

$$\liminf_{t\to\infty} \frac{L_X^*(t)}{(t/\log\log t)^{1/\kappa}} = 4\left(\frac{(\kappa-1)\kappa^2}{8}\right)^{1/\kappa} \qquad \mathbb{P}\text{-}a.s. \quad \text{if } \kappa > 1,$$

where $c_1(\kappa)$ is defined in (65).

Theorem 5 *We have, for any $\varepsilon > 0$,*

$$\liminf_{t\to\infty} \frac{L_X^*(t)}{t/[(\log t)^{1/\kappa}(\log\log t)^{(2/\kappa)+\varepsilon}]} = +\infty \qquad \mathbb{P}\text{-}a.s. \quad \text{if } 0 < \kappa \le 1.$$

In the case $0 < \kappa \le 1$, Theorems 4 and 5 give different bounds, for technical reasons.

We also get the convergence in law under the annealed law $\mathbb{P}$ of $L_X^*(t)$, suitably renormalized, when $\kappa \ge 1$:

Theorem 6 *We have as $t \to +\infty$, under the annealed law $\mathbb{P}$,*

$$\frac{L_X^*(t)}{t/\log t} \xrightarrow{\mathscr{L}} \frac{1}{2\mathscr{E}} \qquad \text{if } \kappa = 1,$$

$$\frac{L_X^*(t)}{t^{1/\kappa}} \xrightarrow{\mathscr{L}} 4\left[\kappa^2(\kappa-1)/8\right]^{1/\kappa}\mathscr{E}^{-1/\kappa} \qquad \text{if } \kappa > 1,$$

where $\mathscr{E}$ denotes an exponential variable with mean 1.

We notice that in the previous theorem, the case $0 < \kappa < 1$ is lacking. Indeed, we did not succeed in obtaining it with the annealed technics of the present paper, because due to (6), $H(r)$ suitably renormalized converges in law but does not converge in probability to a positive constant in this case. This is why we used quenched technics in Andreoletti et al. [5] to prove that $L_X^*(t)/t$ converges in law under $\mathbb{P}$ as $t \to +\infty$ when $0 < \kappa < 1$. To this aim, we used and extended to local time the quenched tools developed in Andreoletti et al. [3] to get the localization of X in this case $0 < \kappa < 1$, combined with some additional tools such as two dimensional Lévy processes and convergence in Skorokhod topology.

So, Theorem 6 completes the results of [4] (see our (3)) and [5], that is, these 3 results give the convergence in law of $L_X^*(t)$ suitably renormalized for any value of $\kappa \in \mathbb{R}$.

In the proof of Theorems 2–5, we will frequently need to use the almost sure asymptotics of the first hitting times $H(\cdot)$. In view of the last part (8) of Theorem 1, we only need to study the case $\kappa \in (0, 1]$.

Theorem 7 *Let $a(\cdot)$ be a positive nondecreasing function. If $0 < \kappa < 1$, then*

$$\sum_{n=1}^{\infty} \frac{1}{na(n)} \begin{cases} < +\infty \\ = +\infty \end{cases} \Longleftrightarrow \limsup_{r\to\infty} \frac{H(r)}{[ra(r)]^{1/\kappa}} = \begin{cases} 0 \\ +\infty \end{cases} \qquad \mathbb{P}\text{-a.s.}$$

If $\kappa = 1$, the statement holds under the additional assumption that

$$\limsup_{r\to+\infty} \frac{\log r}{a(r)} < \infty.$$

Theorem 8 *We have $\mathbb{P}$ a.s. (Γ denotes the usual gamma function)*

$$\liminf_{r\to+\infty} \frac{H(r)}{r^{1/\kappa}/(\log\log r)^{(1/\kappa)-1}} = \frac{8\kappa[\pi\kappa]^{1/\kappa}(1-\kappa)^{\frac{1-\kappa}{\kappa}}}{\left[2\Gamma^2(\kappa)\sin(\pi\kappa)\right]^{1/\kappa}} =: c_2(\kappa) \text{ if } 0 < \kappa < 1, \quad (9)$$

$$\liminf_{r\to+\infty} \frac{H(r)}{r\log r} = 4 \qquad \text{if } \kappa = 1. \quad (10)$$

The following corollary follows immediately from Theorem 7 and gives a negative answer to a question raised in Hu et al. [33, Remark 1.3 p. 3917]:

Corollary 1 *The convergence in probability $H(r)/(r\log r) \to 4$ in Theorem 1 in the case $\kappa = 1$ cannot be strengthened into an almost sure convergence.*

We observe that in the case $0 < \kappa < 1$, the process $H(\cdot)$ has the same almost sure asymptotics as κ-stable subordinators (see Bertoin [7, p. 92]).

Finally, define $\log_1 := \log$ and $\log_k := \log_{k-1} \circ \log$ for $k > 1$. Theorems 7 and 8, and the fact that $X(t)$ is not very far from $\sup_{0\le s\le t} X(s)$ (see Lemma 4 below) lead to

Corollary 2 *Recall that $c_2(\kappa)$ is defined in (9). We have $\mathbb{P}$-a.s. for $k \in \mathbb{N}^*$,*

$$\limsup_{t\to\infty} \frac{X(t)}{t^\kappa (\log\log t)^{1-\kappa}} = \frac{2\Gamma^2(\kappa)\sin(\pi\kappa)}{\pi 8^\kappa \kappa^{\kappa+1}(1-\kappa)^{1-\kappa}} = \frac{1}{[c_2(\kappa)]^\kappa} \quad \text{if } 0<\kappa<1, \tag{11}$$

$$\limsup_{t\to\infty} \frac{X(t)}{t/\log t} = \frac{1}{4} \qquad \text{if } \kappa = 1, \tag{12}$$

$$\begin{cases} \alpha \le 1 \\ \alpha > 1 \end{cases} \iff \liminf_{t\to+\infty} \frac{X(t)}{t^\kappa/[(\log t)\dots(\log_{k-1} t)(\log_k t)^\alpha]} \begin{cases} = 0 \\ = +\infty \end{cases} \quad \text{if } 0<\kappa\le 1, \tag{13}$$

where for $k=1$, $(\log t)\dots(\log_{k-1} t) = 1$ by convention. These results remain true if we replace $X(t)$ by $\sup_{0\le s\le t} X(s)$.

Corresponding results in the recurrent case $\kappa = 0$ are proved by Hu et al. [29], extended later by Singh [45] to some asymptotically stable potentials and following results of Deheuvels et al. [16] for Sinai's walk.

Our proof hinges upon stochastic calculus. In particular, one key ingredient of the proofs of Theorems 2–8 is an approximation of the joint law of the hitting time $H[F(r)]$ of $F(r) \approx r$ by X and the maximum local time $L_X^*[H(F(r))]$ of X at this time, stated in Lemma 2, and proved in Sect. 6. Another important tool is a modification of the Borel-Cantelli lemma, stated in Lemma 3, which, loosely speaking, says that one can chop the real half line $[0,\infty)$ into regions in which the diffusion X behaves in an "independent" way.

The rest of the paper is organized as follows. In Sect. 2.1, we give some preliminaries on local time and Bessel processes. We present in Sect. 2.2 some estimates which will be needed later on; the proof of one key estimate (Lemma 2) is postponed until Sect. 6. Section 3 is devoted to the study of the almost sure asymptotics of $L_X^*[H(r)]$, stated in Theorems 9 and 10. In Sect. 4, we study the Lévy classes for the hitting times $H(r)$ and prove Theorems 7 and 8 and Corollary 2. In Sect. 5, we study $L_X^*[H(r)]/H(r)$ and prove Theorems 2–6. Section 6 is devoted to the proof of Lemma 2. Finally, we prove in Sect. 7 some lemmas dealing with Bessel processes, Jacobi processes and Brownian motion.

Throughout the paper, the letter c with a subscript denotes constants that are finite and positive.

2 Some Preliminaries

We provide in this section some preliminaries on local time, on Bessel processes and on the diffusion X.

2.1 Preliminaries on Local Time and Bessel Processes

We first define, for any Brownian motion $(B(t),\ t \geq 0)$ and $r > 0$, the hitting time

$$\sigma_B(r) := \inf\{t > 0,\ B(t) = r\}.$$

Moreover, we denote by $(L_B(t,x),\ t \geq 0,\ x \in \mathbb{R})$ the local time of B, i.e., the jointly continuous process satisfying $\int_0^t f(B(s))\mathrm{d}s = \int_{-\infty}^{+\infty} f(x)L_B(t,x)\mathrm{d}x$ for any positive measurable function f. We define the inverse local time of B at 0 as

$$\tau_B(a) := \inf\{t \geq 0,\ L_B(t,0) > a\}, \qquad a > 0.$$

Furthermore, for any $\delta \in [0,\infty)$ and $x \in [0,\infty)$, the unique strong solution of the stochastic differential equation

$$Z(t) = x + 2\int_0^t \sqrt{Z(s)}\mathrm{d}\beta(s) + \delta t,$$

where $(\beta(s), s \geq 0)$ is a (one dimensional) Brownian motion, is named a δ-dimensional squared Bessel process starting from x. A Bessel process with dimension δ (or equivalently with order $\delta/2 - 1$) starting from $x \geq 0$ is defined as the (nonnegative) square root of a δ-dimensional squared Bessel process starting from x^2 (see e.g. Borodin et al. [10], 39 p. 73 for a more general definition as a linear diffusion with generator $\frac{1}{2}\frac{d^2}{dx^2} + \frac{\delta-1}{2x}\frac{d}{dx}$ for every $\delta \in \mathbb{R}$; see also Göing-Jaeschke et al. [28, Definition 3 p. 329]). We recall some important results.

Fact 1 (First Ray-Knight Theorem) Consider $r > 0$ and a Brownian motion $(B(t),\ t \geq 0)$. The process $(L_B(\sigma_B(r), r-x),\ x \geq 0)$ is a continuous inhomogeneous Markov process, starting from 0. It is a 2-dimensional squared Bessel process for $x \in [0,r]$ and a 0-dimensional squared Bessel process for $x \geq r$.

Fact 2 (Second Ray-Knight Theorem) Fix $r > 0$, and let $(B(t),\ t \geq 0)$ be a Brownian motion. The process $(L_B(\tau_B(r), x),\ x \geq 0)$ is a 0-dimensional squared Bessel process starting from r.

See e.g. Revuz and Yor [40, Chap. XI] for more details about Ray-Knight theorems and Bessel processes. Following the method used by Hu et al. [33, see Eq. (3.8)], we also need the following well known result:

Fact 3 (Lamperti Representation Theorem, see Yor [51, Eq. (2.e)])
Consider $W_\kappa(x) = W(x) - \kappa x/2$ as in (1) with $\kappa > 0$, where $(W(x),\ x \geq 0)$ is a Brownian motion. There exists a $(2-2\kappa)$-dimensional Bessel process $(\rho(t),\ t \geq 0)$, starting from $\rho(0) = 2$, such that $\exp[W_\kappa(t)/2] = \rho(A(t))/2$ for all $t \geq 0$, where $A(r) := \int_0^r e^{W_\kappa(s)}\mathrm{d}s$, $r \geq 0$.

We also recall the following extension to Bessel processes of Williams' time reversal theorem (see Yor [52, p. 80]; see also Göing-Jaeschke et al. [28, Eq. (34)]).

Fact 4 One has, for $\delta < 2$,

$$(R_\delta(T_0 - s),\ s \le T_0) \overset{\mathcal{L}}{=} (R_{4-\delta}(s),\ s \le \gamma_a),$$

where $\overset{\mathcal{L}}{=}$ denotes equality in law, $(R_\delta(s),\ s \ge 0)$ denotes a δ-dimensional Bessel process starting from $a > 0$, $T_0 := \inf\{s \ge 0,\ R_\delta(s) = 0\}$, $(R_{4-\delta}(s),\ s \ge 0)$ is a $(4-\delta)$-dimensional Bessel process starting from 0, and $\gamma_a := \sup\{s \ge 0,\ R_{4-\delta}(s) = a\}$.

Let S_κ^{ca} be a (positive) completely asymmetric stable variable of index κ for $0 < \kappa < 1$, and C_8^{ca} a (positive) completely asymmetric Cauchy variable of parameter 8. Their characteristic functions are given by:

$$\mathbb{E}e^{itS_\kappa^{ca}} = \exp\left[-|t|^\kappa \left(1 - i\,\mathrm{sgn}(t)\tan\left(\frac{\pi\kappa}{2}\right)\right)\right], \tag{14}$$

$$\mathbb{E}e^{itC_8^{ca}} = \exp\left[-8\left(|t| + it\frac{2}{\pi}\log|t|\right)\right].$$

Throughout the paper, we set $\lambda := 4(1+\kappa)$. If $(B(t),\ t \ge 0)$ denotes, as before, a Brownian motion, we introduce

$$K_\beta(\kappa) := \int_0^{+\infty} x^{1/\kappa - 2} L_\beta(\tau_\beta(\lambda), x)\mathrm{d}x, \qquad 0 < \kappa < 1, \tag{15}$$

$$C_\beta := \int_0^1 \frac{L_\beta(\tau_\beta(8), x) - 8}{x}\mathrm{d}x + \int_1^{+\infty} \frac{L_\beta(\tau_\beta(8), x)}{x}\mathrm{d}x. \tag{16}$$

We have the following equalities in law:

Fact 5 (Biane and Yor [8]) For $0 < \kappa < 1$,

$$C_\beta \overset{\mathcal{L}}{=} 8c_3 + (\pi/2)C_8^{ca}, \qquad K_\beta(\kappa) \overset{\mathcal{L}}{=} (\kappa^{2-1/\kappa} c_4(\kappa)/4) S_\kappa^{ca},$$

where $c_3 > 0$ denotes an unimportant constant, and

$$\psi(\kappa) := \left(\frac{\pi\kappa}{4\Gamma^2(\kappa)\sin(\pi\kappa/2)}\right)^{1/\kappa}, \qquad c_4(\kappa) := 8\psi(\kappa)\lambda^{1/\kappa}\kappa^{-1/\kappa}. \tag{17}$$

This fact is proved in (Biane and Yor [8]); the identity in law related to C_β is given in its paragraph (4.3.2) pp. 64–66 and the one related to $K_\beta(\kappa)$ follows from its (1.a) p. 24.

Finally, the first Ray-Knight theorem leads to the following formula. For $v > 0$ and $y > 0$,

$$\mathbb{P}\Big(\sup_{0\le s\le \tau_\beta(v)} \beta(s) < y\Big) = \mathbb{P}\big[L_\beta(\sigma_\beta(y),0) > v\big] = \mathbb{P}\big(R_2^2(y) > v\big) = \exp\Big(-\frac{v}{2y}\Big), \tag{18}$$

where $(R_2(s),\ s \ge 0)$ is a 2-dimensional Bessel process starting from 0.

2.2 Some Preliminaries on the Diffusion

We assume in the rest of the paper that $\kappa > 0$, and so X is a.s. transient to the right. We start by introducing

$$A(x) := \int_0^x e^{W_\kappa(y)}\mathrm{d}y, \quad x \in \mathbb{R}, \qquad A_\infty := \int_0^\infty e^{W_\kappa(y)}\mathrm{d}y < \infty \text{ a.s.}$$

We recall that A is a scale function of X under the quenched law P_{W_κ} (see e.g. Shi [44, Eq. (2.2)]). That is, if $P_{W_\kappa}^y$ denotes the law of the diffusion X in the potential W_κ, starting from y instead of 0, we have conditionally on the potential W_κ,

$$P_{W_\kappa}^y\big[H(z) < H(x)\big] = \big[A(y) - A(x)\big]/\big[A(z) - A(x)\big], \qquad x < y < z. \tag{19}$$

We observe that, since $\kappa > 0$, $A(x) \to A_\infty < \infty$ a.s. when $x \to +\infty$.

For technical reasons, we have to introduce the random function F as follows. Fix $r > 0$. Since the function $x \mapsto A_\infty - A(x) =: D(x)$ is almost surely continuous and (strictly) decreasing and has limits $+\infty$ and 0 respectively on $-\infty$ and $+\infty$, there exists a unique $F(r) \in \mathbb{R}$, depending only on the process W_κ, such that

$$A_\infty - A(F(r)) = \exp(-\kappa r/2) =: \delta(r). \tag{20}$$

Our first estimate describes how close $F(r)$ is to r, for large r.

Lemma 1 *Let $\kappa > 0$ and $0 < \delta_0 < 1/2$. Define for $r > 0$,*

$$E_1(r) := \big\{\big(1 - 5r^{-\delta_0}/\kappa\big)r \le F(r) \le \big(1 + 5r^{-\delta_0}/\kappa\big)r\big\}. \tag{21}$$

Then for all large r,

$$\mathbb{P}\big[E_1(r)^c\big] \le \exp\big(-r^{1-2\delta_0}\big). \tag{22}$$

As a consequence, for any $\varepsilon > 0$, we have, almost surely, for all large r,

$$(1 - \varepsilon)r \le F(r) \le (1 + \varepsilon)r. \tag{23}$$

Proof of Lemma 1 Let $0 < \delta_0 < 1/2$, and fix $r > 0$. We have

$$\mathbb{P}\big[E_1(r)^c\big] \leq \mathbb{P}\big[F(r) < (1 - 5r^{-\delta_0}/\kappa)r\big] + \mathbb{P}\big[F(r) > (1 + 5r^{-\delta_0}/\kappa)r\big]. \tag{24}$$

Define $s_\pm := (1 \pm 5r^{-\delta_0}/\kappa)r$, and $A_\infty^{(s)} := \int_s^\infty \exp(W_\kappa(u) - W_\kappa(s))\mathrm{d}u$ for $s \geq 0$. Observe that D is strictly decreasing, $D(F(r)) = e^{-\kappa r/2}$ and notice that $D(s_\pm) = A_\infty^{(s_\pm)} \exp(W_\kappa(s_\pm))$. Consequently,

$$\begin{aligned}\mathbb{P}\big[F(r) < (1 - 5r^{-\delta_0}/\kappa)r\big] &\leq \mathbb{P}\big[D(F(r)) > D(s_-)\big]\\ &= \mathbb{P}\big[-\kappa r/2 > \log\big(A_\infty^{(s_-)}\big) + W_\kappa(s_-)\big].\end{aligned}$$

Moreover, $A_\infty^{(s_\pm)} \stackrel{\mathscr{L}}{=} A_\infty \stackrel{\mathscr{L}}{=} 2/\gamma_\kappa$, where γ_κ is a gamma variable of parameter $(\kappa, 1)$ (see Dufresne [24] or Borodin et al. [10, IV.48 p. 78]), that is, γ_κ has density $\frac{1}{\Gamma(\kappa)} e^{-x} x^{\kappa-1} \mathbf{1}_{\mathbb{R}_+}(x)$. Hence

$$\begin{aligned}\mathbb{P}\big[F(r) < (1 - 5r^{-\delta_0}/\kappa)r\big] &\leq \mathbb{P}\big[\log(2/\gamma_\kappa) < -r^{1-\delta_0}\big] + \mathbb{P}\big[W(s_-) < -3r^{1-\delta_0}/2\big]\\ &\leq 2\exp\big(-9r^{1-2\delta_0}/8\big),\end{aligned}$$

for large r, since $\mathbb{P}[W(1) < -x] \leq e^{-x^2/2}$ for $x \geq 1$. Similarly, we have for large r,

$$\begin{aligned}\mathbb{P}\big[F(r) > s_+\big] &\leq \mathbb{P}\big[\log(2/\gamma_\kappa) > r^{1-\delta_0}/2\big] + \mathbb{P}\big[W(s_+) > 2r^{1-\delta_0}\big]\\ &\leq \exp\big(-9r^{1-2\delta_0}/8\big).\end{aligned}$$

This yields (22) in view of (24).

Then $\sum_{n\geq 1} \mathbb{P}\big[E_1(n)^c\big] < \infty$, so (23) follows from the Borel–Cantelli lemma and the monotonicity of $F(\cdot)$. □

In the rest of the paper, we define, for $\delta_1 > 0$ and any $r > 0$,

$$c_5 := 2(\lambda/\kappa)^{\delta_1}, \qquad \psi_\pm(r) := 1 \pm \frac{c_5}{r^{\delta_1}}, \qquad t_\pm(r) := \frac{\kappa\psi_\pm(r)r}{\lambda}. \tag{25}$$

Taking $\psi_\pm(r)$ as defined above instead of simply $1 \pm \varepsilon$ is necessary e.g. in Lemma 5 below. Moreover, if $(\beta(s),\ s \geq 0)$ is a Brownian motion and $v > 0$, we define the Brownian motion $(\beta_v(s),\ s \geq 0)$ by $\beta_v(s) := (1/v)\beta(v^2 s)$, $s \geq 0$.

We prove in Sect. 6 the following approximation of the (annealed) joint law of $\big(L_X^*[H(F(r))], H(F(r))\big)$.

Lemma 2 *Let $\kappa > 0$ and $\varepsilon \in (0, 1)$. For $\delta_1 > 0$ small enough, there exists $c_6 > 0$ and $\alpha > 0$ such that for r large enough, there exist a Brownian motion $(\beta(t),\ t \geq 0)$ such that the following holds:*
(i) *Whenever $\kappa > 0$, we have*

$$\mathbb{P}[E_2(r)] \geq 1 - r^{-\alpha},$$

where

$$E_2(r) := \left\{(1-\varepsilon)\widehat{L}_-(r) \le L_X^*[H(F(r))] \le (1+\varepsilon)\widehat{L}_+(r)\right\}, \tag{26}$$

$$\widehat{L}_\pm(r) := 4[\kappa t_\pm(r)]^{1/\kappa}\left[\sup_{0\le u\le \tau_{\beta_{t_\pm(r)}}(\lambda)} \beta_{t_\pm(r)}(u)\right]^{1/\kappa} = 4\left[\sup_{0\le u\le \tau_\beta(\lambda t_\pm(r))} \kappa\beta(u)\right]^{1/\kappa}. \tag{27}$$

(ii) *If* $0<\kappa\le 1$, *we have*

$$\mathbb{P}[E_3(r)] \ge 1 - r^{-\alpha},$$

where, using the notation introduced in (15) *and* (16),

$$E_3(r) := \left\{(1-\varepsilon)\widehat{I}_-(r) \le H(F(r)) \le (1+\varepsilon)\widehat{I}_+(r)\right\}, \tag{28}$$

$$\widehat{I}_\pm(r) := \begin{cases} 4\kappa^{1/\kappa-2}t_\pm(r)^{1/\kappa}\left[K_{\beta_{t_\pm(r)}}(\kappa) \pm c_6 t_\pm(r)^{1-1/\kappa}\right], & 0<\kappa<1, \\ 4t_\pm(r)\left[C_{\beta_{t_\pm(r)}} + 8\log t_\pm(r)\right], & \kappa = 1. \end{cases} \tag{29}$$

Notice in particular that the Brownian motion β is the same in **(i)** and **(ii)**; this allows to approximate the law of quantities depending on both $L_X^*[H(F(r))]$ and $H(F(r))$, such as $L_X^*[H(F(r))]/H(F(r))$, which is useful in Sect. 5. This is possible because we kept the random function $F(r)$ in the expressions $L_X^*[H(F(r))]$ and $H(F(r))$, in order to have the same Brownian motion β in the left hand side and the right hand side of the inequalities defining $E_2(r)$ and $E_3(r)$.

The proof of Lemma 2 is postponed to Sect. 6.

With an abuse of notation, for $z\ge 0$, we denote by $X\circ\Theta_{H(z)}$ the process $(X(H(z)+t)-z,\ t\ge 0)$. Notice that due to the strong Markov property applied at stopping time $H(z)$ under the quenched law P_{W_κ}, $X\circ\Theta_{H(z)}$ is, conditionally on W_κ, a diffusion in the $(-\kappa/2)$-drifted Brownian potential $W_\kappa\circ\Theta_z := (W_\kappa(x+z)-W_\kappa(z),\ x\in\mathbb{R})$, starting from 0. Define $H_{X\circ\Theta_{H(z)}}(s) = H(z+s)-H(z)$, $s\ge 0$, which is the hitting time of s by $X\circ\Theta_{H(z)}$. In view of (20), we also define $F_{W_\kappa\circ\Theta_z}$ by $\int_{F_{W_\kappa\circ\Theta_z}(r)}^{\infty} e^{W_\kappa\circ\Theta_z(u)}du = \delta(r)$, $r>0$. That is, $F_{W_\kappa\circ\Theta_z}$ plays the same role for $W_\kappa\circ\Theta_z$ (resp. for $X\circ\Theta_{H(z)}$) as F does for W_κ (resp. for X). Similarly, $L^*_{X\circ\Theta_{H(z)}}$ and $(L^*\circ H)_{X\circ\Theta_{H(z)}}$ denote respectively the processes L^* and $L^*\circ H$ for the diffusion $X\circ\Theta_{H(z)}$, with $(L^*)_X := L_X^*$. The following lemma is a modification of the Borel-Cantelli lemma.

Lemma 3 *Let* $\kappa>0$, $\alpha>0$, $r_n := \exp(n^\alpha)$ *and* $Z_n := \sum_{k=1}^n r_k$ *for* $n\ge 1$. *Assume* f *is a continuous function* $(0,+\infty)^2\to\mathbb{R}$ *and* $(\Delta_n)_{n\ge 1}$ *is a sequence of open sets in* $\mathbb{R}$ *such that*

$$\sum_{n\ge 1}\mathbb{P}\{f[(H\circ F)(r_{2n}), (L_X^*\circ H\circ F)(r_{2n})]\in\Delta_n\} = +\infty. \tag{30}$$

Then for any $0 < \varepsilon < 1/2$, $\mathbb{P}$ *almost surely, there exist infinitely many* n *such that for some* $t_n \in [(1-\varepsilon)r_{2n}, (1+\varepsilon)r_{2n}]$,

$$f\big[H_{X\circ\Theta_{H(Z_{2n-1})}}(t_n), (L^*\circ H)_{X\circ\Theta_{H(Z_{2n-1})}}(t_n)\big] \in \Delta_n.$$

The results remain true if $r_n = n^n$ *for every* $n \geq 1$.

Proof (Proof of Lemma 3) We divide $\mathbb{R}_+$ into some regions in which the diffusion X will behave "independently", in order to apply the Borel-Cantelli lemma.

To this aim, let $n \geq 1$ and

$$E_4(n) := \left\{ \inf_{\{t:\, H(Z_{2n-1})\leq t\leq H(Z_{2n}+r_{2n+1}/2)\}} X(t) > Z_{2n-2} + \frac{1}{2}r_{2n-1} \right\}.$$

Define $x_n := r_{2n-1}/2$. For any environment, i.e., for any realization of W_κ, X is a Markov process under P_{W_κ}, and $H(Z_{2n-1})$ is a stopping time. Hence, $P_{W_\kappa}(E_4(n)^c)$ is the probability that the diffusion in the potential W_κ started at Z_{2n-1} hits level $Z_{2n-2}+x_n$ before $Z_{2n}+x_{n+1}$, that is

$$P_{W_\kappa}\big[E_4(n)^c\big] = \left(1 + \frac{\int_{Z_{2n-2}+x_n}^{Z_{2n-1}} e^{W_\kappa(u)}\mathrm{d}u}{\int_{Z_{2n-1}}^{Z_{2n}+x_{n+1}} e^{W_\kappa(u)}\mathrm{d}u}\right)^{-1} \leq \frac{\int_{Z_{2n-1}}^{Z_{2n}+x_{n+1}} e^{W_\kappa(u)}\mathrm{d}u}{\int_{Z_{2n-2}+x_n}^{Z_{2n-1}} e^{W_\kappa(u)}\mathrm{d}u}, \tag{31}$$

where we used (19). Observe that $r_{2n-1}-x_n = x_n$ and define for some $0 < \varepsilon_0 < \kappa/4$,

$$E_5(n) := \left\{ \sup_{0\leq u\leq r_{2n-1}-x_n} \left| W_\kappa(u+Z_{2n-2}+x_n) - W_\kappa(Z_{2n-2}+x_n) + \frac{\kappa}{2}u \right| \leq \varepsilon_0(r_{2n-1}-x_n) \right\}$$

and $E_6(n) := \big\{\sup_{u\geq 0}[W_\kappa(u+Z_{2n-1}) - W_\kappa(Z_{2n-1})] \leq v_n\big\}$, where $v_n := 2(\log n)/\kappa$. Since $\sup_{0\leq u\leq x_n} W(u) \stackrel{\mathcal{L}}{=} |W(x_n)|$ and $\sup_{x\geq 0} W_\kappa(x)$ has an exponential law of parameter κ (see e.g. Borodin et al. [10, 1.1.4 (1) p. 251]), we have for large n,

$$\mathbb{P}\big[E_5(n)^c\big] = \mathbb{P}\Big(\sup_{0\leq u\leq x_n}|W(u)| > \varepsilon_0 x_n\Big) \leq 4\exp\Big[-\frac{\varepsilon_0^2 x_n}{2}\Big],$$

$$\mathbb{P}\big[E_6(n)^c\big] = \exp(-\kappa v_n) = n^{-2}.$$

Moreover by (31), we have for n large enough, on $E_5(n)\cap E_6(n)$,

$$P_{W_\kappa}\big[E_4(n)^c\big]\le \kappa\frac{(r_{2n}+x_{n+1})\exp[v_n+W_\kappa(Z_{2n-1})]}{\exp[W_\kappa(Z_{2n-2}+x_n)-\varepsilon_0(r_{2n-1}-x_n)]}$$
$$\le \kappa(r_{2n}+x_{n+1})\exp[v_n+(2\varepsilon_0-\kappa/2)(r_{2n-1}-x_n)]. \qquad (32)$$

We now integrate (32) over $E_5(n)\cap E_6(n)$. Since $\mathbb{P}[E_5(n)^c]$ and $\mathbb{P}[E_6(n)^c]$ are summable, this yields since $\varepsilon_0<\kappa/4$,

$$\sum_{n=1}^{+\infty}\mathbb{P}\big[E_4(n)^c\big]<\infty. \qquad (33)$$

To complete the proof of Lemma 3, let $0<\varepsilon<1/2$, and define

$$\mathscr{D}_n:=\Big\{\exists t_n\in[(1-\varepsilon)r_{2n},(1+\varepsilon)r_{2n}],$$
$$f\big[H_{X\circ\Theta_{H(Z_{2n-1})}}(t_n),(L^*\circ H)_{X\circ\Theta_{H(Z_{2n-1})}}(t_n)\big]\in\Delta_n\Big\},$$
$$\mathscr{E}_n:=\Big\{\Big(1-5r_{2n}^{-\delta_0}/\kappa\Big)r_{2n}\le F_{W_\kappa\circ\Theta_{Z_{2n-1}}}(r_{2n})\le\Big(1+5r_{2n}^{-\delta_0}/\kappa\Big)r_{2n}\Big\}.$$

Let $\tilde{t}_n:=F_{W_\kappa\circ\Theta_{Z_{2n-1}}}(r_{2n})$. We have uniformly for large n,

$$\mathscr{D}_n\cap E_4(n)\supset\Big\{f\big[H_{X\circ\Theta_{H(Z_{2n-1})}}(\tilde{t}_n),(L^*\circ H)_{X\circ\Theta_{H(Z_{2n-1})}}(\tilde{t}_n)\big]\in\Delta_n\Big\}\cap E_4(n)\cap\mathscr{E}_n. \qquad (34)$$

Due to our assumption (30), $\sum_{n\ge1}\mathbb{P}\{f[H_{X\circ\Theta_{H(Z_{2n-1})}}(\tilde{t}_n),(L^*\circ H)_{X\circ\Theta_{H(Z_{2n-1})}}(\tilde{t}_n)]\in\Delta_n\}=\infty$, since $X\circ\Theta_{H(Z_{2n-1})}$ is a diffusion process in the $(-\kappa/2)$-drifted Brownian potential $W_\kappa\circ\Theta_{Z_{2n-1}}$, which also gives $\mathbb{P}(\mathscr{E}_n)=\mathbb{P}(E_1(r_{2n}))$. In view of (33), (34) and Lemma 1, this yields $\sum_{n\in\mathbb{N}}\mathbb{P}(\mathscr{D}_n\cap E_4(n))=+\infty$.

Define $x\wedge y:=\inf\{x,y\}$, $(x,y)\in\mathbb{R}^2$. Since $\varepsilon r_{2n}\le r_{2n+1}/2$ for large n, the event $\mathscr{D}_n\cap E_4(n)$ is measurable with respect to the σ-field generated by $(W_\kappa(x+Z_{2n-1})-W_\kappa(Z_{2n-1}),-r_{2n-1}/2\le x\le Z_{2n}+r_{2n+1}/2-Z_{2n-1})$ and $\big(X\circ\Theta_{H(Z_{2n-1})}(t),\ 0\le t\le H_{X\circ\Theta_{H(Z_{2n-1})}}(-r_{2n-1}/2)\wedge H_{X\circ\Theta_{H(Z_{2n-1})}}(Z_{2n}+r_{2n+1}/2-Z_{2n-1}\big)$. So, the events $\mathscr{D}_n\cap E_4(n)$, $n\ge1$, are independent by the strong Markov Property, because the intervals $\big[Z_{2n-1}-r_{2n-1}/2,Z_{2n}+r_{2n+1}/2\big)$, $n\ge1$ are disjoint. Hence, Lemma 3 follows by an application of the Borel-Cantelli lemma. □

3 Almost Sure Asymptotics of $L_X^*[H(r)]$

As a warm up, we first prove the following results, which are useful in Sect. 5.

Theorem 9 *Let $\kappa > 0$. For any positive nondecreasing function $a(\cdot)$, we have*

$$\sum_{n=1}^{\infty} \frac{1}{na(n)} \begin{cases} < \infty \\ = +\infty \end{cases} \Longleftrightarrow \limsup_{r\to\infty} \frac{L_X^*[H(r)]}{[ra(r)]^{1/\kappa}} = \begin{cases} 0 \\ +\infty \end{cases} \qquad \mathbb{P}\text{-a.s.}$$

Theorem 10 *For $\kappa > 0$,*

$$\liminf_{r\to+\infty} \frac{L_X^*[H(r)]}{(r/\log\log r)^{1/\kappa}} = 4\left(\frac{\kappa^2}{2}\right)^{1/\kappa} \qquad \mathbb{P}\text{-a.s.}$$

3.1 Proof of Theorem 9

Let $r_n := e^n$ and $Z_n := \sum_{k=1}^n r_k$. Denote by $a(\cdot)$ be a positive nondecreasing function. We begin with the upper bound in Theorem 9.

First, notice that for $\widehat{L}_\pm$ which is defined in (27), and any positive y and r, we have

$$\mathbb{P}\left(\widehat{L}_\pm(r) < (yr)^{1/\kappa}\right) = \mathbb{P}\left[\sup_{0\le u\le \tau_\beta(\lambda t_\pm(r))} \beta(u) < \frac{yr}{4^\kappa \kappa}\right] = \exp\left(-\frac{\kappa^2 4^\kappa \psi_\pm(r)}{2y}\right), \tag{35}$$

by (18) and (25). This together with Lemma 2 gives, for some $\alpha > 0$, $\varepsilon > 0$ and all large r,

$$\begin{aligned}\mathbb{P}\left\{L_X^*[H(F(r))] > \left(ra(e^{-2}r)\right)^{1/\kappa}\right\} &\le 1 - \exp\left(-\frac{(1+\varepsilon)^\kappa \kappa^2 4^\kappa \psi_+(r)}{2a(e^{-2}r)}\right) + r^{-\alpha} \\ &\le \frac{c_7}{a(e^{-2}r)} + r^{-\alpha},\end{aligned} \tag{36}$$

since $1 - e^{-x} \le x$ for all $x \in \mathbb{R}$. Assume $\sum_{n=1}^{+\infty} \frac{1}{na(n)} < \infty$, which is equivalent to $\sum_{n=1}^{+\infty} \frac{1}{a(r_n)} < \infty$. Then (36) leads to $\sum_{n=1}^{+\infty} \mathbb{P}\{L_X^*[H(F(r_n))] > [r_n a(r_{n-2})]^{1/\kappa}\} < \infty$.

So by the Borel–Cantelli lemma, almost surely for all large n, $L_X^*[H(F(r_n))] \le [r_n a(r_{n-2})]^{1/\kappa}$. On the other hand, $r_{n-1} \le F(r_n)$ almost surely for all large n (see (23)). As a consequence, almost surely for all large n, $L_X^*[H(r_{n-1})] \le [r_n a(r_{n-2})]^{1/\kappa}$. Let $r \in [r_{n-2}, r_{n-1}]$, for such large n. Then

$$L_X^*[H(r)] \le L_X^*[H(r_{n-1})] \le [r_n a(r_{n-2})]^{1/\kappa} \le e^{2/\kappa}[ra(r)]^{1/\kappa}.$$

Consequently,

$$\limsup_{r\to+\infty} \frac{L_X^*[H(r)]}{[ra(r)]^{1/\kappa}} \le e^{2/\kappa} \qquad \mathbb{P}\text{-a.s.} \tag{37}$$

Since $\sum_{n=1}^{+\infty} \frac{1}{n\varepsilon a(n)}$ is also finite, (37) holds for $a(\cdot)$ replaced by $\varepsilon a(\cdot)$, $\varepsilon > 0$. Letting $\varepsilon \to 0$ yields the "zero" part of Theorem 9.

Now we turn to the proof of the "infinity" part. Assume $\sum_{n=1}^{+\infty} \frac{1}{na(n)} = +\infty$, that is, $\sum_{n=1}^{+\infty} \frac{1}{a(r_n)} = +\infty$. Observe that we may restrict ourselves to the case $a(x) \to +\infty$ when $x \to +\infty$, since the result in this case yields the result when a is bounded.

By an argument similar to that leading to (36), we have, for some $\alpha > 0$ and all large r,

$$\mathbb{P}\left\{L_X^*[H(F(r))] > \left(ra(e^2 r)\right)^{1/\kappa}\right\} \geq \frac{c_8}{a(e^2 r)} - r^{-\alpha},$$

which implies $\sum_{n=1}^{+\infty} \mathbb{P}\left\{(L_X^* \circ H \circ F)(r_{2n}) > [r_{2n}a(r_{2n+2})]^{1/\kappa}\right\} = +\infty$. Let $0 < \varepsilon < 1/2$ and recall that $Z_n = \sum_{k=1}^{n} r_k$; by Lemma 3, almost surely, there exist infinitely many n such that

$$\sup_{s\in[(1-\varepsilon)r_{2n},(1+\varepsilon)r_{2n}]} (L^* \circ H)_{X\circ\Theta_{H(Z_{2n-1})}}(s) > [r_{2n}a(r_{2n+2})]^{1/\kappa}.$$

For such n, we have $\sup_{s\in[(1-\varepsilon)r_{2n},(1+\varepsilon)r_{2n}]} L_X^*[H(Z_{2n-1}+s)] > [r_{2n}a(r_{2n+2})]^{1/\kappa}$. Consequently,

$$\sup_{s\in[(1-\varepsilon)r_{2n},(1+\varepsilon)r_{2n}]} \frac{L_X^*(H(Z_{2n-1}+s))}{[(Z_{2n-1}+s)a(Z_{2n-1}+s)]^{1/\kappa}} \geq c_9,$$

almost surely for infinitely many n. This gives

$$\limsup_{r\to+\infty} \frac{L_X^*[H(r)]}{[ra(r)]^{1/\kappa}} \geq c_9 \qquad \mathbb{P}\text{-a.s.}$$

Replace $a(\cdot)$ by $a(\cdot)/\varepsilon$, and let $\varepsilon \to 0$. This leads to the "infinity" part of Theorem 9. □

3.2 *Proof of Theorem 10*

We fix $\varepsilon \in (0,1)$. By Lemma 2 and (35), we get for some $\alpha > 0$, for every positive function g and all large r,

$$\mathbb{P}\left[L_X^*[H(F(r))] < [r/g(r)]^{1/\kappa}\right] \leq \exp\left[-\kappa^2 4^\kappa (1-\varepsilon)^\kappa \psi_-(r) g(r)/2\right] + r^{-\alpha}. \qquad (38)$$

We choose $g(r) := \frac{2(1+\varepsilon)}{\kappa^2 4^\kappa (1-\varepsilon)^{\kappa+1}\psi_-(r)} \log\log r$. Let $s_n := \exp(n^{1-\varepsilon})$. It follows from (38) that $\sum_{n=1}^{\infty} \mathbb{P}\{L_X^*[H(F(s_n))] < [s_n/g(s_n)]^{1/\kappa}\} < \infty$. Hence by the Borel-

Cantelli lemma, almost surely for all large n,

$$L_X^*[H(F(s_n))] \geq [s_n/g(s_n)]^{1/\kappa}.$$

On the other hand, by (22) and the Borel-Cantelli lemma, $s_n \geq F(s_{n-1})$ almost surely for all large n, which implies that, for $r \in [s_n, s_{n+1}]$,

$$L_X^*[H(r)] \geq L_X^*[H(F(s_{n-1}))] \geq [s_{n-1}/g(s_{n-1})]^{1/\kappa} \geq (1-\varepsilon)[r/g(r)]^{1/\kappa},$$

since $s_{n-1}/s_{n+1} \to 1$ as $n \to +\infty$. Consequently,

$$\liminf_{r\to\infty} \frac{L_X^*[H(r)]}{(r/\log\log r)^{1/\kappa}} \geq 4\left(\frac{\kappa^2}{2}\right)^{1/\kappa} \qquad \mathbb{P}\text{-a.s.}$$

Now we prove the inequality "$\leq$". Let $\varepsilon \in (0, 1/2)$, $r_n := \exp(n^{1+\varepsilon})$, $Z_n := \sum_{k=1}^n r_k$, $n \geq 1$, and $\tilde{g}(r) := \frac{2(1-\varepsilon)}{\kappa^2 4^\kappa (1+\varepsilon)^{\kappa+1}\psi_+(r)} \log\log r$. By Lemma 2 and (35), for some $\alpha > 0$ and all large r,

$$\mathbb{P}\Big[L_X^*[H(F(r))] < [r/\tilde{g}(r)]^{1/\kappa}\Big] \geq \exp\Big[-\kappa^2 4^\kappa (1+\varepsilon)^\kappa \psi_+(r)\tilde{g}(r)/2\Big] - r^{-\alpha}.$$

Therefore,

$$\sum_{n\geq 1} \mathbb{P}\Big[L_X^*[H(F(r_{2n}))] < \big[r_{2n}/\tilde{g}(r_{2n})\big]^{1/\kappa}\Big] = +\infty.$$

It follows from Lemma 3 that, almost surely, there are infinitely many n such that

$$\inf_{s\in[(1-\varepsilon)r_{2n},(1+\varepsilon)r_{2n}]} (L^* \circ H)_{X\circ\Theta_{H(Z_{2n-1})}}(s) < \big[r_{2n}/\tilde{g}(r_{2n})\big]^{1/\kappa}. \tag{39}$$

On the other hand, an application of Theorem 9 with $a(x) \sim_{x\to+\infty} (\log x)^2$ gives that almost surely for large n, $L_X^*[H(Z_{2n-1})] \leq [Z_{2n-1}\log^2 Z_{2n-1}]^{1/\kappa} \leq \varepsilon\big[r_{2n}/\tilde{g}(r_{2n})\big]^{1/\kappa}$, since $Z_p \leq pr_p \leq p\exp(-p^\varepsilon)r_{p+1}$ for p large enough. Therefore,

$$\inf_{s\in[(1-\varepsilon)r_{2n},(1+\varepsilon)r_{2n}]} L_X^*[H(Z_{2n-1}+s)] \leq (1+\varepsilon)\big[r_{2n}/\tilde{g}(r_{2n})\big]^{1/\kappa}$$

almost surely, for infinitely many n, where we used $L_X^*[H(r+s)] \leq L_X^*[H(r)] + (L^* \circ H)_{X\circ\theta_{H(r)}}(s)$, $r \geq 0$, $s \geq 0$. Hence, for such n,

$$\inf_{s\in[(1-\varepsilon)r_{2n},(1+\varepsilon)r_{2n}]} \frac{L_X^*[H(Z_{2n-1}+s)]}{[(Z_{2n-1}+s)/\log\log(Z_{2n-1}+s)]^{1/\kappa}}$$
$$\leq (1+c_{10}\varepsilon)\left(\frac{\kappa^2 4^\kappa \psi_+(r_{2n})}{2}\right)^{1/\kappa}.$$

This yields

$$\liminf_{r\to+\infty} \frac{L_X^*(H(r))}{(r/\log\log r)^{1/\kappa}} \leq 4\left(\frac{\kappa^2}{2}\right)^{1/\kappa} \qquad \mathbb{P}\text{-a.s.},$$

proving Theorem 10. □

4 Proof of Theorems 7 and 8 and Corollary 2

Recall $\widehat{I}_\pm$ from (29) and $c_4(\kappa)$ from (17). By Fact 5,

$$\widehat{I}_\pm(r) \stackrel{\mathscr{L}}{=} t_\pm(r)^{1/\kappa}\{c_4(\kappa)\, S_\kappa^{ca} \pm c_{11}\, t_\pm(r)^{1-1/\kappa}\}, \qquad 0<\kappa<1, \tag{40}$$

$$\widehat{I}_\pm(r) \stackrel{\mathscr{L}}{=} 4t_\pm(r)[8c_3 + (\pi/2)C_8^{ca} + 8\log t_\pm(r)] \qquad \kappa=1, \tag{41}$$

where $c_{11} > 0$ and $c_3 > 0$ are unimportant constants.

We have now all the ingredients to prove Theorems 7 and 8.

4.1 *Proof of Theorem 7*

This subsection is devoted to the proof of Theorem 7. We start with the case $0 < \kappa < 1$.

4.1.1 Case $0 < \kappa < 1$

We assume $0 < \kappa < 1$. Let $a(\cdot)$ be a positive nondecreasing function. Without loss of generality, we suppose that $a(r) \to \infty$ (as $r \to \infty$).

It is known (see e.g. Samorodnitsky and Taqqu [41, (1.2.8) p. 16]) that

$$\mathbb{P}\big(S_\kappa^{ca} > x\big) \underset{x\to+\infty}{\sim} c_{12}x^{-\kappa},$$

where $f(x) \underset{x\to+\infty}{\sim} g(x)$ means $\lim_{x\to+\infty} f(x)/g(x) = 1$, and $c_{12} > 0$ is a constant depending on κ.

Recall $t_\pm(\cdot)$ from (25). By Lemma 2 and (40), for some $\alpha > 0$, we have for large r,

$$\mathbb{P}\big[H(F(r)) > \big(a(e^{-2}r)t_+(r)\big)^{1/\kappa}\big] \leq \frac{c_{13}}{a(e^{-2}r)} + r^{-\alpha}. \tag{42}$$

As in Sect. 3.1, we define $r_n := e^n$ and $Z_n := \sum_{k=1}^{n} r_k$. Assume $\sum_{n\geq 1} \frac{1}{a(r_n)} < \infty$, which is equivalent to $\sum_{n\geq 1} \frac{1}{na(n)} < \infty$. By the Borel–Cantelli lemma, almost surely for n large enough,

$$H[F(r_n)] \leq [a(r_{n-2})t_+(r_n)]^{1/\kappa}. \tag{43}$$

On the other hand, by Lemma 1, almost surely for all large n, we have $r_{n+1} \leq F(r_{n+2})$, which together with (43) implies that for $r \in [r_n, r_{n+1}]$,

$$H(r) \leq H[F(r_{n+2})] \leq [\psi_+(r_{n+2})\kappa r_{n+2}a(r_n)/\lambda]^{1/\kappa} \leq c_{14}[ra(r)]^{1/\kappa}.$$

Therefore, $\limsup_{r\to+\infty} \frac{H(r)}{[ra(r)]^{1/\kappa}} \leq c_{14}$ $\mathbb{P}$-a.s., implying the "zero" part of Theorem 7, since we can replace $a(\cdot)$ by any constant multiple of $a(\cdot)$.

To prove the "infinity" part, we assume $\sum_{n\geq 1} \frac{1}{na(n)} = +\infty$, and observe that, by an argument similar to that leading to (42), we have, for some $\alpha > 0$ and all r large enough,

$$\mathbb{P}\big[H(F(r)) > \big(a(e^2r)t_-(r)\big)^{1/\kappa}\big] \geq \frac{c_{15}}{a(e^2r)} - r^{-\alpha}. \tag{44}$$

Thanks to Lemma 3, $\sup_{s\in[(1-\varepsilon)r_{2n},(1+\varepsilon)r_{2n}]} H_{X\circ\Theta_{H(Z_{2n-1})}}(s) > [a(r_{2n+2})t_-(r_{2n})]^{1/\kappa}$, almost surely for infinitely many n. Since $H(Z_{2n-1}+s) \geq H_{X\circ\Theta_{H(Z_{2n-1})}}(s)$ for all $s > 0$, this implies, for these n,

$$\sup_{s\in[(1-\varepsilon)r_{2n},(1+\varepsilon)r_{2n}]} H(Z_{2n-1}+s)/[a(Z_{2n-1}+s)(Z_{2n-1}+s)]^{1/\kappa} \geq c_{16}. \tag{45}$$

This gives $\limsup_{r\to+\infty} \frac{H(r)}{[ra(r)]^{1/\kappa}} \geq c_{16}$ $\mathbb{P}$-a.s. , proving the "infinity" part in Theorem 7, in the case $0 < \kappa < 1$ by replacing $a(.)$ by any constant multiple of $a(.)$. □

4.1.2 Case $\kappa = 1$

Let $r_n := e^n$ and $Z_n := \sum_{k=1}^{n} r_k$. We recall that there exists a constant $c_{17} := \frac{16}{\pi}$ such that $\mathbb{P}(C_8^{ca} > x) \underset{x\to+\infty}{\sim} \frac{c_{17}}{x}$ (see e.g. Samorodnitsky et al. [41, Prop. 1.2.15 p. 16]). Hence, by Lemma 2 and (41), for some $\alpha > 0$ and all large r,

$$\mathbb{P}\left\{H(F(r)) > 4t_+(r)(1+\varepsilon)\big[8c_3 + a(e^{-2}r) + 8\log t_+(r)\big]\right\} \leq c_{18}/a(e^{-2}r) + r^{-\alpha}. \tag{46}$$

Assume $\sum_{n\geq 1} \frac{1}{na(n)} < \infty$. Then by the Borel-Cantelli lemma, almost surely, for all large n,

$$H[F(r_n)] \leq 4(1+\varepsilon)t_+(r_n)[8c_3 + a(r_{n-2}) + 8\log(\psi_+(r_n)\kappa r_n/8)].$$

Under the additional assumption $\limsup_{r\to+\infty}(\log r)/a(r) < \infty$, we have, almost surely, for all large n and $r \in [r_n, r_{n+1}]$ (thus $r \le F(r_{n+2})$ by Lemma 1),

$$H(r) \le H[F(r_{n+2})] \le c_{19} r_{n+2}[a(r_n) + \log r_{n+2}] \le c_{20} r a(r).$$

As in the case $0 < \kappa < 1$, this yields the "zero" part of Theorem 7 in the case $\kappa = 1$.

For the "infinity" part, we assume $\sum_{n\ge 1} \frac{1}{na(n)} = +\infty$. As in (46), we have, for some $\alpha > 0$,

$$\mathbb{P}\left\{H(F(r)) > 4t_-(r)(1-\varepsilon)a(e^2 r)\right\} \ge c_{21}/a(e^2 r) - r^{-\alpha},$$

for large r. As in the displays between (44) and (45), this yields the "infinity part" of Theorem 7 in the case $\kappa = 1$. □

4.2 Proof of Theorem 8

This subsection is devoted to the proof of Theorem 8. We start with the case $0 < \kappa < 1$.

4.2.1 Case $0 < \kappa < 1$

We have $\mathbb{E}\left(e^{-tS_\kappa^{ca}}\right) = \exp[-t^\kappa / \cos(\pi\kappa/2)]$, $t \ge 0$, e.g. by Samorodnitsky et al. ([41, Proposition 1.2.12], in the notation of [41], S_κ^{ca} is distributed as $S_\kappa(1, 1, 0)$). So by Bingham et al. [9, Example p. 349],

$$\log \mathbb{P}\left(S_\kappa^{ca} < x\right) \underset{x\to 0,\ x>0}{\sim} -c_{22} x^{-\kappa/(1-\kappa)}, \tag{47}$$

where $c_{22} := (1-\kappa)\kappa^{\kappa/(1-\kappa)}[\cos(\pi\kappa/2)]^{-1/(1-\kappa)}$. By Lemma 2, (40) and (47), for any (strictly) positive function f such that $\lim_{x\to+\infty} f(x) = 0$ and $\varepsilon > 0$ small enough, we have for large r,

$$\begin{aligned} &\mathbb{P}\left[H(F(r)) < t_-(r)^{1/\kappa} f(r)\right] \\ &\le \exp\left[-(c_{22}-\varepsilon)\left(\frac{(1-\varepsilon)c_4(\kappa)}{f(r) + (1-\varepsilon)c_{11} t_-(r)^{1-1/\kappa}}\right)^{\kappa/(1-\kappa)}\right] + r^{-\alpha}. \end{aligned} \tag{48}$$

We define for $\varepsilon > 0$ and $r > 1$,

$$f_\varepsilon^\pm(r) := (1\pm\varepsilon)c_4(\kappa)\left(\frac{(1\pm\varepsilon)(c_{22}\pm\varepsilon)}{(1\mp\varepsilon)\log\log r}\right)^{(1-\kappa)/\kappa} \pm c_{11}(1\pm\varepsilon)t_\pm(r)^{1-1/\kappa}.$$

So, (48) gives

$$\mathbb{P}\big[H(F(r)) < t_-(r)^{1/\kappa} f_\varepsilon^-(r)\big] \le (\log r)^{-(1+\varepsilon)/(1-\varepsilon)} + r^{-\alpha}.$$

With $s_n := \exp(n^{1-\varepsilon})$, this gives $\sum_{n=1}^{+\infty} \mathbb{P}\big[H(F(s_n)) < t_-(s_n)^{1/\kappa} f_\varepsilon^-(s_n)\big] < \infty$, which, by the Borel-Cantelli lemma, implies that, almost surely, for all large n, $H[F(s_n)] \ge t_-(s_n)^{1/\kappa} f_\varepsilon^-(s_n)$.

Recall from Lemma 1 that, almost surely, for all large n, we have $F(s_n) \le (1+\varepsilon)s_n$. Let r be large. There exists n (large) such that $(1+\varepsilon)s_n \le r \le (1+2\varepsilon)s_n$. Then if r is large,

$$H(r) \ge H[F(s_n)] \ge t_-(s_n)^{1/\kappa} f_\varepsilon^-(s_n) \ge t_-^{1/\kappa}\left(\frac{r}{1+2\varepsilon}\right) f_\varepsilon^-\left(\frac{r}{1+\varepsilon}\right).$$

Plugging the value of $t_-(\frac{r}{1+2\varepsilon})$ (defined in (25)), this yields inequality "$\ge$" of (9) with

$$c_2(\kappa) := 8\psi(\kappa)c_{22}^{(1-\kappa)/\kappa} = 8[\pi\kappa]^{1/\kappa}(1-\kappa)^{\frac{1-\kappa}{\kappa}}\kappa/\big[2\Gamma^2(\kappa)\sin(\pi\kappa)\big]^{1/\kappa} \qquad (49)$$

where $c_{22} = c_{22}(\kappa)$ is defined after (47) and ψ and $c_4(\kappa)$ in (17).

To prove the upper bound, let $r_n := \exp(n^{1+\varepsilon})$ and $Z_n := \sum_{k=1}^{n} r_k$. By means of an argument similar to that leading to (48), we have $\sum_{n\ge 1} \mathbb{P}\big[H(F(r_{2n})) < t_+(r_{2n})^{1/\kappa} f_\varepsilon^+(r_{2n})\big] = +\infty$. So by Lemma 3, for $0 < \varepsilon < 1/2$, there exist almost surely infinitely many n such that

$$\inf_{u\in[(1-\varepsilon)r_{2n},(1+\varepsilon)r_{2n}]} H_{X\circ\Theta_{H(Z_{2n-1})}}(u) < [t_+(r_{2n})]^{1/\kappa} f_\varepsilon^+(r_{2n}).$$

Moreover, by Theorem 7, $H(Z_{2n-1}) < \big[Z_{2n-1}\log^2 Z_{2n-1}\big]^{1/\kappa} \le \varepsilon[t_+(r_{2n})]^{1/\kappa} f_\varepsilon^+(r_{2n})$ almost surely for all large n, since $\sum_{n\ge 1} 1/(n\log^2 n) < \infty$ and $Z_p \le p\exp(-p^\varepsilon)r_{p+1}$ for all large p as before. This yields almost surely for large n,

$$\inf_{v\in[Z_{2n-1}+(1-\varepsilon)r_{2n},\, Z_{2n-1}+(1+\varepsilon)r_{2n}]} H(v) < (1+\varepsilon)[t_+(r_{2n})]^{1/\kappa} f_\varepsilon^+(r_{2n}).$$

Consequently,

$$\liminf_{r\to+\infty} \frac{H(r)}{r^{1/\kappa}(\log\log r)^{(\kappa-1)/\kappa}} \le 8\psi(\kappa)c_{22}^{(1-\kappa)/\kappa} = c_2(\kappa) \qquad \mathbb{P}\text{-a.s.}$$

This gives inequality "$\le$" of (9) and thus yields Theorem 8 in the case $0 < \kappa < 1$.

4.2.2 Case $\kappa = 1$

Assume $\kappa = 1$ (thus $\lambda = 8$). By Samorodnitsky et al. [41, Proposition 1.2.12], $\mathbb{E}[\exp(-C_8^{ca})] = 1$ (in the notation of [41], C_8^{ca} is distributed as $S_1(8,1,0)$). Hence,

$$\mathbb{P}[C_8^{ca} \leq -\varepsilon \log r] \leq r^{-\varepsilon}\, \mathbb{E}[\exp(-C_8^{ca})] = r^{-\varepsilon}, \qquad r > 0, \tag{50}$$

for $\varepsilon > 0$. By Lemma 2 and (41), we have if $\varepsilon > 0$ is small enough, for all large r,

$$\begin{aligned}\mathbb{P}\{H[F(r)] \leq 32t_-(r)(1-2\varepsilon)[c_3 + \log t_-(r)]\} &\leq \mathbb{P}\left(C_8^{ca} \leq -\varepsilon \log r\right) + \mathbb{P}[E_3(r)^c] \\ &\leq 2r^{-\varepsilon}.\end{aligned}$$

Let $s_n := \exp(n^{1-\varepsilon})$. Thus, by the Borel-Cantelli lemma, almost surely, for all large n,

$$H[F(s_n)] > 32t_-(s_n)(1-2\varepsilon)[c_3 + \log t_-(s_n)] \geq 4(1-3\varepsilon)s_n \log s_n.$$

In view of the last part of Lemma 1, this yields inequality "$\geq$" in (10) similarly as before (49). The inequality "$\leq$", on the other hand, follows immediately from Theorem 1 (that $H(r)/(r\log r) \to 4$ in probability). Theorem 8 is proved. □

4.3 *Proof of Corollary 2*

First, we need the following lemma, which says that X does not go back too far on the left, and so $X(t)$ is very close from $\sup_{0\leq s\leq t} X(s)$:

Lemma 4 *For every $\kappa > 0$, there exists a constant $c_{23}(\kappa)$ such that $\mathbb{P}$ a.s. for large t,*

$$0 \leq \sup_{0\leq s\leq t} X(s) - X(t) \leq c_{23}(\kappa) \log t. \tag{51}$$

Notice that this is not true in the recurrent case $\kappa = 0$. An heuristic explanation for $0 \leq \kappa < 1$ would be that the valleys of height approximatively $\log t$ have a length of order $(\log t)^2$ in the case $\kappa = 0$, whereas they have a height of order at most $\log t$ in the case $0 < \kappa < 1$, see e.g. Andreoletti et al. [3, Lemma 2.7].

Proof (Proof of Lemma 4) Let $\kappa > 0$. By Kawazu et al. ([35], Theorem p. 79 applied with $c = \kappa/2$ to our $-X$), there exists a constant $c_{24}(\kappa) > 0$ such that $\mathbb{P}\big[\inf_{u\geq 0} X(u) < -c_{24}(\kappa)\log n\big] \leq 1/n^2$ for large n. Since $\inf_{u\geq 0} X(H(n)+u) - n$ has the same law under $\mathbb{P}$ as $\inf_{u\geq 0} X(u)$ due to the strong Markov property as explained before Lemma 3, this gives $\sum_n \mathbb{P}\big[\inf_{u\geq 0} X(H(n)+u) - n < -c_{24}(\kappa)\log n\big] < \infty$. So by the Borel-Cantelli lemma, almost surely for large n,

$$\inf_{u\geq 0} X(H(n)+u) - n \geq -c_{24}(\kappa)\log n. \tag{52}$$

For $t > 0$, there exists $n \in \mathbb{N}$ such that $H(n) \le t < H(n+1)$. We have by (52), almost surely if t is large,

$$\begin{aligned}\sup_{0\le s\le t} X(s) - X(t) &\le \sup_{0\le s\le H(n+1)} X(s) - X(t)\\ &= n+1 - X[H(n) + (t - H(n))]\\ &\le 1 + c_{24}(\kappa)\log n.\end{aligned}$$

Moreover, we have $\log v \le 2\log H(v)$ $\mathbb{P}$ a.s. for large v, by Theorem 1 if $\kappa > 1$ and by Theorem 8 if $0 < \kappa \le 1$. Hence almost surely for large t, with the same notation as before,

$$\sup_{0\le s\le t} X(s) - X(t) \le 1 + c_{24}(\kappa)\log n \le 1 + 2c_{24}(\kappa)\log H(n) \le 1 + 2c_{24}(\kappa)\log t.$$

This proves the second inequality of (51). The first one is clear. □

Proof (Proof of Corollary 2) By Lemma 4,

$$\limsup_{t\to\infty}\left[X(t)/(t/\log t)\right] = \limsup_{t\to\infty}\left[\Big(\sup_{0\le s\le t} X(s)\Big)/(t/\log t)\right].$$

So, (12) is equivalent to (12) with $X(t)$ replaced by $\sup_{0\le s\le t} X(s)$. The same remark also applies to (11) and (13).

Now, we have $\sup_{0\le s\le y} X(s) \ge r \iff H(r) \le y$, $r > 0$, $y > 0$. Consequently (11)–(13) with $X(t)$ replaced by $\sup_{0\le s\le t} X(s)$ follow respectively from (9), (10) and Theorem 7 applied to $a(r) = (\log r)\dots(\log_{k-1} r)(\log_k r)^{\alpha}$. Indeed for (13) when $\kappa = 1$, cases $k = 1$, $\alpha \le 1$ and $k = 2$, $\alpha \le 0$ follow from the case $k = 3$, $\alpha = 1$. This proves Corollary 2. □

5 Proof of Theorems 2–6

Proof (Proof of Theorem 4: Case $\kappa > 1$) Follows from Theorems 10 and 1. □

Proof (Proof of Theorem 3) Follows from Theorems 9 and 1. □

Proof (Proof of Theorem 6) We first notice that for every $\kappa > 0$, thanks to Lemma 2 **(i)**,

$$L_X^*[H(F(r))]/r^{1/\kappa} \xrightarrow{\mathcal{L}} 4\left[\kappa^2/\lambda\right]^{1/\kappa}\left(\sup\nolimits_{0\le u\le \tau_\beta(\lambda)} \beta(u)\right)^{1/\kappa}, \tag{53}$$

where $\xrightarrow{\mathcal{L}}$ denotes convergence in law under $\mathbb{P}$ as $r \to +\infty$.

We now assume $\kappa > 1$. In this case, $H(F(r))/r \to_{r\to+\infty} 4/(\kappa-1)$ $\mathbb{P}$-a.s. by Lemma 1 Eq. (23) and Theorem 1 Eq. (8). This and (53) lead to the convergence in law under $\mathbb{P}$ of $L_X^*(t)/t^{1/\kappa}$ to $4[\kappa^2(\kappa-1)/(4\lambda)]^{1/\kappa}\big(\sup_{0\le u\le\tau_\beta(\lambda)}\beta(u)\big)^{1/\kappa}$. Since $\sup_{0\le u\le\tau_\beta(\lambda)}\beta(u)$ has by (18) the same law as $\lambda/(2\mathscr{E})$, where $\mathscr{E}$ is an exponential variable with mean 1, this proves Theorem 6 when $\kappa > 1$.

We finally assume $\kappa = 1$. In this case, $H(F[t/(4\log t)])/t \to_{t\to+\infty} 1$ in probability under $\mathbb{P}$ by Lemma 1 and Theorem 1 Eq. (7). This, combined with (53) leads to the convergence in law of $L_X^*(t)/(t/\log t)$ to $\lambda^{-1}\sup_{0\le u\le\tau_\beta(\lambda)}\beta(u)$, which proves Theorem 6 when $\kappa = 1$. □

We now assume $0 < \kappa \le 1$, and need to prove Theorems 2, 4 and 5. Unfortunately, it follows immediately from Theorems 7 and 8 that there is no almost sure convergence result for $H(r)$ in this case due to strong fluctuations; hence a joint study of $L_X^*[H(r)]$ and $H(r)$ is useful. In Sect. 5.1, we prove a lemma which will be needed later on. Section 5.2 is devoted to the proof of Theorems 2–5 in the case $0 < \kappa < 1$, whereas Sect. 5.3 to the proof of Theorems 4 and 5 in the case $\kappa = 1$.

5.1 A Lemma

In this section we assume $0 < \kappa \le 1$. Let $\delta_1 > 0$ and recall the definitions of $t_\pm(r)$ from (25) and $\widehat{L}_\pm(r)$ from (27).

Lemma 5 *Define* $E_7(r) := \{\widehat{L}_-(r) = \widehat{L}_+(r)\}$. *For all* $\delta_2 \in (0,\delta_1)$ *and all large* r, *we have*

$$\mathbb{P}\big[E_7(r)^c\big] \le r^{-\delta_2}.$$

Proof Let $\delta_2 \in (0,\delta_1)$. Observe that

$$1 \le \left(\frac{\widehat{L}_+(r)}{\widehat{L}_-(r)}\right)^\kappa \le \max\left(1, \frac{\sup_{0\le u\le\tau_{\tilde\beta}\{[\psi_+(r)-\psi_-(r)]\kappa r\}}\tilde\beta(u)}{\sup_{0\le u\le\tau_\beta(\psi_-(r)\kappa r)}\beta(u)}\right), \tag{54}$$

where $\tilde\beta(u) := \beta[u+\tau_\beta(\psi_-(r)\kappa r)]$, $u \ge 0$, is a Brownian motion independent of the random variable $\sup_{0\le u\le\tau_\beta(\psi_-(r)\kappa r)}\beta(u)$. By (18) and the usual inequality $1-e^{-x}\le x$ (for $x \ge 0$),

$$\mathbb{P}\left(\sup_{0\le u\le\tau_{\tilde\beta}\{[\psi_+(r)-\psi_-(r)]\kappa r\}}\tilde\beta(u) > [\psi_+(r)-\psi_-(r)]\kappa r^{1+\delta_2}\right) \le \frac{1}{2r^{\delta_2}},$$

$$\mathbb{P}\left(\sup_{0\le u\le\tau_\beta(\psi_-(r)\kappa r)}\beta(u) < \frac{\psi_-(r)\kappa r}{4\delta_2\log r}\right) = \frac{1}{r^{2\delta_2}} \le \frac{1}{2r^{\delta_2}},$$

for large r. By definition, $\psi_\pm(r) = 1 \pm c_5 r^{-\delta_1}$ (see (25)). Therefore, we have for large r, with probability greater than $1 - r^{-\delta_2}$,

$$\frac{\sup_{0\le u\le \tau_{\tilde\beta}\{[\psi_+(r)-\psi_-(r)]\kappa r\}} \tilde\beta(u)}{\sup_{0\le u\le \tau_\beta(\psi_-(r)\kappa r)} \beta(u)} \le \frac{[\psi_+(r)-\psi_-(r)]\kappa r^{1+\delta_2}}{\psi_-(r)\kappa r/(4\delta_2\log r)} = \frac{8c_5\delta_2 r^{-(\delta_1-\delta_2)}\log r}{1-c_5 r^{-\delta_1}} < 1.$$

This, combined with (54), yields the lemma. □

5.2 *Case* $0 < \kappa < 1$

This section is devoted to the proof of Theorems 2, 4 and 5 in the case $0 < \kappa < 1$.

For any Brownian motion $(\beta(u),\ u \ge 0)$, let

$$N_\beta := \frac{\int_0^{+\infty} x^{1/\kappa-2} L_\beta(\tau_\beta(\lambda), x)\mathrm{d}x}{\left[\sup_{0\le u\le \tau_\beta(\lambda)} \beta(u)\right]^{1/\kappa}}.$$

So, in the notation of (15), (25) and (27), $N_{\beta_{t_\pm(r)}} = 4[\kappa t_\pm(r)]^{1/\kappa} K_{\beta_{t_\pm(r)}}(\kappa)/\widehat{L}_\pm(r)$, $r > 0$.

On $E_2(r)\cap E_3(r)\cap E_7(r)$ (the events $E_2(r)$ and $E_3(r)$ are defined in Lemma 2, whereas $E_7(r)$ in Lemma 5), we have, for some constant c_{25}, $\varepsilon > 0$ small enough and all large r,

$$\begin{aligned}\frac{H(F(r))}{L_X^*[H(F(r))]} &\ge \frac{4(1-\varepsilon)\kappa^{1/\kappa-2} t_-(r)^{1/\kappa}\{K_{\beta_{t_-(r)}}(\kappa) - c_6 t_-(r)^{1-1/\kappa}\}}{(1+\varepsilon)\widehat{L}_-(r)}\\ &\ge (1-3\varepsilon)\kappa^{-2} N_{\beta_{t_-(r)}} - c_{25} t_-(r)/\widehat{L}_-(r).\end{aligned} \tag{55}$$

Similarly, on $E_2(r)\cap E_3(r)\cap E_7(r)$, for some constant c_{26} and all large r,

$$\frac{H(F(r))}{L_X^*[H(F(r))]} \le (1+3\varepsilon)\kappa^{-2} N_{\beta_{t_+(r)}} + c_{26}\frac{t_+(r)}{\widehat{L}_+(r)}. \tag{56}$$

Define $E_8(r) := \left\{c_{25} t_-(r)/\widehat{L}_-(r) \le \varepsilon,\ c_{26} t_+(r)/\widehat{L}_+(r) \le \varepsilon\right\}$. By (35), $\mathbb{P}[E_8(r)^c] \le 1/r^2$ for large r. Thus $\mathbb{P}[E_2(r)\cap E_3(r)\cap E_7(r)\cap E_8(r)] \ge 1 - r^{-\alpha_1}$ for some $\alpha_1 > 0$ and all large r by Lemmas 2 and 5. In view of (55) and (56), we have, for some $\alpha_1 > 0$ and all large r,

$$\mathbb{P}\left((1-3\varepsilon)\kappa^{-2} N_{\beta_{t_-(r)}} - \varepsilon \le \frac{H(F(r))}{L_X^*[H(F(r))]} \le (1+3\varepsilon)\kappa^{-2} N_{\beta_{t_+(r)}} + \varepsilon\right) \ge 1 - \frac{1}{r^{\alpha_1}}. \tag{57}$$

We now proceed to the study of the law of N_β. By the second Ray-Knight theorem (Fact 2), there exists a 0-dimensional Bessel process $(U(x),\ x \geq 0)$, starting from $\sqrt{\lambda}$, such that

$$(L_\beta(\tau_\beta(\lambda), x), x \geq 0) = \big(U^2(x), x \geq 0\big), \tag{58}$$

$$\sup_{0 \leq u \leq \tau_\beta(\lambda)} \beta(u) = \inf\{x \geq 0,\ U(x) = 0\} =: \zeta_U, \tag{59}$$

$$N_\beta = \zeta_U^{-1/\kappa} \int_0^{\zeta_U} x^{1/\kappa - 2} U^2(x) \mathrm{d}x. \tag{60}$$

By Williams' time reversal theorem (Fact 4), there exists a 4-dimensional Bessel process $(R(s),\ s \geq 0)$, starting from 0, such that

$$(U(\zeta_U - s),\ s \leq \zeta_U) \stackrel{\mathcal{L}}{=} (R(s),\ s \leq \gamma_a),\ a := \sqrt{\lambda},\ \gamma_a := \sup\{s \geq 0,\ R(s) = \sqrt{\lambda}\}. \tag{61}$$

Therefore,

$$N_\beta \stackrel{\mathcal{L}}{=} \gamma_a^{-1/\kappa} \int_0^{\gamma_a} x^{1/\kappa-2} R^2(\gamma_a - x) \mathrm{d}x = \int_0^1 (1-v)^{1/\kappa-2} \left(\frac{R(\gamma_a v)}{\sqrt{\gamma_a}}\right)^2 \mathrm{d}v.$$

Recall (Yor [52], p. 52) that for any bounded measurable functional G,

$$\mathbb{E}\left[G\left(\frac{R(\gamma_a u)}{\sqrt{\gamma_a}}, u \leq 1\right)\right] = \mathbb{E}\left(\frac{2}{R^2(1)} G\big(R(u), u \leq 1\big)\right). \tag{62}$$

In particular, for $x > 0$,

$$\mathbb{P}\big(N_\beta > x\big) = \mathbb{E}\left(\frac{2}{R^2(1)} \mathbf{1}_{\{\int_0^1 (1-v)^{1/\kappa-2} R^2(v) \mathrm{d}v > x\}}\right). \tag{63}$$

5.2.1 Proof of Theorem 5 (Case $0 < \kappa < 1$)

Fix $y > 0$. By (63), for $r > 1$,

$$\begin{aligned} \mathbb{P}(N_\beta > y \log\log r) &\leq \mathbb{E}\left(\frac{2}{R^2(1)} \mathbf{1}_{\{\int_0^1 (1-v)^{1/\kappa-2} R^2(v) \mathrm{d}v > y \log\log r,\ R^2(1) \leq 1\}}\right) \\ &\quad + 2\mathbb{P}\left(\int_0^1 (1-v)^{1/\kappa-2} R^2(v) \mathrm{d}v > y \log\log r\right) \\ &:= \Pi_1(r) + \Pi_2(r) \end{aligned} \tag{64}$$

with obvious notation.

We first consider $\Pi_2(r)$. Let $\mathscr{H} := \{(t \in [0,1] \mapsto \int_0^t f(s)\mathrm{d}s),\ f \in L^2([0,1],\mathbb{R}^4)\}$. As R is the Euclidean norm of a 4-dimensional Brownian motion $(\gamma(t),\ t \geq 0)$, we have by Schilder's theorem (see e.g. Dembo and Zeitouni [17], Theorem 5.2.3),

$$\begin{aligned}
&\lim_{r\to+\infty} \frac{1}{y\log\log r}\log \mathbb{P}\left(\int_0^1 (1-v)^{1/\kappa-2}R^2(v)\mathrm{d}v > y\log\log r\right)\\
&= -\inf\left\{\frac{1}{2}\int_0^1 \|\phi'(v)\|^2\mathrm{d}v :\quad \phi\in\mathscr{H},\ \int_0^1 (1-v)^{1/\kappa-2}\|\phi(v)\|^2\mathrm{d}v \geq 1\right\}\\
&=: -c_1(\kappa),
\end{aligned} \tag{65}$$

where $\|\cdot\|$ denotes the Euclidean norm. For $\phi\in\mathscr{H}$, $\|\phi(v)\|^2 = \left\|\int_0^v \phi'(u)\mathrm{d}u\right\|^2 \leq v\int_0^1 \|\phi'(u)\|^2\mathrm{d}u$, where we applied Cauchy-Schwarz to each coordinate; thus $\int_0^1 (1-v)^{1/\kappa-2}\|\phi(v)\|^2\mathrm{d}v \leq \left[\int_0^1 (1-v)^{1/\kappa-2}v\mathrm{d}v\right]\int_0^1 \|\phi'(v)\|^2\mathrm{d}v$. So, $c_1(\kappa)\in(0,\infty)$.

By (65), for $0<\varepsilon<1$ and large r,

$$\Pi_2(r) \leq (\log r)^{-(1-\varepsilon)yc_1(\kappa)}. \tag{66}$$

Now, we consider $\Pi_1(r)$. As R is the Euclidean norm of a 4-dimensional Brownian motion $(\gamma(t),\ t\geq 0)$, we have

$$\Pi_1(r) = \mathbb{E}\left(\frac{2}{\|\gamma(1)\|^2}\mathbf{1}_{\{\|\gamma(1)\|\leq 1\}}\mathbf{1}_{\{\int_0^1 (1-v)^{1/\kappa-2}\|\gamma(v)\|^2\mathrm{d}v > y\log\log r\}}\right).$$

By the triangular inequality, for any finite positive measure μ on $[0,1]$,

$$\sqrt{\int_0^1 \|\gamma(v)\|^2\mathrm{d}\mu(v)} \leq \sqrt{\int_0^1 \|\gamma(v)-v\gamma(1)\|^2\mathrm{d}\mu(v)} + \sqrt{\int_0^1 v^2\mathrm{d}\mu(v)}\ \|\gamma(1)\|.$$

Therefore, applying this to $\mathrm{d}\mu(v) = (1-v)^{1/\kappa-2}\mathrm{d}v$, we have for large r,

$$\begin{aligned}
\Pi_1(r) &\leq \mathbb{E}\left(\frac{2}{\|\gamma(1)\|^2}\mathbf{1}_{\{\int_0^1 (1-v)^{1/\kappa-2}\|\gamma(v)-v\gamma(1)\|^2\mathrm{d}v > (\sqrt{y\log\log r}-c_{27})^2\}}\right)\\
&:= \mathbb{E}\left(\frac{2}{\|\gamma(1)\|^2}\mathbf{1}_E\right),
\end{aligned}$$

where $c_{27} := \sqrt{\int_0^1 v^2(1-v)^{1/\kappa-2}\mathrm{d}v}$. By the independence of $\gamma(1)$ and $(\gamma(v)-v\gamma(1),\ v\in[0,1])$, the expectation on the right hand side is $= \mathbb{E}(\frac{2}{\|\gamma(1)\|^2})\mathbb{P}(E) = \mathbb{P}(E)$ (the last identity being a consequence of (62) by taking $G=1$ there). Therefore, $\Pi_1(r) \leq \mathbb{P}(E)$.

Again, by the independence of $\gamma(1)$ and $(\gamma(v) - v\gamma(1),\ v \in [0,1])$, we see that, by writing $c_{28} := 1/\mathbb{P}(\|\gamma(1)\| \leq 1)$, $\Pi_1(r) \leq c_{28}\,\mathbb{P}(E,\ \|\gamma(1)\| \leq 1)$. By another application of the triangular inequality, this leads to, for large r:

$$\Pi_1(r) \leq c_{28}\,\mathbb{P}\left(\int_0^1 (1-v)^{1/\kappa-2}\|\gamma(v)\|^2 \mathrm{d}v > \left(\sqrt{y\log\log r} - 2c_{27}\right)^2\right).$$

In view of (65), we have, for all large r, $\Pi_1(r) \leq (\log r)^{-(1-\varepsilon)yc_1(\kappa)}$. Plugging this into (64) and (66) yields that, for any $y > 0$, $\varepsilon > 0$ and all large r,

$$\mathbb{P}(N_\beta > y\log\log r) \leq 2(\log r)^{-(1-\varepsilon)yc_1(\kappa)}. \tag{67}$$

Let $0 < \varepsilon < 1/2$, and $s_n := \exp(n^{1-\varepsilon})$. We get

$$\sum_{n=1}^{+\infty} \mathbb{P}\Big\{\frac{H(F(s_n))}{L_X^*[H(F(s_n))]} > \frac{(1+4\varepsilon)\log\log s_n}{(1-\varepsilon)^3\kappa^2 c_1(\kappa)}\Big\} < \infty$$

due to (57) and (67). By the Borel–Cantelli lemma, almost surely, for all large n,

$$\frac{H(F(s_n))}{L_X^*[H(F(s_n))]} \leq \frac{1+4\varepsilon}{(1-\varepsilon)^3\kappa^2 c_1(\kappa)}\log\log s_n. \tag{68}$$

We now bound $\frac{H(F(s_{n+1}))}{H(F(s_n))}$. Observe that for large n, $s_{n+1} - s_n \leq n^{-\varepsilon}s_n$. By Lemma 1, almost surely for all large n,

$$\begin{aligned}
&H[F(s_{n+1})] - H[F(s_n)] \\
&\leq H\big[\big(1 + 5s_{n+1}^{-\delta_0}/\kappa\big)s_{n+1}\big] - H\big[\big(1 - 5s_n^{-\delta_0}/\kappa\big)s_n\big] \\
&\leq H\big[\big(1 - 5s_n^{-\delta_0}/\kappa\big)s_n + (2-\varepsilon)n^{-\varepsilon}s_n\big] - H\big[\big(1 - 5s_n^{-\delta_0}/\kappa\big)s_n\big] \\
&= \inf\Big\{u \geq 0 :\ \widehat{X}_n(u) > (2-\varepsilon)n^{-\varepsilon}s_n\Big\},
\end{aligned} \tag{69}$$

where $\big(\widehat{X}_n(u),\ u \geq 0\big)$ is, conditionally on W_κ, a diffusion process in the random potential $\widehat{W}_\kappa(x) := W_\kappa\big[x + \big(1 - \frac{5}{\kappa}s_n^{-\delta_0}\big)s_n\big] - W_\kappa\big[\big(1 - \frac{5}{\kappa}s_n^{-\delta_0}\big)s_n\big]$, $x \in \mathbb{R}$, starting from 0. We denote by $\widehat{H}_n(r)$ the hitting time of $r \geq 0$ by $\widehat{X}_n$, so that

$$\inf\Big\{u \geq 0 :\ \widehat{X}_n(u) > (2-\varepsilon)n^{-\varepsilon}s_n\Big\} = \widehat{H}_n\big[(2-\varepsilon)n^{-\varepsilon}s_n\big]. \tag{70}$$

Note that for any $r > 0$, under $\mathbb{P}$, $\widehat{H}_n(r)$ is distributed as $H(r)$. Therefore, applying (42) and Lemma 1 to $r = 2n^{-\varepsilon}s_n$ yields that, for any $0 < \delta_0 < \frac{1}{2}$,

$$\sum_n \mathbb{P}\Big[\widehat{H}_n\left(\big[1 - 5(2n^{-\varepsilon}s_n)^{-\delta_0}/\kappa\big]2n^{-\varepsilon}s_n\right) > \big[n(\log n)^{1+\varepsilon}t_+(2n^{-\varepsilon}s_n)\big]^{1/\kappa}\Big] < \infty.$$

Since $\left[1 - \frac{5}{\kappa}(2n^{-\varepsilon}s_n)^{-\delta_0}\right]2n^{-\varepsilon}s_n \geq (2-\varepsilon)n^{-\varepsilon}s_n$ (for large n), it follows from the Borel–Cantelli lemma that, almost surely for all large n, $\widehat{H}_n[(2-\varepsilon)n^{-\varepsilon}s_n] \leq \left[n(\log n)^{1+\varepsilon}t_+(2n^{-\varepsilon}s_n)\right]^{1/\kappa}$. This, together with (69) and (70), yields that, almost surely for all large n,

$$H[F(s_{n+1})] - H[F(s_n)] \leq \left[n(\log n)^{1+\varepsilon}t_+(2n^{-\varepsilon}s_n)\right]^{1/\kappa} \leq c_{29}\left[n^{1-\varepsilon}(\log n)^{1+\varepsilon}s_n\right]^{1/\kappa}.$$

Recall from Lemma 1 and Theorem 8 that, almost surely, for all large n, $H[F(s_n)] \geq H[(1-\varepsilon)s_n] \geq \frac{c_{30}s_n^{1/\kappa}}{(\log\log s_n)^{1/\kappa-1}}$, which yields

$$\frac{H[F(s_{n+1})]}{H[F(s_n)]} \leq 1 + \frac{c_{29}[n^{1-\varepsilon}(\log n)^{1+\varepsilon}s_n]^{1/\kappa}}{c_{30}s_n^{1/\kappa}/(\log\log s_n)^{1/\kappa-1}} \leq c_{31}(\log s_n)^{1/\kappa}(\log\log s_n)^{(2+\varepsilon)/\kappa-1}.$$

In view of (68), this gives, almost surely, for large n and $t \in [H(F(s_n)),\, H(F(s_{n+1}))]$,

$$\frac{t}{L_X^*(t)} \leq \frac{H[F(s_n)]}{L_X^*[H(F(s_n))]}\,\frac{H[F(s_{n+1})]}{H[F(s_n)]} < c_{32}(\log s_n)^{1/\kappa}(\log\log s_n)^{(2+\varepsilon)/\kappa}.$$

Since, almost surely for all large n, $\log H[F(s_n)] \geq \log H[(1-\varepsilon)s_n] \geq \frac{1-\varepsilon}{\kappa}\log s_n$ (this is seen first by Lemma 1, and then by Theorem 8), we have proved that

$$\liminf_{t\to+\infty} \frac{L_X^*(t)}{t(\log t)^{-1/\kappa}(\log\log t)^{-(2+\varepsilon)/\kappa}} \geq c_{33} \qquad \mathbb{P}\text{-a.s.}$$

Since $\varepsilon \in (0, \frac{1}{2})$ is arbitrary, this proves Theorem 5 in the case $0 < \kappa < 1$. □

5.2.2 Proof of Theorem 4 (Case $0 < \kappa < 1$)

By (63), for any $s > 0$ and $u > 0$,

$$\begin{aligned}\mathbb{P}(N_\beta > s) &\geq \frac{2}{u}\,\mathbb{P}\left(\int_0^1 (1-v)^{1/\kappa-2}R^2(v)\mathrm{d}v > s,\ R^2(1) \leq u\right)\\ &\geq \frac{2}{u}\,\mathbb{P}\left(\int_0^1 (1-v)^{1/\kappa-2}R^2(v)\mathrm{d}v > s\right) - \frac{2}{u}\,\mathbb{P}\left(R^2(1) > u\right).\end{aligned}$$

The first probability term on the right hand side is taken care of by (65), whereas for the second, we have $\frac{1}{u}\log\mathbb{P}(R^2(1) > u) \to -\frac{1}{2}$, for $u \to \infty$, since $R^2(1)$ has a chi-squared distribution with 4 degrees of freedom. Taking $u := \exp(\sqrt{\log\log r})$ leads to: for any $y > 0$,

$$\liminf_{r\to\infty} \frac{\log\mathbb{P}(N_\beta > y\log\log r)}{\log\log r} \geq -yc_1(\kappa).$$

Plugging this into (57) yields that, for $r_n := \exp(n^{1+\varepsilon})$,

$$\sum_{n\geq 1} \mathbb{P}\left(\frac{(H \circ F)(r_{2n})}{(L_X^* \circ H \circ F)(r_{2n})} > \frac{(1-3\varepsilon)\log\log r_{2n}}{\kappa^2 c_1(\kappa)(1+\varepsilon)^3} - \varepsilon \right) = +\infty.$$

Let $Z_n := \sum_{k=1}^n r_k$. By Lemma 3 (in its notation), almost surely, for infinitely many n,

$$\sup_{u\in[(1-\varepsilon)r_{2n},(1+\varepsilon)r_{2n}]} \frac{H_{X\circ\Theta_{H(Z_{2n-1})}}(u)}{(L^*\circ H)_{X\circ\Theta_{H(Z_{2n-1})}}(u)} > \frac{(1-8\varepsilon)\log\log r_{2n}}{\kappa^2 c_1(\kappa)}, \tag{71}$$

if $\varepsilon > 0$ is small enough. Observe that

$$(L^*\circ H)_{X\circ\Theta_{H(Z_{2n-1})}}(u) = \sup_{x\in\mathbb{R}} L_{\widetilde{X}_n}\big(\widetilde{H}_n(u), x\big) =: L^*_{\widetilde{X}_n}\big(\widetilde{H}_n(u)\big), \tag{72}$$

where $\big(\widetilde{X}_n(v),\ v\geq 0\big)$ is a diffusion process in the random potential $W_\kappa(x+Z_{2n-1})-W_\kappa(Z_{2n-1})$, $x\in\mathbb{R}$, $\big(L_{\widetilde{X}_n}(t,x),\ t\geq 0,\ x\in\mathbb{R}\big)$ is its local time and $\widetilde{H}_n(r) := \inf\big\{t > 0,\ \widetilde{X}_n(t) > r\big\}$, $r > 0$. Hence, for any $u > 0$, under $\mathbb{P}$, the left hand side of (72) is distributed as $L_X^*(H(u))$. Applying (38) and Lemma 1 to $\tilde{r}_{2n} := (1-\varepsilon)^2 r_{2n}$, there exists $c_{34} > 0$ such that

$$\sum_n \mathbb{P}\left[L^*_{\widetilde{X}_n}\big(\widetilde{H}_n\big[\big(1+5(\tilde{r}_{2n})^{-\delta_0}/\kappa\big)\tilde{r}_{2n}\big]\big) < c_{34}[r_{2n}/\log\log r_{2n}]^{1/\kappa}\right] < \infty.$$

Since $\big(1+\frac{5}{\kappa}(\tilde{r}_{2n})^{-\delta_0}\big)\tilde{r}_{2n} \leq (1-\varepsilon)r_{2n}$ for large n, the Borel-Cantelli lemma gives that, almost surely, for all large n,

$$c_{34}\big[r_{2n}/\log\log r_{2n}\big]^{1/\kappa} \leq L^*_{\widetilde{X}_n}\big(\widetilde{H}_n[(1-\varepsilon)r_{2n}]\big) \leq L^*_{\widetilde{X}_n}\big(\widetilde{H}_n(u)\big) \tag{73}$$

for any $u \in [(1-\varepsilon)r_{2n}, (1+\varepsilon)r_{2n}]$. Applying Theorem 9, we have almost surely for large n,

$$L_X^*[H(Z_{2n-1})] \leq [Z_{2n-1}\log^2 Z_{2n-1}]^{1/\kappa} \leq \varepsilon[r_{2n}/\log\log r_{2n}]^{1/\kappa} \leq (\varepsilon/c_{34})L^*_{\widetilde{X}_n}\big(\widetilde{H}_n(u)\big)$$

for $u \in [(1-\varepsilon)r_{2n}, (1+\varepsilon)r_{2n}]$, since $Z_k \leq k\exp(-k^\varepsilon)r_{k+1}$ for large k. Hence,

$$L_X^*[H(Z_{2n-1}+u)] \leq (1+\varepsilon/c_{34})L^*_{\widetilde{X}_n}\big(\widetilde{H}_n(u)\big). \tag{74}$$

On the other hand, we have by Theorem 7, almost surely, for all large n,

$$\log\log r_{2n} \geq (1-\varepsilon)\log\log H(Z_{2n-1}+u), \qquad u\in[(1-\varepsilon)r_{2n}, (1+\varepsilon)r_{2n}].$$

Consequently, almost surely for infinitely many n, by (74) and (71),

$$\inf_{v\in[Z_{2n-1}+(1-\varepsilon)r_{2n},Z_{2n-1}+(1+\varepsilon)r_{2n}]}\frac{L_X^*[H(v)]}{H(v)/\log\log H(v)}$$

$$\leq (1+c_{35}\varepsilon)\inf_{u\in[(1-\varepsilon)r_{2n},(1+\varepsilon)r_{2n}]}\frac{(L^*\circ H)_{X\circ\Theta_{H(Z_{2n-1})}}(u)}{H_{X\circ\Theta_{H(Z_{2n-1})}}(u)/\log\log r_{2n}}\leq(1+c_{36}\varepsilon)\kappa^2c_1(\kappa),$$

proving Theorem 4 in the case $0<\kappa<1$. □

5.2.3 Proof of Theorem 2

Assume $0<\kappa<1$. Fix $x>0$, and let $r_n:=\exp(n^{1+\varepsilon})$. Since $\mathbb{P}(N_\beta<x)>0$, (57) implies $\sum_{n\in\mathbb{N}}\mathbb{P}\left(\frac{(H\circ F)(r_{2n})}{(L_X^*\circ H\circ F)(r_{2n})}<\frac{(1+3\varepsilon)x}{\kappa^2}+\varepsilon\right)=+\infty$. By Lemma 3, for small $\varepsilon>0$, almost surely for infinitely many n,

$$\inf_{u\in[(1-\varepsilon)r_{2n},\ (1+\varepsilon)r_{2n}]}\frac{H_{X\circ\Theta_{H(Z_{2n-1})}}(u)}{(L^*\circ H)_{X\circ\Theta_{H(Z_{2n-1})}}(u)}<\frac{(1+3\varepsilon)x}{\kappa^2}+\varepsilon. \tag{75}$$

With the same notation as in (72), $H_{X\circ\Theta_{H(Z_{2n-1})}}(u)=H(Z_{2n-1}+u)-H(Z_{2n-1})$ is the hitting time $\widetilde{H}_n(u)$ of u by the diffusion $\widetilde{X}_n$. For any u, under $\mathbb{P}$, it has the same distribution as $H(u)$. Hence, applying (48) and Lemma 1 to $\tilde{r}_{2n}=(1-\varepsilon)^2r_{2n}$ leads to (for $0<\delta_0<1/2$)

$$\sum_n\mathbb{P}\left[\widetilde{H}_n\left(\left(1+\frac{5}{\kappa}(\tilde{r}_{2n})^{-\delta_0}\right)\tilde{r}_{2n}\right)<r_{2n}^{1/\kappa}/\log r_{2n}\right]<\infty.$$

Since $\left(1+\frac{5}{\kappa}(\tilde{r}_{2n})^{-\delta_0}\right)\tilde{r}_{2n}<(1-\varepsilon)r_{2n}$ for large n, it follows from the Borel-Cantelli lemma that, almost surely, for all large n,

$$\frac{r_{2n}^{1/\kappa}}{\log r_{2n}}\leq\widetilde{H}_n[(1-\varepsilon)r_{2n}]\leq\inf_{u\in[(1-\varepsilon)r_{2n},(1+\varepsilon)r_{2n}]}H_{X\circ\Theta_{H(Z_{2n-1})}}(u). \tag{76}$$

On the other hand, by Theorem 7, $H(Z_{2n-1})\leq[Z_{2n-1}\log^2Z_{2n-1}]^{1/\kappa}\leq\varepsilon\frac{r_{2n}^{1/\kappa}}{\log r_{2n}}$ almost surely, for all large n. This and (76) give, for $u\in[(1-\varepsilon)r_{2n},\ (1+\varepsilon)r_{2n}]$, $H(Z_{2n-1}+u)\leq(1+\varepsilon)H_{X\circ\Theta_{H(Z_{2n-1})}}(u)$. Plugging this into (75) yields that, almost surely, for infinitely many n,

$$\inf_{u\in[(1-\varepsilon)r_{2n},\ (1+\varepsilon)r_{2n}]}\frac{H(Z_{2n-1}+u)}{L_X^*(H(Z_{2n-1}+u))}<\frac{(1+\varepsilon)(1+3\varepsilon)x}{\kappa^2}+\varepsilon(1+\varepsilon).$$

Hence $\limsup_{t\to+\infty}\frac{L_X^*(t)}{t}\geq\frac{\kappa^2}{x}$, a.s. Sending $x\to 0$ completes the proof of Theorem 2. □

5.3 *Case* $\kappa = 1$

This section is devoted to the proofs of Theorems 4 and 5 in the case $\kappa = 1$ (thus $\lambda = 8$; since $\lambda = 4(1+\kappa)$). Let

$$N_\beta(t) := \frac{1}{\sup_{0\leq u\leq \tau_\beta(8)}\beta(u)}\left[\int_0^1 \frac{L_\beta(\tau_\beta(8),x)-8}{x}\mathrm{d}x + \int_1^{+\infty}\frac{L_\beta(\tau_\beta(8),x)}{x}\mathrm{d}x + 8\log t\right].$$

Exactly as in (57), we have, for some $\alpha_1>0$, any $\varepsilon\in(0,1/3)$, and all large r,

$$\mathbb{P}\left((1-3\varepsilon)N_{\beta_{t_-(r)}}[t_-(r)]\leq\frac{H(F(r))}{L_X^*[H(F(r))]}\leq(1+3\varepsilon)N_{\beta_{t_+(r)}}[t_+(r)]\right)\geq 1-\frac{1}{r^{\alpha_1}}, \tag{77}$$

where $t_\pm(\cdot)$ are defined in (25), and C_β in (16). (Compared to (57), we no longer have the extra "$\pm\varepsilon$" terms, since they are already taken care of by the presence of $8\log t$ in the definition of $N_\beta(t)$).

With the same notation as in (58) and (59), the second Ray-Knight theorem (Fact 2) gives

$$N_\beta(t) = \frac{1}{\zeta_U}\left[\int_0^1\frac{U^2(x)-8}{x}\mathrm{d}x + \int_1^{+\infty}\frac{U^2(x)}{x}\mathrm{d}x + 8\log t\right] \tag{78}$$

$$= \frac{1}{\zeta_U}\left[\int_0^{\zeta_U}\frac{U^2(x)-8}{x}\mathrm{d}x + 8\log\zeta_U + 8\log t\right], \tag{79}$$

since $U(x)=0$ for every $x\geq\zeta_U$.

5.3.1 Proof of Theorem 5 (Case $\kappa = 1$)

We have $\lambda = 8$ in the case $\kappa = 1$. Since $\sup_{x>0}\frac{\log x}{x}<\infty$, we have

$$N_\beta(t)\leq c_{37}+\frac{1}{\zeta_U}\int_0^{\zeta_U}\frac{|U^2(x)-8|}{x}\mathrm{d}x+\frac{8\log t}{\zeta_U}. \tag{80}$$

We claim that for some constant $c_{38} > 0$,

$$\limsup_{y\to+\infty} \frac{1}{y} \log \mathbb{P}\left(\frac{1}{\zeta_U}\int_0^{\zeta_U} \frac{|U^2(x)-8|}{x}\mathrm{d}x > y\right) \le -c_{38}. \tag{81}$$

Indeed, $\zeta_U = \sup_{0\le u\le \tau_\beta(8)} \beta(u)$ by definition (see (59)), which, in view of (18), implies that $\mathbb{P}(\zeta_U > z) = 1 - e^{-4/z} \le 4/z$ for $z > 0$. Therefore, if we write $p(y)$ for the probability expression at (81), we have, for any $z > 0$,

$$p(y) \le \frac{4}{z} + \mathbb{P}\left(\frac{1}{\zeta_U}\int_0^{\zeta_U} \frac{|U^2(x)-8|}{x}\mathrm{d}x > y,\ \zeta_U \le z\right).$$

In the notation of (61)–(62), this yields

$$\begin{aligned} p(y) &\le \frac{4}{z} + \mathbb{P}\left(\frac{1}{\gamma_a}\int_0^1 \frac{|R^2(\gamma_a v)-8|}{1-v}\mathrm{d}v > y,\ \gamma_a \le z\right)\\ &= \frac{4}{z} + \mathbb{E}\left(\frac{2}{R^2(1)}\mathbf{1}_{\{\int_0^1 \frac{|R^2(v)-R^2(1)|}{1-v}\mathrm{d}v > y,\ R^2(1)\ge 8/z\}}\right)\\ &\le \frac{4}{z} + \frac{z}{4}\mathbb{P}\left(\int_0^1 \frac{|R^2(v)-R^2(1)|}{1-v}\mathrm{d}v > y\right). \end{aligned} \tag{82}$$

In order to apply Schilder's theorem as in (65), let $\phi \in \mathscr{H}$. As before between (65) and (66), we have $\|\phi(t)\| \le \sqrt{t}\big[\int_0^1 \|\phi'(s)\|^2\mathrm{d}s\big]^{1/2}$. Similarly, $\big|\|\phi(u)\| - \|\phi(1)\|\big| \le \|\phi(u)-\phi(1)\| \le \sqrt{1-u}\big[\int_0^1 \|\phi'(s)\|^2\mathrm{d}s\big]^{1/2}$. Hence,

$$\begin{aligned} \int_0^1 \frac{\big|\|\phi(u)\|^2 - \|\phi(1)\|^2\big|}{1-u}\mathrm{d}u &= \int_0^1 \frac{\big|\|\phi(u)\| - \|\phi(1)\|\big|}{1-u}\big[\|\phi(u)\| + \|\phi(1)\|\big]\mathrm{d}u\\ &\le 2\left(\int_0^1 \frac{\mathrm{d}u}{\sqrt{1-u}}\right)\int_0^1 \|\phi'(s)\|^2\mathrm{d}s. \end{aligned}$$

Consequently,

$$c_{39} := \inf\left\{\frac{1}{2}\int_0^1 \|\phi'(u)\|^2\mathrm{d}u :\ \phi\in\mathscr{H},\ \int_0^1 \frac{\big|\|\phi(u)\|^2 - \|\phi(1)\|^2\big|}{1-u}\mathrm{d}u > 1\right\} \in (0,\infty).$$

Applying Schilder's theorem gives $\limsup_{y\to+\infty} \frac{1}{y}\log \mathbb{P}\left(\int_0^1 \frac{|R^2(v)-R^2(1)|}{1-v}\mathrm{d}v > y\right) \le -c_{39}$. Plugging this into (82), and taking $z = \exp(\frac{c_{39}}{2}y)$ there, we obtain the claimed inequality in (81), with $c_{38} := c_{39}/2$.

On the other hand, by (18) and (59),

$$\mathbb{P}\left(\frac{8\log t}{\zeta_U} > 2(1+2\varepsilon)(\log t)\log\log t\right) = \frac{1}{(\log t)^{1+2\varepsilon}}.$$

This, combined with (80) and (81) gives, for all large t,

$$\mathbb{P}\left\{N_\beta(t) > 2(1+3\varepsilon)(\log t)\log\log t\right\} \le \frac{2}{(\log t)^{1+2\varepsilon}}.$$

Let $s_n := \exp(n^{1-\varepsilon})$. By (77), we have

$$\sum_{n=1}^{+\infty} \mathbb{P}\left(\frac{H(F(s_n))}{L_X^*[H(F(s_n))]} > 2(1+3\varepsilon)^2(\log s_n)\log\log s_n\right) < \infty,$$

which, by means of the Borel–Cantelli lemma, implies that, almost surely, for all large n,

$$\frac{H(F(s_n))}{L_X^*[H(F(s_n))]} \le 2(1+3\varepsilon)^2(\log s_n)\log\log s_n. \tag{83}$$

Now we give an upper bound for $\frac{H(F(s_{n+1}))}{H(F(s_n))}$. By Lemma 1, almost surely for n large enough, $F(s_n) \ge (1-\varepsilon)s_n$. An application of Theorem 8 yields that, almost surely, for large n,

$$H[F(s_n)] \ge H[(1-\varepsilon)s_n] \ge 4(1-2\varepsilon)s_n \log s_n. \tag{84}$$

With the same notation and the same arguments as in (69) and (70), almost surely for all large n, $H[F(s_{n+1})] - H[F(s_n)] \le \widehat{H}_n[(2-\varepsilon)n^{-\varepsilon}s_n]$. Moreover, $\widehat{H}_n(r)$ is distributed as $H(r)$ under $\mathbb{P}$ for any $r > 0$. Hence, applying Lemma 1 and (46) to $r = \tilde{s}_n := 2n^{-\varepsilon}s_n$ and $a(e^{-2}\tilde{s}_n) = 8n(\log n)^{1+\varepsilon}$ for $0 < \delta_0 < \frac{1}{2}$, we get

$$\sum_n \mathbb{P}\left[\widehat{H}_n\left((1-5(\tilde{s}_n)^{-\delta_0}/\kappa)\widetilde{s_n}\right) > 32(1+\varepsilon)t_+(\tilde{s}_n)\left[c_3 + n(\log n)^{1+\varepsilon} + \log t_+(\tilde{s}_n)\right]\right]$$
$$< \infty.$$

Since $[1 - \frac{5}{\kappa}(\tilde{s}_n)^{-\delta_0}]\tilde{s}_n \ge (2-\varepsilon)n^{-\varepsilon}s_n$ (for large n), the Borel-Cantelli lemma yields that

$$\widehat{H}_n\left((2-\varepsilon)n^{-\varepsilon}s_n\right) \le 32(1+\varepsilon)t_+(2n^{-\varepsilon}s_n)[c_3 + n(\log n)^{1+\varepsilon} + \log t_+(2n^{-\varepsilon}s_n)],$$

almost surely for large n. Hence, $H[F(s_{n+1})] - H[F(s_n)] \le c_{39}s_n(\log s_n)(\log n)^{1+\varepsilon}$. Hence, by (84), we have, almost surely, for all large n,

$$H[F(s_{n+1})]/H[F(s_n)] \le c_{40}(\log\log s_n)^{1+\varepsilon}.$$

Let $t \in [H(F(s_n)), H(F(s_{n+1}))]$. By (83), if t is large enough,

$$\frac{t}{L_X^*(t)} \leq \frac{H[F(s_n)]}{L_X^*[H(F(s_n))]} \frac{H[F(s_{n+1})]}{H[F(s_n)]} < 3c_{40}(\log s_n)(\log\log s_n)^{2+\varepsilon}.$$

Since almost surely for large n, $\log H[F(s_n)] \geq \log H[(1-\varepsilon)s_n] \geq \log s_n$ (by Lemma 1 and Theorem 8), this yields

$$\liminf_{t\to+\infty} \frac{L_X^*(t)}{t/[(\log t)(\log\log t)^{2+\varepsilon}]} \geq \frac{1}{3c_{40}} \qquad \mathbb{P}\text{-a.s.}$$

Theorem 5 is proved in the case $\kappa = 1$. □

5.3.2 Proof of Theorem 4 (Case $\kappa = 1$)

Again, $\lambda = 8$. Let $0 < \varepsilon < 1/2$. Recall that $\zeta_U = \sup_{0\leq u\leq \tau_\beta(8)} \beta(u)$, and that $N_\beta(t) = \frac{1}{\zeta_U}\left[\int_0^{\zeta_U} \frac{U^2(x)-8}{x}\mathrm{d}x + 8\log\zeta_U + 8\log t\right]$ (see (59) and (79)). This time, we need to bound $N_\beta(t)$ from below. By (18) for large z,

$$\mathbb{P}\left(8\frac{\log\zeta_U}{\zeta_U} < -z\right) \leq \mathbb{P}\left(-\frac{1}{\zeta_U^2} < -z\right) = \mathbb{P}\left(\zeta_U < \frac{1}{\sqrt{z}}\right) = \exp\left(-4\sqrt{z}\right).$$

By (18) again,

$$\mathbb{P}\left(\frac{8\log t}{\zeta_U} > 2(1-\varepsilon)(\log t)\log\log t\right) = \frac{1}{(\log t)^{1-\varepsilon}}.$$

On the other hand, for all large y, $\mathbb{P}\big(\frac{1}{\zeta_U}\int_0^{\zeta_U} \frac{|U^2(x)-8|}{x}\mathrm{d}x > y\big) \leq e^{-c_{41}y}$ (see (81)). Assembling these pieces yields that, for all large t,

$$\mathbb{P}[N_\beta(t) > 2(1-2\varepsilon)(\log t)\log\log t] \geq \frac{1}{2(\log t)^{1-\varepsilon}}.$$

Let $r_n := \exp(n^{1+\varepsilon})$. In view of (77) and Lemma 3, we get almost surely for infinitely many n,

$$\sup_{u\in[(1-\varepsilon)r_{2n},\,(1+\varepsilon)r_{2n}]} \frac{H_{X\circ\Theta_{H(Z_{2n-1})}}(u)}{(L^*\circ H)_{X\circ\Theta_{H(Z_{2n-1})}}(u)} > 2(1-2\varepsilon)(1-3\varepsilon)(\log r_{2n})\log\log r_{2n}. \tag{85}$$

The expression on the left hand side of (85) is "close to" $H(r_{2n})/L_X^*[H(r_{2n})]$, but we need to prove this rigorously. With the same argument as in the displays between (72) and (73), we get that there exists $c_{42} > 0$ such that, almost surely for large n,

$$\inf_{u\in[(1-\varepsilon)r_{2n},(1+\varepsilon)r_{2n}]}(L^*\circ H)_{X\circ\Theta_{H(Z_{2n-1})}}(u) \geq c_{42}r_{2n}/\log\log r_{2n}.$$

Observe that $Z_k \leq k\exp(-k^{\varepsilon})r_{k+1}$ for large k, as in the paragraph after (39). Exactly as in the case $0 < \kappa < 1$, we apply Theorem 9, to see that almost surely for large n,

$$L_X^*[H(Z_{2n-1})] \leq \varepsilon r_{2n}/\log\log r_{2n} \leq (\varepsilon/c_{42})\inf_{u\in[(1-\varepsilon)r_{2n},(1+\varepsilon)r_{2n}]}(L^*\circ H)_{X\circ\Theta_{H(Z_{2n-1})}}(u),$$

which implies, for all $u \in [(1-\varepsilon)r_{2n}, (1+\varepsilon)r_{2n}]$,

$$L_X^*[H(Z_{2n-1}+u)] \leq (1+\varepsilon/c_{42})(L^*\circ H)_{X\circ\Theta_{H(Z_{2n-1})}}(u). \tag{86}$$

By Theorem 7, almost surely for all large n, $\sup_{u\in[(1-\varepsilon)r_{2n},(1+\varepsilon)r_{2n}]}\log H(Z_{2n-1}+u) \leq (1+\varepsilon)\log r_{2n}$. In view of (86) and then (85), there are almost surely infinitely many n such that

$$\begin{aligned}
&\inf_{v\in[Z_{2n-1}+(1-\varepsilon)r_{2n},Z_{2n-1}+(1+\varepsilon)r_{2n}]}\frac{L_X^*[H(v)]}{H(v)/[(\log H(v))\log\log H(v)]}\\
&\leq (1+c_{43}\varepsilon)\inf_{u\in[(1-\varepsilon)r_{2n},(1+\varepsilon)r_{2n}]}\frac{(L^*\circ H)_{X\circ\Theta_{H(Z_{2n-1})}}(u)}{H_{X\circ\Theta_{H(Z_{2n-1})}}(u)[(\log r_{2n})\log\log r_{2n}]^{-1}}\\
&\leq (1+c_{44}\varepsilon)/2.
\end{aligned}$$

This proves Theorem 4 in the case $\kappa = 1$. □

Remark 1 Assume $\kappa = 1$. We also prove that in this case, $\mathbb{P}$ almost surely,

$$\limsup_{t\to+\infty} L_X^*(t)/t \geq 8/[c_{17}\pi] = 1/2. \tag{87}$$

This is in agreement with Theorem 1.1 of Gantert and Shi [26] for RWRE. However, we could not prove whether this lim sup is finite or not, contrarily to the cases $\kappa \in (0,1)$ and $\kappa > 1$, and to the case of RWRE, for which the maximum local time at time t is clearly less than $t/2$.

We now prove (87). With the same notation as in (58) and (59), let $\widehat{C}_U := \int_0^1 \frac{U^2(x)-8}{x}\mathrm{d}x + \int_1^{+\infty}\frac{U^2(x)}{x}\mathrm{d}x$, $\varepsilon \in (0,1/3)$ and $y := (1+\varepsilon)^2 c_{17}\pi/[8(1-\varepsilon)]$. We have for $z > 0$, by (78),

$$\mathbb{P}[N_\beta(t) < y] = \mathbb{P}\big[\widehat{C}_U + 8\log t < y\zeta_U\big] \geq \mathbb{P}\big[(z+8)\log t < y\zeta_U, \widehat{C}_U \leq z\log t\big].$$

Notice that $\widehat{C}_U \overset{\mathcal{L}}{=} C_\beta \overset{\mathcal{L}}{=} 8c_3 + (\pi/2)C_8^{ca}$ first by (58) and (16), then by Fact 5. So, $\mathbb{P}\big[\widehat{C}_U > z\log t\big] = \mathbb{P}\big[C_8^{ca} > (2z/\pi)\log t - 16c_3/\pi\big] \sim_{t\to+\infty} \pi c_{17}/(2z\log t)$ (see before (46)). Moreover, $\mathbb{P}[(z+8)\log t < y\zeta_U] \sim_{t\to+\infty} 4y/[(z+8)\log t]$ by (59) and (18). Thus,

$$\begin{aligned}\mathbb{P}[N_\beta(t) < y] &\geq \mathbb{P}\big[(z+8)\log t < y\zeta_U\big] - \mathbb{P}\big[\widehat{C}_U > z\log t\big]\\ &\geq [4(1-\varepsilon)y/(z+8) - (1+\varepsilon)\pi c_{17}/(2z)]/\log t\\ &= (1+\varepsilon)c_{17}\pi[(1+\varepsilon)/(z+8) - 1/z]/(2\log t).\end{aligned}$$

So we can choose z so that $\mathbb{P}[N_\beta(t) < y] \geq c_{45}/\log t$ for some constant $c_{45} > 0$. We now set $r_k := k^k$, $k \in \mathbb{N}^*$. This and (77) give for some $\alpha_1 > 0$,

$$\begin{aligned}\mathbb{P}\left[\frac{L_X^*[H(F(r_{2n}))]}{H(F(r_{2n}))} > [(1+3\varepsilon)y]^{-1}\right] &\geq \mathbb{P}\Big[N_\beta[t_+(r_{2n})] < y\Big] - \frac{1}{r_{2n}^{\alpha_1}}\\ &\geq \frac{c_{45}}{2\log r_{2n}} \geq \frac{c_{45}}{5n\log n}\end{aligned}$$

for large n. Hence by Lemma 3 in its notation, $\mathbb{P}$ almost surely, there exist infinitely many n such that for some $t_n \in [(1-\varepsilon)r_{2n}, (1+\varepsilon)r_{2n}]$,

$$\frac{(L^*\circ H)_{X\circ\Theta_{H(Z_{2n-1})}}(t_n)}{H_{X\circ\Theta_{H(Z_{2n-1})}}(t_n)} > [(1+3\varepsilon)y]^{-1}. \tag{88}$$

Notice that $Z_{2n-1} = \sum_{k=1}^{2n-1} k^k \leq (2n-1)^{2n-1} + \sum_{k=1}^{2n-2}(2n-2)^{2n-2} \leq 2(2n)^{2n-1} = r_{2n}/n$. We have by Theorem 7, almost surely for all large n,

$$H(Z_{2n-1}) \leq Z_{2n-1}(\log Z_{2n-1})(\log\log Z_{2n-1})^2 \leq \varepsilon r_{2n}\log r_{2n}. \tag{89}$$

On the other hand, first by Lemmas 1 and 2, then by (41) and since $\mathbb{E}[\exp(-C_8^{ca})] = 1$ as before (50), for every $\varepsilon > 0$ small enough,

$$\begin{aligned}&\mathbb{P}\big[H_{X\circ\Theta_{H(Z_{2n-1})}}((1-\varepsilon)r_{2n}) < (1-10\varepsilon)32t_-(r_{2n})[\log t_-(r_{2n}) + c_3]\big]\\ &\leq \mathbb{P}\big[(1-\varepsilon)\widehat{I}_-((1-2\varepsilon)r_{2n}) < (1-10\varepsilon)32t_-(r_{2n})[\log t_-(r_{2n}) + c_3]\big] + 2r_{2n}^{-\alpha_1}\\ &\leq \mathbb{P}\big[C_8^{ca} < -\varepsilon(16/\pi)\log r_{2n}\big] + 2r_{2n}^{-\alpha_1} \leq 2r_{2n}^{-16\varepsilon/\pi}.\end{aligned}$$

Hence, thanks to the Borel Cantelli lemma, almost surely for large n,

$$\begin{aligned}H_{X\circ\Theta_{H(Z_{2n-1})}}(t_n) &\geq H_{X\circ\Theta_{H(Z_{2n-1})}}((1-\varepsilon)r_{2n})\\ &\geq (1-10\varepsilon)32t_-(r_{2n})[\log t_-(r_{2n}) + c_3] \geq (1-11\varepsilon)4r_{2n}\log r_{2n}.\end{aligned}$$

This together with (88) and (89) gives $\mathbb{P}$ almost surely for infinitely many n,

$$\frac{L_X^*[H(Z_{2n-1}+t_n)]}{H(Z_{2n-1}+t_n)} \geq \frac{(L^*\circ H)_{X\circ\Theta_{H(Z_{2n-1})}}(t_n)}{H_{X\circ\Theta_{H(Z_{2n-1})}}(t_n)} \frac{H_{X\circ\Theta_{H(Z_{2n-1})}}(t_n)}{H(Z_{2n-1})+H_{X\circ\Theta_{H(Z_{2n-1})}}(t_n)}$$
$$\geq [(1+3\varepsilon)y]^{-1}(1+\varepsilon)^{-1} \geq (1-10\varepsilon)8/[c_{17}\pi],$$

for small ε. As before, let $t \to +\infty$, and then $\varepsilon \to 0$. This proves (87) since $c_{17} = \frac{16}{\pi}$ as before (46).

6 Proof of Lemma 2

This section is devoted to the proof of Lemma 2. The basic idea goes back to Hu et al. [33], but requires considerable refinements due to the complicated nature of the process $x \mapsto L_X(t,x)$ and to the fact that we are interested in the joint law of $(L_X^*[H(.)], H(.))$. Throughout the proof we consider the annealed probability $\mathbb{P}$.

Let $\kappa > 0$ and $\varepsilon \in (0,1)$. We fix $r > 1$. Recall that $A(x) = \int_0^x e^{W_\kappa(u)}du$, and $A_\infty = \lim_{x\to+\infty} A(x) < \infty$, a.s. As in Brox [11, Eq. (1.1)], the general diffusion theory leads to

$$X(t) = A^{-1}[B(T^{-1}(t))], \qquad t \geq 0, \tag{90}$$

where $(B(s),\ s \geq 0)$ is a Brownian motion independent of W, and for $0 \leq u < \sigma_B(A_\infty)$, $T(u) := \int_0^u \exp\{-2W_\kappa[A^{-1}(B(s))]\}ds$ (A^{-1} and T^{-1} denote respectively the inverses of A and T). The local time of X can be written as (see Shi [43, Eq. (2.5)])

$$L_X(t,x) = e^{-W_\kappa(x)} L_B\big(T^{-1}(t), A(x)\big), \qquad t \geq 0,\ x \in \mathbb{R}. \tag{91}$$

As in (5), $H(\cdot)$ denotes the first hitting time of X. Then as in Shi [44, Eqs. (4.3)–(4.6)],

$$H(u) = T[\sigma_B(A(u))] = \int_{-\infty}^0 + \int_0^{A(u)} e^{-2W_\kappa[A^{-1}(x)]} L_B(\sigma_B[A(u)], x)dx$$
$$=: H_-(u) + H_+(u) \tag{92}$$

for $u \geq 0$. Recall F from (20) and notice that $F(r) > 0$ on $E_1(r)$ if r is large enough. By scaling since W_κ and then $A(F(r))$ are independent of B, and then by the first Ray-Knight theorem (Fact 1), there exists a squared Bessel process of dimension 2, starting from 0 and denoted by $\big(R_2^2(s),\ s \geq 0\big)$, independent of W_κ, such that

$$\left(\frac{L_B\{\sigma_B[A(F(r))], A(F(r)) - sA(F(r))\}}{A(F(r))},\ s \in [0,1]\right) = \big(R_2^2(s),\ s \in [0,1]\big).$$

Hence, it is more convenient to study $L_X^*[H(.)]$ instead of $L_X^*(t)$. We consider

$$L_X^+[H(u)] := \sup_{x\geq 0} L_X(H(u), x) = \sup_{0\leq x\leq u} \{e^{-W_\kappa(x)} L_B[\sigma_B(A(u)), A(x)]\}, \qquad u \geq 0.$$

In particular,

$$L_X^+[H(F(r))] = \sup_{x\in[0,F(r)]} \left\{e^{-W_\kappa(x)} A(F(r)) R_2^2\left[\frac{A(F(r)) - A(x)}{A(F(r))}\right]\right\}.$$

Moreover, by Lamperti's representation theorem (Fact 3), there exists a Bessel process $\rho = (\rho(t),\ t \geq 0)$, of dimension $(2 - 2\kappa)$, starting from $\rho(0) = 2$, such that for all $t \geq 0$, $e^{W_\kappa(t)/2} = \frac{1}{2}\rho(A(t))$. Now, let

$$\widetilde{R}_{2+2\kappa}(t) := \rho(A_\infty - t), \qquad 0 \leq t \leq A_\infty.$$

By Williams' time reversal theorem (Fact 4), $\widetilde{R}_{2+2\kappa}$ is a Bessel process of dimension $(2 + 2\kappa)$, starting from 0. Since W_κ and $A(F(r))$ are independent of R_2, $u \mapsto \sqrt{A(F(r))}R_2(u/A(F(r)))$ is a 2-dimensional Bessel process, starting from 0 and independent of $\widetilde{R}_{2+2\kappa}$. We still denote by R_2 this new Bessel process. We obtain

$$\begin{aligned} L_X^+[H(F(r))] &= 4 \sup_{x\in[0,F(r)]} \frac{R_2^2[A(F(r)) - A(x)]}{\widetilde{R}_{2+2\kappa}^2[A_\infty - A(x)]} \\ &= 4 \sup_{v\in[0,A(F(r))]} \frac{R_2^2(v)}{\widetilde{R}_{2+2\kappa}^2[A_\infty - A(F(r)) + v]}. \end{aligned}$$

Doing the same transformations on $H_+(F(r))$ (see (92)) and recalling that $A_\infty - A(F(r)) = \delta(r) = \exp(-\kappa r/2)$ and so is deterministic thanks to the random function F, we obtain

$$\begin{aligned} &\big(L_X^+[H(F(r))], H_+[F(r)]\big) \\ &= \left(4 \sup_{v\in[0,A(F(r))]} \frac{R_2^2(v)}{\widetilde{R}_{2+2\kappa}^2[\delta(r) + v]}, 16\int_0^{A(F(r))} \frac{R_2^2(s)}{\widetilde{R}_{2+2\kappa}^4[\delta(r) + s]} \mathrm{d}s\right) \\ &= \left(4 \sup_{u\in[0,\delta(r)^{-1}A(F(r))]} \frac{R_2^2[\delta(r)u]}{\widetilde{R}_{2+2\kappa}^2[\delta(r)(1+u)]}, 16\int_0^{\delta(r)^{-1}A(F(r))} \frac{R_2^2[\delta(r)u]\delta(r)\mathrm{d}u}{\widetilde{R}_{2+2\kappa}^4[\delta(r)(1+u)]}\right). \end{aligned}$$

We still denote by R_2 the 2-dimensional Bessel process $u \mapsto \frac{1}{\sqrt{\delta(r)}}R_2(\delta(r)u)$. We define

$$\widehat{R}_{2+2\kappa}(u) = \frac{1}{\sqrt{\delta(r)}}\widetilde{R}_{2+2\kappa}[\delta(r)(1+u)], \qquad u \geq 0. \tag{93}$$

Notice that $\widehat{R}_{2+2\kappa}(u)$ is a $(2+2\kappa)$-dimensional Bessel process, starting from $\widetilde{R}_{2+2\kappa}(\delta(r))/\sqrt{\delta(r)}$ and independent of R_2.

Recall (see e.g. Karlin and Taylor [34, p. 335]) that a Jacobi process $(Y(t),\ t \geq 0)$ of dimensions (d_1, d_2) is a solution of the stochastic differential equation

$$\mathrm{d}Y(t) = 2\sqrt{Y(t)(1-Y(t))}\,\mathrm{d}\hat{\beta}(t) + \big[d_1 - (d_1+d_2)Y(t)\big]\mathrm{d}t, \tag{94}$$

where $\big(\hat{\beta}(t),\ t \geq 0\big)$ is a standard Brownian motion.

Due to Warren and Yor [50, p. 337], there exists a Jacobi process $(Y(t),\ t \geq 0)$ of dimensions $(2, 2+2\kappa)$, starting from 0, independent of $\big(R_2^2(t) + \widehat{R}_{2+2\kappa}^2(t),\ t \geq 0\big)$, such that

$$\forall u \geq 0, \qquad \frac{R_2^2(u)}{R_2^2(u) + \widehat{R}_{2+2\kappa}^2(u)} = Y \circ \Lambda_Y(u), \quad \Lambda_Y(u) := \int_0^u \frac{\mathrm{d}s}{R_2^2(s) + \widehat{R}_{2+2\kappa}^2(s)}. \tag{95}$$

In particular, $(\Lambda_Y(t),\ t \geq 0)$ is independent of Y. As a consequence, for all $r \geq 0$,

$$\begin{aligned}
&(L_X^+[H(F(r))], H_+[F(r)]) \\
&= \left(4 \sup_{u \in [0, \delta(r)^{-1}A(F(r))]} \frac{Y \circ \Lambda_Y(u)}{1 - Y \circ \Lambda_Y(u)}, 16 \int_0^{\delta(r)^{-1}A(F(r))} \frac{[Y \circ \Lambda_Y(u)]\Lambda_Y'(u)\mathrm{d}u}{[1 - Y \circ \Lambda_Y(u)]^2}\right) \\
&= \left(4 \sup_{u \in [0, \varphi(r)]} \frac{Y(u)}{1 - Y(u)}, 16 \int_0^{\varphi(r)} \frac{Y(u)}{(1 - Y(u))^2}\mathrm{d}u\right),
\end{aligned}$$

where

$$\varphi(r) := \Lambda_Y\big[\delta(r)^{-1}A(F(r))\big]. \tag{96}$$

Define $\alpha_\kappa := 1/(4+2\kappa)$ and let $T_Y(\alpha_\kappa) := \inf\{t > 0, Y(t) = \alpha_\kappa\}$ be the hitting time of α_κ by Y. We introduce

$$\overline{L}(r) := 4 \sup_{u \in [0, T_Y(\alpha_\kappa)]} \frac{Y(u)}{1 - Y(u)}, \qquad \overline{H}(r) := 16 \int_0^{T_Y(\alpha_\kappa)} \frac{Y(u)}{(1 - Y(u))^2}\mathrm{d}u, \tag{97}$$

$$L_0(r) := 4 \sup_{u\in[T_Y(\alpha_\kappa),\varphi(r)]} \frac{Y(u)}{1-Y(u)}, \qquad H_0(r) := 16 \int_{T_Y(\alpha_\kappa)}^{\varphi(r)} \frac{Y(u)}{(1-Y(u))^2} du.$$

We have on $E_9 := \{T_Y(\alpha_\kappa) \le 64 \log r \le \kappa r/(2\lambda) \le \varphi(r)\}$,

$$\left(L_X^+[H(F(r))],\ H_+(F(r))\right) = \left(\max\{\overline{L}(r), L_0(r)\},\ \overline{H}(r) + H_0(r)\right), \tag{98}$$

$$\overline{L}(r) \le \frac{4\alpha_\kappa}{1-\alpha_\kappa} \quad \text{and} \quad \overline{H}(r) \le \frac{16\alpha_\kappa}{(1-\alpha_\kappa)^2} T_Y(\alpha_\kappa) \le \frac{2^{10}\alpha_\kappa}{(1-\alpha_\kappa)^2} \log r. \tag{99}$$

Observe that $S(y) := \int_{\alpha_\kappa}^{y} \frac{dx}{x(1-x)^{1+\kappa}}$, $y \in (0,1)$ is a scale function of Y, as in Hu et al. [33, Eq. (2.27)]. Hence $t \mapsto S[Y(t + T_Y(\alpha_\kappa))]$ is a continuous local martingale, so by Dubins-Schwarz theorem, there exists a Brownian motion $(\beta(t),\ t \ge 0)$ such that for all $t \ge 0$,

$$Y[t + T_Y(\alpha_\kappa)] = S^{-1}\{\beta[U_Y(t)]\}, \tag{100}$$

$$U_Y(t) := 4 \int_0^t \frac{ds}{Y[s + T_Y(\alpha_\kappa)]\{1 - Y[s + T_Y(\alpha_\kappa)]\}^{1+2\kappa}}. \tag{101}$$

The rest of the proof of Lemma 2 requires some more estimates, stated as Lemmas 6–9 below. Lemmas 6–8 deal only with Bessel processes, Jacobi processes and Brownian motion, and may be of independent interest, whereas Lemma 9 gives an upper bound for the total time spent by X on $(-\infty, 0)$, and for the maximum local time of X in $(-\infty, 0)$. We defer the proofs of Lemmas 6–8 to Sect. 7, and we complete the proof of Lemma 2.

Lemma 6 *Let $(R(t),\ t \ge 0)$ be a Bessel process of dimension $d > 4$, starting from $R_0 \stackrel{\mathcal{L}}{=} \widetilde{R}_{d-2}(1)$, where $\left(\widetilde{R}_{d-2}(t),\ t \in [0,1]\right)$ is $(d-2)$-dimensional Bessel process. For any $\delta_3 \in (0, \frac{1}{2})$ and all large t,*

$$\mathbb{P}\left\{\left|\frac{1}{\log t}\int_0^t \frac{ds}{R^2(s)} - \frac{1}{d-2}\right| > \frac{1}{(\log t)^{(1/2)-\delta_3}}\right\} \le \exp\left(-c_{46}\,(\log t)^{2\delta_3}\right).$$

Lemma 7 *Let $\delta_1 > 0$ and define*

$$E_{10} := \left\{\tau_\beta\left[\left(1 - v^{-\delta_1}\right)\lambda v\right] \le U_Y(v) \le \tau_\beta\left[\left(1 + v^{-\delta_1}\right)\lambda v\right]\right\}. \tag{102}$$

If δ_1 is small enough, then for all large v, $\mathbb{P}(E_{10}^c) \le \frac{1}{v^{1/4-5\delta_1}}$.

In the two previous lemmas, taking respectively $\frac{1}{(\log t)^{(1/2)-\delta_3}}$ and $v^{-\delta_1}$ instead of simply some fixed $\tilde{\varepsilon} > 0$ is necessary to obtain Lemma 2 with $\psi_\pm(r)$ instead of simply $1 \pm \tilde{\varepsilon}$ in the definition of $\widehat{L}_\pm(r)$ and $\widehat{I}_\pm(r)$, which itself is necessary for example to prove Lemma 5.

Lemma 8 *Let $(\beta(s),\ s \geq 0)$ be a Brownian motion, and $\lambda = 4(1+\kappa)$ as before. We define*

$$J_\beta(\kappa, t) := \int_0^1 y(1-y)^{\kappa-2} L_\beta\big[\tau_\beta(\lambda), S(y)/t\big]dy, \qquad 0 < \kappa \leq 1,\ t \geq 0. \tag{103}$$

Let $0 < d < 1$ and recall that $0 < \varepsilon < 1$.

(i) *Case $0 < \kappa < 1$: recall $K_\beta(\kappa)$ from (15). There exist $c_{47} > 0$ and $c_{48} > 0$ such that for all large t, on an event E_{11} of probability greater than $1 - c_{47}/t^d$, we have*

$$(1-\varepsilon)K_\beta(\kappa) - c_{48}t^{1-1/\kappa} \leq \kappa^{2-1/\kappa} t^{1-1/\kappa} J_\beta(\kappa, t) \leq (1+\varepsilon)K_\beta(\kappa) + c_{48}t^{1-1/\kappa}. \tag{104}$$

(ii) *Case $\kappa = 1$: recall C_β from (16). There exists $c_{47} > 0$ such that for t large enough, on an event E_{11} of probability greater than $1 - c_{47}/t^d$,*

$$(1-\varepsilon)[C_\beta + 8\log t] \leq J_\beta(1, t) \leq (1+\varepsilon)[C_\beta + 8\log t]. \tag{105}$$

Lemma 9 *Let $\kappa > 0$ and define*

$$L_X^{*-}(+\infty) := \sup_{r\geq 0}\ \sup_{x<0} L_X(H(r), x) = \sup_{t\geq 0}\ \sup_{x<0} L_X(t, x), \quad H_-(+\infty) := \lim_{r\to+\infty} H_-(r).$$

There exist $c_{49} > 0$ and $c_{50} > 0$ such that for all large z,

$$\mathbb{P}\big[L_X^{*-}(+\infty) > z\big] \leq c_{49} z^{-\kappa/(\kappa+2)}, \tag{106}$$

$$\mathbb{P}\big[H_-(+\infty) > z\big] \leq c_{50}[(\log z)/z]^{\kappa/(\kappa+2)}. \tag{107}$$

Proof (Proof of· Lemma 9) This lemma is proved in Andreoletti et al. ([3, Lemma 3.5], which is true for every $\kappa > 0$). More precisely, (107) is proved in [3, Eq. (3.29)], whereas (106) is proved in [3, Eq. (3.31)]. □

By admitting Lemmas 6–8, we can now complete the proof of Lemma 2.

Proof (Proof of Lemma 2: Part (i)) Notice that

$$S(y) \underset{y\to 1}{\sim} \int_{\alpha_\kappa}^{y} \frac{ds}{(1-s)^{1+\kappa}} \underset{y\to 1}{\sim} \frac{1}{\kappa}\frac{1}{(1-y)^\kappa}. \qquad \frac{y}{1-y} \underset{y\to 1}{\sim} [\kappa S(y)]^{1/\kappa}. \tag{108}$$

Define $\widetilde{L}_0(r) := 4\big[\sup_{u\in[T_Y(\alpha_\kappa),\varphi(r)]} \kappa S(Y(u))\big]^{1/\kappa}$. We have,

$$\widetilde{L}_0(r) = 4\Big[\sup_{u\in[0,\varphi(r)-T_Y(\alpha_\kappa)]} \kappa\beta\big(U_Y(u)\big)\Big]^{1/\kappa} = 4\Big[\sup_{s\in[0,U_Y(\varphi(r)-T_Y(\alpha_\kappa))]} \kappa\beta(s)\Big]^{1/\kappa}. \tag{109}$$

Recall L_0 from (97). By (108), there exists a constant $c_{51} > 0$ depending on ε such that

$$\left\{\widetilde{L}_0(r) > c_{51}\right\} \subset \left\{(1-\varepsilon)\widetilde{L}_0(r) \le L_0(r) \le (1+\varepsilon)\widetilde{L}_0(r)\right\}. \tag{110}$$

We look for an estimate of $U_Y[\varphi(r) - T_Y(\alpha_\kappa)]$ appearing in the expression of $\widetilde{L}_0(r)$ in the right hand side of (109). Recall (see Dufresne [24], or Borodin et al. [10, IV.48 p. 78]) that $A_\infty \overset{\mathcal{L}}{=} 2/\gamma_\kappa$, where γ_κ is a gamma variable of parameter $(\kappa, 1)$, with density $e^{-x}x^{\kappa-1}\mathbf{1}_{(0,\infty)}(x)/\Gamma(\kappa)$. Since $A(F(r)) \le A_\infty$, we have

$$\mathbb{P}\left[A(F(r)) > r^{2/\kappa}\right] \le \mathbb{P}\left[\gamma_\kappa < 2r^{-2/\kappa}\right] \le 2^\kappa r^{-2}/[\kappa\Gamma(\kappa)].$$

On the other hand, by definition, $A(F(r)) = A_\infty - \delta(r) = A_\infty - e^{-\kappa r/2}$ (see (20)), which implies

$$\mathbb{P}\left[A(F(r)) < \frac{1}{2\log r}\right] = \mathbb{P}\left[\frac{2}{\gamma_\kappa} < \frac{1}{2\log r} + \delta(r)\right] \le \frac{1}{r^2}$$

for large r. Consequently,

$$\mathbb{P}\left\{\kappa r/2 - 2\log\log r \le \log\left[\delta(r)^{-1}A(F(r))\right] \le \kappa r/2 + (2/\kappa)\log r\right\} \ge 1 - c_{52}r^{-2}.$$

Recall that $\varphi(r) = \Lambda_Y[\delta(r)^{-1}A(F(r))]$ by (96). Thus, for large r,

$$\begin{aligned}&\mathbb{P}\left\{\Lambda_Y\left[\exp(\kappa r/2 - 2\log\log r)\right] \le \varphi(r) \le \Lambda_Y\left[\exp(\kappa r/2 + (2/\kappa)\log r)\right]\right\}\\ &\ge 1 - c_{52}r^{-2}.\end{aligned}$$

By definition, $\Lambda_Y(u) = \int_0^u \frac{\mathrm{d}s}{R_2^2(s)+\widehat{R}_{2+2\kappa}^2(s)}$. Notice that $\left(R_2^2(t) + \widehat{R}_{2+2\kappa}^2(t),\ t \ge 0\right)$ is a $(4+2\kappa)$-dimensional squared Bessel process starting from $\widetilde{R}_{2+2\kappa}^2[\delta(r)]/\delta(r)$ by the additivity property of squared Bessel processes (see e.g. Revuz et al. [40, XI th. 1.2]). So, it follows from Lemma 6 applied with $d = 4 + 2\kappa$ and $\delta_3 = 1/4$, that there exist constants $c_{53} > 0$ and $c_{54} > 0$, such that

$$\mathbb{P}\left\{\kappa r/\lambda - c_{53}r^{1/2+\delta_3} \le \varphi(r) \le \kappa r/\lambda + c_{53}r^{1/2+\delta_3}\right\} \ge 1 - c_{54}r^{-2}, \tag{111}$$

for large r, where $\lambda = 4(1+\kappa)$, as before.

In order to study $T_Y(\alpha_\kappa)$, we go back to the stochastic differential equation in (94) satisfied by the Jacobi process $Y(\cdot)$, with $d_1 = 2$ and $d_2 = 2 + 2\kappa$. Note that $Y(s) \in (0,1)$ for any $s > 0$. By the Dubins–Schwarz theorem, there exists a Brownian motion $\left(\widehat{B}(s),\ s \ge 0\right)$ such that

$$Y(t) = \widehat{B}\left(4\int_0^t Y(s)(1-Y(s))\mathrm{d}s\right) + \int_0^t [2 - (4+2\kappa)Y(s)]\mathrm{d}s, \qquad t \ge 0.$$

Recall that $\alpha_\kappa = 1/(4+2\kappa)$, and let $t \geq 2\alpha_\kappa$. We have, on the event $\{T_Y(\alpha_\kappa) \geq t\}$,

$$\inf_{0\leq s\leq 4t} \widehat{B}(s) \leq \widehat{B}\left(4\int_0^t Y(s)(1-Y(s))\mathrm{d}s\right) \leq \alpha_\kappa - t \leq -\frac{t}{2},$$

since $Y(s) \leq \alpha_\kappa \leq 1$ if $0 \leq s \leq t \leq T_Y(\alpha_\kappa)$. As a consequence, for $t \geq 2\alpha_\kappa$,

$$\mathbb{P}[T_Y(\alpha_\kappa) > t] \leq \mathbb{P}\left(\inf_{0\leq s\leq 4t} \widehat{B}(s) \leq -\frac{t}{2}\right) = \mathbb{P}\left(\left|\widehat{B}(4t)\right| \geq \frac{t}{2}\right) \leq 2\exp\left(-\frac{t}{32}\right). \tag{112}$$

In particular, $\mathbb{P}[T_Y(\alpha_\kappa) > 64\log r] \leq 2r^{-2}$ for large r. Plug this into (111), let $c_{55} > c_{53}$ and define $\underline{\varphi} = \underline{\varphi}(r) := \kappa r/\lambda - c_{55}r^{1/2+\delta_3}$ and $\overline{\varphi} = \overline{\varphi}(r) := \kappa r/\lambda + c_{53}r^{1/2+\delta_3}$. This gives,

$$\mathbb{P}\left\{U_Y(\underline{\varphi}) \leq U_Y[\varphi(r) - T_Y(\alpha_\kappa)] \leq U_Y(\overline{\varphi})\right\} \geq 1 - c_{56}r^{-2}$$

for large r. By Lemma 7, for small $\delta_1 > 0$ and all large r,

$$\mathbb{P}\left\{\tau_\beta\left[\left(1-(\underline{\varphi})^{-\delta_1}\right)\lambda\underline{\varphi}\right] \leq U_Y[\varphi(r) - T_Y(\alpha_\kappa)] \leq \tau_\beta\left[\left(1+(\overline{\varphi})^{-\delta_1}\right)\lambda\overline{\varphi}\right]\right\}$$
$$\geq 1 - r^{-c_{57}}.$$

We choose δ_1 such that $\delta_1 < 1/2 - \delta_3$. Then for large r, we have $\left(1-(\underline{\varphi})^{-\delta_1}\right)\lambda\underline{\varphi} \geq \left[1-2(\frac{\lambda}{\kappa})^{\delta_1}r^{-\delta_1}\right]\kappa r = \lambda t_-(r)$, and $\left(1+(\overline{\varphi})^{-\delta_1}\right)\lambda\overline{\varphi} \leq \left[1+2(\frac{\lambda}{\kappa})^{\delta_1}r^{-\delta_1}\right]\kappa r = \lambda t_+(r)$ (see (25)). Thus,

$$\mathbb{P}\left\{\tau_\beta[\lambda t_-(r)] \leq U_Y[\varphi(r) - T_Y(\alpha_\kappa)] \leq \tau_\beta[\lambda t_+(r)]\right\} \geq 1 - r^{-c_{57}}. \tag{113}$$

With $\widehat{L}_\pm(r) = 4\left[\sup_{s\in[0,\tau_\beta(\lambda t_\pm(r))]}\kappa\beta(s)\right]^{1/\kappa}$ (see (27)), (113) and (109) give for large r,

$$\mathbb{P}\left\{\widehat{L}_-(r) \leq \widetilde{L}_0(r) \leq \widehat{L}_+(r)\right\} \geq 1 - r^{-c_{57}}. \tag{114}$$

By (35), $\mathbb{P}\left\{\widehat{L}_-(r) > r^{(1-\delta_1)/\kappa}\right\} \geq 1 - r^{-1}$ for large r. Applying (110) and (114), this yields

$$\mathbb{P}\left\{(1-\varepsilon)r^{(1-\delta_1)/\kappa} < (1-\varepsilon)\widehat{L}_-(r) \leq L_0(r) \leq (1+\varepsilon)\widehat{L}_+(r)\right\} \geq 1 - r^{-c_{58}}.$$

Recall that $\mathbb{P}[T_Y(\alpha_\kappa) > 64\log r] \leq 2r^{-2}$ for large r, which together with (111) gives $\mathbb{P}(E_9^c) \leq (c_{54}+2)r^{-2}$. In view of (98) and (99), for large r,

$$\mathbb{P}\left\{(1-\varepsilon)r^{(1-\delta_1)/\kappa} < (1-\varepsilon)\widehat{L}_-(r) \leq L_X^+[H(F(r))] \leq (1+\varepsilon)\widehat{L}_+(r)\right\} \geq 1 - r^{-c_{59}}. \tag{115}$$

On the other hand, applying Lemma 9 to $z = r^{(1-2\delta_1)/\kappa}$ gives

$$\mathbb{P}\Big[\sup_{x<0} L_X[H(F(r)),x] > r^{(1-2\delta_1)/\kappa}\Big] \le \mathbb{P}[L_X^{*-}(+\infty) > r^{(1-2\delta_1)/\kappa}] \le \frac{c_{49}}{r^{(1-2\delta_1)/(\kappa+2)}}$$

for large r. This implies

$$\mathbb{P}\left\{(1-\varepsilon)\widehat{L}_-(r) \le L_X^*[H(F(r))] \le (1+\varepsilon)\widehat{L}_+(r)\right\} \ge 1 - \frac{1}{r^{c_{59}}} - \frac{c_{49}}{r^{(1-2\delta_1)/(\kappa+2)}},$$

proving the first part of Lemma 2. □

Proof (Proof of Lemma 2: Part (ii)) In this part, we assume $0 < \kappa \le 1$.

Recall that $H_0(r) = 16\int_0^{\varphi(r)-T_Y(\alpha_\kappa)} \frac{Y[u+T_Y(\alpha_\kappa)]}{(1-Y[u+T_Y(\alpha_\kappa)])^2}du$ and that $Y[t+T_Y(\alpha_\kappa)] = S^{-1}\{\beta[U_Y(t)]\}$, see (97) and (100). As in Hu et al. ([33] p. 3923, calculation of Υ_μ), this and (101) lead to:

$$\begin{aligned} H_0(r) &= 4\int_0^{\varphi(r)-T_Y(\alpha_\kappa)} (Y[u+T_Y(\alpha_\kappa)])^2(1-Y[u+T_Y(\alpha_\kappa)])^{2\kappa-1}\mathrm{d}U_Y(u) \\ &= 4\int_0^1 x(1-x)^{\kappa-2}L_\beta[U_Y(\varphi(r)-T_Y(\alpha_\kappa)),S(x)]\mathrm{d}x. \end{aligned} \tag{116}$$

Recall that $t_\pm(r) = \left[1 \pm 2(\frac{\lambda}{\kappa})^{\delta_1} r^{-\delta_1}\right]\frac{\kappa}{\lambda}r$, $\beta_v(s) = \beta(v^2 s)/v$ and let J_β be as in (103). We have,

$$\begin{aligned} &\int_0^1 x(1-x)^{\kappa-2}L_\beta\{\tau_\beta[\lambda t_\pm(r)],S(x)\}\mathrm{d}x \\ &= t_\pm(r)\int_0^1 x(1-x)^{\kappa-2}L_{\beta_{t_\pm(r)}}\{\tau_{\beta_{t_\pm(r)}}(\lambda),\frac{S(x)}{t_\pm(r)}\}\mathrm{d}x \\ &= t_\pm(r)J_{\beta_{t_\pm(r)}}[\kappa,t_\pm(r)]. \end{aligned}$$

By (113) and (116), we have for large r,

$$\mathbb{P}\left[4t_-(r)J_{\beta_{t_-(r)}}[\kappa,t_-(r)] \le H_0(r) \le 4t_+(r)J_{\beta_{t_+(r)}}[\kappa,t_+(r)]\right] \ge 1 - r^{-c_{57}}.$$

Now, apply Lemma 8 to $d = 1/2$. So there exist $c_6 > 0$ and $c_{60} > 0$ such that for large r,

$$\mathbb{P}\left\{(1-\varepsilon)\widehat{I}_-(r) \le H_0(r) \le (1+\varepsilon)\widehat{I}_+(r)\right\} \ge 1 - r^{-c_{60}}, \tag{117}$$

where $\widehat{I}_\pm(r)$ is defined in (29).

In the case $0 < \kappa < 1$, we know that $\mathbb{P}(E_9^c) \le (c_{54}+2)r^{-2}$ for large r as proved before (115), so by (99), $\mathbb{P}\big[\overline{H}(r) \le c_{61}\log r\big] \ge \mathbb{P}(E_9) \ge 1-(c_{54}+2)r^{-2}$ for some c_{61} and all large r. On the other hand, by Lemma 9, $\mathbb{P}[H_-(F(r)) \le \varepsilon r] \ge \mathbb{P}[H_-(+\infty) \le \varepsilon r] \ge 1 - c_{62}r^{-(1-\delta_1)\kappa/(\kappa+2)}$, for all large r. Consequently, by (117) and (98), for large r,

$$\mathbb{P}\{(1-\varepsilon)\widehat{I}_-(r) \le H(F(r)) \le (1+\varepsilon)\widehat{I}_+(r) + (4\varepsilon\lambda/\kappa)t_+(r)\} \ge 1 - r^{-c_{63}}.$$

Changing the value of c_6, this proves Lemma 2 (ii) in the case $0 < \kappa < 1$.

Now we consider the case $\kappa = 1$. As before, $\mathbb{P}[H_-(F(r)) + \overline{H}(r) \le 2\varepsilon r] \ge 1 - r^{-c_{64}}$ (for large r). Moreover, $\mathbb{P}\big[C_{\beta_{t_\pm(r)}} > -\pi\log r\big] \ge 1 - r^{-2}$ by Fact 5 and (50). Therefore, (29) gives $\mathbb{P}\big\{\widehat{I}_+(r) \ge 16t_+(r)\log r\big\} \ge 1 - r^{-2}$. Consequently, for large r,

$$\mathbb{P}\big[0 \le H_-(F(r)) + \overline{H}(r) \le \varepsilon\widehat{I}_+(r)\big] \ge 1 - r^{-c_{65}},$$

which, in view of (117), yields that, for large r,

$$\mathbb{P}\big[(1-\varepsilon)\widehat{I}_-(r) \le H(F(r)) \le (1+2\varepsilon)\widehat{I}_+(r)\big] \ge 1 - r^{-c_{66}}.$$

This proves Lemma 2 (ii) in the case $\kappa = 1$. □

7 Proof of Lemmas 6–8

This section is devoted to the proof of Lemmas 6–8. For the sake of clarity, the proofs of these lemmas are presented in separated subsections.

7.1 Proof of Lemma 6

First, notice that we can not apply Talet [48, Lemma 3.2 Eq. (3.4)] since her constant c_3 depends on her (fixed) δ, whereas we would like to take her $\delta = (\log t)^{\delta_3 - 1/2} \to_{t\to+\infty} 0$, which is necessary for example for our Lemma 5. A similar remark applies for Talet [48, Prop. 5.1] and our Lemma 7. So we need different estimates than in her paper.

Let $d > 4$ and $R_0 \stackrel{\mathscr{L}}{=} \widetilde{R}_{d-2}(1)$, where $\widetilde{R}_{d-2}$ is a $(d-2)$-dimensional Bessel process. We consider a d-dimensional Bessel process R, starting from R_0. We introduce $\theta(t) := \int_0^t R^{-2}(s)\mathrm{d}s$. Itô's formula gives $\log R(t) = \log R_0 + M(t) + \frac{d-2}{2}\theta(t)$, where $M(t) := \int_0^t R(s)^{-1}\mathrm{d}\hat{\beta}(s)$ and $\big(\hat{\beta}(t),\ t \ge 0\big)$ is a Brownian motion. By the Dubins–Schwarz theorem, there exists a Brownian motion $\big(\tilde{\beta}(t),\ t \ge 0\big)$ such that $M(t) = \tilde{\beta}(\theta(t))$ for all $t \ge 0$. Accordingly,

$$(d-2)\theta(t)/2 = \log R(t) - \log R_0 - \tilde{\beta}(\theta(t)), \qquad t \ge 0. \tag{118}$$

Let $\delta_3 \in (0, \frac{1}{2})$, $0 < \varepsilon < 1$, and $x = x(t) := \frac{d-2}{6} \frac{1}{(\log t)^{(1/2)-\delta_3}}$. We have (see e.g. Göing-Jaeschke et al. [28, Eq. (50)]), $\mathbb{P}\big(R_0^2 \in du\big) = u^{d/2-2}e^{-u/2}\mathbf{1}_{(0,\infty)}(u)/\big[\Gamma(d/2-1)2^{d/2-1}\big]$. So for large t,

$$\mathbb{P}\left(\left|\frac{\log R_0}{\log t}\right| > x\right) = \mathbb{P}\left(\frac{\log R_0}{\log t} > x\right) + \mathbb{P}\left(\frac{\log R_0}{\log t} < -x\right) \le \exp\left(-(1-\varepsilon)\frac{t^{2x}}{2}\right) + \frac{c_{67}}{t^{x(d/2-1)}}. \tag{119}$$

Denote by $n := \lceil d \rceil$ the smallest integer such that $n \ge d$. Since an n-dimensional Bessel process can be realized as the Euclidean modulus of an $\mathbb{R}^n$-valued Brownian motion, it follows from the triangular inequality that $R(t) \le_{\mathscr{L}} R_0 + \widehat{R}_n(t)$, where $(\widehat{R}_n(t),\ t \ge 0)$ is an n-dimensional Bessel process starting from 0. Consequently, for large t, $\mathbb{P}\big(R(t) > t^{(1/2)+x}\big) \le \mathbb{P}\big(\widehat{R}_n(t) > t^{(1/2)+x}/2\big) + \mathbb{P}\big(R_0 > t^x\big) \le 2\exp(-(1-\varepsilon)t^{2x}/8)$, and $\mathbb{P}\big(R(t) < t^{(1/2)-x}\big) \le 2t^{-x}$, e.g. since $\big|\tilde{\beta}(t)\big| \le_{\mathscr{L}} R(t)$. Therefore, for large t,

$$\mathbb{P}\left(\left|\frac{\log R(t)}{\log t} - \frac{1}{2}\right| > x\right) \le 2\exp\left(-(1-\varepsilon)\frac{t^{2x}}{8}\right) + 2t^{-x}. \tag{120}$$

Define $E_{12} := \left\{\left|\frac{\log R(t)}{\log t} - \frac{1}{2}\right| \le x\right\} \cap \left\{\left|\frac{\log R_0}{\log t}\right| \le x\right\}$ and

$$E_{13} := \left\{\frac{d-2}{2}\theta(t) < \log t\right\}, \qquad E_{14} := \left\{\sup_{0 \le s \le 2(\log t)/(d-2)} \big|\tilde{\beta}(s)\big| \le x \log t\right\}.$$

By (119) and (120), we have for large t,

$$\mathbb{P}(E_{12}^c) \le 3\exp\big[-(1-\varepsilon)t^{2x}/8\big] + 3t^{-x}. \tag{121}$$

We now estimate $\mathbb{P}(E_{12} \cap E_{13}^c)$. We first observe that on E_{12}, we have, by (118),

$$\big|\tilde{\beta}(\theta(t)) + (d-2)\theta(t)/2 - (\log t)/2\big| \le 2x \log t.$$

We claim that $E_{12} \cap E_{13}^c \subset \big\{\big|\tilde{\beta}(\theta(t))\big| > \frac{d-2}{6}\theta(t)\big\}$ for large t. Indeed, on the event $E_{12} \cap E_{13}^c \cap \big\{\big|\tilde{\beta}(\theta(t))\big| \le \frac{d-2}{6}\theta(t)\big\}$,

$$(d-2)\theta(t)/2 \le (2x + 1/2)\log t - \tilde{\beta}(\theta(t)) \le (2x + 1/2)\log t + (d-2)\theta(t)/6,$$

which implies $\frac{d-2}{2}\theta(t) \le (\frac{3}{4} + 3x)\log t$. This, for large t, contradicts $\frac{d-2}{2}\theta(t) \ge \log t$ on E_{13}^c. Therefore, $E_{12} \cap E_{13}^c \subset \big\{\big|\tilde{\beta}(\theta(t))\big| > \frac{d-2}{6}\theta(t)\big\}$ holds for all large t, from which it follows that

$$\begin{aligned}\mathbb{P}\big(E_{12}\cap E_{13}^c\big) &\le \mathbb{P}\bigg(\sup_{s\ge 2(\log t)/(d-2)}\frac{|\tilde\beta(s)|}{s}>\frac{d-2}{6}\bigg)\\ &=\mathbb{P}\bigg(\sup_{u\ge 1}\frac{|\tilde\beta(u)|}{u}>\sqrt{\frac{(d-2)\log t}{18}}\bigg)\\ &=\mathbb{P}\bigg(\sup_{0\le v\le 1}\big|\tilde\beta(v)\big|>\sqrt{(d-2)(\log t)/18}\bigg)\\ &\le 4\exp\big[-(d-2)(\log t)/36\big],\end{aligned}$$

because $u\mapsto u\tilde\beta(1/u)$ is a Brownian motion and $\sup_{0\le v\le 1}\tilde\beta(v)\overset{\mathcal{L}}{=}\big|\tilde\beta(1)\big|$. Since $\mathbb{P}\big(E_{14}^c\big)\le 4\exp\big[-\frac{d-2}{4}x^2\log t\big]$ for large t, this and (121) give for large t,

$$\begin{aligned}\mathbb{P}\big(E_{12}^c\cup E_{13}^c\cup E_{14}^c\big) &\le \mathbb{P}\big(E_{12}^c\big)+\mathbb{P}\big(E_{12}\cap E_{13}^c\big)+P\big(E_{12}\cap E_{13}\cap E_{14}^c\big)\\ &\le \exp\big(-c_{68}\,x^2\log t\big).\end{aligned}$$

Since $E_{12}\cap E_{13}\cap E_{14}\subset\big\{\big|\frac{\theta(t)}{\log t}-\frac{1}{d-2}\big|\le\frac{6x}{d-2}\big\}$ by (118), this completes the proof of Lemma 6. □

7.2 *Proof of Lemma 7*

Let $v>0$. Recall that for every $x\ge 0$, $\beta_v(x)=(1/v)\beta(v^2x)$, and notice that $v^2\tau_{\beta_v}(x)=\tau_\beta(xv)$ almost surely. Then,

$$E_{10}=\big\{\tau_{\beta_v}\big[\big(1-v^{-\delta_1}\big)\lambda\big]\le U_Y(v)/v^2\le\tau_{\beta_v}\big[\big(1+v^{-\delta_1}\big)\lambda\big]\big\}. \tag{122}$$

For $\delta_1>0$, define $E_{15}:=\big\{\sup_{0\le s\le\tau_{\beta_v}(2\lambda)}|\varepsilon_1(v,s)|<v^{-\delta_1}\big\}$, where

$$\varepsilon_1=\varepsilon_1(v,s):=\frac{1}{4}\int_0^1(1-x)^\kappa\left[L_{\beta_v}\left(s,\frac{S(x)}{v}\right)-L_{\beta_v}(s,0)\right]\mathrm{d}x,\qquad s\ge 0.$$

By Hu et al. [33, Eq. (2.34) p. 3924], $E_{15}\subset E_{10}$. Thus it remains to prove that for δ_1 small enough, $\mathbb{P}(E_{15}^c)\le 1/v^{1/4-5\delta_1}$ for large v. Notice that for $s\ge 0$,

$$|\varepsilon_1|\le\left(\int_{\{S(x)>\sqrt{v}\}}+\int_{\{S(x)<-\sqrt{v}\}}+\int_{\{|S(x)|\le\sqrt{v}\}}\right)$$

$$\frac{(1-x)^{\kappa}}{4}\left|L_{\beta_v}\left(s, \frac{S(x)}{v}\right) - L_{\beta_v}(s,0)\right| \mathrm{d}x$$
$$=: \varepsilon_2(v,s) + \varepsilon_3(v,s) + \varepsilon_4(v,s). \tag{123}$$

Since $S(y) = \int_{\alpha_\kappa}^{y} \frac{\mathrm{d}x}{x(1-x)^{1+\kappa}}$, we have $1 - S^{-1}(u) \underset{u\to+\infty}{\sim} (\kappa u)^{-1/\kappa}$. So, we have

$$\sup_{0\le s\le \tau_{\beta_v}(2\lambda)} \varepsilon_2(v,s) \le \frac{1}{4}\int_{1-\left(\frac{2}{\kappa\sqrt{v}}\right)^{1/\kappa}}^{1} (1-x)^{\kappa} \sup_{0\le s\le \tau_{\beta_v}(2\lambda)} \sup_{u\ge 0}\left[L_{\beta_v}(s,u) + L_{\beta_v}(s,0)\right]\mathrm{d}x$$
$$\le [2/(\kappa\sqrt{v})]^{\frac{1}{\kappa}+1} \sup_{u\ge 0}\left[L_{\beta_v}(\tau_{\beta_v}(2\lambda),u) + 2\lambda\right],$$

for all large v. Thanks to the second Ray-Knight theorem (Fact 2), we know that $Q := (L_{\beta_v}(\tau_{\beta_v}(2\lambda),u),\ u\ge 0)$ is a 0-dimensional squared Bessel process starting from 2λ. Moreover, $x \mapsto x$ is a scale function of Q (see e.g. Revuz et al. [40, p. 442]). Hence, for large v,

$$\mathbb{P}\left(\sup_{0\le s\le \tau_{\beta_v}(2\lambda)} \varepsilon_2(v,s) \ge \left[2/(\kappa\sqrt{v})\right]^{1/\kappa+1}\sqrt{v}\right) \le \mathbb{P}\left(\sup_{u\ge 0} Q(u) \ge \frac{\sqrt{v}}{2}\right) = \frac{4\lambda}{\sqrt{v}}. \tag{124}$$

Similarly (this time, using $S(x) \sim \log x$, $x \to 0$), we have, for large v,

$$\mathbb{P}\left[\sup_{0\le s\le \tau_{\beta_v}(2\lambda)} \varepsilon_3(v,s) \ge \exp(-\sqrt{v}/2)\sqrt{v}\right] \le 4\lambda/\sqrt{v}. \tag{125}$$

To estimate $\varepsilon_4(v,s)$, we note that

$$\varepsilon_4(v,s) \le \sup_{|u|\le 1/\sqrt{v}} \left|L_{\beta_v}(s,u) - L_{\beta_v}(s,0)\right|. \tag{126}$$

Let $\varepsilon \in (0,1/2)$, $t_v > 0$, $\gamma \ge 1$ and define $(M)_t^* := \sup_{0\le s\le t}|M(s)|$ for $t > 0$ and any Brownian motion $(M(s),\ s\ge 0)$. Applying Barlow and Yor [6, (ii) p. 199] to the continuous martingale $\beta_v(.\wedge t_v)$ and its jointly continuous local time $(L_{\beta_v}(u\wedge t_v, a),\ u\ge 0,\ a\in\mathbb{R})$, we see that for some constant $C_{\gamma,\varepsilon} > 0$,

$$\left\|\sup_{0\le s\le t_v, a\ne b} \frac{|L_{\beta_v}(s,b) - L_{\beta_v}(s,a)|}{|b-a|^{1/2-\varepsilon}}\right\|_{\gamma} \le C_{\gamma,\varepsilon}\left\|[(\beta_v)_{t_v}^*]^{1/2+\varepsilon}\right\|_{\gamma} = C_{\gamma,\varepsilon}\left\|[(\beta)_{t_v}^*]^{1/2+\varepsilon}\right\|_{\gamma},$$

where $\|.\|_\gamma = \mathbb{E}(|.|^\gamma)^{1/\gamma}$. Then, by Chebyshev's inequality and a change of scale, for $\alpha > 0$,

$$\mathbb{P}\left(\sup_{0\le s\le t_v, a\ne b} \frac{|L_{\beta_v}(s,b) - L_{\beta_v}(s,a)|}{|b-a|^{1/2-\varepsilon}} \ge \alpha\right) \le \frac{(\sqrt{t_v})^{(1/2+\varepsilon)\gamma}}{\alpha^\gamma}\left[C_{\gamma,\varepsilon}\left\|[(\beta)_1^*]^{1/2+\varepsilon}\right\|_{\gamma}\right]^{\gamma}. \tag{127}$$

On $E_{16} := \left\{ \sup_{0\leq s\leq \tau_{\beta_v}(2\lambda), a\neq b} \frac{|L_{\beta_v}(s,b)-L_{\beta_v}(s,a)|}{|b-a|^{1/2-\varepsilon}} \leq v^{\frac{1}{2}(\frac{1}{2}-2\varepsilon)} \right\}$, we have by (126),

$$\sup_{0\leq s\leq \tau_{\beta_v}(2\lambda)} \varepsilon_4(v,s) \leq v^{-\frac{1}{2}(\frac{1}{2}-\varepsilon)} v^{\frac{1}{2}(\frac{1}{2}-2\varepsilon)} = v^{-\varepsilon/2}. \tag{128}$$

We now choose $\gamma := 2$ and $t_v := v^{\frac{1/4-\varepsilon}{1/2+\varepsilon}}$. Since $\mathbb{P}[\tau_{\beta_v}(2\lambda) > t_v] = \mathbb{P}[L_{\beta_v}(t_v, 0) < 2\lambda] = \mathbb{P}[\sup_{0\leq s\leq t_v} \beta(s) < 2\lambda] = \mathbb{P}[|\beta(t_v)| < 2\lambda] \leq 4\lambda/\sqrt{t_v}$ by Lévy's theorem (see e.g. Revuz et al. [40, VI th. 2.3]), we get for all large v (if ε is small enough),

$$\begin{aligned}\mathbb{P}\big[E_{16}(v)^c\big] \leq & \mathbb{P}\big[\tau_{\beta_v}(2\lambda) > t_v\big] + \mathbb{P}\left(\sup_{0\leq s\leq t_v, a\neq b} \frac{|L_{\beta_v}(s,b) - L_{\beta_v}(s,a)|}{|b-a|^{1/2-\varepsilon}} \geq v^{\frac{1}{2}(\frac{1}{2}-2\varepsilon)}\right) \\ \leq & 4\lambda\, v^{\frac{\varepsilon-1/4}{1+2\varepsilon}} + c_{69} v^{-1/4+\varepsilon} \leq v^{-1/4+2\varepsilon}/2.\end{aligned}$$

Combining this with (123)–(125) and (128), we obtain that, for $\varepsilon > 0$ small enough,

$$\mathbb{P}\left(\sup_{0\leq s\leq \tau_{\beta_v}(2\lambda)} |\varepsilon_1(v,s)| \geq 2v^{-\varepsilon/2}\right) \leq v^{-1/4+2\varepsilon}.$$

This gives, with the choice of $\delta_1 := 2\varepsilon/5$, $\mathbb{P}(E_{10}^c) \leq \mathbb{P}(E_{15}^c) \leq v^{-1/4+5\delta_1}$ (for large v). □

7.3 *Proof of Lemma 8*

Assume $0 < \kappa \leq 1$. Consider $0 < d < 1$, $\varepsilon \in (0, 1/2)$ such that $d(1/2+\varepsilon) + (\varepsilon - 1)(1/2-\varepsilon) < 0$, $M_\varepsilon > 0$, and a Brownian motion $(\beta(t),\ t \geq 0)$. We can write for $t > 0$,

$$\begin{aligned} & J_\beta(\kappa, t) \\ & = \left(\int_0^{S^{-1}(-t^\varepsilon)} + \int_{S^{-1}(-t^\varepsilon)}^{\alpha_\kappa} + \int_{\alpha_\kappa}^{S^{-1}(M_\varepsilon)} + \int_{S^{-1}(M_\varepsilon)}^1\right) L_\beta\left(\tau_\beta(\lambda), \frac{S(y)}{t}\right) \frac{y dy}{(1-y)^{2-\kappa}} \\ & := J_1 + J_2 + J_3 + J_4. \end{aligned}$$

We begin by estimating J_1. Since $S(x) \sim_{x\to 0} \log x$, we have, for large t, $J_1 \leq \exp(-t^\varepsilon/2) \sup_{s\geq 0} \tilde{Q}(s)$ where $\tilde{Q}$ is a 0-dimensional squared Bessel process starting from λ (by the second Ray-Knight theorem stated in Fact 2, applied to $-\beta$). Hence, we get $\mathbb{P}\left[J_1 \geq \exp(-t^\varepsilon/2)t^d\right] \leq \lambda/t^d$.

Fix a large constant $\gamma > 0$ such that $d(1/2 + \varepsilon + 1/\gamma) + (\varepsilon - 1)(1/2 - \varepsilon) < 0$, and define

$$E_{17} := \{\tau_\beta(\lambda) \le t^{2d}\},$$
$$E_{18} := \left\{ \sup_{0 \le s \le t^{2d}, a \ne b} \frac{|L_\beta(s,b) - L_\beta(s,a)|}{|b-a|^{1/2-\varepsilon}} \le t^{d(1/2+\varepsilon+1/\gamma)} \right\}.$$

Recall that $S(\alpha_\kappa) = 0$. To estimate J_2, we note that, on $E_{17} \cap E_{18}$, uniformly for all large t,

$$J_2 \le \left[\int_0^{\alpha_k} \frac{y\mathrm{d}y}{(1-y)^{2-\kappa}}\right] \sup_{-t^{\varepsilon-1} \le b \le 0} L_\beta(\tau_\beta(\lambda), b)$$
$$\le \frac{\alpha_\kappa\left[\lambda + t^{d(1/2+\varepsilon+1/\gamma)}(t^{\varepsilon-1})^{\frac{1}{2}-\varepsilon}\right]}{(1-\alpha_k)^{2-\kappa}} \le \frac{2\alpha_\kappa\lambda}{(1-\alpha_k)^{2-\kappa}}.$$

Notice that $\mathbb{P}(E_{17}^c) \le 2\lambda t^{-d}$ as proved after (128), and that $\mathbb{P}(E_{18}^c) \le c_{70} t^{-d}$ (by (127) with t^{2d} instead of t_v). Therefore, there exists $c_{71} > 0$ such that for large t,

$$\mathbb{P}\left(J_2 \le c_{71}\right) \ge \mathbb{P}(E_{17} \cap E_{18}) \ge 1 - c_{72} t^{-d}. \qquad (129)$$

We now turn to J_3. As already noticed after (123), we have $1 - S^{-1}(u) \underset{u \to +\infty}{\sim} (\kappa u)^{-1/\kappa}$. Therefore, we can choose $M_\varepsilon > 0$ such that

$$\forall u \ge M_\varepsilon, \qquad \frac{[1 - S^{-1}(u)]^{2\kappa-1}}{(\kappa u)^{1/\kappa-2}} \in (1-\varepsilon, 1+\varepsilon) \quad \text{and} \quad S^{-1}(u) \ge 1 - \varepsilon. \qquad (130)$$

On the event $E_{17} \cap E_{18}$, uniformly for all large t,

$$J_3 \le \sup_{0 \le x \le M_\varepsilon/t} L_\beta(\tau_\beta(\lambda), x) \int_{\alpha_\kappa}^{S^{-1}(M_\varepsilon)} y(1-y)^{\kappa-2}\mathrm{d}y$$
$$\le c_{73}\left[\lambda + t^{d(1/2+\varepsilon+1/\gamma)}(M_\varepsilon/t)^{\frac{1}{2}-\varepsilon}\right] \le 2\lambda c_{73}.$$

Consequently, $\mathbb{P}[J_3 \le 2\lambda c_{73}] \ge \mathbb{P}(E_{17} \cap E_{18}) \ge 1 - c_{72} t^{-d}$ for large t.

Now we write

$$J_4 = \kappa^{1/\kappa-2}\, t^{1/\kappa-1} \int_{M_\varepsilon/t}^{+\infty} \left[S^{-1}(tx)\right]^2 \frac{\left[1 - S^{-1}(tx)\right]^{2\kappa-1}}{(\kappa t)^{1/\kappa-2}} L_\beta(\tau_\beta(\lambda), x)\mathrm{d}x.$$

Therefore, (130) leads to

$$(1-\varepsilon)^3\int_{M_\varepsilon/t}^{+\infty} x^{1/\kappa-2}L_\beta(\tau_\beta(\lambda),x)\mathrm{d}x \le \kappa^{2-1/\kappa}t^{1-1/\kappa}J_4$$
$$\le (1+\varepsilon)\int_{M_\varepsilon/t}^{+\infty} x^{1/\kappa-2}L_\beta(\tau_\beta(\lambda),x)\mathrm{d}x. \tag{131}$$

Proof (Proof of Lemma 8: Part (i)) We first assume $0<\kappa<1$.

On $E_{17}\cap E_{18}$, for large t, we have $\int_0^{M_\varepsilon/t} x^{1/\kappa-2}L_\beta(\tau_\beta(\lambda),x)\mathrm{d}x \le c_{74}t^{1-1/\kappa}$. Recall K_β from (15). It follows from (131) and (129) that, for large t,

$$\mathbb{P}\left[(1-\varepsilon)^3K_\beta(\kappa)-(1-\varepsilon)^3c_{74}t^{1-1/\kappa}\le\kappa^{2-1/\kappa}t^{1-1/\kappa}J_4\le(1+\varepsilon)K_\beta(\kappa)\right]$$
$$\ge 1-c_{72}t^{-d}.$$

Since $J_\beta(\kappa,t)=J_1+J_2+J_3+J_4$, we get for large t,

$$\mathbb{P}\left\{(1-\varepsilon)^3K_\beta(\kappa)-\frac{c_{48}}{t^{1/\kappa-1}}\le\kappa^{2-1/\kappa}t^{1-1/\kappa}J_\beta(\kappa,t)\le(1+\varepsilon)K_\beta(\kappa)+\frac{c_{48}}{t^{1/\kappa-1}}\right\}$$
$$\ge 1-c_{75}t^{-d},$$

for some $c_{48}>0$, proving the lemma in the case $0<\kappa<1$. □

Proof of Lemma 8: Part (ii) We assume $\kappa=1$, thus $\lambda=8$.

By the definition of C_β (see (16)), we have

$$\int_{M_\varepsilon/t}^{\infty}\frac{L_\beta(\tau_\beta(8),x)}{x}\mathrm{d}x=C_\beta-\int_0^{M_\varepsilon/t}\frac{L_\beta(\tau_\beta(8),x)-8}{x}\mathrm{d}x+8\log t-8\log M_\varepsilon.$$

On $E_{17}\cap E_{18}$, for large t,

$$\int_0^{M_e/t}\frac{|L_\beta(\tau_\beta(8),x)-8|}{x}\mathrm{d}x\le\int_0^{M_e/t}\frac{t^{d(1/2+\varepsilon+1/\gamma)}x^{1/2-\varepsilon}}{x}\mathrm{d}x\le\varepsilon.$$

As in (50), $\mathbb{P}(C_\beta+8\log t<\log t)\le t^{-7}$. Therefore, by (131) and (129), we have for large t,

$$\mathbb{P}\left\{(1-\varepsilon)^4[C_\beta+8\log t]\le J_4\le(1+\varepsilon)^2[C_\beta+8\log t]\right\}\ge 1-c_{76}t^{-d}.$$

Since $J_\beta(1,t)=J_1+J_2+J_3+J_4$, we get for large t,

$$\mathbb{P}\left\{(1-\varepsilon)^4[C_\beta+8\log t]\le J_\beta(1,t)\le(1+\varepsilon)^3[C_\beta+8\log t]\right\}\ge 1-c_{77}t^{-d}.$$

This proves the lemma in the case $\kappa=1$. □

Acknowledgements I would like to thank Zhan Shi for many helpful discussions. I am also grateful to an anonymous referee for a very careful reading of the first version of the paper and for valuable comments which helped improve the presentation of the paper, including a recommendation to merge the first version of this paper with [19].

This research was partially supported by the French ANR project MEMEMO2 2010 BLAN 0125.

References

1. P. Andreoletti, On the concentration of Sinai's walk. Stoch. Process. Appl. **116**, 1377–1408 (2007)
2. P. Andreoletti, Almost sure estimates for the concentration neighborhood of Sinai's walk. Stoch. Process. Appl. **117**, 1473–1490 (2007)
3. P. Andreoletti, A. Devulder, Localization and number of visited valleys for a transient diffusion in random environment. Electron. J. Probab. **20**, 1–58 (2015)
4. P. Andreoletti, R. Diel, Limit law of the local time for Brox's diffusion. J. Theor. Probab. **24**, 634–656 (2011)
5. P. Andreoletti, A. Devulder, G. Véchambre, Renewal structure and local time for diffusions in random environment. ALEA, Lat. Am. J. Probab. Math. Stat. **13**, 863–923 (2016)
6. M.T. Barlow, M. Yor, Semimartingale inequalities via the Garsia-Rodemich-Rumsey lemma, and applications to local times. J. Funct. Anal. **49**, 198–229 (1982)
7. J. Bertoin, *Lévy Processes* (Cambridge University Press, Cambridge, 1996)
8. P. Biane, M. Yor, Valeurs principales associées aux temps locaux browniens. Bull. Sci. Math. **111**, 23–101 (1987)
9. N.H. Bingham, J.L. Teugels, Duality for regularly varying functions. Q. J. Math. Oxford Ser. (2) **26**, 333–353 (1975)
10. A.N. Borodin, P. Salminen, *Handbook of Brownian Motion—Facts and Formulae*, 2nd edn. (Birkhäuser, Boston, 2002)
11. T. Brox, A one-dimensional diffusion process in a Wiener medium. Ann. Probab. **14**, 1206–1218 (1986)
12. P. Carmona, The mean velocity of a Brownian motion in a random Lévy potential. Ann. Probab. **25**, 1774–1788 (1997)
13. D. Cheliotis, One-dimensional diffusion in an asymmetric random environment. Ann. Inst. Henri Poincaré Probab. Stat. **42**, 715–726 (2006)
14. D. Cheliotis, Localization of favorite points for diffusion in a random environment. Stoch. Proc. Appl. **118**, 1159–1189 (2008)
15. F. Comets, M. Falconnet, O. Loukianov, D. Loukianova, C. Matias, Maximum likelihood estimator consistency for a ballistic random walk in a parametric random environment. Stoch. Proc. Appl. **124**, 268–288 (2014)
16. P. Deheuvels, P. Révész, Simple random walk on the line in random environment. Probab. Theory Relat. Fields **72**, 215–230 (1986)
17. A. Dembo, O. Zeitouni, *Large Deviations Techniques and Applications*, 2nd edn. (Springer, New York, 1998)
18. A. Dembo, N. Gantert, Y. Peres, Z. Shi, Valleys and the maximum local time for random walk in random environment. Probab. Theory Relat. Fields **137**, 443–473 (2007)
19. A. Devulder, Almost sure asymptotics for a diffusion process in a drifted Brownian potential. Preprint arXiv:math/0511053 (2005)
20. A. Devulder, Some properties of the rate function of quenched large deviations for random walk in random environment. Markov Process. Relat. Fields **12**, 27–42 (2006)
21. A. Devulder, Persistence of some additive functionals of Sinai's walk. Ann. Inst. Henri Poincaré Probab. Stat. **52**, 1076–1105 (2016)

22. R. Diel, Almost sure asymptotics for the local time of a diffusion in Brownian environment. Stoch. Proc. Appl. **121**, 2303–2330 (2011)
23. R. Diel, G. Voisin, Local time of a diffusion in a stable Lévy environment. Stochastics **83**, 127–152 (2011)
24. D. Dufresne, Laguerre series for Asian and other options. Math. Finance **10**, 407–428 (2000)
25. G. Faraud, Estimates on the speedup and slowdown for a diffusion in a drifted Brownian potential. J. Theor. Probab. **24**, 194–239 (2009)
26. N. Gantert, Z. Shi, Many visits to a single site by a transient random walk in random environment. Stoch. Process. Appl. **99**, 159–176 (2002)
27. N. Gantert, Y. Peres, Z. Shi, The infinite valley for a recurrent random walk in random environment. Ann. Inst. Henri Poincaré Probab. Stat. **46**, 525–536 (2010)
28. A. Göing-Jaeschke, M. Yor, A survey and some generalizations of Bessel processes. Bernoulli **9**, 313–349 (2003)
29. Y. Hu, Z. Shi, The limits of Sinai's simple random walk in random environment. Ann. Probab. **26**, 1477–1521 (1998)
30. Y. Hu, Z. Shi, The local time of simple random walk in random environment. J. Theor. Probab. **11**, 765–793 (1998)
31. Y. Hu, Z. Shi, The problem of the most visited site in random environment. Probab. Theory Relat. Fields **116**, 273–302 (2000)
32. Y. Hu, Z. Shi, Moderate deviations for diffusions with Brownian potential. Ann. Probab. **32**, 3191–3220 (2004)
33. Y. Hu, Z. Shi, M. Yor, Rates of convergence of diffusions with drifted Brownian potentials. Trans. Am. Math. Soc. **351**, 3915–3934 (1999)
34. S. Karlin, H.M. Taylor, *A Second Course in Stochastic Processes* (Academic, New York, 1981)
35. K. Kawazu, H. Tanaka, On the maximum of a diffusion process in a drifted Brownian environment, in *Séminaire de Probabilités XXVII.* Lecture Notes in Mathematics, vol. 1557 (Springer, Berlin, 1993), pp. 78–85
36. K. Kawazu, H. Tanaka, A diffusion process in a Brownian environment with drift. J. Math. Soc. Japan **49**, 189–211 (1997)
37. P. Le Doussal, C. Monthus, D. Fisher, Random walkers in one-dimensional random environments: exact renormalization group analysis. Phys. Rev. E 3 **59**, 4795–4840 (1999)
38. P. Mathieu, On random perturbations of dynamical systems and diffusions with a Brownian potential in dimension one. Stoch. Process. Appl. **77**, 53–67 (1998)
39. P. Révész, *Random Walk in Random and Non-random Environments* (World Scientific, Singapore, 1990)
40. D. Revuz, M. Yor, *Continuous Martingales and Brownian Motion*, 3rd edn. (Springer, Berlin, 1999)
41. G. Samorodnitsky, M.S. Taqqu, *Stable Non-Gaussian Random Processes* (Chapman and Hall, New York, 1994)
42. S. Schumacher, Diffusions with random coefficients. Contemp. Math. **41**, 351–356 (1985)
43. Z. Shi, A local time curiosity in random environment. Stoch. Process. Appl. **76**, 231–250 (1998)
44. Z. Shi, Sinai's walk via stochastic calculus, in *Panoramas et Synthèses*, vol. 12, ed. by F. Comets, E. Pardoux (Société Mathématique de Francen, Paris, 2001), pp. 53–74
45. A. Singh, Limiting behavior of a diffusion in an asymptotically stable environment. Ann. Inst. Henri Poincaré Probab. Stat. **43**, 101–138 (2007)
46. A. Singh, Rates of convergence of a transient diffusion in a spectrally negative lévy potential. Ann. Probab. **36**, 279–318 (2008)
47. M. Taleb, Large deviations for a Brownian motion in a drifted Brownian potential. Ann. Probab. **29**, 1173–1204 (2001)
48. M. Talet, Annealed tail estimates for a Brownian motion in a drifted Brownian potential. Ann. Probab. **35**, 32–67 (2007)

49. H. Tanaka, Limit distribution for 1-dimensional diffusion in a reflected Brownian medium, in *Séminaire de Probabilités XXI*. Lecture Notes in Mathematics, vol. 1247 (Springer, Berlin, 1987), pp. 246–261
50. J. Warren, M. Yor, The Brownian burglar: conditioning Brownian motion by its local time process, in *Séminaire de Probabilités XXXII*. Lecture Notes in Mathematics, vol. 1686 (Springer, Berlin, 1998), pp. 328–342
51. M. Yor, Sur certaines fonctionnelles exponentielles du mouvement brownien réel. J. Appl. Probab. **29**, 202–208 (1992)
52. M. Yor, *Local Times and Excursions for Brownian Motion: A Concise Introduction*. Lecciones en Mathemàticas, vol. 1 (Universidad Central de Venuezuela, Caracas, 1995)
53. O. Zeitouni, Lecture notes on random walks in random environment, in *École d'été de Probabilités de Saint-Flour 2001*. Lecture Notes in Mathematics, vol. 1837 (Springer, Berlin, 2004), pp. 189–312
54. O. Zindy, Upper limits of Sinai's walk in random scenery. Stoch. Process. Appl. **118**, 981–1003 (2008)

A Link Between Bougerol's Identity and a Formula Due to Donati-Martin, Matsumoto and Yor

Mátyás Barczy and Peter Kern

Dedicated to the memory of Marc Yor 1949–2014

Abstract We point out an easy link between two striking identities on exponential functionals of the Wiener process and the Wiener bridge originated by Bougerol, and Donati-Martin, Matsumoto and Yor, respectively. The link is established using a continuous one-parameter family of Gaussian processes known as α-Wiener bridges or scaled Wiener bridges, which in case $\alpha = 0$ coincides with a Wiener process and for $\alpha = 1$ is a version of the Wiener bridge.

1 Introduction

Our starting point is Bougerol's identity in [5] which states that

$$\sinh(B_t) \stackrel{\mathrm{d}}{=} W_{A_t} \qquad \text{for every fixed } t \geq 0, \tag{1}$$

where $(B_t)_{t\geq 0}$ and $(W_t)_{t\geq 0}$ are independent standard Wiener processes, $\stackrel{\mathrm{d}}{=}$ denotes equality in distribution, and

$$A_t = \int_0^t \exp(2\,B_s)\,\mathrm{d}s \qquad \text{for } t \geq 0.$$

M. Barczy (✉)
Faculty of Informatics, University of Debrecen, Pf. 12, H-4010 Debrecen, Hungary
e-mail: barczy.matyas@inf.unideb.hu

P. Kern
Mathematical Institute, Heinrich-Heine-University Düsseldorf, Universitätsstr. 1, 40225 Düsseldorf, Germany
e-mail: kern@math.uni-duesseldorf.de

C. Donati-Martin et al. (eds.), *Séminaire de Probabilités XLVIII*, Lecture Notes in Mathematics 2168, DOI 10.1007/978-3-319-44465-9_6

In fact there is also a generalization of Bougerol's identity with equality in law for stochastic processes due to Alili et al. [1, Proposition 2]; cf. also [13, formula (69)] or [15, page 200]. Recently, there has been a renewed interest in generalizations of Bougerol's identity (1). Bertoin et al. [3] presented a two-dimensional extension of (1) that involves some exponential functional and the local time at 0 of a standard Wiener process. For another two-dimensional extension of (1), and even a three-dimensional one we refer to Vakeroudis [13, Sects. 4.2 and 4.3].

We are only interested in the following particular case of the identity (1) presented in [13, 14]. Bougerol's identity (1) is equivalent to the equality of the corresponding continuous Lebesgue densities, which yields

$$\frac{1}{\sqrt{(1+x^2)t}}\exp\left(-\frac{\mathrm{Arsinh}^2(x)}{2t}\right) = \mathbb{E}\left[\frac{1}{\sqrt{A_t}}\exp\left(-\frac{x^2}{2A_t}\right)\right]$$

for all $t > 0$ and $x \in \mathbb{R}$, see, e.g., [14, formula (1.e)]. Especially, for $x = 0$, by the $1/2$-self-similarity of a standard Wiener process and a change of variables $r = (4/\beta^2)s$ for some $\beta > 0$ we get

$$\begin{aligned} t^{-1/2} &= \mathbb{E}\left[\left(\int_0^t \exp(2B_s)\,\mathrm{d}s\right)^{-1/2}\right] = \mathbb{E}\left[\left(\int_0^t \exp(\beta B_{(4/\beta^2)s})\,\mathrm{d}s\right)^{-1/2}\right] \\ &= \frac{2}{\beta}\cdot\mathbb{E}\left[\left(\int_0^{(4/\beta^2)t} \exp(\beta B_r)\,\mathrm{d}r\right)^{-1/2}\right]. \end{aligned} \tag{2}$$

Hence, setting $t = \beta^2/4$ we get for every $\beta > 0$

$$\mathbb{E}\left[\left(\int_0^1 \exp(\beta B_s)\,\mathrm{d}s\right)^{-1/2}\right] = 1.$$

This formula is a consequence of Bougerol's identity (1) which obviously holds for $\beta = 0$ and also remains true for $\beta < 0$, since $(-B_t)_{t\geq 0}$ is a Wiener process, i.e.,

$$\mathbb{E}\left[\left(\int_0^1 \exp(\beta B_s)\,\mathrm{d}s\right)^{-1/2}\right] = 1 \quad \text{for every } \beta \in \mathbb{R}. \tag{3}$$

A similar identity due to Donati-Martin, Matsumoto and Yor [7, 8] holds when replacing the Wiener process $(B_t)_{t\geq 0}$ by a Wiener bridge $(B_t^\circ = B_t - tB_1)_{t\in[0,1]}$, a zero mean Gaussian process with covariance function $\mathrm{Cov}(B_s^\circ, B_t^\circ) = s(1-t)$ for $0 \leq s \leq t \leq 1$. Namely, this identity states that

$$\mathbb{E}\left[\left(\int_0^1 \exp(\beta B_s^\circ)\,\mathrm{d}s\right)^{-1}\right] = 1 \quad \text{for every } \beta \in \mathbb{R}. \tag{4}$$

Hobson [9] provides a simple proof of (4) using a relationship between a Wiener bridge and a Wiener excursion obtained by Biane [4]. A further elementary proof of (4) is given in [7, Proposition 2.1].

Donati-Martin et al. [7] already pointed out how to obtain a link between the two identities (3) and (4) in the sense that the identity (3) follows from the identity (4) as a consequence of a formula combining exponential functionals of the Wiener process and the Wiener bridge, for details we refer to [7, Proposition 3.2].

Our aim is to give a different link between the two identities (3) and (4) using so-called α-Wiener bridges (also known as scaled Wiener bridges). These processes build a one-parameter family of Gaussian processes for parameter $\alpha \in \mathbb{R}$. They have been first considered by Brennan and Schwartz [6] and later have been investigated by Mansuy [11] and Barczy and Pap [2]. For our purposes an α-Wiener bridge $(X_t^{(\alpha)})_{t\in[0,1)}$ can be defined as a (weak) solution of the stochastic differential equation (SDE)

$$\mathrm{d}X_t^{(\alpha)} = -\frac{\alpha}{1-t} X_t^{(\alpha)}\,\mathrm{d}t + \mathrm{d}B_t, \qquad t \in [0,1), \tag{5}$$

with initial condition $X_0^{(\alpha)} = 0$. Barczy and Pap [2] have shown that $(X_t^{(\alpha)})_{t\in[0,1)}$ is a bridge in the sense that $X_t^{(\alpha)} \to 0 =: X_1^{(\alpha)}$ as $t \uparrow 1$ almost surely if and only if $\alpha > 0$. Moreover, for $\alpha \geq 0$ it is shown in [2] that $(X_t^{(\alpha)})_{t\in[0,1]}$ is a zero mean Gaussian process with covariance function

$$\mathrm{Cov}(X_s^{(\alpha)}, X_t^{(\alpha)}) = \begin{cases} \frac{(1-s)^\alpha(1-t)^\alpha}{1-2\alpha}\left(1-(1-s)^{1-2\alpha}\right) & \text{if } \alpha \neq \frac{1}{2} \\ \sqrt{(1-s)(1-t)}\,\log\left(\frac{1}{1-s}\right) & \text{if } \alpha = \frac{1}{2} \end{cases} \tag{6}$$

for $0 \leq s \leq t \leq 1$. Note that for fixed $0 \leq s \leq t \leq 1$, (6) is continuous in $\alpha \geq 0$, which for $\alpha \to \frac{1}{2}$ can be easily seen by l'Hospital's rule. The unique strong solution of the SDE (5) with initial condition $X_0^{(\alpha)} = 0$ is given by

$$X_t^{(\alpha)} = \int_0^t \left(\frac{1-t}{1-s}\right)^\alpha \mathrm{d}B_s \quad \text{for } t \in [0,1), \tag{7}$$

and shows that $(X_t^{(0)})_{t\in[0,1]} = (B_t)_{t\in[0,1]}$ and $(X_t^{(1)})_{t\in[0,1]} \stackrel{\mathrm{d}}{=} (B_t^\circ)_{t\in[0,1]}$. The latter is due to the fact that both sides of the equation are zero mean Gaussian processes with the same covariance function. Hence, variation of the parameter $\alpha \in [0,1]$ continuously connects the Wiener process for $\alpha = 0$ with the Wiener bridge for $\alpha = 1$ in the sense that for $\alpha, \alpha_0 \geq 0$, the finite dimensional distributions of $(X^{(\alpha)})_{t\in[0,1]}$ converge weakly to those of $(X^{(\alpha_0)})_{t\in[0,1]}$ as $\alpha \to \alpha_0$. This follows directly from the continuity in α of the covariance function (6) and is the key observation for our link between the identities (3) and (4).

The paper is organized as follows. We will first show that certain space-time rescalings of an α-Wiener bridge either coincide in law with a usual Wiener bridge for $\alpha > \frac{1}{2}$ or with the Wiener process for $0 \le \alpha < \frac{1}{2}$, see Proposition 1. Then an application of these space-time rescalings to the identity (4) and (3), respectively, yields two new identities for certain transformations of exponential functionals of α-Wiener bridges which coincide when $\alpha = \frac{1}{2}$, see Theorem 1. We further show that a $\frac{1}{2}$-Wiener bridge can be scaled to both, a Wiener bridge and a standard Wiener process, see Proposition 2. As a consequence, we present another two identities for certain transformations of exponential functionals of $\frac{1}{2}$-Wiener bridges in Theorem 2.

2 Link Between the Identities

In the sequel, $\stackrel{\mathcal{D}}{=}$ denotes equality in law for stochastic processes on the space of continuous functions $C([0,1])$ or $C([0,\infty))$, respectively.

Proposition 1 *(a) For $\alpha > \frac{1}{2}$ we have*

$$\left(\sqrt{2\alpha-1}\, t^{\frac{\alpha-1}{2\alpha-1}} X^{(\alpha)}_{1-t^{1/(2\alpha-1)}}\right)_{t\in[0,1]} \stackrel{\mathcal{D}}{=} (X^{(1)}_t)_{t\in[0,1]}.$$

(b) For $0 \le \alpha < \frac{1}{2}$ we have

$$\left(\sqrt{1-2\alpha}\,(1-t)^{-\frac{\alpha}{1-2\alpha}} X^{(\alpha)}_{1-(1-t)^{1/(1-2\alpha)}}\right)_{t\in[0,1]} \stackrel{\mathcal{D}}{=} (X^{(0)}_t)_{t\in[0,1]}.$$

Proof We will first prove that the processes under consideration are zero mean Gaussian processes having almost surely continuous trajectories, which is not obvious for the left-hand sides as $t \downarrow 0$ for $\alpha \in (\frac{1}{2}, 1)$ in (a), and as $t \uparrow 1$ in (b), respectively. Once we know this, it remains to show the equality of covariance functions.

(a) Let us introduce a continuous martingale $(M_t)_{t\in[0,1)}$ related to the process $X^{(\alpha)}$ given by (7). Namely, let

$$M_t := \frac{X^{(\alpha)}_t}{(1-t)^\alpha} = \int_0^t \frac{1}{(1-s)^\alpha}\,\mathrm{d}B_s \quad \text{for } t \in [0,1)$$

with quadratic variation $\langle M\rangle_t = (1-(1-t)^{1-2\alpha})/(1-2\alpha) \to \infty$ as $t \uparrow 1$ for $\alpha > \frac{1}{2}$ as obtained in [2, formula (3.1)]. Then, similarly to the proof of [2, Lemma 3.1], for the increasing function $[1,\infty) \ni x \mapsto f(x) = x^{3/4}$ with $\int_1^\infty (f(x))^{-2}\,\mathrm{d}x < \infty$, an application of [10, Theoreme 1] or Exercise 1.16 in Chapter V of [12] gives $M_t/f(\langle M\rangle_t) \to 0$ a.s. as $t \uparrow 1$. Letting $t = 1-s^{1/(2\alpha-1)} \uparrow$

1 as $s \downarrow 0$ this shows

$$\frac{s^{\frac{-\alpha}{2\alpha-1}} X^{(\alpha)}_{1-s^{1/(2\alpha-1)}}}{\left((1-s^{-1})/(1-2\alpha)\right)^{3/4}} \to 0 \quad \text{a.s. as } s \downarrow 0.$$

To obtain $s^{\frac{\alpha-1}{2\alpha-1}} X^{(\alpha)}_{1-s^{1/(2\alpha-1)}} \to 0$ a.s. as $s \downarrow 0$ it suffices to see that for $s \downarrow 0$ we have

$$s^{\frac{\alpha-1}{2\alpha-1}} s^{\frac{\alpha}{2\alpha-1}} (s^{-1}-1)^{3/4} = s(s^{-1}-1)^{3/4} = s^{1/4}(1-s)^{3/4} \to 0.$$

Hence the centered Gaussian processes under consideration almost surely have continuous sample paths on $[0, 1]$ starting in the origin. Thus it remains to show the equality of their covariance functions for $0 < s \leq t \leq 1$. Using (6) and the fact that the function $(0, 1] \ni t \mapsto 1 - t^{1/(2\alpha-1)}$ is decreasing, we get for $0 < s \leq t \leq 1$

$$\operatorname{Cov}\left(X^{(\alpha)}_{1-s^{1/(2\alpha-1)}}, X^{(\alpha)}_{1-t^{1/(2\alpha-1)}}\right) = \frac{s^{\frac{\alpha}{2\alpha-1}} t^{\frac{\alpha}{2\alpha-1}}}{1-2\alpha} (1-t^{-1})$$

$$= \frac{s^{\frac{\alpha}{2\alpha-1}} t^{\frac{\alpha}{2\alpha-1}-1}}{2\alpha-1} (1-t) = \frac{s^{\frac{1-\alpha}{2\alpha-1}} t^{\frac{1-\alpha}{2\alpha-1}}}{2\alpha-1} s(1-t)$$

from which the assertion easily follows.

(b) In case $\alpha = 0$ the identity is trivially fulfilled. For $0 < \alpha < \frac{1}{2}$ it is shown in the proof of [2, Lemma 3.1] that $\lim_{t\uparrow 1}(1-t)^{-\alpha} X^{(\alpha)}_t$ exists in $\mathbb{R}$ almost surely and has a normal distribution as a limit of normally distributed random variables. Letting $t = 1 - (1-s)^{1/(1-2\alpha)} \uparrow 1$ as $s \uparrow 1$ we have

$$\lim_{s\uparrow 1} (1-s)^{-\frac{\alpha}{1-2\alpha}} X^{(\alpha)}_{1-(1-s)^{1/(1-2\alpha)}} \quad \text{exists a.s.,}$$

which shows that the centered Gaussian processes under consideration almost surely have continuous sample paths on $[0, 1]$ starting in the origin. Thus it remains to show the equality of their covariance functions. Using (6) and the fact that the function $[0, 1] \ni t \mapsto 1 - (1-t)^{1/(1-2\alpha)}$ is increasing, we get for $0 \leq s \leq t \leq 1$

$$\operatorname{Cov}\left(X^{(\alpha)}_{1-(1-s)^{1/(1-2\alpha)}}, X^{(\alpha)}_{1-(1-t)^{1/(1-2\alpha)}}\right) = \frac{(1-s)^{\frac{\alpha}{1-2\alpha}} (1-t)^{\frac{\alpha}{1-2\alpha}}}{1-2\alpha} s$$

from which again the assertion easily follows. □

Theorem 1

(a) For $\alpha > \frac{1}{2}$ and any $\beta \in \mathbb{R}$ we have

$$\mathbb{E}\left[\left(\int_0^1 \exp\left(\frac{\beta}{(1-s)^{1-\alpha}} X_s^{(\alpha)}\right) \frac{ds}{(1-s)^{2(1-\alpha)}}\right)^{-1}\right] = 2\alpha - 1.$$

(b) For $0 \le \alpha < \frac{1}{2}$ and any $\beta \in \mathbb{R}$ we have

$$\mathbb{E}\left[\left(\int_0^1 \exp\left(\frac{\beta}{(1-s)^{\alpha}} X_s^{(\alpha)}\right) \frac{ds}{(1-s)^{2\alpha}}\right)^{-1/2}\right] = \sqrt{1-2\alpha}.$$

(c) For $\alpha = \frac{1}{2}$ and any $\beta \in \mathbb{R}$ both identities in (a) and (b) hold.

Remark 1 For the $\frac{1}{2}$-Wiener bridge the two identities in (a) and (b) of Theorem 1 are valid by part (c) and are in fact equivalent, since both identities show that for any $\beta \in \mathbb{R}$ the non-negative random variable

$$Y(\beta) := \left(\int_0^1 \exp\left(\frac{\beta}{\sqrt{1-s}} X_s^{(1/2)}\right) \frac{ds}{1-s}\right)^{-1/2} = 0 \quad \text{almost surely.}$$

Hence the version of the Bougerol identity in (b) represents the mean $\mathbb{E}[Y(\beta)] = 0$, whereas the formula (a), as a version of the identity due to Donati-Martin, Matsumoto and Yor, represents the second moment $\mathbb{E}[(Y(\beta))^2] = 0$.

Proof of Theorem 1

(a) An application of Proposition 1 (a) to (4) together with a change of variables $s = 1 - t^{\frac{1}{2\alpha-1}}$ yields for any $\beta \in \mathbb{R}$

$$\begin{aligned}
1 &= \mathbb{E}\left[\left(\int_0^1 \exp(\beta X_t^{(1)})\, dt\right)^{-1}\right] \\
&= \mathbb{E}\left[\left(\int_0^1 \exp\left(\beta\sqrt{2\alpha-1}\, t^{\frac{\alpha-1}{2\alpha-1}} X^{(\alpha)}_{1-t^{1/(2\alpha-1)}}\right) dt\right)^{-1}\right] \\
&= \mathbb{E}\left[\left(\int_0^1 \exp\left(\frac{\tilde{\beta}}{(1-s)^{1-\alpha}} X_s^{(\alpha)}\right) \cdot (2\alpha-1) \frac{ds}{(1-s)^{2(1-\alpha)}}\right)^{-1}\right],
\end{aligned}$$

where $\tilde{\beta} = \beta\sqrt{2\alpha-1} \in \mathbb{R}$ is arbitrary.

(b) For $\alpha = 0$ the identity is a restatement of (3). For $0 < \alpha < \frac{1}{2}$ an application of Proposition 1 (b) to (3) together with a change of variables $s = 1-(1-t)^{1/(1-2\alpha)}$

yields for any $\beta \in \mathbb{R}$

$$\begin{aligned}
1 &= \mathbb{E}\left[\left(\int_0^1 \exp(\beta X_t^{(0)})\,\mathrm{d}t\right)^{-1/2}\right] \\
&= \mathbb{E}\left[\left(\int_0^1 \exp\left(\beta\sqrt{1-2\alpha}\,(1-t)^{-\frac{\alpha}{1-2\alpha}} X^{(\alpha)}_{1-(1-t)^{1/(1-2\alpha)}}\right)\mathrm{d}t\right)^{-1/2}\right] \\
&= \mathbb{E}\left[\left(\int_0^1 \exp\left(\frac{\tilde{\beta}}{(1-s)^{\alpha}} X_s^{(\alpha)}\right)\cdot(1-2\alpha)\,\frac{\mathrm{d}s}{(1-s)^{2\alpha}}\right)^{-1/2}\right],
\end{aligned}$$

where $\tilde{\beta} = \beta\sqrt{1-2\alpha} \in \mathbb{R}$ is arbitrary.

(c) For $\alpha = \frac{1}{2}$ the process $(M_t)_{t\in[0,1)}$ with $M_t = (1-t)^{-1/2}X_t^{(1/2)} = \int_0^t (1-s)^{-1/2}\mathrm{d}B_s$ is a centered continuous martingale with quadratic variation $\langle M\rangle_t = -\log(1-t) \to \infty$ as $t \uparrow 1$; see formulas (3.1) and (3.2) in [2]. Hence by the Dambis, Dubins-Schwarz theorem there exists a Wiener process $(\tilde{B}_t)_{t\geq 0}$ such that $(M_t)_{t\in[0,1)} = (\tilde{B}_{\langle M\rangle_t})_{t\in[0,1)}$ almost surely; see Theorem 1.6 in Chapter V of [12]. It follows by a change of variables $t = \langle M\rangle_s = -\log(1-s)$ and monotone convergence that for $\beta \neq 0$

$$\begin{aligned}
&\mathbb{E}\left[\left(\int_0^1 \exp\left(\frac{\beta}{\sqrt{1-s}} X_s^{(1/2)}\right)\frac{\mathrm{d}s}{1-s}\right)^{-1/2}\right] \\
&\quad = \mathbb{E}\left[\left(\int_0^1 \exp\left(\beta\tilde{B}_{-\log(1-s)}\right)\frac{\mathrm{d}s}{1-s}\right)^{-1/2}\right] = \mathbb{E}\left[\left(\int_0^\infty \exp\left(\beta\tilde{B}_t\right)\mathrm{d}t\right)^{-1/2}\right] \\
&\quad = \lim_{T\to\infty}\mathbb{E}\left[\left(\int_0^T \exp\left(\beta\tilde{B}_t\right)\mathrm{d}t\right)^{-1/2}\right] = \lim_{T\to\infty} T^{-1/2} = 0,
\end{aligned}$$

where the last but one equality follows by setting $t = \beta^2 T/4$ in (2). Since in case $\beta = 0$ the expectation is obviously vanishing, this shows that the identity in (b) is fulfilled for $\alpha = \frac{1}{2}$. In particular it shows that a non-negative random variable has zero expectation and thus is equal to zero almost surely. Hence also its second moment vanishes, which proves the identity in (a) for $\alpha = \frac{1}{2}$. □

In case $\alpha = \frac{1}{2}$ it is possible to link the $\frac{1}{2}$-Wiener bridge $(X_t^{(1/2)})_{t\in[0,1]}$ to both identities (4) and (3) with non-vanishing expectation by either introducing an additional log-term in the integrand or by integrating over a smaller domain as follows. We first present the corresponding space-time scalings, which might be of independent interest.

Proposition 2 *We have*

$$\left(t\sqrt{\exp(t^{-1}-1)}X^{(1/2)}_{1-\exp(1-t^{-1})}\right)_{t\in[0,1]} \stackrel{\mathcal{D}}{=} (X^{(1)}_t)_{t\in[0,1]}. \tag{8}$$

$$\left(\mathrm{e}^{t/2}X^{(1/2)}_{1-\exp(-t)}\right)_{t\geq 0} \stackrel{\mathcal{D}}{=} (X^{(0)}_t)_{t\geq 0}. \tag{9}$$

Proof We first show that as $t \downarrow 0$ we have

$$t\sqrt{\exp(t^{-1}-1)}X^{(1/2)}_{1-\exp(1-t^{-1})} \to 0 \quad \text{a.s.} \tag{10}$$

From the proof of part (c) of Theorem 1 we know that there exists a Wiener process $(\tilde{B}_t)_{t\geq 0}$ such that $((1-s)^{-1/2}X^{(1/2)}_s)_{s\in[0,1)} = (\tilde{B}_{-\log(1-s)})_{s\in[0,1)}$ almost surely. Letting $s = 1-\exp(1-t^{-1})$ we get

$$\left(\sqrt{\exp(t^{-1}-1)}X^{(1/2)}_{1-\exp(1-t^{-1})}\right)_{t\in(0,1]} = \left(\tilde{B}_{t^{-1}-1}\right)_{t\in(0,1]} \qquad \text{a.s.}$$

from which (10) follows by the strong law of large numbers for Brownian motion, since almost surely

$$t\sqrt{\exp(t^{-1}-1)}X^{(1/2)}_{1-\exp(1-t^{-1})} = t\tilde{B}_{t^{-1}-1} = (1-t)\,\frac{t}{1-t}\,\tilde{B}_{\frac{1-t}{t}} \to 0$$

as $t \downarrow 0$. Hence the centered Gaussian processes under consideration in (8) almost surely have continuous sample paths on $[0,1]$ starting in the origin. Thus it remains to show the equality of their covariance functions for $0 < s \leq t \leq 1$. Using (6) and the fact that the function $(0,1] \ni t \mapsto 1-\exp(1-t^{-1})$ is decreasing, we get for any $0 < s \leq t \leq 1$,

$$\begin{aligned}\operatorname{Cov}\left(X^{(1/2)}_{1-\exp(1-s^{-1})}, X^{(1/2)}_{1-\exp(1-t^{-1})}\right) &= \sqrt{\exp(1-s^{-1})}\sqrt{\exp(1-t^{-1})}(t^{-1}-1) \\ &= \frac{\sqrt{\exp(1-s^{-1})}\sqrt{\exp(1-t^{-1})}}{s\cdot t}\, s(1-t),\end{aligned}$$

from which (8) easily follows. Similarly, for any $0 \leq s \leq t$ we get using (6)

$$\operatorname{Cov}\left(X^{(1/2)}_{1-\exp(-s)}, X^{(1/2)}_{1-\exp(-t)}\right) = \mathrm{e}^{-s/2}\mathrm{e}^{-t/2}\,s$$

from which (9) easily follows. □

Theorem 2 *For any $\beta \in \mathbb{R}$ we have*

$$\mathbb{E}\left[\left(\int_0^1 \exp\left(\frac{\beta}{\sqrt{1-s}\,(1-\log(1-s))}X^{(1/2)}_s\right)\frac{\mathrm{d}s}{(1-s)\,(1-\log(1-s))^2}\right)^{-1}\right] = 1$$

and

$$\mathbb{E}\left[\left(\int_0^{1-e^{-1}} \exp\left(\frac{\beta}{\sqrt{1-s}} X_s^{(1/2)}\right) \frac{ds}{1-s}\right)^{-1/2}\right] = 1.$$

Proof Applying (8) to (4) together with a change of variables $s = 1 - e^{-(t^{-1}-1)}$ yields for any $\beta \in \mathbb{R}$

$$\begin{aligned}
1 &= \mathbb{E}\left[\left(\int_0^1 \exp(\beta X_t^{(1)})\,dt\right)^{-1}\right] \\
&= \mathbb{E}\left[\left(\int_0^1 \exp\left(\beta t\sqrt{\exp(t^{-1}-1)} X^{(1/2)}_{1-\exp(1-t^{-1})}\right) dt\right)^{-1}\right] \\
&= \mathbb{E}\left[\left(\int_0^1 \exp\left(\frac{\beta}{\sqrt{1-s}\,(1-\log(1-s))} X_s^{(1/2)}\right) \frac{ds}{(1-s)\,(1-\log(1-s))^2}\right)^{-1}\right]
\end{aligned}$$

which proves the first identity. Similarly, an application of (9) to (3) together with a change of variables $s = 1 - e^{-t}$ yields for any $\beta \in \mathbb{R}$

$$\begin{aligned}
1 &= \mathbb{E}\left[\left(\int_0^1 \exp(\beta X_t^{(0)})\,dt\right)^{-1/2}\right] \\
&= \mathbb{E}\left[\left(\int_0^1 \exp\left(\beta\, e^{t/2} X^{(1/2)}_{1-e^{-t}}\right) dt\right)^{-1/2}\right] \\
&= \mathbb{E}\left[\left(\int_0^{1-e^{-1}} \exp\left(\frac{\beta}{\sqrt{1-s}} X_s^{(1/2)}\right) \frac{ds}{1-s}\right)^{-1/2}\right]
\end{aligned}$$

which proves the second identity. □

Remark 2 Motivated by the identities (3) and (4), one can formulate the open question whether there exists a (continuous) function $p : [0,1] \to (-\infty, 0)$ such that

$$\mathbb{E}\left[\left(\int_0^1 \exp\left(\beta X_t^{(\alpha)}\right) dt\right)^{p(\alpha)}\right] = 1 \qquad \text{for every } \beta \in \mathbb{R}.$$

References

1. L. Alili, D. Dufresne, M. Yor, Sur l'identité de Bougerol pour les functionelles du mouvement brownien avec drift, in *Exponential Functionals and Principal Values related to Brownian Motion*, ed. by M. Yor (Biblioteca Revista Matemática Iberoamericana, Madrid, 1997), pp. 3–14
2. M. Barczy, G. Pap, Alpha-Wiener bridges: singularity of induced measures and sample path properties. Stoch. Anal. Appl. **28**, 447–466 (2010)
3. J. Bertoin, D. Dufresne, M. Yor, Some two-dimensional extensions of Bougerol's identity in law for the exponential functional of linear Brownian motion. Rev. Mat. Iberoam. **29**, 1307–1324 (2013)
4. P. Biane, Relations entre pont et excursion du mouvement Brownian réel. Ann. Inst. Henri Poincaré Ser. B **22**, 1–7 (1986)
5. P. Bougerol, Exemples de thèorémes locaux sur les groupes résolubles. Ann. Inst. Henri Poincaré **19**, 369–391 (1983)
6. M.J. Brennan, E.S. Schwartz, Arbitrage in stock index futures. J. Bus. **63**, S7–S31 (1990)
7. C. Donati-Martin, H. Matsumoto, M. Yor, On striking identities about the exponential functionals of the Brownian bridge and Brownian motion. Period. Math. Hung. **41**, 103–119 (2000)
8. C. Donati-Martin, H. Matsumoto, M. Yor, The law of geometric Brownian motion and its integrals, revisited; application to conditional moments, in *Mathematical Finance, Bachelier Congress 2000*, ed. by H. Geman, et al. (Springer, Berlin, 2002), pp. 221–243
9. D. Hobson, A short proof of an identity for a Brownian bridge due to Donati-Martin, Matsumoto and Yor. Stat. Probab. Lett. **77**, 148–150 (2007)
10. D. Lépingle, Sur les comportement asymptotique des martingales locales. Semin. Probab. XII, Lect. Notes Math. **649**, 148–161 (1978)
11. R. Mansuy, On a one-parameter generalization of the Brownian bridge and associated quadratic functionals. J. Theor. Probab. **17**, 1021–1029 (2004)
12. D. Revuz, M. Yor, *Continuous Martingales and Brownian Motion*, 3rd edn., corr. 2nd printing (Springer, Berlin, 2001)
13. S. Vakeroudis, Bougerol's identity in law and extensions. Probab. Surv. **9**, 411–437 (2012)
14. M. Yor, On some exponential functionals of Brownian motion. Adv. Appl. Probab. **24**, 509–531 (1992)
15. M. Yor, *Exponential Functionals of Brownian Motion and Related Processes* (Springer, Berlin, 2001)

Large Deviation Principle for Bridges of Sub-Riemannian Diffusion Processes

Ismaël Bailleul

Abstract We prove that bridges of subelliptic diffusions on a compact manifold, with distinct ends, satisfy a large deviation principle in the space of Hölder continuous functions, with a good rate function, when the travel time tends to 0. This leads to the identification of the deterministic first order asymptotics of the distribution of the bridge under generic conditions on the endpoints of the bridge.

1 Introduction

Let M be a compact, connect and oriented m-dimensional smooth manifold, embedded in some ambient Euclidean space $\left(\mathbb{R}^d, |\cdot|_d\right)$. Let $V_1, \ldots, V_\ell$ be smooth vector fields on M, whose Lie algebra has maximal rank everywhere. Given another vector field V on M, set

$$\mathcal{L} = \frac{1}{2}\sum_{i=1}^{\ell} V_i^2 + V. \tag{1}$$

The semi-group associated with $\mathcal{L}$ has a smooth positive fundamental solution $p_t(z, z')$ with respect to any smooth volume measure VOL on M. Given any two points x and y in M, denote by $\Omega^{x,y}$ the set of continuous paths $\omega : [0,1] \to M$ with $\omega_0 = x$ and $\omega_1 = y$. For $\epsilon > 0$, we define uniquely a probability measure $\mathbb{P}_\epsilon^{x,y}$ on $\Omega^{x,y}$ defining $\mathbb{P}_\epsilon^{x,y}\left(\omega_{t_1} \in A_1, \ldots, \omega_{t_k} \in A_k\right)$ for all $k \geqslant 1$, $0 < t_1 < \cdots < t_k < 1$ and any Borel sets $A_1, \ldots, A_k$ of M, by the formula

$$\frac{1}{p_\epsilon(x,y)} \int \left(\prod_{j=1}^{k} \left(p_{t_j\epsilon - t_{j-1}\epsilon}(x_{j-1}, x_j) \mathbf{1}_{A_j}(x_j) \right) \right) p_{\epsilon - t_k\epsilon}(x_k, y)\, \mathrm{VOL}(dx_1) \cdots \mathrm{VOL}(dx_k) \tag{2}$$

I. Bailleul (✉)
IRMAR, 263 Av. du General Leclerc, 35042 Rennes, France
e-mail: ismael.bailleul@univ-rennes1.fr

C. Donati-Martin et al. (eds.), *Séminaire de Probabilités XLVIII*, Lecture Notes in Mathematics 2168, DOI 10.1007/978-3-319-44465-9_7

where $t_0 = 0$ and $x_0 = x$. This formula describes the law of the diffusion process associated with $\epsilon\mathcal{L}$, conditioned on having position x at time 0 and position y at time 1.

We are interested in this work in proving a large deviation result for the family of measures $\left(\mathbb{P}_\epsilon^{x,y}\right)_{0<\epsilon\leqslant 1}$, seen as measures supported on some space of Hölder functions. Such kind of results were stated without proofs in Takanobu and Watanabe's famous paper [1], and proved for the first time in a recent paper by Inahama [2] in as general a framework as in the present paper. See also an earlier work of his [3] where similar results are proved under some stronger ellipticity assumptions. His analysis rests on the dynamic description of the diffusion associated with $\mathcal{L}$, given by the stochastic differential equation $dx_t = V(x_t)dt + \sum_{i=1}^{\ell} V_i(x_t)\circ dB_t^i$, or rather on its rough path counterpart. By using quasi-sure analysis, he is able to lift the measures $\mathbb{P}_\epsilon^{x,y}$ to some measures $\mathbf{P}_\epsilon^{x,y}$ on the space of geometric rough paths, which requires the quasi-sure existence of the Brownian rough path. The large deviation principle for $\mathbb{P}_\epsilon^{x,y}$ is then obtained as a consequence of a subtle large deviation principle for $\mathbf{P}_\epsilon^{x,y}$ involving Watanabe's theory of Donsker's Delta function, via the continuity of the Ito-Lyons map. Our proof is more analytic, in that its essential ingredients are the heat kernel estimates of Léandre and Sanchez-Calle. We also use the machinery of rough paths as a convenient tool for proving the exponential tightness of the family of probability measures $\left(\mathbb{P}_\epsilon^{x,y}\right)_{0<\epsilon\leqslant 1}$ on $\mathcal{C}_{x,y}^{\alpha}\big([0,1], M\big)$. This way of proceeding is probably simpler than Inahama's approach, and much shorter; however, Inahama's approach has the advantage to prove a large deviation result on a rough paths space, so his result on the rough bridge measure can be propagated to other measures constructed as the image measure of the former via any Itô-Lyons map.

We need a little bit of notation to state our result. Write H_0^1 for the set of $\mathbb{R}^\ell$-valued paths h over the time interval $[0,1]$, with starting point 0; its H^1-norm is denoted by $\|h\|$. Given $h \in H_0^1$, we define a path γ^h by solving the differential equation

$$\dot{\gamma}_t^h = \sum_{i=1}^{\ell} V_i\big(\gamma_t^h\big)\dot{h}_t^i, \tag{3}$$

for $0 \leqslant t \leqslant 1$, given any specified starting point. The Lie bracket condition ensures that one defines a metric topology identical to the manifold topology setting for any pair of points (a, b) in M

$$d(a, b) = \inf \int_0^1 |\dot{h}_s|_\ell ds$$

where the infimum is over the non-empty set of H_0^1-controls h such that $\gamma_0^h = a$ and $\gamma_1^h = b$. It is called the sub-Riemannian distance associated with $\mathcal{L}$. The notation $|\cdot|_\ell$ stands here for the Euclidean norm on $\mathbb{R}^\ell$. We define an $[0,\infty]$-valued function J on $\Omega^{x,y}$ setting

$$J(\gamma) = \frac{1}{2}\Big(\inf\big\{\|h\|^2\,;\,\gamma^h = \gamma\big\} - d(x,y)^2\Big), \tag{4}$$

with the convention $\inf \emptyset = \infty$. The above infimum is called the **energy of the path** γ, classically denoted by $2I(\gamma)$. Given any $0 < \alpha < 1$, denote by $\|x\|_\alpha$ the α-Hölder norm of a path x from $[0,1]$ to the ambient space $\mathbb{R}^d$. Write $\mathcal{C}^\alpha_{x,y}\big([0,1],M\big)$ for the set of all M-valued paths with finite α-Hölder norm, with endpoints x and y; it is equipped with the topology associated with $\|\cdot\|_\alpha$.

Theorem 1 (Large Deviation Principle for Bridges of Degenerate Diffusion Processes)

(i) Given any $\frac{1}{3} < \alpha < \frac{1}{2}$, the probabilities $\mathbb{P}_\epsilon^{x,y}$ are supported on $\mathcal{C}^\alpha_{x,y}\big([0,1],M\big)$.
(ii) The family $\big(\mathbb{P}_\epsilon^{x,y}\big)_{0<\epsilon<1}$ satisfies a large deviation principle in $\mathcal{C}^\alpha_{x,y}\big([0,1],M\big)$, with good rate function J.

This statement calls for a few remarks.

1. *The above definition of the space $\mathcal{C}^\alpha_{x,y}\big([0,1],M\big)$ rests on the ambient Euclidean metric. It is straightforward to see that it coincides with the set of M-valued paths which are α-Hölder for any choice of Riemannian metric on M, so $\mathcal{C}^\alpha_{x,y}\big([0,1],M\big)$ is intrinsically defined.*
2. *It seems possible however to trace back the large deviation upper bound to some works of Gao* [4] *and Gao and Ren* [5] *on large deviation principles for stochastic flows in the framework of capacities on Wiener space. They prove in these works a Freidlin-Wentzell estimate/large deviation principle for (r,p)-capacities on Wiener space. Denote by X^ϵ the solution to the stochastic differential equation*

$$dX_t^\epsilon = \epsilon\, V(X_t^\epsilon)dt + \epsilon^{\frac{1}{2}} \sum_{i=1}^{\ell} V_i(X_t^\epsilon)\circ dw_t^i,$$

for a Brownian motion $w = (w^1,\dots,w^\ell)$. As the probability measure $\mathbb{P}_\epsilon^{x,y}$ has finite energy [6], *a theorem of Sugita, theorem 4.2 in* [7], *ensures that we have*

$$\big\{\mathbb{P}_\epsilon^{x,y}(A)\big\}^p \leqslant c\, \mathrm{C}_p^r(X^\epsilon \in A),$$

for some positive constant c and all Borel sets A in Wiener space; so a large deviation upper bound for C_p^r implies a corresponding result for $\mathbb{P}_\epsilon^{x,y}(\cdot)$. It does not seem possible to get the large deviation lower bound by these methods. Theorem 1 was also proved by different methods in the recent work [8] *of the author, in possibly unbounded manifolds—this setting is not covered by*

Inahama's result in so far as unbounded manifolds do not necessary have immersed versions in an ambient Euclidean space unlike closed manifolds, by Whitney's theorem.

3. *We shall see in Sect. 3 that the large deviation principle stated in Theorem 1 leads directly to the identification of the first order asymptotics of* $\mathbb{P}_\epsilon^{x,y}$ *under some mild conditions on* (x, y)*, in the sense that* $\mathbb{P}_\epsilon^{x,y}$ *converges weakly to a Dirac mass on some particular path* γ *from* x *to* y*. It is natural in that setting to push further the analysis and try and get a second order asymptotics. This is done in the work* [8] *where it is proved that the fluctuation process around the deterministic limit* γ *is a Gaussian process whose covariance involves the (non-constant rank) sub-Riemannian geometry associated with the operator* $\mathcal{L}$*. This requires that the pair* (x, y) *lies outside some intrinsic cutlocus associated with* $\mathcal{L}$
4. *See the two works* [9, 10] *of Baldi, Caramellino and Rossi for recent results related to the above theorem., on large deviation results for the probability of exit of a domain for a bridge of a diffusion process.*

2 Proof of the Large Deviation Principle

The proof of Theorem 1 follows the pattern of proof devised by Hsu in [11] to prove a similar large deviation principle in a Riemannian setting where $\mathcal{L}$ is the Laplacian of some Riemannian metric on M. Our reasoning relies crucially on Léandre's logarithmic estimate [12, 13]

$$\lim_{\epsilon \searrow 0} \epsilon \log p_\epsilon(z, z') = -\frac{d^2(z, z')}{2}, \tag{5}$$

which holds uniformly with respect to $(z, z') \in M^2$, as well as on Sanchez-Calle's estimate

$$p_t(z, z') \leqslant c\, t^{-m}, \tag{6}$$

which holds for some positive constant c and all $z, z' \in M$ and $t > 0$, see [14].

Write Ω^x for the set of continuous paths $\omega : [0, 1] \to M$ started from x; we equip Ω^x and $\Omega^{x,y}$ with the metric of uniform convergence inherited from the ambient space. Fix $\alpha \in \left(\frac{1}{3}, \frac{1}{2}\right)$.

2.1 Exponential Tightness of the Family of Probability Measures $(\mathbb{P}^{x,y}_\epsilon)_{0<\epsilon\leqslant 1}$ on $\mathcal{C}^\alpha_{x,y}([0,1],M)$

Given $n = n(N) \geqslant 7$ and $K = K(N)$, to be fixed later as functions of some parameter N, we define a compact subset C_N both of $\Omega^{x,y}$ and $\mathcal{C}^\alpha_{x,y}([0,1],M)$ setting

$$C_N = \left\{\omega \in \Omega^{x,y};\ \sup_{0<t-s\leqslant\frac{1}{n}} \frac{|\omega_t-\omega_s|_d}{|t-s|^\alpha} \leqslant K\right\}.$$

The above supremum is over the set of all times $s,t \in [0,1]$. We first work on the time interval $[0,2/3]$ to evaluate the $\mathbb{P}^{x,y}_\epsilon$-probability of C_N, to avoid the difficulties coming from the singularities of the drift at time 1, in the classical dynamical description of the bridge as the solution to a stochastic differential equation. Set

$$\begin{aligned}
(\star) &:= \mathbb{P}^{x,y}_\epsilon\left(\sup_{s,t\in[0,2/3],\ 0<t-s\leqslant\frac{1}{n}} \frac{|\omega_t-\omega_s|_d}{|t-s|^\alpha} > K\right)\\
&\leqslant \frac{n}{2}\sup_{0\leqslant r\leqslant 2/3} \mathbb{P}^{x,y}_\epsilon\left(\sup_{r\leqslant s<t\leqslant r+2/n} \frac{|\omega_t-\omega_s|_d}{|t-s|^\alpha} > K\right).
\end{aligned}$$

Using (2) and the Markov property provides the upper bound

$$\begin{aligned}
(\star) &\leqslant \sup_{0\leqslant r\leqslant\frac{2\epsilon}{3}} \mathbb{E}^x\left[\frac{p_{(\epsilon-r-\frac{2\epsilon}{n})}\big(\omega_{\epsilon(r+\frac{2\epsilon}{n})},y\big)}{p_\epsilon(x,y)};\ \sup_{r\leqslant s<t\leqslant r+\frac{2\epsilon}{n}} \frac{|\omega_t-\omega_s|_d}{|t-s|^\alpha} \geqslant K\right]\\
&\leqslant \frac{c\epsilon^{-m}}{p_\epsilon(x,y)}\sup_{z\in M}\mathbb{P}^z\left(\sup_{0\leqslant s<t\leqslant\frac{2\epsilon}{n}} \frac{|\omega_t-\omega_s|_d}{|t-s|^\alpha} \geqslant K\right).
\end{aligned} \tag{7}$$

By Lyons' universal limit theorem, as stated for instance under the form given in Theorem 11 in [15], there exists universal controls on the oscillation of solutions of stochastic differential equations in terms of the oscillations of Brownian motion and its Lévy area. More precisely, there exists positive constants a_i, b_i, depending only on the vector fields V, V_i, such that

$$\begin{aligned}
\sup_{z\in M}\mathbb{P}^z\left(\sup_{0\leqslant s<t\leqslant\frac{2\epsilon}{n}} \frac{|\omega_t-\omega_s|}{|t-s|^\alpha} \geqslant K\right) &\leqslant a_1\left\{\mathbf{P}\Big(\big\|\mathbf{B}_{[0,(2\epsilon)/n]}\big\| \geqslant b_1 K\Big) + \mathbf{P}\Big(\big\|\mathbf{B}_{[0,(2\epsilon)/n]}\big\|^3 \geqslant K\wedge\frac{n}{3}\Big)\right\},\\
&\leqslant a_2\mathbf{P}\Big(\big\|\mathbf{B}_{[0,(2\epsilon)/n]}\big\| \geqslant b_2(K\wedge n)^{1/3}\Big)
\end{aligned} \tag{8}$$

where $\mathbf{B}_{[0,(2\epsilon)/n]}$ is the Brownian $\frac{1}{\alpha}$-rough path on the time interval $\big[0,\frac{2\epsilon}{n}\big]$, defined on some probability space $(\boldsymbol{\Omega},\mathcal{F},\mathbf{P})$, and $\big\|\mathbf{B}_{[0,(2\epsilon)/n]}\big\|$ stands for the homogeneous rough path norm of $\mathbf{B}_{[0,(2\epsilon)/n]}$; see for instance Chap. 10.1 of [16]. It follows from the equality in law $\left\|\mathbf{B}_{[0,(2\epsilon)/n]}\right\| = \sqrt{\frac{2\epsilon}{n}}\left\|\mathbf{B}_{[0,1]}\right\|$, the Gaussian character of $\left\|\mathbf{B}_{[0,1]}\right\|$ under $\mathbf{P}$, and Léandre's estimate (5) for $p_\epsilon(x,y)$, that

$$\epsilon \log \mathbb{P}_\epsilon^{x,y}\left(\sup_{s,t\in[0,2/3],\,0<t-s\leqslant\frac{1}{n}} \frac{|\omega_t-\omega_s|}{|t-s|^\alpha} > K\right) \leqslant \frac{d^2(x,y)}{2} + o_\epsilon(1) - \frac{n(K\wedge n)^{2/3}}{2} b_2^2,$$

so we have

$$\overline{\lim_{\epsilon\searrow 0}}\, \epsilon \log \mathbb{P}_\epsilon^{x,y}\left(\sup_{s,t\in[0,2/3],\,0<t-s\leqslant\frac{1}{n}} \frac{|\omega_t-\omega_s|}{|t-s|^\alpha} > K\right) \leqslant -N \tag{9}$$

by choosing $n = n(N)$ and $K = K(N)$ big enough.

To get a similar estimate when working on the whole time interval $[0,1]$, remark that since M is compact and the operator $\mathcal{L}$ is hypoelliptic, it has a smooth positive invariant measure by Hörmander's theorem; positivity is [17]. If we use this measure as our reference measure Vol, then $\hat{p}_t(z,z') = p_t(z',z)$ is the heat kernel of another operator $\widehat{\mathcal{L}}$ which satisfies the same conditions as $\mathcal{L}$. Write $\widehat{\mathbb{P}}_\epsilon^{z,z'}$ for the law of the associated bridge. So the class of measures $\left(\mathbb{P}_\epsilon^{z,z'}\right)_{z\neq z'\in M}$ constructed from hypoelliptic operators $\mathcal{L}$ as in (1), satisfying the Lie bracket assumption, is preserved under time reversal. Applying inequality (9) to the measure $\widehat{\mathbb{P}}_\epsilon^{y,x}$ on $\Omega^{y,x}$ obtained by time-reversal of $\mathbb{P}_\epsilon^{x,y}$, we conclude with (9) that

$$\overline{\lim_{\epsilon\searrow 0}}\, \epsilon \log \mathbb{P}_\epsilon^{x,y}(C_N^c) \leqslant -N.$$

So the family $\left(\mathbb{P}_\epsilon^{x,y}\right)_{0<\epsilon<1}$ of probabilities on $C_\alpha^{x,y}\big([0,1],M\big)$ is exponentially tight, which proves in particular point (i). As the inclusion of $C_\alpha^{x,y}\big([0,1],M\big)$ into $\big(\Omega^{x,y}, \|\cdot\|_\infty\big)$ is continuous, it suffices, by the inverse contraction principle, to prove that $\left(\mathbb{P}_\epsilon^{x,y}\right)_{0<\epsilon<1}$ satisfies a large deviation principle in $\big(\Omega^{x,y}, \|\cdot\|_\infty\big)$, with good rate function J, to prove point (ii) of the theorem, in so far as J is also a good rate function on $C_\alpha^{x,y}\big([0,1],M\big)$. We follow closely Hsu's work [11] to prove that fact.

2.2 *Large Deviation Upper Bound for* $(\mathbb{P}_\epsilon^{x,y})_{0<\epsilon\leqslant 1}$

We first prove the upper bound for a compact subset C of $\Omega^{x,y}$. For $0<a<1$, set

$$C^a = \big\{\omega \in \Omega^{x,y}\,;\, \exists\,\rho\in C \text{ such that } \omega_s = \rho_s, \text{ for } 0\leqslant s\leqslant 1-a\big\}$$

and

$$C_*^a = \{\omega \in \Omega^x\,;\, \exists\, \rho \in C \text{ such that } \omega_s = \rho_{(1-a)s}, \text{ for all } 0 \leqslant s \leqslant 1\}.$$

The set C^a is closed in both Ω^x and $\Omega^{x,y}$, and $C \subset C^a$. Using (2) and the Markov property, we get as in (7) the inequality

$$\begin{aligned}\mathbb{P}_\epsilon^{x,y}(C) \leqslant \mathbb{P}_\epsilon^{x,y}(C^a) &\leqslant \mathbb{E}_\epsilon^x\left[\frac{p_{a\epsilon}(\omega_1, y)}{p_\epsilon(x,y)}\mathbf{1}_{\omega\in C_*^a}\right]\\ &\leqslant \frac{c\epsilon^{-m}}{p_\epsilon(x,y)}\mathbb{P}_\epsilon^x(C_*^a).\end{aligned}$$

As C_*^a is closed in Ω^x, we have by the classical Freidlin-Wentzell large deviation principle for $\mathbb{P}_\epsilon^x$

$$\limsup_{\epsilon\searrow 0} \epsilon \log \mathbb{P}_\epsilon^{x,y}(C) \leqslant \frac{d^2(x,y)}{2} - \frac{1}{1-a}\inf_{\omega\in C_*^a} I(\omega).$$

Using the lower semicontinuity of I on Ω^x, it is straightforward to use the compactness of C to see that $\limsup_{a\searrow 0}\inf_{\omega\in C_*^a} I(\omega) \geqslant \inf_{\omega\in C} I(\omega)$, as done in [11, p. 112]. This proves the upper bound

$$\limsup_{\epsilon\searrow 0} \epsilon \log \mathbb{P}_\epsilon^{x,y}(C) \leqslant -\inf_C J,$$

for a compact set C; it is classical that the exponential tightness proved in Sect. 2.1 implies in that case the upper bound for any closed set.

2.3 Large Deviation Lower Bound for $(\mathbb{P}_\epsilon^{x,y})_{0<\epsilon\leqslant 1}$

We use the notation $\|f\|_{[a,b]}$ to denote the uniform norm of some function f defined on some time interval $[a,b]$. Given an open set U in $\Omega^{x,y}$,we aim at proving that we have

$$\liminf_{\epsilon\searrow 0} \epsilon \log \mathbb{P}_\epsilon^{x,y}(U) \geqslant -J(\gamma) \tag{10}$$

for any $\gamma \in U$ with finite energy $I(\gamma)$. Pick such a path $\gamma \in U$ and $b > 0$ small enough for the ball in Ω^x with center γ and radius b to be included in U. Set for $0 < a < 1$

$$U^{a,b} = \{\omega \in \Omega^{x,y}\,;\, \|\omega-\gamma\|_{[0,1-a]} < b\}, \quad F^{a,b} = \{\omega \in \Omega^{x,y}\,;\, \|\omega-\gamma\|_{[1-a,1]} \geqslant b\}$$

and $U_*^{a,b} = \{\omega_* \in \Omega^x\,;\, \exists\,\omega \in U \text{ such that } \omega_*(s) = \omega_{(1-a)s}, \text{ for all } 0 \leqslant s \leqslant 1\}$. We have $U^{a,b} \subset (U \cup F^{a,b})$, so $\mathbb{P}_\epsilon^{x,y}(U) \geqslant \mathbb{P}_\epsilon^{x,y}(U^{a,b}) - \mathbb{P}_\epsilon^{x,y}(F^{a,b})$. We prove (10) by showing that $\liminf_{\epsilon\searrow 0} \epsilon \log \mathbb{P}_\epsilon^{x,y}(U^{a,b}) \geqslant -J(\gamma)$, and $\liminf_{\epsilon\searrow 0} \epsilon \log \mathbb{P}_\epsilon^{x,y}(F^{a,b}) = -\infty$.

Given $\lambda > 0$, write $B_\lambda(y)$ for the sub-Riemannian open ball in M, with center y and radius λ. Using the Markov property as above, we have

$$\begin{aligned}
\mathbb{P}_\epsilon^{x,y}(U^{a,b}) &= \mathbb{E}_\epsilon^{x,y}\Big[\mathbb{P}_{\epsilon(1-a)}^{x,X_{1-a}}(U_*^{a,b})\Big] \geqslant \int \mathbb{P}_{\epsilon(1-a)}^{x,z}(U_*^{a,b}) \frac{p_{\epsilon(1-a)}(x,z)p_{\epsilon a}(z,y)}{p_\epsilon(x,y)} \mathbf{1}_{z\in B_\lambda(y)}\, dz \\
&\geqslant \frac{\min_{z\in B_\lambda(y)} p_{\epsilon a}(z,y)}{p_\epsilon(x,y)} \int \mathbb{P}_{\epsilon(1-a)}^{x,z}(U_*^{a,b}) \mathbf{1}_{z\in B_\lambda(y)} p_{\epsilon(1-a)}(x,z)\, dz \\
&\geqslant \frac{\min_{z\in B_\lambda(y)} p_{\epsilon a}(z,y)}{p_\epsilon(x,y)} \mathbb{P}_{\epsilon(1-a)}^{x}\big(U_*^{a,b} \cap \{\omega_1 \in B_\lambda(y)\}\big).
\end{aligned}$$

Define $\gamma_a(s) = \gamma_{(1-a)s}$ for all $0 \leqslant s \leqslant 1$. As γ has finite energy, one can pick some control $h \in H_0^1$ such that $\gamma^h = \gamma$; we have $d(\gamma_a(1), y) \leqslant \int_{1-a}^1 |\dot h_s|_\ell\, ds \leqslant \sqrt{a \int_{1-a}^1 |\dot h_s|_\ell^2\, ds}$. The choice of $\lambda = \lambda(a) = 2\sqrt{a \int_{1-a}^1 |\dot h_s|_\ell^2\, ds}$ ensures that the open set $U_*^{a,b} \cap \{\omega_1 \in B_\lambda(y)\}$ contains γ_a, so it is nonempty; also, $\frac{\lambda(a)^2}{a} \to 0$ as a tends to 0. Using the classical Freidlin-Wentzell large deviation theory and the uniform character of Léandre's estimate (5), the above lower bound for $\mathbb{P}_\epsilon^{x,y}(U^{a,b})$ gives

$$\liminf_{\epsilon\searrow 0} \epsilon \log \mathbb{P}_\epsilon^{x,y}(U^{a,b}) \geqslant \frac{-I(\gamma_a)}{1-a} + \frac{d(x,y)^2}{2} - \frac{\lambda(a)^2}{2a},$$

from which the inequality $\liminf_{\epsilon\searrow 0} \epsilon \log \mathbb{P}_\epsilon^{x,y}(U^{a,b}) \geqslant -J(\gamma)$ follows, since $I(\gamma_a) \to I(\gamma)$ and $\frac{\lambda(a)^2}{a} \to 0$ as a tends to 0.

We now deal with the term $\mathbb{P}_\epsilon^{x,y}(F^{a,b})$. Set $\overline{\gamma}_s = \gamma_{1-s}$, for $0 \leqslant s \leqslant 1$, and choose a small enough to have $\|\overline{\gamma} - y\|_{[0,a]} \leqslant \frac{b}{2}$. We use the same time reversal trick and notations as above to estimate $\mathbb{P}_\epsilon^{x,y}(F^{a,b})$. Write

$$\begin{aligned}
\mathbb{P}_\epsilon^{x,y}(F^{a,b}) &= \widehat{\mathbb{P}}_\epsilon^{y,x}\big(\|\omega - \overline{\gamma}\|_{[0,a]} \geqslant b\big) \leqslant \widehat{\mathbb{P}}_\epsilon^{y,x}\Big(\|\omega - y\|_{[0,a]} \geqslant \frac{b}{2}\Big) \\
&\leqslant \frac{c\epsilon^{-m}}{p_\epsilon(y,x)} \widehat{\mathbb{P}}_\epsilon^{y}\Big(\|\omega - y\|_{[0,a]} \geqslant \frac{b}{2}\Big).
\end{aligned}$$

Léandre's estimate (5) and the classical large deviation results for $\widehat{\mathbb{P}}_\epsilon^{y}$ give the existence of a positive constant c such that we have

$$\liminf_{\epsilon\searrow 0} \epsilon \log \mathbb{P}_\epsilon^{x,y}(F^{a,b}) \leqslant \frac{d(x,y)^2}{2} - \frac{c}{a};$$

this upper bound tends to $-\infty$ as a tends to 0. Sections 2.1, 2.2 and 2.3 all together prove Theorem 1.

3 First Order Asymptotics for Bridges of Degenerate Diffusion Processes

Theorem 1 provides a straightforward mean for investigating the first order asymptotics of $\mathbb{P}_\epsilon^{x,y}$ as ϵ tends to 0, for x and y in generic positions.

Theorem 2 (First Order Asymptotics of $\mathbb{P}_\epsilon^{x,y}$) *If there exists a unique path γ with minimal energy from x to y, then $\mathbb{P}_\epsilon^{x,y}$ converges weakly in $\left(\Omega^{x,y}, \|\cdot\|_\infty\right)$ to a Dirac mass on γ as ϵ tends to 0.*

The proof of this result follows the proof of Lemma 3.1 in [11]. Since the family $(\mathbb{P}_\epsilon^{x,y})_{0<\epsilon\leqslant 1}$ is tight by point (a) in Sect. 2, let $\mathbb{Q}$ be any limit measure. Given $b > 0$, set

$$C_N^b = C_N \cap \{\omega \in \Omega^{x,y}\,;\, \|\omega - \gamma\|_\infty > b\};$$

then $\inf_{\omega \in C_N^b} J(\omega) > 0$. Indeed, since the paths of C_N^b are equicontinuous, if the infimum were null, we could extract from any sequence of paths $(\omega_n)_{n\geqslant 0}$ such that $J(\omega_n)$ converges to 0 a uniformly converging subsequence with limit $\omega \in \overline{C_N^b}$, say. We should then have $J(\omega) = 0$, by the lower semicontinuity of J, that is $\omega = \gamma$, since there is a unique path from x to y with minimal energy, in contradiction with the fact that elements of $\overline{C_N^b}$ satisfy the inequality $\|\omega - \gamma\|_\infty \geqslant b > 0$. As a consequence, the above large deviation upper bound implies

$$\mathbb{Q}\big(C_N^b\big) \leqslant \liminf_{\epsilon \searrow} \mathbb{P}_\epsilon^{x,y}\big(\overline{C_N^b}\big) = 0\,;$$

sending N tend to infinity, it follows that

$$\mathbb{Q}\big(\omega \in \Omega^{x,y}\,;\, \|\omega - \gamma\|_\infty > b\big) = 0.$$

As this holds for all $b > 0$, we have $\mathbb{Q} = \delta_\gamma$, from which the convergence of $\mathbb{P}_\epsilon^{x,y}$ to δ_γ follows.

Note that the set of pairs of points $(x, y) \in M^2$ such that x and y are joined by a unique path of minimal energy is dense in M^2.

A different proof of this result is given in the work [8] of the author, where we also study the limit law of the fluctuations of the bridge process around this deterministic limit. It happens to be a Gaussian process, whose covariance is explicitly determined by the bicharacteristic flow in the cotangent bundle of the manifold.

References

1. S. Takanobu, S. Watanabe, Asymptotic expansion formulas of the Schilder type for a class of conditional Wiener functional integrations, in *Asymptotic Problems in Probability Theory: Wiener Functionals and Asymptotics (Sanda/Kyoto, 1990)*. Pitman Research Notes in Mathematics Series, vol. 284 (Longman Scientific and Technical, Harlow, 1993), pp. 194–241
2. Y. Inahama, Large deviations for rough path lifts of Wanatabe's pullbacks of delta functions. International Mathematics Research Notices (Oxford University Press, Oxford, 2015), p. rnv349
3. Y. Inahama, Large deviations of Freidlin-Wentzel type for pinned diffusion processes. Trans. Am. Math. Soc. **367**, 8107–8137 (2015)
4. F. Gao, Large deviations of (r, p)-capacities for diffusion processes. Adv. Math. (China) **25**(6), 500–509 (1996)
5. F. Gao, J. Ren, Large deviations for stochastic flows and their applications. Sci. China Ser. A **44**(8), 1016–1033 (2001)
6. H. Airault, P. Malliavin, Intégration géométrique sur l'espace de Wiener. Bull. Sci. Math. **112**(1), 3–52 (1988)
7. H. Sugita, Positive generalized Wiener functions and potential theory over abstract Wiener space. Osaka J. Math. **25**(3), 665–696 (1988)
8. I. Bailleul, L. Mesnager, J. Norris, Small time fluctuations for the bridge of a sub-Riemannian diffusion. arXiv:1505.03464 (2015)
9. P. Baldi, L. Caramellino, M. Rossi, On sharp large deviations for the bridge of a general diffusion. arXiv:1410.0863 (2014)
10. P. Baldi, L. Caramellino, M. Rossi, Large deviation asymptotics for the exit from a domain of the bridge of a general diffusion. arXiv:1410.0863 (2014)
11. P. Hsu, Brownian bridges on Riemannian manifolds. Prob. Theory Relat. Fields **84**(1), 103–118 (1990)
12. R. Léandre, Majoration en temps petit de la densité d'une diffusion dégénérée. Probab. Theory Relat. Fields **74**(2), 289–294 (1987)
13. R. Léandre, Minoration en temps petit de la densité d'une diffusion dégénérée. J. Funct. Anal. **74**(2), 399–414 (1987)
14. A. Sanchez-Calle, Fundamental solution and geometry of the sum of squares of vector fields. Inv. Math. **78**, 143–160 (1984)
15. I. Bailleul, Flows driven by rough paths. Rev. Mat. Iberoam. **31**(3), 901–934 (2015)
16. P. Friz, N. Victoir, *Multidimensional Stochastic Processes as Rough Paths* (Cambridge University Press, Cambridge, 2010)
17. S. Aida, S. Kusuoka, D. Stroock, On the support of Wiener functionals, in *Asymptotic Problems in Probability Theory: WIener Functionals and Asymptotics (Sanda/Kyoto, 1990)*. Pitman Research Notes in Mathematics Series, vol. 284 (Longman Scientific and Technical, Harlow, 1993), pp. 3–34

Dévissage of a Poisson Boundary Under Equivariance and Regularity Conditions

Jürgen Angst and Camille Tardif

Abstract We present a method that allows, under suitable equivariance and regularity conditions, to determine the Poisson boundary of a diffusion starting from the Poisson boundary of a sub-diffusion of the original one. We then give two examples of application of this dévissage method. Namely, we first recover the classical result that the Poisson boundary of Brownian motion on a rotationally symmetric manifolds is generated by its escape angle, and we then give an "elementary" probabilistic proof of the delicate result of Bailleul (Probab Theory Relat Fields 141(1–2):283-329, 2008), i.e. the determination of the Poisson boundary of the relativistic Brownian motion in Minkowski space-time.

MSC Classification: 60J45, 60J60, 53C30, 58D19

1 Introduction

The Poisson boundary of a Markov process is a measure space which reflects precisely its long-time asymptotic behavior. In the same time, it can be seen as a random compactification of the state space since it gives some salient information on its geometry at infinity. Finally, it provides a nice representation of bounded harmonic functions associated to the generator of the process, see for example [3, 9] for nice introductions to the topic.

Focusing on the case of Brownian motion on Lie groups or Riemannian manifolds, the Poisson boundary can be computed explicitly in a number of examples: semi-simple groups [13], constant curvature Riemannian manifolds and pinched Cartan-Hadamard manifolds [15] etc. Nevertheless, the explicit determination of the

J. Angst (✉)
IRMAR, Université de Rennes 1, Rennes, France
e-mail: jurgen.angst@univ-rennes1.fr

C. Tardif
LPMA, Université Pierre et Marie Curie, Paris, France
e-mail: camille.tardif@upmc.fr

C. Donati-Martin et al. (eds.), *Séminaire de Probabilités XLVIII*, Lecture Notes in Mathematics 2168, DOI 10.1007/978-3-319-44465-9_8

Poisson boundary of the Brownian on a general Riemannian manifold is largely out of reach. Indeed, even in the case of Cartan-Hadamard manifolds, the question of its triviality (which is a priori a much simpler problem than its explicit determination) is equivalent to the Green-Wu conjecture on the existence of bounded harmonic functions, see [2] and the references therein.

In this paper, we present a so-called dévissage method that allows, under equivariance and regularity conditions, to determine the Poisson boundary of a diffusion starting from the Poisson boundary of a sub-diffusion of the original one. Namely, if the state space E can be written as $E = X \times G$ in an appropriate coordinate system, where X is a differentiable manifold and G is a finite dimensional connected Lie group, and with standard notations recalled in Sect. 2.1 below, we prove the following result:

Theorem A.1 (Theorems 1–2 Below) *Let $(x_t, g_t)_{t\geq 0}$ be a diffusion process with values in $X \times G$, starting from $(x, g) \in X \times G$ and satisfying Hypotheses 1–4 of Sect. 2.2 below. In particular the first projection $(x_t)_{t\geq 0}$ is itself a diffusion process with values in X and when t goes to infinity, the second projection $(g_t)_{t\geq 0}$ converges $\mathbb{P}_{(x,g)}$-almost surely to a random element g_∞ of G. Then, the invariant sigma field $Inv((x_t, g_t)_{t\geq 0})$ of the full diffusion coincides up to $\mathbb{P}_{(x,g)}$-negligeable sets with $Inv((x_t)_{t\geq 0}) \vee \sigma(g_\infty)$.*

Under some natural extra hypothesis, the above theorem can be extended to the case where the group G is replaced by a finite dimensional co-compact homogeneous space $Y := G/K$.

Theorem B.1 (Theorem 3 Below) *Let $(x_t, y_t)_{t\geq 0}$ be a diffusion process with values in $X \times Y$, starting from $(x, y) \in X \times Y$ and satisfying Hypothesis 5 of Sect. 2.2 below. In particular the first projection $(x_t)_{t\geq 0}$ is itself a diffusion process with values in X and when t goes to infinity, the second projection $(y_t)_{t\geq 0}$ converges $\mathbb{P}_{(x,y)}$-almost surely to a random element y_∞ of Y. Then, the two sigma fields $Inv((x_t, y_t)_{t\geq 0})$ and $Inv((x_t)_{t\geq 0}) \vee \sigma(y_\infty)$ coincide up to $\mathbb{P}_{(x,y)}$-negligeable sets.*

The plan of the paper is the following: in the next Sect. 2, we specify the geometric and probabilistic backgrounds and then the equivariance and regularity conditions under which the dévissage method can be applied. Section 3 is devoted to the proofs of the results stated above: we first consider the case where $\text{Inv}((x_t)_{t\geq 0})$ is trivial and G is a finite dimensional Lie group (Theorem 1 of Sect. 3.1), then the case where $\text{Inv}((x_t)_{t\geq 0})$ is non-trivial but G is still a finite dimensional Lie group (Theorem 2 of Sect. 3.2). Finally, we extend this result to case where $Y = G/K$ is a finite dimensional co-compact homogeneous space (Theorem 3 of Sect. 3.3). In Sect. 4.1, we first apply the dévissage method to recover the classical result that the Poisson boundary of the standard Brownian motion on a rotationally symmetric manifold is generated by its limit escape angle, see e.g. [1, 12]. To conclude, in Sect. 4.2, we give an "elementary" probabilistic proof of the main result of [4] i.e. the determination of the Poisson boundary of the relativistic Brownian motion in Minkowski space-time.

2 The Dévissage Method Framework

2.1 Geometric and Probabilistic Background

Let X be a differentiable manifold and G a finite dimensional connected Lie group, in particular G carries a right invariant Haar measure μ. As usual, let us denote by $C^\infty(X \times G, \mathbb{R})$ the set of smooth functions from $X \times G$ to the real line $\mathbb{R}$. From the natural left action of G on itself

$$\begin{aligned} G \times G &\to G \\ (g,h) &\mapsto g.h := gh \end{aligned},$$

we deduce a left action of G on $C^\infty(X \times G, \mathbb{R})$, namely:

$$\begin{aligned} G \times C^\infty(X \times G, \mathbb{R}) &\to C^\infty(X \times G, \mathbb{R}) \\ (g,f) &\mapsto g \cdot f := ((x,h) \mapsto f(x, g.h)) \end{aligned}.$$

In this context, let $(x_t, g_t)_{t\geq 0}$ be a diffusion process with values in $X \times G$ and with infinite lifetime. We denote by $\mathscr{L}$ its infinitesimal generator acting on $C^\infty(X \times G, \mathbb{R})$. Without loss of generality, we can suppose that the process $(x_t, g_t)_{t\geq 0}$ is defined on the canonical space $(\Omega, \mathscr{F})$ where $\Omega := C(\mathbb{R}^+, X \times G)$ is the paths space and $\mathscr{F}$ is its standard Borel sigma field. A generic element $\omega = (\omega_t)_{t\geq 0} \in \Omega$ can be written $\omega = (\omega^X, \omega^G)$ where $\omega^X = (\omega_t^X)_{t\geq 0} \in C(\mathbb{R}^+, X)$ and $\omega^G = (\omega_t^G)_{t\geq 0} \in C(\mathbb{R}^+, G)$. The law of a sample path $(x_t, g_t)_{t\geq 0}$ starting from (x, g) will be denoted by $\mathbb{P}_{(x,g)}$ and $\mathbb{E}_{(x,g)}$ will denote the associated expectation. Note that we have again a natural left action of G on Ω:

$$\begin{aligned} G \times \Omega &\to \Omega \\ (g,\omega) &\mapsto g.\omega := (\omega^X, g.\omega^G) \end{aligned},$$

where $g.\omega^G := (g.\omega_t^G)_{t\geq 0} \in C(\mathbb{R}^+, G)$. Without loss of generality, we can also suppose that $(x_t, g_t)_{t\geq 0}$ is the coordinate process, namely: $x_t(\omega) = \omega_t^X$, $g_t(\omega) = \omega_t^G$, for all $t \geq 0$. With standard notations, we introduce tail sigma field associated to the diffusion: $\mathscr{F}^\infty := \bigcap_{t\geq 0} \sigma((x_s, g_s),\ s \geq t)$, and we consider the classical shift operators $(\theta_s)_{s\geq 0}$ on Ω:

$$\begin{aligned} \theta_s : \quad \Omega &\to \Omega \\ \omega = (\omega_t)_{t\geq 0} &\mapsto \theta_s\omega := (\omega_{t+s})_{t\geq 0} \end{aligned}.$$

Recall that, by definition, the invariant sigma field $\mathrm{Inv}((x_t, g_t)_{t\geq 0})$ associated to the diffusion process $(x_t, g_t)_{t\geq 0}$ is the sub-sigma field of $\mathscr{F}^\infty$ composed of invariant events, that is events $A \in \mathscr{F}^\infty$ such that $\theta_s^{-1}A = A$ for all $s > 0$.

If K is a compact subgroup of G, we will denote by Y the associated homogeneous space i.e. $Y := G/K$ and by π the canonical projection $\pi : G \to G/K$. As above, given a diffusion process $(x_t, y_t)_{t\geq 0}$ on $X \times Y$, we will denote by $\mathbb{P}_{(x,y)}$ the law of the path starting from $(x, y) \in X \times Y$, which we realize as probability measure on the canonical space $\pi(\Omega) = C(\mathbb{R}^+, X \times Y)$.

2.2 *Dévissage, Convergence, Equivariance and Regularity Conditions*

In the case of a group i.e. given a diffusion process $(x_t, g_t)_{t\geq 0}$ with values in $X \times G$, the dévissage method can be applied under the following set of hypotheses:

Hypothesis 1 (Dévissage Condition) The first projection $(x_t)_{t\geq 0}$ is a sub-diffusion of the full process $(x_t, g_t)_{t\geq 0}$. Its own invariant sigma field $\mathrm{Inv}((x_t)_{t\geq 0})$ is either trivial or generated by a random variable ℓ_∞ with values in a separable measure space $(S, \mathscr{G}, \lambda)$ and the law of ℓ_∞ is absolutely continuous with respect to λ.

Hypothesis 2 (Convergence Condition) For any starting point $(x, g) \in X \times G$, the process $(g_t)_{t\geq 0}$ converges $\mathbb{P}_{(x,g)}$-almost surely when t goes to infinity to a random variable g_∞ in G.

Hypothesis 3 (Equivariance Condition) The infinitesimal generator $\mathscr{L}$ of the diffusion is equivariant under the action of G on $C^\infty(X \times G, \mathbb{R})$, i.e. $\forall f \in C^\infty(X \times G, \mathbb{R})$, we have

$$\mathscr{L}(g \cdot f) = g \cdot (\mathscr{L} f).$$

Hypothesis 4 (Regularity Condition) All bounded $\mathscr{L}$-harmonic functions are continuous on the state space $X \times G$.

In the homogeneous case, i.e. given a diffusion process $(x_t, y_t)_{t\geq 0}$ on $X \times Y$ where $Y = G/K$ is a co-compact homogeneous space, our hypothesis can be formulated as follows:

Hypothesis 5 (Homogeneous Case) There exists a K-right equivariant diffusion $(x_t, g_t)_{t\geq 0}$ in $X \times G$ satisfying Hypotheses 1–4 above such that under $\mathbb{P}_{(x,y)}$ the process $(x_t, y_t)_{t\geq 0}$ has the same law as $(x_t, \pi(g_t))_{t\geq 0}$ under $\mathbb{P}_{(x,g)}$ for $g \in \pi^{-1}(\{y\})$.

2.3 *Comments on the Assumptions*

Let us first remark that Hypotheses 1 and 2 ensure that the two sigma fields $\mathrm{Inv}((x_t)_{t\geq 0})$ and $\sigma(g_\infty)$ appearing in Theorem A.1 are well defined.

2.3.1 On the Dévissage Condition

The starting point of the dévissage method is that the state space E of the original diffusion can be written as $X \times G$ (resp. $X \times G/K$) in an appropriate coordinate system, where the corresponding first projection $(x_t)_{t\geq 0}$ is a sub-diffusion of $(x_t, g_t)_{t\geq 0}$ (resp. $(x_t, y_t)_{t\geq 0}$). This "splitting property" occurs in a large number of situations, in particular when considering diffusion processes on manifolds that show some symmetries.

For example, any left invariant diffusion $(z_t)_{t\geq 0}$ with values in a semi-simple Lie group H can be decomposed in Iwasawa coordinates as $z_t = n_t a_t k_t$ where $n_t \in N$, $a_t \in A$, $k_t \in K$ take values in Lie subgroups and $(k_t)_{t\geq 0}$ and $(a_t, k_t)_{t\geq 0}$ are sub-diffusions. In other words, the state space can be decomposed as the product of $X = A \times K$ and $G = N$. Under some regularity conditions (see e.g. [11]), it can be shown that the Poisson boundary of the sub-diffusion (a_t, k_t) is trivial and that n_t converges almost-surely to a random variable $n_\infty \in N$ when t goes to infinity. Thus, our results ensure that the Poisson boundary of the full diffusion $(z_t)_{t\geq 0}$ is generated by the single random variable n_∞.

Another typical situation where the dévissage condition is fulfilled is the case of standard Brownian motion on a Riemannian manifold with a warped product structure, a very representative example being the classical hyperbolic space $\mathbb{H}^d$ seen in polar coordinates $(r, \theta) \in \mathbb{R}_+^* \times \mathbb{S}^{d-1}$, i.e. $X = \mathbb{R}_+^*$ and $G/K = SO(d)/SO(d-1)$. In that case, the radial component $(r_t)_{t\geq 0}$ is a one-dimensional transient sub-diffusion whose Poisson boundary is trivial and the angular component $(\theta_t)_{t\geq 0}$ is a time-changed spherical Brownian motion on $\mathbb{S}^{d-1}$ that converges almost surely to a random variable $\theta_\infty \in \mathbb{S}^{d-1}$. Again, the dévissage method ensures that the Poisson boundary of the full diffusion is generated by the single random variable θ_∞. This example generalizes to the case of a standard Brownian motion on a rotationally symmetric manifold, see Sect. 4.1.

The hypothesis that the first projection $(x_t)_{t\geq 0}$ is a sub-diffusion of the full diffusion $(x_t, g_t)_{t\geq 0}$ (resp. $(x_t, y_t)_{t\geq 0}$) is convenient and easy to check when considering examples. Nevertheless it is not necessary in the sense that there are cases where the couple $(x_t, g_t)_{t\geq 0}$ does not a priori satisfy the dévissage condition, but where a simple change of coordinates allows to implement the method, see Remark 1 below for such an example.

Finally, remark that the absolute continuity condition required when $\mathrm{Inv}((x_t)_{t\geq 0})$ is non-trivial, is ensured for example if the infinitesimal generator of the diffusion process $(x_t)_{t\geq 0}$ is hypoelliptic. Moreover, without loss of generality, we can suppose in that case that the measure λ on $(S, \mathscr{G})$ is a probability measure, see [10].

2.3.2 On the Equivariance Condition

The main hypothesis that allows to implement the dévissage scheme is the third one i.e. the equivariance condition. To emphasize its role, let us first consider the following example where the diffusion process $(x_t, g_t)_{t\geq 0}$ with values $X \times G = \mathbb{R} \times \mathbb{R}$

is solution of the stochastic differential equations system:

$$dx_t = dt + e^{-x_t^2} dB_t, \quad dg_t = e^{-x_t} dt, \quad (x_0, g_0) \in \mathbb{R} \times \mathbb{R}, \tag{1}$$

where $(B_t)_{t \geq 0}$ is a standard real Brownian motion. The infinitesimal generator $\mathscr{L}$ of the diffusion is hypoelliptic, so that Hypothesis 4 is fulfilled. Naturally, the process $(x_t)_{t \geq 0}$ is a one dimensional sub-diffusion of $(x_t, g_t)_{t \geq 0}$ and from the Lemma 1 below, Hypothesis 1 is also fulfilled.

Lemma 1 *There exists a process $(u_t)_{t \geq 0}$ that converges $\mathbb{P}_{(x,g)}$-almost surely to a random variable u_∞ in $\mathbb{R}$ when t goes to infinity such that for all $t \geq 0$*

$$x_t = x_0 + t + u_t.$$

Moreover, the invariant sigma field $Inv((x_t)_{t \geq 0})$ is trivial.

Proof For all $t \geq 0$, we have $x_t = x_0 + t + u_t$, where $u_t := \int_0^t e^{-x_s^2} dB_s$. The martingale u_t satisfies $\langle u \rangle_t = \int_0^t e^{-2x_s^2} ds \leq t$ so that from the law of iterated logarithm, we have almost surely $x_t \geq t/2$ for t sufficiently large. In particular, $\langle u \rangle_\infty < +\infty$ almost surely and u_t is convergent. Since x_t goes almost surely to infinity with t, standard shift-coupling arguments apply and we deduce that $\mathrm{Inv}((x_t)_{t \geq 0})$ is trivial. Note however that the tail sigma field of $(x_t)_{t \geq 0}$ i.e. the invariant sigma field of the space-time process $\mathrm{Inv}((t, x_t)_{t \geq 0})$ is not trivial. Indeed, $x_0 + u_\infty = \lim_{t \to +\infty} (x_t - t)$ is a non-trivial shift invariant random variable. □

From Lemma 1 again, the second projection $g_t = g_0 + \int_0^t e^{-x_s} ds$ converges $\mathbb{P}_{(x,g)}$-almost surely when t goes to infinity to a random variable g_∞ in $\mathbb{R}$ and Hypothesis 2 is satisfied. Finally, considering the action of $G = (\mathbb{R}, +)$ on itself by translation, Hypothesis 3 is also satisfied since, for $f \in C(\mathbb{R} \times \mathbb{R}, \mathbb{R})$ and $(x, g, h) \in \mathbb{R}^3$ we have

$$\begin{aligned}
\mathscr{L}(h \cdot f)(x, g) &= (\partial_x f)(x, g + h) + \frac{1}{2} e^{-x^2} (\partial_x^2 f)(x, g + h) + e^{-x} (\partial_g f)(x, g + h) \\
&= h \cdot (\mathscr{L} f)(x, g).
\end{aligned}$$

Hence, from Theorem A, the invariant sigma field $\mathrm{Inv}((x_t, g_t)_{t \geq 0})$ coincide with $\mathrm{Inv}((x_t)_{t \geq 0}) \vee \sigma(g_\infty) = \sigma(g_\infty)$ up to $\mathbb{P}_{(x,g)}$-negligeable sets i.e. the dévissage scheme applies. Let us now consider a very similar process, namely the diffusion process $(x_t, g_t)_{t \geq 0}$ with values $X \times G = \mathbb{R} \times \mathbb{R}$ which is solution of the new following stochastic differential equations system:

$$dx_t = dt + e^{-x_t^2} dB_t, \quad dg_t = -g_t dt, \quad (x_0, y_0) \in \mathbb{R} \times \mathbb{R}, \tag{2}$$

where $(B_t)_{t \geq 0}$ is again a standard real Brownian motion. With a view to apply the dévissage method, the context seems favorable because the infinitesimal generator $\mathscr{L}$ of the diffusion is hypoelliptic, $(x_t)_{t \geq 0}$ is a one dimensional sub-diffusion of

$(x_t, g_t)_{t\geq 0}$, and $g_t = g_0 e^{-t}$ converges (deterministically) to $g_\infty = 0$ when t goes to infinity. In particular, the sigma field $\mathrm{Inv}((x_t)_{t\geq 0}) \vee \sigma(g_\infty)$ is trivial. Nevertheless, we have the following proposition:

Proposition 1 *Let $(x, g) \in \mathbb{R} \times \mathbb{R}$ with $g \neq 0$, then the sigma field $Inv((x_t, g_t)_{t\geq 0})$ differs from $Inv((x_t)_{t\geq 0}) \vee \sigma(g_\infty)$ by a $\mathbb{P}_{(x,g)}$-non-negligeable set.*

Proof If $g \neq 0$, the sigma field $\mathrm{Inv}((x_t, g_t)_{t\geq 0})$ is not trivial under $\mathbb{P}_{(x,g)}$ because the process $x_t + \log(|g_t|)$ converges $\mathbb{P}_{(x,g)}$-almost surely to $x_0 + \log(|g_0|) + u_\infty$ which, from the proof of Lemma 1, is a non-trivial invariant random variable. □

The reason for which the dévissage method does not apply in this last example is that Hypothesis 3 i.e. the equivariance condition is not fulfilled. Indeed, the generator of the full diffusion writes

$$\mathscr{L} = \partial_x + \frac{1}{2} e^{-x^2} \partial_x^2 - g\partial_g,$$

and in general, for $f \in C(\mathbb{R} \times \mathbb{R}, \mathbb{R})$ and $(x, g, h) \in \mathbb{R}^3$ we have

$$\begin{aligned} \mathscr{L}(h \cdot f)(x, g) &= (\partial_x f)(x, g + h) + \frac{1}{2} e^{-x^2} (\partial_x^2 f)(x, g + h) - g(\partial_g f)(x, g + h) \\ &\neq \\ h \cdot (\mathscr{L} f)(x, g) &= (\partial_x f)(x, g + h) + \frac{1}{2} e^{-x^2} (\partial_x^2 f)(x, g + h) - (g + h)(\partial_g f)(x, g + h). \end{aligned}$$

Remark 1 The equivariance condition is relatively strong and forces $(x_t)_{t\geq 0}$ to be a sub-diffusion (which is already supposed in Hypothesis 1). Indeed, since a function $f : X \times G \to \mathbb{R}$ does not depend on its second variable if and only if $g \cdot f = f$ for all $g \in G$, the equivariance condition implies that $\mathscr{L}$ maps $C^\infty(X)$ onto $C^\infty(X)$ (and thus (x_t) is a sub-diffusion). Nevertheless, some cases where this assumption is not fulfilled can be solved by the dévissage method. For example, consider the diffusion process $(x_t, g_t)_{t\geq 0}$ solution of following system of stochastic differential equations

$$\begin{cases} dx_t = \left(\frac{x_t g_t^2}{x_t^2 + g_t^2} + g_t \right) dt + g_t dB_t, \\ \\ dg_t = \frac{g_t^3}{x_t^2 + g_t^2} dt, \end{cases} \tag{3}$$

where, clearly, there is no equivariance. It is yet possible to show that, almost surely, x_t escapes to infinity with t, g_t converges to a random variable g_∞ and that $\mathrm{Inv}((x_t, g_t)_{t\geq 0}) = \sigma(g_\infty)$ almost surely. Indeed the invariant sigma-field of (x_t, g_t) coincides with the one of $(u_t, v_t) := (x_t/g_t, \log(g_t))$ (since the map is bijective). But now (u_t, v_t) solves the system

$$\begin{cases} du_t = dt + dB_t, \\ dv_t = \frac{1}{1+u_t^2} dt, \end{cases} \tag{4}$$

and one can easily check that, for this new diffusion, all the hypotheses of the dévissage method are now fulfilled. Therefore, applying Theorem A, we can conclude that $\mathrm{Inv}((x_t, g_t)_{t\geq 0}) = \mathrm{Inv}((u_t, v_t)_{t\geq 0}) = \sigma(v_\infty) = \sigma(g_\infty)$.

2.3.3 On the Regularity Condition

As already noticed in the examples of the last section, the regularity condition is automatically satisfied for a large class of diffusion processes, namely when the infinitesimal generator $\mathscr{L}$ is elliptic or hypoelliptic. The role of this assumption will be clear at the end of the proof of Theorems 1 and 2, since it allows to go to the limit in the regularization procedure. In a more heuristical way, the regularity condition can be seen as a mixing hypothesis which prevents pathologies that may occur when considering foliated dynamics.

To be more precise on the kind of pathologies we have in mind, consider the following discrete and deterministic example that was suggested to us by S. Gouëzel. The underlying space is the product space $X \times Y = \mathbb{S}^1 \times \mathbb{S}^1$ where $\mathbb{S}^1$ is identified to $\mathbb{R}/\mathbb{Z}$. Fix $\alpha \notin \mathbb{Q}$, and define the transformation $T : X \times Y \to X \times Y$ such that $T(x, y) := (x+\alpha, y)$. Now let $X(x, y) := x$ and $Y(x, y) := y$ be the first and second projections and for $n \geq 0$ define $X_n := X \circ T^n$ i.e. $X_n(x, y) = (x + n\alpha, y)$ and $Y_n := Y \circ T^n \equiv Y$. In this discrete time context, the resulting sequence $(X_n, Y_n)_{n\geq 0}$ plays the role of $(x_t, y_t)_{t\geq 0}$ in the framework described in Sect. 2.1. The dynamics of (X_n) does not depend on (Y_n), which is constant, and thus converges when n goes to infinity. It is thus natural to ask if the devissage method applies or not in this context. The answer is negative in general. To see this, for $y \in \mathbb{S}^1$, consider the probability measure

$$\nu_y := C \sum_{n\in\mathbb{Z}} \frac{1}{1+n^2} \delta_{y+n\alpha},$$

where C is a normalizing constant and define a measure $\mathbb{P}$ on $X \times Y$ such that

$$\int_{X\times Y} h(x, y)\mathbb{P}(dx, dy) := \int_{y\in\mathbb{S}^1} \left[\int_{x\in\mathbb{S}^1} h(x, y) \left(\frac{1}{2}\nu_y(dx) + \frac{1}{2}\nu_{y+1/2}(dx) \right) \right] dy.$$

Note that the first marginal $\mathbb{P}_X(\cdot) = \int_Y \mathbb{P}(\cdot, dy)$ of $\mathbb{P}$ is the Lebesgue measure hence the invariant sigma field $\mathrm{Inv}((X_n)_{n\geq 0})$ is trivial under $\mathbb{P}$. Since Y is T-invariant, the invariant sigma field $\mathrm{Inv}((Y_n)_{n\geq 0})$ is composed of events that do not depend on the first coordinate x. Thus, under $\mathbb{P}$, the sigma field $\mathrm{Inv}((X_n)_{n\geq 0}) \vee \mathrm{Inv}((Y_n)_{n\geq 0})$ is only composed of events that do not depend on the first coordinate. Now consider the sets $A := \{(y + n\alpha, y), y \in \mathbb{S}^1, n \in \mathbb{Z}\}$ and $B := \{(y + 1/2 + n\alpha, y), y \in \mathbb{S}^1, n \in \mathbb{Z}\}$. Both sets are invariant by the dynamics but they do depend on the first coordinate. Hence, the global invariant sigma field $\mathrm{Inv}((X_n, Y_n)_{n\geq 0})$ differs from $\mathrm{Inv}((Y_n)_{n\geq 0})$ by some $\mathbb{P}$-non negligeable events and the dévissage method does not apply here:

$$\mathrm{Inv}((X_n, Y_n)_{n\geq 0}) \neq \mathrm{Inv}((X_n)_{n\geq 0}) \vee \mathrm{Inv}((Y_n)_{n\geq 0}).$$

3 Proof of the Main Result

We now give the proof of our results. To highlight the main ideas behind the proof, we first consider the simplest case when the invariant sigma field of $(x_t)_{t\geq 0}$ is trivial and when $Y = G$ is a finite dimensional Lie group. Then, we extend the result in the case where the invariant sigma field of $(x_t)_{t\geq 0}$ is non-trivial and finally, we consider the homogeneous case.

3.1 *Starting from a Trivial Poisson Boundary*

Let us first prove the following result:

Theorem 1 *Suppose that the full diffusion $(x_t, g_t)_{t\geq 0}$ satisfies Hypotheses 1–4. Suppose moreover that for all $(x, g) \in X \times G$, the invariant sigma field $Inv((x_t)_{t\geq 0})$ is trivial for the measure $\mathbb{P}_{(x,g)}$. Then the two sigma fields*

$$Inv((x_t, g_t)_{t\geq 0}) \;\; and \;\; \sigma(g_\infty)$$

coincide up to $\mathbb{P}_{(x,g)}$-negligeable sets. Equivalently, if H is a bounded $\mathscr{L}$-harmonic function, then there exists a bounded measurable function ψ on G such that H can be written as $H(x, g) = \mathbb{E}_{(x,g)}[\psi(g_\infty)]$, for all $(x, g) \in X \times G$.

Proof (Proof of Theorem 1) The first step of the proof is the following lemma, which is valid under Hypotheses 1–4 (the triviality of $\mathrm{Inv}((x_t)_{t\geq 0})$ is not required here). From Hypothesis 2, for all $(x, g) \in X \times G$, the process $(g_t)_{t\geq 0}$ converges $\mathbb{P}_{(x,g)}$-almost surely to a random variable $g_\infty = g_\infty(\omega)$ in G.

Lemma 2 *Under Hypotheses 1–4, and for all starting points $(x, g) \in X \times G$ and $h \in G$, the law of the process $h.(x_t, g_t)_{t\geq 0} = (x_t, h.g_t)_{t\geq 0}$ under $\mathbb{P}_{(x,g)}$ coincides with the law of $(x_t, g_t)_{t\geq 0}$ under $\mathbb{P}_{(x,h.g)}$. In particular,*

1. *the law of the limit g_∞ under $\mathbb{P}_{(x,h.g)}$ is the law of $h.g_\infty$ under $\mathbb{P}_{(x,g)}$;*
2. *for all $(g, g') \in G^2$, the push-forward measures of both $\mathbb{P}_{(x,g)}$ and $\mathbb{P}_{(x,g')}$ under the measurable map $\omega = (\omega^X, \omega^G) \mapsto hg_\infty^{-1}.\omega = (\omega^X, hg_\infty^{-1}(\omega).\omega^G)$ coincide.*

Proof (Proof of Lemma 2) The result is an direct consequence of the equivariance Hypothesis 3. Indeed, if $f \in C^\infty(X \times G, \mathbb{R})$ is compactly supported, from Itô's formula, under $\mathbb{P}_{(x,g)}$ we have for all $h \in G$:

$$f(x_t, h.g_t) = (h \cdot f)(x_t, g_t) = (h \cdot f)(x, g) + \int_0^t \mathcal{L}\,(h \cdot f)\,(x_s, g_s)ds + M_t$$

$$= f(x, h.g) + \int_0^t h \cdot (\mathcal{L} f)\,(x_s, g_s)ds + M_t$$

$$= f(x, h.g) + \int_0^t (\mathcal{L} f)\,(x_s, h.g_s)ds + M_t,$$

where M_t is a martingale vanishing at zero. Otherwise, under $\mathbb{P}_{(x,h.g)}$ we have:

$$f(x_t, g_t) = f(x, h.g) + \int_0^t (\mathcal{L} f)(x_s, g_s)ds + N_t,$$

where N_t is again a martingale vanishing at zero. In other words, under $\mathbb{P}_{(x,g)}$ and $\mathbb{P}_{(x,h.g)}$ respectively, both processes $h.(x_t, g_t)_{t\geq 0}$ and $(x_t, g_t)_{t\geq 0}$ solve the same martingale problem, hence their laws coincide. □

Let us go back to the proof of Theorem 1. From Hypothesis 2, for all starting points $(x, g) \in X \times G$, the process $(g_t)_{t\geq 0}$ converges $\mathbb{P}_{(x,g)}$-almost surely to a random variable $g_\infty = g_\infty(\omega)$ in G. We define

$$\Omega_0^{(x,g)} := \{\omega \in \Omega, \lim_{t\to+\infty} g_t(\omega) \text{ exists}\},$$

and consider $\tilde{g}_\infty$ such that $\tilde{g}_\infty := g_\infty$ on $\Omega_0^{(x,g)}$ and $\tilde{g}_\infty := \mathrm{Id}_G$ on $\Omega \backslash \Omega_0^{(x,g)}$. Let H be a bounded $\mathcal{L}$-harmonic function. By the standard duality between bounded invariant random variables and bounded harmonic functions, see e.g. Proposition 3.4 p. 423 of [14], there exists a bounded variable $Z : \Omega \to \mathbb{R}$ which is measurable with respect to $\mathrm{Inv}((x_t, g_t)_{t\geq 0})$, i.e. Z is $\mathcal{F}^\infty$-measurable and satisfies $Z(\theta_s \omega) = Z(\omega)$ for all $\omega \in \Omega$, such that for all $(x, g) \in X \times G$:

$$H(x, g) = \mathbb{E}_{(x,g)}[Z].$$

Moreover, $(x, g) \in X \times G$ being fixed, for $\mathbb{P}_{(x,g)}$-almost all paths ω, we have:

$$Z(\omega) = \lim_{t\to+\infty} H(x_t(\omega), g_t(\omega)).$$

The first idea here is to use the Lie group structure to condition the diffusion to escape at a prescribed point in G. Remark that standard conditioning methods such as Doob h-transform can not be implemented here since the law of the limit g_∞ is not known a priori. For $h \in G$, consider the random variable

$$Z^h(\omega) := Z(h.\tilde{g}_\infty^{-1}.\omega) = Z(\omega^X, h\tilde{g}_\infty(\omega)^{-1}\omega^G).$$

This new variable Z^h can be seen as modification of the initial variable Z so that the value of $Z^h(\omega)$ is the value of Z but conditioned by the event that the G-valued component ω^G of the sample path ω does not exit at the random point $g_\infty(\omega)$ but at the fixed point h. This variable is again $\mathrm{Inv}((x_t, g_t)_{t\geq 0})$-measurable. Indeed, since the constant function equal to h and the random variable Z are shift-invariant, we have

$$Z(h.\tilde{g}_\infty^{-1}(\theta_s\omega).\theta_s\omega) = Z(\theta_s(h.\tilde{g}_\infty^{-1}.\omega)) = Z(h.\tilde{g}_\infty^{-1}.\omega).$$

Since Z^h is bounded and is measurable with respect to $\mathrm{Inv}((x_t, g_t)_{t\geq 0})$, the function $(x, g) \mapsto \mathbb{E}_{(x,g)}[Z^h]$ is also a bounded $\mathscr{L}$-harmonic function. But from the second point of Lemma 2, for all starting points $(x, g, g') \in X \times G^2$, we have

$$\mathbb{E}_{(x,g)}[Z^h] = \mathbb{E}_{(x,g')}[Z^h].$$

In other words, the harmonic function $(x, g) \mapsto \mathbb{E}_{(x,g)}[Z^h]$ is constant in g and its restriction to X is $\mathscr{L}^X$-harmonic, where $\mathscr{L}^X$ denotes the infinitesimal generator of the sub-diffusion $(x_t)_{t\geq 0}$. Since $\mathrm{Inv}((x_t)_{t\geq 0})$ is supposed to be trivial, we deduce that the function $(x, g) \mapsto \mathbb{E}_{(x,g)}[Z^h]$ is constant. In the sequel, we will denote by $\psi(h)$ the value of this constant. Note that $h \mapsto \psi(h)$ is a bounded measurable function since $h \mapsto Z^h$ is. The resulting function ψ is precisely the one appearing in the statement of Theorem 1. By construction, $\psi(h)$ is the common value, for all starting points $(x, g) \in X \times G$, of $\mathbb{E}_{(x,g)}[Z^h]$, i.e. the expectation of Z "conditioned" by the event that ω^G exit in h instead of $g_\infty(\omega)$. The second step of the proof consists in considering a "smooth version" of the map $h \mapsto Z^h$, that will allow us to deal with non-countable union of negligeable sets, which is necessary if we want to mimic the above approach replacing h by $g_\infty(\omega)$. So let us introduce an approximate unity $(\rho_n)_{n\geq 0}$ on G, fix $\mathbf{g} \in G$, $n \in \mathbb{N}$ and consider the "conditioned and regularized" version Z, namely:

$$Z^{\mathbf{g},n}(\omega) := \int_G Z^h(\omega)\rho_n(\mathbf{g}h^{-1})\mu(dh).$$

The exact same reasoning as above shows that $Z^{\mathbf{g},n}$ is bounded and measurable with respect to $\mathrm{Inv}((x_t, g_t)_{t\geq 0})$ so that the function $(x, g) \mapsto \mathbb{E}_{(x,g)}[Z^{\mathbf{g},n}]$ is constant. Hence, for all $\mathbf{g} \in G$, $n \in \mathbb{N}$ and $(x, g) \in X \times G$, there exists a set $\Omega^{\mathbf{g},n,(x,g)} \subset \Omega$ such that $\mathbb{P}_{(x,g)}(\Omega^{\mathbf{g},n,(x,g)}) = 1$ and such that for all paths ω in $\Omega^{\mathbf{g},n,(x,g)}$, we have:

$$Z^{\mathbf{g},n}(\omega) = \lim_{t\to\infty} \mathbb{E}_{(x_t(\omega),g_t(\omega))}[Z^{\mathbf{g},n}] = \mathbb{E}_{(x_0(\omega),g_0(\omega))}[Z^{\mathbf{g},n}] = \mathbb{E}_{(x,g)}[Z^{\mathbf{g},n}].$$

Let D be a countable dense set in G and consider the intersection

$$\Omega^{(x,g)} := \bigcap_{\mathbf{g}\in D, n\in\mathbb{N}} \Omega^{\mathbf{g},n,(x,g)}.$$

We have naturally $\mathbb{P}_{(x,g)}(\Omega^{(x,g)}) = 1$ and for $\omega \in \Omega^{(x,g)}$:

$$\forall \mathbf{g} \in D,\ n \in \mathbb{N},\quad Z^{\mathbf{g},n}(\omega) = \mathbb{E}_{(x,g)}[Z^{\mathbf{g},n}].$$

Since the above expressions are continuous in $\mathbf{g}$, we deduce that the last inequality is true for all $\mathbf{g} \in G$. In other words, we have shown that for all $\mathbf{g} \in G$ and for all ω in $\Omega^{(x,g)}$:

$$Z^{\mathbf{g},n}(\omega) = \mathbb{E}_{(x,g)}[Z^{\mathbf{g},n}] = \int_G \psi(h)\rho_n(\mathbf{g}h^{-1})\mu(dh).$$

In particular, taking $\mathbf{g} = \tilde{g}_\infty(\omega)$, we obtain that for all $\omega \in \Omega^{(x,g)}$ and for all $n \in \mathbb{N}$:

$$Z^{\tilde{g}_\infty(\omega),n}(\omega) = \int_G \psi(h)\rho_n(\tilde{g}_\infty(\omega)h^{-1})\mu(dh). \tag{5}$$

Recall that the Haar measure μ is right invariant so that

$$Z^{\tilde{g}_\infty(\omega),n}(\omega) = \int_G Z^h(\omega)\rho_n(g_\infty(\omega)h^{-1})\mu(dh) = \int_G Z(h.\omega))\rho_n(h^{-1})\mu(dh),$$

and

$$\int_G \psi(h)\rho_n(\tilde{g}_\infty(\omega)h^{-1})\mu(dh) = \int_G \psi(h\tilde{g}_\infty(\omega))\rho_n(h^{-1})\mu(dh).$$

Thus, Eq. (5) is equivalent to

$$\int_G Z(h.\omega)\rho_n(h^{-1})\mu(dh) = \int_G \psi(h\tilde{g}_\infty(\omega))\rho_n(h^{-1})\mu(dh).$$

Taking the integral in ω with respect to $\mathbb{P}_{(x,g)}$ on $\Omega^{(x,g)}$, we deduce that for all $n \in \mathbb{N}$:

$$\int_G \mathbb{E}_{(x,g)}[Z(h.\omega)]\rho_n(h^{-1})\mu(dh) = \int_G \mathbb{E}_{(x,g)}[\psi(h\tilde{g}_\infty)]\rho_n(h^{-1})\mu(dh),$$

which, from Lemma 2 yields

$$\int_G H(x, hg)\rho_n(h^{-1})\mu(dh) = \int_G \mathbb{E}_{(x,hg)}[\psi(\tilde{g}_\infty)]\rho_n(h^{-1})\mu(dh).$$

From Hypothesis 4, bounded $\mathscr{L}$-harmonic functions are continuous, hence we can let n go to infinity in the above expressions to get the desired result, namely:

$$H(x,g) = \mathbb{E}_{(x,g)}[\psi(g_\infty)].$$

□

3.2 Starting from a Non-trivial Poisson Boundary

Let us now consider the general case when $\text{Inv}((x_t)_{t\geq 0})$ is not trivial but generated by a random variable ℓ_∞ with values in a separable measure space $(S, \mathscr{G})$. We will prove the following result:

Theorem 2 *Suppose that the full diffusion $(x_t, g_t)_{t\geq 0}$ satisfies Hypotheses 1–4. Then, for all starting points $(x,g) \in X \times G$, the two sigma fields*

$$\textit{Inv}((x_t, g_t)_{t\geq 0}) \ \textit{and}\ \sigma(\ell_\infty, g_\infty)$$

coincide up to $\mathbb{P}_{(x,g)}$-negligeable sets. Equivalently, if H is a bounded $\mathscr{L}$-harmonic function, there exists a bounded measurable function ψ on $S \times G$ such that H can be written as $H(x,g) = \mathbb{E}_{(x,g)}[\psi(\ell_\infty, g_\infty)]$ for all $(x,g) \in X \times G$.

Proof The proof is very similar to the one of Theorem 1, but it requires an extra argument to ensure the measurability of the function ψ. So let H be a bounded $\mathscr{L}$-harmonic function and $Z : \Omega \to \mathbb{R}$ the associated bounded random variable which is measurable with respect to $\text{Inv}((x_t, g_t)_{t\geq 0})$. For $\mathbf{g}, h \in G$ and $n \in \mathbb{N}$, we consider the random variables

$$Z^h(\omega) := Z(h.\tilde{g}_\infty^{-1}.\omega), \qquad Z^{\mathbf{g},n}(\omega) := \int_G Z^h(\omega)\rho_n(\mathbf{g}h^{-1})\mu(dh).$$

As in the proof of Theorem 1, the element h being fixed, the variable Z^h is bounded and $\text{Inv}((x_t, g_t)_{t\geq 0})$-measurable, so that the function $(x,g) \mapsto \mathbb{E}_{(x,g)}[Z^h]$ is bounded and $\mathscr{L}$-harmonic. From Lemma 2, this function is constant in g and its restriction to X is thus $\mathscr{L}^X$-harmonic. Hence, there exists a bounded measurable function $\psi_h : S \to \mathbb{R}$ such that

$$\forall (x,g) \in X \times G, \ \ \mathbb{E}_{(x,g)}[Z^h] = \mathbb{E}_{(x,g)}[\psi_h(\ell_\infty)]. \tag{6}$$

By Hypothesis 1, the random variable ℓ_∞ admits a density k with respect to the reference probability measure λ on $(S, \mathscr{G})$, so that the last equation can be written

$$\forall (x,g) \in X \times G, \ \ \mathbb{E}_{(x,g)}[Z^h] = \int \psi_h(\ell)k(x,\ell)\lambda(d\ell).$$

The difficulty here is that, a priori, the function $(h,\ell) \mapsto \psi_h(\ell)$ is not measurable in both variables. To deal with this difficulty, note that for any $A \in \mathscr{G}$, we have also:

$$\mathbb{E}_{x,g}[\mathbf{1}_{\ell_\infty \in A} Z^h] = \mathbb{E}_{x,g}[\mathbf{1}_{\ell_\infty \in A} \psi_h(\ell_\infty)] = \int \mathbf{1}_A(\ell)\psi_h(\ell)k(x,\ell)\lambda(d\ell). \tag{7}$$

Let us fix $x_0 \in X$ and define

$$\mathbb{Q}_h(d\ell) := \frac{\psi_h(\ell)}{\mathbb{E}_{(x_0,g)}[Z^h]} k(x_0,\ell)\lambda(d\ell).$$

For each h, the measure $\mathbb{Q}_h$ is absolutely continuous with respect to $k(x_0,\ell)\lambda(d\ell)$ and, by Eq. (7), the one parameter family $(\mathbb{Q}_h)_h$ is a measurable family of probability measures. Recall that by Hypothesis 1, the measurable space $(S,\mathscr{G})$ is separable, thus Theorem 58 p. 57 of [6] applies and there exists a measurable map $X : S \times G \to \mathbb{R}$ such that $X(.,h)$ is a density of $\mathbb{Q}_h$ with respect to $k(x_0,\ell)\lambda(d\ell)$, i.e. for all $h \in G$

$$X(\ell,h) = \frac{\psi_h(\ell)}{\mathbb{E}_{(x_0,g)}[Z^h]} \quad \text{for } \lambda-\text{almost all } \ell.$$

The map $h \mapsto \mathbb{E}_{(x_0,g)}[Z^h]$ being measurable, the function

$$\tilde{\psi}(\ell,h) := X(\ell,h)\mathbb{E}_{(x_0,g)}[Z^h]$$

is also measurable and for all $(x,g) \in X \times G$, we have $\mathbb{E}_{(x,g)}[Z^h] = \mathbb{E}_{(x,g)}[\tilde{\psi}(\ell_\infty,h)]$. For all $\mathbf{g} \in G$, $n \in \mathbb{N}$ and $(x,g) \in X \times G$, we thus have:

$$\begin{aligned}\mathbb{E}_{(x,g)}[Z^{\mathbf{g},n}] &= \int_G \mathbb{E}_{(x,g)}[Z^h]\rho_n(\mathbf{g}h^{-1})\mu(dh) = \int_G \mathbb{E}_{(x,g)}[\tilde{\psi}(\ell_\infty,h)]\rho_n(\mathbf{g}h^{-1})\mu(dh)\\ &= \mathbb{E}_{(x,g)}\left[\int_G \tilde{\psi}(\ell_\infty,h)\rho_n(\mathbf{g}h^{-1})\mu(dh)\right].\end{aligned}$$

Hence, $(x,g) \in X \times G$ being fixed, we obtain that $\mathbb{P}_{(x,g)}$-almost surely

$$\begin{aligned}Z^{\mathbf{g},n} &= \lim_{t\to+\infty} \mathbb{E}_{(x_t,g_t)}[Z^{\mathbf{g},n}]\\ &= \lim_{t\to+\infty} \mathbb{E}_{(x_t,g_t)}\left[\int_G \tilde{\psi}(\ell_\infty,h)\rho_n(\mathbf{g}h^{-1})\mu(dh)\right] = \int_G \tilde{\psi}(\ell_\infty,h)\rho_n(\mathbf{g}h^{-1})\mu(dh).\end{aligned}$$

In other words, $(\mathbf{g}, n, (x,g))$ being fixed, there exists a set $\Omega^{\mathbf{g},n,(x,g)} \subset \Omega$ of full measure i.e. $\mathbb{P}_{(x,g)}(\Omega^{\mathbf{g},n,(x,g)}) = 1$ such that for all $\omega \in \Omega^{\mathbf{g},n,(x,g)}$

$$Z^{\mathbf{g},n}(\omega) = \int_G \tilde{\psi}(\ell_\infty(\omega),h)\rho_n(\mathbf{g}h^{-1})\mu(dh).$$

If D a countable dense set in G, we get that for all $\omega \in \Omega^{(x,g)} := \bigcap_{\mathbf{g}\in D, n\in\mathbb{N}} \Omega^{\mathbf{g},n,(x,g)}$:

$$\forall \mathbf{g} \in D, n \in \mathbb{N}, \quad Z^{\mathbf{g},n}(\omega) = \int_G \tilde{\psi}(\ell_\infty(\omega), h)\rho_n(\mathbf{g}h^{-1})\mu(dh).$$

The above expressions being continuous in $\mathbf{g}$, we can take $\mathbf{g} = \tilde{g}_\infty(\omega)$ to get

$$\begin{aligned}\forall \omega \in \Omega^{(x,g)}, \forall n \in \mathbb{N}, \quad Z^{\tilde{g}_\infty(\omega),n}(\omega) &= \int_G \tilde{\psi}(\ell_\infty(\omega), h)\rho_n(\tilde{g}_\infty(\omega)h^{-1})\mu(dh) \\ &= \int_G \tilde{\psi}(\ell_\infty(\omega), h\tilde{g}_\infty(\omega))\rho_n(h^{-1})\mu(dh).\end{aligned}$$

Taking the expectation under $\mathbb{P}_{(x,g)}$, the left hand side gives :

$$\begin{aligned}\mathbb{E}_{(x,g)}[Z^{\tilde{g}_\infty,n}] &= \int_\Omega \left[\int_G Z^h(\omega)\rho_n(\tilde{g}_\infty(\omega)h^{-1})\mu(dh)\right]\mathbb{P}_{(x,g)}(d\omega) \\ &= \int_\Omega \left[\int_G Z^{h\tilde{g}_\infty(\omega)}(\omega)\rho_n(h^{-1})\mu(dh)\right]\mathbb{P}_{(x,g)}(d\omega) \\ &= \int_\Omega \left[\int_G Z(h.\omega)\rho_n(h^{-1})\mu(dh)\right]\mathbb{P}_{(x,g)}(d\omega) \\ &= \int_\Omega \left[\int_G Z(\omega)\rho_n(h^{-1})\mu(dh)\right]\mathbb{P}_{(x,hg)}(d\omega) \\ &= \int_G \mathbb{E}_{(x,hg)}[Z]\rho_n(h^{-1})\mu(dh)\end{aligned}$$

and the right hand side

$$\begin{aligned}\mathbb{E}_{(x,g)}\left[\int_G \tilde{\psi}(\ell_\infty, h\tilde{g}_\infty)\rho_n(h^{-1})\mu(dh)\right] &= \int_G \mathbb{E}_{(x,g)}\left[\tilde{\psi}(\ell_\infty, h\tilde{g}_\infty)\right]\rho_n(h^{-1})\mu(dh) \\ &= \int_G \mathbb{E}_{(x,hg)}\left[\tilde{\psi}(\ell_\infty, \tilde{g}_\infty)\right]\rho_n(h^{-1})\mu(dh).\end{aligned}$$

Since $\mathscr{L}$-harmonic functions are continuous, letting n go to infinity, we deduce

$$\mathbb{E}_{(x,g)}[Z] = \mathbb{E}_{(x,g)}[\tilde{\psi}(\ell_\infty, \tilde{g}_\infty)].$$

□

3.3 Extension to Homogeneous Manifolds

Finally, we give the proof of Theorem B, i.e. we extend the previous results to the homogeneous case. Consider K a compact sub-group of G, denote by $Y := G/K$ the homogenous space associated and $\pi : G \to G/K$ the canonical projection. Let (x_t, y_t) be a diffusion on $X \times Y$.

Theorem 3 *Suppose that the full diffusion $(x_t, y_t)_{t\geq 0}$ satisfies Hypothesis 5 of Sect. 2.2, then for all starting points $(x, y) \in X \times Y$, the two sigma fields*

$$Inv((x_t, y_t)_{t\geq 0}) \text{ and } Inv((x_t)_{t\geq 0}) \vee \sigma(y_\infty)$$

coincide up to $\mathbb{P}_{(x,y)}$-negligeable sets.

Proof We consider here the case where $\mathrm{Inv}((x_t)_{t\geq 0})$ is generated by a random variable ℓ_∞ with values in a separable measure space $(S, \mathcal{G}, \lambda)$. The case where $\mathrm{Inv}((x_t)_{t\geq 0})$ is trivial can be treated is a very similar way. Let us fix $(x, y) \in X \times Y$ and $g \in \pi^{-1}(\{y\})$. By Hypothesis 5, there exists a K-right equivariant diffusion $(x_t, g_t)_{t\geq 0}$ on $X\times G$ such that, under $\mathbb{P}_{(x,g)}$, the process $(x_t, \pi(g_t))_{t\geq 0}$ has the same law as the process $(x_t, y_t)_{t\geq 0}$ under $\mathbb{P}_{(x,\pi(g))}$. Moreover, $(x_t, g_t)_{t\geq 0}$ satisfies Hypotheses 1–4, in particular g_t converges $\mathbb{P}_{(x,g)}$-almost surely to g_∞ when t goes to infinity. Hence, y_t converges $\mathbb{P}_{(x,\pi(g))}$-almost surely to the asymptotic random variable

$$y_\infty : \omega \in \pi(\Omega) \mapsto \begin{cases} \lim_{t\to+\infty} y_t(\omega) \text{ if exists,} \\ \pi(\mathrm{Id}) \text{ else.} \end{cases}$$

Moreover, for any g in $\pi^{-1}(\{y\})$, the law of y_∞ under $\mathbb{P}_{(x,y)}$ is the same as the law of $\pi(g_\infty)$ under $\mathbb{P}_{(x,g)}$. So let us consider $Z : \pi(\Omega) \to \mathbb{R}$ a bounded θ_t-invariant random variable, for any g in $\pi^{-1}(\{y\})$, we have:

$$\mathbb{E}_{(x,y)}[\dot{Z}] = \mathbb{E}_{(x,g)}[Z \circ \pi] = \int_K \mathbb{E}_{(x,gk)}[Z \circ \pi]\mathrm{Haar}(dk).$$

Since $Z \circ \pi$ is $\mathrm{Inv}((x_t, g_t)_{t\geq 0})$-measurable and bounded, by Theorem 2 applied to $(x_t, g_t)_{t\geq 0}$, there exists a bounded measurable function $(\ell, g) \mapsto \widetilde{H}(\ell, g)$ such that:

$$\forall (x, g) \in X \times G,\ \forall k \in K, \quad \mathbb{E}_{(x,gk)}[Z \circ \pi] = \mathbb{E}_{(x,gk)}[\widetilde{H}(\ell_\infty, g_\infty)].$$

Then, using the K-right equivariance of $(x_t, g_t)_{t\geq 0}$ we obtain for $g \in \pi^{-1}(\{y\})$:

$$\begin{aligned} \mathbb{E}_{(x,y)}[Z] &= \int_K \mathbb{E}_{(x,gk)}[\widetilde{H}(\ell_\infty, g_\infty)]\mathrm{Haar}(dk) \\ &= \int_K \mathbb{E}_{(x,g)}[\widetilde{H}(\ell_\infty, g_\infty k)]\mathrm{Haar}(dk). \end{aligned}$$

Now introduce $\mathscr{S} : Y \to G$ a measurable section of π. Then $g_\infty = \mathscr{S}(\pi(g_\infty))k'$ for some random $k' \in K$ and we have

$$\begin{aligned}\mathbb{E}_{(x,y)}[Z] &= \mathbb{E}_{(x,g)}\left[\int_K \widetilde{H}(\ell_\infty, \mathscr{S}(\pi(g_\infty))k'k)\text{Haar}(dk)\right] \\ &= \mathbb{E}_{(x,g)}\left[\int_K \widetilde{H}(\ell_\infty, \mathscr{S}(\pi(g_\infty))k)\text{Haar}(dk)\right].\end{aligned}$$

Finally denoting by H the bounded measurable function on $S \times Y$ defined by

$$H(\ell, y) := \int_K \widetilde{H}(\ell, \mathscr{S}(y)k)\text{Haar}(dk),$$

we have

$$\mathbb{E}_{(x,y)}[Z] = \mathbb{E}_{(x,y)}[H(\ell_\infty, y_\infty)].$$

□

4 Examples of Application

In this last section, we give two examples of application of the dévissage method. The first one, which concerns the asymptotic behavior of the standard Brownian motion on a rotationally symmetric manifold, is a direct consequence of Theorem 3. The second one, which characterizes the Poisson boundary of the relativistic Brownian motion in Minkowski space is an application of Theorem 1, after a suitable change of coordinates.

4.1 *Brownian Motion on Rotationally Invariant Models*

Let us consider a rotationally invariant model (M, g) i.e. a differentiable Riemannian manifold (M, g), diffeomorphic to the Euclidean space $\mathbb{R}^n$ and such that there exists a point $o \in M$ called the center of the manifold, such that $M \backslash \{o\}$ admits a global polar coordinate system $(r, \theta) \in \mathbb{R}_+^* \times \mathbb{S}^{n-1}$, in which the metric g takes the form

$$g = dr^2 + f^2(r)d\theta^2,$$

where the warping function f is smooth and positive on $\mathbb{R}_+^*$. The Laplace-Beltrami operator Δ^M on M is then given by the formula

$$\Delta^M = \partial_r^2 + (n-1)\frac{f'}{f}(r)\partial_r + \frac{1}{f(r)^2}\Delta_\theta^{\mathbb{S}^{n-1}},$$

where $\Delta_\theta^{\mathbb{S}^{n-1}}$ is the classical Laplace operator on the round sphere $\mathbb{S}^{n-1}$. In this context, let X be a Brownian motion on (M,g) starting from $X_0 = x_0 \neq o$ which is decomposed according to $M\backslash\{o\} = \mathbb{R}_+^* \times \mathbb{S}^{n-1}$ into its radial and angular process, namely $X_t = (r_t, \theta_t)$. The process (r_t, θ_t) then solves the following system of stochastic differential equations:

$$dr_t = dW_t + \frac{n-1}{2}\frac{f'}{f}(r_t)dt, d\theta_t = \frac{1}{f(r_t)}d\Theta_t,$$

where W_t and Θ_t are independent Brownian motions on $\mathbb{R}$ and $\mathbb{S}^{n-1}$ respectively. Note that, whatever the warping function is, the radial component r_t is a one dimensional sub-diffusion of X_t and the angular component θ_t is time-changed spherical Brownian motion parametrized by the clock

$$C_t := \int_0^t \frac{1}{f(r_s)^2}ds,$$

which only depends on the radial component. The general theory of one-dimensional diffusions then ensures that (see e.g. [1])

1. if

$$\int_1^{+\infty} f^{1-n}(r)dr < +\infty,$$

then r_t goes to infinity almost surely;

2. if

$$\int_1^{+\infty} f^{n-1}(r)\left(\int_r^{+\infty} f^{1-n}(\rho)d\rho\right)dr = +\infty,$$

then the lifetime of r_t is almost surely infinite;

3. if

$$\int_1^{+\infty} f^{n-3}(r)\left(\int_r^{+\infty} f^{1-n}(\rho)d\rho\right)dr < +\infty,$$

then the clock C_t is almost surely convergent i.e.

$$C_\infty = \int_0^{+\infty} \frac{1}{f(r_s)^2}ds < +\infty.$$

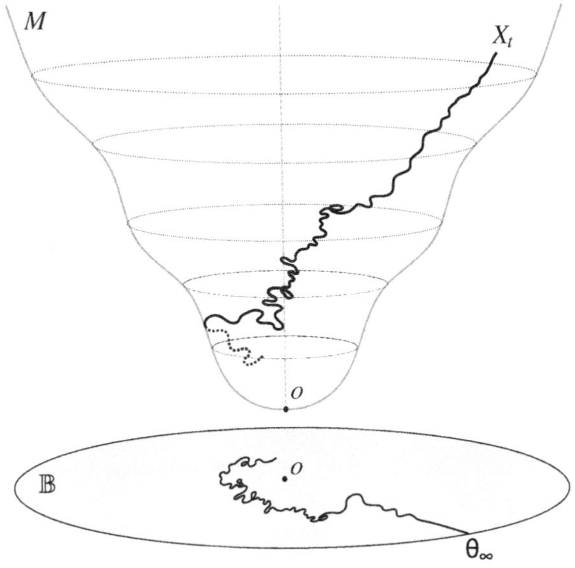

Fig. 1 Asymptotic behavior of X_t under conditions 1–3

Thus, under these integrability conditions on f, the process X_t is transient, does not explode and its angular part θ_t converges almost surely to some asymptotic random variable $\theta_\infty \in \mathbb{S}^{n-1}$. In other words, X_t goes to infinity in a random preferred direction or equivalently, the model M being seen as the interior of the unit ball $\mathbb{B}$, its converges to a random point θ_∞ of its visual boundary, see Fig. 1 below.

Thus, choosing the natural polar coordinates on M, and considering the sphere $\mathbb{S}^{n-1}$ as a $SO(n)$-homogeneous space, we are in position to apply Theorem 3 to obtain:

Theorem 4 *Under conditions 1–3 on the warping function f, the Poisson boundary of the Brownian motion $X_t = (r_t, \theta_t)$ on M is generated by its escape angle θ_∞.*

Proof Let us denote by $\mathcal{L}_r$ the infinitesimal generator of the radial sub-diffusion (r_t). The devissage and regularity conditions are fulfilled and since r_t goes almost surely to infinity with t, by shift coupling, it is easy to see that bounded $\mathcal{L}_r$-harmonic functions are constant, or equivalently that the invariant sigma field of the radial diffusion is trivial. Indeed, let h be a bounded $\mathcal{L}_r$-harmonic function and fix two points $r_0^1 \neq r_0^2$ in $\mathbb{R}^+$. Without loss of generality, we can suppose that $r_0^2 < r_0^1$. Let (r_t^2) be a version of the radial process starting from r_0^2. Almost surely, the process r_t^2 goes to infinity with t so that the stopping time $T := \inf\{t > 0, r_t^2 = r_0^1\}$ is finite almost surely. By the stopping time theorem, we obtain that $h(r_0^1) = \mathbb{E}[h(r_T^2)] = h(r_0^2)$, hence the function h is constant. It remains to check that Hypothesis 5, concerning the homogeneous case, is fulfilled. Let us denote by

$$\pi : SO(n) \longrightarrow \mathbb{S}^{n-1} = SO(n)/SO(n-1)$$

the canonical projection which makes $SO(n)$ the orthonormal frame bundle over $\mathbb{S}^{n-1}$. We can lift horizontally the $\mathbb{S}^{n-1}$-Brownian motion $(\Theta_t)_{t\geq 0}$ into a left invariant diffusion living in $SO(n)$. Namely, denoting by $(H_i)_{i=1...n-1}$ the canonical horizontal vector fields on $SO(n)$ (which are moreover left invariant) we have

$$\forall \phi \in C^2(\mathbb{S}^{n-1}, \mathbb{R}), \quad \Delta(\phi) \circ \pi = \left(\sum_{i=1}^{n-1} H_i^2 \right) (\phi \circ \pi).$$

Thus denoting by (r_t, g_t) the diffusion on $]0, +\infty[\times SO(n)$ generated by

$$\mathscr{L} := \partial_r^2 + (n-1)\frac{f'}{f}(r)\partial_r + \frac{1}{f(r)^2} \sum_{i=1}^{n-1} H_i^2,$$

we obtain that $(r_t, \pi(g_t))$ is a Brownian motion on M. Moreover, according that $\int^{+\infty} f(r_s)^{-2} ds$ is almost surely finite, the process g_t converges almost surely to some $SO(n)$-valued asymptotic random variable g_∞. Note that, $(H_i)_{i=1...n-1}$ being left-invariant, the equivariance condition is satisfied, and finally Hypothesis 5 is fulfilled. □

4.2 *Relativistic Brownian Motion in Minkowski Space*

Both Euclidean Brownian motion B_t and Langevin process are standard models for the physical Brownian motion. They are non relativistic models because, in both cases, the reference frame in which the fluid is at rest plays a specific role (taking into account the fluid viscosity). For instance the dynamics of those processes change when a constant drift is added to the frame. So, in both models, there is no Galilean covariance and a fortiori no Lorentzian covariance neither. Nevertheless, it is remarkable that when the viscosity coefficient of the fluid is null the Langevin process simply writes $(B_t, \int^t B_s ds)$ and it shows a Galilean covariant dynamics. In 1966, Dudley introduced in [7] a Lorentzian analogue to this process, more precisely he proved that there exists a unique diffusion process, taking values in the Minkowskian phase space and having a Lorentzian covariant dynamics with time-like C^1 trajectories. In [4], Bailleul characterized the long-time asymptotic behavior of this relativistic diffusion by computing its Poisson boundary. He showed in particular that it corresponds to the causal boundary to Minkowski space-time. We propose in this section to use Theorem 1 to provide a direct proof of his result.

4.2.1 Dudley Diffusion in Minkowski Space-Time

We denote by $\mathbb{R}^{1,d}$ the Minkowski space-time $\mathbb{R}^{d+1}$ endowed with the Lorentz quadratic form q

$$q(\xi) := \left(\xi^0\right)^2 - \sum_{i=1}^{d} \left(\xi^i\right)^2.$$

The canonical basis is denoted by $(e_0, \dots, e_d)$. Let denote by $\mathbb{H}^d$ the half pseudo unit sphere

$$\mathbb{H}^d := \{\xi \in \mathbb{R}^{1,d},\ q(\xi) = 1\ \xi^0 > 0\}.$$

The restriction of the quadratic form $-q$ to $T\mathbb{H}^d$ makes $\mathbb{H}^d$ a Riemannian manifold of constant negative curvature, so $\mathbb{H}^d$ is the hyperboloïd model of the hyperbolic space of dimension d.

Via the following polar coordinates

$$\begin{array}{ccc} \mathbb{R}_+^* \times \mathbb{S}^{d-1} & \longrightarrow & \mathbb{H}^d \setminus \{e_0\} \\ (r, \theta) & \longmapsto & \cosh(r) e_0 + \sinh(r) \sum_{i=1}^{d} \theta^i e_i \end{array}$$

the hyperbolic space $\mathbb{H}^d$ is a rotationally invariant model centered at e_0 with metric $ds^2 = dr^2 + \sinh(r)^2 d\theta^2$, and the Laplacian is given by

$$\Delta_{r,\theta}^{\mathbb{H}^d} = \partial_r^2 + (d-1)\coth(r)\partial_r + \frac{1}{\sinh(r)^2}\Delta_\theta^{\mathbb{S}^{d-1}}.$$

Let now define Dudley's diffusion introduced in [7].

Definition 1 Dudley's diffusion is the diffusion process $(\dot{\xi}_t, \xi_t)$ with values in the phase space $\mathbb{H}^d \times \mathbb{R}^{1,d}$ and generated by

$$\mathscr{L} := \frac{\sigma^2}{2}\Delta_{\dot{\xi}}^{\mathbb{H}^d} + \dot{\xi} \cdot \partial_\xi,$$

thus $\dot{\xi}_t$ is a classical Riemannian Brownian motion in $\mathbb{H}^d$ and $\xi_t = \xi_0 + \int_0^t \dot{\xi}_s ds$.

Note that the paths ξ_t are C^1 and time-like (since $q(\dot{\xi}_t) = 1$). Moreover since Lorentz linear transforms act by isometry on $\mathbb{H}^d$ and the $\mathbb{H}^d$-Brownian motion $\dot{\xi}_t$ has isometries equivariant dynamics, it follows that Dudley's process has Lorentz equivariant dynamics.

4.2.2 Asymptotic Random Variables

Since the warping function $f := \sinh$ satisfies the integrability condition 1–3 of Sect. 4.1, the angular process θ_t of $\dot{\xi}_t$ converges almost surely to a random variable $\theta_\infty \in \mathbb{S}^{d-1}$. There is another asymptotic random variable associated to ξ_t. We have indeed that $q(\xi_t, e_0+\theta_\infty)$ converges almost surely to some real random variable R_∞. Geometrically this asymptotic random variable R_∞ defines the position of some asymptotic affine hyperplan, whose direction is q-orthogonal to $e_0 + \theta_\infty$, see Fig. 2 below.

We refer to [4] for a detailed proof. Briefly, it follows from the decomposition

$$\begin{aligned} q(\xi_t, e_0 + \theta_\infty) &= q(\xi_0, e_0 + \theta_\infty) + \int_0^t (\cosh(r_s) - \sinh(r_s)\langle \theta_s, \theta_\infty \rangle)\, ds \\ &= q(\xi_0, e_0 + \theta_\infty) + \int_0^t \left(e^{-r_s} - \sinh(r_s) d(\theta_s, \theta_\infty)^2\right) ds, \end{aligned}$$

where $d(\cdot,\cdot)$ is the Riemannian distance on $\mathbb{S}^{d-1}$ and from the fact

$$\frac{r_t}{t} \underset{t\to\infty}{\longrightarrow} \frac{d-1}{2}\sigma^2, \qquad \limsup_{t\to+\infty} \log(d(\theta_t, \theta_\infty)) \le -\frac{d-1}{2}\sigma^2.$$

In [4], Bailleul proved that the two variables θ_∞ and R_∞ are the only asymptotic variables associated to Dudley's diffusion. Namely, using coupling techniques, almost-coupling techniques, uniform continuity estimates for harmonic functions obtained via delicate Harnack inequalities, he proved that

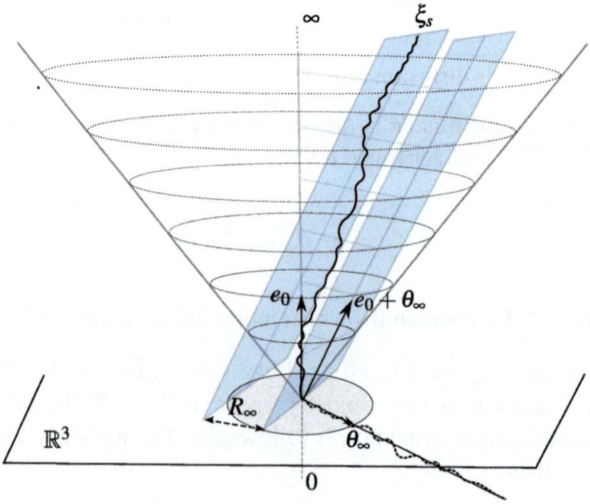

Fig. 2 Typical behavior of the relativistic diffusion in Minkowski space

Theorem 5 (Bailleul) *The invariant σ-field of Dudley's diffusion $(\dot{\xi}_t, \xi_t)_{t\geq 0}$ is generated by the couple $(\theta_\infty, R_\infty) \in \mathbb{S}^{d-1} \times \mathbb{R}_+^*$.*

We propose here to use the devissage method to recover this result. For that we write the dynamics of the diffusion in a new coordinate system $(\alpha_t, \beta_t, \gamma_t, h_t, \delta_t)$ which makes appear a decomposition of the original diffusion into a sub-diffusion $(\alpha_t, \beta_t, \gamma_t)$ with values in $\mathbb{R} \times \mathbb{R} \times \mathbb{R}^{d-1}$ and a process (h_t, δ_t) with values in the group $\mathbb{R}^{d-1} \times \mathbb{R}$ and which has equivariant dynamics. Then, we show that this $\mathbb{R}^{d-1} \times \mathbb{R}$-valued process converges almost surely to some asymptotic random variable $(h_\infty, \delta_\infty) \in \mathbb{R}^{d-1} \times \mathbb{R}$. We conclude the proof by checking that the sub-diffusion $(\alpha_t, \beta_t, \gamma_t)$ has a trivial Poisson boundary. In Remark 4 below, we explicit the link between the limit variables $(h_\infty, \delta_\infty)$ and $(\theta_\infty, R_\infty)$, namely h_∞ is a stereographical projection of θ_∞ and δ_∞ is proportional to R_∞. Our approach is inspired by Bailleul and Raugi's work [5] where the authors use Raugi's results on random walks on Lie groups to find the Poisson boundary of the Dudley diffusion.

4.2.3 New System of Coordinates in $\mathbb{H}^d \times \mathbb{R}^{1,d}$

We first exhibit a new coordinates in which Dudley's diffusion splits up in a sub-diffusion and a process with values in the group $\mathbb{R}^{d-1} \times \mathbb{R}$. For this, let us introduce the Iwasawa coordinates on $\mathbb{H}^d$ (see for instance [8] where the authors consider the equivalent decomposition $\mathscr{T}_h^t \mathscr{D}_\alpha(e_0)$)

$$\begin{array}{ccc} \mathbb{R} \times \mathbb{R}^{d-1} & \longrightarrow & \mathbb{H}^d \\ (\alpha, h) & \longmapsto & \begin{pmatrix} \frac{e^\alpha}{2}(1 + |h|^2) + \frac{e^{-\alpha}}{2} \\ \frac{e^\alpha}{2}(1 - |h|^2) - \frac{e^{-\alpha}}{2} \\ e^\alpha h \end{pmatrix} = \mathscr{T}_h \mathscr{D}_\alpha(e_0), \end{array}$$

where $\mathscr{T}_h$ and $\mathscr{D}_\alpha$ are the following matrix of q-isometries of $\mathbb{R}^{1,d}$:

$$\mathscr{T}_h := \exp \begin{pmatrix} 0 & 0 & h^t \\ 0 & 0 & -h^t \\ h & h & 0 \end{pmatrix} = \begin{pmatrix} 1 + \frac{|h|^2}{2} & \frac{|h|^2}{2} & h^t \\ -\frac{|h|^2}{2} & 1 - \frac{|h|^2}{2} & -h^t \\ h & h & \mathrm{Id} \end{pmatrix},$$

and

$$\mathscr{D}_\alpha := \exp \begin{pmatrix} 0 & \alpha & 0 \\ \alpha & 0 & 0 \\ 0 & 0 & 0 \end{pmatrix} = \begin{pmatrix} \cosh(\alpha) & \sinh(\alpha) & 0 \\ \sinh(\alpha) & \cosh(\alpha) & 0 \\ 0 & 0 & \mathrm{Id} \end{pmatrix}.$$

Note that we have the product relations $\mathscr{T}_{h+h'} = \mathscr{T}_h \mathscr{T}_{h'}$ and $\mathscr{D}_{\alpha+\alpha'} = \mathscr{D}_\alpha \mathscr{D}_{\alpha'}$ and that the coordinates $(y := e^{-\alpha}, h)$ are the classical half-space coordinates of the hyperbolic space $\mathbb{H}^d$. For $\xi \in \mathbb{R}^{1,d}$ denote by

$$(\xi)^+ := q(\xi, e_0 - e_1), \quad (\xi)^- := q(\xi, e_0 + e_1), \quad (\xi)^\perp := (q(\xi, e_i))_{i=2,\dots,d} \in \mathbb{R}^{d-1}.$$

We have then the following q-orthogonal decomposition of ξ on the eigenspaces of the matrices $\mathcal{D}_\alpha$

$$\xi = (\xi)^+ \frac{e_0 + e_1}{2} + (\xi)^- \frac{e_0 - e_1}{2} + \sum_{i=2}^{d} (\xi)_i^\perp e_i.$$

Now let us consider the new system of coordinates in $\mathbb{H}^d \times \mathbb{R}^{1,d}$ given by the following diffeomorphism

$$\begin{array}{ccc} \mathbb{H}^d \times \mathbb{R}^{1,d} & \longrightarrow & \mathbb{R} \times \mathbb{R} \times \mathbb{R}^{d-1} \times \mathbb{R}^{d-1} \times \mathbb{R} \\ (\mathcal{T}_h \mathcal{D}_\alpha(e_0), \xi) & \longmapsto & (\alpha, \beta, \gamma, h, \delta) \\ & & \parallel \\ & & (\alpha, (\mathcal{T}_h^{-1}\xi)^+, (\mathcal{T}_h^{-1}\xi)^\perp, h, (\mathcal{T}_h^{-1}\xi)^-) \end{array}.$$

Remark 2 In Proposition 2 below, we recover the fact that the $\mathbb{H}^d$-component $\dot{\xi}_t = \mathcal{T}_{h_t}\mathcal{D}_{\alpha_t}(e_0)$ of Dudley diffusion escapes at infinity (α_t goes to infinity) and have an asymptotic angle (the stereographical projection h_t converges to h_∞). Thus, one can naively imagine that the $\mathbb{R}^{1,d}$-component ξ_t of Dudley diffusion asymptotically rotates together with its derivative $\dot{\xi}_t$ in the asymptotic direction given by h_∞. In order to catch some extra invariant information, it is then natural to consider the process $\mathcal{T}_{h_t}^{-1}\xi_t$ rather than ξ_t and to decompose it on the eigenspaces of the matrices $\mathcal{D}_\alpha$. This explains why we introduced the previous diffeomorphism.

Moreover, this new system of coordinates in $\mathbb{H}^d \times \mathbb{R}^{1,d}$ corresponds to coordinates in the Poincaré group introduced by Bailleul and Raugi in [5] where they study the asymptotic behavior of random walks in the Poincaré group lifting Dudley diffusion.

The following lemma shows that the Dudley diffusion, written in those new coordinates, splits up in a tower of sub-diffusions.

Lemma 3 *In this new system of coordinates, Dudley's diffusion* $(\alpha_t, \beta_t, \gamma_t, h_t, \delta_t)$ *satisfies the following system of stochastic differential equations:*

$$\begin{cases} d\alpha_t = \sigma dW_t + \frac{1}{2}\sigma^2(d-1)dt, \\ d\beta_t = e^{\alpha_t}dt, \\ d\gamma_t = \sigma e^{-\alpha_t}\beta_t dB_t, \\ dh_t = \sigma e^{-\alpha_t} dB_t, \\ d\delta_t = \left(e^{-\alpha_t} + \sigma^2(d-1)\beta_t e^{-2\alpha_t}\right) dt + 2\sigma e^{-\alpha_t}\gamma_t \cdot dB_t, \end{cases}$$

where (W_t, B_t) *is a standard Brownian motion on* $\mathbb{R} \times \mathbb{R}^{d-1}$.

Proof Considering Iwasawa coordinates (α, h), the restricted Riemannian metric $-q|_{T\mathbb{H}^d}$ writes $d\alpha^2 + e^{2\alpha}|dh|^2$ and the hyperbolic Laplacian is (see also [8])

$$\Delta^{\mathbb{H}} = e^{-2\alpha}\Delta_h^{\mathbb{R}^{d-1}} + \frac{\partial^2}{\partial\alpha^2} + (d-1)\frac{\partial}{\partial\alpha}. \tag{8}$$

Thus there exists two standard independent Brownian motions W_t and B_t with values in $\mathbb{R}$ and $\mathbb{R}^{d-1}$ respectively such that

$$\begin{cases} d\alpha_t = \sigma dW_t + \frac{\sigma^2}{2}(d-1)dt, \\ dh_t = \sigma e^{-\alpha_t} dB_t. \end{cases}$$

Moreover, recall that by definition we have $\dot{\xi}_t := \mathscr{T}_{h_t}\mathscr{D}_{\alpha_t}(e_0)$ thus

$$\begin{aligned} d\beta_t := d(\mathscr{T}_{h_t}^{-1}\xi_t)^+ &= dq(\mathscr{T}_{h_t}^{-1}\xi_t, e_0 - e_1) \\ &= dq(\xi_t, e_0 - e_1) \quad \text{since } \mathscr{T}_{h_t}(e_0 - e_1) = e_0 - e_1 \\ &= q(\dot{\xi}_t, e_0 - e_1)dt = e^{\alpha_t}dt. \end{aligned}$$

Moreover for $i = 2 \dots d$, we have

$$\begin{aligned} d(\gamma_t)^i := dq(\mathscr{T}_{h_t}^{-1}\xi_t, e_i) &= dq(\xi_t, \mathscr{T}_{h_t}(e_i)) \\ &= q(\dot{\xi}_t, \mathscr{T}_{h_t}(e_i)) + q(\xi_t, \circ d\mathscr{T}_{h_t}(e_i)) \\ &= q(\mathscr{D}_{\alpha_t}(e_0), e_i) + q(\mathscr{T}_{h_t}^{-1}\xi_t, \mathscr{T}_{h_t}^{-1} \circ d\mathscr{T}_{h_t}(e_i)) \\ &= 0 + q(\mathscr{T}_{h_t}^{-1}\xi_t, \circ dh_t^i(e_0 - e_1)) \\ &= \beta_t \circ dh_t^i = \beta_t e^{-\alpha_t} dB_t^i. \end{aligned}$$

Finally, we have

$$\begin{aligned} d\delta_t &= dq(\mathscr{T}_{h_t}^{-1}\xi_t, e_0 + e_1) \\ &= dq(\xi_t, \mathscr{T}_{h_t}(e_0 + e_1)) \\ &= q(\dot{\xi}_t, \mathscr{T}_{h_t}(e_0 + e_1)) + q(\mathscr{T}_{h_t}^{-1}\xi_t, \mathscr{T}_{h_t}^{-1} \circ d\mathscr{T}_{h_t}(e_0 + e_1)) \\ &= e^{-\alpha_t}dt + q(\mathscr{T}_{h_t}^{-1}\xi_t, 2\sum_{i=2}^{d} \circ dh_t^i e_i) \end{aligned}$$

$$= e^{-\alpha_t} dt + 2 \sum_{i=2}^{d} \gamma_t^i \circ dh_t^i$$
$$= e^{-\alpha_t} dt + 2\sigma e^{-\alpha_t} \gamma_t \cdot dB_t + \sigma^2 (d-1) \beta_t e^{-2\alpha_t} dt.$$

□

Remark 3 From Lemma 3, one can check easily that the process $(\alpha_t, \beta_t, \gamma_t)_{t\geq 0}$ with values in $X = \mathbb{R} \times \mathbb{R} \times \mathbb{R}^{d-1}$ is indeed a $(d+1)$-dimensional sub-diffusion of Dudley's diffusion. Moreover, the infinitesimal generator $\mathscr{L}$ of the whole process $(\alpha_t, \beta_t, \gamma_t, h_t, \delta_t)_{t\geq 0}$ is of the form

$$\mathscr{L} = \mathscr{L}_{\alpha\beta\gamma} + F(\alpha)\frac{\partial^2}{\partial h^2} + G(\alpha, \beta)\frac{\partial}{\partial \delta} + H(\alpha, \gamma)\frac{\partial^2}{\partial \delta^2},$$

where $\mathscr{L}_{\alpha\beta\gamma}$ is the generator of the sub-diffusion $(\alpha_t, \beta_t, \gamma_t)_{t\geq 0}$ and F, G, H are smooth functions. Therefore, $\mathscr{L}$ is clearly equivariant under the action by translation of elements (h, δ) of the additive group $G = \mathbb{R}^{d-1} \times \mathbb{R}$.

4.2.4 Asymptotic Behavior in the New Coordinates

We now establish the asymptotic behavior of Dudley's diffusion in the new coordinates system $(\alpha, \beta, \gamma, h, \delta)$. Namely, we show that both processes h_t and δ_t converge almost surely to some asymptotic random variables $h_\infty \in \mathbb{R}^{d-1}$ and $\delta_\infty \in \mathbb{R}$. Moreover α_t and β_t are transient and go almost surely to infinity with t, and finally the process γ_t is recurrent in $\mathbb{R}^{d-1}$. This qualitative asymptotic behavior is illustrated in Fig. 3 below.

Proposition 2 *We have almost surely*

$$\frac{\alpha_t}{t} \underset{t\to+\infty}{\longrightarrow} \frac{1}{2}\sigma^2(d-1), \tag{9}$$

$$h_t \underset{t\to+\infty}{\longrightarrow} h_\infty, \tag{10}$$

$$\delta_t \underset{t\to+\infty}{\longrightarrow} \delta_\infty, \tag{11}$$

where h_∞ and δ_∞ are two asymptotic random variables. Moreover, $(\gamma_t)_{t\geq 0}$ is a local martingale whose quadratic variation satisfies

$$\langle \gamma \rangle_\infty = \int_0^{+\infty} \beta_u^2 e^{-2\alpha_u} du = +\infty \text{ almost surely}, \tag{12}$$

in particular $(\gamma_t)_{t\geq 0}$ is recurrent.

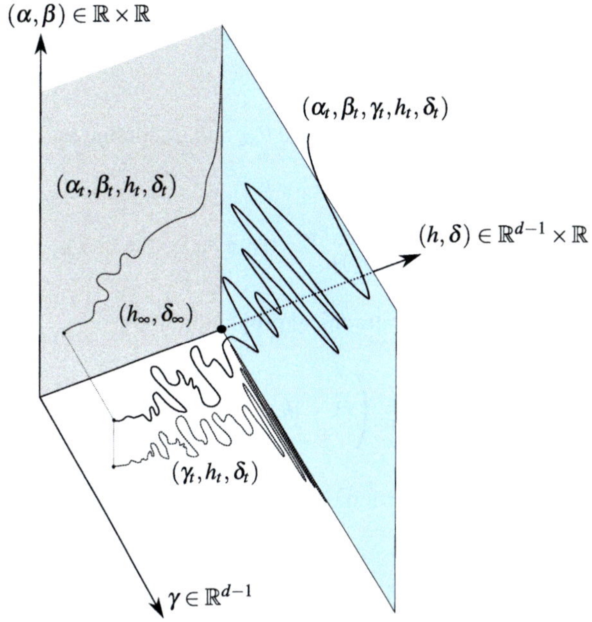

Fig. 3 Asymptotics of Dudley's diffusion in the new coordinates

Note that the asymptotic random variables (h_∞,δ_∞) associated to $(\alpha_t,\beta_t,\gamma_t,h_t,\delta_t)_{t\geq 0}$ are related to the limit variables (θ_∞,R_∞) introduced in Sect. 4.2.2 in simple way which is explained in Remark 4 below.

Proof Since $\alpha_t = \alpha_0+\frac{1}{2}\sigma^2(d-1)t+\sigma W_t$, the convergence (9) follows from the law of iterated logarithm. Hence, the integrants in the expressions defining h_t and δ_t are dominated by $\exp\left(-(\frac{1}{2}\sigma^2(d-1)-\varepsilon)t\right)$ for some $\varepsilon>0$ fixed and for t sufficiently large, so that h_t and δ_t converge almost surely. Let us now check (12) and set $u_t:=\beta_t e^{-\alpha_t}$. Then u_t is a one dimensional diffusion solution of

$$du_t=-\sigma u_t dW_t+\left(1-\frac{d-2}{2}\sigma^2 u_t\right)dt,$$

and it admits the explicit representation

$$u_t=u_0 e^{-\frac{d-1}{2}\sigma^2 t-\sigma W_t}\left[1+\frac{1}{u_0}\int_0^t e^{\frac{d-1}{2}\sigma^2 s+\sigma W_s}ds\right].$$

Then, one easily checks that $(u_t)_{t\geq 0}$ is ergodic in $(0,+\infty)$ with invariant measure

$$\mu(dx):=\frac{1}{Z_\mu}\frac{e^{-\frac{2}{\sigma^2 x}}}{x^d}\mathbf{1}_{(0,+\infty)}(x)dx,$$

where Z_μ is a normalizing constant. In particular, we have $\int_0^{+\infty} u_s^2 ds = +\infty$ almost surely, hence the result.

□

Remark 4 The asymptotic random variable θ_∞ and h_∞ define the same asymptotic line in the light cone. Namely we have

$$\frac{1}{q(e_0, \mathscr{T}_{h_\infty}(e_0 + e_1))} \mathscr{T}_{h_\infty}(e_0 + e_1) = e_0 + \theta_\infty,$$

or more explicitly, h_∞ is a stereographical projection of θ_∞

$$\theta_\infty = \frac{1}{1 + |h_\infty|^2} \left((1 - |h_\infty|^2) e_1 + 2 \sum_{i=2}^{d} h_\infty^i e_i \right).$$

Moreover R_∞ and δ_∞ are proportional

$$R_\infty = \frac{1}{1 + |h_\infty|^2} \delta_\infty.$$

4.2.5 Poisson Boundary

To finally recover Bailleul's result using the devissage method, we have to show that the sub-diffusion $(\alpha_t, \beta_t, \gamma_t)$ has a trivial Poisson boundary. Precisely we find a two steps shift-coupling for that sub-diffusion and obtain the following proposition.

Proposition 3 *The sub-diffusion $(\alpha_t, \beta_t, \gamma_t)$ has a trivial Poisson boundary.*

Proof Fix two initials points $x := (\alpha, \beta, \gamma)$ and $\bar{x} := (\bar{\alpha}, \bar{\beta}, \bar{\gamma})$ in $\mathbb{R} \times \mathbb{R} \times \mathbb{R}^{d-1}$. We will exhibit two random times S and $\bar{S}$, that are almost surely finite, and two copies $x_t := (\alpha_t, \beta_t, \dot{\gamma}_t)$ and $\bar{x}_t := (\bar{\alpha}_t, \bar{\beta}_t, \bar{\gamma}_t)$ starting at x and $\bar{x}$ respectively such that $x_S = \bar{x}_{\bar{S}}$ (Fig. 4).

Let us first consider two copies α_t and $\bar{\alpha}_t$ starting respectively at α and $\bar{\alpha}$

$$\alpha_t := \alpha + \sigma W_t + \frac{d-1}{2}\sigma^2 t \text{ and } \bar{\alpha}_t := \bar{\alpha} + \sigma \bar{W}_t + \frac{d-1}{2}\sigma^2 t,$$

where W_t and $\bar{W}_t$ are two independent Brownian motions. Then, the following increasing processes

$$\beta_t := \beta + \int_0^t e^{\alpha_s} ds \quad \text{and} \quad \bar{\beta}_t := \bar{\beta} + \int_0^t e^{\bar{\alpha}_s} ds$$

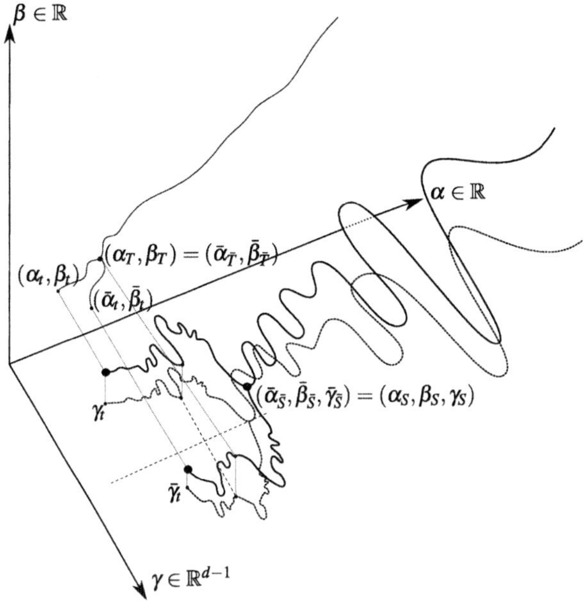

Fig. 4 A two steps shift-coupling

admit inverses denoted respectively by $(\beta_t^{-1})_{t \geq \beta}$ and $(\tilde{\beta}_t^{-1})_{t \geq \tilde{\beta}}$. The time changed processes

$$u_t := \alpha(\beta_t^{-1}) \quad \text{and} \quad \tilde{u}_t := \tilde{\alpha}(\tilde{\beta}_t^{-1})$$

are then two independent copies of a one-dimensional diffusion, and their difference $u_t - \tilde{u}_t$ (for $t \geq \max(\beta, \tilde{\beta})$) is recurrent, in particular the processes will couple automatically at some random and almost surely finite time R. Defining $T := \beta_R^{-1}$ and $\bar{T} := \tilde{\beta}_R^{-1}$ we obtain a first shift-coupling

$$\alpha_T = \tilde{\alpha}_{\bar{T}}, \quad \beta_T = \tilde{\beta}_{\bar{T}}.$$

Let us then define $(\bar{\alpha}_t, \bar{\beta}_t)$ to be $(\tilde{\alpha}_t, \tilde{\beta}_t)$ up to time $\bar{T}$

$$\forall t \in [0, \bar{T}] \quad (\bar{\alpha}_t, \bar{\beta}_t) := (\tilde{\alpha}_t, \tilde{\beta}_t),$$

and coinciding with (α_t, β_t) after time $\bar{T}$

$$\forall s \geq 0, \quad (\bar{\alpha}_{\bar{T}+s}, \bar{\beta}_{\bar{T}+s}) := (\alpha_{T+s}, \beta_{T+s}). \tag{13}$$

By the Markov property $(\bar{\alpha}_t, \bar{\beta}_t)_{t \geq 0}$ is well defined and is a copy of $(\alpha_t, \beta_t)_{t \geq 0}$. Now let define two copies γ_t and $\bar{\gamma}_t$ which will couple at different times. Denote by $(B_u)_{u\geq 0}$ and $(\hat{B}_u)_{u\geq 0}$ two independent Brownian motions on $\mathbb{R}^{d-1}$ starting at 0 and independent of $(W_u)_{u\geq 0}$ and $(\bar{W}_u)_{u\geq 0}$. Set

$$\gamma_t := \gamma + B\left(\int_0^{t\vee T} \sigma^2\beta_s^2 e^{-2\alpha_s} ds\right) + \hat{B}\left(\int_T^{T\wedge t} \sigma^2\beta_s^2 e^{-2\alpha_s} ds\right).$$

Now consider $\check{B}$ the Brownian motion starting at 0 obtained from $\hat{B}$ by reflexion in the linear hyperplan orthogonal to $\gamma_T - \left(\bar{\gamma} + B\left(\int_0^{\bar{T}} \sigma^2\bar{\beta}_s^2 e^{-2\bar{\alpha}_s} ds\right)\right)$ and set

$$\bar{\gamma}_t := \bar{\gamma} + B\left(\int_0^{t\vee \bar{T}} \sigma^2\bar{\beta}_s^2 e^{-2\bar{\alpha}_s} ds\right) + \check{B}\left(\int_{\bar{T}}^{\bar{T}\wedge t} \sigma^2\bar{\beta}_s^2 e^{-2\bar{\alpha}_s} ds\right).$$

By the Markov property $(\bar{\gamma}_t)_{t\geq 0}$ is well defined and is a copy of $(\gamma_t)_{t\geq 0}$. Moreover, by symmetry there exists a finite time U such that $\gamma_T + \hat{B}_U = \bar{\gamma}_T + \check{B}_U$. By (12) of Proposition 2 we know that $\int_T^s \sigma^2\beta_u^2 e^{-2\alpha_u} du$ tends to $+\infty$ when s goes to $+\infty$, thus we can consider the finite time S such that $U = \int_T^S \sigma^2\beta_s^2 e^{-2\alpha_s} ds$. Then define $\bar{S} := S - T + \bar{T}$ and using (13) we obtain

$$\begin{aligned}\int_{\bar{T}}^{\bar{S}} \sigma^2\bar{\beta}_s^2 e^{-2\bar{\alpha}_s} ds &= \int_0^{\bar{S}-\bar{T}} \sigma^2\bar{\beta}_{\bar{T}+s}^2 e^{-2\bar{\alpha}_{\bar{T}+s}} ds \\ &= \int_0^{S-T} \sigma^2\beta_{T+s}^2 e^{-2\alpha_{T+s}} ds = \int_T^S \sigma^2\beta_s^2 e^{-2\alpha_s} ds = U.\end{aligned}$$

Thus we finally obtain

$$\begin{aligned}\gamma_S &= \gamma_T + \hat{B}\left(\int_T^S \sigma^2\beta_s^2 e^{-2\alpha_s} ds\right) \\ &= \gamma_T + \hat{B}_U \\ &= \bar{\gamma}_T + \check{B}_U \\ &= \bar{\gamma}_T + \check{B}\left(\int_{\bar{T}}^{\bar{S}} \sigma^2\bar{\beta}_s^2 e^{-2\bar{\alpha}_s} ds\right) \\ &= \bar{\gamma}_{\bar{S}}.\end{aligned}$$

Moreover, since $\bar{S} - \bar{T} = S - T$ we have also $\alpha_S = \bar{\alpha}_{\bar{S}}$ and $\beta_S = \bar{\beta}_{\bar{S}}$. Thus we have constructed two copies $(\alpha_t, \beta_t, \gamma_t)_{t\geq 0}$ and $(\bar{\alpha}_t, \bar{\beta}_t, \bar{\gamma}_t)_{t\geq 0}$ that couple at the finite times S and $\bar{S}$ and it ends the proof of the proposition. □

We are finally in position to apply the dévissage scheme to the diffusion $(x_t, g_t)_{t\geq 0}$ where $x_t := (\alpha_t, \beta_t, \gamma_t) \in X := \mathbb{R} \times \mathbb{R} \times \mathbb{R}^{d-1}$ and $g_t := (h_t, \delta_t) \in G := \mathbb{R}^{d-1} \times \mathbb{R}$. Namely, from Theorem 1, we can conclude that

Theorem 6 *For all* $(\dot{\xi}, \xi) \in \mathbb{H}^d \times \mathbb{R}^{1,d}$, *the invariant* σ*-algebra of Dudley's diffusion starting from* $(\dot{\xi}, \xi)$ *coincides with* $\sigma(h_\infty, \delta_\infty)$ *up to* $\mathbb{P}_{(\dot{\xi},\xi)}$ *negligible sets.*

Note that by Remark 4, we have $\sigma(h_\infty, \delta_\infty) = \sigma(\theta_\infty, R_\infty)$ i.e. we precisely recover Bailleul's result.

References

1. M. Arnaudon, A. Thalmaier, Brownian motion and negative curvature, in *Random Walks, Boundaries and Spectra*. Progress in Probability, vol. 64, ed. by L. Daniel et al. (Birkhäuser, Basel, 2011), pp. 143–161. Proceedings of the workshop on boundaries, Graz, Austria, June 29–July 3, 2009 and the Alp-workshop, Sankt Kathrein, Austria, July 4–5, 2009
2. M. Arnaudon, A. Thalmaier, S. Ulsamer, Existence of non-trivial harmonic functions on Cartan-Hadamard manifolds of unbounded curvature. Math. Z. **263**, 369–409 (2009)
3. M. Babillot, An introduction to Poisson boundaries of Lie groups, in *Probability Measures on Groups: Recent Directions and Trends* (Tata Institute of Fundamental Research, Mumbai, 2006), pp. 1–90
4. I. Bailleul, Poisson boundary of a relativistic diffusion. Probab. Theory Relat. Fields **141**(1–2), 283–329 (2008)
5. I. Bailleul, A. Raugi, Where does randomness lead in spacetime? ESAIM Probab. Stat. **14**, 16–52 (2010)
6. C. Dellacherie, P.-A. Meyer, *Probabilities and Potential. B*. North-Holland Mathematics Studies, vol. 72 (North-Holland, Amsterdam, 1982). Theory of martingales, Translated from the French by J.P. Wilson
7. R.M. Dudley, Lorentz-invariant Markov processes in relativistic phase space. Ark. Mat. **6**, 241–268 (1966)
8. J. Franchi, Y. Le Jan, *Hyperbolic Dynamics and Brownian Motion*. Oxford Mathematical Monographs (Oxford University Press, Oxford, 2012). An introduction
9. H. Furstenberg, Boundary theory and stochastic processes on homogeneous spaces, in *Harmonic Analysis on Homogeneous Spaces*. Proceedings of Symposia in Pure Mathematics, vol. XXVI, Williams Coll., Williamstown, Mass., 1972 (American Mathematical Society, Providence, RI, 1973), pp. 193–229
10. V.A. Kaimanovich, Measure-theoretic boundaries of Markov chains, 0-2 laws and entropy, in *Harmonic Analysis and Discrete Potential Theory (Frascati, 1991)* (Plenum, New York, 1992), pp. 145–180
11. M. Liao, *Lévy Processes in Lie Groups*. Cambridge Tracts in Mathematics, vol. 162 (Cambridge University Press, Cambridge, 2004)
12. P. March, Brownian motion and harmonic functions on rotationally symmetric manifolds. Ann. Probab. **14**(3), 793–801 (1986)
13. A. Raugi, Fonctions harmoniques sur les groupes localement compacts à base dénombrable. Bull. Soc. Math. France Mém. **54**, 5–118 (1977)
14. D. Revuz, M. Yor, *Continuous Martingales and Brownian Motion*, 3rd edn. (Springer, New York, 1999)
15. D. Sullivan, The Dirichlet problem at infinity for a negatively curved manifold. J. Differ. Geom. **18**(4), 723–732 (1983/1984)

Weitzenböck and Clark-Ocone Decompositions for Differential Forms on the Space of Normal Martingales

Nicolas Privault

Abstract We present a framework for the construction of Weitzenböck and Clark-Ocone formulae for differential forms on the probability space of a normal martingale. This approach covers existing constructions based on Brownian motion, and extends them to other normal martingales such as compensated Poisson processes. It also applies to the path space of Brownian motion on a Lie group and to other geometries based on the Poisson process. Classical results such as the de Rham-Hodge-Kodaira decomposition and the vanishing of harmonic differential forms are extended in this way to finite difference operators by two distinct approaches based on the Weitzenböck and Clark-Ocone formulae.

AMS Classification: 60H07, 60J65, 58J65, 58A10

1 Introduction

Vanishing theorems for harmonic forms on Riemannian manifolds can be proved by the Bochner method, which involves the Weitzenböck formula and relates the Hodge Laplacian on differential n-forms $\Delta_n = \mathrm{d}^{n-1}\mathrm{d}^{(n-1)*} + \mathrm{d}^{n*}\mathrm{d}^n$ to the Bochner Laplacian $L = \nabla^*\nabla$ through a zero order curvature term R_n, i.e.

$$\Delta_n = L + R_n. \tag{1}$$

Here, d^n is the exterior derivative on n-forms with adjoint d^{n*}, ∇ is the covariant derivative with adjoint ∇^*, and R_n is the Weitzenböck curvature which reduces to the usual Ricci tensor on one-forms. In particular, since both Laplacian operators Δ_n and L are non-negative, the identity (1) shows that there are no L^2 harmonic n-forms on a complete manifold when the curvature term R_n is positive.

N. Privault (✉)
Division of Mathematical Sciences, School of Physical and Mathematical Sciences, Nanyang Technological University, 21 Nanyang Link, Singapore 637371, Singapore
e-mail: nprivault@ntu.edu.sg

C. Donati-Martin et al. (eds.), *Séminaire de Probabilités XLVIII*, Lecture Notes in Mathematics 2168, DOI 10.1007/978-3-319-44465-9_9

The Bochner vanishing technique extends to infinite dimension, in particular in the linear case. On abstract Wiener spaces, the de Rham-Hodge decomposition and a Weitzenböck formula have been derived in [20] with $-L$ the Ornstein-Uhlenbeck operator and $R_n = n\mathrm{Id}$, and it has been shown therein that there exist no nontrivial harmonic n-forms for $n \geq 1$. Various other Weitzenböck-type formulae have been established on infinite-dimensional manifolds with curvature, for example, on submanifolds of the Wiener space in [12], on path spaces over Riemannian manifolds [6], and on loop spaces over Lie groups [10], with more complicated curvature terms R_n. On the path spaces over compact Lie groups, the Itô map has been in used in [11] to construct a diffeomorphism which transfers the Weitzenböck formula of [20], and thus the vanishing theorem, from the Wiener space to path groups. The vanishing of harmonic one-forms on loop groups has also been proved in [1] using the Weitzenböck formula [10].

Vanishing theorems on path spaces can also be proved using martingale representation and the Clark-Ocone formula

$$F = \mathbb{E}[F] + \int_0^\infty \mathbb{E}[D_t F \mid \mathscr{F}_t] dB_t,$$

cf. [5, 15], which decomposes a square-integrable function into the sum of a constant and an Itô integral with respect to Brownian motion, where D_t denotes the Malliavin gradient, cf. (6) below. The Clark-Ocone formula has been extended to differential forms in [22] on the Wiener space and in [7] on the path space over a Riemannian manifold, in order to decompose an n-form into the sum of an exact form and a martingale. As the martingale component in the decomposition vanishes when the given n-form is closed, such a formula can be used to show that there exist no non-trivial harmonic n-forms. On the classical Wiener space, these generalised Clark-Ocone formulae [22] provide an alternative proof of the vanishing results of [20]; in addition, they give explicit expression for closed differential forms, while their dual versions apply to the representation of co-closed forms. The vanishing of harmonic one-forms on the path spaces over Riemannian manifolds, has been proved by the Clark-Ocone formula in [7].

Until now, those vanishing techniques have only been applied in the Brownian framework, where the underlying gradient operator satisfies the derivation property. The aim of this paper is to show that they also apply to a large family of stochastic processes, without requiring the derivation property of the gradient operator nor a Gaussian setting. In particular, our argument applies to gradient operators defined by chaos expansion methods with respect to normal martingales. In this way the Weitzenböck and Clark-Ocone decompositions are shown to apply not only on the Wiener space, but also to other normal continuous-time martingales such as the compensated Poisson process, for which the gradient operator can be defined by finite differences. Our approach relies on a direct proof inspired by the arguments of [9] and [10] for the Weitzenböck formula on path and loop groups.

In Sect. 3 we construct a Hodge Laplacian $\Delta_n = \mathrm{d}^{n-1}\mathrm{d}^{(n-1)*} + \mathrm{d}^{n*}\mathrm{d}^n$ on differential n-forms and we derive the de Rham-Hodge decomposition

$$L^2(\Omega; H^{\wedge n}) = \mathrm{Im}\,\mathrm{d}^{n-1} \oplus \mathrm{Im}\,\mathrm{d}^{n*} \oplus \mathrm{Ker}\,\Delta_n, \qquad n \geq 1,$$

cf. (21). Section 4 deals with examples to which our general framework applies, including chaos-based settings and a non-chaos based constructions such as the path space over a Lie group.

In Theorem 1 we prove the Weitzenböck identity

$$\Delta_n = n\mathrm{Id}_{H^{\wedge n}} + \nabla^*\nabla, \qquad n \geq 1,$$

and in Proposition 2 we show the vanishing of harmonic forms $\mathrm{Ker}\,\Delta_n = \{0\}$, from which the de Rham-Hodge decomposition

$$L^2(\Omega; H^{\wedge n}) = \mathrm{Im}\,\mathrm{d}^{n-1} \oplus \mathrm{Im}\,\mathrm{d}^{n*}, \qquad n \geq 1,$$

follows. This result is also derived in Corollary 1 from the Clark-Ocone formula of Theorem 2, showing the complementarity of the two approaches.

It can be shown in addition that this method goes beyond chaos expansions and encompasses other natural geometries in addition to the path space over a Lie group described in Sect. 4, for example on the Poisson space over the half line $\mathbf{R}_+$, cf. [19], in which case the gradient operator has the derivation property.

In [2, 3], n-differential forms on the configuration space over a Riemannian manifold under a Poisson random measure have been constructed in a different way by a lifting of the underlying differential structure on the manifold to the configuration space. We also refer the reader to [4] for a different approach to the construction of the Hodge decomposition on abstract metric spaces.

This paper is organized as follows. Sections 2 and 3 introduce the general framework of differential and divergence operators on functions and differential forms, including duality and commutation relations. Section 4 describes a number of examples to which this framework applies, while Sects. 5 and 6 present the main results on the Weitzenböck identity and the generalised Clark-Ocone formulae, respectively, including the vanishing of harmonic forms. The examples of Sect. 4, which range from normal martingales to Lie-group valued Brownian motion, are revisited one by one in the frameworks of Sects. 5 and 6. The appendix contains the proofs of the main results.

2 Differential Forms and Exterior Derivative

In this section we introduce an abstract gradient and divergence framework based on a probability space $(\Omega, \mathscr{F}, \mathbb{P})$ and an algebra $\mathscr{S} \subset L^2(\Omega)$ for the pointwise product of random variables, dense in $L^2(\Omega)$.

Deterministic Forms

We fix a measure space (X, σ) and we consider a linear space H of $\mathbf{R}^d$-valued functions, dense in $L^2(X, \mathbf{R}^d)$, $d \geq 1$, and endowed with the inner product induced from $L^2(X, \mathbf{R}^d)$. Denote by $H^{\otimes n}$ the n-th tensor power of H, and by $H^{\circ n}$, resp. $H^{\wedge n}$, its subspaces of symmetric, resp. skew-symmetric tensors, completed using the inherited Hilbert space cross norm. The exterior product $\wedge$ is defined as

$$h_1 \wedge \cdots \wedge h_n := \mathscr{A}_n(h_1 \otimes \cdots \otimes h_n), \quad h_1, \ldots, h_n \in H, \tag{2}$$

where $\mathscr{A}_n$ denotes the antisymmetrization map on n-tensors given by

$$\mathscr{A}_n(h_1 \otimes \cdots \otimes h_n) = \sum_{\sigma \in \Sigma_n} \operatorname{sign}(\sigma)(h_{\sigma(1)} \otimes \cdots \otimes h_{\sigma(n)}), \tag{3}$$

and the summation is over $n!$ elements of the symmetric group Σ_n consisting of all permutations of $\{1, \ldots, n\}$. We also equip $H^{\wedge n}$ with the inner product

$$\begin{aligned}
\langle f_n, g_n \rangle_{H^{\wedge n}} &:= \frac{1}{n!} \int_X \cdots \int_X \langle f_n(x_1, \ldots, x_n), g_n(x_1, \ldots, x_n) \rangle_{(\mathbf{R}^d)^{\otimes n}} \sigma(dx_1) \cdots \sigma(dx_n) \\
&= \frac{1}{n!} \langle f_n, g_n \rangle_{H^{\otimes n}}, \qquad f_n, g_n \in H^{\wedge n},
\end{aligned}$$

so that we have in particular

$$\begin{aligned}
&\langle h_1 \wedge \cdots \wedge h_n, k_1 \wedge \cdots \wedge k_n \rangle_{H^{\wedge n}} \\
&:= \frac{1}{n!} \int_X \cdots \int_X \langle \mathscr{A}_n(h_1 \otimes \cdots \otimes h_n), \mathscr{A}_n(k_1 \otimes \cdots \otimes k_n) \rangle_{(\mathbf{R}^d)^{\otimes n}} \sigma(dx_1) \cdots \sigma(dx_n) \\
&= \frac{1}{n!} \int_X \cdots \int_X \Big\langle \sum_{\sigma \in \Sigma_n} \operatorname{sign}(\sigma)(h_{\sigma(1)} \otimes \cdots \otimes h_{\sigma(n)}), \\
&\quad \sum_{\eta \in \Sigma_n} \operatorname{sign}(\eta)(k_{\eta(1)} \otimes \cdots \otimes k_{\eta(n)}) \Big\rangle_{(\mathbf{R}^d)^{\otimes n}} \sigma(dx_1) \cdots \sigma(dx_n) \\
&= \sum_{\sigma \in \Sigma_n} \operatorname{sign}(\sigma) \int_X \cdots \int_X \langle h_{\sigma(1)} \otimes \cdots \otimes h_{\sigma(n)}, k_1 \otimes \cdots \otimes k_n \rangle_{(\mathbf{R}^d)^{\otimes n}} \sigma(dx_1) \cdots \sigma(dx_n) \\
&= \sum_{\sigma \in \Sigma_n} \operatorname{sign}(\sigma) \int_X \cdots \int_X \langle h_{\sigma(1)}, k_1 \rangle_{\mathbf{R}^d} \cdots \langle h_{\sigma(n)}, k_n \rangle_{\mathbf{R}^d} \sigma(dx_1) \cdots \sigma(dx_n) \\
&= \det((\langle h_i, k_j \rangle_H)_{1 \leq i,j \leq n}),
\end{aligned}$$

$h_1, \ldots, h_n, k_1, \ldots, k_n \in H$.

Covariant Derivative

In the sequel we will use a covariant derivative operator

$$\begin{aligned}\nabla : H &\longrightarrow H \otimes H \\ h &\longmapsto \nabla h = (\nabla_x h)_{x \in X},\end{aligned}$$

where $\nabla_x h \in \mathbf{R}^d \otimes H$, $x \in X$, is defined from the relation

$$\langle \nabla_x h, k \rangle_H := \langle \nabla h, k \rangle_H (x) \in \mathbf{R}^d, \qquad x \in X, \quad h, k \in H.$$

We will extend the definition of ∇ to an operator

$$\nabla : H^{\wedge n} \longrightarrow H \otimes H^{\wedge n}$$

on differential forms in $H^{\wedge n}$, by the following steps.

(i) Let

$$\nabla_x^{(l)}(h_1 \wedge \cdots \wedge h_n) \in \underbrace{H \wedge \cdots \wedge H}_{\text{l-1 times}} \wedge (\mathbf{R}^d \otimes H) \wedge \underbrace{H \wedge \cdots \wedge H}_{\text{n-l times}}$$

as

$$\nabla_x^{(l)}(h_1 \wedge \cdots \wedge h_n) := h_1 \wedge \cdots \wedge h_{l-1} \wedge \nabla_x h_l \wedge h_{l+1} \wedge \cdots \wedge h_n, \qquad x \in X,$$

$l = 1, \ldots, n$.

(ii) We define $\nabla_x(h_1 \wedge \cdots \wedge h_n)$ in $\mathbf{R}^d \otimes H^{\wedge n}$, $x \in X$, as

$$\begin{aligned}\nabla_x(h_1 \wedge \cdots \wedge h_n) &:= \sum_{j=1}^{n} \nabla_x^{(j)}(h_1 \wedge \cdots \wedge h_n) \\ &= \sum_{j=1}^{n} h_1 \wedge \cdots \wedge h_{j-1} \wedge \nabla_x h_j \wedge h_{j+1} \wedge \cdots \wedge h_n,\end{aligned} \tag{4}$$

by canonically identifying the space

$$\underbrace{H \wedge \cdots \wedge H}_{\text{l-1 times}} \wedge (\mathbf{R}^d \otimes H) \wedge \underbrace{H \wedge \cdots \wedge H}_{\text{n-l times}}$$

to $\mathbf{R}^d \otimes H^{\wedge n}$, for $l = 1, \ldots, n$.

Given $g \in H$ we also define $\nabla_g(h_1 \wedge \cdots \wedge h_n) \in H^{\wedge n}$ by

$$\nabla_g(h_1 \wedge \cdots \wedge h_n) = \int_X \langle g(x), \nabla_x(h_1 \wedge \cdots \wedge h_n) \rangle_{\mathbf{R}^d} \sigma(dx)$$

$$= \sum_{j=1}^{n} \int_X \langle g(x), (h_1 \wedge \cdots \wedge h_{j-1} \wedge \nabla_x h_j \wedge h_{j+1} \wedge \cdots \wedge h_n) \rangle_{\mathbb{R}^d} \sigma(dx)$$

$$= \sum_{j=1}^{n} (h_1 \wedge \cdots \wedge h_{j-1} \wedge \nabla_g h_j \wedge h_{j+1} \wedge \cdots \wedge h_n).$$

Exterior Derivative–Deterministic Forms

We now define the exterior derivative on n-forms $u_n \in H^{\wedge n}$ by

$$\langle \mathrm{d}^n u_n, h_1 \wedge \cdots \wedge h_{n+1} \rangle_{H^{\wedge(n+1)}} := \sum_{k=1}^{n+1} (-1)^{k-1} \langle \nabla_{h_k} u_n, h_1 \wedge \cdots \wedge h_{k-1} \wedge h_{k+1} \wedge \cdots \wedge h_{n+1} \rangle_{H^{\wedge n}}, \tag{5}$$

where $h_1, \ldots, h_{n+1} \in H$, i.e. $\mathrm{d}^n = \frac{1}{n!} \mathscr{A}_{n+1} \nabla$, and the $(n+1)$-form $\mathrm{d}^n(h_1 \wedge \cdots \wedge h_n)$ is given by

$$\mathrm{d}^n_{x_{n+1}}((h_1 \wedge \cdots \wedge h_n)(x_1, \ldots, x_n)) = \mathrm{d}^n(h_1 \wedge \cdots \wedge h_n)(x_1, \ldots, x_{n+1})$$

$$= \sum_{j=1}^{n+1} (-1)^{j-1} \nabla_{x_j} (h_1 \wedge \cdots \wedge h_n)(x_1, \ldots, x_{j-1}, x_{j+1}, \ldots, x_{n+1})$$

$$= \sum_{j=1}^{n+1} (-1)^{j-1} \sum_{i=1}^{n} \nabla^{(i)}_{x_j} (h_1 \wedge \cdots \wedge h_n)(x_1, \ldots, x_{j-1}, x_{j+1}, \ldots, x_{n+1})$$

$$= \sum_{j=1}^{n+1} (-1)^{j-1} \sum_{i=1}^{n} (h_1 \wedge \cdots \wedge \nabla_{x_j} h_i \wedge \cdots \wedge h_n)(x_1, \ldots, x_{j-1}, x_{j+1}, \ldots, x_{n+1}).$$

Random Forms

In the sequel we will need a linear gradient operator

$$\begin{aligned} D : \mathscr{S} &\longrightarrow L^2(\Omega; H) \\ F &\longmapsto DF = (D_x F)_{x \in X} \end{aligned} \tag{6}$$

acting on random variables in $\mathscr{S}$.

We work on the space $\mathscr{S} \otimes H^{\wedge n}$ of elementary (random) n-forms that can be written as linear combinations of terms for the form

$$u_n = F \otimes h \in \mathscr{S} \otimes H^{\wedge n}, \qquad F \in \mathscr{S}, \quad h \in H^{\wedge n}. \tag{7}$$

The operator D is extended to $u_n \in \mathscr{S} \otimes H^{\wedge n}$ as in (7) by the pointwise equality

$$D_x u_n := (D_x F) \otimes (h_1 \wedge \cdots \wedge h_n), \qquad x \in X, \tag{8}$$

i.e. $Du_n \in \mathcal{S} \otimes H \otimes H^{\wedge n}$. We also extend ∇ to random forms $u_n = F \otimes f_n \in \mathcal{S} \otimes H^{\wedge n}$ by defining $\nabla u_n \in \mathcal{S} \otimes H \otimes H^{\wedge n}$ as

$$\begin{aligned}\nabla_y(u_n(x_1,\ldots,x_n)) &:= (D_y F) \otimes f_n(x_1,\ldots,x_n) + F \otimes \nabla_y f_n(x_1,\ldots,x_n) \\ &= (D_y F) \otimes f_n(x_1,\ldots,x_n) + F \otimes \sum_{l=1}^{n} \nabla_y^{(l)} f_n(x_1,\ldots,x_n),\end{aligned}$$

$x_1,\ldots,x_n,y \in X$. In particular for $n = 1$, ∇ extends to stochastic processes (or one-forms) as

$$\nabla_x(u_1(y)) = \nabla_x(F \otimes f_1(y)) = (D_x F) \otimes f_1(y) + F \otimes \nabla_x f_1(y),$$

with $u_1 = F \otimes f_1 \in \mathcal{S} \otimes H$.

Lie Bracket and Vanishing of Torsion

The Lie bracket $\{f,g\}$ of $f,g \in H$, is defined to be the unique element w of H satisfying

$$(D_f D_g - D_g D_f)F = D_w F, \qquad F \in \mathcal{S}, \tag{9}$$

where

$$D_f F := \langle f, DF \rangle_H, \qquad f \in H, \quad F \in \mathrm{Dom}(D),$$

and is extended to $u,v \in \mathcal{S} \otimes H$ by

$$\{Fu, Gv\}(x) = FG\{u,v\}(x) + v(x)FD_u G - u(x)GD_v F, \qquad x \in X,$$

$u,v \in H$, $F,G \in \mathcal{S}$. In the sequel we will make the following assumption.

(A1) Vanishing of torsion. The connection defined by ∇ has a vanishing torsion, i.e. we have

$$\{u,v\} = \nabla_u v - \nabla_v u, \qquad u,v \in \mathcal{S} \otimes H. \tag{\textbf{A1}}$$

From (9) the vanishing of torsion Assumption (A1) can be written as

$$\begin{aligned}&\int_X \int_X \langle (D_x D_y F - D_y D_x F), f(x) \otimes g(y) \rangle_{\mathbf{R}^d \otimes \mathbf{R}^d} \sigma(dx)\sigma(dy) \qquad (10)\\ &= \int_X \int_X \langle \nabla_y g(x), D_x F \otimes f(y) \rangle_{\mathbf{R}^d \otimes \mathbf{R}^d} \sigma(dx)\sigma(dy) \\ &- \int_X \int_X \langle \nabla_y f(x), D_x F \otimes g(y) \rangle_{\mathbf{R}^d \otimes \mathbf{R}^d} \sigma(dx)\sigma(dy), \qquad F \in \mathcal{S}, \quad f,g \in H.\end{aligned}$$

Exterior Derivative–Random Forms
From Assumption (A1) we may now define the $(n+1)$-form $\mathrm{d}^n u_n$ as

$$\begin{aligned}
\mathrm{d}^n u_n(x_1,\ldots,x_{n+1}) &:= \mathrm{d}^n_{x_{n+1}}(u_n(x_1,\ldots,x_n)) \\
&= \frac{1}{n!}\mathscr{A}_{n+1}(\nabla.u_n)(x_1,\ldots,x_{n+1}) \\
&= (D.F\wedge f_n)(x_1,\ldots,x_{n+1}) + \frac{1}{n!}F\otimes\mathscr{A}_{n+1}(\nabla.f_n)(x_1,\ldots,x_{n+1}),
\end{aligned}$$

which is also equal to

$$\sum_{j=1}^{n+1}(-1)^{j-1}\nabla_{x_j}u_n(x_1,\ldots,x_{j-1},x_{j+1},\ldots,x_{n+1})$$

in $H^{\wedge(n+1)}$, for $u_n\in\mathscr{S}\otimes H^{\wedge n}$ of the form (7). In other words, on elementary forms we have

$$\begin{aligned}
\mathrm{d}^n_{x_{n+1}}(F\otimes h_1\wedge\cdots\wedge h_n(x_1,\ldots,x_n)) &= \mathrm{d}^n(F\otimes h_1\wedge\cdots\wedge h_n)(x_1,\ldots,x_{n+1}) \qquad (11)\\
&= ((D.F)\wedge h_1\wedge\cdots\wedge h_n)(x_1,\ldots,x_{n+1}) + F\otimes\mathrm{d}^n(h_1\wedge\cdots\wedge h_n)(x_1,\ldots,x_{n+1}).
\end{aligned}$$

In particular, for $n=0$ we have

$$\mathrm{d}^0_x F = \nabla_x F = D_x F, \qquad F\in\mathscr{S}\otimes H^{\wedge 0}=\mathscr{S}.$$

and for $n=1$,

$$\mathrm{d}^1(F\otimes h_1)(x_1,x_2) = \mathrm{d}^1_{x_2}(F\otimes h_1(x_1)) = F\otimes\mathrm{d}^1_{x_1}h_1(x_2)+(D_{x_1}F)\otimes h(x_2)-D_{x_2}F\otimes h(x_1).$$

We also note that

$$\mathrm{d}^n(\mathscr{S}\otimes H^{\wedge n})\subset\mathrm{Dom}(\mathrm{d}^{n+1}), \qquad n\in\mathbb{N}. \tag{12}$$

Assumption (A1) and the invariant formula for differential forms (see e.g. Proposition 3.11 page 36 of [13]) also show that we have

$$\mathrm{d}^{n+1}\mathrm{d}^n = 0, \qquad n\in\mathbb{N}, \tag{13}$$

which implies

$$\mathrm{Im}\,\mathrm{d}^n\subset\mathrm{Ker}\,\mathrm{d}^{n+1}\subset\mathrm{Dom}(\mathrm{d}^{n+1}), \qquad n\in\mathbb{N}. \tag{14}$$

3 Divergence of n-Forms and Duality

In this section we consider a divergence operator

$$\delta : \mathscr{S} \otimes H \longrightarrow L^2(\Omega),$$
$$u = (u(x))_{x \in X} \longmapsto \delta(u)$$

acting on stochastic processes, and extended to elementary n-forms by letting

$$\delta(h_1 \wedge \cdots \wedge h_n)$$
$$:= \frac{1}{n} \sum_{j=1}^{n} (-1)^{j-1} \delta(h_j) \otimes (h_1 \wedge \cdots \wedge h_{j-1} \wedge h_{j+1} \wedge \cdots \wedge h_n) \in \mathscr{S} \otimes H^{\wedge(n-1)},$$

and to simple elements $u \in \mathscr{S} \otimes H^{\wedge n}$ of the form (7) by

$$\begin{aligned} \delta(u_n)(x_1, \ldots, x_{n-1}) &:= \delta(u_n(\cdot, x_1, \ldots, x_{n-1})) \\ &= \delta(F \otimes (h_1 \wedge \cdots \wedge h_n))(x_1, \ldots, x_{n-1}) \\ &= \frac{1}{n} \sum_{j=1}^{n} (-1)^{j-1} \delta(F \otimes h_j) \otimes (h_1 \wedge \cdots \wedge h_{j-1} \wedge h_{j+1} \wedge \cdots \wedge h_n)(x_1, \ldots, x_{n-1}). \end{aligned} \tag{15}$$

Divergence of Random Forms
The divergence operator d^{n*} on $(n+1)$-forms $u_{n+1} = F \otimes f_{n+1} \in \mathscr{S} \otimes H^{\wedge(n+1)}$ of the form (7) is defined by

$$\mathrm{d}^{n*} u_{n+1}(x_1, \ldots, x_n) := \delta(F \otimes f_{n+1}(\cdot, x_1, \ldots, x_n)) - F \mathrm{trace}(\nabla . f_{n+1}(\cdot, x_1, \ldots, x_n))$$

where

$$\mathrm{trace}(\nabla . f_{n+1}(\cdot, x_1, \ldots, x_n)) := \int_0^\infty \mathrm{Tr}\, \nabla_x f_{n+1}(x, x_1, \ldots, x_n) \sigma(dx)$$

and Tr denotes the trace on $\mathbf{R}^d \otimes \mathbf{R}^d$, i.e. we have

$$\begin{aligned} &\mathrm{d}^{n*} u_{n+1}(x_1, \ldots, x_n) \\ &\quad = \delta(F \otimes f_{n+1}(\cdot, x_1, \ldots, x_n)) - F \int_0^\infty \mathrm{Tr}\, \nabla_x f_{n+1}(x, x_1, \ldots, x_n) \sigma(dx), \end{aligned} \tag{16}$$

which belongs to $\mathscr{S} \otimes H^{\wedge n}$ from (15), $n \geq 1$, and

$$\mathrm{d}^{0*} u_1 = \delta(u_1), \qquad u_1 \in \mathscr{S} \otimes H, \tag{17}$$

when $n = 0$ since $\nabla_x f_1(x) = 0$.

Duality Relations

We will make the following assumptions on δ, D and ∇:

(A2) The operators D and δ satisfy the duality relation

$$\mathbb{E}[\langle DF, u\rangle_H] = \mathbb{E}[F\delta(u)], \qquad F \in \mathrm{Dom}(D), \quad u \in \mathrm{Dom}(\delta). \tag{A2}$$

The above duality condition (A2) implies that the operators D and δ are closable, cf. Proposition 3.1.2 of [18], and the operators D and δ are extended to their respective closed domains $\mathrm{Dom}(D)$ and $\mathrm{Dom}(\delta)$.

(A3) The operator ∇ satisfies the condition

$$\int_{X^{n+1}} \langle g_{n+1}(x_1, \ldots, x_{n+1}), \nabla_{x_1} f_n(x_2, \ldots, x_{n+1})\rangle_{(\mathbf{R}^d)^{\otimes(n+1)}} \sigma(dx_1)\cdots\sigma(dx_{n+1}) =$$

$$- \int_{X^{n+1}} \langle \mathrm{Tr}\, \nabla_x g_{n+1}(x, x_1, \ldots, x_n), f_n(x_1, \ldots, x_n)\rangle_{(\mathbf{R}^d)^{\otimes n}} \sigma(dx_1)\cdots\sigma(dx_n)\sigma(dx), \tag{A3}$$

$f_n \in H^{\otimes n}$, $g_{n+1} \in H^{\otimes(n+1)}$, $n \geq 1$.

The compatibility condition (A3) is weaker than the usual compatibility of ∇ with the metric $\langle \cdot, \cdot \rangle_H$, which reads

$$\langle \nabla_x f_n, g_n\rangle_{H^{\otimes n}} = -\langle f_n, \nabla_x g_n\rangle_{H^{\otimes n}}, \qquad x \in X, \tag{18}$$

$f_n, g_n \in H^{\otimes n}$.. Indeed, when applying (18) to $f_{n+1}(x, \cdot) \in H^{\otimes n}$ and $g_n \in H^{\otimes n}$, $x \in X$, $n \geq 1$, we get

$$\int_{X^{n+1}} \langle \mathrm{Tr}\, \nabla_x f_{n+1}(x, x_1, \ldots, x_n), g_n(x_1, \ldots, x_n)\rangle_{(\mathbf{R}^d)^{\otimes n}} \sigma(dx_1)\cdots\sigma(dx_n)\sigma(dx)$$

$$= \sum_{k=1}^{d} \int_{X^{n+1}} \langle (\nabla_x f_{n+1}^{(k)}(x, x_1, \ldots, x_n))^{(k)}, g_n(x_1, \ldots, x_n)\rangle_{(\mathbf{R}^d)^{\otimes n}} \sigma(dx_1)\cdot\sigma(dx_n)\sigma(dx)$$

$$= -\sum_{k=1}^{d} \int_{X^{n+1}} \langle f_{n+1}^{(k)}(x, x_1, \ldots, x_n), (\nabla_x g_n(x_1, \ldots, x_n))^{(k)}\rangle_{(\mathbf{R}^d)^{\otimes n}} \sigma(dx_1)\cdot\sigma(dx_n)\sigma(dx)$$

$$= -\int_{X^{n+1}} \langle f_{n+1}(x, x_1, \ldots, x_n), \nabla_x g_n(x_1, \ldots, x_n)\rangle_{(\mathbf{R}^d)^{\otimes(n+1)}} \sigma(dx_1)\cdot\sigma(dx_n)\sigma(dx),$$

where $f_{n+1}^{(k)}(x, x_1, \ldots, x_n)$ denotes the k-th component in $\mathbf{R}^d$ of the first component of $f_{n+1}(x, x_1, \ldots, x_n)$ in $(\mathbf{R}^d)^{\otimes(n+1)}$. In this sense, Assumption (A3) is automatically satisfied in all settings which incorporate the compatibility (18) with $\langle \cdot, \cdot \rangle$.

Proposition 1 (Duality) *Under Assumptions (A1)–(A3), for any $u_n \in \mathscr{S} \otimes H^{\wedge n}$ and $v_{n+1} \in \mathscr{S} \otimes H^{\wedge(n+1)}$ we have*

$$\langle \mathrm{d}^n u_n, v_{n+1} \rangle_{L^2(\Omega, H^{\wedge(n+1)})} = \langle u_n, \mathrm{d}^{n*} v_{n+1} \rangle_{L^2(\Omega, H^{\wedge n})}. \tag{19}$$

As above we note that the duality (19) implies the closability of both d^n and $\mathrm{d}^{n*} v_{n+1}$, which are extended to their closed domains $\mathrm{Dom}(\mathrm{d}^n)$ and $\mathrm{Dom}(\mathrm{d}^{n*})$, $n \in \mathbb{N}$, by the same argument as in Proposition 3.1.2 of [18]. When $n = 0$, the statement of Proposition 1 reduces to (A2).

The proof of Proposition 1 is postponed to the appendix. In the case of one-forms it reads

$$\begin{aligned}
&\langle \mathrm{d}^1_{t_2}(Ff_1(t_1)), Gg_2(t_1, t_2) \rangle_{L^2(\Omega, H^{\wedge 2})} \\
&= \left\langle \nabla_{t_1} \left(\frac{F}{2} f_1(t_2) \right), Gg_2(t_1, t_2) \right\rangle_{L^2(\Omega, H^{\otimes 2})} - \left\langle \nabla_{t_2} \left(\frac{F}{2} f_1(t_1) \right), Gg_2(t_1, t_2) \right\rangle_{L^2(\Omega, H^{\otimes 2})} \\
&= \left\langle D_{t_1} \frac{F}{2} f_1(t_2), Gg_2(t_1, t_2) \right\rangle_{L^2(\Omega, H^{\otimes 2})} - \left\langle D_{t_2} \frac{F}{2} f_1(t_1), Gg_2(t_1, t_2) \right\rangle_{L^2(\Omega, H^{\otimes 2})} \\
&\quad + \left\langle \frac{F}{2} \nabla_{t_1} f_1(t_2), Gg_2(t_1, t_2) \right\rangle_{L^2(\Omega, H^{\otimes 2})} - \left\langle \frac{F}{2} \nabla_{t_2} f_1(t_1), Gg_2(t_1, t_2) \right\rangle_{L^2(\Omega, H^{\otimes 2})} \\
&= \left\langle \frac{F}{2} f_1(t_2), \delta(Gg_2(\cdot, t_2)) \right\rangle_{L^2(\Omega, H)} - \left\langle \frac{F}{2} f_1(t_1), \delta(Gg_2(t_1, \cdot)) \right\rangle_{L^2(\Omega, H)} \\
&\quad - \left\langle \frac{F}{2} f_1(t_2), G \int_0^\infty \mathrm{Tr}\, \nabla_t g_2(t, t_2) dt \right\rangle_{L^2(\Omega, H)} - \left\langle \frac{F}{2} f_1(t_1), G \int_0^\infty \mathrm{Tr}\, \nabla_t g_2(t_1, t) dt \right\rangle_{L^2(\Omega, H)} \\
&= \left\langle \frac{F}{2} f_1(t_2), \delta(Gg_2(\cdot, t_2)) \right\rangle_{L^2(\Omega, H)} + \left\langle \frac{F}{2} f_1(t_1), \delta(Gg_2(\cdot, t_1)) \right\rangle_{L^2(\Omega, H)} \\
&\quad - \left\langle \frac{F}{2} f_1(t_2), G \int_0^\infty \mathrm{Tr}\, \nabla_t g_2(t, t_2) dt \right\rangle_{L^2(\Omega, H)} + \left\langle \frac{F}{2} f_1(t_1), G \int_0^\infty \mathrm{Tr}\, \nabla_t g_2(t, t_1) dt \right\rangle_{L^2(\Omega, H)} \\
&= \langle Ff_1(t_1), \delta(Gg_2(\cdot, t_1)) \rangle_{L^2(\Omega, H)} - \left\langle Ff_1(t_1), G \int_0^\infty \mathrm{Tr}\, \nabla_t g_2(t, t_1) dt \right\rangle_{L^2(\Omega, H)} \\
&= \langle Ff_1(t_1), \mathrm{d}^{1*}(Gg_2)(t_1) \rangle_{L^2(\Omega, H)}, \qquad F, G \in \mathscr{S}, \quad f_1 \in H, \quad g_2 \in H^{\wedge 2}.
\end{aligned}$$

As in (A2) above, the duality (19) shows that d^n extends to a closed operator

$$\mathrm{d}^n : \mathrm{Dom}(\mathrm{d}^n) \longrightarrow L^2(\Omega; H^{\wedge(n+1)})$$

with domain $\mathrm{Dom}(\mathrm{d}^n) \subset L^2(\Omega; H^{\wedge n})$, and d^{n*}, $n \in \mathbb{N}$, extends to a closed operator

$$\mathrm{d}^{n*} : \mathrm{Dom}(\mathrm{d}^{n*}) \longrightarrow L^2(\Omega; H^{\wedge n})$$

with domain $\mathrm{Dom}(\mathrm{d}^{n*}) \subset L^2(\Omega; H^{\wedge(n+1)})$, by the same argument as in Proposition 3.1.2 of [18].

In addition, by the coboundary condition (13) and the duality (19) we find

$$\mathrm{d}^{n*}\mathrm{d}^{(n+1)*} = 0, \qquad n \in \mathbb{N}.$$

Based on (14) we define the Hodge Laplacian on differential n-forms as

$$\Delta_n = \mathrm{d}^{n-1}\mathrm{d}^{(n-1)*} + \mathrm{d}^{n*}\mathrm{d}^n, \tag{20}$$

and call harmonic n-forms the elements of the kernel $\mathrm{Ker}\,\Delta_n$ of Δ_n. By (13)–(14) and Proposition 1 we have the de Rham-Hodge decomposition

$$L^2(\Omega; H^{\wedge n}) = \mathrm{Im}\,\mathrm{d}^{n-1} \oplus \mathrm{Im}\,\mathrm{d}^{n*} \oplus \mathrm{Ker}\,\Delta_n, \qquad n \geq 1. \tag{21}$$

Indeed, the spaces of exact and co-exact forms $\mathrm{Im}\,\mathrm{d}^{n-1}$ and $\mathrm{Im}\,\mathrm{d}^{n*}$ are mutually orthogonal by (13) and the duality of Proposition 1. Moreover, the orthogonal complement $(\mathrm{Ker}\,\mathrm{d}^{(n-1)*}) \cap (\mathrm{Ker}\,\mathrm{d}^n)$ of $\mathrm{Im}\,\mathrm{d}^{n-1} \oplus \mathrm{Im}\,\mathrm{d}^{n*}$ in $L^2(\Omega; H^{\wedge n})$ is made of n-forms u_n that are both closed ($d^n u_n = 0$) and co-closed ($d^{(n-1)*}u_n = 0$), hence it is contained in (and equal to) $\mathrm{Ker}\,\Delta_n$ by (20).

Intertwining Relations

The statements and proofs of both the Weitzenböck identity and Clark-Ocone formula in the sequel will also require the following conditions. We assume that

(A4) the operator ∇ satisfies

$$\nabla_x f(y) \cdot \nabla_y f(x) = 0, \qquad \sigma(dx)\sigma(dy) - a.e., \qquad f \in H, \tag{A4}$$

i.e. $\sigma(dx)\sigma(dy) - a.e.$ we have $\nabla_x f(y) = 0$ or $\nabla_y f(x) = 0$.

(A5) Intertwining relation. For all $u \in \mathcal{S} \otimes H$ of the form $u = F \otimes f$ we have

$$\langle g, D\delta(u)\rangle_H = \langle g, u\rangle_H + \delta(\nabla_g u) + \langle DF, \nabla_f g\rangle_H, \qquad g \in H. \tag{A5}$$

We make the following remarks.

Remark 1

(i) When $\nabla = 0$ on H, Assumption (A5) reads

$$D_x\delta(u) = u(x) + \delta(D_x u), \qquad x \in X,$$

for all $u = F \otimes f \in \mathcal{S} \otimes H$.

(ii) Under the torsion free Assumption (A1), Relation (10) shows that (A5) reads

$$D_x\delta(u) = u(x)+\delta(\nabla_x u)+\langle DD_xF, f\rangle_H-\langle D_xDF, f\rangle_H+\langle D.F, \nabla_x f(\cdot)\rangle_H, \quad (22)$$

$u = F \otimes f \in \mathcal{S} \otimes H, x \in X$.

(iii) As a consequence of (22), Assumption (A5) simplifies to

$$D_x\delta(h) = h(x) + \delta(\nabla_x h), \qquad x \in X, \quad h \in H. \qquad \textbf{(A5')}$$

when D satisfies the Leibniz rule of derivation

$$D_x(FG) = FD_xG + GD_xF, \qquad F, G \in \mathcal{S}, \quad x \in X. \quad (23)$$

This will be the case in examples where ∇ does not vanish on H.

(iv) When D has the derivation property (23), Relation (16) rewrites as the divergence formula

$$\begin{aligned} &\mathrm{d}^{n*} u_{n+1}(x_1, \ldots, x_n) \\ &= F\delta(f_{n+1}(\cdot, x_1, \ldots, x_n)) - \int_0^\infty \mathrm{Tr}\,\nabla_x(Ff_{n+1}(x, x_1, \ldots, x_n))\sigma(dx), \end{aligned}$$

$u_{n+1} = F \otimes f_{n+1} \in \mathcal{S} \otimes H^{\wedge(n+1)}$.

Proof We only prove (iii) and (iv).

(iii) First, we note that the duality condition (A2) and the Leibniz rule (23) imply

$$\delta(Fh) = F\delta(h) - \langle DF, h\rangle_H, \qquad F \in \mathcal{S}, \quad h \in H. \quad (24)$$

Hence by (A5') and (23) we have

$$\begin{aligned} D_x\delta(Fh) &= D_x(F\delta(h) - \langle DF, h\rangle_H) \\ &= \delta(h)D_xF + FD_x\delta(h) - D_x\langle DF, h\rangle_H \\ &= \delta(h)D_xF + Fh(x) + F\delta(\nabla_x h) - D_x\langle DF, h\rangle_H \\ &= Fh(x) + \delta(hD_xF) + \langle DD_xF, h\rangle_H + \delta(F\nabla_x h) + \langle D.F, \nabla_x h(\cdot)\rangle_H \\ &\quad - \langle D_xDF, h\rangle_H \\ &= u_1(x) + \delta(\nabla_x u_1) + \langle DD_xF, h\rangle_H - \langle D_xDF, h\rangle_H + \langle D.F, \nabla_x h(\cdot)\rangle_H. \end{aligned}$$

The converse statement is immediate.

(iv) This is a consequence of (16) and the divergence formula (24).

4 Examples

In this section we consider examples of frameworks satisfying Assumptions (A1)–(A5).

Commutative Examples–Chaos Expansions

We start by considering a family of examples based on chaos expansions, in which we take $\nabla h = 0$ for all $h \in H$. Here (X, σ) is a measure space, we take $H = L^2(X, \sigma)$ and $d \geq 1$ and we assume that the chaos decomposition holds, i.e. every $F \in L^2(\Omega, \mathscr{F}, P)$ can be decomposed into a series

$$F = \sum_{n=0}^{\infty} I_n(f_n), \qquad f_n \in H^{\circ n},$$

of multiple stochastic integrals, where $I_0(f_0) = \mathbb{E}[F]$ and for all $n \geq 1$, the multiple stochastic integral $I_n : H^{\circ n} \to L^2(\Omega)$ satisfies the isometry condition

$$\langle I_n(f_n), I_m(g_m) \rangle_{L^2(\Omega)} = n! \mathbf{1}_{\{n=m\}} \langle f_n, g_m \rangle_{H^{\circ n}}, \qquad n, m \geq 1.$$

In this case the space $\mathscr{S}$ is made of the finite chaos expansions

$$\mathscr{S} = \left\{ C + \sum_{k=1}^{n} I_k(f_k), \qquad f_k \in L^2(X)^{\circ k}, \quad k = 1, \ldots, n, \quad n \geq 1, \ C \in \mathbf{R} \right\},$$

the operator

$$D : \mathrm{Dom}(D) \longrightarrow L^2(\Omega \times X, dP \times \sigma(dx))$$

is defined by

$$D_x I_n(f_n) := n I_{n-1}(f_n(*, x)), \qquad dP \times \sigma(dx) - a.e., \quad n \in \mathbb{N}. \tag{25}$$

On the other hand,

$$\delta : \mathrm{Dom}(\delta) \longrightarrow L^2(\Omega),$$

is defined on processes of the form $(I_n(f_{n+1}(*, t)))_{t \in \mathbf{R}_+}$ as

$$\delta(I_n(f_{n+1}(*, \cdot))) := I_{n+1}(\tilde{f}_{n+1}), \qquad n \in \mathbb{N}, \tag{26}$$

where $\tilde{f}_{n+1}$ denotes the symmetrization of $f_{n+1} \in H^{\circ n} \otimes H$ in $(n+1)$ variables. In this chaos expansion framework we have $\nabla = D$ and $\nabla f = 0$ for all $f \in H$, hence Assumptions (A1) and (A3)–(A4) are obviously satisfied and the exterior derivative d is defined by the skew-symmetrisation of D, i.e. (11) becomes

$$\mathrm{d}^n_{x_{n+1}}(F \otimes h_1 \wedge \cdots \wedge h_n(x_1, \ldots, x_n)) = ((D.F) \wedge h_1 \wedge \cdots \wedge h_n)(x_1, \ldots, x_{n+1}), \qquad (27)$$

$x_1, \ldots, x_{n+1} \in X$. As for Assumptions (A2) and (A5) we have the following:

(A2) the duality relation holds in the general framework of chaos expansions; see, e.g., Proposition 4.1.3 of [18];
(A5) given that $\nabla = D$, the commutation relation (A5) holds for $u \in \mathscr{S} \otimes H$, see, e.g., Proposition 4.1.4 of [18].

Note that here, Relations (13)–(14) hold by the definition (27) of the exterior derivative d and the symmetry of second derivative.

Next, we consider some specific examples based on chaos expansions.

Example 1.1—Poisson random measures
On the probability space of a Poisson random measure $\omega(dx)$ with σ-finite intensity measure $\sigma(dx)$ on X, $I_n(f)$ is the multiple compensated Poisson stochastic integral

$$I_n(f_n) := \int_{\Delta_n} f_n(x_1, \ldots, x_n)(\omega(dx_1) - \sigma(dx_1)) \cdots (\omega(dx_n) - \sigma(dx_n))$$

of the symmetric function $f_n \in H^{\circ n}$ with respect to $\omega(dx)$, where

$$\Delta_n = \{(x_1, \ldots, x_n) \in X^n \ : \ x_i \neq x_j, \quad 1 \leq i < j \leq n\}.$$

Here the operator D_x defined in (25) acts by finite differences and addition of a configuration point at $x \in X$, i.e.

$$D_x F(\omega) = F(\omega \cup \{x\}) - F(\omega), \qquad x \in X,$$

where $\omega \cup \{x\}$ represents the addition of the point x to the point configuration ω, see e.g., Proposition 6.4.7 of [18]. Being a finite difference operator, D does not have the derivation property. We refer the reader to Sect. 6.5 of [18] and references therein for the expression of δ defined in (26) in this setting.

Example 1.2—Normal martingales
When $X = \mathbf{R}_+$, chaos-based examples satisfying the above conditions (A1)–(A5) include normal martingales having the chaos representation property (CRP). An $(\mathscr{F}_t)_{t\in\mathbf{R}_+}$-martingale $(M_t)_{t\in\mathbf{R}_+}$ on a filtered probability space $(\Omega, (\mathscr{F}_t)_{t\in\mathbf{R}_+}, P)$ is called a normal martingale under P if

$$\mathbb{E}[(M_t - M_s)^2 \mid \mathscr{F}_s] = t - s, \qquad 0 \leq s \leq t,$$

see, e.g., [18] and references therein. Here we also assume that $(M_t)_{t\in\mathbf{R}_+}$ has the chaos representation property (CRP), and the multiple stochastic integral $I_n(f_n)$ of

$f_n \in L^2(\mathbf{R}_+)^{\circ n}$ with respect to $(M_t)_{t\in\mathbf{R}_+}$ is given by

$$I_n(f_n) = n! \int_0^\infty \int_0^{t_n} \cdots \int_0^{t_2} f_n(t_1, \ldots, t_n) dM_{t_1} \cdots dM_{t_n}, \qquad n \geq 1.$$

Examples of normal martingales satisfying the CRP include the Brownian motion and the compensated Poisson process, both of which we discuss in more details below, as well as certain processes with non-independent increments such as the Azéma martingales, for which the explicit expression of the gradient D is generally unknown; see [8] and Sect. 2.10 of [18].

Example 1.2-a)—Brownian motion
When $X = \mathbf{R}_+$ and $(M_t)_{t\in\mathbf{R}_+}$ is the standard Brownian motion with respect to its own filtration $(\mathscr{F}_t)_{t\in\mathbf{R}_+}$, it is usual to take $\mathscr{S}$ as the space of smooth cylindrical functionals of the form

$$F = f(I_1(h_1), \ldots, I_1(h_n)), \qquad h_1, \ldots, h_n \in H, \quad f \in C_b^\infty(\mathbf{R}^n),$$

on which the gradient D is defined by

$$DF = \sum_{i=1}^n h_i \partial_i f(I_1(h_1), \ldots, I_1(h_n)), \tag{28}$$

e.g., see Definition 1.2.1 in [14]. Here, D is a derivation, whose adjoint δ is also called the Skorohod integral, the multiple integrals I_n are the well-known multiple Itô integrals, and $(\mathscr{F}_t)_{t\in\mathbf{R}_+}$ is the standard Brownian filtration.

Example 1.2-b)—Standard Poisson process
In the special case $X = \mathbf{R}_+$ we can define a standard compensated Poisson process $(M_t)_{t\in\mathbf{R}_+}$ as $(M_t)_{t\in\mathbf{R}_+} := (N_t - t)_{t\in\mathbf{R}_+}$, which is a martingale with respect to its own filtration $(\mathscr{F}_t)_{t\in\mathbf{R}_+}$, and D_t becomes a finite difference operator whose action is given by addition of a Poisson jump at time $t \in \mathbf{R}_+$, i.e.

$$D_t F(N.) = F(N. + \mathbf{1}_{[t,\infty)}(\cdot)) - F(N.), \qquad t \in \mathbf{R}_+, \tag{29}$$

which does not have the derivation property. The construction in [19] also applies to the standard Poisson process, via a different construction using differential operators on the Poisson space.

Example 1.2-c)—Discrete-time chaos expansions
Let $\Omega = \{-1, 1\}^{\mathbb{N}}$ with $X = \mathbb{N}$, and consider the family $(Y_k)_{k\geq 1}$ of independent $\{-1, 1\}$-valued Bernoulli random variables constructed from the canonical projections on Ω under P. That is, with $\mathscr{F}_{-1} = \{\emptyset, \Omega\}$ and $\mathscr{F}_n = \sigma(Y_0, \ldots, Y_n)$ for $n \in \mathbb{N}$, the conditional probabilities $p_n := P(Y_n = 1 \mid \mathscr{F}_{n-1})$ and $q_n := P(Y_n = -1 \mid \mathscr{F}_{n-1})$ are given by

$$p_n = P(Y_n = 1) \quad \text{and} \quad q_n = P(Y_n = -1),$$

respectively. We take $X = \mathbb{N}$ and $H = \ell^2(\mathbb{N}, \sigma)$, where σ is the counting measure on $\mathbb{N}$, and

$$\mathscr{S} = \left\{F = f(Y_0, \dots, Y_n), \quad f : \mathbb{N}^{n+1} \longrightarrow \mathbf{R} \text{ bounded}, \ n \in \mathbb{N}\right\},$$

As in the continuous-time case, every $F \in L^2(\Omega, \mathscr{F}, P)$ can be decomposed into a series of discrete-time multiple stochastic integrals, which here take the form

$$I_n(f_n) = \sum_{k_1 \neq \dots \neq k_n \geq 0} f_n(k_1, \dots, k_n) Z_{k_1} \cdots Z_{k_n}, \tag{30}$$

where the sequence $Z_k := \mathbf{1}_{\{Y_k=1\}} \sqrt{\frac{q_k}{p_k}} - \mathbf{1}_{\{Y_k=-1\}} \sqrt{\frac{p_k}{q_k}}$, $k \in \mathbb{N}$, defines a normalized i.i.d. sequence of centered random variables with unit variance; see, e.g., Chap. 1 of [18]. The gradient D is given by

$$D_k I_n(f_n) = n I_{n-1}(f_n(*, k) \mathbf{1}_{\{k \notin *\}}), \qquad k \in \mathbb{N},$$

and more explicitly it satisfies, for $F \in \mathscr{S}$ and $k \in \mathbb{N}$,

$$D_k F(\omega) = \sqrt{p_k q_k} (F((\omega_i \mathbf{1}_{\{i \neq k\}} + \mathbf{1}_{\{i=k\}})_{i \in \mathbb{N}}) - F((\omega_i \mathbf{1}_{\{i \neq k\}} - \mathbf{1}_{\{i=k\}})_{i \in \mathbb{N}})). \tag{31}$$

The divergence δ is defined as in (26), and again we have $\nabla = D$, hence Assumptions (A1) and (A3)–(A4) are automatically satisfied. Similarly, the duality relation (A2) is known to hold in the discrete-time case by e.g. Proposition 1.8.2 of [18].

Note however that here the operators D and δ do *not* satisfy the commutation relation (A5) in this discrete-time setting, due to the exclusion of diagonals in the construction (30) of multiple stochastic integrals. For this reason, the framework of Sect. 3 and the subsequent sections do not cover this discrete-time setting.

Noncommutative Example

Here we consider an example which is not based on chaos (or multiple stochastic integral) expansions, and for which ∇ does not vanish on $H^{\wedge n}$, $n \geq 1$, with $X = \mathbf{R}_+$. In this case we need to show that Assumptions (A1)–(A5) are satisfied. A different noncommutative example, based on the standard Poisson process, is given in [19].

Example 1.3—The Lie-Wiener path space

Take $X = \mathbf{R}_+$ and let G be a compact connected m-dimensional Lie group, with identity e and whose Lie algebra $\mathscr{G}$, with orthonormal basis $(e_1, \dots, e_m)$ and Lie bracket $[\cdot, \cdot]$, is identified to $\mathbf{R}^m$ and equipped with an Ad-invariant, left invariant metric $\langle \cdot, \cdot \rangle$.

Brownian motion $(\gamma(t))_{t\in\mathbf{R}_+}$ on G is constructed from a standard m-dimensional Brownian motion $(B_t)_{t\in\mathbf{R}_+}$ via the Stratonovich differential equation

$$\begin{cases} d\gamma(t) = \gamma(t) \circ dB_t \\ \gamma(0) = \mathrm{e}, \end{cases} \tag{32}$$

with the image measure of the Wiener measure by the mapping $I : (B_t)_{t\in\mathbf{R}_+} \longmapsto (\gamma(t))_{t\in\mathbf{R}_+}$. Here we take $H = L^2(\mathbf{R}_+; \mathscr{G})$ with the inner product induced by $\mathscr{G}$, and let

$$\mathscr{S} = \{F = f(\gamma(t_1), \ldots, \gamma(t_n)) \ : \ f \in \mathscr{C}_b^\infty(\mathrm{G}^n)\}.$$

Next is the definition of the right derivative operator D, cf. [9].

Definition 1 For F of the form

$$F = f(\gamma(t_1), \ldots, \gamma(t_n)) \in \mathscr{S}, \qquad f \in \mathscr{C}_b^\infty(\mathrm{G}^n), \tag{33}$$

we let $DF \in L^2(\Omega \times \mathbf{R}_+; \mathscr{G})$ be defined as

$$\langle DF, h\rangle_H := \frac{d}{d\varepsilon} f\left(\gamma(t_1)e^{\varepsilon\int_0^{t_1} h_s ds}, \ldots, \gamma(t_n)e^{\varepsilon\int_0^{t_n} h_s ds}\right)_{|\varepsilon=0}, \quad h \in L^2(\mathbf{R}_+, \mathscr{G}).$$

Given F of the form (33) we also have

$$D_t F = \sum_{i=1}^n \partial_i f(\gamma(t_1), \ldots, \gamma(t_n)) \mathbf{1}_{[0,t_i]}(t), \qquad t \geq 0. \tag{34}$$

The covariant derivative operator $\nabla : \mathscr{S} \otimes H \to L^2(\Omega; H \otimes H)$ is defined as

$$\nabla_s u(t) = D_s u(t) + \mathbf{1}_{[0,t]}(s) \mathrm{ad} u(t) \in \mathscr{G} \otimes \mathscr{G}, \qquad s, t \in \mathbf{R}_+, \tag{35}$$

where $\mathrm{ad}(u)v = [u, v]$, $u, v \in \mathscr{G}$, $(\mathrm{ad}u)(t) := \mathrm{ad}(u(t))$, $t \in \mathbf{R}_+$, for $u \in \mathscr{S} \otimes H$, and $\mathrm{ad}u$ is the linear operator defined on $\mathscr{G}$ by

$$\langle e_i \otimes e_j, \mathrm{ad}u\rangle_{\mathscr{G}\otimes\mathscr{G}} = \langle e_j, \mathrm{ad}(e_i)u\rangle_{\mathscr{G}} = \langle e_j, [e_i, u]\rangle_{\mathscr{G}}, \quad i, j = 1, \ldots, m, \quad u \in \mathscr{G}.$$

We now check that all required assumptions are satisfied in the present setting.

(A1) The vanishing of torsion is satisfied from Theorem 2.3-(i) of [9].
(A2) The operator D admits an adjoint δ that satisfies the duality relation

$$E[F\delta(v)] = E[\langle DF, v\rangle_H], \quad F \in \mathscr{S}, \ v \in L^2(\mathbf{R}_+; \mathscr{G}), \tag{36}$$

cf. e.g. [9], which shows that (A2) is satisfied.

(A3) We note that ad$u$ is skew-adjoint as the inner product in $\mathscr{G}$ is chosen Ad-invariant, hence the connection from ∇ is torsion free and (18) is satisfied.

(A4) Assumption (A4) clearly holds by the definition (35) which shows that

$$\nabla_t f(s) = 0, \qquad 0 < s < t, \quad f \in H.$$

This also applies in the setting of loop groups [10].

(A5) By Theorem 2.4-(*ii*) of [9] it is known that D and ∇ satisfy the commutation relation (A5) for $f \in H$, hence by Remark 1, Assumption (A5) is satisfied for $u \in H \otimes \mathscr{S}$ since by (34) the operator D satisfies the chain rule of derivation (23). Here, (13)–(14) also hold from Corollary 2 of [11] which is proved using a mapping of ∇ on the path group to D on the Wiener space by the Itô map.

5 Weitzenböck Identities for n-Forms

In this section we will need the following additional assumption:

(B1) For all $n \geq 1$ the covariant derivative operator satisfies

$$d_{x_n}^{n-1} \nabla_x f_n(x, x_1, \ldots, x_{n-1}) = \nabla_x d_{x_n}^{n-1} f_n(x, x_1, \ldots, x_{n-1}), \qquad \textbf{(B1)}$$

$x_1, \ldots, x_n \in X, x \in X, f_n \in H^{\wedge n}$.

Assumption (B1) is straightforwardly satisfied in all examples of Sect. 4, except for the discrete-time Example 1.2-c), however it requires a specific proof in the Poisson derivation case of [19]. Note that Assumption (B1) differs from the usual vanishing of curvature condition, which reads

$$\nabla_u \nabla_v - \nabla_v \nabla_u = \nabla_{\{u,v\}}$$

where $\{u, v\}$ is the Lie bracket of two vector fields u, v, cf. Theorem 2.3-(*ii*) of [9].

Lemma 1 (Intertwining Relation) *Under Assumptions (A1)–(A5) and (B1), for any $u_n \in \mathscr{S} \otimes H^{\wedge n}$ of the form $u_n = F \otimes f_n$, with $F \in \mathscr{S}$ and $f_n, g_n \in H^{\wedge n}$, we have*

$$\begin{aligned}
&\langle d^{n-1} d^{(n-1)*} u_n(x_1, \ldots, x_n), g_n(x_1, \ldots, x_n)\rangle_{H^{\wedge n}} = nF\langle f_n(x_1, \ldots, x_n), g_n(x_1, \ldots, x_n)\rangle_{H^{\wedge n}} \\
&\quad + \sum_{j=1}^{n} \langle \delta(\nabla_{x_j}(F \otimes f_n(x_1, \ldots, x_{j-1}, \cdot, x_{j+1}, \ldots, x_n))), g_n(x_1, \ldots, x_n)\rangle_{H^{\wedge n}} \\
&\quad - \left\langle d_{x_n}^{n-1}(F \otimes \operatorname{trace} \nabla_{\cdot} f_n(\cdot, x_1, \ldots, x_{n-1})), g_n(x_1, \ldots, x_n)\right\rangle_{H^{\wedge n}} \\
&\quad + \sum_{j=1}^{n} \int_X \langle D_x F \otimes f_n(x_1, \ldots, x_n), \nabla_{x_j}^{(j)} g_n(x_1, \ldots, x_{j-1}, x, x_{j+1}, \ldots, x_n)\rangle_{H^{\wedge n}} \sigma(dx).
\end{aligned} \qquad (37)$$

The proof of Lemma 1 is deferred to the appendix. Here we verify that for $n = 1$, Lemma 1 coincides with Assumption (A5) since $\nabla_x f_1(x) = 0$ and

$$
\begin{aligned}
\mathrm{d}^0_{x_1} \mathrm{d}^{0*} u_1 &= D_{x_1} \delta(F f_1) \\
&= \delta(f_1 D_{x_1} F) + F f_1(x_1) + \delta(F \nabla_{x_1} f_1) \\
&\quad + \langle D D_{x_1} F, f_1 \rangle_H - \langle D_{x_1} D F, f_1 \rangle_H + \langle D.F, \nabla_{x_1} f_1(\cdot) \rangle_H \\
&= u_{x_1} + \delta(\nabla_{x_1} u_1) + \langle D D_{x_1} F, f_1 \rangle_H - \langle D_{x_1} D F, f_1 \rangle_H + \langle D.F, \nabla_{x_1} f_1(\cdot) \rangle_H,
\end{aligned}
$$

for any $u_1, v_1 \in \mathscr{S} \otimes H$ of the form $u_1 = F \otimes f_1$ with $F \in \mathscr{S}$, and $f_1, g_1 \in H$. By the vanishing of torsion Assumption (A1), this yields

$$
\begin{aligned}
&\langle \mathrm{d}^0_{x_1} \mathrm{d}^{0*} u_1(x_1), g_1(x_1) \rangle_H = F \langle f(x_1), g_1(x_1) \rangle_H + \langle \delta(\nabla_{x_1}(F f_1)), g_1(x_1) \rangle_H \\
&\quad + \langle \langle D D_{x_1} F, f_1 \rangle_H - \langle D_{x_1} D F, f_1 \rangle_H, g_1(x_1) \rangle_H + \langle \langle D.F, \nabla_{x_1} f_1(\cdot) \rangle_H, g_1(x_1) \rangle_H \\
&= F \langle f(x_1), g_1(x_1) \rangle_H + \langle \delta(\nabla_{x_1}(F f_1)), g_1(x_1) \rangle_H \\
&\quad + \int_X \int_X \langle \nabla_y g_1(x), D_x F \otimes f_1(y) \rangle_{\mathbf{R}^d \otimes \mathbf{R}^d} \sigma(dx) \sigma(dy) \\
&\quad - \int_X \int_X \langle \nabla_y f_1(x), D_x F \otimes g_1(y) \rangle_{\mathbf{R}^d \otimes \mathbf{R}^d} \sigma(dx) \sigma(dy) \\
&\quad + \langle \langle D.F, \nabla_{x_1} f_1(\cdot) \rangle_H, g_1(x_1) \rangle_H \\
&= F \langle f(x_1), g_1(x_1) \rangle_H + \langle \delta(\nabla_{x_1}(F \otimes f_1)), g_1(x_1) \rangle_H + \langle \langle D.F, \nabla_{x_1} g_1(\cdot) \rangle_H, f_1(x_1) \rangle_H.
\end{aligned}
$$

which coincides with Lemma 1 since $\nabla_x f_1(x) = 0$ when $n = 1$.

Weitzenböck Identity

Recall that the de Rham-Hodge Laplacian (21) is given on n-forms by

$$
\Delta_n = \mathrm{d}^{n-1} \mathrm{d}^{(n-1)*} + \mathrm{d}^{n*} \mathrm{d}^n, \qquad n \geq 1.
$$

Theorem 1 *Under the Assumptions (A1)–(A5) and (B1) we have the Weitzenböck identity*

$$
\Delta_n = n \mathrm{Id}_{H^{\wedge n}} + \nabla^* \nabla, \qquad u_n \in \mathscr{S} \otimes H^{\wedge n}, \quad n \geq 1. \tag{38}
$$

By duality (38) shows that

$$
\begin{aligned}
n! \|\mathrm{d}^{(n-1)*} u_n\|^2_{L^2(\Omega, H^{\wedge(n-1)})} &+ n! \|\mathrm{d}^n u_n\|^2_{L^2(\Omega, H^{\wedge(n+1)})} \\
&= n n! \|u_n\|^2_{L^2(\Omega, H^{\wedge n})} + \|\nabla u_n\|^2_{L^2(\Omega, H^{\otimes(n+1)})},
\end{aligned} \tag{39}
$$

$u_n \in \mathscr{S} \otimes H^{\wedge n}$, $n \geq 1$. We first check that the Weitzenböck identity (39) holds for one-forms, i.e.

$$\|\mathrm{d}^{0*} u_1\|^2_{L^2(\Omega)} + \frac{1}{2}\|\mathrm{d}^1 u_1\|^2_{L^2(\Omega,H^{\otimes 2})} = \|u_1\|^2_{L^2(\Omega,H)} + \|\nabla u_1\|^2_{L^2(\Omega,H^{\otimes 2})}, \tag{40}$$

and we refer to the appendix for the proof in the case of n-forms. By Assumption (A5) and the commutation relation (10) we have, taking $u_1 = F \otimes f_1$ and following the argument of [9],

$$\begin{aligned}
\langle \mathrm{d}^0_{x_1} \mathrm{d}^{0*} u_1, u_1(x_1)\rangle_{L^2(\Omega,H)} &= \langle D_{x_1}\delta(F\otimes f_1), Ff_1(x_1)\rangle_{L^2(\Omega,H)} \\
&= \langle Ff_1(x_1) + \delta(\nabla_{x_1}(F\otimes f_1)) + \langle DD_{x_1}F, f_1\rangle_H - \langle D_{x_1}DF, f_1\rangle_H \\
&\quad + \langle D.F, \nabla_{x_1} f_1(\cdot)\rangle_H, Ff_1(x_1)\rangle_{L^2(\Omega,H)} \\
&= \langle u_1(x_1), u_1(x_1)\rangle_{L^2(\Omega,H)} + \langle \nabla_{x_1} u_1(\cdot), D.u_1(x_1)\rangle_{L^2(\Omega,H^{\otimes 2})} \\
&\quad + \langle F\langle DD_{x_1}F, f_1\rangle_H - F\langle D_{x_1}DF, f_1\rangle_H + F\langle D.F, \nabla_{x_1} f_1(\cdot)\rangle_H, f_1(x_1)\rangle_{L^2(\Omega,H)} \\
&= \langle u_1(x_1), u_1(x_1)\rangle_{L^2(\Omega,H)} + \langle \nabla_{x_1} u_1(\cdot), D.u_1(x_1)\rangle_{L^2(\Omega,H^{\otimes 2})} \\
&\quad + \langle \langle D_{x_1}F, F\nabla.f_1(x_1)\rangle_H, f_1(\cdot)\rangle_{L^2(\Omega,H)} \\
&= \langle u_1(x_1), u_1(x_1)\rangle_{L^2(\Omega,H)} + \langle \nabla_{x_1} u_1(x_2), \nabla_{x_2} u_1(x_1)\rangle_{L^2(\Omega,H^{\otimes 2})} \qquad (41)\\
&= \langle u_1(x_1), u_1(x_1)\rangle_{L^2(\Omega,H)} - \frac{1}{2}\langle \mathrm{d}^1 u_1, \mathrm{d}^1 u_1\rangle_{L^2(\Omega,H^{\otimes 2})} \\
&\quad + \langle \nabla_{x_2} u_1(x_1), \nabla_{x_2} u_1(x_1)\rangle_{L^2(\Omega,H^{\otimes 2})}, \qquad (42)
\end{aligned}$$

where we used Assumption (A4) to reach (41). This implies (40) and

$$\|\delta(u_1)\|^2_{L^2(\Omega,H)} + \frac{1}{2}\|\mathrm{d}^1 u_1\|^2_{L^2(\Omega,H^{\wedge 2})} = \|u_1\|^2_{L^2(\Omega,H)} + \|\nabla u_1\|^2_{L^2(\Omega,H^{\otimes 2})}, \quad u_1 \in \mathscr{S}\otimes H.$$

Theorem 1 shows that the Bochner Laplacian $L = -\nabla^*\nabla$ and the Hodge Laplacian Δ_n have same closed domain $\mathrm{Dom}(\Delta_n)$ on the random n-forms and that all eigenvalues λ_n of the Bochner Laplacian L satisfy $\lambda_n \geq n$. Indeed, if w_n is an eigenvector of Δ_n with eigenvalue λ_n, by rewriting (38) as

$$L = n\mathrm{Id}_{H^{\wedge n}} - \Delta_n,$$

we find that L and Δ_n share the same eigenvectors and that $\lambda_n \geq n \geq 1$ since

$$\begin{aligned}
0 &\leq -\langle Lw_n, w_n\rangle_{L^2(\Omega,H^{\wedge n})} \\
&= \langle (\Delta_n - n\mathrm{Id}_{H^{\wedge n}})w_n, w_n\rangle_{L^2(\Omega,H^{\wedge n})} \\
&= \langle \Delta_n w_n, w_n\rangle_{L^2(\Omega,H^{\wedge n})} - n\langle w_n, w_n\rangle_{L^2(\Omega,H^{\wedge n})} \\
&= (\lambda_n - n)\langle w_n, w_n\rangle_{L^2(\Omega,H^{\wedge n})}, \qquad n \geq 1. \qquad (43)
\end{aligned}$$

Proposition 2 *Under Assumptions (A1)–(A5) and (B1), the de Rham-Hodge-Kodaira decomposition* (21) *rewrites as*

$$L^2(\Omega; H^{\wedge n}) = \operatorname{Im} \mathrm{d}^{n-1} \oplus \operatorname{Im} \mathrm{d}^{n*}, \qquad n \geq 1. \tag{44}$$

Proof By (43) the operator $\Delta_n = n\mathrm{Id}_{H^{\otimes n}} - L$ becomes invertible for all $n \geq 1$, and the space Ker Δ_n of harmonic forms for the de Rham Laplacian Δ_n is equal to $\{0\}$. i.e. any harmonic form for the de Rham Laplacian Δ_n has to vanish, and we conclude by (21).

Next, we consider a number of examples to which the framework and results of this section can be applied.

Commutative Examples–Chaos Expansions

All commutative examples of Sect. 3 satisfy Assumption (B1) since in this case, ∇ vanishes on $H^{\wedge n}$, $n \geq 1$, as in all chaos-based examples.

Example 2.1—Poisson random measures
In the Poisson case the semi-group $(P_t)_{t\in\mathbb{R}_+}$ associated to the Bochner Laplacian $L := -\nabla^*\nabla$, cf. [21], admits an integral representation, cf. e.g. Lemma 6.8.1 of [18]. Proposition 2 shows here that any harmonic form for the de Rham Laplacian has to vanish.

Example 2.2—Normal martingales

Example 2.2-a)—Brownian and Poisson cases
In the Brownian case, Theorem 1 covers Proposition 3.1 of [20] on the Weitzenböck decomposition, and Proposition 2 is known to hold also from [20].

Example 2.2-b)—Discrete-time case
Proposition 2 holds in this discrete-time setting as the semi-group $(P_t)_{t\in\mathbb{R}_+}$ is contractive, cf. Proposition 1.9.3 and Lemma 1.9.4 of [18]. However, Theorem 1 does not hold here as (A5) is not satisfied.

Noncommutative Example
Example 2.3—Lie-Wiener path space
We need to check the following condition, which immediately holds because the operation ad in (35) commutes with itself.

(B1) In other words, we can write $\mathrm{ad}\, u$ as

$$\begin{aligned}
\mathrm{ad}\, u &= \sum_{k=1}^{m} \langle u, e_k\rangle_{\mathscr{G}}\, \mathrm{ad}\, e_k \\
&= \sum_{i,j,k=1}^{m} \langle u, e_k\rangle_{\mathscr{G}} (e_i \otimes e_j) \langle e_i \otimes e_j, \mathrm{ad}\, e_k\rangle_{\mathscr{G}\otimes\mathscr{G}} \\
&= \sum_{i,j,k=1}^{m} \langle u, e_k\rangle_{\mathscr{G}} (e_i \otimes e_j) A_{i,j,k},
\end{aligned}$$

where the matrix $A = (A_{i,j,k})_{1 \le i,j,k \le m}$ is the 3-tensor given by

$$A_{i,j,k} = \langle e_i \otimes e_j, \mathrm{ad}\, e_k \rangle_{\mathscr{G} \otimes \mathscr{G}} = \langle e_j, \mathrm{ad}(e_i) e_k \rangle_{\mathscr{G}} = \langle e_j, [e_i, e_k] \rangle_{\mathscr{G}},$$

$1 \le i, j, k \le m$.

Note that Assumption (B1) differs from the vanishing of curvature in e.g. Theorem 2.3-(*ii*) of [9] in the path group case.

6 Clark-Ocone Representation Formula

In this section we take $d = 1$ and $X = \mathbf{R}_+$, and we consider a normal martingale $(M_t)_{t \in \mathbf{R}_+}$ generating a filtration $(\mathscr{F}_t)_{t \in \mathbf{R}_+}$ on the probability space $(\Omega, \mathscr{F}, P)$. We assume that D satisfies the following Assumptions (C1) and (C2), in addition to (A1)–(A5).

(C1) The operator D satisfies the Clark-Ocone formula

$$F = E[F \mid \mathscr{F}_t] + \int_t^\infty E[D_r F \mid \mathscr{F}_r] dM_r, \qquad t \in \mathbf{R}_+, \tag{C1}$$

for $F \in \mathrm{Dom}(D)$.

(C2) The operator D satisfies the commutation relation

$$D_s E[F \mid \mathscr{F}_t] = \mathbf{1}_{[0,t]}(s) E[D_s F \mid \mathscr{F}_t], \qquad s, t \in \mathbf{R}_+, \tag{C2}$$

for $F \in \mathrm{Dom}(D)$,

(A4') The operator ∇ satisfies the condition

$$\nabla_s f(t) = 0, \qquad 0 \le t < s, \qquad f \in H, \tag{A4'}$$

We note that (A4') is stronger than (A4), and that (C2) implies

$$\nabla_s E[u(t) \mid \mathscr{F}_t] = \mathbf{1}_{[0,t]}(s) E[\nabla_s u(t) \mid \mathscr{F}_t], \qquad s, t \in \mathbf{R}_+,$$

for $u \in \mathrm{Dom}(\nabla)$. In addition, under the duality assumption (A2), Assumption (C1) is equivalent to stating that $(M_t)_{t \in \mathbf{R}_+}$ has the predictable representation property and δ coincides with the stochastic integral with respect to $(M_t)_{t \in \mathbf{R}_+}$ on the square-integrable predictable processes, cf. Corollary 3.2.8 and Propositions 3.3.1 and 3.3.2 of [18]. Also it is sufficient to assume that (C1) holds for $t = 0$, cf. Proposition 3.2.3 of [18].

Clark-Ocone Formula for *n*-Forms

In this section extend the Clark-Ocone formula for differential forms of [22] to the general framework of this paper.

Theorem 2 *Under the Assumptions (A1)–(A5) and (C1)–(C2), for* $u_n \in \mathrm{Dom}(\mathrm{d}^n)$, *we have, for a.e.* $t_1, \ldots, t_n \in \mathbf{R}_+$,

$$u_n(t_1, \ldots, t_n) = \mathrm{d}_{t_n}^{n-1} \int_{t_1 \vee \cdots \vee t_{n-1}}^{\infty} E[u_n(r, t_1, \ldots, t_{n-1}) \mid \mathscr{F}_r] dM_r \qquad (45)$$

$$+ \int_{t_1 \vee \cdots \vee t_n}^{\infty} E[\mathrm{d}_{t_n}^n u_n(r, t_1, \ldots, t_{n-1}) \mid \mathscr{F}_r] dM_r.$$

In particular, Theorem 2 shows that any closed form $u_n \in \mathrm{Dom}(\mathrm{d}^n)$ can be written as

$$u_n(t_1, \ldots, t_n) = \mathrm{d}_{t_n}^{n-1} \int_{t_1 \vee \cdots \vee t_{n-1}}^{\infty} E[u_n(r, t_1, \ldots, t_{n-1}) \mid \mathscr{F}_r] dM_r,$$

$t_1, \ldots, t_n \in \mathbf{R}_+$. As a consequence of Theorem 2 the range of the exterior derivative d^n is closed, and similarly for its adjoint d^{n*}, for all $n \geq 1$. In this way we recover the fact that the Hodge Laplacian Δ_n has a closed range as well, so it has a spectral gap, cf. Theorem 6.6 and Corollary 6.7 of [7]. However this does not yield an explicit Poincaré inequality and lower bound for the spectral gap, unlike for the classical Clark-Ocone formula cf. e.g. Proposition 3.2.7 of [18]. Note that the Weitzeböck formula (43) also shows that all eigenvalues λ_n of the Bochner Laplacian $L = -\nabla^* \nabla$ on n-forms satisfy $\lambda_n \geq n$, $n \geq 1$. A quick proof of the identity (45) for one-forms is instructive while we delay the proof for general n-forms to the appendix. When $n = 1$, for $u \in \mathscr{S} \otimes H$, we have, by (A4') and the Clark-Ocone formula (C1) for $t \in \mathbf{R}_+$,

$$u(t) = E[u(t) \mid \mathscr{F}_t] + \int_t^{\infty} E[D_r u(t) \mid \mathscr{F}_r] dM_r$$

$$= E[u(t) \mid \mathscr{F}_t] + \int_t^{\infty} E[\nabla_r u(t) \mid \mathscr{F}_r] dM_r,$$

and by (A5) and (C2) we find

$$D_t \int_0^{\infty} E[u(r) \mid \mathscr{F}_r] dM_r = E[u(t) \mid \mathscr{F}_t] + \int_0^{\infty} \nabla_t E[u(r) \mid \mathscr{F}_r] dM_r \qquad (46)$$

$$= E[u(t) \mid \mathscr{F}_t] + \int_t^{\infty} E[\nabla_t u(r) \mid \mathscr{F}_r] dM_r,$$

hence

$$u(t) = D_t \int_0^{\infty} E[u(r) \mid \mathscr{F}_r] dM_r + \int_t^{\infty} E[\nabla_r u(t) - \nabla_t u(r) \mid \mathscr{F}_r] dM_r$$

$$= D_t \int_0^{\infty} E[u(r) \mid \mathscr{F}_r] dM_r + \int_t^{\infty} E[\mathrm{d}_t^1 u(r) \mid \mathscr{F}_r] dM_r.$$

Note that in (46) above we applied Assumption (A5) to an adapted process v, in which case the condition simply reads

$$D_t\delta(v) = v(t) + \delta(\nabla_t v), \qquad t \in \mathbf{R}_+, \qquad \textbf{(A5")}$$

since when $v(\cdot) = \mathbf{1}_{[t,\infty)}(\cdot)F \otimes a$ is a simple adapted process, where $a \in \mathbf{R}^d$ and F is $\mathscr{F}_t$-measurable, we have $\mathbf{1}_{[t,\infty)}(r)D_rF = 0$, $r \in \mathbf{R}_+$, as follows from (C2), i.e.

$$D_rF = D_rE[F \mid \mathscr{F}_t] = \mathbf{1}_{[0,t]}(r)E[D_rF \mid \mathscr{F}_t], = 0, \qquad r \geq t.$$

The Clark-Ocone formula Theorem 2 allows us in particular to recover the de Rham-Hodge-Kodaira decomposition (44).

Corollary 1 *We have* $\mathrm{Im}\, \mathrm{d}^n = \mathrm{Ker}\, \mathrm{d}^{n+1}$, $n \in \mathbb{N}$, *and the de Rham-Hodge-Kodaira decomposition* (21) *reads*

$$L^2(\Omega; H^{\wedge n}) = \mathrm{Im}\, \mathrm{d}^{n-1} \oplus \mathrm{Im}\, \mathrm{d}^{n*}, \qquad n \geq 1.$$

Proof By Theorem 2 we have $\mathrm{Im}\, \mathrm{d}^n \supset \mathrm{Ker}\, \mathrm{d}^{n+1}$, which shows by (14) that $\mathrm{Im}\, \mathrm{d}^n = \mathrm{Ker}\, \mathrm{d}^{n+1}$, $n \in \mathbb{N}$.

As a consequence of Corollary 1, we also get the exactness of the sequence

$$\mathrm{Dom}\,(\mathrm{d}^n) \xrightarrow{\mathrm{d}^n} \mathrm{Im}\,(\mathrm{d}^n) = \mathrm{Ker}\,(\mathrm{d}^{n+1}) \xrightarrow{\mathrm{d}^{n+1}} \mathrm{Im}\,(\mathrm{d}^{n+1}), \qquad n \in \mathbb{N}, \tag{47}$$

as in Theorem 3.2 of [20]. By duality of (47) we also find by Corollary 1 that

$$\mathrm{Im}\, \mathrm{d}^{(n+1)*} = \mathrm{Ker}\, \mathrm{d}^{n*}, \qquad n \in \mathbb{N},$$

and the following sequence is also exact:

$$\mathrm{Im}\,(\mathrm{d}^{n*}) \xleftarrow{\mathrm{d}^{n*}} \mathrm{Ker}\,(\mathrm{d}^{n*}) = \mathrm{Im}\,(\mathrm{d}^{(n+1)*}) \xleftarrow{\mathrm{d}^{(n+1)*}} \mathrm{Dom}\,(\mathrm{d}^{(n+1)*}), \qquad n \in \mathbb{N}.$$

Next, we consider some examples to which the above framework applies.

Commutative Examples–Chaos Expansions

As written at the beginning of this section, we take $X = \mathbf{R}_+$ in all cases due to the need of a time scale in order to state the Clark-Ocone formula.

Example 3.1-a)—Normal martingales

As in Section "Commutative Examples–Chaos Expansions" we have $X = \mathbf{R}_+$ and $\nabla = 0$ on H, i.e. $\nabla = D$ and Assumption (A4') is automatically satisfied. Let us check that Assumptions (C1) and (C2) are satisfied in the framework of normal martingales that have the chaos representation property (CRP).

(C1) Since the normal martingale $(M_t)_{t\in\mathbf{R}_+}$ has the chaos representation property, the Clark-Ocone formula holds for any $F \in \mathrm{Dom}(D) \subset L^2(\Omega, \mathscr{F}, P)$ as

$$F = \mathbb{E}[F] + \int_0^\infty \mathbb{E}[D_t F \mid \mathscr{F}_t] dM_t, \tag{48}$$

cf. Proposition 4.2.3 of [18] for a proof via the chaos expansion of F.

(C2) This condition is satisfied from the definition (25) of D and e.g. Lemma 2.7.2 page 88 of [18] or Proposition 1.2.8 page 34 of [14] in the Wiener case.

Example 3.1-b)—Discrete-time chaos expansions
The Clark-Ocone formula (C1) holds in the discrete-time case as

$$F = \mathbb{E}[F] + \sum_{k=0}^{\infty} \mathbb{E}[D_k F \mid \mathscr{F}_{k-1}] Z_k, \tag{49}$$

cf. Proposition 1.7.1 of [18] and references therein, hence Assumption (C2) is also satisfied here, as it is satisfied for normal martingales. However, Theorem 2 does not hold here because (A5) is not satisfied.

Noncommutative Example
Example 3.2)—Lie-Wiener path space (Example 1.3 continued)

(C1) On the classical Wiener space, when $(u(t))_{t\in\mathbf{R}_+}$ is square-integrable and adapted to the Brownian filtration $(\mathscr{F}_t)_{t\in\mathbf{R}_+}$, $\delta(u)$ coincides with the Itô integral of $u \in L^2(\Omega; H)$ with respect to the underlying Brownian motion $(B_t)_{t\in\mathbf{R}_+}$, i.e.

$$\delta(u) = \int_0^\infty u(t) dB_t, \tag{50}$$

and this shows that Assumption (C1) is satisfied, cf. e.g. Proposition 3.3.2 of [18].

(C2) Assumption (C2) is satisfied here as in the case of normal martingales as in e.g. Lemma 2.7.2 page 88 of [18], or for Brownian motion as in Proposition 1.2.8 page 34 of [14].
On the Lie-Wiener path space we note that we have the relation

$$\langle DF, h\rangle_H = \langle \hat{D}F, h\rangle_H + \hat{\delta}\left(\int_0^{\cdot} \mathrm{ad}(h(s)) ds \hat{D}.F\right), \qquad F \in \mathscr{S}, \tag{51}$$

where $\hat{D}$ and $\hat{\delta}$ denote here the gradient and divergence appearing in (28) on the underlying standard Wiener space with Brownian motion $(B_t)_{t\in\mathbf{R}_+}$ in (32), cf. e.g. Lemma 4.1 of [17] and references therein, or Corollary 5.2.1 of [16] for the more general setting of Riemannian manifolds. Relation (51) shows that

$$E[D_s F \mid \mathscr{F}_r] = E[\hat{D}_s F \mid \mathscr{F}_r], \qquad 0 \le r \le s, \tag{52}$$

cf. also Relation (5.7.5) page 191 of [18] on Riemannian manifolds, hence (C1) is satisfied for D because it holds for $\hat{D}$ as noted above.

Consequently, (C2) holds on the Lie-Wiener path space since we have

$$\begin{aligned}
\langle DE[F \mid \mathscr{F}_t], h\rangle_H &= \langle \hat{D}E[F \mid \mathscr{F}_t], h\rangle_H + \hat{\delta}\left(\int_0^{\cdot} \mathrm{ad}(h(s))ds\hat{D}.E[F \mid \mathscr{F}_t]\right) \\
&= \langle \mathbf{1}_{[0,t]}(\cdot)E[\hat{D}.F \mid \mathscr{F}_t], h\rangle_H + \hat{\delta}\left(\int_0^{\cdot} \mathrm{ad}(h(s))ds\mathbf{1}_{[0,t]}(\cdot)E[\hat{D}.F \mid \mathscr{F}_t]\right) \\
&= \langle \mathbf{1}_{[0,t]}(\cdot)E[\hat{D}.F \mid \mathscr{F}_t], h\rangle_H + E\left[\hat{\delta}\left(\int_0^{\cdot} \mathbf{1}_{[0,t]}(s)\mathrm{ad}(h(s))ds\hat{D}.F\right)\Big|\mathscr{F}_t\right] \\
&= \langle \mathbf{1}_{[0,t]}(\cdot)E[D.F \mid \mathscr{F}_t], h\rangle_H, \qquad t \in \mathbf{R}_+.
\end{aligned}$$

Assumption (A4') is also clearly satisfied by the definition (35) of ∇.

Hence Theorems 2 covers Theorems 3.1 of [22] on the Wiener space as well as its extension to the path space using the diffeomorphism approach of [11].

Acknowledgements Numerous comments and suggestions by Yuxin Yang on this paper are gratefully acknowledged. This work was supported by Singapore MOE Tier 2 grant MOE2012-T2-2-033 ARC 3/13.

Appendix

In this section we state the proofs of Proposition 1, Theorems 1 and 2, by extension of the original arguments of [10, 11, 20] and [22] to our framework.

Proof of Proposition 1 (Duality Relation) Assuming that $u_n \in \mathscr{S}\otimes H^{\wedge n}$ and $v_{n+1} \in \mathscr{S} \otimes H^{\wedge(n+1)}$ have the form (7) and using the definition (5) of d^n and the duality assumption (A2) we have, using the antisymmetry of g_{n+1},

$$\begin{aligned}
&\langle \mathrm{d}^n_{t_{n+1}}(Ff_n(t_1,\ldots,t_n)), Gg_{n+1}(t_1,\ldots,t_{n+1})\rangle_{L^2(\Omega,H^{\wedge(n+1)})} \\
&= \sum_{j=1}^{n+1}\frac{(-1)^{j-1}}{(n+1)!}\langle \nabla_{t_j}(Ff_n(t_1,\ldots,t_{j-1},t_{j+1},\ldots,t_{n+1})), Gg_{n+1}(t_1,\ldots,t_{n+1})\rangle_{L^2(\Omega,H^{\otimes(n+1)})} \\
&= \sum_{j=1}^{n+1}\frac{(-1)^{j-1}}{(n+1)!}\langle D_{t_j}Ff_n(t_1,\ldots,t_{j-1},t_{j+1},\ldots,t_{n+1}), Gg_{n+1}(t_1,\ldots,t_{n+1})\rangle_{L^2(\Omega,H^{\otimes(n+1)})} \\
&\quad + \sum_{j=1}^{n+1}\frac{(-1)^{j-1}}{(n+1)!}\langle F\nabla_{t_j}f_n(t_1,\ldots,t_{j-1},t_{j+1},\ldots,t_{n+1}), Gg_{n+1}(t_1,\ldots,t_{n+1})\rangle_{L^2(\Omega,H^{\otimes(n+1)})}
\end{aligned}$$

$$= \frac{1}{(n+1)!}\sum_{j=1}^{n+1}(-1)^{j-1}\langle Ff_n(t_1,\ldots,t_{j-1},t_{j+1},\ldots,t_{n+1}),$$

$$\delta(Gg_{n+1}(t_1,\ldots,t_{j-1},\cdot,t_{j+1},\ldots,t_{n+1}))\rangle_{L^2(\Omega,H^{\otimes n})}$$

$$-\frac{1}{(n+1)!}\sum_{j=1}^{n+1}(-1)^{j-1}\langle Ff_n(t_1,\ldots,t_{j-1},t_{j+1},\ldots,t_{n+1}),$$

$$G\int_0^\infty \operatorname{Tr}\nabla_t g_{n+1}(t_1,\ldots,t_{j-1},t,t_{j+1},\ldots,t_{n+1})dt\rangle_{L^2(\Omega,H^{\otimes n})}$$

$$= \frac{1}{(n+1)!}\sum_{j=1}^{n+1}\langle Ff_n(t_1,\ldots,t_{j-1},t_{j+1},\ldots,t_{n+1}),$$

$$\delta(Gg_{n+1}(\cdot,t_1,\ldots,t_{j-1},t_{j+1},\ldots,t_{n+1}))\rangle_{L^2(\Omega,H^{\otimes n})}$$

$$-\frac{1}{(n+1)!}\sum_{j=1}^{n+1}\langle Ff_n(t_1,\ldots,t_{j-1},t_{j+1},\ldots,t_{n+1}),$$

$$G\int_0^\infty \operatorname{Tr}\nabla_t g_{n+1}(t,t_1,\ldots,t_{j-1},t_{j+1},\ldots,t_{n+1})dt\rangle_{L^2(\Omega,H^{\otimes n})}$$

$$= \langle Ff_n(t_1,\ldots,t_n),\mathrm{d}^{n*}(Gg_{n+1})(t_1,\ldots,t_n)\rangle_{L^2(\Omega,H^{\wedge n})},$$

where we applied the antisymmetry condition (A3) and the definition (16) of d^{n*}.

□

Proof of Lemma 1 (Intertwining Relation) When $n = 1$, by (A5) or (22) we have

$$\begin{aligned}\mathrm{d}^0_{x_1}\mathrm{d}^{0*}u_1 &= D_{x_1}\delta(u_1)\\ &= u_1(x_1)+\delta(\nabla_{x_1}u_1)+\langle DD_{x_1}F,f_1\rangle_H-\langle D_{x_1}DF,f_1\rangle_H+\langle D.F,\nabla_{x_1}f_1(\cdot)\rangle_H,\end{aligned}$$

for $u_1 = F\otimes f_1 \in \dot{\mathscr{S}}\otimes H$. Next, by the definition (5) of d^n and (A5) or (22) we have

$$\begin{aligned}&\mathrm{d}^{n-1}_{t_n}\delta(F\otimes f_n(\cdot,x_1,\ldots,x_{n-1}))\\ &= \sum_{j=1}^{n}(-1)^{j-1}\nabla_{x_j}\delta(F\otimes f_n(\cdot,x_1,\ldots,x_{j-1},x_{j+1},\ldots,x_n))\\ &= \sum_{j=1}^{n}(-1)^{j-1}F\otimes f_n(x_j,x_1,\ldots,x_{j-1},x_{j+1},\ldots,x_n)\\ &\quad+\sum_{j=1}^{n}(-1)^{j-1}\delta(\nabla_{x_j}(F\otimes f_n(\cdot,x_1,\ldots,x_{j-1},x_{j+1},\ldots,x_n)))\end{aligned}$$

$$+\sum_{j=1}^{n}(-1)^{j-1}\langle D.D_{x_j}F - D_{x_j}D.F, f_n(\cdot, x_1, \ldots, x_{j-1}, x_{j+1}, \ldots, x_n)\rangle_H$$

$$+\sum_{j=1}^{n}(-1)^{j-1}\langle D.F, \nabla_{x_j}^{(1)} f_n(\cdot, x_2, \ldots, x_{j-1}, x_{j+1}, \ldots, x_n)\rangle_H$$

$$= nF \otimes f_n(x_1, \ldots, x_n)$$

$$+\sum_{j=1}^{n}\delta(\nabla_{x_j}(F \otimes f_n(x_1, \ldots, x_{j-1}, \cdot, x_{j+1}, \ldots, x_n)))$$

$$+\sum_{j=1}^{n}\int_X \langle D_x F \otimes f_n(x_1, \ldots, x_n), \nabla_{x_j}^{(j)} g_n(x_1, \ldots, x_{j-1}, x, x_{j+1}, \ldots, x_n)\rangle_{H^{\wedge n}} \sigma(dx).$$

where we applied (10). We conclude by the definition (16) of d^{n*} which states that

$$\mathrm{d}^{(n-1)*}(F \otimes f_n)(x_1, \ldots, x_{n-1}) = \delta(F \otimes f_n(\cdot, x_1, \ldots, x_{n-1})) - F\int_X \mathrm{Tr}\, \nabla_x f_n(x, x_1, \ldots, x_{n-1})\sigma(dx).$$

□

Proof of Theorem 1 (Weitzenböck Identity) We will show that

$$\|\mathrm{d}^{(n-1)*}u_n\|^2_{L^2(\Omega, H^{\otimes(n-1)})} + \frac{1}{n+1}\|\mathrm{d}^n u_n\|^2_{L^2(\Omega, H^{\otimes(n+1)})} = n\|u_n\|^2_{L^2(\Omega, H^{\otimes n})} + \|\nabla u_n\|^2_{L^2(\Omega, H^{\otimes(n+1)})},$$

for $u_n \in \mathscr{S} \otimes H^{\wedge n}$. By the intertwining relation of Lemma 1 combined with the use of Assumption (B1) to reach (53) below, we have

$$\langle \mathrm{d}^{n-1}\mathrm{d}^{(n-1)*}u_n(x_1, \ldots, x_n), g_n(x_1, \ldots, x_n)\rangle_{H^{\wedge n}} = nF\langle f_n(x_1, \ldots, x_n), g_n(x_1, \ldots, x_n)\rangle_{H^{\wedge n}}$$

$$+\sum_{j=1}^{n}\langle \delta((D_{x_j}F) \otimes f_n(x_1, \ldots, x_{j-1}, \cdot, x_{j+1}, \ldots, x_n)), g_n(x_1, \ldots, x_n)\rangle_{H^{\wedge n}}$$

$$+\sum_{j=1}^{n}\langle \delta(F \otimes \nabla_{x_j} f_n(x_1, \ldots, x_{j-1}, \cdot, x_{j+1}, \ldots, x_n), g_n(x_1, \ldots, x_n)\rangle_{H^{\wedge n}}$$

$$-\sum_{j=1}^{n}(-1)^{j-1} \tag{53}$$

$$\sum_{l=2}^{n}\left\langle (D_{x_j}F)\otimes\int_X \mathrm{Tr}\,\nabla_x^{(l)} f_n(x,x_1,\ldots,x_{j-1},x_{j+1},\ldots,x_n)\sigma(dx), g_n(x_1,\ldots,x_n)\right\rangle_{H^{\wedge n}}$$

$$-F\left\langle \int_X \mathrm{Tr}\,\nabla_x \mathrm{d}_{x_n}^{n-1} f_n(x,x_1,\ldots,x_{n-1})\sigma(dx), g_n(x_1,\ldots,x_n)\right\rangle_{H^{\wedge n}} \tag{54}$$

$$+\sum_{j=1}^{n}\int_X \langle (D_xF)\otimes f_n(x_1,\ldots,x_n), \nabla_{x_j}^{(j)} g_n(x_1,\ldots,x_{j-1},x,x_{j+1},\ldots,x_n)\rangle_{H^{\wedge n}}\sigma(dx).$$

Hence, applying Assumption (A4) from (53)–(56), and Assumption (A4) from (56) to (57), we find

$$\langle \mathrm{d}_{x_n}^{n-1}\mathrm{d}^{(n-1)*}u_n(\cdot,x_1,\ldots,x_{n-1}), v_n(x_1,\ldots,x_n)\rangle_{L^2(\Omega,H^{\wedge n})}$$

$$= n\langle u_n(x_1,\ldots,x_n), v_n(x_1,\ldots,x_n)\rangle_{L^2(\Omega,H^{\wedge n})}$$

$$+\sum_{j=1}^{n}\langle \delta((D_{x_j}F)\otimes f_n(x_1,\ldots,x_{j-1},\cdot,x_{j+1},\ldots,x_n)), G\otimes g_n(x_1,\ldots,x_n)\rangle_{L^2(\Omega,H^{\wedge n})}$$

$$+\sum_{j=1}^{n}\langle \delta(F\otimes \nabla_{x_j} f_n(x_1,\ldots,x_{j-1},\cdot,x_{j+1},\ldots,x_n)), G\otimes g_n(x_1,\ldots,x_n)\rangle_{L^2(\Omega,H^{\wedge n})}$$

$$+\frac{1}{n}\sum_{j=1}^{n}(-1)^{n-j}\sum_{l=1}^{n-1}\int_X\int_X \langle (D_yF)\otimes f_n(x_1,\ldots,x_{j-1},x,x_{j+1},\ldots,x_n), \tag{55}$$

$$G\otimes \nabla_x^{(l)} g_n(x_1,\ldots,x_{j-1},x_{j+1},\ldots,x_n,y)\rangle_{L^2(\Omega,H^{\wedge(n-1)})}\sigma(dx)\sigma(dy)$$

$$+\frac{1}{n}\sum_{l=1}^{n}\sum_{\substack{j=1\\ j\neq l}}^{n}\int_X\int_X \langle F\otimes \nabla_y^{(j)} f_n(x_1,\ldots,x_{l-1},x,x_{l+1},\ldots,x_n), \tag{56}$$

$$\cdot \quad G\otimes \nabla_x^{(l)} g_n(x_1,\ldots,x_{j-1},y,x_{j+1},\ldots,x_n)\rangle_{L^2(\Omega,H^{\wedge(n-1)})}\sigma(dx)\sigma(dy)$$

$$+\sum_{l=1}^{n}(-1)^{n-j}\int_X \big\langle (D_yF)\otimes f_n(x_1,\ldots,x_n),$$

$$G\otimes \nabla_{x_l}^{(n)} g_n(x_1,\ldots,x_{l-1},x_{l+1},\ldots,x_n,y)\big\rangle_{L^2(\Omega,H^{\wedge n})}\sigma(dy)$$

$$= n\langle u_n(x_1,\ldots,x_n), v_n(x_1,\ldots,x_n)\rangle_{L^2(\Omega,H^{\wedge n})}$$

$$+\frac{1}{n}\sum_{j=1}^{n}\int_X\int_X \langle (D_xF)\otimes f_n(x1,\ldots,x_{j-1},y,x_{j+1},\ldots,x_n),$$

$$(D_yG)\otimes g_n(x_1,\ldots,x_{j-1},x,x_{j+1},\ldots,x_n)\rangle_{L^2(\Omega,H^{\wedge(n-1)})}\sigma(dx)\sigma(dy)$$

$$+\frac{1}{n}\sum_{j=1}^{n}\int_X\int_X \langle F\otimes \nabla_x f_n(x_1,\ldots,x_{j-1},y,x_{j+1},\ldots,x_n),$$

$$(D_yG) \otimes g_n(x_1, \ldots, x_{j-1}, x, x_{j+1}, \ldots, x_n)\rangle_{L^2(\Omega, H^{\wedge(n-1)})}\sigma(dx)\sigma(dy)$$

$$+ \frac{1}{n}\sum_{j=1}^{n}(-1)^{n-j}\sum_{l=1}^{n}\int_X\int_X\langle(D_yF)\otimes f_n(x_1, \ldots, x_{j-1}, x, x_{j+1}, \ldots, x_n),$$

$$G \otimes \nabla_x^{(l)} g_n(x_1, \ldots, x_{j-1}, x_{j+1}, \ldots, x_n, y)\rangle_{L^2(\Omega, H^{\wedge(n-1)})}\sigma(dx)\sigma(dy)$$

$$+ \frac{1}{n}\sum_{j=1}^{n}\sum_{l=1}^{n}\int_X\int_X\langle F\otimes\nabla_y^{(j)}f_n(x_1, \ldots, x_{l-1}, x, x_{l+1}, \ldots, x_n), \tag{57}$$

$$G \otimes \nabla_x^{(l)} g_n(x_1, \ldots, x_{j-1}, y, x_{j+1}, \ldots, x_n)\rangle_{L^2(\Omega, H^{\wedge(n-1)})}\sigma(dx)\sigma(dy)$$

$$= n\langle u_n(x_1, \ldots, x_n), v_n(x_1, \ldots, x_n)\rangle_{L^2(\Omega, H^{\wedge n})}$$

$$+ \frac{1}{n}\sum_{l=1}^{n}\int_X\int_X\langle(D_xF)\otimes f_n(x_1, \ldots, x_{l-1}, y, x_{l+1}, \ldots, x_n),$$

$$(D_yG) \otimes g_n(x_1, \ldots, x_{j-1}, x, x_{j+1}, \ldots, x_n)\rangle_{L^2(\Omega, H^{\wedge(n-1)})}\sigma(dx)\sigma(dy)$$

$$+ \frac{1}{n}\sum_{j=1}^{n}\int_X\int_X\langle F\otimes\nabla_x f_n(x_1, \ldots, x_{j-1}, y, x_{j+1}, \ldots, x_n),$$

$$(D_yG) \otimes g_n(x_1, \ldots, x_{j-1}, x, x_{j+1}, \ldots, x_n)\rangle_{L^2(\Omega, H^{\wedge(n-1)})}\sigma(dx)\sigma(dy)$$

$$+ \frac{1}{n}\sum_{j=1}^{n}(-1)^{n-j}\sum_{l=1}^{n}\int_X\int_X\langle(D_yF)\otimes f_n(x_1, \ldots, x_{j-1}, x, x_{j+1}, \ldots, x_n),$$

$$G \otimes \nabla_x^{(l)} g_n(x_1, \ldots, x_{j-1}, x_{j+1}, \ldots, x_n, y)\rangle_{L^2(\Omega, H^{\wedge(n-1)})}\sigma(dx)\sigma(dy)$$

$$+ \frac{1}{n}\sum_{j=1}^{n}(-1)^{n-j}\sum_{l=1}^{n}\int_X\int_X\langle F\otimes\nabla_y^{(j)}f_n(x_1, \ldots, x_{l-1}, x, x_{l+1}, \ldots, x_n),$$

$$G \otimes \nabla_x^{(l)} g_n(x_1, \ldots, x_{j-1}, x_{j+1}, \ldots, x_n, y)\rangle_{L^2(\Omega, H^{\wedge(n-1)})}\sigma(dx)\sigma(dy)$$

$$= n\langle u_n(x_1, \ldots, x_n), v_n(x_1, \ldots, x_n)\rangle_{L^2(\Omega, H^{\wedge n})}$$

$$+ \frac{1}{n!}\Big\langle(D_{x_{n+1}}F)\otimes f_n(x_1, \ldots, x_n) + F\otimes\nabla_{x_{n+1}}f_n(x_1, \ldots, x_n),$$

$$\sum_{j=1}^{n}(-1)^{n-j}(D_{x_j}G)\otimes g_n(x_1, \ldots, x_{j-1}, x_{j+1}, \ldots, x_n, x_{n+1})$$

$$+ \sum_{j=1}^{n}(-1)^{n-j}\sum_{l=1}^{n} G\otimes\nabla_{x_j}^{(l)} g_n(x_1, \ldots, x_{j-1}, x_{j+1}, \ldots, x_n, x_{n+1})\Big\rangle_{L^2(\Omega, H^{\otimes(n+1)})}$$

$$= n\langle u_n(x_1, \ldots, x_n), v_n(x_1, \ldots, x_n)\rangle_{L^2(\Omega, H^{\wedge n})}$$

$$- \frac{(-1)^n}{n!}\langle(D_{x_{n+1}}F)\otimes f_n(x_1, \ldots, x_n) + F\otimes\nabla_{x_{n+1}}f_n(x_1, \ldots, x_n),$$

$$\sum_{j=1}^{n+1}(-1)^{j-1}(D_{x_j}G)\otimes g_n(x_1,\ldots,x_{j-1},x_{j+1},\ldots,x_{n+1})$$

$$+\sum_{j=1}^{n+1}(-1)^{j-1}\sum_{l=1}^{n}G\otimes\nabla_{x_j}^{(l)}g_n(x_1,\ldots,x_{j-1},x_{j+1},\ldots,x_{n+1})\Big\rangle_{L^2(\Omega,H^{\otimes(n+1)})}$$

$$+\frac{1}{n!}\Big\langle(D_{x_{n+1}}F)\otimes f_n(x_1,\ldots,x_n)+F\otimes\nabla_{x_{n+1}}f_n(x_1,\ldots,x_n),$$

$$(D_{x_{n+1}}G)\otimes g_n(x_1,\ldots,x_n)+G\otimes\sum_{l=1}^{n}\nabla_{x_{n+1}}^{(l)}g_n(x_1,\ldots,x_n)\Big\rangle_{L^2(\Omega,H^{\otimes(n+1)})}$$

$$=n\langle u_n(x_1,\ldots,x_n),v_n(x_1,\ldots,x_n)\rangle_{L^2(\Omega,H^{\wedge n})}$$

$$-\frac{(-1)^n}{n!}\Big\langle(D_{x_{n+1}}F)\otimes f_n(x_1,\ldots,x_n)+F\otimes\nabla_{x_{n+1}}f_n(x_1,\ldots,x_n),$$

$$\sum_{j=1}^{n+1}(-1)^{j-1}(D_{x_j}G)\otimes g_n(x_1,\ldots,x_{j-1},x_{j+1},\ldots,x_{n+1})$$

$$+\sum_{j=1}^{n+1}(-1)^{j-1}G\otimes\nabla_{x_j}g_n(x_1,\ldots,x_{j-1},x_{j+1},\ldots,x_{n+1})\Big\rangle_{L^2(\Omega,H^{\otimes(n+1)})}$$

$$+\frac{1}{n!}\Big\langle(D_{x_{n+1}}F)\otimes f_n(x_1,\ldots,x_n)+F\otimes\nabla_{x_{n+1}}f_n(x_1,\ldots,x_n),$$

$$(D_{x_{n+1}}G)\otimes g_n(x_1,\ldots,x_n)+G\otimes\nabla_{x_{n+1}}g_n(x_1,\ldots,x_n)\Big\rangle_{L^2(\Omega,H^{\otimes(n+1)})}$$

$$=n\langle u_n(x_1,\ldots,x_n),v_n(x_1,\ldots,x_n)\rangle_{L^2(\Omega,H^{\wedge n})}$$

$$-\frac{1}{(n+1)!}\Big\langle\sum_{j=1}^{n+1}(-1)^{j-1}(D_{x_j}F)\otimes f_n(x_1,\ldots,x_{j-1},x_{j+1},\ldots,x_{n+1})$$

$$+\sum_{j=1}^{n+1}(-1)^{j-1}F\otimes\nabla_{x_j}f_n(x_1,\ldots,x_{j-1},x_{j+1},\ldots,x_{n+1}),$$

$$\sum_{j=1}^{n+1}(-1)^{j-1}(D_{x_j}G)\otimes g_n(x_1,\ldots,x_{j-1},x_{j+1},\ldots,x_{n+1})$$

$$+\sum_{j=1}^{n+1}(-1)^{j-1}G\otimes\nabla_{x_j}g_n(x_1,\ldots,x_{j-1},x_{j+1},\ldots,x_{n+1})\Big\rangle_{L^2(\Omega,H^{\otimes(n+1)})}$$

$$+\frac{1}{n!}\Big\langle(D_{x_{n+1}}F)\otimes f_n(x_1,\ldots,x_n)+F\otimes\nabla_{x_{n+1}}f_n(x_1,\ldots,x_n),$$

$$(D_{x_{n+1}}G)\otimes g_n(x_1,\ldots,x_n)+G\otimes\nabla_{x_{n+1}}g_n(x_1,\ldots,x_n)\Big\rangle_{L^2(\Omega,H^{\otimes(n+1)})}$$

$$= n\langle u_n(x_1,\dots,x_n), v_n(x_1,\dots,x_n)\rangle_{L^2(\Omega,H^{\wedge n})} - \langle \mathrm{d}^n u_n, \mathrm{d}^n v_n\rangle_{L^2(\Omega,H^{\wedge(n+1)})}$$

$$+ \frac{1}{n!}\langle \nabla_{x_{n+1}} u_n(x_1,\dots,x_n), \nabla_{x_{n+1}} v_n(x_1,\dots,x_n)\rangle_{L^2(\Omega,H^{\wedge(n+1)})},$$

where we used (A4). Hence we have

$$\langle \mathrm{d}^{(n-1)*} u_n, \mathrm{d}^{(n-1)*} v_n\rangle_{L^2(\Omega,H^{\wedge n})} + \langle \mathrm{d}^n u_n, \mathrm{d}^n v_n\rangle_{L^2(\Omega,H^{\wedge(n+1)})}$$
$$= n\langle u_n, v_n\rangle_{L^2(\Omega,H^{\wedge n})} + \frac{1}{n!}\langle \nabla u_n, \nabla v_n\rangle_{L^2(\Omega,H^{\otimes(n+1)})},$$

i.e.

$$\langle \mathrm{d}^{(n-1)*} u_n, \mathrm{d}^{(n-1)*} v_n\rangle_{L^2(\Omega,H^{\otimes n})} + \frac{1}{n+1}\langle \mathrm{d}^n u_n, \mathrm{d}^n v_n\rangle_{L^2(\Omega,H^{\otimes(n+1)})}$$
$$= n\langle u_n, v_n\rangle_{L^2(\Omega,H^{\otimes n})} + \langle \nabla u_n, \nabla v_n\rangle_{L^2(\Omega,H^{\otimes(n+1)})},$$

and applying the duality

$$\langle \nabla u_n, \nabla v_n\rangle_{L^2(\Omega,H^{\otimes(n+1)})} = \langle \nabla^* \nabla u_n, v_n\rangle_{L^2(\Omega,H^{\wedge n})}, \qquad u_n, v_n \in \mathscr{S} \otimes H^{\wedge n},$$

we get

$$\mathrm{d}^{n-1}\mathrm{d}^{(n-1)*} + \mathrm{d}^{n*}\mathrm{d}^n = n\mathrm{I}_{H^{\wedge n}} + \nabla^*\nabla.$$

□

Proof of Theorem 2 (Clark-Ocone Formula) By the Clark-Ocone formula (C1) and Assumption (A4') we have

$$u_n(t_1,\dots,t_n) = E[u_n(t_1,\dots,t_n) \mid \mathscr{F}_{t_1\vee\cdots\vee t_n}] + \int_{t_1\vee\cdots\vee t_n}^{\infty} E[D_r u_n(t_1,\dots,t_n) \mid \mathscr{F}_r] dM_r$$

$$= E[u_n(t_1,\dots,t_n) \mid \mathscr{F}_{t_1\vee\cdots\vee t_n}] + \int_{t_1\vee\cdots\vee t_n}^{\infty} E[\nabla_r u_n(t_1,\dots,t_n) \mid \mathscr{F}_r] dM_r, \tag{58}$$

$t_1,\dots,t_n \in \mathbf{R}_+$. Next, by the definition (5) of d^n and (22) applied to adapted processes we have

$$\mathrm{d}_{t_n}^{n-1} \int_{t_1\vee\cdots\vee t_{n-1}}^{\infty} E[u_n(r,t_1,\dots,t_{n-1}) \mid \mathscr{F}_r] dM_r$$

$$= \sum_{j=1}^{n} (-1)^{j-1} \nabla_{t_j} \int_{t_1\vee\cdots\vee t_{j-1}\vee t_{j+1}\vee\cdots\vee t_n}^{\infty} E[u_n(r,t_1,\dots,t_{j-1},t_{j+1},\dots,t_n) \mid \mathscr{F}_r] dM_r$$

$$= \sum_{j=1}^{n}(-1)^{j-1}\mathbf{1}_{[t_1\vee\cdots\vee t_n,\infty)}(t_j)E[u_n(t_j,t_1,\ldots,t_{j-1},t_{j+1},\ldots,t_n)\mid\mathscr{F}_{t_j}]$$

$$+\sum_{j=1}^{n}(-1)^{j-1}\int_{t_1\vee\cdots\vee t_{j-1}\vee t_{j+1}\cdots\vee t_n}^{\infty}\nabla_{t_j}E[u_n(t_{n+1},t_1,\ldots,t_{j-1},t_{j+1},\ldots,t_n)\mid\mathscr{F}_{t_{n+1}}]dM_{t_{n+1}}$$

$$= \sum_{j=1}^{n}\mathbf{1}_{[t_1\vee\cdots\vee t_n,\infty)}(t_j)E[u_n(t_1,\ldots,t_n)\mid\mathscr{F}_{t_j}]$$

$$+\sum_{j=1}^{n}(-1)^{j-1}\int_{t_1\vee\cdots\vee t_{j-1}\vee t_{j+1}\cdots\vee t_n}^{\infty}\nabla_{t_j}E[u_n(t_{n+1},t_1,\ldots,t_{j-1},t_{j+1},\ldots,t_n)\mid\mathscr{F}_{t_{n+1}}]dM_{t_{n+1}}$$

$$= E[u_n(t_1,\ldots,t_n)\mid\mathscr{F}_{t_1\vee\cdots\vee t_n}]$$

$$+\sum_{j=1}^{n}(-1)^{j-1}\int_{t_1\vee\cdots\vee t_n}^{\infty}E[\nabla_{t_j}u_n(t_{n+1},t_1,\ldots,t_{j-1},t_{j+1},\ldots,t_n)\mid\mathscr{F}_{t_{n+1}}]dM_{t_{n+1}}, \qquad (59)$$

where on the last line we used the fact that by (A4), $\nabla_{t_j}u_n(t_{n+1},t_1,\ldots,t_{j-1},t_{j+1},\ldots,t_n)$ vanishes when $t_1\vee\cdots\vee t_{j-1}\vee t_{j+1}\cdots\vee t_n < t_{n+1} < t_j$, hence by taking the difference of (58) and (59) we find

$$u_n(t_1,\ldots,t_n) = \mathrm{d}_{t_n}^{n-1}\int_{t_1\vee\cdots\vee t_{n-1}}^{\infty}E[u_n(r,t_1,\ldots,t_{n-1})\mid\mathscr{F}_r]dM_r$$

$$-\sum_{j=1}^{n}(-1)^{j-1}\int_{t_1\vee\cdots\vee t_n}^{\infty}E[\nabla_{t_j}u_n(t_1,\ldots,t_{j-1},t_{j+1},\ldots,t_{n+1})\mid\mathscr{F}_{t_{n+1}}]dM_{t_{n+1}}$$

$$+\int_{t_1\vee\cdots\vee t_n}^{\infty}E[\nabla_r u_n(t_1,\ldots,t_n)\mid\mathscr{F}_r]dM_r$$

$$= \mathrm{d}_{t_n}^{n-1}\int_{t_1\vee\cdots\vee t_{n-1}}^{\infty}E[u_n(r,t_1,\ldots,t_{n-1})\mid\mathscr{F}_r]dM_r$$

$$+\int_{t_1\vee\cdots\vee t_n}^{\infty}E[\mathrm{d}_{t_{n+1}}^{n}u_n(t_1,\ldots,t_n)\mid\mathscr{F}_{t_{n+1}}]dM_{t_{n+1}},$$

$t_1,\ldots,t_n \in \mathbf{R}_+$. □

References

1. S. Aida, Vanishing of one-dimensional L^2-cohomologies of loop groups. J. Funct. Anal. **261**(8), 2164–2213 (2011)
2. S. Albeverio, A. Daletskii, E. Lytvynov, De Rham cohomology of configuration spaces with Poisson measure. J. Funct. Anal. **185**(1), 240–273 (2001)
3. S. Albeverio, A. Daletskii, E. Lytvynov, Laplace operators on differential forms over configuration spaces. J. Geom. Phys. **37**(1–2), 15–46 (2001)
4. L. Bartholdi, T. Schick, N. Smale, S. Smale, Hodge theory on metric spaces. Found. Comput. Math. **12**(1), 1–48 (2012)
5. J.M.C. Clark, The representation of functionals of Brownian motion by stochastic integrals. Ann. Math. Stat. **41**, 1281–1295 (1970)
6. A.B. Cruzeiro, P. Malliavin, Renormalized differential geometry on path space: structural equation, curvature. J. Funct. Anal. **139**, 119–181 (1996)
7. K.D. Elworthy, Y. Yang, The vanishing of L^2 harmonic one-forms on based path spaces. J. Funct. Anal. **264**(5), 1168–1196 (2013)
8. M. Émery, On the Azéma martingales, in *Séminaire de Probabilités XXIII*. Lecture Notes in Mathematics, vol. 1372 (Springer, New York, 1990), pp. 66–87
9. S. Fang, J. Franchi, Flatness of Riemannian structure over the path group and energy identity for stochastic integrals. C. R. Acad. Sci. Paris Sér. I Math. **321**(10), 1371–1376 (1995)
10. S. Fang, J. Franchi, De Rham-Hodge-Kodaira operator on loop groups. J. Funct. Anal. **148**(2), 391–407 (1997)
11. S. Fang, J. Franchi, A differentiable isomorphism between Wiener space and path group, in *Séminaire de Probabilités, XXXI*. Lecture Notes in Mathematics, vol. 1655 (Springer, Berlin, 1997), pp. 54–61
12. T. Kazumi, I. Shigekawa, Differential calculus on a submanifold of an abstract Wiener space, II: Weitzenböck formula, in *Proceedings of the Conference on Dirichlet Forms and Stochastic Processes* (Walter de Gruyter and Co., Hawthorne, 1993), pp. 235–251
13. S. Kobayashi, K. Nomizu, *Foundations of Differential Geometry*, vol. I (Interscience Publishers, a division of John Wiley and Sons, New York/London, 1963)
14. D. Nualart, *The Malliavin Calculus and Related Topics*. Probability and Its Applications, 2nd edn. (Springer, Berlin, 2006)
15. D. Ocone, Malliavin's calculus and stochastic integral representations of functionals of diffusion processes. Stochastics **12**(3–4), 161–185 (1984)
16. J.J. Prat, N. Privault, Explicit stochastic analysis of Brownian motion and point measures on Riemannian manifolds. J. Funct. Anal. **167**, 201–242 (1999)
17. N. Privault, Quantum stochastic calculus applied to path spaces over Lie groups, in *Proceedings of the International Conference on Stochastic Analysis and Applications* (Kluwer, Dordrecht, 2004), pp. 85–94
18. N. Privault, *Stochastic Analysis in Discrete and Continuous Settings with Normal Martingales*. Lecture Notes in Mathematics, vol. 1982 (Springer, Berlin, 2009)
19. N. Privault, De Rham-Hodge decomposition and vanishing of harmonic forms by derivation operators on the Poisson space. Preprint, 36 pp, to appear in Infinite Dimensional Analysis, Quantum Probability and Related Topics, 2015
20. I. Shigekawa, de Rham-Hodge-Kodaira's decomposition on an abstract Wiener space. J. Math. Kyoto Univ. **26**(2), 191–202 (1986)
21. D. Surgailis, On multiple Poisson stochastic integrals and associated Markov semi-groups. Probab. Math. Stat. **3**, 217–239 (1984)
22. Y. Yang, Generalised Clark-Ocone formulae for differential forms. Commun. Stoch. Anal. **6**(2), 323–337 (2012)

On the Range of Exponential Functionals of Lévy Processes

Anita Behme, Alexander Lindner, and Makoto Maejima

Abstract We characterize the support of the law of the exponential functional $\int_0^\infty e^{-\xi_{s-}}\, d\eta_s$ of two one-dimensional independent Lévy processes ξ and η. Further, we study the range of the mapping Φ_ξ for a fixed Lévy process ξ, which maps the law of η_1 to the law of the corresponding exponential functional $\int_0^\infty e^{-\xi_{s-}}\, d\eta_s$. It is shown that the range of this mapping is closed under weak convergence and in the special case of positive distributions several characterizations of laws in the range are given.

1 Introduction

Given a bivariate Lévy process $(\xi, \eta)^T = ((\xi_t, \eta_t)^T)_{t\geq 0}$, its *exponential functional* is defined as

$$V := \int_0^\infty e^{-\xi_{s-}} d\eta_s, \tag{1}$$

A. Behme (✉)
Institut für Mathematische Stochastik, Technische Universität Dresden, 01062 Dresden, Germany

Zentrum Mathematik, Technische Universität München, Boltzmannstraße 3, 85748 Garching bei München, Germany
e-mail: a.behme@tum.de

A. Lindner
Institut für Mathematische Stochastik, Technische Universität Braunschweig, 38106 Braunschweig, Germany

Institute of Mathematical Finance, Ulm University, 89081 Ulm, Germany
e-mail: alexander.lindner@uni-ulm.de

M. Maejima
Department of Mathematics, Keio University, Hiyoshi, Yokohama 223-8522, Japan
e-mail: maejima@math.keio.ac.jp

C. Donati-Martin et al. (eds.), *Séminaire de Probabilités XLVIII*, Lecture Notes in Mathematics 2168, DOI 10.1007/978-3-319-44465-9_10

provided that the integral converges almost surely. Exponential functionals of Lévy processes appear as stationary distributions of generalized Ornstein-Uhlenbeck (GOU) processes. In particular, if ξ and η are independent and ξ_t tends to $+\infty$ as $t \to \infty$ almost surely, then the law of V defined in (1) is the stationary distribution of the GOU process

$$V_t = e^{-\xi_t}\left(\int_0^t e^{\xi_{s-}} d\eta_s + V_0\right), \quad t \geq 0, \tag{2}$$

where V_0 is a starting random variable, independent of $(\xi, \eta)^T$, on the same probability space (cf. [22, Theorem 2.1]). Hence, when V_0 is chosen to have the same distribution as V, then the process $(V_t)_{t\geq 0}$ is strictly stationary.

Unless $\xi_t = at$ with $a > 0$, the distribution of V is known only in a few special cases. See e.g. Bertoin and Yor [7] for a survey on exponential functionals of the form $V = \int_0^\infty e^{-\xi_{s-}} ds$, or Gjessing and Paulsen [15], who determine the distribution of $\int_0^\infty e^{-\xi_{s-}} d\eta_s$ for some cases. A thorough study of distributions of the form $\int_0^\infty e^{-\xi_{s-}} d\eta_s$ when η is a Brownian motion is carried out in Kuznetsov et al. [20]. We state the following example due to Dufresne (e.g. [7, Eq. (16)]) of an exponential functional whose distribution has been determined and to which we will refer later. Here and in the following we write "$\stackrel{d}{=}$" to denote equality in distribution of random variables.

Example 1 For $(\xi_t, \eta_t) = (\sigma B_t + at, t)$ with $\sigma > 0, a > 0$ and a standard Brownian motion $(B_t)_{t\geq 0}$ it holds

$$V = \int_0^\infty e^{-(\sigma B_t + at)} dt \stackrel{d}{=} \frac{2}{\sigma^2 \Gamma_{\frac{2a}{\sigma^2}}},$$

where Γ_r denotes a standard Gamma random variable with shape parameter r, i.e. with density

$$P(\Gamma_r \in dx) = \frac{x^{r-1}}{\Gamma(r)} e^{-x} \mathbb{1}_{(0,\infty)}(x) dx.$$

Denote by $\mathscr{L}(X)$ the law of a random variable X and let $\xi = (\xi_t)_{t\geq 0}$ be a one-dimensional Lévy process drifting to $+\infty$. In this paper we will consider the mapping

$$\Phi_\xi : D_\xi \to \mathscr{P}(\mathbb{R}) := \text{the set of probability distributions on } \mathbb{R},$$

$$\mathscr{L}(\eta_1) \mapsto \mathscr{L}\left(\int_0^\infty e^{-\xi_{s-}} d\eta_s\right),$$

defined on

$$D_\xi := \{\mathscr{L}(\eta_1) : \eta = (\eta_t)_{t\geq 0} \text{ one-dimensional Lévy process independent of } \xi$$
$$\text{such that } \int_0^\infty e^{-\xi_{s-}}\, d\eta_s \text{ converges a.s.}\}.$$

An explicit description of D_ξ in terms of the characteristic triplets [cf. (3)] of ξ and η follows from Theorem 2 in Erickson and Maller [14]. Denote the range of Φ_ξ by

$$R_\xi := \Phi_\xi(D_\xi).$$

Although the domain Φ_ξ can be completely characterized by Erickson and Maller [14], much less is known about the range R_ξ and properties of the mapping Φ_ξ. In the case that $\xi_t = at, a > 0$ is deterministic, it is well known that $D_\xi = \mathrm{ID}_{\log}(\mathbb{R})$, the set of real-valued infinitely divisible distributions with finite $\log^+$-moment, and that Φ_ξ is an algebraic isomorphism between $\mathrm{ID}_{\log}(\mathbb{R})$ and $R_\xi = L(\mathbb{R})$, the set of real-valued selfdecomposable distributions [17, Proposition 3.6.10].

For general ξ, the mapping Φ_ξ has already been studied in [4], where it has been shown that Φ_ξ is injective in many cases, while injectivity cannot be obtained if ξ and η are allowed to exhibit a dependence structure. Further in [4] conditions for continuity (in a weak sense) of Φ_ξ are given. These results were then used to obtain some information on the range R_ξ. In particular it has been shown that centered Gaussian distributions can only be obtained in the setting of (classical) OU processes, namely, for ξ being deterministic and η being a Brownian motion.

In this paper we take up the subject of studying properties of the mapping Φ_ξ and of distributions in R_ξ, and start in Sect. 2 with a classification of possible supports of the laws in R_ξ. Section 3 is devoted to show closedness of the range R_ξ under weak convergence. It also follows that the inverse mapping Φ_ξ^{-1} is continuous if it is well-defined, i.e. if Φ_ξ is injective. In Sects. 4 and 5 we specialize on positive distributions in R_ξ. Section 4 gives a general criterion for positive distributions to belong to R_ξ. In Sect. 5 we use this criterion to obtain further results in the case that ξ is a Brownian motion with drift. We derive a differential equation for the Laplace exponent of a positive distribution in R_ξ and from this we gain concrete conditions in terms of Lévy measure and drift for some distributions to be in R_ξ. We end up studying the special case of positive stable distributions in R_ξ.

For an $\mathbb{R}^d$-valued Lévy process $X = (X_t)_{t\geq 0}$, the *characteristic exponent* is given by its Lévy-Khintchine formula (e.g. [28, Theorem 8.1])

$$\begin{aligned}\log \phi_X(u) &:= \log E\left[e^{i\langle u, X_1\rangle}\right] \\ &= i\langle \gamma_X, u\rangle - \frac{1}{2}\langle u, A_X u\rangle + \int_{\mathbb{R}^d} (e^{i\langle u,x\rangle} - 1 - i\langle u, x\rangle \mathbb{1}_{|x|\leq 1})\nu_X(dx), \quad u \in \mathbb{R},\end{aligned} \tag{3}$$

where (γ_X, A_X, ν_X) is the *characteristic triplet* of the Lévy process X. In case that X is real-valued we will usually replace A_X by σ_X^2. In the special case of subordinators in $\mathbb{R}$, i.e. nondecreasing Lévy processes, we will also use the Laplace transform

$$\mathbb{L}_X(u) := E[e^{-uX_1}] = e^{\psi_X(u)}, \quad u \geq 0,$$

of X and call $\psi_X(u)$ the Laplace exponent of the Lévy process X. We refer to [28] for further information on Lévy processes. In the following, when the symbol X is regarded as a real-valued random variable, we also use the notation $\phi_X(u)$ and $\mathbb{L}_X(u)$ for its characteristic function and Laplace transform, respectively. The Fourier transform of a finite measure μ on $\mathbb{R}$ is written as $\hat{\mu}(u) = \int_{\mathbb{R}} e^{iux}\,\mu(dx)$. We write "$\xrightarrow{d}$" to denote convergence in distribution of random variables, and "$\xrightarrow{w}$" to denote weak convergence of probability measures. We use the abbreviation "i.i.d." for "independent and identically distributed". The set of all twice continuously differentiable functions $f : \mathbb{R} \to \mathbb{R}$ which are bounded will be denoted by $C_b^2(\mathbb{R})$, and the subset of all $f : \mathbb{R} \to \mathbb{R}$ which have additionally compact support by $C_c^2(\mathbb{R})$.

2 On the Support of the Exponential Functional

In this section we shall give the support of the distribution of the exponential functional $V = \int_0^\infty e^{-\xi_{s-}}\,d\eta_s$ when ξ and η are independent Lévy processes. In particular it turns out that the support will always be a closed interval. A similar result does not hold for solutions of arbitrary random recurrence equations, or for exponential functionals of Lévy processes with dependent ξ and η, as we shall show in Remark 1.

For ξ being spectrally negative, it is well known (e.g. [8]) that V has a selfdecomposable and hence infinitely divisible distribution. In [28, Theorem 24.10] a characterization of the support of infinitely divisible distributions is given in terms of the Lévy triplet. In particular the support of a selfdecomposable distribution on $\mathbb{R}$ is either a single point, $\mathbb{R}$ itself or a one-sided unbounded interval. Unfortunately the characteristic triplet of V is not known in general and also, for not spectrally negative ξ this result can not be applied.

Before we characterize the support of the law of $V = \int_0^\infty e^{-\xi_{s-}}\,d\eta_s$ when ξ and η are general independent Lévy processes, we treat the special case when $\eta_t = t$ in the following lemma. Much attention has been paid to this case, and in particular, it has been shown that the stationary solution has a density under various conditions, see e.g. Pardo et al. [25] or Carmona at al. [12]. Haas and Rivero [16, Theorem 1.4, Lemma 2.1] gave a characterization when this law is bounded and obtained that this is the case if and only if ξ is a subordinator with strictly positive drift, and derived the support then. So parts of the following lemma follow already from results in [16], nevertheless we have decided to give a detailed proof.

Lemma 1 *Let ξ be a Lévy process drifting to $+\infty$ and set $V = \int_0^\infty e^{-\xi_s} ds$. Then*

$$\operatorname{supp} \mathscr{L}(V) = \begin{cases} \left\{\frac{1}{b}\right\}, & \text{if } \xi_t = bt \text{ with } b > 0, \\ \left[0, \frac{1}{b}\right], & \text{if } \xi \text{ is a non-deterministic subordinator with drift } b > 0, \\ \left[\frac{1}{b}, \infty\right), & \text{if } \xi \text{ is non-deterministic and of finite variation,} \\ & \text{with drift } b > 0 \text{ and } \nu_\xi((0,\infty)) = 0, \\ [0, \infty), & \text{otherwise.} \end{cases}$$

Proof The claim is clear if ξ is deterministic, while it follows from Remark 1 if ξ is a Brownian motion with drift, so suppose that $\nu_\xi \not\equiv 0$. Suppose first that $\nu_\xi((0,\infty)) > 0$, and let $x_0 \in \operatorname{supp} \mathscr{L}(V) \cap (0, \infty)$. Let $c \in \operatorname{supp} \nu_\xi \cap (0, \infty)$ and $y_0 \in (e^{-c} x_0, x_0)$. We shall show that also $y_0 \in \operatorname{supp} \mathscr{L}(V)$, so that by induction $\operatorname{supp} \mathscr{L}(V)$ must be an interval with lower endpoint 0 if $\nu_\xi((0,\infty)) > 0$. To see this, define $z_0 \in (0, y_0)$ so that

$$z_0 + e^{-c}(x_0 - z_0) = y_0.$$

Let $\varepsilon \in (0, \frac{x_0 - z_0}{2})$ and define

$$A = A_\varepsilon := \left\{\omega \in \Omega : \int_0^\infty e^{-\xi_s(\omega)} \, ds \in (x_0 - \varepsilon, x_0 + \varepsilon)\right\}.$$

Then $P(A) > 0$ since $x_0 \in \operatorname{supp} \mathscr{L}(V)$. Define the stopping time $T_1 \in [0, \infty]$ by

$$T_1(\omega) := \inf\left\{t \geq 0 : \int_0^t e^{-\xi_s(\omega)} \, ds = z_0\right\}.$$

Since $t \mapsto \int_0^t e^{-\xi_s(\omega)} \, ds$ is continuous, T_1 is finite on A. Let $\delta_1 \in (0, \frac{x_0 - z_0}{2})$ and $\delta_2 \in (0, c)$. Then $\nu_\eta((c - \delta_2, c + \delta_2)) > 0$, and since $P(A) > 0$, there are a (sufficiently large) constant $K = K(\varepsilon, \delta_1, \delta_2) > 0$ and a (sufficiently small) constant $\delta = \delta(\varepsilon, \delta_1, \delta_2) > 0$ such that $\delta < 1$ and

$$B := B_{\varepsilon,\delta_1,\delta_2,\delta,K} := A \cap \Big\{T_1 \leq K, \int_{T_1}^{T_1+\delta} e^{-\xi_s} \, ds \leq \delta_1, \\ \Delta\xi_s \notin (c - \delta_2, c + \delta_2), \ \forall\, s \in (T_1, T_1 + \delta]\Big\}$$

has a positive probability. Now define the set $C = C_{\varepsilon,\delta_1,\delta_2,\delta,K}$ to be the set of all $\omega \in \Omega$, for which there exists an $\omega' \in B$, some time $t(\omega') \in (T_1 \wedge K, (T_1 \wedge K) + \delta]$ and some $\alpha(\omega') \in (c - \delta_2, c + \delta_2)$ such that

$$(\xi_t(\omega))_{t \geq 0} = (\xi_t(\omega') + \alpha(\omega') \mathbb{1}_{[t(\omega'),\infty)})_{t \geq 0},$$

namely, the set of ω whose paths behave exactly like a sample path from the set B, but with the exception that additionally exactly one jump of size in $(c-\delta_2, c+\delta_2)$ occurs in the interval $(T_1 \wedge K, (T_1 \wedge K)+\delta]$. Since $T_1 \wedge K$ is a finite stopping time, it follows from the strong Markov property of ξ and from $P(B)>0$ that also $P(C)>0$. But for $\omega \in C$, with $\omega' \in B$ and $\alpha = \alpha(\omega') \in (c-\delta_2, c+\delta_2)$ as in the definition of C, we obtain

$$\begin{aligned}
&\int_0^\infty e^{-\xi_s(\omega)}\,ds \\
&= \int_0^{T_1(\omega')} e^{-\xi_s(\omega')}\,ds + \int_{T_1(\omega')}^{T_1(\omega')+\delta} e^{-\xi_s(\omega)}\,ds + e^{-\alpha}\int_{T_1(\omega')+\delta}^{\infty} e^{-\xi_s(\omega')}\,ds \\
&\in \Big[z_0 + \int_{T_1(\omega')}^{T_1(\omega')+\delta} e^{-\xi_s(\omega)}\,ds + e^{-\alpha}\Big(x_0 - \varepsilon - z_0 - \int_{T_1(\omega')}^{T_1(\omega')+\delta} e^{-\xi_s(\omega')}\,ds\Big), \\
&\qquad z_0 + \int_{T_1(\omega')}^{T_1(\omega')+\delta} e^{-\xi_s(\omega)}\,ds + e^{-\alpha}\Big(x_0 + \varepsilon - z_0 - \int_{T_1(\omega')}^{T_1(\omega')+\delta} e^{-\xi_s(\omega')}\,ds\Big)\Big] \\
&\subset \Big[z_0 - \delta_1 + e^{-c}(x_0 - z_0 - \varepsilon) + (e^{-c-\delta_2} - e^{-c})(x_0 - z_0 - \varepsilon) - e^{-c+\delta_2}\delta_1, \\
&\qquad z_0 + \delta_1 + e^{-c}(x_0 - z_0 + \varepsilon) + (e^{-c+\delta_2} - e^{-c})(x_0 - z_0 + \varepsilon) + e^{-c+\delta_2}\delta_1\Big].
\end{aligned}$$

Since $y_0 = z_0 + e^{-c}(x_0 - z_0)$, we see that $y_0 \in \operatorname{supp}\mathscr{L}(V)$ by choosing ε, δ_1 and δ_2 sufficiently small. So we have shown that $\operatorname{supp}\mathscr{L}(V)$ is an interval with 0 as its lower endpoint if $\nu_\xi((0,\infty))>0$.

By a similar reasoning, one can show that $\operatorname{supp}\mathscr{L}(V)$ is an interval with $+\infty$ as its upper endpoint if $\nu_\xi((-\infty,0))>0$.

It follows that $\operatorname{supp}\mathscr{L}(V) = [0,\infty)$ if $\nu_\xi((0,\infty))>0$ and $\nu_\xi((-\infty,0))>0$. Now suppose that ξ is of infinite variation with $\nu_\xi((0,\infty))>0$ (but $\nu_\xi((-\infty,0)) = 0$), or $\nu_\xi((-\infty,0))>0$ (but $\nu_\xi((0,\infty)) = 0$). Then there is $\alpha>0$ such that for each $t_1, t_0 > 0$ with $t_1 > t_0$ and $K>0$ the event

$$\{\xi_s \geq -2,\ \forall s \in [0,t_0],\quad \xi_s \geq K,\ \forall s \in [t_0,t_1],\quad \xi_s \geq \alpha s,\ \forall s \geq t_1\}$$

has a positive probability, since $\lim_{t\to\infty} t^{-1}\xi_t$ exists almost surely in $(0,\infty]$ by Doney and Maller [13, Theorems 4.3 and 4.4]) and since $\operatorname{supp}\mathscr{L}(\xi_t) = \mathbb{R}$ for all $t>0$ (cf. [28, Theorem 24.10]). Choosing t_0 small enough and t_1, K big enough, it follows that $0 \in \operatorname{supp}\mathscr{L}(V)$ since $\operatorname{supp}\mathscr{L}(V)$ is closed. On the other hand, since also the event

$$\{\xi_s \leq 2,\ \forall s \in [0,t_2]\}$$

has positive probability for each $t_2 > 0$ as a consequence of the infinite variation of ξ, it follows that $\operatorname{supp} \mathscr{L}(V)$ is unbounded, hence showing that $\operatorname{supp} \mathscr{L}(V) = [0, \infty)$ if ξ is of infinite variation.

Now assume that ξ is of finite variation with drift $b \in \mathbb{R}$, $\nu_\xi((0,\infty)) > 0$ and $\nu_\xi((-\infty, 0)) = 0$. We already know that $0 \in \operatorname{supp} \mathscr{L}(V)$. If $b \leq 0$, then the event $\{\xi_s \leq 2, \ \forall s \in [0, t_2]\}$ has a positive probability for each $t_2 > 0$, and hence $\operatorname{supp} \mathscr{L}(V)$ is unbounded. If $b > 0$, then for each $\varepsilon > 0$ and $t_2 > 0$, the event $\{\xi_s \leq (b+\varepsilon)s, \ \forall s \in [0, t_2]\}$ has a positive probability by Shtatland's result (cf. [28, Theorem 43.20]), so that $\sup \operatorname{supp} \mathscr{L}(V) \geq \int_0^{t_2} e^{-(b+\varepsilon)s}\, ds$ for each $t_2 > 0$ and $\varepsilon > 0$, and hence $\sup \operatorname{supp} \mathscr{L}(V) \geq 1/b$. On the other hand, in that case $V = \int_0^\infty e^{-\xi_s}\, ds \leq \int_0^\infty e^{-bs}\, ds = 1/b$, so that $\operatorname{supp} \mathscr{L}(V) = [0, 1/b]$.

Finally, assume that ξ is of finite variation with drift $b > 0$, $\nu_\xi((0,\infty)) = 0$ and $\nu_\xi((-\infty, 0)) > 0$. Then $\operatorname{supp} \mathscr{L}(V)$ is unbounded and by arguments similar to above, using that $\lim_{t\to\infty} t^{-1}\xi_t = E[\xi_1] \in (0, b)$, we see that $\inf \operatorname{supp} \mathscr{L}(V) = 1/b$, so that $\operatorname{supp} \mathscr{L}(V) = [1/b, \infty)$. This finishes the proof. □

Now we can characterize the support of $\mathscr{L}\left(\int_0^\infty e^{-\xi_{s-}}\, d\eta_s\right)$ when ξ and η are independent Lévy processes. Observe that Theorem 1 below together with Lemma 1 provides a complete characterization of all possible cases.

Theorem 1 *Let ξ and η be two independent Lévy processes such that $V := \int_0^\infty e^{-\xi_{s-}}\, d\eta_s$ converges almost surely.*

(i) Suppose η is of infinite variation, or that $\nu_\eta((0,\infty)) > 0$ and $\nu_\eta((-\infty,0)) > 0$. Then $\operatorname{supp} \mathscr{L}(V) = \mathbb{R}$.

(ii) Suppose η is of finite variation with drift a, $\nu_\eta((0,\infty)) > 0$ and $\nu_\eta((-\infty,0)) = 0$. Then for $a \geq 0$

$$\operatorname{supp} \mathscr{L}(V) = \begin{cases} \left[\frac{a}{b}, \infty\right), & \text{if } \xi \text{ is of finite variation with drift } b > 0 \\ & \text{and } \nu_\xi((0,\infty)) = 0, \\ [0, \infty), & \text{otherwise,} \end{cases}$$

and for $a < 0$

$$\operatorname{supp} \mathscr{L}(V) = \begin{cases} \left[\frac{a}{b}, \infty\right), & \text{if } \xi \text{ is a subordinator with drift } b > 0, \\ \mathbb{R}, & \text{otherwise.} \end{cases}$$

(iii) Suppose η is of finite variation with drift a, $\nu_\eta((0,\infty)) = 0$ and $\nu_\eta((-\infty,0)) > 0$. Then for $a > 0$

$$\operatorname{supp} \mathscr{L}(V) = \begin{cases} \left(-\infty, \frac{a}{b}\right], & \text{if } \xi \text{ is a subordinator with drift } b > 0, \\ \mathbb{R}, & \text{otherwise,} \end{cases}$$

and for $a \le 0$

$$\operatorname{supp} \mathscr{L}(V) = \begin{cases} \left(-\infty, \frac{a}{b}\right], & \textit{if } \xi \textit{ is of finite variation with drift } b > 0 \\ & \textit{and } \nu_\xi((0,\infty)) = 0, \\ (-\infty, 0], & \textit{otherwise.} \end{cases}$$

Proof Denote by $D([0,\infty), \mathbb{R})$ the set of all real valued càdlàg functions on $[0,\infty)$. Since ξ and η are independent, we can condition on $\xi = f$ with $f \in D([0,\infty), \mathbb{R})$ and it follows that, for P_ξ-almost every $f \in D([0,\infty), \mathbb{R})$,

$$V_f := \int_0^\infty e^{-f(s-)}\, d\eta_s = \lim_{T\to\infty} \int_0^T e^{-f(s-)}\, d\eta_s$$

converges almost surely. Hence we can apply the results in [27] for such f, and obtain that V_f is infinitely divisible with Gaussian variance

$$A_f = A_\eta \int_0^\infty e^{-2f(s)}\, ds$$

and Lévy measure ν_f, given by

$$\nu_f(B) = \int_0^\infty ds \int_{\mathbb{R}} \mathbb{1}_B(e^{-f(s)}x)\, \nu_\eta(dx) \quad \text{for } B \in \mathscr{B}(\mathbb{R}^d) \text{ with } 0 \notin B$$

(cf. [27, Theorem 3.10]). In particular, $A_f > 0$ if and only if $A_\eta > 0$, $\nu_f((0,\infty)) > 0$ if and only if $\nu_\eta((0,\infty)) > 0$, and $\nu_f((-\infty,0)) > 0$ if and only if $\nu_\eta((-\infty,0)) > 0$. Further, since $\lim_{s\to\infty} f(s) = +\infty$ P_ξ-a.s.(f), for any $\varepsilon > 0$ we conclude that

$$\nu_f((-\varepsilon,\varepsilon) \setminus \{0\}) = \int_0^\infty \nu_\eta((-e^{f(s)}\varepsilon, e^{f(s)}\varepsilon) \setminus \{0\}) ds = \infty$$

provided that $\nu_\eta \not\equiv 0$. This shows that $0 \in \operatorname{supp} \nu_f$, P_ξ-a.s.(f). It then follows from [28, Theorem 24.10] that

$$\operatorname{supp} \mathscr{L}(V_f) = \mathbb{R}, \quad P_\xi - \text{a.s.}(f)$$

if $A_\eta > 0$, or if $\nu_\eta((0,\infty)) > 0$ and $\nu_\eta((-\infty,0)) > 0$.

Hence in that case $P(V_f \in B | \xi = f) > 0$ P_ξ-a.s.(f) for any open set $B \neq \emptyset$, so that $P(V \in B) = \int P(V_f \in B|\xi = f)\, dP_\xi(f) > 0$. Thus $\operatorname{supp} \mathscr{L}(V) = \mathbb{R}$, which shows (i).

To show (ii), suppose η is of finite variation with drift a, and $\nu_\eta((0,\infty)) > 0$ and $\nu_\eta((-\infty,0)) = 0$. Then, for P_ξ-a.e. f, $V_f \ge a \int_0^\infty e^{-f(s)}\, ds > -\infty$ and hence V_f is of finite variation. It then follows from [27, Theorem 3.15] that V_f has drift

$a\int_0^\infty e^{-f(s)}\,ds$ and [28, Theorem 24.10] gives

$$\operatorname{supp}\mathscr{L}(V_f) = \left[a\int_0^\infty e^{-f(s)}\,ds, \infty\right).$$

Since $P(V \in B) = \int P(V_f \in B|\xi = f)\,dP_\xi(f)$, the assertion (ii) follows from Lemma 1. Finally, (iii) follows from (ii) by replacing η by $-\eta$. □

The following result is now immediate.

Corollary 1 *Let ξ be a Lévy process drifting to $+\infty$, and η another Lévy process, independent of ξ such that $\mathscr{L}(\eta_1) \in D_\xi$. Then $V = \int_0^\infty e^{-\xi_{s-}}\,d\eta_s \geq 0$ a.s. if and only if η is a subordinator.*

Remark 1

(i) Let ξ and η be two independent Lévy processes such that $V = \int_0^\infty e^{-\xi_{s-}}\,d\eta_s$ converges almost surely and consider the associated GOU process $(V_t)_{t\geq 0}$ defined by (2). Then it is easy to see that $V_n = A_n V_{n-1} + B_n$ for each $n \in \mathbb{N}$, where $((A_n, B_n)^T)_{n\in\mathbb{N}}$ is an i.i.d. sequence of bivariate random vectors given by

$$(A_n, B_n)^T = \left(e^{-(\xi_n-\xi_{n-1})}, e^{-(\xi_n-\xi_{n-1})}\int_{(n-1,n]} e^{\xi_{s-}-\xi_{n-1}}\,d\eta_s\right)^T$$

(e.g. [22, Lemma 6.2]). Further, if V_0 is chosen to be independent of $(\xi,\eta)^T$, then $(V_0,\ldots,V_{n-1})^T$ is independent of $((A_k,B_k)^T)_{k\geq n}$ for each n. Since $\mathscr{L}(V)$ is the stationary marginal distribution of the GOU process, it is also the stationary marginal distribution of the random recurrence equation $V_n = A_n V_{n-1} + B_n$, $n \in \mathbb{N}$. We have seen in particular, that the support of $\mathscr{L}(V)$ was always an interval. Hence it is natural to ask if stationary solutions to arbitrary random recurrence equations will always have an interval as its support. We will see that this is not the case. To be more precise, let $\left((A_n,B_n)^T\right)_{n\in\mathbb{N}}$ be a given i.i.d. sequence of bivariate random vectors. Suppose that $(X_n)_{n\in\mathbb{N}_0}$ is a strictly stationary sequence which satisfies the random recurrence equation

$$X_n = A_n X_{n-1} + B_n, \quad n \in \mathbb{N}, \tag{4}$$

such that $(X_0,\ldots,X_{n-1})$ is independent of $\left((A_k,B_k)^T\right)_{k\geq n}$ (provided that such a solution exists) for every $n \in \mathbb{N}$. Then the support of $\mathscr{L}(X_0)$ does not need to be an interval, even if A_n is constant and hence A_n and B_n are independent. To see this, let $A_n = 1/3$ and let $(B_n)_{n\in\mathbb{Z}}$ be an i.i.d. sequence such that $P(B_n = 0) = P(B_n = 2) = \frac{1}{2}$. Then

$$X_n = \sum_{k=0}^{\infty} 3^{-k} B_{n-k}, \quad n \in \mathbb{N}_0, \tag{5}$$

defines a stationary solution of (4), which is unique in distribution. Obviously, the support of $\mathscr{L}(X_0)$ is given by the Cantor set

$$\left\{ \sum_{n=0}^{\infty} 3^{-n} z_n : z_n \in \{0, 2\},\ \forall\, n \in \mathbb{N}_0 \right\},$$

which is totally disconnected and not an interval.

(ii) The stationary solution constructed in (5) is a $1/3$-decomposable distribution (see [28, Definition 64.1] for the definition). By Proposition 6.2 in [4], there exists a bivariate Lévy process $(\xi, \eta)^T$ such that $\xi_t = (\log 3) N_t$ for a Poisson process $(N_t)_{t \geq 0}$ and such that

$$\int_0^\infty e^{-\xi_{s-}}\, d\eta_s = \int_0^\infty 3^{-N_{s-}}\, d\eta_s$$

has the same distribution as X_0 from (5). In particular, its support is not an interval. Hence a similar statement to Theorem 1 does not hold under dependence.

3 Closedness of the Range

This section is devoted to show that, as in the well-known case of a deterministic process ξ, the range $R_\xi = \Phi_\xi(D_\xi)$ is closed under weak convergence. On the contrary, closedness of R_ξ under convolution does not hold any more as will be shown in Corollary 2 below.

It will also follow that the inverse mapping $(\Phi_\xi)^{-1}$ is continuous, provided that Φ_ξ is injective. Recall that Φ_ξ is injective if, for instance, ξ is spectrally negative (cf. [4, Theorem 5.3]). Further, for any ξ drifting to $+\infty$, Φ_ξ is always injective when restricted to positive measures $\mathscr{L}(\eta_1)$ [4, Remark 5.4]. Thus, although Φ_ξ need not be continuous (which follows by an argument similar to [4, Example 7.1]), the inverse of Φ_ξ restricted to positive measures will turn out to be always continuous.

We start with the following proposition, which shows that the mapping Φ_ξ is closed.

Proposition 1 *Let ξ be a Lévy process drifting to $+\infty$. Then the mapping Φ_ξ is closed in the sense that if $\mathscr{L}(\eta_1^{(n)}) \in D_\xi$, $\eta_1^{(n)} \xrightarrow{d} \eta_1$ and $\Phi_\xi(\mathscr{L}(\eta_1^{(n)})) \xrightarrow{w} \mu$ for some random variable η_1 and probability measure μ as $n \to \infty$, then $\mathscr{L}(\eta_1) \in D_\xi$ and $\Phi_\xi(\mathscr{L}(\eta_1)) = \mu$.*

Proof For $n \in \mathbb{N}$, let $W^{(n)}$ be a random variable such that

$$W^{(n)} \stackrel{d}{=} \int_0^\infty e^{-\xi_{s-}}\, d\eta_s^{(n)} \quad \text{and } W^{(n)} \text{ is independent of } (\xi, \eta^{(n)})^T,$$

where $\eta^{(n)}$ is a Lévy process induced by $\eta_1^{(n)}$ independent of ξ. Then the limit $\mathscr{L}(\eta_1)$ is infinitely divisible by Sato [28, Lemma 7.8]) and we can define η as a Lévy process induced by η_1, independent of ξ. Let W be a random variable with distribution μ, independent of $(\xi, \eta)^T$. The proof of Behme and Lindner [4, Theorem 7.3], more precisely the part leading to Eq. (7.12) there, then shows that for every $t > 0$ we have

$$\left(e^{-\xi_t}, \int_0^t e^{\xi_{s-}}\, d\eta_s^{(n)}\right)^T \xrightarrow{d} \left(e^{-\xi_t}, \int_0^t e^{\xi_{s-}}\, d\eta_s\right)^T, \quad n \to \infty.$$

Due to independence this yields

$$\left(W^{(n)}, e^{-\xi_t}, \int_0^t e^{\xi_{s-}}\, d\eta_s^{(n)}\right)^T \xrightarrow{d} \left(W, e^{-\xi_t}, \int_0^t e^{\xi_{s-}}\, d\eta_s\right)^T, \quad n \to \infty,$$

and since $\mathscr{L}(W^{(n)})$ is the invariant distribution of the GOU process driven by $(\xi, \eta^{(n)})^T$, this implies

$$W^{(n)} \stackrel{d}{=} e^{-\xi_t}\left(W^{(n)} + \int_0^t e^{\xi_{s-}}\, d\eta_s^{(n)}\right) \xrightarrow{d} e^{-\xi_t}\left(W + \int_0^t e^{\xi_{s-}}\, d\eta_s\right), \quad n \to \infty.$$

Since also $W^{(n)} \xrightarrow{d} W$ as $n \to \infty$, this shows that

$$W \stackrel{d}{=} e^{-\xi_t}\left(W + \int_0^t e^{\xi_{s-}}\, d\eta_s\right)$$

for any $t > 0$, so that $\mu = \mathscr{L}(W)$ is an invariant distribution of the GOU process driven by $(\xi, \eta)^T$. By Lindner and Maller [22, Theorem 2.1], or alternatively [6, Theorem 2.1 (a)], this shows that $\int_0^\infty e^{-\xi_{s-}}\, d\eta_s$ converges a.s., i.e. $\mathscr{L}(\eta_1) \in D_\xi$, and that

$$\mu = \mathscr{L}(W) = \mathscr{L}\left(\int_0^\infty e^{-\xi_{s-}}\, d\eta_s\right) = \Phi_\xi(\mathscr{L}(\eta_1)),$$

giving the claim. □

In order to show that R_ξ is closed, we shall first show in Proposition 2 below that if a sequence $(\Phi_\xi(\mathscr{L}(\eta_1^{(n)})))_{n\in\mathbb{N}}$ is tight, then $(\eta_1^{(n)})_{n\in\mathbb{N}}$ is tight. To achieve this, observe first that as a consequence of Kallenberg [19, Lemma 15.15] and Prokhorov's theorem, a sequence $(\mathscr{L}(\eta_1^{(n)}))_{n\in\mathbb{N}}$ of infinitely divisible distributions on $\mathbb{R}$ with characteristic triplets $(\gamma_n, \sigma_n^2, \nu_n)$ is tight if and only if

$$\sup_{n\in\mathbb{N}} \left|\gamma_n + \int_{\mathbb{R}} x\left(\frac{1}{1+x^2} - \mathbb{1}_{|x|\le 1}\right) \nu_n(dx)\right| < \infty$$

and the sequence $(\tilde{\nu}_n)_{n\in\mathbb{N}}$ of finite positive measures on $\mathbb{R}$ with

$$\tilde{\nu}_n(dx) = \sigma_n^2\,\delta_0(dx) + \frac{x^2}{1+x^2}\nu_n(dx)$$

is weakly relatively compact (in particular, this implies that $\sup_{n\in\mathbb{N}} \tilde{\nu}_n(\mathbb{R}) < \infty$). Using Prokhorov's theorem for finite measures (e.g. [1, Theorem 7.8.7]), it is easy to see that this is equivalent to

$$\sup_{n\in\mathbb{N}} \sigma_n^2 < \infty, \tag{6}$$

$$\sup_{n\in\mathbb{N}} \int_{[-1,1]} x^2\,\nu_n(dx) < \infty, \tag{7}$$

$$\sup_{n\in\mathbb{N}} \nu_n(\mathbb{R}\setminus[-r,r]) < \infty, \quad \forall\, r > 0, \tag{8}$$

$$\lim_{r\to\infty} \sup_{n\in\mathbb{N}} \nu_n(\mathbb{R}\setminus[-r,r]) = 0, \quad \text{and} \tag{9}$$

$$\sup_{n\in\mathbb{N}} |\gamma_n| < \infty. \tag{10}$$

The following lemma gives direct uniform estimates for $\mu([-r,r])$ in terms of the Lévy measure or Gaussian variance of an infinitely divisible distribution μ which will be needed to prove Proposition 2.

Lemma 2 *Let μ be an infinitely divisible distribution on $\mathbb{R}$ with characteristic triplet (γ, σ^2, ν). For $\varepsilon \in (0,1)$ denote by I_ε the set*

$$I_\varepsilon := \{z \in \mathbb{R} : 1 - \cos z \geq \varepsilon\}.$$

Then for any $p \in (0,1)$ and $a > 0$, there is some $\varepsilon = \varepsilon(a,p) \in (0,1)$ such that

$$\frac{\lambda^1(I_\varepsilon \cap [-y,y])}{\lambda^1([-y,y])} \geq 1 - p, \quad \forall\, y \geq a, \tag{11}$$

where λ^1 denotes the Lebesgue measure on $\mathbb{R}$. For $\delta > 0$, denote by

$$\|\nu\|_\delta := \nu(\mathbb{R}\setminus[-\delta,\delta])$$

the total mass of $\nu_{|\mathbb{R}\setminus[-\delta,\delta]}$ and

$$M(\nu) := \int_{[-1,1]} x^2\,\nu(dx).$$

Further, let $c > 0$ be a constant such that

$$\cos(t) - 1 \leq -ct^2, \quad \forall\, t \in [-1, 1].$$

Then

$$\mu([-r, r]) \leq 4(e^{-\varepsilon(\delta/r, p)\|\nu\|_\delta}(1-p) + p), \quad \forall\, p \in (0,1), r, \delta > 0, \tag{12}$$

$$\mu([-r, r]) \leq 1 - \min\{e^{-\|\nu\|_{2r}}, 1 - e^{-\|\nu\|_{2r}/2}\}, \quad \forall\, r > 0, \tag{13}$$

$$\mu([-r, r]) \leq 2r \int_{-1/r}^{1/r} e^{-M(\nu)ct^2}\, dt, \quad \forall\, r \geq 1, \tag{14}$$

and

$$\mu([-r, r]) \leq 2r \int_{-1/r}^{1/r} e^{-\sigma^2 t^2/2}\, dt, \quad \forall\, r > 0. \tag{15}$$

Proof Equation (11) is clear. Let $r > 0$. Then an application of Kallenberg [19, Lemma 5.1] shows

$$\mu([-r, r]) \leq 2r \int_{-1/r}^{1/r} |\hat{\mu}(t)|\, dt = 2r \int_{-1/r}^{1/r} \exp\left(-\sigma^2 t^2/2 + \int_{\mathbb{R}} (\cos(xt) - 1)\nu(dx)\right) dt \tag{16}$$

which immediately gives (15). Let $\delta > 0$. Equation (12) is trivial when $\|\nu\|_\delta = 0$, and for $\|\nu\|_\delta > 0$ observe that by (16) and Jensen's inequality we can estimate

$$\begin{aligned}
\mu([-r, r]) &\leq 2r \int_{-1/r}^{1/r} \exp\left(\int_{|x|>\delta} (\cos(xt) - 1)\|\nu\|_\delta \frac{\nu(dx)}{\|\nu\|_\delta}\right) dt \\
&\leq 2r \int_{-1/r}^{1/r} \left(\int_{|x|>\delta} e^{(\cos(xt)-1)\|\nu\|_\delta} \frac{\nu(dx)}{\|\nu\|_\delta}\right) dt \\
&= \int_{|x|>\delta} \left(\frac{2r}{|x|} \int_{-|x|/r}^{|x|/r} e^{(\cos z - 1)\|\nu\|_\delta}\, dz\right) \frac{\nu(dx)}{\|\nu\|_\delta}.
\end{aligned} \tag{17}$$

By (11) we estimate for $|x| \geq \delta$ and $p \in (0,1)$ with $\varepsilon = \varepsilon(\delta/r, p)$

$$\begin{aligned}
&\frac{2r}{|x|} \int_{-|x|/r}^{|x|/r} e^{(\cos z - 1)\|\nu\|_\delta}\, dz \\
&\leq \frac{4}{\lambda^1([-\frac{|x|}{r}, \frac{|x|}{r}])} \left(e^{-\varepsilon\|\nu\|_\delta} \lambda^1\left([-\frac{|x|}{r}, \frac{|x|}{r}] \cap I_\varepsilon\right) + \lambda^1\left([-\frac{|x|}{r}, \frac{|x|}{r}] \setminus I_\varepsilon\right)\right) \\
&\leq 4(e^{-\varepsilon\|\nu\|_\delta}(1-p) + p),
\end{aligned}$$

which together with (17) results in (12). Similarly, (14) is trivial when $M(\nu) = 0$, while for $M(\nu) > 0$ define the finite measure ρ on $[-1, 1]$ by $\rho(dx) = x^2\nu(dx)$. We then estimate with (16) and Jensen's inequality, for $r \geq 1$,

$$\begin{aligned}
\mu([-r,r]) &\leq 2r \int_{-1/r}^{1/r} \exp\left(\int_{[-1,1]} \frac{\cos(xt)-1}{x^2} M(\nu) \frac{\rho(dx)}{M(\nu)} \right) dt \\
&\leq 2r \int_{-1/r}^{1/r} \left(\int_{[-1,1]} \exp\left(\frac{\cos(xt)-1}{x^2} M(\nu) \right) \frac{\rho(dx)}{M(\nu)} \right) dt \\
&\leq 2r \int_{-1/r}^{1/r} \left(\int_{[-1,1]} e^{-ct^2 M(\nu)} \frac{\rho(dx)}{M(\nu)} \right) dt,
\end{aligned}$$

which gives (14). Finally, let us prove Eq. (13). This is again trivial when $\|\nu\|_{2r} = 0$, so assume $\|\nu\|_{2r} > 0$. By symmetry, we can assume without loss of generality that

$$\nu((-\infty, -2r)) \geq \|\nu\|_{2r}/2 > 0.$$

Let $(X_t)_{t\geq 0}$ be a Lévy process with $\mathscr{L}(X_1) = \mu$, and define

$$Y_t := \sum_{0<s\leq t, \Delta X_s < -2r} \Delta X_s \quad \text{and} \quad Z_t := X_t - Y_t, \quad t \in \mathbb{R},$$

where $\Delta X_s := X_s - X_{s-}$ denotes the jump size of X at time s. Then $(Y_t)_{t\geq 0}$ and $(Z_t)_{t\geq 0}$ are two independent Lévy processes, and $(Y_t)_{t\geq 0}$ is a compound Poisson process with Lévy measure $\nu_{|(-\infty,-2r)}$. Denote by $(N_t)_{t\geq 0}$ the underlying Poisson process in $(Y_t)_{t\geq 0}$ which counts the number of jumps of $(Y_t)_{t\geq 0}$. Then

$$\begin{aligned}
\mu(\mathbb{R} \setminus [-r,r]) &= P(|Y_1 + Z_1| > r) \\
&\geq P(|Z_1| > r, Y_1 = 0) + P(|Z_1| \leq r, Y_1 < -2r) \\
&= P(|Z_1| > r)\, P(N_1 = 0) + P(|Z_1| \leq r)\, P(N_1 \geq 1) \\
&= P(|Z_1| > r) e^{-\nu((-\infty,-2r))} + (1 - P(|Z_1| > r))(1 - e^{-\nu((-\infty,-2r))}) \\
&\geq \min\{e^{-\nu((-\infty,-2r))}, 1 - e^{-\nu((-\infty,-2r))}\} \\
&\geq \min\{e^{-\|\nu\|_{2r}}, 1 - e^{-\|\nu\|_{2r}/2}\},
\end{aligned}$$

which implies (13). □

The next result is the key step in proving closedness of R_ξ.

Proposition 2 *Let ξ be a Lévy process drifting to $+\infty$ and $(\mathscr{L}(\eta_1^{(n)}))_{n\in\mathbb{N}}$ be a sequence in D_ξ such that $(\mu_n := \Phi_\xi(\mathscr{L}(\eta_1^{(n)})))_{n\in\mathbb{N}}$ is tight. Then also $(\eta_1^{(n)})_{n\in\mathbb{N}}$ is tight.*

Proof Denote by $(\gamma_n, \sigma_n^2, \nu_n)$ the characteristic triplet of $\eta_1^{(n)}$. We have to show that conditions (6)–(10) are satisfied. Let $n \in \mathbb{N}$. Since $\int_0^\infty e^{-\xi_{s-}}\, d\eta_s^{(n)}$ converges almost surely and since $\eta^{(n)}$ and ξ are independent, conditioning on $\xi = f$ shows that

$$\left(\int_0^\infty e^{-\xi_{s-}}\, d\eta_s^{(n)} \middle| \xi = f\right) = \int_0^\infty e^{-f(s-)}\, d\eta_s^{(n)}$$

for P_ξ-almost every $f \in D([0,\infty), \mathbb{R})$, where the integral $\int_0^\infty e^{-f(s-)}\, d\eta_s^{(n)}$ converges almost surely for each such f. Further, since $\sup_{s\in[0,1]} |\xi_s| < \infty$ a.s. by the càdlàg paths of ξ, there are $0 < D_1 \le 1 \le D_2 < \infty$ such that

$$P\left(D_1 \le e^{-\xi_s} \le D_2 \;\forall\, s \in [0,1]\right) \ge 1/2.$$

Consequently there are some measurable sets $A_n \subset D([0,\infty), \mathbb{R})$ with $P_\xi(A_n) \ge 1/2$ such that

$$D_1 \le e^{-f(s)} \le D_2 \quad \forall\, f \in A_n,\; s \in [0,1],$$

and

$$\int_0^\infty e^{-f(s-)}\, d\eta_s^{(n)} \text{ converges a.s., } \forall f \in A_n.$$

Further, we obtain

$$\begin{aligned} \mu_n(\mathbb{R} \setminus [-r,r]) &\ge \int_{A_n} P\left(\left|\int_0^\infty e^{-f(s-)}\, d\eta_s^{(n)}\right| > r\right) P_\xi(df) \\ &\ge \frac{1}{2}\left(1 - \sup_{f\in A_n} P\left(\left|\int_0^\infty e^{-f(s-)}\, d\eta_s^{(n)}\right| \le r\right)\right). \end{aligned} \tag{18}$$

For fixed $f \in A_n$ the distribution of $\int_0^\infty e^{-f(s-)}\, d\eta_s^{(n)}$ is infinitely divisible with Gaussian variance

$$\sigma_{f,n}^2 = \sigma_n^2 \int_0^\infty (e^{-f(s)})^2\, ds \ge D_1^2 \sigma_n^2 \tag{19}$$

and Lévy measure $\nu_{f,n}$ satisfying

$$\nu_{f,n}(B) = \int_0^\infty ds \int_{\mathbb{R}} \mathbb{1}_B(e^{-f(s)}x)\, \nu_n(dx) \tag{20}$$

for any Borel set $B \subset \mathbb{R} \setminus \{0\}$ (cf. [27, Theorem 3.10]). In particular, for $f \in A_n$ and any $\delta > 0$,

$$\nu_{f,n}(\mathbb{R} \setminus [-\delta, \delta]) \geq \int_0^1 ds \int_{\mathbb{R}} \mathbb{1}_{\mathbb{R}\setminus[-\delta e^{f(s)}, \delta e^{f(s)}]}(x)\, \nu_n(dx)$$
$$\geq \nu_n(\mathbb{R} \setminus [-\delta/D_1, \delta/D_1]). \tag{21}$$

From (20) we obtain

$$\int_{[-1,1]} t^2 \nu_{f,n}(dt) = \int_0^\infty ds \int_{\mathbb{R}} \left(e^{-f(s)}x\right)^2 \mathbb{1}_{\{|e^{-f(s)}x| \leq 1\}}(x)\, \nu_n(dx),$$

for $f \in A_n$, hence

$$\int_{[-1,1]} t^2 \nu_{f,n}(dt) \geq D_1^2 \int_{[-1,1]} x^2 \mathbb{1}_{\{|D_2 x| \leq 1\}}(x) \nu_n(dx)$$
$$= D_1^2 \int_{[-1/D_2, 1/D_2]} x^2 \nu_n(dx). \tag{22}$$

Now suppose (6) were violated. Then by (18), (15) and (19) we conclude that

$$\sup_{n\in\mathbb{N}} \{\mu_n(\mathbb{R} \setminus [-r, r])\} \geq \frac{1}{2} \sup_{n\in\mathbb{N}} \left\{ 1 - 2r \int_{-1/r}^{1/r} e^{-D_1^2 \sigma_n^2 t^2/2}\, dt \right\} = \frac{1}{2}$$

for every $r > 0$, contradicting tightness of $(\mu_n)_{n\in\mathbb{N}}$. Hence (6) must be true.

Now suppose that (8) were violated, so that there is some $\delta > 0$ such that $\sup_{n\in\mathbb{N}} \|\nu_n\|_\delta = \infty$ with the notions of Lemma 2. Let $p \in (0, 1/4)$ be arbitrary. Then by (12) and (21), we have for every $f \in A_n$, with $\epsilon = \epsilon(D_1\delta/r, p)$ as defined in Lemma 2, that

$$P\left(\left| \int_0^\infty e^{-f(s-)}\, d\eta_s^{(n)} \right| \leq r \right) \leq 4 \left(e^{-\varepsilon(D_1\delta/r, p)\|\nu_{f,n}\|_{D_1\delta}} (1-p) + p \right)$$
$$\leq 4 e^{-\varepsilon(D_1\delta/r,p)\|\nu_n\|_\delta} (1-p) + 4p. \tag{23}$$

From (18) we then obtain that

$$\sup_{n\in\mathbb{N}} \{\mu_n(\mathbb{R} \setminus [-r, r])\} \geq \frac{1}{2}(1 - 4p) > 0, \quad \forall\, r > 0,$$

which again contradicts tightness of $(\mu_n)_{n\in\mathbb{N}}$ so that (8) must hold.

Now suppose that (9) were violated. Then there is some $a > 0$ and a sequence $(\delta_k)_{k\in\mathbb{N}}$ of positive real numbers tending to $+\infty$ and an index $n(k) \in \mathbb{N}$ for each k

such that

$$\|\nu_{n(k)}\|_{2\delta_k/D_1} \geq a, \quad \forall\, k \in \mathbb{N}.$$

Let $p \in (0, 1/4)$ be arbitrary and choose $\varepsilon = \varepsilon(D_1, p)$ as in Lemma 2. Let $b > 0$ be such that

$$b_1 := 4\left(e^{-\varepsilon(D_1,p)b}(1-p)+p\right) < 1.$$

Let $f \in A_n$. Then if $\|\nu_{f,n(k)}\|_{D_1\delta_k} \geq b$ we have

$$P\left(\left|\int_0^\infty e^{-f(s-)}\, d\eta_s^{(n(k))}\right| \leq \delta_k\right) \leq b_1 < 1$$

by (23), while if $\|\nu_{f,n(k)}\|_{D_1\delta_k} < b$ we obtain from (13) and (21) that

$$\begin{aligned} P\left(\left|\int_0^\infty e^{-f(s-)}\, d\eta_s^{(n(k))}\right| \leq \delta_k\right) &\leq 1 - \min\{e^{-\|\nu_{f,n(k)}\|_{2\delta_k}}, 1 - e^{-\|\nu_{f,n(k)}\|_{2\delta_k}/2}\} \\ &\leq 1 - \min\{e^{-b}, 1 - e^{-\|\nu_{n(k)}\|_{2\delta_k/D_1}/2}\} \\ &\leq 1 - \min\{e^{-b}, 1 - e^{-a/2}\}. \end{aligned}$$

From (18) we then conclude

$$\mu_{n(k)}(\mathbb{R} \setminus [-\delta_k, \delta_k]) \geq \frac{1}{2}\left(1 - \max\{b_1, 1-e^{-b}, e^{-a/2}\}\right) > 0 \quad \forall\, k \in \mathbb{N}.$$

In particular,

$$\limsup_{r\to\infty} \sup_{n\in\mathbb{N}} \{\mu_n(\mathbb{R} \setminus [-r, r])\} \geq \frac{1}{2}\left(1 - \max\{b_1, 1-e^{-b}, e^{-a/2}\}\right) > 0,$$

which again contradicts tightness of $(\mu_n)_{n\in\mathbb{N}}$. We conclude that also (9) must be valid.

Now suppose that (7) were violated, but (8) holds. Then by (18), (14), (22) and with c from Lemma 2 we have for every $r \geq 1$

$$\begin{aligned} &\sup_{n\in\mathbb{N}} \{\mu_n(\mathbb{R} \setminus [-r, r])\} \\ &\geq \frac{1}{2} \sup_{n\in\mathbb{N}} \left\{1 - 2r\int_{-1/r}^{1/r} \exp\left(-D_1^2 c t^2 \int_{[-1/D_2, 1/D_2]} x^2\, \nu_n(dx)\right) dt\right\} = \frac{1}{2}, \end{aligned}$$

where we have used that (8) together with $\sup_{n\in\mathbb{N}} \int_{[-1,1]} x^2\, \nu_n(dx) = \infty$ imply $\sup_{n\in\mathbb{N}} \int_{[-1/D_2,1/D_2]} x^2\, \nu_n(dx) = \infty$. This again contradicts tightness of $(\mu_n)_{n\in\mathbb{N}}$ so that (7) must hold.

Finally, suppose that (10) were violated but that (6)–(9) hold. Then there is a subsequence of $(\gamma_n)_{n\in\mathbb{N}}$ which diverges to $+\infty$ or $-\infty$, and without loss of generality assume that this is $(\gamma_n)_{n\in\mathbb{N}}$. Since $(\mu_n)_{n\in\mathbb{N}}$ is tight by assumption, there is a subsequence of $(\mu_n)_{n\in\mathbb{N}}$ which converges weakly, and for the convenience of notation assume again that $(\mu_n)_{n\in\mathbb{N}}$ converges weakly to some distribution μ. Let the Lévy process U with characteristic triplet $(\gamma_U, \sigma_U^2, \nu_U)$ be related to ξ by $\mathscr{E}(U)_t = e^{-\xi_t}$, where $\mathscr{E}(U)$ denotes the stochastic exponential of U. Then it follows from [4, Corollary 3.2 and Eq. (4.1)] that

$$\begin{aligned}
\gamma_n \int_{\mathbb{R}} f'(x)\, \mu_n(dx) = & -\frac{1}{2}\sigma_n^2 \int_{\mathbb{R}} f''(x)\, \mu_n(dx) \\
& - \int_{\mathbb{R}} \mu_n(dx) \int_{\mathbb{R}} \left(f(x+y) - f(x) - f'(x)y\mathbb{1}_{|y|\le 1}\right) \nu_n(dy) \\
& - \gamma_U \int_{\mathbb{R}} f'(x)x\, \mu_n(dx) - \frac{1}{2}\sigma_U^2 \int_{\mathbb{R}} f''(x)x^2\, \mu_n(dx) \\
& - \int_{\mathbb{R}} \mu_n(dx) \int_{\mathbb{R}} \left(f(x+xy) - f(x) - f'(x)xy\mathbb{1}_{|y|\le 1}\right) \nu_U(dy)
\end{aligned}$$

for every function $f \in C_c^2(\mathbb{R})$. Consider the right hand side of this equation. The first summand remains bounded in n by (6) and weak convergence of μ_n, and the second remains bounded in n by (7) and (8), since

$$|f(x+y) - f(x) - f'(x)\mathbb{1}_{|y|\le 1}| \le 2\|f\|_\infty \mathbb{1}_{|y|>1} + \|f''\|_\infty y^2 \mathbb{1}_{|y|\le 1}$$

(cf. [4, Proof of Lemma 4.2]), where $\|\cdot\|_\infty$ denotes the supremum norm. The third and fourth summands converge by weak convergence of μ_n, and the fifth summand remains bounded in n by Behme and Lindner [4, Eq. (3.6)] (actually, the fifth summand can be shown to converge). We conclude also that $\gamma_n \int_{\mathbb{R}} f'(x)\mu_n(dx)$ must be bounded in n for every $f \in C_c^2(\mathbb{R})$. Choosing $f \in C_c^2(\mathbb{R})$ such that $\int_{\mathbb{R}} f'(x)\, \mu(dx) \neq 0$, we obtain that $(\gamma_n)_{n\in\mathbb{N}}$ must be bounded and hence the desired contradiction. Summing up, we have verified (6)–(10) so that $(\eta_1^{(n)})_{n\in\mathbb{N}}$ must be tight. □

Now define

$$\begin{aligned}
D_\xi^+ &:= \{\mathscr{L}(\eta_1) \in D_\xi : \eta_1 \ge 0 \text{ a.s.}\}, \\
\Phi_\xi^+ &:= (\Phi_\xi)|_{D_\xi^+},
\end{aligned}$$

and

$$R_\xi^+ := \Phi_\xi(D_\xi^+) = \Phi_\xi^+(D_\xi^+).$$

By Corollary 1,

$$R_\xi^+ = R_\xi \cap \{\mu \in \mathscr{P}(\mathbb{R}) : \operatorname{supp}\mu \subset [0,\infty)\}.$$

We now show closedness of R_ξ under weak convergence and that the inverse of Φ_ξ (provided that it exists) is continuous.

Theorem 2 *Let $\xi = (\xi_t)_{t\geq 0}$ be a Lévy process drifting to $+\infty$.*

(i) Then R_ξ and R_ξ^+ are closed under weak convergence.
(ii) If Φ_ξ is injective, then the inverse $\Phi_\xi^{-1} : R_\xi \to D_\xi$ is continuous with respect to the topology induced by weak convergence.
(iii) The inverse $(\Phi_\xi^+)^{-1} : R_\xi^+ \to D_\xi^+$ is continuous.

Proof

(i) Let $(\mu_n = \Phi_\xi(\mathscr{L}(\eta_1^{(n)})))_{n\in\mathbb{N}}$ be a sequence in R_ξ which converges weakly to some $\mu \in \mathscr{P}(\mathbb{R})$. Then $(\mu_n)_{n\in\mathbb{N}}$ is tight, and by Proposition 2, $(\eta_1^{(n)})_{n\in\mathbb{N}}$ must be tight, too. Hence there is a subsequence $(\eta_1^{(n_k)})_{k\in\mathbb{N}}$ which converges weakly to some random variable η_1. It then follows from Proposition 1 that also $\mathscr{L}(\eta_1) \in D_\xi$ and that $\Phi_\xi(\mathscr{L}(\eta_1)) = \mu$. Hence $\mu \in R_\xi$ so that R_ξ is closed. Since $\{\mu \in \mathscr{P}(\mathbb{R}) : \operatorname{supp}\mu \subset [0,\infty)\}$ is closed, this gives also closedness of R_ξ^+.

(ii) Let $(\mu_n = \Phi_\xi(\mathscr{L}(\eta_1^{(n)})))_{n\in\mathbb{N}}$ be a sequence in R_ξ which converges weakly to some μ. By Proposition 2, $(\eta_1^{(n)})_{n\in\mathbb{N}}$ is tight. Let $(\eta_1^{(k_n)})_{k\in\mathbb{N}}$ be a subsequence which converges weakly to some η_1, say. Then $\mathscr{L}(\eta_1) \in D_\xi$ and $\Phi_\xi(\mathscr{L}(\eta_1)) = \mu$ by Proposition 1, and since Φ_ξ is injective we have $\mathscr{L}(\eta_1) = \Phi_\xi^{-1}(\mu)$. Since the convergent subsequence was arbitrary, this shows that $\mathscr{L}(\eta_1^{(n)}) = \Phi_\xi^{-1}(\mu_n)$ converges weakly to $\Phi_\xi^{-1}(\mu)$ as $n \to \infty$ (cf. [9, Corollary to Theorem 25.10]). Hence Φ_ξ is continuous.

(iii) This can be proved in complete analogy to (ii). □

Remark 2 Closedness of R_ξ^+ under weak convergence and continuity of $(\Phi_\xi^+)^{-1}$ can also be proved in a simpler way by circumventing Proposition 2 but using a formula for the Laplace transforms of $\eta_1^{(n)}$ and μ_n (cf. [4, Remark 4.5], or Theorem 3 below), and showing that $\mu^{(n)} \xrightarrow{w} \mu$ implies convergence of the Laplace transforms of $\eta_1^{(n)}$. A similar approach for showing closedness of R_ξ is not evident since there is not a similarly convenient formula for the Fourier transforms available, but only one in terms of suitable two-sided limits (cf. [4, Eq. (4.7)]).

As a consequence of Theorem 2, we can now show that R_ξ will not be closed under convolution if ξ is non-deterministic and satisfies a suitable moment condition. We conjecture that R_ξ will never be closed under convolution unless ξ is deterministic.

Corollary 2 *Let $\xi = (\xi_t)_{t\geq 0}$ be a non-deterministic Lévy process drifting to $+\infty$ such that $E[(e^{-2\xi_1})] < 1$. Then R_ξ is not closed under convolution.*

Proof Let $(\eta_t)_{t\geq 0}$ be a symmetric compound Poisson process with Lévy measure $\nu = \delta_{-1} + \delta_1$, where δ_a denotes the Dirac measure at a. Then $\mathscr{L}(\eta_1) \in D_\xi$ and $V := \int_0^\infty e^{-\xi_{s-}} d\eta_s$ is symmetric, too, and since by Behme [3, Theorem 3.3] we have $E[V^2] < \infty$, this yields $E[V] = 0$. Now let $(V_i)_{i\in\mathbb{N}}$ be an i.i.d. family of independent copies of V. Then by the Central Limit Theorem,

$$\mathscr{L}\left(n^{-\frac{1}{2}}(V_1 + \cdots + V_n)\right) \to \mathscr{N}(0, \operatorname{Var}(V)), \quad n \to \infty,$$

with $\operatorname{Var}(V) \neq 0$. If the range R_ξ was closed under convolution, we consequently had $\mathscr{L}(n^{-\frac{1}{2}}(V_1 + \cdots + V_n)) \in R_\xi$ and due to closedness of R_ξ under weak convergence this gave $\mathscr{N}(0, \operatorname{Var}(V)) \in R_\xi$. This contradicts [4, Theorem 6.4]. □

4 A General Criterion for a Positive Distribution to be in the Range

From this section on, we restrict ourselves to positive distributions in R_ξ, i.e. we only consider Φ_ξ^+ and R_ξ^+ as defined in Sect. 3. We start by giving a general criterion to decide whether a positive distribution is in the range R_ξ^+ of Φ_ξ^+ for a given Lévy process ξ.

Theorem 3 *Let ξ be a Lévy process drifting to $+\infty$. Let $\mu = \mathscr{L}(V)$ be a probability measure on $[0,\infty)$ with Laplace exponent ψ_V. Then $\mu \in R_\xi^+$ if and only if the function*

$$g_\mu : (0,\infty) \to \mathbb{R}$$

$$g_\mu(u) := (\gamma_\xi - \frac{\sigma_\xi^2}{2})u\psi_V'(u) - \frac{\sigma_\xi^2}{2}u^2\left(\psi_V''(u) + (\psi_V'(u))^2\right)$$
$$- \int_{\mathbb{R}} \left(e^{\psi_V(ue^{-y}) - \psi_V(u)} - 1 + u\psi_V'(u)y\mathbb{1}_{|y|\leq 1}\right)\nu_\xi(dy), \quad u > 0, \tag{24}$$

defines the Laplace exponent of some subordinator η, i.e. if there is some subordinator η such that

$$E[e^{-\eta_1 u}] = e^{g_\mu(u)}, \quad \forall\, u > 0. \tag{25}$$

In that case, $\Phi_\xi(\mathscr{L}(\eta_1)) = \mu$.

Using a Taylor expansion for $|y| \leq 1$, it is easy to see that the integral defining g_μ converges for every distribution μ on $[0,\infty)$.

Proof Observe first that

$$-E\left[Ve^{-uV}\right] = \mathbb{L}_V'(u) = \psi_V'(u)e^{\psi_V(u)} \tag{26}$$

$$E\left[V^2e^{-uV}\right] = \mathbb{L}_V''(u) = \psi_V''(u)e^{\psi_V(u)} + (\psi_V'(u))^2e^{\psi_V(u)} \tag{27}$$

for $u > 0$. Hence

$$\begin{aligned} g_\mu(u)\mathbb{L}_V(u) &= -u\gamma_\xi E\left[Ve^{-uV}\right] - \frac{\sigma_\xi^2}{2}\left(E\left[V^2e^{-uV}\right]u^2 - E\left[Ve^{-uV}\right]u\right) \\ &\quad - \int_{(-1,\infty)} \left(\mathbb{L}_V(ue^{-y}) - \mathbb{L}_V(u) - uE\left[Ve^{-uV}\right]y\mathbb{1}_{|y|\leq 1}\right)\nu_\xi(dy), \ \forall\, u > 0. \end{aligned} \tag{28}$$

Now if $\mu = \mathscr{L}(V) \in R_\xi^+$, let $\mathscr{L}(\eta_1) \in D_\xi^+$ such that $\mu = \mathscr{L}(V) = \Phi_\xi(\mathscr{L}(\eta_1))$. Then $g_\mu = \log \mathbb{L}_\eta$ by Remark 4.5 of Behme and Lindner [4], so that (25) is satisfied.

Conversely, suppose that $V \geq 0$, and let η be a subordinator such that (25) is true. Define the Lévy process U by $e^{-\xi_t} = \mathscr{E}(U)_t$, where U denotes the stochastic exponential of U. Then by Behme and Lindner [4, Remark 4.5] and (28), (25) is equivalent to

$$\begin{aligned} \log \mathbb{L}_\eta(u)\,\mathbb{L}_V(u) &= u\gamma_U E\left[Ve^{-uV}\right] - \frac{\sigma_U^2u^2}{2}E\left[V^2e^{-uV}\right] \\ &\quad - \int_{(-1,\infty)} \left(\mathbb{L}_V(u(1+y)) - \mathbb{L}_V(u) + uE\left[Ve^{-uV}\right]y\mathbb{1}_{|y|\leq 1}\right)\nu_U(dy), \\ &\quad \forall\, u > 0, \end{aligned}$$

and a direct computation using (26) and (27) shows that this in turn is equivalent to

$$\begin{aligned} 0 &= \int_{[0,\infty)} \left(f'(x)(x\gamma_U + \gamma_\eta^0) + \frac{1}{2}f''(x)x^2\sigma_U^2\right)\mu(dx) \\ &\quad + \int_{[0,\infty)} \mu(dx) \int_{(-1,\infty)} \left(f(x+xy) - f(x) - f'(x)xy\mathbb{1}_{|y|\leq 1}\right)\nu_U(dy) \\ &\quad + \int_{[0,\infty)} \mu(dx) \int_{[0,\infty)} \left(f(x+y) - f(x)\right)\nu_\eta(dy) \end{aligned} \tag{29}$$

for all functions $f \in \mathscr{G} := \{h \in C_b^2(\mathbb{R}) : \exists\, u > 0 \text{ such that } h(x) = e^{-ux},\ \forall x \geq 0\}$, where γ_η^0 denotes the drift of η. Observe that (29) is also trivially true for $f \equiv 1$.

Denote by

$$\mathbf{R}[\mathscr{G}] := \{h \in C_b^2(\mathbb{R}) : \exists n \in \mathbb{N}_0, \exists \lambda_1, \dots, \lambda_n \in \mathbb{R}, \exists u_1, \dots, u_n \geq 0$$
$$\text{such that } h(x) = \sum_{k=1}^{n} \lambda_k e^{-u_k x} \ \forall\, x \geq 0\}$$

the algebra generated by $\mathscr{G}$. By linearity, (29) holds true also for all $f \in \mathbf{R}[\mathscr{G}]$. Since $\mathscr{G}$ is strongly separating and since for each $x \in \mathbb{R}$ there exists $h \in \mathscr{G}$ such that $g'(x) \neq 0$, the set $\mathscr{G}$ satisfies condition (N) of Llavona [23, Definition 1.4.1], and hence $\mathbf{R}[\mathscr{G}]$ is dense in $\mathscr{S}^2(\mathbb{R})$ by Llavona [23, Corollary. 1.4.10], where

$$\mathscr{S}^2(\mathbb{R}) := \{h \in C^2(\mathbb{R}) : \lim_{|x|\to\infty} (1+|x|^2)^k(|h(x)|+|h'(x)|+|h''(x)|) = 0, \ \forall\, k \in \mathbb{N}_0\}$$

is the space of rapidly decreasing functions of order 2, endowed with the usual topology (cf. [23, Definition 0.1.8]). In particular, for every $f \in C_c^2(\mathbb{R}) \subset \mathscr{S}^2(\mathbb{R})$ there exists a sequence $(f_n)_{n\in\mathbb{N}}$ in $\mathbf{R}[\mathscr{G}]$ such that

$$\lim_{n\to\infty} \sup_{x\in\mathbb{R}} \left[(1+|x|^2)\left(|f_n(x)-f(x)| + |f_n'(x)-f'(x)| + |f_n''(x)-f''(x)|\right)\right] = 0.$$

Since (29) holds for each f_n, an application of Lebesgue's dominated convergence theorem shows that (29) also holds for $f \in C_c^2(\mathbb{R})$; remark that Lebesgue's theorem can be applied by Eqs. (3.5) and (3.6) in [4] for the integral with respect to ν_U and μ and by observing that

$$|f(x+y)-f(x)| \leq 2\|f\|_\infty \mathbb{1}_{y>1} + \|f'\|_\infty y \mathbb{1}_{0<y\leq 1}$$

for the integral with respect to ν_η and μ.

Since $C_c^2(\mathbb{R})$ is a core for the Feller process

$$W_t^x = x + \int_0^t W_{s-}^x \, dU_s + \eta_t \tag{30}$$

with generator

$$\begin{aligned} A^W f(x) &= f'(x)(x\gamma_U + \gamma_\eta^0) + \frac{1}{2} f''(x) x^2 \sigma_U^2 \\ &\quad + \int_{(-1,\infty)} (f(x+xy) - f(x) - f'(x)xy\mathbb{1}_{|y|\leq 1})\nu_U(dy) \\ &\quad + \int_{[0,\infty)} (f(x+y) - f(x))\nu_\eta(dy) \end{aligned}$$

for $f \in C_c^2(\mathbb{R})$ (cf. [4, Theorem 3.1 and Corollary 3.2] and [28, Eq. (8.6)]), we have that $\int_{\mathbb{R}} A^W f(x)\,\mu(dx) = 0$ for all f from a core, and hence $\mu = \mathscr{L}(V)$ is an invariant measure for the GOU process (30) by Liggett [21, Theorem 3.37]. By Behme et al. [6, Theorem 2.1(a)], this implies that $\int_0^\infty e^{-\xi_{s-}}\,d\eta_s$ converges a.s. and that $\mathscr{L}(\int_0^\infty e^{-\xi_{s-}}\,d\eta_s) = \mu$, so that $\mathscr{L}(\eta_1) \in D_\xi^+$ and $\Phi_\xi(\mathscr{L}(\eta_1)) = \mu$, completing the proof. □

Remark 3 To obtain a similar handy criteria for a non-positive distribution to be in the range D_ξ seems harder. A general *necessary* condition in this vein for a distribution $\mu = \mathscr{L}(V)$ to be in the range R_ξ, where ξ is a Lévy process with characteristic triplet $(\gamma_\xi, \sigma_\xi^2, \nu_\xi)$, can be derived from Eq. (4.7) in [4]. If further $E[V^2] < \infty$, and $\log \phi_\eta(u)$ denotes the characteristic exponent of a Lévy process η such that $E[e^{iu\eta_1}] = \phi_\eta(u)$, $u \in \mathbb{R}$, then by Eq. (4.8) in [4],

$$\phi_V(u) \log \phi_\eta(u) = \gamma_\xi u \phi_V'(u) - \frac{\sigma_\xi^2}{2}\left(u^2 \phi_V''(u) + u\phi_V'(u)\right)$$
$$- \int_{\mathbb{R}} (\phi_V(ue^{-y}) - \phi_V(u) + uy\phi_V'(u)\mathbb{1}_{|y|\le 1})\,\nu_\xi(du). \quad (31)$$

In [5, Example 3.2], this equation has been derived using the theory of symbols. Hence, a necessary condition for V with $E[V^2] < \infty$ to be in R_ξ is that there is a Lévy process η, such that the right-hand side of (31) can be expressed as $\phi_V(u) \log \phi_\eta(u)$, $u \in \mathbb{R}$. In Example 4.3 of Behme and Schnurr [5] it has been shown that the existence of some Lévy process η such that the right-hand side of (31) can be expressed as $\phi_V(u) \log \phi_\eta(u)$ is also *sufficient* for $\mu = \mathscr{L}(V)$ with $E[V^2] < \infty$ to be in R_ξ, hence this is a *necessary and sufficient* condition for $\mathscr{L}(V)$ with $E[V^2] < \infty$ to be in R_ξ, similar to Theorem 3. Without the assumption $EV^2 < \infty$, a necessary and sufficient condition is not established at the moment.

We conclude this section with the following results:

Lemma 3 *Let ξ be a spectrally negative Lévy process of infinite variation, drifting to $+\infty$. Then every element in R_ξ^+ is selfdecomposable and of finite variation with drift* 0.

Proof That any element in R_ξ^+ must be selfdecomposable has been shown in [8], since ξ is spectrally negative. Since every element in R_ξ^+ is positive, it must be of finite variation, and it follows from Theorem 1 and [28, Theorem 24.10] that the drift must be 0. □

Remark 4 It is well known that a selfdecomposable distribution cannot have finite non-zero Lévy measure, in particular it cannot be a compound Poisson distribution, which follows for instance immediately from [28, Corollary 15.11]. This applies in particular to exponential functionals of Lévy processes with spectrally negative ξ. However, even if ξ is not spectrally negative, and $(\xi, \eta)^T$ is a bivariate (possibly dependent) Lévy process, then $\int_0^\infty e^{-\xi_{s-}}\,d\eta_s$ (provided it converges) still cannot be a non-trivial compound Poisson distribution, with or without drift. For if c denotes the drift of a non-trivial compound Poisson distribution with drift, then this distribution must have an atom at c. However, e.g. by Bertoin et al. [8, Theorem 2.2], $\mathscr{L}(\int_0^\infty e^{-\xi_{s-}}\, d\eta_s)$ must be continuous unless constant. In other words, if $\int_0^\infty e^{-\xi_{s-}}\, d\eta_s$ is infinitely divisible, non-constant and has no Gaussian part, then its Lévy measure must be infinite. In particular, it follows that if η is a subordinator and $\int_0^\infty e^{-\xi_{s-}}\, d\eta_s$ is infinitely divisible and non-constant, then its Lévy measure must be infinite.

5 Some Results on R_ξ^+ When ξ is a Brownian Motion

It is particularly interesting to study the distributions $\int_0^\infty e^{-\xi_{s-}}\, d\eta_s$ when one of the independent Lévy processes ξ or η is a Brownian motion with drift. While the paper [20] focuses on the case when η is a Brownian motion with drift, in this section we specialise to the case $\xi_t = \sigma B_t + at$, $t \geq 0$, with $\sigma, a > 0$ and $(B_t)_{t\geq 0}$ a standard Brownian motion. Then by Lemma 3, R_ξ^+ is a subset of $L(\mathbb{R}_+)$, the class of selfdecomposable distributions on $\mathbb{R}_+$. Recall that a distribution $\mu = \mathscr{L}(V)$ on $\mathbb{R}_+$ is selfdecomposable if and only if it is infinitely divisible with non-negative drift and its Lévy measure has a Lévy density of the form $(0,\infty) \to [0,\infty)$, $x \mapsto x^{-1}k(x)$ with a non-increasing function $k = k_V : (0,\infty) \to [0,\infty)$ (cf. [28, Corollary 15.11]). Further, to every distribution $\mu = \mathscr{L}(V) \in L(\mathbb{R}_+)$ there exists a subordinator $X = (X_t)_{t\geq 0} = (X_t(\mu))_{t\geq 0}$, unique in distribution, such that

$$\mu = \mathscr{L}\left(\int_0^\infty e^{-t}\, dX_t\right), \tag{32}$$

(cf. [18, 31]). The Laplace exponents of V and X are related by

$$\psi_X(u) = u\psi_V'(u), \quad u > 0 \tag{33}$$

[e.g. [2, Remark 4.3]; alternatively, (33) can be deduced from (24)]. Denoting the drifts of V and X by b_V and b_X, respectively, it is easy to see that

$$b_V = b_X \int_0^\infty e^{-t} dt = b_X. \tag{34}$$

Since the negative of the Laplace exponent of any infinitely divisible positive distribution is a Bernstein function and these are concave (cf. [29, Definition 3.1 and Theorem 3.2]) it holds $u\psi'(u) \geq \psi(u)$ for any such Laplace exponent. Together with the above we observe that $\psi_X(u) \geq \psi_V(u)$ and hence

$$\int_{(0,\infty)} (1 - e^{-ut})\, \nu_X(dt) \leq \int_{(0,\infty)} (1 - e^{-ut})\, \nu_V(dt), \quad \forall\, u \geq 0.$$

Finally, the Lévy density $x^{-1}k(x)$ of V with k non-increasing and the Lévy measure ν_X are related by

$$k(x) = \nu_X((x,\infty)), \quad x > 0 \tag{35}$$

(e.g. [2, Eq. (4.17)]). In particular, the condition $k(0+) < \infty$ is equivalent to $\nu_X(\mathbb{R}_+) < \infty$, and the derivative of $-k$ is the Lévy density of ν_X.

5.1 Differential Equation, Necessary Conditions, and Nested Ranges

In the next result we give the differential equation for the Laplace transform of V, which has to be satisfied if $\mathscr{L}(V)$ is in the range D_ξ^+. In the special case when η is a compound Poisson process with non-negative jumps, this differential equation (36) below has already been obtained by Nilsen and Paulsen [24, Proposition 2]. We then rewrite this differential equation in terms of ψ_X, which turns out to be very useful for the further investigations.

Theorem 4 *Let $\xi_t = \sigma B_t + at$, $t \geq 0$, $\sigma, a > 0$ for some standard Brownian motion $(B_t)_{t\geq 0}$. Let $\mu = \mathscr{L}(V) \in L(\mathbb{R}_+)$ have drift b_V and Lévy density given by $x^{-1}k(x)$, $x > 0$, where $k : (0,\infty) \to [0,\infty)$ is non-increasing. Then the following are true:*

(i) $\mu \in R_\xi^+$ if and only if there is some subordinator η such that

$$\frac{1}{2}\sigma^2 u^2 \mathbb{L}_V''(u) + \left(\frac{\sigma^2}{2} - a\right) u \mathbb{L}_V'(u) + \psi_\eta(u)\mathbb{L}_V(u) = 0, \quad u > 0, \tag{36}$$

in which case $\mu = \mathscr{L}(V) = \Phi_\xi(\mathscr{L}(\eta_1))$. In particular, if η is a subordinator, then the Laplace transform of V satisfies (36) with $\mathbb{L}_V(0) = 1$, and if V is not constant 0, then $\lim_{u\to\infty} \mathbb{L}_V(u) = 0$.

(ii) Let the subordinator $X = X(\mu)$ be related to μ by (32). Then $\mu \in R_\xi^+$ if and only if the function

$$(0,\infty) \to \mathbb{R}, \quad u \mapsto a\psi_X(u) - \frac{\sigma^2}{2} u \psi_X'(u) - \frac{\sigma^2}{2}(\psi_X(u))^2$$

defines the Laplace exponent $\psi_\eta(u)$ of some subordinator η. In that case

$$\Phi_\xi(\mathscr{L}(\eta_1)) = \mathscr{L}\left(\int_0^\infty e^{-t}\, dX_t\right) = \mu. \tag{37}$$

Proof

(i) By Theorem 3, $\mu = \mathscr{L}(V) \in R_\xi^+$ if and only if

$$\psi_\eta(u) = \left(a - \frac{\sigma^2}{2}\right) u\psi_V'(u) - \frac{\sigma^2}{2} u^2 \left(\psi_V''(u) + (\psi_V'(u))^2\right), \quad u > 0, \tag{38}$$

for some subordinator η, in which case $\mu = \Phi_\xi(\mathscr{L}(\eta_1))$. Using (26) and (27), it is easy to see that this is equivalent to (36). That $\mathbb{L}_V(0) = 1$ is clear. If V is not constant 0, then it cannot have an atom at 0 (e.g. [8, Theorem 2.2]), hence $\lim_{u\to\infty} \mathbb{L}_V(u) = 0$.

(ii) If $\mathscr{L}(V) = \mathscr{L}\left(\int_0^\infty e^{-t}\, dX_t\right) \in L(\mathbb{R}_+)$ for some subordinator X, then by (33) $\psi_V'(u) = u^{-1}\psi_X(u)$ and $\psi_V''(u) = u^{-1}\psi_X'(u) - u^{-2}\psi_X(u)$. Inserting this into (38) yields the condition

$$\psi_\eta(u) = a\psi_X(u) - \frac{\sigma^2}{2} u\psi_X'(u) - \frac{\sigma^2}{2}(\psi_X(u))^2, \quad u > 0, \tag{39}$$

which gives the claim. □

Remark 5

(i) Since $u\psi_X'(u) \geq \psi_X(u)$ as observed after Eq. (34), it follows from (39) that

$$\psi_\eta(u) \leq \left(a - \frac{\sigma^2}{2}\right)\psi_X(u) - \frac{\sigma^2}{2}(\psi_X(u))^2, \quad u > 0,$$

when the subordinators X and η are related by (37).

(ii) Equation (39) is a Riccati equation for ψ_X. Using the transformation $y(u) = \exp(\int_1^u \frac{\psi_X(v)}{v}\, dv) = C\,\mathbb{L}_V(u)$ for $u > 0$ by (33), it is easy to see that it reduces to the linear equation (36). Unfortunately, in general it does not seem possible to solve (36) in a closed form.

(iii) Since for any subordinator η, $\psi_\eta(u)$ has a continuous continuation to $\{z \in \mathbb{C} : \Re(z) \geq 0\}$ which is analytic in $\{z \in \mathbb{C} : \Re(z) > 0\}$ (e.g. [29, Proposition 3.6]), for any fixed $u_0 > 0$ Eq. (36) can be solved in principle on $(0, 2u_0)$ by the power series method (e.g. [11, Sect. 2.8, Theorem 7, p. 190]). In particular when ν_η is such that $\int_{(1,\infty)} e^{ux}\nu_\eta(dx) < \infty$ for every $u > 0$ (e.g. if ν_η has compact support), then $\psi_\eta(z) = -b_\eta z + \int_{(0,\infty)}(e^{-zx} - 1)\,\nu_\eta(dx)$, $z \in \mathbb{C}$, is an analytic continuation of ψ_η in the complex plane. Hence it admits a power series expansion of the form $\psi_\eta(z) = \sum_{n=0}^\infty f_n z^n$, $z \in \mathbb{C}$, with $f_0 = 0$ and Eq. (36) may be solved by the Frobenius method (e.g. [11, Sect. 2.8, Theorem 8, p. 215]). To exemplify this, assume for simplicity that $2a/\sigma^2$ is not an integer. Equation (36) has a weak singularity at 0. Its so-called indicial polynomial is given by

$$r \mapsto r(r-1) + \left(1 - \frac{2a}{\sigma^2}\right) r = r\left(r - \frac{2a}{\sigma^2}\right).$$

The exponents of singularity are the zeros of this polynomial, i.e. 0 and $2a/\sigma^2$, and since we have assumed that $2a/\sigma^2$ is not an integer, the general real solution of (36) is given by

$$\mathbb{L}_V(u) = C_1 u^{2a/\sigma^2} \sum_{n=0}^{\infty} c_n u^n + C_2 \sum_{n=0}^{\infty} d_n u^n, \quad u > 0,$$

where $C_1, C_2 \in \mathbb{R}$, $c_0 = d_0 = 1$, the coefficients c_n, d_n are defined recursively by

$$c_n := \frac{-1}{n(n+2a/\sigma^2)} \sum_{k=0}^{n-1} c_k f_{n-k}, \quad d_n = \frac{-1}{n(n-2a/\sigma^2)} \sum_{k=0}^{n-1} d_k f_{n-k}, \quad n \in \mathbb{N},$$

(e.g. [11, Sect. 2.8, Eq. (14), p. 209]) and the power series $\sum_{n=0}^{\infty} c_n u^n$ and $\sum_{n=0}^{\infty} d_n u^n$ converge in $u \in \mathbb{C}$. Since $\mathbb{L}_V(0) = 1$, we even conclude that $C_2 = 1$.

Next, we show that the ranges of Φ_ξ, when $\xi_t = \sigma B_t + at$, are nested when σ and a vary over all positive parameters.

Theorem 5 *Let $B = (B_t)_{t\geq 0}$ be a standard Brownian motion. For $a, \sigma > 0$ let $\xi^{(a,\sigma)} := (\xi_t^{(a,\sigma)})_{t\geq 0} := (\sigma B_t + at)_{t\geq 0}$. Then $R^+_{\xi^{(a,\sigma)}} = R^+_{\xi^{(a/\sigma^2,1)}}$.*

Further, for $a, \sigma, a', \sigma' > 0$ such that $a/\sigma^2 \leq a'/\sigma'^2$ we have $R^+_{\xi^{(a,\sigma)}} \subset R^+_{\xi^{(a',\sigma')}}$.

In particular, for fixed $\sigma > 0$, the family $R^+_{\xi^{(a,\sigma)}}$, $a > 0$, is nested and non-decreasing in a, and for fixed $a > 0$ the family $R^+_{\xi^{(a,\sigma)}}$, $\sigma > 0$, is nested and non-increasing in σ.

Proof Since $(\sigma B_t + at)_{t\geq 0}$ has the same distribution as $(B_{t\sigma^2} + at)_{t\geq 0}$, we obtain for a Lévy process $\eta = (\eta_t)_{t\geq 0}$ such that $\mathcal{L}(\eta_1) \in D_{\xi^{(a,\sigma)}}$ and η is independent of B,

$$\int_0^\infty e^{-(\sigma B_t + at)}\, d\eta_t \overset{d}{=} \int_0^\infty e^{-(B_{t\sigma^2}+at)}\, d\eta_t = \int_0^\infty e^{-(B_t + (a/\sigma^2)t)}\, d\eta_{t/\sigma^2}.$$

Hence $\mathcal{L}(\eta_{1/\sigma^2}) \in D_{\xi^{(a/\sigma^2,1)}}$ and $\Phi_{\xi^{(a,\sigma)}}(\mathcal{L}(\eta_1)) = \Phi_{\xi^{(a/\sigma^2,1)}}(\mathcal{L}(\eta_{1/\sigma^2}))$. In particular, $R^+_{\xi^{(a,\sigma)}} \subset R^+_{\xi^{(a/\sigma^2,1)}}$. Similarly, $R^+_{\xi^{(a,\sigma)}} \supset R^+_{\xi^{(a/\sigma^2,1)}}$ so that $R^+_{\xi^{(a,\sigma)}} = R^+_{\xi^{(a/\sigma^2,1)}}$. For the second assertion, it is hence sufficient to assume $\sigma = 1$. Now if $a < a'$ and $\mu \in R^+_{\xi^{(a,1)}}$, let the subordinator X be related to μ by (32). Then

$$a\psi_X(u) - \frac{1}{2}u\psi_X'(u) - \frac{1}{2}(\psi_X(u))^2 = \psi_\eta(u), \quad u > 0,$$

by Theorem 4 (ii), hence

$$a'\psi_X(u) - \frac{1}{2}u\psi_X'(u) - \frac{1}{2}(\psi_X(u))^2 = \psi_\eta(u) + (a'-a)\psi_X(u), \quad u > 0,$$

defines the Laplace exponent of a subordinator by Schilling et al. [29, Corollary 3.8 (i)]. Hence $\mu \in R^+_{\xi(a',1)}$ again by Theorem 4 (ii). The remaining assertions are clear. □

Remark 6 Although $R^+_{\xi(1,\sigma)} \subset R^+_{\xi(1,\sigma')}$ for $0 < \sigma' < \sigma$, and $\sigma B_t + t$ converges pointwise to t when $\sigma \to 0$, we do not have $\bigcup_{\sigma>0} R^+_{\xi(1,\sigma)} = R^+_{\xi_t=t}\ (= L(\mathbb{R}_+))$. For example, a positive 3/4-stable distribution is in $L(\mathbb{R}_+)$ but not in $\bigcup_{\sigma>0} R^+_{\xi(1,\sigma)}$, as follows from Example 2 or Corollary 3 below.

While it is difficult to solve Eqs. (36) and (39) for given ψ_η, they still allow to obtain results about the qualitative structure of the range. The following gives a simple necessary condition in terms of the Lévy density $x^{-1}k(x)$ for μ to be in R^+_ξ, and to calculate the drift b_η of $(\Phi^+_\xi)^{-1}(\mu)$ when $\mu \in R^+_\xi$.

Theorem 6 *Let $\xi_t = \sigma B_t + at$, $t \geq 0$, for $\sigma, a > 0$ and some standard Brownian motion $B = (B_t)_{t\geq 0}$. Let $\mu = \mathcal{L}(V) \in L(\mathbb{R}_+)$ with drift b_V and Lévy density $x^{-1}k(x)$. Let the subordinator X be related to μ by (32) and denote its drift by b_X.*

(i) If $\mu \in R^+_\xi$, then $b_X = 0$ and $\lim_{u\to\infty} u^{-1/2}|\psi_X(u)| = \lim_{u\to\infty} u^{1/2}|\psi_V'(u)|$ exists and is finite. If $\mu = \Phi_\xi(\mathcal{L}(\eta_1))$ for some subordinator η with drift b_η, then b_η and ψ_X are related by

$$b_\eta = \frac{\sigma^2}{2}\lim_{u\to\infty} u^{-1}(\psi_X(u))^2 = \frac{\sigma^2}{2}\lim_{u\to\infty} u(\psi_V'(u))^2. \tag{40}$$

(ii) If $\mu \in R^+_\xi$ has Lévy density $x^{-1}k(x)$, then it holds $\limsup_{x\downarrow 0} x^{-1/2}\int_0^x k(s)\,ds < \infty$ and $b_V = 0$. In particular, if $\mu = \Phi_\xi(\mathcal{L}(\eta_1))$ for some subordinator η with drift b_η, then $b_\eta > 0$ if and only if $\limsup_{x\downarrow 0} x^{-1/2}\int_0^x k(s)\,ds > 0$.

Proof

(i) Suppose that $\mu = \mathcal{L}(V) = \Phi_\xi(\mathcal{L}(\eta_1)) \in R^+_\xi$. Then $b_V = 0$ by Lemma 3 and hence $b_X = 0$ by (34). Since $\psi_X'(u) = -\int_{(0,\infty)} e^{-ux} x\,\nu_X(dx)$ we conclude that $\lim_{u\to\infty}\psi_X'(u) = 0$ by dominated convergence. Since $b_X = 0$ and $\lim_{u\to\infty} u^{-1}\psi_X(u) = -b_X = 0$ and $\lim_{u\to\infty} u^{-1}\psi_\eta(u) = -b_\eta$ by Schilling et al. [29, Remark 3.3 (iv)], (40) as well as the necessity of the stated condition follow from (39) and (33).

(ii) Since $k(x) = \nu_X((x,\infty))$ by (35), it follows from [29, Lemma 3.4] that

$$\frac{e-1}{e} \leq \frac{|\psi_X(u)|}{u\int_0^{1/u} k(s)\,ds} \leq 1, \quad u > 0.$$

Hence (ii) is an immediate consequence of (i) and Lemma 3. □

Example 2 Let $\xi_t = \sigma B_t + at$ be as in Theorem 6. Let $\mu \in L(\mathbb{R}_+)$ with Lévy density $x^{-1}k(x)$. Then $\int_0^1 k(x)\,dx < \infty$.

If $\liminf_{s\downarrow 0} k(s)s^{1/2} = +\infty$, then $\liminf_{x\downarrow 0} x^{-1/2}\int_0^x k(s)\,ds = +\infty$. Hence $\mu \notin R_\xi^+$. In particular, a non-degenerate positive α-stable distribution with $\alpha > 1/2$ cannot be in R_ξ^+. A more detailed result will be given in Corollary 3 below.

5.2 Selfdecomposable Distributions with $k(0+) < \infty$

In this subsection we specialize to selfdecomposable distributions with $k(0+) < \infty$ and give a characterization when they are in the range R_ξ^+ for ξ a Brownian motion with drift.

Theorem 7 *Let $\xi_t = \sigma B_t + at$, $t \geq 0$, $\sigma, a > 0$ for some standard Brownian motion $(B_t)_{t\geq 0}$. Let $\mu = \mathscr{L}(V) \in L(\mathbb{R}_+)$ have drift b_V and Lévy density $x^{-1}k(x)$, $x > 0$, where $k = k_V : (0,\infty) \to [0,\infty)$ is non-increasing. Let the subordinator $X = X(\mu)$ be related to μ by (32). Assume that $k(0+) < \infty$, equivalently that $\nu_X(\mathbb{R}_+) < \infty$.*

(i) Then $\mu \in R_\xi^+$ if and only if $b_X = 0$ and ν_X has a density g on $(0,\infty)$ such that

$$\lim_{t\to\infty} tg(t) = \lim_{t\to 0} tg(t) = 0 \tag{41}$$

and such that the function

$$G : (0,\infty) \to [0,\infty), \tag{42}$$

$$t \mapsto (a + \sigma^2 \nu_X(\mathbb{R}_+)) \int_0^t g(v)\,dv + \frac{\sigma^2}{2} tg(t) - \frac{\sigma^2}{2}\int_0^t (g * g)(v)\,dv$$

is non-decreasing. If these conditions are satisfied, then

$$\Phi_\xi(\mathscr{L}(\eta_1)) = \mu,$$

where η is the subordinator with drift 0 and finite Lévy measure $\nu_\eta(dx) = dG(x)$.

(ii) Equivalently, $\mu = \mathscr{L}(V) \in R_\xi^+$ if and only if $b_V = 0$ and $-k : (0,\infty) \to (-\infty, 0]$ is absolutely continuous with derivative g on $(0,\infty)$ satisfying (41) and such that G defined by (42) is non-decreasing. In that case, $\Phi_\xi(\mathscr{L}(\eta_1)) = \mu$, where η is a subordinator with drift 0 and finite Lévy measure $\nu_\eta(dx) = dG(x)$.

Proof

(i) Assume that $\nu_X(\mathbb{R}_+) < \infty$. Suppose first that $\mu \in R_\xi^+$, and let $(\eta_t)_{t\geq 0}$ be a subordinator such that $\Phi_\xi(\mathscr{L}(\eta_1)) = \mu$. Then $b_X = 0$ by Theorem 6 (i), and by Theorem 4 (ii), we have (39) with

$$\psi_\eta(u) = -b_\eta u - \int_{(0,\infty)} (1 - e^{-ut})\, \nu_\eta(dt)$$

$$\text{and} \quad \psi_X(u) = -\int_{(0,\infty)} (1 - e^{-ut})\, \nu_X(dt), \quad u \geq 0.$$

Since $\mathbb{L}_{\nu_X}(u)^2 = \mathbb{L}_{\nu_X * \nu_X}(u)$ and $(\nu_X * \nu_X)(\mathbb{R}_+) = \nu_X(\mathbb{R}_+)^2$, where $\mathbb{L}_{\nu_X}$ denotes the Laplace transform of the finite measure ν_X, we conclude

$$\begin{aligned}
\psi_X(u)^2 &= \left(\int_{(0,\infty)} (1 - e^{-ut})\, \nu_X(dt) \right)^2 \\
&= \nu_X(\mathbb{R}_+)^2 - 2\nu_X(\mathbb{R}_+) \int_{(0,\infty)} e^{-ut}\, \nu_X(dt) + \int_{(0,\infty)} e^{-ut} (\nu_X * \nu_X)(dt) \\
&= \int_{(0,\infty)} (1 - e^{-ut})\, (2\nu_X(\mathbb{R}_+)\nu_X - \nu_X * \nu_X)(dt).
\end{aligned}$$

Hence, from (39), on the one hand

$$\frac{\sigma^2}{2} u \psi_X'(u) = b_\eta u + \int_{(0,\infty)} (1 - e^{-ut})\, \rho_1(dt) - \int_{(0,\infty)} (1 - e^{-ut})\, \rho_2(dt), \tag{43}$$

where

$$\rho_1 := \nu_\eta + \frac{\sigma^2}{2} \nu_X * \nu_X \quad \text{and} \quad \rho_2 := (a + \sigma^2 \nu_X(\mathbb{R}_+))\nu_X.$$

On the other hand, $u\psi_X'(u) = -u \int_{(0,\infty)} e^{-ut} t\, \nu_X(dt)$, and rewriting the integral $\int_{(0,\infty)} (1 - e^{-ut}) \rho_i(dt) = \int_0^\infty u e^{-ut} \rho_i((t,\infty))\, dt$ by Fubini's theorem as in [29, Remark 3.3(ii)], (43) gives

$$\frac{\sigma^2}{2} u \int_{(0,\infty)} e^{-ut} t \nu_X(dt) = -b_\eta u + u \int_0^\infty e^{-ut} \left(\rho_2((t,\infty)) - \rho_1((t,\infty))\right) dt, \quad u > 0.$$

Dividing by u, the uniqueness theorem for Laplace transforms then shows $b_\eta = 0$ and that ν_X has a density g, given by

$$g(t) = \frac{2}{\sigma^2 t} \left(\rho_2((t,\infty)) - \rho_1((t,\infty))\right), \quad t > 0. \tag{44}$$

From this we conclude that $\lim_{t\to\infty} tg(t) = 0$ and that the limit $\lim_{t\to 0} tg(t) = \frac{2}{\sigma^2}(\rho_2(\mathbb{R}_+) - \rho_1(\mathbb{R}_+))$ exists in $[-\infty, \infty)$ since $\rho_2(\mathbb{R}_+) < \infty$. But since $g \geq 0$, the limit must be in $[0, \infty)$, hence $\rho_1(\mathbb{R}_+) < \infty$ so that $\nu_\eta(\mathbb{R}_+) < \infty$, and since

$\int_0^1 \frac{tg(t)}{t}\,dt = \int_0^1 g(t)\,dt < \infty$, we also have $\lim_{t\to 0} tg(t) = 0$. Further, by (44), the total variation of $t \mapsto tg(t)$ over $(0,\infty)$ is finite. Knowing now that ν_X has a density g with $\lim_{t\to\infty} tg(t) = \lim_{t\to 0} tg(t) = 0$, we can write using partial integration

$$\begin{aligned} u\psi_X'(u) &= \int_0^\infty \left(\frac{d}{dt}e^{-ut}\right) tg(t)\,dt = \int_0^\infty tg(t)d\left(e^{-ut}\right) \\ &= tg(t)e^{-ut}\Big|_{t=0}^{t=\infty} - \int_0^\infty e^{-ut}\,d(tg(t)) = \int_0^\infty (1-e^{-ut})\,d(tg(t)). \end{aligned}$$

Inserting this in (43), we obtain by uniqueness of the representation of Bernstein functions (cf. [29, Theorem 3.2]) that

$$\frac{\sigma^2}{2}d(tg(t)) = \nu_\eta(dt) + \frac{\sigma^2}{2}(g*g)(t)\,dt - (a+\sigma^2\nu_X(\mathbb{R}_+))g(t)\,dt,$$

or equivalently

$$\nu_\eta(dt) = (a+\sigma^2\nu_X(\mathbb{R}_+))g(t)\,dt + \frac{\sigma^2}{2}d(tg(t)) - \frac{\sigma^2}{2}(g*g)(t)\,dt. \tag{45}$$

Since ν_η is a positive (and finite) measure, so is the right-hand side of (45), and hence G is non-decreasing with $\nu_\eta(dt) = dG(t)$, finishing the proof of the "only if"-assertion. The converse follows by reversing the calculations above, by defining a subordinator η with drift 0 and Lévy measure $\nu_\eta(dt) := dG(t)$, observing that $t \mapsto tg(t)$ is of finite total variation on $(0,\infty)$ by (41) and (42), and then showing that ν_η satisfies (43) and hence that ψ_η satisfies (39).

(ii) This follows immediately from (i), (34) and (35). □

Remark 7 Let $\xi_t = \sigma B_t + at$, $t \geq 0$, with $\sigma, a > 0$ and $(B_t)_{t\geq 0}$ a standard Brownian motion.

(i) If $\mu \in R_\xi^+$ and X is a subordinator such that (32) holds and such that $\nu_X(\mathbb{R}_+) < \infty$, then the Lévy density g of ν_X cannot have negative jumps, since by (42) this would contradict non-decreasingness of G.

(ii) Let X be a subordinator with $\nu_X(\mathbb{R}_+) < \infty$ and $b_X = 0$, and suppose that ν_X has a density g such that there is $r \geq 0$ with $g(t) = 0$ for $t \in (0,r]$ and g is differentiable on (r,∞) (the case $r = 0$ is allowed). Then $\mathscr{L}(\int_0^\infty e^{-t}\,dX_t) \in R_\xi^+$ if and only if g satisfies (41) and

$$\left(a+\sigma^2\nu_X(\mathbb{R}_+)+\frac{\sigma^2}{2}\right)g(t)+\frac{\sigma^2}{2}tg'(t)-\frac{\sigma^2}{2}(g*g)(t) \geq 0, \quad \forall\, t > r. \tag{46}$$

This follows immediately from Theorem 4 (iii) since the right-hand side of (46) is the derivative of the function G defined by (42).

The following gives an example for a distribution in R_ξ^+ such that $\nu_X(\mathbb{R}_+) < \infty$.

Example 3 Let $r \geq 0$ and let $g : [0, \infty) \to [0, \infty)$ be a function such that $g(t) = 0$ for all $t \in (0, r)$ (a void assumption if $r = 0$), $g_{|[r,\infty)}$ is continuously differentiable with derivative g', such that g is strictly positive on $[r, \infty)$, $\lim_{t\to\infty} g(t) = 0$ and such that $-g'$ is regularly varying at ∞ with index $\beta < -2$ (in particular, $g'(t) < 0$ for large enough t). Then g defines a Lévy density of a subordinator X with drift 0 such that $\nu_X(\mathbb{R}_+) < \infty$ and $\mathcal{L}(\int_0^\infty e^{-t}\, dX_t) \in R^+_{\sigma B_t + at}$ for large enough a (but σ fixed).

Proof Since $-g'$ is regularly varying with index β and $\lim_{t\to\infty} g(t) = 0$, g is regularly varying at ∞ with index $\beta + 1 < -1$ and $\lim_{t\to\infty} \frac{-tg'(t)}{g(t)} = -\beta - 1$ by Karamata's Theorem (e.g. [10, Theorem 1.5.11]). In particular, $\lim_{t\to\infty} tg(t) = 0$, further $\lim_{t\to 0} tg(t) = 0$ since $g(0) < \infty$, and g is a density of a finite measure. Next, observe that

$$\frac{(g*g)(t)}{g(t)} = \int_r^{t/2} \frac{g(t-x)}{g(t)} g(x)\, dx + \int_{t/2}^{t-r} \frac{g(x)}{g(t)} g(t-x)\, dx, \quad t \geq 2r.$$

But for any $\varepsilon > 0$, when $t \geq t_\varepsilon$ is large enough, we have $g(t-x)/g(t) \leq 2^{-\beta-1} + \varepsilon$ for $x \in (r, t/2]$, and $g(x)/g(t) \leq 2^{-\beta-1} + \varepsilon$ for $x \in [t/2, t-r]$ by the uniform convergence theorem for regularly varying functions (e.g. [10, Theorem 1.5.2]). As $\int_0^\infty g(t)\, dt < \infty$, this shows that $\limsup_{t\to\infty} \frac{(g*g)(t)}{g(t)} < \infty$. Since also $g * g$ as well as $|g'|$ are bounded on $[r, \infty)$, it follows that (46) is satisfied for all $t \geq r$ for large enough a, and for $t \in (0, r)$ it is trivially satisfied. Hence $\mathcal{L}(\int_0^\infty e^{-t}\, dX_t) \in R^+_{\sigma B_t + at}$ for large enough a. □

Next we give some examples of selfdecomposable distributions which are not in R_ξ^+.

Example 4 Let $\xi_t = \sigma B_t + at$, $t \geq 0$, with a standard Brownian motion B and parameters $\sigma, a > 0$.

(i) A selfdecomposable distribution with Lévy density $c\mathbb{1}_{(0,1)}(x)x^{-1}$ and $c > 0$ is not in R_ξ^+ by Theorem 7, since $k(x) = \mathbb{1}_{(0,1)}(x)$ satisfies $k(0+) < \infty$ but is not continuous.
(ii) If X is a subordinator with non-trivial Lévy measure ν_X such that ν_X has compact support, then $\mathcal{L}(\int_0^\infty e^{-t}\, dX_t)$ is not in R_ξ^+ by Theorem 7, since if it were then ν_X had a density g, and if x_g denotes the right end point of the support of g, then $2x_g$ is the right endpoint of the support of $g * g$, showing that the function G defined by (42) cannot be non-decreasing on $(0, \infty)$.
(iii) If X is a subordinator with finite Lévy measure and non-trivial Lévy density g which is a step function (with finitely or infinitely many steps), then $\mathcal{L}(\int_0^\infty e^{-t}\, dX_t)$ is not in R_ξ^+ by Remark 7 (i), since g must have at least one negative jump as a consequence of $\int_0^\infty g(t)\, dt < \infty$.

5.3 Positive Stable Distributions

In this subsection we characterize when a positive stable distribution is in the range R_ξ^+. We also consider (finite) convolutions of positive stable distributions, i.e. distributions of the form $\mathscr{L}(\sum_{k=1}^n X_i)$, where $n \in \mathbb{N}$ and $X_1, \ldots, X_n$ are independent positive stable distributions.

Theorem 8 *Set $\xi_t = \sigma B_t + at$, $t \geq 0$, $a, \sigma > 0$ for some standard Brownian motion $(B_t)_{t\geq 0}$. Let $0 < \alpha_1 < \cdots < \alpha_n < 1$ for some $n \in \mathbb{N}$ and $b_i \geq 0$, $i = 1, \ldots, n$ and let μ be the distribution of $\sum_{i=1}^n X_i$ where the X_i are independent and each X_i is non-trivial and positive α_i-stable with drift b_i. Then if μ is in R_ξ^+ it holds $b_i = 0$, $i = 0, \ldots, n$, $\alpha_1 \leq (\frac{2a}{\sigma^2} \wedge \frac{1}{2})$ and $\alpha_n \leq \frac{1}{2}$. Conversely, if $b_i = 0$, $i = 0, \ldots, n$ and $\alpha_n \leq (\frac{2a}{\sigma^2} \wedge \frac{1}{2})$, then μ is in R_ξ^+.*

Proof Assume $\mu = \mathscr{L}(V) = \mathscr{L}(\int_0^\infty e^{-\xi_{s-}} d\eta_s) \in R_\xi^+$ for some subordinator η. Since $\psi_V(u) = \sum_{i=1}^n \psi_{X_i}(u)$, the drift of V is $\sum_{i=1}^n b_i$. By Lemma 3, this implies $\sum_{i=1}^n b_i = 0$ and hence $b_i = 0$ for all i. Since each X_i is positive α_i-stable with drift 0 and non-trivial, we know from [28, Remarks 14.4 and 21.6] that the Laplace exponent of X_i is given by

$$\psi_{X_i}(u) = \int_{(0,\infty)} (e^{-ux} - 1)\nu_{X_i}(dx) = \int_0^\infty (e^{-ux} - 1)c_i x^{-1-\alpha_i} dx$$

with $c_i > 0$. Hence

$$\psi_V(u) = \sum_{i=1}^n \int_0^\infty (e^{-ux} - 1)c_i x^{-1-\alpha_i} dx,$$

such that

$$\psi_V'(u) = -\sum_{i=1}^n c_i u^{\alpha_i - 1}\Gamma(1-\alpha_i) \quad \text{and} \quad \psi_V''(u) = \sum_{i=1}^n c_i u^{\alpha_i-2}\Gamma(2-\alpha_i), \quad u > 0.$$

Hence (38) reads

$$\begin{aligned}\psi_\eta(u) = -\sum_{i=1}^n \Bigg[\left(\left(a - \frac{\sigma^2}{2}\right) c_i\,\Gamma(1-\alpha_i) + \frac{\sigma^2}{2} c_i\,\Gamma(2-\alpha_i)\right) u^{\alpha_i} \\ + \sigma^2 \sum_{j=1}^{i-1} c_i c_j \Gamma(1-\alpha_i)\Gamma(1-\alpha_j) u^{\alpha_i+\alpha_j} + \frac{\sigma^2}{2} c_i^2 (\Gamma(1-\alpha_i))^2 u^{2\alpha_i}\Bigg] \\ =: -\sum_{i=1}^n \left(A_i u^{\alpha_i} + \sum_{j=1}^{i-1} B_{i,j} u^{\alpha_i+\alpha_j} + C_i u^{2\alpha_i}\right) =: -f(u), \quad u > 0. \qquad (47)\end{aligned}$$

Observe that $A_i \in \mathbb{R}$, and $B_{i,j}, C_i > 0$ for all i,j. As the left hand side of (47) is the Laplace exponent of a subordinator it is the negative of a Bernstein function [29, Theorem 3.2] and thus $f(u)$, $u \geq 0$, has to be a Bernstein function if a solution to (47) exists. By Schilling et al. [29, Corollary 3.8 (viii)] a Bernstein function cannot grow faster than linearly, which yields directly that $\alpha_i \in (0, 1/2]$, $i = 1, \ldots, n$. As by Schilling et al. [29, Definition 3.1] the first derivative of a Bernstein function is completely monotone, considering $\lim_{u\to 0} f'(u) \geq 0$ we further conclude that necessarily $A_1 \geq 0$, which is equivalent to $\alpha_1 \leq \frac{2a}{\sigma^2}$.

Conversely, let V be a non-trivial finite convolution of positive α_i-stable distributions with drift 0 and $0 < \alpha_1 < \cdots < \alpha_n \leq (\frac{2a}{\sigma^2} \wedge \frac{1}{2})$. Then $A_i \geq 0$ for all i and the preceding calculations show that the right hand side of (38) is given by $f(u)$, which is the Laplace exponent of a subordinator, namely an independent sum of positive α_i-stable subordinators (for each $A_i \geq 0$), $(\alpha_i + \alpha_j)$-stable subordinators (for each $B_{i,j}$), $2\alpha_i$-stable subordinators (for each C_i with $\alpha_i < \frac{1}{2}$) and possibly a deterministic subordinator (if $\alpha_n = 1/2$). Hence $\mathcal{L}(V) \in R_\xi^+$ by Theorem 3. □

As a consequence of the above theorem, we can characterize which positive α-stable distributions are in R_ξ^+:

Corollary 3 *Let $\xi_t = \sigma B_t + at$, $t \geq 0$, $a, \sigma > 0$ for some standard Brownian motion $(B_t)_{t\geq 0}$. Then a non-degenerate positive α-stable distribution μ is in R_ξ^+ if and only if its drift is 0 and $\alpha \in (0, \frac{2a}{\sigma^2} \wedge \frac{1}{2}]$. If this condition is satisfied and μ has Lévy density $x \mapsto cx^{-1-\alpha}$ on $(0,\infty)$ with $c > 0$, then $\mu = \Phi_\xi(\mathcal{L}(\eta_1))$, where in the case $\alpha < 1/2$, η is a subordinator with drift 0 and Lévy density on $(0,\infty)$ given by*

$$x \mapsto c\alpha\left(a - \frac{\sigma^2}{2}\alpha\right)x^{-\alpha-1} + \sigma^2 c^2 \frac{\alpha(\Gamma(1-\alpha))^2}{\Gamma(1-2\alpha)} x^{-2\alpha-1},$$

and in the case $\alpha = 1/2 = 2a/\sigma^2$, η is a deterministic subordinator with drift $\sigma^2 c^2 (\Gamma(1-\alpha))^2/2$.

Proof The equivalence is immediate from Theorem 8. Further, by (47), we have $\Phi_\xi(\mathcal{L}(\eta_1)) = \mu$ where the Laplace exponent of η is given by

$$\psi_\eta(u) = -\left(\left(a - \frac{\sigma^2}{2}\right)c\,\Gamma(1-\alpha) + \frac{\sigma^2}{2}c\,\Gamma(2-\alpha)\right)u^\alpha - \frac{\sigma^2}{2}c^2(\Gamma(1-\alpha))^2 u^{2\alpha}.$$

The case $\alpha = 1/2 = 2a/\sigma^2$ now follows immediately, and for $\alpha < 1/2$ observe that

$$\int_0^\infty (e^{-ux} - 1)x^{-1-\beta}\,dx = \int_0^u \left(\frac{d}{dv}\int_0^\infty (e^{-vx} - 1)x^{-1-\beta}\,dx\right)dv$$
$$= -\int_0^u v^{\beta-1}\Gamma(1-\beta)\,dv = -\frac{\Gamma(1-\beta)}{\beta}u^\beta$$

for $\beta \in (0, 1)$ and $u > 0$, which gives the desired form of the drift and Lévy density of η also in this case. □

Example 5 Reconsider Example 1, namely,

$$V = \int_0^\infty e^{-(\sigma B_t + at)} dt \overset{d}{=} \frac{2}{\sigma^2 \Gamma_{\frac{2a}{\sigma^2}}},$$

where V has the law of a scaled inverse Gamma distributed random variable with parameter $\frac{2a}{\sigma^2}$. In the case that $\frac{2a}{\sigma^2} = \frac{1}{2}$, or equivalently $a = \sigma^2/4$ this is a so called Lévy distribution and it is $1/2$-stable (cf. [30, p. 507]). Reassuringly, by Corollary 3, $\mathscr{L}(V)$ is a $1/2$-stable distribution if $a = \sigma^2/4$.

Corollary 4 *Let $\xi_t = \sigma B_t + at$, $t \geq 0$, $\sigma, a > 0$ for some standard Brownian motion $(B_t)_{t\geq 0}$. Then R_ξ^+ contains the closure of all finite convolutions of positive α-stable distributions with drift* 0 *and $\alpha \in (0, \frac{2a}{\sigma^2} \wedge \frac{1}{2}]$, which is characterized as the set of infinitely divisible distributions μ with Laplace exponent*

$$\psi(u) = \int_{(0, \frac{2a}{\sigma^2} \wedge \frac{1}{2}]} m(d\alpha) \int_0^\infty (e^{-ux} - 1) x^{-1-\alpha} dx \tag{48}$$

where m is a measure on $(0, \frac{2a}{\sigma^2} \wedge \frac{1}{2}]$ such that

$$\int_{(0, \frac{2a}{\sigma^2} \wedge \frac{1}{2}]} \alpha^{-1} m(d\alpha) < \infty. \tag{49}$$

Proof Denote by M_1 the class of all finite convolutions of positive α-stable distributions with drift 0 and $\alpha \in (0, \frac{2a}{\sigma^2} \wedge \frac{1}{2}]$, by M_2 its closure with respect to weak convergence, and by M_3 the class of all positive distributions on $\mathbb{R}$ whose characteristic exponent can be represented in the form (48) with m subject to (49). We show that $M_2 = M_3$, then since $M_2 \subset R_\xi^+$ by Theorems 8 and 2 (i), this implies the statement. To see $M_2 \subset M_3$, denote by $L_\infty(\mathbb{R})$ the closure of all finite convolutions of stable distributions on $\mathbb{R}$ (cf. [26, Theorem 3.5], where $L_\infty(\mathbb{R})$ is defined differently, but shown to be equivalent to this definition). Using the fact that $L_\infty(\mathbb{R})$ is closed, it then follows easily from [26, Theorem 4.1] that also M_3 is closed under weak convergence. Since obviously $M_1 \subset M_3$ (take m to be a measure supported on a finite set), we also have $M_2 \subset M_3$. Conversely, $M_3 \subset M_2$ can be shown in complete analogy to the proof of Sato [26, Theorem 3.5]. □

Remark 8 From the proof of Theorem 8 it is possible to obtain a necessary and sufficient condition for a finite convolution of positive, stable distributions to be in R_ξ^+. Indeed if the X_i are such that $\psi_{X_i}(u) = -c_i u^{\alpha_i}$ with $c_i > 0$ and $\alpha_i \in (0, 1)$, then $\mu = \mathscr{L}(\sum_{i=1}^n X_i)$ is in R_ξ^+ if and only if the function f defined by (47) is a Bernstein function. After ordering the indices, the function f can be written as $\sum_{i=1}^m D_i u^{\gamma_i}$ with

$0 < \gamma_1 < \cdots \gamma_m < 2$ and coefficients $D_i \in \mathbb{R} \setminus \{0\}$. Since

$$\sum_{i=1,\ldots,m;\gamma_i<1} D_i u^{\gamma_i} = \int_0^\infty (1 - e^{-ux}) \sum_{i=1,\ldots,m;\gamma_i<1} \frac{D_i \gamma_i}{\Gamma(1-\gamma_i)} x^{-1-\gamma_i} \, dx$$

as seen in the proof of Corollary 3, it follows from [29, Corollary 3.8(viii)] and [28, Example 12.3] that f is a Bernstein function if and only if $\gamma_m \leq 1$, $D_m \geq 0$ and

$$\sum_{i=1,\ldots,m;\gamma_i<1} \frac{D_i \gamma_i}{\Gamma(1-\gamma_i)} x^{-1-\gamma_i} \geq 0, \quad \forall\, x > 0.$$

Acknowledgements We would like to thank the anonymous referee for valuable suggestions which helped to improve the exposition of the manuscript. Makoto Maejima's research was partially supported by JSPS Grand-in-Aid for Science Research 22340021.

References

1. R.B. Ash, C.A. Dolèans-Dade, *Probability & Measure*, 2nd edn. (Academic, New York, 2000)
2. O.E. Barndorff-Nielsen, N. Shephard, Modelling by Lévy processes for financial econometrics, in *Lévy Processes: Theory and Applications*, ed. by O.E. Barndorff-Nielsen, T. Mikosch, S. Resnick (Birkhäuser, Boston, 2001), pp. 283–318
3. A. Behme, Distributional properties of solutions of $dV_t = V_{t-}dU_t + dL_t$ with Lévy noise. Adv. Appl. Probab. **43**, 688–711 (2011)
4. A. Behme, A. Lindner, On exponential functionals of Lévy processes. J. Theor. Probab. (2013). doi:10.1007/s10959-013-0507-y
5. A. Behme, A. Schnurr, A criterion for invariant measures of Itô processes based on the symbol. Bernoulli **21**(3), 1697–1718 (2015)
6. A. Behme, A. Lindner, R. Maller, Stationary solutions of the stochastic differential equation $dV_t = V_{t-}dU_t + dL_t$ with Lévy noise. Stoch. Process. Appl. **121**, 91–108 (2011)
7. J. Bertoin, M. Yor, Exponential functionals of Lévy processes. Probab. Surv. **2**, 191–212 (2005)
8. J. Bertoin, A. Lindner, R. Maller, On continuity properties of the law of integrals of Lévy processes, in *Séminaire de Probabilités XLI*, ed. by C. Donati-Martin, M. Émery, A. Rouault, C. Stricker. Lecture Notes in Mathematics, vol. 1934 (Springer, Berlin, 2008), pp. 137–159
9. P. Billingsley, *Probability and Measure*, 3rd edn. Wiley Series in Probability and Mathematical Statistics (Wiley, New York, 1995)
10. N.H. Bingham, C.M. Goldie, J.L. Teugels, *Regular Variation*. Encyclopedia of Mathematics and its Applications, vol. 27 (Cambridge University Press, Cambridge, 1989)
11. M. Braun, *Differential Equations and Their Applications*, 4th edn. (Springer, New York, 1993)
12. P. Carmona, F. Petit, M. Yor, On the distribution and asymptotic results for exponential functionals of Lévy processes, in *Exponential Functionals and Principal Values Related to Brownian Motion*. Bibl. Rev. Mat. Iberoamericana (Rev. Mat. Iberoamericana, Madrid, 1997), pp. 73–130
13. R.A. Doney, R.A. Maller, Stability and attraction to normality for Lévy processes at zero and infinity. J. Theor. Probab. **15**, 751–792 (2002)
14. K.B. Erickson, R.A. Maller, Generalised Ornstein-Uhlenbeck processes and the convergence of Lévy integrals, in *Séminaire de Probabilités XXXVIII*, ed. by M. Emery, M. Ledoux, M., Yor. Lecture Notes in Mathematics, vol. 1857 (Springer, Berlin, 2005), pp. 70–94

15. H.K. Gjessing, J. Paulsen, Present value distributions with applications to ruin theory and stochastic equations. Stoch. Process. Appl. **71**, 123–144 (1997)
16. B. Haas, V. Rivero, Quasi-stationary distributions and Yaglom limits of self-similar Markov processes. Stoch. Process. Appl. **122**, 4054–4095 (2012)
17. Z.J. Jurek, J.D. Mason, *Operator-Limit Distributions in Probability Theory* (Wiley, New York, 1993)
18. Z.J. Jurek, W. Vervaat, An integral representation for self-decomposable Banach space valued random variables. Z. Wahrsch. Verw. Gebiete **62**, 247–262 (1983)
19. O. Kallenberg, *Foundations of Modern Probability*, 2nd edn. (Springer, Berlin, 2001)
20. A. Kuznetsov, J.C. Pardo, M. Savov, Distributional properties of exponential functionals of Lévy processes. Electron. J. Probab. **17**, 1–35 (2012)
21. T. Liggett, *Continuous Time Markov Processes. An Introduction.* AMS Graduate Studies in Mathematics, vol. 113 (American Mathematical Society, Providence, RI, 2010)
22. A. Lindner, R. Maller, Lévy integrals and the stationarity of generalised Ornstein-Uhlenbeck processes. Stoch. Process. Appl. **115**, 1701–1722 (2005)
23. J.G. Llavona, *Approximation of Continuously Differentiable Functions.* Mathematics Studies, vol. 130 (North-Holland, Amsterdam, 1986)
24. T. Nilsen, J. Paulsen, On the distribution of a randomly discounted compound Poisson process. Stoch. Process. Appl. **61**, 305–310 (1996)
25. J.C. Pardo, V. Rivero, K. van Schaik, On the density of exponential functionals of Lévy processes. Bernoulli **19**(5A), 1938–1964 (2013)
26. K. Sato, Class L of multivariate distributions and its subclasses. J. Multivar. Anal. **10**, 207–232 (1980)
27. K. Sato, Transformations of infinitely divisible distributions via improper stochastic integrals. ALEA **3**, 67–110 (2007)
28. K. Sato, *Lévy Processes and Infinitely Divisible Distributions*, revised edn. (Cambridge University Press, Cambridge, 2013)
29. R.L. Schilling, R. Song, Z. Vondracek, *Bernstein Functions. Theory and Applications*, 2nd edn. De Gruyter Studies in Mathematics, vol. 37 (Walter de Gruyter, Berlin, 2012)
30. F.W. Steutel, K. van Harn, *Infinite Divisibility of Probability Distributions on the Real Line* (Dekker, New York, 2003)
31. S.J. Wolfe, On a continuous analogue of the stochastic difference equation $X_n = \rho X_{n-1} + B_n$. Stoch. Process. Appl. **12**, 301–312 (1982)

t-Martin Boundary of Killed Random Walks in the Quadrant

Cédric Lecouvey and Kilian Raschel

Abstract We compute the t-Martin boundary of two-dimensional small steps random walks killed at the boundary of the quarter plane. We further provide explicit expressions for the (generating functions of the) discrete t-harmonic functions. Our approach is uniform in t, and shows that there are three regimes for the Martin boundary.

1 Introduction and Main Results

1.1 Aim of This Paper

This work is concerned with discrete t-harmonic functions associated to Laplacian operators with Dirichlet conditions in the quarter plane. Let $\{p_{k,\ell}\}$ be non-negative numbers summing to 1. Consider the associated discrete t-Laplacian, acting on functions f defined on the quarter plane $\mathbf{N}^2 = \{0, 1, 2, \ldots\}^2$ by

$$L_t(f)(i,j) = \sum_{k,\ell} p_{k,\ell} f(i+k, j+\ell) - t \cdot f(i,j), \qquad \forall i,j \geq 1.$$

Our aim is to characterize the functions $f = \{f(i,j)\}_{i,j\geq 0}$ which are

1. t-harmonic in the interior of the quarter plane, i.e., $L_t(f)(i,j) = 0$ for all i, $j \geq 1$;
2. positive in the interior of the quarter plane: for all $i,j \geq 1, f(i,j) > 0$;

C. Lecouvey
Laboratoire de Mathématiques et Physique Théorique, Université de Tours, Parc de Grandmont, 37200 Tours, France
e-mail: Cedric.Lecouvey@lmpt.univ-tours.fr

K. Raschel (✉)
CNRS & Pacific Institute for the Mathematical Sciences, Vancouver, BC, Canada

Laboratoire de Mathématiques et Physique Théorique, Université de Tours, Parc de Grandmont, 37200 Tours, France
e-mail: Kilian.Raschel@lmpt.univ-tours.fr

C. Donati-Martin et al. (eds.), *Séminaire de Probabilités XLVIII*, Lecture Notes in Mathematics 2168, DOI 10.1007/978-3-319-44465-9_11

3. zero on the boundary and at the exterior of the quarter plane: for all $(i,j) \in \mathbf{Z}^2$ such that $i \leq 0$ and/or $j \leq 0, f(i,j) = 0$.

The probabilistic counterpart of this potential theory viewpoint is the following: t-harmonic functions satisfying to (1)–(3) are t-harmonic for random walks (whose increments have the law $\{p_{k,\ell}\}$) killed at the boundary of the quarter plane.

1.2 Literature

In general, it is a difficult problem to determine the Martin boundary (essentially, the set of harmonic functions) of a given class of Markov chains, especially for non-homogeneous processes (in our case, the inhomogeneity comes from the boundary of the quadrant). The t-Martin boundary plays a crucial role to determine the Martin boundary (i.e., the t-Martin boundary with $t = 1$) of products of transition kernels [9, 15]. Moreover, via the procedure of Doob h-transform, discrete harmonic functions have also applications to defining random processes conditioned on staying in given domains of $\mathbf{Z}^d$ (the latter processes arise a great interest in the mathematical community, as they appear in several distinct domains: quantum random walks [2, 3], random matrices, non-colliding random walks [13]). Further details and motivations of considering the t-Martin boundary can be found in [9, Introduction].

For non-zero drift random walks in cones, the Martin boundary has essentially been found for very particular cones, as half spaces $\mathbf{N} \times \mathbf{Z}^{d-1}$ and orthants $\mathbf{N}^d$. In [10, Corollary 1.1] it has been found in the case of $\mathbf{N}^2$ for random walks with exponential moments, using ratio limit theorems for local processes and large deviation techniques. The Martin boundary was proved to be homeomorphic to $[0, \pi/2]$. In [12], under the small steps and non-degeneration hypotheses, namely,

(a) the $p_{k,\ell}$ are 0 as soon as $|k| > 1$ and/or $|\ell| > 1$,
(b) in the (clockwise) list $p_{1,1}, p_{1,0}, p_{1,-1}, p_{0,-1}, p_{-1,-1}, p_{-1,0}, p_{-1,1}, p_{0,1}$, there are no three consecutive zeros,

the exact asymptotics of Green functions was obtained, and a similar result as in [10] on the Martin boundary was derived. In [10, 12], no explicit expressions for the harmonic functions were provided.

For random walks with zero drift, the results are rarer, and typically require a strong underlying structure: the random walks are cartesian products in [15], they are associated with Lie algebras in [2, 3], etc. Last but not least, knowing the harmonic functions for zero drift random walks in $\mathbf{N}^{d-1}$ is necessary for constructing harmonic functions of walks with drift in $\mathbf{N}^d$, see [8]. The first systematical result was obtained in [17]: under (a)–(c), where (c) is the zero drift hypothesis

(c) $\sum_{k,\ell} k p_{k,\ell} = \sum_{k,\ell} \ell p_{k,\ell} = 0$, it was proved that there is a unique discrete harmonic function (up to multiplicative factors). In [17], there is also an explicit expression for the generating function

$$H(x,y)=\sum_{i,j\geq 1} f(i,j)x^{i-1}y^{j-1} \tag{1}$$

of the values of the harmonic function. Finally, this uniqueness result is extended in [4] to a much larger class of transition probabilities and dimension.

The are less examples of studies of t-Martin boundary. One of them is [9], for reflected random walks in half-spaces.

1.3 Main Results

Our main results are on the structure of the t-Martin boundary (Theorem 1) and on the explicit expression of the t-harmonic functions (Theorem 2).

Define t_0 by

$$t_0 = \min_{a\in\mathbf{R}^2} \phi(a), \tag{2}$$

where we have noted

$$\phi(a)=\phi(a_1,a_2)=\sum_{k,\ell} p_{k,\ell}e^{ka_1}e^{\ell a_2}. \tag{3}$$

Notice that $t_0 \in (0,1]$, and $t_0 = 1$ if and only if (c) holds (Fig. 1).

Theorem 1 *For any random walk satisfying to (a)–(b), the t-Martin boundary is,*

(i) *for $t > t_0$, homeomorphic to a segment S_t (with non-empty interior);*
(ii) *for $t = t_0$, reduced to one point;*
(iii) *for $t < t_0$, empty.*

For $t = 1$ and non-zero drift, Theorem 1 (i) is proved in [10, 12]; for $t = 1$ and zero drift, Theorem 1 (ii) is obtained in [17, Theorem 12]. Theorem 1 (iii) follows from general results on Markov kernels, see, e.g., [16].

Fig. 1 Location of t_0 in the non-zero drift case (*left*) and in the zero drift case (*right*). There are three regimes for the t-Martin boundary: empty (*green, left*), reduced to one point (*red, middle*), homeomorphic to a segment (*blue, right*), see Theorem 1

We shall give two proofs of Theorem 1. The first one is based on a functional equation satisfied by the generating function (1) of any t-harmonic function [see (6)]. This method (solving the functional equation via complex analysis) was introduced in [17] for the case $t = 1$ and zero drift. As we shall see, it has the following advantages: it works for any t; the critical value t_0 appears very naturally; finally, it provides an expression for the t-harmonic functions (see our Theorem 2).

The second proof is based on an exponential change of measure, which allows to reduce the general case to the case $t = 1$. In particular, with this second method, Theorem 1 can be extended to a much larger class than those satisfying to (a)–(b): namely, Theorem 1 (i) to the class of random walks whose increments have exponential moments (thanks to [10, Corollary 1.1]), and Theorem 1 (ii) to random walks with bounded symmetric jumps (thanks to [4, Theorem 1]).

Our second theorem provides an explicit expression for the generating function (1). We recall that a t-harmonic function $f > 0$ is said to be minimal if for any t-harmonic function $g > 0$, the inequality $g \leq f$ implies the equality $g = c \cdot f$, for some $c > 0$. Notice that

$$H(x,0) = \sum_{i\geq 1} f(i,1)x^{i-1}, \qquad H(0,y) = \sum_{j\geq 1} f(1,j)y^{j-1} \tag{4}$$

are the generating functions of the values of the harmonic function above/on the right of the coordinate axes. Introduce the second order (in x and y) polynomial, called the kernel,

$$L(x,y) = xy\left(\sum_{-1\leq k,\ell\leq 1} p_{k,\ell}x^{-k}y^{-\ell} - t\right). \tag{5}$$

The kernel is fully characterized by the jumps $\{p_{k,\ell}\}$.

Theorem 2 *Let $\{p_{k,\ell}\}$ be any jumps satisfying to (a)–(b), and let S_t be the segment in Theorem 1.*

In case (i) ($t \in (t_0,\infty)$), there exists a universal function w [see (19) *and* (20)*], i.e., a function depending only on the kernel L (and therefore also on t), such that for any minimal t-harmonic function $\{f(i,j)\}$, there exist $p \in S_t$ and two constants α, β [see* (13) *and* (14)*], with*

$$H(x,0) = \frac{1}{L(x,0)}\left(\frac{\alpha}{w(x) - w(p)} + \beta\right).$$

In case (ii) ($t = t_0$), the t-harmonic function is unique, up to multiplicative factors. Its expression can be obtained either as the limit of the above expression when $t \to t_0$, or directly with Eqs. (19) *and* (23).

A similar expression holds for $H(0,y)$, and finally the functional equation (6) gives the announced expression for $H(x,y)$. Theorem 2 will be stated in full details in Sect. 3.

1.4 Organization of the Paper

In Sect. 2 we state the functional equation (6), we introduce some notation, we compute the growth of t-harmonic functions (Lemmas 2 and 3), and we finally show that the generating function (4) satisfies a simple boundary value problem (Lemma 4). In Sect. 3 we solve this boundary value problem, by introducing the notion of conformal gluing functions (Definition 1). In Sect. 4 we extend our Theorem 1 to a larger class of jumps $\{p_{k,\ell}\}$, by making an exponential change of jumps (Corollary 1). In Sect. 5 we propose some remarks and a conjecture around our results. In Appendix we give an explicit expression for the conformal gluing function w of Theorem 2.

Our paper is self-contained. However, for some technical aspects of our work, specially those concerning random walks in the quarter plane, we decided to state the results without proof, referring the readers to the large existing literature (see, e.g., [5, 6, 11, 12, 17]).

2 Boundary Value Problem for the Generating Functions of Harmonic Functions

Our approach extends the one in [17], and consists in using the generating function $H(x,y)$ [see (1)] of the harmonic function. The key point is that this function $H(x,y)$ satisfies the functional equation

$$L(x,y)H(x,y) = L(x,0)H(x,0) + L(0,y)H(0,y) - L(0,0)H(0,0), \tag{6}$$

where L is defined in (5).

The proof of (6) simply comes from multiplying the relation $L_t(f)(i,j) = 0$ by $x^i y^j$ and then from summing w.r.t. $i,j \geq 1$. In (6), the variables x and y can be seen as formal variables, but they will mostly be used as complex variables.

This section is organized as follows: we first study important properties of the kernel (5). Then we are interested in the regularity (as complex functions) of $H(x,0)$ and $H(0,y)$, which is related to the exponential growth of harmonic functions. Then we state a boundary value problem satisfied by these generating functions.

2.1 Notations

The kernel $L(x, y)$ in (5) can also be written

$$L(x,y) = \alpha(x)y^2 + \beta(x)y + \gamma(x) = \tilde{\alpha}(y)x^2 + \tilde{\beta}(y)x + \tilde{\gamma}(y), \tag{7}$$

where (without loss of generality, we assume that $p_{0,0} = 0$)

$$\begin{cases} \alpha(x) = p_{-1,-1}x^2 + p_{0,-1}x + p_{1,-1}, \\ \beta(x) = p_{-1,0}x^2 - tx + p_{1,0}, \\ \gamma(x) = p_{-1,1}x^2 + p_{0,1}x + p_{1,1}, \\ \tilde{\alpha}(y) = p_{-1,-1}y^2 + p_{-1,0}y + p_{-1,1}, \\ \tilde{\beta}(y) = p_{0,-1}y^2 - ty + p_{0,1}, \\ \tilde{\gamma}(y) = p_{1,-1}y^2 + p_{1,0}y + p_{1,1}. \end{cases}$$

We also define

$$\delta(x) = \beta(x)^2 - 4\alpha(x)\gamma(x), \qquad \tilde{\delta}(y) = \tilde{\beta}(y)^2 - 4\tilde{\alpha}(y)\tilde{\gamma}(y), \tag{8}$$

which are the discriminants of the polynomial $L(x, y)$ as a function of y and x, respectively. The following facts regarding the polynomial δ are proved in [6, Chap. 2] for $t = 1$, their proof for general values of $t \geq t_0$ [t_0 being defined in (2)] would be similar: under (a)–(b), δ has degree (in x) three or four. We denote its roots by $\{x_\ell\}_{1\leq\ell\leq4}$, with

$$|x_1| \leq |x_2| \leq |x_3| \leq |x_4|, \tag{9}$$

and $x_4 = \infty$ if δ has degree three. We have $x_1 \in [-1, 1)$, $x_4 \in (1, \infty) \cup \{\infty\} \cup (-\infty, -1]$, and $x_2, x_3 > 0$. Further $\delta(x)$ is negative on $\mathbf{R}$ if and only if $x \in (x_1, x_2) \cup (x_3, x_4)$. The polynomial $\tilde{\delta}$ in (8) and its roots $\{y_\ell\}_{1\leq\ell\leq4}$ satisfy similar properties.

In what follows, we define the algebraic functions $X(y)$ and $Y(x)$ by $L(X(y), y) = 0$ and $L(x, Y(x)) = 0$. With (7) and (8) we have the obvious expressions

$$X(y) = \frac{-\tilde{\beta}(y) \pm \sqrt{\tilde{\delta}(y)}}{2\tilde{\alpha}(y)}, \qquad Y(x) = \frac{-\beta(x) \pm \sqrt{\delta(x)}}{2\alpha(x)}. \tag{10}$$

The functions $X(y)$ and $Y(x)$ both have two branches, called X_0, X_1 and Y_0, Y_1, which are meromorphic on the cut planes $\mathbf{C} \setminus ([y_1, y_2] \cup [y_3, y_4])$ and $\mathbf{C} \setminus ([x_1, x_2] \cup [x_3, x_4])$, respectively. The numbering of the branches can be chosen so as to satisfy $|X_0(y)| \leq |X_1(y)|$ (resp. $|Y_0(x)| \leq |Y_1(x)|$) on the whole of the cut planes, see [6, Theorem 5.3.3].

Note that except $\alpha, \gamma, \tilde{\alpha}, \tilde{\gamma}$, all quantities defined above depend on t.

2.2 Growth of t-Harmonic Functions

By definition, the exponential growth of a sequence $\{u_i\}$ of positive real numbers is $\limsup_{i\to\infty} u_i^{1/i}$. We first identify (Lemma 1) the exponential growth of $\{f(i,1)\}$ and $\{f(1,j)\}$ for $t = 1$, and then (Lemma 2) we treat the general case in t.

2.2.1 First Case: $t = 1$

Consider on $\mathbf{R}^2$ the function ϕ defined by (3), and define the set $D_1 = \{a \in \mathbf{R}^2 : \phi(a) \leq 1\}$ and its boundary $\partial D_1 = \{a \in \mathbf{R}^2 : \phi(a) = 1\}$. If the drift is zero (hypothesis (c)), the set D_1 is reduced to $\{0\}$, see [7, Proposition 4.3]. If not, it is homeomorphic to the unit disc. More precisely, for $a \in \partial D_1$, let $q(a) = \frac{\nabla\phi(a)}{|\nabla\phi(a)|} \in \mathbf{S}^1$ (the unit circle). If the drift is non-zero, the function q is a homeomorphism between ∂D_1 and $\mathbf{S}^1$, see [7, Proposition 4.4] or [10, Introduction]. Define finally $\mathbf{S}^1_+ = \mathbf{S}^1 \cap \mathbf{R}^2_+$ as well as $\Gamma_1^+ = \{a \in \partial D_1 : q(a) \in \mathbf{S}^1_+\}$. The following result is proved in [10].

Lemma 1 ([10]) *For any non-zero minimal* 1*-harmonic function* f*, there exists* $a = (a_1, a_2) \in \Gamma_1^+$ *such that the exponential growth of* $\{f(i,1)\}$ *(resp.* $\{f(1,j)\}$*) is* a_1 *(resp.* a_2*).*

(And reciprocally, any $a \in \Gamma_1^+$ is the growth of a minimal 1-harmonic function.) Lemma 1 follows from Eq. (1.3) in [10], which gives the structure of any minimal harmonic function. It holds a priori only in the case of a non-zero drift, but it turns out to be also true in the zero drift case, as there is then no exponential growth, i.e., $a_1 = a_2 = 0$, which is guaranteed by Raschel [17, Lemma 2].

2.2.2 General Case in t

We introduce $D_t = \{a \in \mathbf{R}^2 : \phi(a) \leq t\}$ as well as (with obvious notation) ∂D_t and Γ_t^+.

The function ϕ is strictly convex on $\mathbf{R}^2$, and due to the hypothesis (b) it admits a global minimum on $\mathbf{R}^2$. Let t_0 as in (2). Note that $t_0 \leq 1$ (evaluate ϕ at 0) and that $t_0 = 1$ if and only if the drift is zero (see [7, Proposition 4.3]). The following result extends Lemma 1 to t-harmonic functions.

Lemma 2 *Let* $t \geq t_0$*. For any non-zero minimal* t*-harmonic function* f*, there exists* $a \in \Gamma_t^+$ *such that the exponential growth of* $\{f(i,1)\}$ *(resp.* $\{f(1,j)\}$*) is* a_1 *(resp.* a_2*).*

(And reciprocally, any $a \in \Gamma_t^+$ is the growth of a minimal t-harmonic function.) Lemma 2 could be proved along the same lines as Lemma 1, but it can also be obtained thanks to the exponential change of the parameters $\{p_{i,j}\}$ presented in Sect. 4. As Lemma 1, Lemma 2 holds for any value of the drift.

2.2.3 Reformulation in Terms of the Kernel

Lemma 2 can be reformulated as follows, in terms of quantities related to the kernel (5). This will be more convenient for our analysis.

Lemma 3 *Let $t \geq t_0$. For any non-zero minimal t-harmonic function f, there exists $p \in [x_2, X(y_2)]$ (resp. $p' \in [y_2, Y(x_2)]$) with $p' = Y_0(p)$ (or $p = X_0(p')$), such that the exponential growth of $\{f(i,1)\}$ (resp. $\{f(1,j)\}$) is $1/p$ (resp. $1/p'$).*

Proof We first notice that $\phi(a) = t$ if and only if $L(1/e^{a_1}, 1/e^{a_2}) = 0$, see (3) and (5). Moreover, the real and positive points of $\{(x,y) \in \mathbf{C}^2 : L(x,y) = 0\}$ are (see [6, 11])

$$\mathscr{P} = \{(x, Y_0(x)) : x \in [x_2, x_3]\} \cup \{(x, Y_1(x)) : x \in [x_2, x_3]\}.$$

The fact that $a \in \Gamma_t^+$ implies that $x \in [x_2, X(y_2)]$, since the normal to the curve $\mathscr{P}$ at x_2 (resp. $X(y_2)$) is $(-1, 0)$ (resp. $(0, -1)$). □

As a consequence of Lemmas 2 and 3, we obtain a proof of Theorem 1 (i). We shall give more details on the proof of Theorem 1 in Sect. 3.

2.3 A Boundary Value Problem

In this section we prove that the function $L(x,0)H(x,0)$ satisfies a simple boundary value problem. A boundary value problem is composed of a boundary condition (Lemma 4) and a regularity condition (Lemma 5).

With the previous notation we introduce

$$\mathscr{M} = X([y_1, y_2]) = X_0([y_1, y_2]) \cup X_1([y_1, y_2]).$$

This curve is symmetrical w.r.t. the real axis, since $\tilde{\delta}$ is non-positive on $[y_1, y_2]$, and hence the two branches X_0 and X_1 are complex conjugate on that interval. See Fig. 2 for an example of curve $\mathscr{M}$.

Denote by $\overline{x}$ the complex conjugate of $x \in \mathbf{C}$.

Lemma 4 *We have the boundary condition: for all x in $\mathscr{M}$,*

$$L(x,0)H(x,0) - L(\overline{x},0)H(\overline{x},0) = 0.$$

We have a similar equation for $L(0,y)H(0,y)$ on the curve $\mathscr{L} = Y([x_1, x_2])$.

Proof Lemma 4 is classical; see [17, Sect. 2.6] for the original proof in the zero drift case. The main idea is to evaluate (6) at $(X_0(y), y)$, and then to make the difference of the two equations obtained by letting y go to $[y_1, y_2]$ from above and below in $\mathbf{C}$ (i.e., with y having a positive and then a negative imaginary part). □

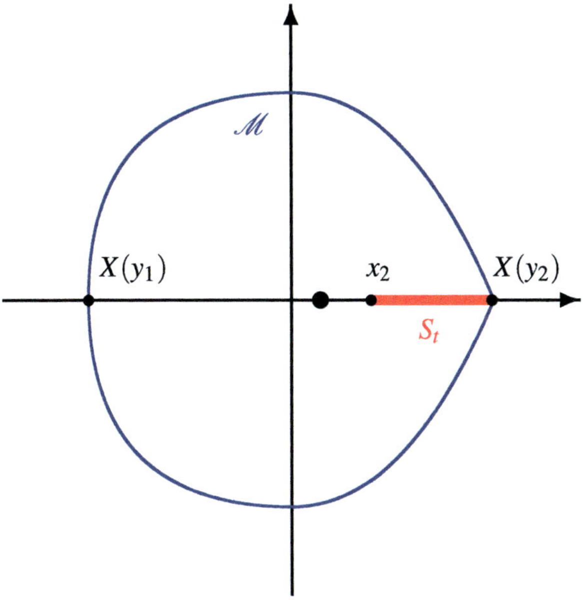

Fig. 2 The curve $\mathscr{M} = X([y_1, y_2])$ is symmetrical w.r.t. the real axis. It is smooth everywhere except at $X(y_2)$, where it may have a corner point (if and only if $t = t_0$). Any $1/p$, with p point of the *red* segment S_t, is the exponential growth of a harmonic function

Lemma 3 implies that: to any minimal t-harmonic function $\{f(i,j)\}$ we can associate a number $p \in [x_2, X(y_2)]$ such that $1/p$ (resp. $1/p'$, with $p' = Y_0(p)$) is the exponential growth of $\{f(i, 1)\}$ (resp. $\{f(1,j)\}$). We write $\{f_p(i,j)\}$ and $H_p(x, y)$ to emphasize this exponential growth.

Let $\mathscr{D}_{\mathscr{M}}$ be the interior domain delimited by the curve $\mathscr{M}$ (containing x_2 on Fig. 2) and $\overline{\mathscr{D}}_{\mathscr{M}} = \mathscr{D}_{\mathscr{M}} \cup \mathscr{M}$ be its closure. The lemma hereafter gives the regularity of the complex function $H_p(x, 0)$ (a similar result holds for $H_p(0, y)$).

Lemma 5 *For any $t \in [t_0, \infty)$, the generating function $H_p(x, 0)$ is meromorphic in $\mathscr{D}_{\mathscr{M}}$, and has in $\overline{\mathscr{D}}_{\mathscr{M}}$ a unique singularity, at p. The singularity is on the boundary $\mathscr{M}$ if and only if $t \neq t_0$ and $p = X(y_2)$ or if $t = t_0$. In case (i) ($t \neq t_0$), the singularity is polar. In case (ii) ($t = t_0$), it can be polar or not polar.*

Proof The case $t = t_0$ is rather special. In the case of a zero drift ($t_0 = 1$), it has been proved in [17, Lemma 3]. For other values of $t = t_0$, the proof would be completely similar.

We therefore assume that $t \neq t_0$. It follows from Lemma 3 that the function $H_p(x, 0)$ is analytic in the open disc $\mathscr{D}(0, p)$ centered at 0 and of radius p. The same holds for $H_p(0, y)$ in $\mathscr{D}(0, p')$. Consider the identity

$$L(x, 0)H_p(x, 0) + L(0, Y_0(x))H_p(0, Y_0(x)) - L(0, 0)H_p(0, 0) = 0, \tag{11}$$

which is the functional equation (6) evaluated at $(x, Y_0(x))$. The fact that (11) holds on a non-empty set is not clear a priori, and follows from Lemma 6, with x on the circle of radius p. A consequence of (11) is that $H_p(x,0)$ can be continued on the whole of $\mathscr{D}_{\mathscr{M}}$. Indeed, writing

$$\mathscr{D}_{\mathscr{M}} = \mathscr{D}(0,p) \cup (\mathscr{D}_{\mathscr{M}} \setminus \mathscr{D}(0,p)),$$

the generating function is defined through its power series in the first domain, and thanks to $H_p(0, Y_0(x))$ in the complementary domain. Further, we have that $\lim_{x\to p} |H_p(x,0)| = \infty$ (independently of the way that $x \to p$), so that p is indeed a pole, and not an essential singularity. □

The following result has been used in the proof of Lemma 5. For the proof we refer to [12, Lemma 28], which is a very close statement.

Lemma 6 *Let $x \in [x_2, X(y_2)]$. Then for all $|u| = x$, we have $|Y_0(u)| \leq Y_0(|u|)$. Furthermore, the inequality is strict, except for $u = x$.*

3 Resolution of the Boundary Value Problem: Proof of Theorems 1 and 2

3.1 Conformal Gluing Functions

Lemmas 4 and 5 imply that the function $L(x,0)H(x,0)$ belongs to the set of functions f which are meromorphic in $\mathscr{D}_{\mathscr{M}}$ and satisfy on $\mathscr{M}$ the equality $f(x) = f(\overline{x})$. This set of functions is too large to work on: for instance, $P \circ f$ still belongs to this set for any polynomial P. The good idea is to impose a minimality condition on f, and to introduce the notion of conformal gluing functions (our general reference for this is the book of Litvinchuk [14], and more specifically its second chapter).

Definition 1 A conformal gluing function w for $\mathscr{D}_{\mathscr{M}}$ is a function meromorphic and injective on $\mathscr{D}_{\mathscr{M}}$, continuous on $\overline{\mathscr{D}_{\mathscr{M}}}$ except at a finite number of points, and such that $w(x) = w(\overline{x})$ for $x \in \mathscr{M}$.

As stated in the lemma below, conformal gluing functions exist. They must have a unique singularity on $\overline{\mathscr{D}_{\mathscr{M}}}$ (of order 1 if the singularity is in the interior $\mathscr{D}_{\mathscr{M}}$), and are essentially characterized by the location of this singularity.

Lemma 7 ([14]) *Let $p \in \overline{\mathscr{D}_{\mathscr{M}}}$. Up to additive and multiplicative constants, there exists a unique conformal gluing function w for $\mathscr{D}_{\mathscr{M}}$ with a pole at p. Further, for any two conformal gluing functions w_1 and w_2, there exist $a, b, c, d \in \mathbf{C}$ with $ad - bc \neq 0$ such that $w_2 = \frac{aw_1+b}{cw_1+d}$.*

3.2 Complete Statement of Theorem 2

Let w be a conformal mapping as in Lemma 7 with a pole at $x_0 \in (X(y_1), x_2) \setminus \{0\}$ (this reference point x_0 is arbitrary), see Fig. 2 for its location. Subtracting by $w(0)$, we may assume that $w(0) = 0$.

The singularity of $L(x, 0)H_p(x, 0)$ is not located anywhere in $\overline{\mathscr{D}_{\mathscr{M}}}$, but on the segment $[x_2, X(y_2)]$, see Lemma 5. Let us call $\frac{\alpha}{w-w(p)} + \beta$ the class of conformal gluing functions with a pole at $p \in [x_2, X(y_2)]$, see Lemma 7. Our Theorem 2 will be restated as:

$$L(x, 0)H_p(x, 0) = \frac{\alpha}{w(x) - w(p)} + \beta. \tag{12}$$

In other words, the conformal gluing functions parametrized by $p \in [x_2, X(y_2)]$ offer a complete solution to our problem. Notice that expressions for the constants α and β will follow from a one or two term(s) expansion of the equality (12).

We now state the theorem in full details. Define

$$\alpha = -f(1,1) \times \begin{cases} \dfrac{p_{0,1} w(p)^2}{w'(0)} & \text{if } p_{1,1} = 0 \text{ and } p_{0,1} \neq 0, \\ \dfrac{2p_{-1,1} w(p)^2}{w''(0)} & \text{if } p_{1,1} = 0 \text{ and } p_{0,1} = 0, \\ \dfrac{(w(X_0(0)) - w(p))w(p)}{w(X_0(0))} & \text{if } p_{1,1} \neq 0, \end{cases} \tag{13}$$

and

$$\beta = p_{1,1} f(1,1) + \frac{\alpha}{w(p)}. \tag{14}$$

Theorem 3 (Complete version of Theorem 2) *Let α and β be defined in* (13) *and* (14). *We have*

$$H_p(x, 0) = \frac{1}{L(x,0)} \left(\frac{\alpha}{w(x) - w(p)} + \beta \right),$$

$$H_p(0, y) = \frac{p_{1,1} f(1,1) - L(X_0(y), 0) H_p(X_0(y), 0)}{L(0, y)}.$$

3.3 Proof of Theorem 3

Proof The proof of Theorem 3 is based on the following remark (see [17, Lemma 4], taken from [14]): if a function f is analytic on $\mathscr{D}_{\mathscr{M}}$, continuous on $\overline{\mathscr{D}_{\mathscr{M}}}$ and satisfies $f(x) = f(\overline{x})$ for $x \in \mathscr{M}$, then it must be a constant function.

Let us begin with the case $t > t_0$. Lemma 5 implies that $L(x,0)H_p(x,0)$ has a unique pole at p. Let us assume for a while that this pole is of order 1 (which will always be the case, except if $p = X(y_2)$, in which case the pole has order 2). Then by choosing suitably the value of α, the function

$$L(x,0)H_p(x,0) - \frac{\alpha}{w(x)-w(p)} - \beta \tag{15}$$

has no pole in $\mathscr{D}_{\mathscr{M}}$ (since w is injective in $\mathscr{D}_{\mathscr{M}}$, see Definition 1, the function $\frac{1}{w(x)-w(p)}$ has a pole of order 1 at p, as soon as $p \in \mathscr{D}_{\mathscr{M}}$) and is continuous on $\overline{\mathscr{D}_{\mathscr{M}}}$. The function (15) also satisfies the condition $f(x) = f(\overline{x})$ on the boundary. Hence we can use the above remark to conclude that (15) is a constant function. The value of β can be adapted so as to have that (15) is 0. To compute the exact values of the constants α and β, a series expansion of (12) around 0 is enough.

In the case $t > t_0$ but $p = X(y_2)$, the pole of $\frac{1}{w(x)-w(p)}$ at p is of order 2 (indeed, around $p \in \{X(y_1), X(y_2)\}$, the equality $w(x) = w(\overline{x})$ yields $w'(p) = 0$), and the same expression as (15) can be obtained.

The fact that $L(x,0)H_p(x,0)$ has a pole of order 1 or 2 at p follows essentially from that we are looking for positive harmonic functions $\{f(i,j)\}$. If the pole were of higher order, then the solution would have negative coefficients in its expansion near 0, see [17, Lemma 11], which is impossible.

See [17, Sects. 3.1–3.3] for the proof of Theorem 3 in the case $t = t_0 = 1$, which can be immediately adapted to the case $t = t_0 \neq 1$. □

3.4 Proof of Theorem 1

Proof In the small steps case [assumptions (a)–(b)], (i) and (ii) of Theorem 1 follow independently from Theorem 2 or from Lemma 2 (or its reformulation Lemma 3). Theorem 1 (iii) is a consequence of classical results, see [16]. □

4 Extension and Second Proof of Theorem 1

In this section we consider weights $\{p_{k,\ell}\}$ having exponential moments. This assumption implies that the function ϕ introduced in (3) is well defined on $\mathbf{R}^2$.

We note $\{f[p_{k,\ell}]\}$ the set of 1-harmonic functions, once the jumps $\{p_{k,\ell}\}$ have been fixed. For any a such that $\phi(a) = t$, we define new weights as follows (with $\langle\cdot,\cdot\rangle$ denoting the standard scalar product in $\mathbf{R}^2$)

$$p^a_{k,\ell} = p_{k,\ell}e^{\langle a,(k,\ell)\rangle}t^{-1}. \tag{16}$$

The identity $\phi(a) = t$ implies that $\sum_{k,\ell} p^a_{k,\ell} = 1$, and thus the $\{p^a_{k,\ell}\}$ can be interpreted as transition probabilities. Our main result in Sect. 4 is the following:

Proposition 1 *Assume that the $\{p_{k,\ell}\}$ have all exponential moments. Then the set of t-harmonic functions is equal to*

$$\{(i,j) \mapsto e^{\langle a_t,(i,j)\rangle} f[p^{a_t}_{k,\ell}](i,j)\},$$

for any a_t such that $\phi(a_t) = t$.

Proposition 1 is a direct consequence of the following simple correspondence between t-harmonic and 1-harmonic functions. Hereafter, we shall denote by f^a $(a \in \mathbf{R}^2)$ the function

$$f^a(i,j) = f(i,j)e^{-\langle a,(i,j)\rangle}. \tag{17}$$

Lemma 8 *For any a_t such that $\phi(a_t) = t$ and any t-harmonic function f, f^{a_t} is 1-harmonic w.r.t. the weights $\{p^{a_t}_{k,\ell}\}$.*

As a consequence of Proposition 1 and Lemma 8, we can reprove and extend Theorem 1.

Corollary 1 *Theorem 1, initially proved for small steps random walks, can be generalized as follows:*

- *Theorem 1 (i) to random walks whose increments have exponential moments,*
- *Theorem 1 (ii) to random walks with bounded symmetric jumps.*

Proof We first assume that the equation $\phi(a_t) = t$ has a unique solution. In this case a_t is the global minimizer of ϕ on $\mathbf{R}^2$ and the new weights $\{p^{a_t}_{k,\ell}\}$ have zero drift (this corresponds to $\phi'(a_t) = 0$). For random walks with bounded symmetric jumps, there exists a unique $f[p^{a_t}_{k,\ell}]$ which is 1-harmonic (up to multiplicative factors), see [4, Theorem 1]. Corollary 1 follows in this case.

We now suppose that the equation $\phi(a_t) = t$ has more than one solution (and then in fact, infinitely many). In this case the $\{p^{a_t}_{k,\ell}\}$ have non-zero drift (independently of a_t). For any choice of a_t, we can use the result [10, Corollary 1.1] for $t = 1$ (valid for random walks whose increments have exponential moments), and then with (17) and Lemma 8 we transfer it to other values of $t > t_0$. □

We now present some remarks and consequences of Proposition 1:

- Proposition 1 is independent of the choice of a_t.
- Proposition 1 is not only a result on the structure of the Martin boundary, it also provides an expression of the t-harmonic functions in terms of the 1-harmonic functions.

- The exponential factor in (17) does not affect the fact that on the boundary of the quadrant, the functions f and f^a are 0. Incidentally, this explains that the simple exponential change (17) cannot be used in other situations than killed random walks, like reflected random walks on a quadrant (see [9] for the study of the t-Martin boundary of reflected random walks on a half-space).

5 Miscellaneous

5.1 *Stable Martin Boundaries*

According to [15, Definition 2.4], the Martin boundary is stable if the Martin compactification does not depend on the eigenvalue t (with a possible exception at the critical value) and if the Martin kernels are jointly continuous w.r.t. space variable and eigenvalue.

The first item is clearly satisfied in our context (see our Theorem 1). As for the second one, it does not formally come from our results. However, it is most probably true (in this direction, see Sect. 5.5, where we show that the harmonic functions are continuous w.r.t. the eigenvalue t). For small steps random walks and $t = 1$, it is proved in [12, Remark 29] that the Martin kernel is continuous w.r.t. the space variable.

5.2 *Transformations of the Step Set and Consequences on Harmonic Functions*

It is natural to make some transformations of the step set, as $\{p_{k,\ell}\} \to \{p_{\pm k,\pm \ell}\}$, and to see the effect on the harmonic functions $\{f(i,j)\}$. In fact, the consequence will be simpler to read on the generating functions $H(x,0)$ and $H(0,y)$, without obvious implications on the coefficients $\{f(i,j)\}$.

The starting point of all our approach is the functional equation (6), and the difference between two functional equations associated with different jumps is all contained in the kernel (5).

Consider first the transformation $\{p_{k,\ell}\} \to \{p_{-k,\ell}\}$. The new kernel is $x^2L(1/x,y)$, with new branch points in x equal to the $1/x_\ell$, while the y_ℓ remain the same. The new roots of the kernel are $1/X(y)$ and $Y(1/x)$. The curve $\mathscr{L}$ is the same, and the new $\mathscr{M}$ is obtained by an inversion.

The new conformal mapping is an algebraic function of w. To find it we can proceed as in the proof of Theorem 3, by compensating the poles of w in the new curve $\mathscr{M}$ [see (15)]. Changing accordingly the values of the constants α and β and replacing $L(x,0)$ by $x^2L(1/x,0)$ yields the correct statement of Theorem 3 for the step set $\{p_{-k,\ell}\}$.

Regarding the transformation $\{p_{k,\ell}\} \to \{p_{k,-\ell}\}$, w takes the same value but $L(x,0)$, which is equal to $\gamma(x)$ in the case $\{p_{k,\ell}\}$, should be $\alpha(x)$.

Similar facts can be obtained for other transformations, or for the symmetry $\{p_{k,\ell}\} \to \{p_{\ell,k}\}$.

5.3 *Simple Random Walks*

If $p_{0,1} + p_{1,0} + p_{0,-1} + p_{-1,0} = 1$, the minimal t-harmonic functions take the form (with $p \in [x_2, X(y_2)]$ and $p' = Y_0(p)$)

$$f_p(i,j) = \begin{cases} \{(\frac{1}{p})^i - (\frac{p_{-1,0}}{p_{1,0}}p)^i\} j (\frac{1}{p'})^j & \text{if } p = x_2, \\ \{(\frac{1}{p})^i - (\frac{p_{-1,0}}{p_{1,0}}p)^i\}\{(\frac{1}{p'})^j - (\frac{p_{0,-1}}{p_{0,1}}p')^j\} & \text{if } p \in (x_2, X(y_2)), \\ i(\frac{1}{p})^i \{(\frac{1}{p'})^j - (\frac{p_{0,-1}}{p_{0,1}}p')^j\} & \text{if } p = X(y_2). \end{cases} \quad (18)$$

In the particular case $t = 1$, Eq. (18) is obtained in [12, Sect. 5.1]. By using techniques coming from representation theory, the authors of Lecouvey et al. [13] have obtained an explicit expression for one 1-harmonic function, the one equal to the probability of never hitting the cone.

The explicit expression (18) could also be obtained from Sect. 5.5 (via the computation of the generating functions $H_p(x,y)$), where we derive an expression for the function w.

5.4 *Generating Functions of Discrete Harmonic Functions as Tutte's Invariants*

The equality $L(x,0)H(x,0) = L(\overline{x},0)H(\overline{x},0)$ for $x \in \mathscr{M}$ (Lemma 4) implies that, in the terminology of Bernardi et al. [1], $L(x,0)H(x,0)$ is a Tutte's invariant (these invariants were introduced in the 1970s, when studying properly q-colored triangulations, see [18]).

In [1] the authors identity some models of quadrant walks such that their generating function can be expressed in terms of such Tutte's invariants. This illustrates that our results do not only concern Martin boundary theory, but also combinatorial problems as the enumeration of walks in the quarter plane.

5.5 A Conjecture

Our conjecture is that Theorem 1, a priori valid only for a subclass of jumps $\{p_{k,\ell}\}$ with exponential moments, can be extended as follows:

Conjecture 1 Theorem 1 is valid for any $\{p_{k,\ell}\}$ such that $\sum_{k,\ell}(k^2+\ell^2)p_{k,\ell}<\infty$ (i.e., with moments of order 2).

In the particular case of zero drift jumps $\{p_{k,\ell}\}$ (i.e., $t=t_0=1$), this conjecture is stated in [17, Conjecture 1].

Acknowledgements We would like to thank Irina Ignatiouk-Robert for very interesting discussions, in particular for suggesting Proposition 1 and Lemma 8. We acknowledge support from the project MADACA of the Région Centre. The second author would like to thank Gerold Alsmeyer and the Institut für Mathematische Statistik (Universität Münster, Germany), where the project has started. We finally thank a referee for useful comments.

Appendix: Explicit Expressions for *w*

It turns out that a suitable expression for w has been found in [6, Sect. 5.5.2]. Let us briefly explain why such a function appears in [6]. The main goal of Fayolle et al. [6] is to develop a theory for solving a functional equation satisfied by the stationary probabilities generating function of reflected random walks in the quarter plane. The functional equation in [6] is closed to ours [compare [6, Eq. (1.3.6)] with (6)]. Roughly speaking, the general solution of Fayolle et al. [6] can be expressed as

$$\int f(y)\frac{w'(y)}{w(x)-w(y)}\mathrm{d}y$$

for some function f, with the same function w as ours. Our situation is therefore simpler, since we can express the solutions directly in terms of w, without any integral.

In this section we simply state the expression of w, and we refer to [6, Chap. 5] or to [12, Sect. 3] for the details. The expression is

$$w(x)=\frac{u(x_0)}{u(x)-u(x_0)}-\frac{u(x_0)}{u(0)-u(x_0)} \tag{19}$$

(the second term $\frac{u(x_0)}{u(0)-u(x_0)}$ in (19) is to ensure that $w(0)=0$), where the function u is different in the two cases $t\in(t_0,\infty)$ and $t=t_0$.

Case $t \in (t_0, \infty)$

In that case, u can be expressed in terms of Weiertrass elliptic functions, with the formula

$$u(x) = \wp_{1,3}(s^{-1}(x) - \omega_2/2), \tag{20}$$

where

- $\wp_{1,3}$ is the Weierstrass elliptic function associated with the periods ω_1 and ω_3 defined in (21), i.e.,

$$\wp_{1,3}(\omega) = \frac{1}{\omega^2} + \sum_{n_1,n_3 \in \mathbf{Z}} \left\{ \frac{1}{(\omega - n_1\omega_1 - n_3\omega_3)^2} - \frac{1}{(n_1\omega_1 + n_3\omega_3)^2} \right\},$$

- ω_1 and ω_2 are defined as below [with δ as in (8)]:

$$\omega_1 = i\int_{x_1}^{x_2} \frac{\mathrm{d}x}{\sqrt{-\delta(x)}}, \qquad \omega_2 = \int_{x_2}^{x_3} \frac{\mathrm{d}x}{\sqrt{\delta(x)}}, \qquad \omega_3 = \int_{X(y_1)}^{x_1} \frac{\mathrm{d}x}{\sqrt{\delta(x)}}, \tag{21}$$

- $s(\omega) = g^{-1}(\wp_{1,2}(\omega))$, where $\wp_{1,2}$ is the Weierstrass elliptic function associated with the periods ω_1 and ω_2, and g^{-1} is the reciprocal function of

$$g(x) = \begin{cases} \dfrac{\delta''(x_4)}{6} + \dfrac{\delta'(x_4)}{x - x_4} & \text{if } x_4 \neq \infty, \\ \dfrac{\delta''(0)}{6} + \dfrac{\delta'''(0)x}{6} & \text{if } x_4 = \infty, \end{cases} \tag{22}$$

- $x_0 \in (X(y_1), x_2) \setminus \{0\}$ is arbitrary.

Case $t = t_0$

We have

$$u(x) = \left(\frac{\pi}{\omega_3}\right)^2 \left\{ \sin\left(\frac{\pi}{\theta}\left[\arcsin\left(\frac{1}{\sqrt{\frac{1}{3} - \frac{2g(x)}{\delta''(1)}}}\right) - \frac{\pi}{2}\right]\right)^{-2} - \frac{1}{3} \right\}, \tag{23}$$

with g as in (22) and

$$\theta = \arccos\left(-\frac{\sum_{-1\leq i,j\leq 1} ijp_{i,j}x_2^i y_2^j}{2\sqrt{\alpha(x_2)\tilde{\alpha}(y_2)}}\right).$$

Remarks

It can be shown that:

- The expressions given in (20) and (23) are a priori complicated, but it may happen that for some $\{p_{k,\ell}\}$, they become much simpler. If $p_{0,1}+p_{1,0}+p_{0,-1}+p_{-1,0}=1$ for instance, the function u is rational. More generally, if $\omega_2/\omega_3 \in \mathbf{Q}$, then u is an algebraic function. See [12, Proposition 15 and Remark 16] for further remarks on u.
- The function u is continuous w.r.t. the eigenvalue $t \in [t_0, \infty)$, see [5, Sect. 2.2].
- At $t = t_0$ we have $x_2 = x_3$ and $y_2 = y_3$ (in fact $t_0 = \inf\{t > 0 : x_2 = x_3\} = \inf\{t > 0 : y_2 = y_3\}$).

References

1. O. Bernardi, M. Bousquet-Mélou, K. Raschel, Counting quadrant walks via Tutte's invariant method. 1–13 (2015). Preprint. arXiv:1511.04298
2. P. Biane, Quantum random walk on the dual of SU(n). Probab. Theory Relat. Fields **89**, 117–129 (1991)
3. P. Biane, Minuscule weights and random walks on lattices, in *Quantum Probability & Related Topics* (World Scientific, River Edge, NJ, 1992), pp. 51–65
4. A. Bouaziz, S. Mustapha, M. Sifi, Discrete harmonic functions on an orthant in $\mathbf{Z}^d$. Electron. Commun. Probab. **20**, 1–13 (2015)
5. G. Fayolle, K. Raschel, Random walks in the quarter-plane with zero drift: an explicit criterion for the finiteness of the associated group. Markov Process. Relat. Fields **17**, 619–636 (2011)
6. G. Fayolle, R. Iasnogorodski, V. Malyshev, *Random Walks in the Quarter Plane* (Springer, Berlin, 1999)
7. P.-L. Hennequin, Processus de Markoff en cascade. Ann. Inst. H. Poincaré **18**, 109–195 (1963)
8. I. Ignatiouk-Robert, Martin boundary of a killed random walk on $\mathbf{Z}^d$. 1–49 (2009). Preprint. arXiv:0909.3921
9. I. Ignatiouk-Robert, t-Martin boundary of reflected random walks on a half-space. Electron. Commun. Probab. **15**, 149–161 (2010)
10. I. Ignatiouk-Robert, C. Loree, Martin boundary of a killed random walk on a quadrant. Ann. Probab. **38**, 1106–1142 (2010)
11. I. Kurkova, V. Malyshev, Martin boundary and elliptic curves. Markov Process. Relat. Fields **4**, 203–272 (1998)
12. I. Kurkova, K. Raschel, Random walks in $\mathbf{Z}_+^2$ with non-zero drift absorbed at the axes. Bull. Soc. Math. Fr. **139**, 341–387 (2011)
13. C. Lecouvey, E. Lesigne, M. Peigné, Random walks in Weyl chambers and crystals. Proc. Lond. Math. Soc. **104**, 323–358 (2012)

14. G. Litvinchuk, *Solvability Theory of Boundary Value Problems and Singular Integral Equations with Shift* (Kluwer, Dordrecht, 2000)
15. M. Picardello, W. Woess, Martin boundaries of Cartesian products of Markov chains. Nagoya Math. J. **128**, 153–169 (1992)
16. W. Pruitt, Eigenvalues of nonnegative matrices. Ann. Math. Stat. **35**, 1797–1800 (1964)
17. K. Raschel, Random walks in the quarter plane, discrete harmonic functions and conformal mappings. Stoch. Process. Appl. **124**, 3147–3178 (2014). With an appendix by S. Franceschi
18. W. Tutte, Chromatic sums for rooted planar triangulations. V. Special equations. Can. J. Math. **26**, 893–907 (1974)

On the Harmonic Measure of Stable Processes

Christophe Profeta and Thomas Simon

Abstract Using three hypergeometric identities, we evaluate the harmonic measure of a finite interval and of its complementary for a strictly stable real Lévy process. This gives a simple and unified proof of several results in the literature, old and recent. We also provide a full description of the corresponding Green functions. As a by-product, we compute the hitting probabilities of points and describe the non-negative harmonic functions for the stable process killed outside a finite interval.

1 Introduction and Statement of the Results

Let $L = \{L_t, \, t \geq 0\}$ be a real strictly α-stable Lévy process, with characteristic exponent

$$\Psi(\lambda) \;=\; \log(\mathbb{E}[e^{\mathrm{i}\lambda L_1}]) \;=\; -\,(\mathrm{i}\lambda)^{\alpha} e^{-\mathrm{i}\pi\alpha\rho\,\mathrm{sgn}(\lambda)}, \qquad \lambda \in \mathbb{R}. \tag{1}$$

Above, $\alpha \in (0, 2]$ is the self-similarity parameter and $\rho = \mathbb{P}[L_1 \geq 0]$ is the positivity parameter. Recall that when $\alpha = 2$, one has $\rho = 1/2$ and $\Psi(\lambda) = -\lambda^2$, so that L is a rescaled Brownian motion. When $\alpha = 1$, one has $\rho \in (0, 1)$ and L is a Cauchy process with a linear drift. When $\alpha \in (0, 1) \cup (1, 2)$ the characteristic exponent reads

$$\Psi(\lambda) \;=\; -\,\kappa_{\alpha,\rho}|\lambda|^{\alpha}(1 - \mathrm{i}\beta\tan(\pi\alpha/2)\,\mathrm{sgn}(\lambda)),$$

C. Profeta (✉)
Laboratoire de mathématiques et modélisation d'Evry, Université d'Evry-Val d'Essonne, F-91037 Evry Cedex, France
e-mail: christophe.profeta@univ-evry.fr

T. Simon
Laboratoire Paul Painlevé, Université Lille 1, F-59655 Villeneuve d'Ascq Cedex, France

Laboratoire de physique théorique et modèles statistiques, Université Paris-Sud, F-91405 Orsay Cedex, France
e-mail: simon@math.univ-lille1.fr

C. Donati-Martin et al. (eds.), *Séminaire de Probabilités XLVIII*, Lecture Notes in Mathematics 2168, DOI 10.1007/978-3-319-44465-9_12

where $\beta \in [-1, 1]$ is the asymmetry parameter, whose connection with the positivity parameter is given by Zolotarev's formula:

$$\rho = \frac{1}{2} + \frac{1}{\pi\alpha} \arctan(\beta \tan(\pi\alpha/2)),$$

and $\kappa_{\alpha,\rho} = \cos(\pi\alpha(\rho - 1/2)) > 0$ is a scaling constant. We refer e.g. to Chap. VIII in [2] for more details on this parametrization. One has $\rho \in [0, 1]$ if $\alpha < 1$ and $\rho \in [1 - 1/\alpha, 1/\alpha]$ if $\alpha > 1$. When $\alpha > 1, \rho = 1/\alpha$ or $\alpha < 1, \rho = 0$, the process L has no positive jumps, whereas it has no negative jumps when $\alpha > 1, \rho = 1 - 1/\alpha$ or $\alpha < 1, \rho = 1$.

Set $\hat{L} = -L$ for the dual process and $\hat{\rho} = 1 - \rho$ for its positivity parameter. Throughout, it will be implicitly assumed that all quantities enhanced with a hat refer to the same quantities for the dual process, that is with ρ and $\hat{\rho}$ switched. We denote by $\mathbb{P}_x$ the law of L starting from $x \in \mathbb{R}$. Introduce the harmonic measures

$$H_x(dy) = \mathbb{P}_x[L_T \in dy,\ T < \infty] \quad \text{and} \quad H^*_x(dy) = \mathbb{P}_x[L_{T^*} \in dy,\ T^* < \infty],$$

where $T = \inf\{t > 0,\ |L_t| > 1\}$ and $T^* = \inf\{t > 0,\ |L_t| < 1\}$. Observe that by spatial homogeneity and the scaling relationship

$$(\{kL_t,\ t \geq 0\}, \mathbb{P}_x) \stackrel{d}{=} (\{L_{k^\alpha t},\ t \geq 0\}, \mathbb{P}_{kx}), \qquad k > 0, \tag{2}$$

we can deduce from H_x the expression of the harmonic measure of the complementary of any closed bounded interval, whereas the knowledge of H^*_x gives that of the harmonic measure of any open bounded interval. Introduce the following notation

$$x_+ = \max(x, 0), \quad c_{\alpha,\rho} = \frac{\sin(\pi\alpha\rho)}{\pi} \quad \text{and} \quad \psi_{\alpha,\rho}(t) = (t-1)^{\alpha\hat{\rho}-1}(t+1)^{\alpha\rho-1}.$$

In the remainder of this section it will be implicitly assumed that L has jumps of both signs. The corresponding results where L has one-sided jumps, which are simpler, will be briefly described in the last section.

Theorem A

(a) For any $x \in (-1, 1)$, the measure $H_x(dy)$ has density

$$h(x, y) = c_{\alpha,\rho}\,(1+x)^{\alpha\hat{\rho}}(1-x)^{\alpha\rho}(1+y)^{-\alpha\hat{\rho}}(y-1)^{-\alpha\rho}(y-x)^{-1}$$

if $y > 1$ and $h(x, y) = \hat{h}(-x, -y)$ if $y < -1$.

(b) For any $x \in [-1,1]^c$, the measure $H_x^*(dy)$ has density

$$h^*(x,y) = c_{\alpha,\hat\rho}\,(1+y)^{-\alpha\rho}(1-y)^{-\alpha\hat\rho}\left((x+1)^{\alpha\rho}(x-1)^{\alpha\hat\rho}(x-y)^{-1}\right.$$
$$\left. -(\alpha-1)_+\int_1^x \psi_{\alpha,\rho}(t)\,dt\right)$$

if $x > 1$, and $h^*(x,y) = \hat h^*(-x,-y)$ if $x < -1$.

In the symmetric case, these computations date back to [5]—see Theorems A–C therein. Notice that the results of [5], which rely on Kelvin's transformation and the principle of unicity of potentials, deal with the more general rotation invariant stable processes on Euclidean space. In the general case, Part (a) of the above theorem was proved in Theorem 1 of [17], whereas Part (b) was recently obtained in Theorem 1 of [13]. Both methods used in [17] (coupled integral equations) and in [13] (Lamperti's representation and the Wiener-Hopf factorization) are complicated. In this paper we show that the original method of [5] works in the asymmetric case as well, thanks to elementary considerations on the hypergeometric function

$${}_2F_1\left[\begin{matrix}a,b\\c\end{matrix};z\right] = \sum_{n\geq 0}\frac{(a)_n(b)_n}{(c)_n}\frac{z^n}{n!}.$$

More precisely, we use three basic identities for the latter function, due respectively to Euler, Pfaff and Gauss, allowing to perform a simple potential analysis of the function

$$\varphi(t) = (1-t)^{-\alpha\rho}(1+t)^{-\alpha\hat\rho} \tag{3}$$

and to obtain the required generalization of the key Lemma 3.1 in [5].

Define next the killed potential measures

$$G_x(dy) = \mathbb{E}_x\left[\int_0^T \mathbf{1}_{\{L_t\in dy\}}\,dt\right] \qquad \text{and} \qquad G_x^*(dy) = \mathbb{E}_x\left[\int_0^{T^*} \mathbf{1}_{\{L_t\in dy\}}\,dt\right].$$

It is easy to see from the absolute continuity of the two killed semi-groups with respect to the original stable semigroup, that both these measures are absolutely continuous. We denote by $g(x,y)$ and $g^*(x,y)$ their respective densities on $(-1,1)$ and $[-1,1]^c$, the so-called Green functions. These functions are of central interest because they allow to invert the stable infinitesimal generator on $(-1,1)$ and on $[-1,1]^c$—see e.g. Formula (1.42) in [6] in the symmetric case. Observe that they are related to the harmonic measure and to the density of the Lévy measure of L:

$$\nu(y) = \Gamma(\alpha+1)|y|^{-\alpha-1}\left(c_{\alpha,\rho}\mathbf{1}_{\{y>0\}} + c_{\alpha,\hat\rho}\mathbf{1}_{\{y<0\}}\right), \tag{4}$$

through the integral formulæ

$$h(x,y) = \int_{(-1,1)} g(x,v)\nu(y-v)\,dv \quad \text{resp.} \quad h^*(x,y) = \int_{[-1,1]^c} g^*(x,v)\nu(y-v)\,dv$$

for all $x \in (-1,1)$ and $y \in [-1,1]^c$ resp. for all $x \in [-1,1]^c$ and $y \in (-1,1)$, which are both instances of a general formula by Ikeda-Watanabe—see Theorem 1 in [9]. For this reason, the density of the harmonic measure coincides with that of the Poisson kernel—see [6, pp. 16–17]. The closed expression of the Poisson kernel and the Green function for $(-1,1)$ in the symmetric case, and more generally for the open unit ball in the rotation invariant case, are classic results dating back to Riesz [15, 16]. We refer to [6, pp. 18–19] for more details and references, and to the whole monograph [6] for several extensions, all in the rotation invariant framework.

Theorem B *Set* $z = z(x,y) = \left|\frac{1-xy}{x-y}\right|$ *for every* $x \neq y$.

1. *For every* $x \in (-1,1)$, *one has*

$$g(x,y) = \frac{1}{\Gamma(\alpha\rho)\Gamma(\alpha\hat{\rho})} \left(\frac{y-x}{2}\right)^{\alpha-1} \int_1^z \psi_{\alpha,\rho}(t)\,dt$$

if $y \in (x,1)$, *and* $g(x,y) = \hat{g}(y,x)$ *if* $y \in (-1,x)$.

2. *For every* $x > 1$, *one has*

$$g^*(x,y) = \frac{2^{1-\alpha}}{\Gamma(\alpha\rho)\Gamma(\alpha\hat{\rho})} \left((y-x)^{\alpha-1} \int_1^z \psi_{\alpha,\rho}(t)dt - (\alpha-1)_+ \int_1^x \psi_{\alpha,\rho}(t)dt \int_1^y \psi_{\alpha,\hat{\rho}}(t)\,dt\right)$$

if $y \in (x,\infty)$, $g(x,y) = \hat{g}(y,x)$ *if* $y \in (1,x)$, *and*

$$g^*(x,y) = \frac{c_{\alpha,\hat{\rho}}\, 2^{1-\alpha}}{c_{\alpha,\rho}\,\Gamma(\alpha\rho)\Gamma(\alpha\hat{\rho})} \left((x-y)^{\alpha-1} \int_1^z \psi_{\alpha,\rho}(t)dt - (\alpha-1)_+ \int_1^x \psi_{\alpha,\rho}(t)\,dt \int_1^{|y|} \psi_{\alpha,\rho}(t)\,dt\right)$$

if $y < -1$.

Observe that in Part (b) of the above result, the condition $x > 1$ is no restriction since by duality we have $g^*(x,y) = \hat{g}^*(-x,-y)$ for every $x < -1$ and $y \in [-1,1]^c$. Part (a) was obtained as Corollary 4 of [5] in the symmetric case, and as Theorem 1 of [12] in the general case. Part (b) was proved as Theorem 4 in [13], in the only cases $\alpha \leq 1$ and $x, y > 1$. The methods of [12, 13], relying on the Lamperti transformation and an analysis of the reflected process, are complicated. In this paper, we observe that all formulæ of Theorem B can be quickly obtained from the Désiré André equation and one of the two simple lemmas leading to the proof of Theorem A.

The explicit knowledge of the Green function has a number of classical consequences. In this paper we will focus on two of them. The first one deals, in the relevant case $\alpha > 1$, with the hitting probability $\rho(x,y) = \mathbb{P}_x[T_y < T]$, where $T_y = \inf\{t > 0,\, L_t = y\}$.

Corollary 1 *Assume $\alpha > 1$ and set $z = \left|\frac{1-xy}{x-y}\right|$. For every $x, y \in (-1,1)$, one has*

$$\rho(x,y) \;=\; (\alpha-1)\left(\frac{x-y}{1-y^2}\right)^{\alpha-1}\int_1^z \psi_{\alpha,\hat\rho}(t)\,dt$$

if $x > y$, and $\rho(x,y) = \hat\rho(-x,-y)$ if $x < y$.

Observe that the above formula extends by continuity on the diagonal, with the expected property that $\mathbb{P}_x[T_x < T] = 1$. Of course, this follows from the fact that $\{x\}$ is regular for x in the case $\alpha > 1$. When $\alpha \to 2$, Corollary 1 amounts to the very standard Brownian formula

$$\mathbb{P}_x[T_y < T_1] \;=\; \frac{1-x}{1-y}\cdot$$

By the Markov property, one can deduce from Corollary 1 the harmonic measure of the set $\{y\} \cup [-1,1]^c$. Using one of our three hypergeometric identities, it is also possible to derive the asymptotic behaviour of $\mathbb{P}_x[T_y < T]$ when $x \to y$, which is fractional. Last, by spatial homogeneity and scaling, we can quickly recover the statement of Theorem 1.5 in [13]. See Remark 6 below for more detail.

We next consider non-negative harmonic functions on $(-1,1)$, which are the non-negative solutions to

$$\mathcal{L}_{\alpha,\rho} u \;\equiv\; 0$$

on $(-1,1)$, where $\mathcal{L}_{\alpha,\rho}$ is the infinitesimal generator of L. As in the Brownian case, an equivalent characterization—see e.g. [6, p. 20] in the symmetric case—is the mean-value property, which reads $\mathbb{E}_x[u(L_{\tau_U})] = u(x)$ for every open set U whose closure belongs to $(-1,1)$, where $\tau_U = \inf\{t > 0,\, L_t \notin U\}$.

Corollary 2 *The non-negative harmonic functions on $(-1,1)$ which vanish on $[-1,1]^c$ are of the type*

$$x \;\mapsto\; \lambda(1-x)^{\alpha\rho}(1+x)^{\alpha\hat\rho-1} \;+\; \mu(1+x)^{\alpha\hat\rho}(1-x)^{\alpha\rho-1}$$

with $\lambda, \mu \geq 0$.

This result might be already known—compare e.g. with Theorem 10, p. 569 in [18], although we could not find it written down explicitly in the literature. Recall that in order to obtain all non-negative harmonic functions on $(-1,1)$, one needs—see e.g. Theorem 2.6 in [6] in the symmetric case—to add to the above functions the integral of the Poisson kernel $h(x,y)$ along some suitably integrable measure on $[-1,1]^c$.

Both Corollaries 1 and 2 could be obtained for the process killed inside the interval $(-1,1)$, with analogous computations relying on Part (b) of Theorem B. But the formulæ have a rather lengthy aspect, so that we prefer leaving them to the interested reader. The remainder of the paper is as follows. In the three next sections we prove Theorem A, Theorem B, and the two Corollaries. In the last section we gather, for the sake of completeness, the corresponding formulæ in the cases of semi-finite intervals and of one-sided jumps.

2 Proof of Theorem A

As mentioned in the introduction, the argument hinges upon three classical hypergeometric identities, to be found in Theorem 2.2.1, Formula (2.2.6) and Formula (2.3.12) of [1], and which will be henceforth referred to as Euler, Pfaff and Gauss[1] formula respectively.

2.1 Proof of Part (b)

2.1.1 The Case $\alpha < 1$

We reason along the same lines as in Theorem B of [5]. Set $p_t(x)$ for the transition density of L. The following computation, left to the reader, is a well-known consequence of (1), Fourier inversion and the Fresnel integral: one has

$$\int_0^\infty p_t(z)\,dt \;=\; \Gamma(1-\alpha)\,c_{\alpha,\rho}\,z^{\alpha-1}$$

for every $z > 0$. Observe that by duality, one also has

$$\int_0^\infty p_t(z)\,dt \;=\; \int_0^\infty \hat{p}_t(-z)\,dt \;=\; \Gamma(1-\alpha)\,c_{\alpha,\hat{\rho}}\,|z|^{\alpha-1}$$

[1] Among of course many others. This one is a simple consequence of the two-dimensional structure of the space of solutions to the hypergeometric equation. Notice that it can also be obtained by Mellin-Barnes inversion. See the end of the article *Calculs asymptotiques* in Encyclopedia Universalis.

for every $z < 0$. Applying the Désiré André equation (2.1) in [5] and letting $s \to 0$ therein shows that

$$\int_{-1}^{1} u(t,y)\, H_x^*(dt) \;=\; c_{\alpha,\hat{\rho}}\, |x-y|^{\alpha-1} \tag{5}$$

for every $x > 1$ and $y \in (-1,1)$, where we have set

$$u(t,y) \;=\; \left(c_{\alpha,\rho}\, \mathbf{1}_{\{y>t\}} + c_{\alpha,\hat{\rho}}\, \mathbf{1}_{\{y<t\}}\right) |t-y|^{\alpha-1}.$$

In the symmetric case, this Abelian integral equation with constant boundary terms is solved in Sect. 3 of [5], following the method of [15]. See also [7] for the original solution, with a more general term on the left-hand side. After proving the following lemma, which remains valid for $\alpha \in (1,2)$, we will see that the pole-seeking method of [15] applies in the asymmetric case as well.

Lemma 1 *The unique positive measure on* $(-1,1)$ *satisfying*

$$\int_{-1}^{1} \hat{u}(t,y)\, \mu(dt) \;=\; 1, \qquad y \in (-1,1) \tag{6}$$

has the density $\varphi(t)$ *given in* (3).

Proof The fact that there is a unique measure solution of (6) is a standard fact in potential theory—see e.g. Theorem 1 in [14] or Proposition VI.1.15 in [4]. In our concrete context, this unicity can also be obtained by a straightforward adaptation of Lemma 4.1 in [5]. To show the lemma, we compute by a change of variable

$$\int_{-1}^{1} \hat{u}(t,y)(1-t)^{-\alpha\rho}(1+t)^{-\alpha\hat{\rho}}\, dt = c_{\alpha,\hat{\rho}} \int_0^{1+y} t^{\alpha-1}(1-y+t)^{-\alpha\rho}(1+y-t)^{-\alpha\hat{\rho}}\, dt$$
$$+ c_{\alpha,\rho} \int_0^{1-y} t^{\alpha-1}(1-y-t)^{-\alpha\rho}(1+y+t)^{-\alpha\hat{\rho}}\, dt.$$

Using two further changes of variable, the Euler formula, and the complement formula for the Gamma function, we transform the expression on the right-hand side into

$$\frac{\Gamma(\alpha)}{\Gamma(\alpha\hat{\rho})\Gamma(1+\alpha\rho)} \left(\frac{1+y}{1-y}\right)^{\alpha\rho} \left({}_2F_1\left[\begin{matrix}\alpha\rho, \alpha \\ 1+\alpha\rho\end{matrix}; \frac{y+1}{y-1}\right]\right.$$
$$\left. + \frac{\rho}{\hat{\rho}} \left(\frac{1-y}{1+y}\right)^{\alpha} {}_2F_1\left[\begin{matrix}\alpha\hat{\rho}, \alpha \\ 1+\alpha\hat{\rho}\end{matrix}; \frac{y-1}{y+1}\right]\right),$$

and then, using the notation

$$z = \frac{y+1}{y-1},$$

into

$$\frac{\Gamma(\alpha)(-z)^{\alpha\rho}}{\Gamma(\alpha\hat{\rho})\Gamma(1+\alpha\rho)}\left({}_2F_1\left[\begin{matrix}\alpha\rho,\alpha\\1+\alpha\rho\end{matrix};z\right]+\frac{\rho(-z)^{-\alpha}}{\hat{\rho}}\,{}_2F_1\left[\begin{matrix}\alpha\hat{\rho},\alpha\\1+\alpha\hat{\rho}\end{matrix};\frac{1}{z}\right]\right) = 1,$$

where the last equality follows from the Gauss formula.

Remark 1 The solution to (6) in the symmetric case was obtained in Lemma 3.1 of [5], via a reflection argument. Alternatively, the non-symmetric solution can be deduced in a constructive way, following the approach of [15, pp. 41–42] or that of [7]. Observe that the above argument is significantly shorter than in these three references.

We can now finish the proof. Introduce the changes of variables

$$t = x + \frac{1-x^2}{x-s} \quad \text{and} \quad y = x + \frac{1-x^2}{x-z}, \tag{7}$$

and observe that they map $(-1, 1)$ onto $(-1, 1)$, in a decreasing way. Plugging these changes of variables into (6) implies after some computation that

$$(x+1)^{\alpha\rho}(x-1)^{\alpha\hat{\rho}}\int_{-1}^{1}(1+s)^{-\alpha\rho}(1-s)^{-\alpha\hat{\rho}}(x-s)^{-1}\,u(s,z)\,ds = |x-z|^{\alpha-1}$$

for every $x > 1$ and $z \in (-1, 1)$. Multiplying both sides by $c_{\alpha,\hat{\rho}}$ shows the required solution to (5), which is unique by Lemma 1 and the changes of variables (7). □

Remark 2 In the following, we shall make a repeated use of the changes of variables (7), which may be written formally :

$$|1+x|^{\alpha\rho}|1-x|^{\alpha\hat{\rho}}\int|y-t|^{\alpha-1}\frac{|1+t|^{-\alpha\rho}|t-1|^{-\alpha\hat{\rho}}}{|x-t|}\,dt = |x-y|^{\alpha-1}\int|z-s|^{\alpha-1}\varphi(s)ds.$$

The interest of this change of variable is to transform an Abelian integral with two inside parameters into an integral of the hypergeometric type, with one parameter inside.

2.1.2 The Case $\alpha > 1$

We follow the method of Theorem C in [5]. Recall that since L a.s. hits points in finite time, the measure $H_x^*(dt)$ has total mass one. We will need the evaluation

$$\int_0^\infty (p_t(z) - p_t(0))\, dt \;=\; \Gamma(1-\alpha)\, c_{\alpha,\rho}\, z^{\alpha-1}$$

for every $z > 0$, which is easy and classical—see the introduction of [14]. This implies

$$\left(\int_0^\infty e^{-st} p_t(z)\, dt \;-\; p_1(0)\Gamma(1-1/\alpha)s^{\frac{1}{\alpha}-1}\right) \downarrow \Gamma(1-\alpha)\, c_{\alpha,\rho}\, z^{\alpha-1}$$

as $s \to 0$, for every $z > 0$. Proceeding as in [5, pp. 544–545] shows that

$$c_{\alpha,\hat\rho}\, |x-y|^{\alpha-1} \;=\; \int_{-1}^1 u(t,y)\, H_x^*(dt) \;+\; \kappa^*_{\alpha,\rho}(x) \tag{8}$$

for every $x > 1$ and $y \in (-1, 1)$, where

$$\kappa^*_{\alpha,\rho}(x) \;=\; \frac{p_1(0)\Gamma(1-1/\alpha)}{\Gamma(1-\alpha)} \times \lim_{\lambda\to 0} \lambda^{1/\alpha-1}\left(\mathbb{E}_x\left[e^{-\lambda T^*}\right] - 1\right)$$

is a non-negative function which will be determined in the same way as in (4.1) of [5]. Multiplying both sides of (8) by $\varphi(y)$ and integrating on $(-1, 1)$ shows by Lemma 1 that

$$\kappa^*_{\alpha,\rho}(x) \;=\; \left(\int_{-1}^1 \varphi(y)\, dy\right)^{-1} \left(c_{\alpha,\hat\rho}\int_{-1}^1 (x-y)^{\alpha-1}\varphi(y)\, dy \;-\; 1\right)$$

for every $x > 1$. The next lemma, generalizing the second part of Lemma 3.1 in [5], allows to compute the right-hand side.

Lemma 2 *With the above notation, one has*

$$c_{\alpha,\hat\rho}\int_{-1}^1 (x-y)^{\alpha-1}\varphi(y)\, dy \;=\; 1 \;-\; \frac{\Gamma(1-\alpha\rho)2^{1-\alpha}}{\Gamma(\alpha\hat\rho)\Gamma(1-\alpha)}\int_1^x \psi_{\alpha,\rho}(t)\, dt$$

for every $x > 1$.

Proof As in Lemma 1, a change of variable and the Euler formula show first that

$$\sin(\pi\alpha\hat{\rho})\int_{-1}^{1}(x-y)^{\alpha-1}\varphi(y)\,dy$$

$$= \frac{\Gamma(1-\alpha\rho)}{\Gamma(\alpha\hat{\rho})\Gamma(2-\alpha)}\left(\frac{x+1}{2}\right)^{\alpha-1}{}_2F_1\left[\begin{matrix}1-\alpha,1-\alpha\hat{\rho}\\2-\alpha\end{matrix};\frac{2}{x+1}\right]$$

$$= \frac{\Gamma(1-\alpha\rho)}{\Gamma(\alpha\hat{\rho})\Gamma(2-\alpha)}\left(\frac{x-1}{2}\right)^{\alpha-1}{}_2F_1\left[\begin{matrix}1-\alpha,1-\alpha\rho\\2-\alpha\end{matrix};\frac{2}{1-x}\right],$$

where the second equality follows from the Pfaff formula. Using now the Gauss formula, we next transform

$$(-z)^{\alpha-1}{}_2F_1\left[\begin{matrix}1-\alpha,1-\alpha\rho\\2-\alpha\end{matrix};\frac{1}{z}\right]$$

$$= \frac{\Gamma(\alpha\hat{\rho})\Gamma(2-\alpha)}{\Gamma(1-\alpha\rho)} + \frac{(\alpha-1)}{\alpha\hat{\rho}}(-z)^{\alpha\hat{\rho}}{}_2F_1\left[\begin{matrix}1-\alpha\rho,\alpha\hat{\rho}\\1+\alpha\hat{\rho}\end{matrix};z\right]$$

with the notation $z=(1-x)/2$. Putting everything together and applying again the Euler formula completes the proof.

We can now finish the proof of the case $\alpha > 1$. From Lemma 2 and an easy computation, we first deduce

$$\kappa^*_{\alpha,\rho}(x) = c_{\alpha,\hat{\rho}}\,2^{\alpha-1}\frac{1}{\alpha\hat{\rho}}\left(\frac{x-1}{2}\right)^{\alpha\hat{\rho}}{}_2F_1\left[\begin{matrix}1-\alpha\rho,\alpha\hat{\rho}\\1+\alpha\hat{\rho}\end{matrix};\frac{1-x}{2}\right].$$

Coming back to (8) and reasoning as in [5, p. 552], we finally see from Lemma 1 that $H^*_x(dy)$ has density

$$c_{\alpha,\hat{\rho}}\,(x+1)^{\alpha\rho}(x-1)^{\alpha\hat{\rho}}(1+y)^{-\alpha\rho}(1-y)^{-\alpha\hat{\rho}}(x-y)^{-1} - \kappa^*_{\alpha,\rho}(x)\hat{\varphi}(y).$$

To conclude the proof, it suffices to observe by the Euler formula and a change of variable that

$$\kappa^*_{\alpha,\rho}(x)\hat{\varphi}(y)$$

$$= c_{\alpha,\hat{\rho}}\,(1+y)^{-\alpha\rho}(1-y)^{-\alpha\hat{\rho}}(\alpha-1)\,2^{\alpha-1}\frac{1}{\alpha\hat{\rho}}\left(\frac{x-1}{2}\right)^{\alpha\hat{\rho}}{}_2F_1\left[\begin{matrix}1-\alpha\rho,\alpha\hat{\rho}\\1+\alpha\hat{\rho}\end{matrix};\frac{1-x}{2}\right]$$

$$= c_{\alpha,\hat{\rho}}\,(1+y)^{-\alpha\rho}(1-y)^{-\alpha\hat{\rho}}\,(\alpha-1)\int_1^x \psi_{\alpha,\rho}(t)\,dt.$$

□

Remark 3 Since $\kappa^*_{\alpha,\rho}(x)$ is finite and positive, we can deduce from Karamata's Tauberian theorem that

$$\mathbb{P}_x[T^* > t] \;\sim\; -\frac{\Gamma(1-\alpha)\sin(\pi/\alpha)}{\pi p_1(0)}\,\kappa^*_{\alpha,\rho}(x)\, t^{1/\alpha-1} \qquad \text{as } t \to +\infty.$$

This asymptotic is given in Corollary 3 of [5] in the symmetric case, and in Theorem 2 of [14] in the asymmetric case, with a more general formulation. Notice that T^* has infinite expectation.

2.1.3 The Case $\alpha = 1$

This case is known to be more subtle from the computational point of view, because it involves logarithmic kernels. The transition density of L_t is

$$p_t(x) \;=\; \frac{c_{1,\rho}\, t}{t^2 + 2tx\cos\pi\rho + x^2}.$$

The process L does not hit points a.s. but it is recurrent, so that $H^*_x(dt)$ has total mass one. After some computation, one finds

$$\int_0^\infty (p_t(1) - p_t(x))\, dt \;=\; c_{1,\rho}\, \log|x|.$$

See also [14, p. 391]. With this formula, it is possible to finish the proof as in the case $\alpha > 1$, but the computations are lengthy and we hence prefer to invoke a simple argument relying on the Skorokhod topology. Fix $\rho \in (-1, 1)$ and let $\alpha \downarrow 1$. It follows from Corollary VII.3.6 in [10] that

$$\mathcal{L}(L^{\alpha,\rho}) \;\Rightarrow\; \mathcal{L}(L^{1,\rho})$$

with obvious notation for $L^{1,\rho}$ and $L^{\alpha,\rho}$, and where $\Rightarrow$ means weak convergence in the classical Skorokhod space. Using Remark VI.3.8 and Proposition VI.2.12 in [10], it is then easy to deduce that

$$L^{\alpha,\rho}_T \;\xrightarrow{d}\; L^{1,\rho}_T.$$

The conclusion follows from pointwise convergence of the densities $h^*(x, y)$ as $\alpha \downarrow 1$, and Scheffé's lemma. □

Remark 4 The above argument relying on a.s. continuity for the Skorokhod topology will be used repeatedly in the sequel, under the denomination Skorokhod continuity argument.

2.2 Proof of Part (a)

By the Skorokhod continuity argument, it is enough to consider the case $\alpha \neq 1$. Fixing $x \in (-1, 1)$ and proceeding as in [5, pp. 544–545], the harmonic measure $H_x(dt)$ is seen to be the unique solution of the equation

$$u(x, y) = \int_{(-1,1)^c} u(t, y)\, H_x(dt) \tag{9}$$

for every $y \in [-1, 1]^c$. In the case $\alpha < 1$, this is indeed an immediate consequence of the Markov property, leading to the corresponding equation (5). And in the case $\alpha > 1$, the well-known fact—see Lemma 4.1 in [19]—that the tail distribution of T is exponentially small at infinity implies that the perturbative term $\kappa_{\alpha,\rho}$ is zero in the corresponding equation (8). Define

$$\mu_x(dt) = \begin{cases} c_{\alpha,\rho}\, \hat{\varphi}(t)\, dt & \text{if } t \leq x, \\ c_{\alpha,\hat{\rho}}\, \varphi(t)\, dt & \text{if } t > x. \end{cases}$$

We shall deal with the two cases $y > 1$ and $y < -1$ separately.

(*i*) Let $\nu \in (-1, x)$. Applying Lemma 1 with ρ and $\hat{\rho}$ interchanged, we get

$$c_{\alpha,\rho} \int_{-1}^{\nu} |\nu - t|^{\alpha-1} \mu_x(dt) + c_{\alpha,\hat{\rho}} \int_{\nu}^{x} |\nu - t|^{\alpha-1} \mu_x(dt) + c_{\alpha,\rho} \int_{x}^{1} |\nu - t|^{\alpha-1} \mu_x(dt)$$

$$= c_{\alpha,\rho} \left(\int_{-1}^{\nu} c_{\alpha,\rho} |\nu - t|^{\alpha-1} \hat{\varphi}(t) dt + \int_{\nu}^{1} c_{\alpha,\hat{\rho}} |\nu - t|^{\alpha-1} \varphi(t) dt \right) = c_{\alpha,\rho}.$$

The changes of variable (7) implies after some rearrangement

$$\int_{[-1,1]^c} |y - t|^{\alpha-1}\, H_x(dt) = c_{\alpha,\rho} (y - x)^{\alpha-1}$$

for every $y > 1$ with the required expression for $H_x(dt)$, which is Eq. (9).

(*ii*) Take now $\nu \in (x, 1)$. Applying again Lemma 1, we have

$$c_{\alpha,\hat{\rho}} \int_{-1}^{x} |\nu - t|^{\alpha-1} \mu_x(dt) + c_{\alpha,\rho} \int_{x}^{\nu} |\nu - t|^{\alpha-1} \mu_x(dt) + c_{\alpha,\rho} \int_{\nu}^{1} |\nu - t|^{\alpha-1} \mu_x(dt)$$

$$= c_{\alpha,\hat{\rho}} \left(\int_{-1}^{\nu} c_{\alpha,\rho} |\nu - t|^{\alpha-1} \hat{\varphi}(t) dt + \int_{\nu}^{1} c_{\alpha,\hat{\rho}} |\nu - t|^{\alpha-1} \hat{\varphi}(t) dt \right) = c_{\alpha,\hat{\rho}}.$$

The same changes of variables (7) gives

$$\int_{[-1,1]^c} |y-t|^{\alpha-1}\, H_x(dt) \;=\; c_{\alpha,\hat\rho}(x-y)^{\alpha-1}$$

for every $y < -1$, which is again Eq. (9). □

Remark 5

(a) The behaviour at infinity of the distribution function of T is more mysterious than that of T^*. In the non-subordinator case it is known—see Proposition VIII.3 in [2]—that there exists κ_x positive and finite such that

$$-\log \mathbb{P}_x[T > t] \;\sim\; -\kappa_x\, t \qquad \text{as } t \to +\infty,$$

but the exact value of κ_x is unknown except in the completely asymmetric case—see [3]. We refer to Chap. 4 in [6] for more on this topic in the rotation invariant case. Notice that the result of Theorem B (a) allows to compute the expectation of T:

$$\mathbb{E}_x[T] \;=\; \int_{-1}^{1} g(x,y)\, dy \;=\; \frac{(1-x)^{\alpha\rho}(1+x)^{\alpha\hat\rho}}{\Gamma(\alpha+1)}.$$

(b) With our computations, we can also check the values of the total masses $H_x(-1,1)^c$ and $H_x^*(-1,1)$. On the one hand, Lemma 1 and the changes of variables (7) imply

$$\int_{(-1,1)^c} H_x(dt) \;=\; c_{\alpha,\rho}\int_{-1}^{x} (x-z)^{\alpha-1}\hat\varphi(z)dz \;+\; c_{\alpha,\hat\rho}\int_{x}^{1} (z-x)^{\alpha-1}\hat\varphi(z)dz \;=\; 1.$$

On the other hand, in the case $\alpha > 1$, (7) and Lemma 2 show that

$$\begin{aligned}\int_{-1}^{1} H_x^*(dt) &= c_{\alpha,\hat\rho}\int_{-1}^{1} (x-z)^{\alpha-1}\varphi(z)\, dz \;-\; \kappa^*_{\alpha,\rho}(x)\int_{-1}^{1}\hat\varphi(y)\, dy\\ &= 1 - \left(\frac{\Gamma(1-\alpha\rho)2^{1-\alpha}}{\Gamma(\alpha\hat\rho)\Gamma(1-\alpha)} + (\alpha-1)\, c_{\alpha,\hat\rho}\int_{-1}^{1}\hat\varphi(y)dy\right)\int_{1}^{x}\psi_{\alpha,\rho}(t)dt = 1,\end{aligned}$$

because

$$\int_{-1}^{1}\hat\varphi(y)dy \;=\; 2^{1-\alpha}\,\mathrm{B}(1-\alpha\rho, 1-\alpha\hat\rho) \;=\; \frac{2^{1-\alpha}\Gamma(1-\alpha\rho)}{(1-\alpha)c_{\alpha,\hat\rho}\Gamma(\alpha\hat\rho)\Gamma(1-\alpha)}.$$

In the case $\alpha = 1$, the measure H_x^* has also total mass one by continuity. In the case $\alpha < 1$, we find

$$\int_{-1}^{1} H_x^*(dt) = 1 - \mathbb{P}_x[T^* = \infty] = 1 - \frac{\Gamma(1-\alpha\rho)2^{1-\alpha}}{\Gamma(\alpha\hat{\rho})\Gamma(1-\alpha)} \int_1^x \psi_{\alpha,\rho}(t)dt,$$

in accordance with Corollary 2 of [5] and Corollary 1.2 of [13].

3 Proof of Theorem B

3.1 Proof of Part (a)

It is enough to consider the case $y > x$, the case $x > y$ following from Hunt's switching identity—see e.g. Theorem II.5 in [2]. By the Skorokhod continuity argument, it is also enough to consider $\alpha \neq 1$. Reasoning as above, the Désiré André equation yields

$$\begin{aligned} g(x,y) &= c_\alpha \left(c_{\alpha,\rho}\,(y-x)^{\alpha-1} - \int_{(-1,1)^c} u(t,y) H_x(dt) \right) \\ &= c_\alpha \left(c_{\alpha,\rho}(y-x)^{\alpha-1} - c_{\alpha,\rho} \int_{-\infty}^{-1} (y-t)^{\alpha-1} H_x(dt) \right. \\ &\quad \left. - c_{\alpha,\hat{\rho}} \int_1^{+\infty} (t-y)^{\alpha-1} H_x(dt) \right) \end{aligned}$$

with $c_\alpha = \Gamma(1-\alpha)$. Changing the variables as in (7), we deduce

$$\begin{aligned} g(x,y) &= \Gamma(1-\alpha)\, c_{\alpha,\rho}\,(y-x)^{\alpha-1} \left(1 - c_{\alpha,\hat{\rho}} \int_{-1}^{1} (z+t)^{\alpha-1} \hat{\varphi}(t) dt \right) \\ &= \Gamma(1-\alpha)\, c_{\alpha,\rho}\,(y-x)^{\alpha-1} \left(1 - c_{\alpha,\hat{\rho}} \int_{-1}^{1} (z-s)^{\alpha-1} \varphi(s) ds \right) \end{aligned}$$

and the result follows from Lemma 2, since $z > 1$. □

3.2 Proof of Part (b) in the Case $\alpha < 1$

3.2.1 The Case $y > 1$

Hunt's switching identity shows again that it is enough to consider the case $y > x$. As above, the Désiré André equation and the changes of variables (7) imply

$$g^*(x,y) = \Gamma(1-\alpha)\, c_{\alpha,\rho}\left((y-x)^{\alpha-1} - \int_{-1}^{1}(y-t)^{\alpha-1}H_x^*(dt)\right)$$
$$= \Gamma(1-\alpha)\, c_{\alpha,\rho}\,(y-x)^{\alpha-1}\left(1 - c_{\alpha,\hat{\rho}}\int_{-1}^{1}(z-u)^{\alpha-1}\varphi(u)du\right)$$

with $z > x > 1$, and we can conclude by Lemma 2.

3.2.2 The Case $y < -1$

Still using (7), we now have

$$g^*(x,y) = \Gamma(1-\alpha)\, c_{\alpha,\hat{\rho}}\left((x-y)^{\alpha-1} - \int_{-1}^{1}(t-y)^{\alpha-1}H_x^*(dt)\right)$$
$$= \Gamma(1-\alpha)\, c_{\alpha,\hat{\rho}}\,(x-y)^{\alpha-1}\left(1 - \int_{-1}^{1}(z-u)^{\alpha-1}\varphi(u)du\right)$$

with $z \in (1,x)$, and we again conclude by Lemma 2. □

3.3 Proof of Part (b) in the Case $\alpha > 1$

We only consider the case $y > x$. The case $x > y > 1$ is obtained by Hunt's switching identity and the case $y < -1$ by analogous computations. Proceeding as for Eq. (8), we first deduce

$$g^*(x,y) \;=\; \Gamma(1-\alpha)\, c_{\alpha,\rho}\left((y-x)^{\alpha-1} - \int_{-1}^{1}(y-t)^{\alpha-1}H_x^*(dt)\right) - \Gamma(1-\alpha)\kappa^*_{\alpha,\rho}(x).$$

Using Theorem A and the computations of the case $\alpha < 1$, the expression may be transformed into

$$\frac{1}{\Gamma(\alpha\rho)\Gamma(\alpha\hat{\rho})}\left(\frac{y-x}{2}\right)^{\alpha-1}\int_1^z \psi_{\alpha,\rho}(t)\,dt$$
$$-\ \Gamma(1-\alpha)\,\kappa^*_{\alpha,\rho}(x)\left(1 - c_{\alpha,\rho}\int_{-1}^{1}(y-t)^{\alpha-1}\hat{\varphi}(t)dt\right).$$

The result follows from the hat version of Lemma 2 and the expression of $\kappa^*_{\alpha,\rho}(x)$.

□

4 Proof of the Corollaries

4.1 *Proof of Corollary 1*

By duality, it is enough to consider the case $x > y$. From Part (a) of Theorem B and a change of variable, we see that $g(x,y)$ extends by continuity on the diagonal, with

$$g(y,y) = \frac{1}{(\alpha-1)\Gamma(\alpha\rho)\Gamma(\alpha\hat{\rho})}\left(\frac{1-y^2}{2}\right)^{\alpha-1}.$$

Moreover, it is clear that g vanishes on the boundary $\{|x| = 1\} \cup \{|y| = 1\}$ and is hence bounded on $(-1,1)\times(-1,1)$. By Proposition VI.4.11, Exercise VI.4.18 and Formula V.3.16 in [4], we deduce

$$\mathbb{P}_x[T_y < T] = \frac{g(x,y)}{g(y,y)}$$

and the conclusion follows by Theorem B. □

Remark 6

(a) In the case $\alpha \le 1$, the process L does not hit points, so that the problem is irrelevant. In general, one can ask for an evaluation of the probability $\mathbb{P}_x[T_I < T]$ where I is a closed subinterval of $(-1,1)$ not containing x, and T_I is its first hitting time. In the transient case $\alpha < 1$, this problem is solved theoretically as a particular instance of the so-called condenser problem—see Formula (3.4) in [8]. It is an interesting open problem to find out an explicit formula in the real stable framework.

(b) By the Markov property, we can write down the following expression for the harmonic measure $H_x^{\{y\}}(dt)$ of the set $\{y\} \cup [-1,1]^c$:

$$H_x^{\{y\}}(dt) = \rho(x,y)(\delta_{\{y\}}(dt) - H_y(dt)) + H_x(dt). \tag{10}$$

In particular, for every $x, y \in (-1, 1)$, one has

$$\mathbb{P}_x[L_T \in dt,\, T < T_y] \;=\; H_x(dt) - \rho(x,y)H_y(dt).$$

(c) It is interesting to mention that using the Gauss formula, we can deduce the asymptotic behaviour of $\mathbb{P}_x[T_y > T]$ when $x \to y$, which is fractional. For instance, if $y = 0$, one has

$$\mathbb{P}_x[T_0 > T] \underset{x\to 0+}{\sim} \frac{\Gamma(2-\alpha)\Gamma(\alpha\rho)}{\Gamma(1-\alpha\hat{\rho})}\,(2x)^{\alpha-1}$$

$$\text{and} \quad \mathbb{P}_x[T_0 > T] \underset{x\to 0-}{\sim} \frac{\Gamma(2-\alpha)\Gamma(\alpha\hat{\rho})}{\Gamma(1-\alpha\rho)}\,|2x|^{\alpha-1}.$$

(d) By (2) and spatial homogeneity, it is easy to deduce from Corollary 1 the following expression of $\tilde{\rho}(x,y) = \mathbb{P}_x[T_y < \tau]$ where $\tau = \inf\{t > 0,\, L_t > 1\}$: one finds

$$\tilde{\rho}(x,y) \;=\; (\alpha-1)\left|\frac{x-y}{1-y}\right|^{\alpha-1}\int_0^{\left|\frac{1-x}{x-y}\right|} t^{\alpha\rho-1}(t+1)^{\alpha\hat{\rho}-1}\,dt$$

if $x > y$, and $\tilde{\rho}(x,y) = \hat{\tilde{\rho}}(-x,-y)$ if $x < y$. When $y = 0$, this is Theorem 1.5 in [13], correcting a misprint (the $1 - 1/x$ in the second integral should be $-1/x$) therein. Notice that Corollary 1.6 in [13] is also analogously recovered from (10).

4.2 *Proof of Corollary 2*

By the general theory of Martin boundary—see e.g. Theorem 1 in [11], we need to compute the Martin kernels

$$M_1(x) \;=\; \lim_{y\to 1}\frac{g(x,y)}{g(0,y)} \qquad \text{and} \qquad M_{-1}(x) \;=\; \lim_{y\to -1}\frac{g(x,y)}{g(0,y)}.$$

Part (a) of Theorem B and a straightforward asymptotic analysis show that these Martin kernels exist and equal respectively

$$M_1(x) \;=\; (1-x)^{\alpha\rho-1}(1+x)^{\alpha\hat{\rho}} \qquad \text{and} \qquad M_{-1}(x) \;=\; (1+x)^{\alpha\hat{\rho}-1}(1-x)^{\alpha\rho},$$

whence the result. □

5 Final Remarks

In this section, we briefly describe the analogues of the above results in the case of semi-finite intervals and in the spectrally one-sided situation.

5.1 *The Case of Semi-finite Intervals*

By scaling and spatial homogeneity, one can deduce from Theorem A—either its Part (a) or its Part (b)—the following expression of the density of L_τ under $\mathbb{P}_x$, where $x < 1$ and $\tau = \inf\{t > 0,\, L_t > 1\}$. One finds

$$f_{L_\tau}(y) = \frac{c_{\alpha,\rho}(1-x)^{\alpha\rho}}{(y-1)^{\alpha\rho}(y-x)}.$$

This expression has been found by several authors and can be obtained in different ways (see Exercise VIII.3 in [2] and the references therein). Observe that it serves as a starting formula in [17] in order to prove Part (a) of Theorem A. Notice last that the expression extends to the case with no negative jumps, by the Skorokhod continuity argument. In the relevant case with no positive jumps $\alpha > 1, \rho = 1/\alpha$, the law of L_τ is a Dirac mass at one.

The Green function is

$$g_\tau(x,y) = \frac{(y-x)^{\alpha-1}}{\Gamma(\alpha\rho)\Gamma(\alpha\hat{\rho})} \int_0^{\frac{1-y}{y-x}} \psi_{\alpha,\rho}(t)\, dt$$

if $x < y < 1$ and $g_\tau(x,y) = \hat{g}_\tau(y,x)$ if $y < x < 1$. In the case $\alpha > 1$, the analogue of Corollary 1 which is already given in Remark 6 (d) above, can then be recovered. Finally, one finds that the non-negative harmonic functions vanishing on $(1,+\infty)$ are of the type

$$\lambda(1-x)^{\alpha\rho} + \mu(1-x)^{\alpha\rho-1}$$

with $\lambda, \mu \geq 0$, in accordance with Theorem 4 in [18] and the paragraph thereafter.

5.2 *The Case of Stable Processes with One-Sided Jumps*

By duality, it is enough to consider the two cases $\alpha < 1, \rho = 1$ and $\alpha > 1, \rho = 1/\alpha$.

5.2.1 The Case $\alpha < 1, \rho = 1$

It follows readily from the above paragraph that

$$h(x,y) = \frac{c_{\alpha,1}(1-x)^{\alpha}}{(y-1)^{\alpha}(y-x)} \mathbf{1}_{\{y>1\}}$$

for all $x \in (-1,1)$. See also Example 3 in [9] and the references therein for the expression of the density of (L_{T-}, L_T) under $\mathbb{P}_x$. Similarly, one has

$$h^*(x,y) = \frac{c_{\alpha,1}|1+x|^{\alpha}}{(1+y)^{\alpha}(y-x)} \mathbf{1}_{\{|y|<1\}}$$

for all $x < -1$ and $h^*(x,y) = 0$ for all $x > 1$. In accordance with the fact that L is a subordinator, the Green function is

$$g(x,y) = \frac{(y-x)^{\alpha-1}}{\Gamma(\alpha)} \mathbf{1}_{\{x<y\}}$$

for all $x, y \in (-1,1)$,

$$g^*(x,y) = \frac{(y-x)^{\alpha-1}}{\Gamma(\alpha)} \left(\mathbf{1}_{\{x<y<-1\}} + c_{\alpha,1} \left(\int_0^{\frac{|1+x|(y-1)}{2}} t^{\alpha-1}(1+t)^{-1}\, dt \right) \mathbf{1}_{\{y>1\}} \right)$$

for all $x < -1$, and $g^*(x,y) = g(x,y)$ for all $x > 1$. The problem of Corollary 1 is irrelevant. Finally, the non-negative harmonic functions on $(-1,1)$ vanishing on $[-1,1]^c$ are constant multiples of $(1-x)^{\alpha-1}$.

5.2.2 The Case $\alpha > 1, \rho = 1/\alpha$

Using Skorokhod continuity in Theorem A (a) and the absence of positive jumps, one has

$$H_x(dy) = c_{\alpha,1-1/\alpha}(1-x)\frac{(1+x)^{\alpha-1}|y+1|^{1-\alpha}}{(1-y)(x-y)} \mathbf{1}_{\{y<-1\}}\, dy + \mathbb{P}_x[T_1 < T]\, \delta_1(dy).$$

Either taking the limit in Remark 6 (d) or integrating the first term, we can compute the weight of the Dirac mass, and find

$$H_x(dy) = c_{\alpha,1-1/\alpha}(1-x)\frac{(1+x)^{\alpha-1}|y+1|^{1-\alpha}}{(1-y)(x-y)} \mathbf{1}_{\{y<-1\}}\, dy + \left(\frac{x+1}{2}\right)^{\alpha-1} \delta_1(dy).$$

The corresponding Green function is

$$g(x,y) = \frac{1}{\Gamma(\alpha)}\left(\left(\frac{(1-y)(1+x)}{2}\right)^{\alpha-1} - (x-y)^{\alpha-1}\mathbf{1}_{\{x>y\}}\right).$$

The hitting probabilities are

$$\mathbb{P}_x[T_y < T] = \left(\frac{1+x}{1+y}\right)^{\alpha-1}$$

for every $x \le y$, which is also a consequence of a well-known result on scale functions—see e.g. Theorem VII.8 in [2], and

$$\mathbb{P}_x[T_y < T] = \left(\frac{1+x}{1+y}\right)^{\alpha-1} - \left(\frac{2(x-y)}{1-y^2}\right)^{\alpha-1}$$

for every $x > y$. Finally, the non-negative harmonic functions on $(-1,1)$ which vanish on $[-1,1]^c$ are of the type $\lambda(1-x)^{\alpha-1}(1+x)^{\alpha-2} + \mu(1+x)^{\alpha-1}$ with $\lambda, \mu \ge 0$.

It is clear that $H_x^*(dy) = \delta_{-1}(dy)$ for all $x < -1$. To compute $H_x^*(dy)$ for $x > 1$, let us introduce $\tau^* = \inf\{t > 0,\, L_t < 1\}$. The absence of positive jumps and the formula for semi-finite intervals imply after some computation

$$\begin{aligned} H_x^*(dy) &= \mathbf{1}_{\{|y|<1\}}\mathbb{P}_x[L_{\tau^*} \in dy] + \mathbb{P}_x[L_{\tau^*} < -1]\delta_{-1}(dy) \\ &= c_{\alpha,1-1/\alpha}\left(\frac{(x-1)^{\alpha-1}(1-y)^{1-\alpha}}{x-y}\mathbf{1}_{\{|y|<1\}}dy \right. \\ &\quad \left. + \left(\int_0^{\frac{x-1}{x+1}} z^{\alpha-2}(1-z)^{1-\alpha}dz\right)\delta_{-1}(dy)\right), \end{aligned}$$

in accordance with Remark 3 in [14]—see also Proposition 1.3 in [13].

Acknowledgements Nous savons gré à Jean Jacod d'un chat instructif sur la distance de Skorokhod. C. P. a bénéficié du support de la Chaire *Marchés en Mutation*, Fédération Bancaire Française. Travail dédié à l'association Laplace-Gauss.

References

1. G.E. Andrews, R. Askey, R. Roy, *Special Functions* (Cambridge University Press, Cambridge, 1999)
2. J. Bertoin, *Lévy Processes* (Cambridge University Press, Cambridge, 1996)
3. J. Bertoin, On the first exit time of a completely asymmetric stable process from a finite interval. Bull. Lond. Math. Soc. **28**(5), 514–520 (1996)

4. R.M. Blumenthal, R.K. Getoor, *Markov Processes and Potential Theory* (Academic, New York, 1968)
5. R.M. Blumenthal, R.K. Getoor, D.B. Ray, On the distribution of first hits for the symmetric stable processes. Trans. Am. Math. Soc. **99**, 540–554 (1961)
6. K. Bogdan, T. Byczkowski, T. Kulczycki, M. Ryznar, R. Song, Z. Vondraček, *Potential Analysis of Stable Processes and Its Extensions*. Lecture Notes in Mathematics, vol. 1980 (Springer, Berlin, 2009)
7. T. Carleman, Über die Abelsche Integralgleichung mit konstanten Integrationsgrenzen. Math. Z. **15**, 111–120 (1922)
8. K.L. Chung, R.K. Getoor, The condenser problem. Ann. Probab. **5**(1), 82–86 (1977)
9. N. Ikeda, S. Watanabe, On some relations between the harmonic measure and the Lévy measure for a certain class of Markov processes. J. Math. Kyoto Univ. **2**, 79–95 (1962)
10. J. Jacod, A.N. Shiryaev, *Limit Theorems for Stochastic Processes* (Springer, Berlin, 1987)
11. H. Kunita, T. Watanabe, Markov processes and Martin boundaries. Bull. Am. Math. Soc. **69**, 386–391 (1963)
12. A.E. Kyprianou, A.R. Watson, Potential of stable processes. Séminaire de Probabilités **XLVI**, 333–343 (2014)
13. A.E. Kyprianou, J.-C. Pardo, A.R. Watson, Hitting distributions of α-stable processes via path-censoring and self-similarity. Ann. Probab. **42**(1), 398–430 (2014)
14. S.C. Port, Hitting times and potentials for recurrent stable processes. J. Anal. Math. **20**, 371–395 (1967)
15. M. Riesz, Intégrales de Riemann-Liouville et potentiels. Acta Sci. Math. Szeged. **9**, 1–42 (1938)
16. M. Riesz, Rectification au travail "Intégrales de Riemann-Liouville et potentiels". Acta Sci. Math. Szeged. **9**, 116–118 (1938)
17. B.A. Rogozin, The distribution of the first hit for stable and asymptotically stable walks on an interval. Theory Probab. Appl. **17**, 332–338 (1972)
18. M.L. Silverstein, Classification of coharmonic and coinvariant functions for a Lévy process. Ann. Probab. **8**(3), 539–575 (1980)
19. S.J. Taylor, Sample path properties of a transient stable process. J. Math. Mech. **16**, 1229–1246 (1967)

On High Moments of Strongly Diluted Large Wigner Random Matrices

Oleskiy Khorunzhiy

Abstract We consider a dilute version of the Wigner ensemble of $n \times n$ random real symmetric matrices $H^{(n,\rho)}$, where ρ denotes the average number of non-zero elements per row. We study the asymptotic properties of the moments $M_{2s}^{(n,\rho)} = \mathbf{E}\,\mathrm{Tr}(H^{(n,\rho)})^{2s}$ in the limit when n, s and ρ tend to infinity. Our main result is that the sequence $M_{2s_n}^{(n,\rho_n)}$ with $s_n = \lfloor \chi \rho_n \rfloor$, $\chi > 0$ and $\rho_n = o(n^{1/5})$ is asymptotically close to a sequence of numbers $n\hat{m}_{s_n}^{(\rho_n)}$, where $\{\hat{m}_s^{(\rho)}\}_{s\geq 0}$ are determined by an explicit recurrence that involves the second and the fourth moments of the random variables $(H^{(n,\rho)})_{ij}$, V_2 and V_4, respectively. This recurrent relation generalizes the one that determines the moments of the Wigner's semicircle law given by $m_s = \lim_{\rho\to\infty} \hat{m}_s(\rho)$, $s \in \mathbb{N}$. It shows that the spectral properties of random matrices at the edge of the limiting spectrum in the asymptotic regime of the strong dilution essentially differ from those observed in the case of the weak dilution, where the dependence on the fourth moment V_4 does not intervene.

1 Introduction, Main Results and Discussion

Spectral theory of high dimensional random matrices represents an intensively developing branch of modern mathematical physics that reveals deep links between probability theory, analysis, combinatorics and other various fields of mathematics (see monographs [1, 16]). The first studies of spectral properties of random matrices of infinitely increasing dimensions were started by E. Wigner (see e.g. [25]), where the ensemble of real symmetric matrices of the form

$$(A^{(n)})_{ij} = \frac{1}{\sqrt{n}} a_{ij} \tag{1}$$

was introduced and the limiting eigenvalue distribution of $A^{(n)}$, $n \to \infty$ was determined explicitly. The random matrix entries of $A^{(n)}$ (1) are given by jointly

O. Khorunzhiy (✉)
Université de Versailles - Saint-Quentin, 45, Avenue des Etats-Unis, 78035 Versailles, France
e-mail: oleksiy.khorunzhiy@uvsq.fr

C. Donati-Martin et al. (eds.), *Séminaire de Probabilités XLVIII*, Lecture Notes in Mathematics 2168, DOI 10.1007/978-3-319-44465-9_13

independent random variables $\{a_{ij}, i \le j\}$ that have all moments finite and the odd moments zero. At present, this ensemble is referred to as the Wigner ensemble of random matrices. It was proved in [25] that the eigenvalue distribution of $A^{(n)}$ converges in average as $n \to \infty$ to a non-random limit with the density of the semi-circle form. At present this convergence is widely known as the semicircle (or Wigner) law for random matrix ensembles.

The semicircle law was generalized in several directions. One group of generalizations concerns the properties of the probability distributions of elements a_{ij}, another one is related with the studies of the spectral norm of $A^{(n)}$ and other local properties of the eigenvalue distribution at the border of the limiting spectrum or inside of it.

A large number of works is related with various generalizations of the Wigner ensemble that involve modifications of the random matrix entries. In the present paper we study one of such generalizations given by the ensemble of dilute random matrices. We consider a family of real symmetric random matrices $\{H^{(n,\rho)}\}$ whose elements are determined by equality

$$\left(H^{(n,\rho)}\right)_{ij} = a_{ij}\, b_{ij}^{(n,\rho)}, \quad 1 \le i \le j \le n, \tag{2}$$

where $\mathfrak{A} = \{a_{ij}, 1 \le i \le j\}$ is an infinite family of jointly independent identically distributed random variables and $\mathfrak{B}_n = \{b_{ij}^{(n,\rho)}, 1 \le i \le j \le n\}$ is a family of jointly independent between themselves random variables that are also independent from $\mathfrak{A}$. We denote by $\mathbf{E} = \mathbf{E}_n$ the mathematical expectation with respect to the measure $\mathbf{P} = \mathbf{P}_n$ generated by random variables $\{\mathfrak{A}, \mathfrak{B}_n\}$. We assume that the probability distribution of random variables a_{ij} is symmetric and denote their even moments by

$$V_{2l} = \mathbf{E}(a_{ij})^{2l}, \quad l = 1, 2, 3, \ldots$$

Random variables $b_{ij}^{(n,\rho)}$ are proportional to the Bernoulli ones,

$$b_{ij}^{(n,\rho)} = \frac{1}{\sqrt{\rho}} \begin{cases} 1 - \delta_{ij}, & \text{with probability } \rho/n, \\ 0, & \text{with probability } 1 - \rho/n, \end{cases} \tag{3}$$

where $\delta_{ij} = \delta_{i,j}$ is the Kronecker δ-symbol. In the case when the dilution parameter ρ is equal to n, one gets the Wigner ensemble of real symmetric random matrices A_n (1). Let us note that the random matrix $B^{(n,\rho)}$ with the entries $\sqrt{\rho}\, b_{ij}$ (3) can be regarded as the adjacency matrix of the Erdős-Rényi random graph [3]. In this interpretation, the dilution parameter ρ represents the average degree of a given vertex of the graph.

The initial interest in the dilute versions of Wigner ensemble was motivated by theoretical physics studies (see for instance, the pioneering works [17, 18] and the review [14] for more references), where the spectral properties of large systems with a number of broken interactions were considered. This kind of random matrices is also important in the studies of various mathematical models, such as random graphs [4, 5, 7, 15] and many others.

In the present paper we study the asymptotic behavior of the moments of $H^{(n,\rho)}$ given by expression

$$M_{2s}^{(n,\rho)} = E\left(\sum_{i=1}^{n} (H^{(n,\rho)})_{ii}^{2s}\right) = E\left(\mathrm{Tr}\,(H^{(n,\rho)})^{2s}\right).$$

The moment method represents an effective tool of the spectral theory. It is used in the studies of the spectral properties of large random matrices since the pioneering works of E. Wigner [25]. In particular, the semicircle law was proved initially by the convergence of the moments $M_{2s}^{(n,n)}$ in the limit of infinite n and given s. The principal idea of the Wigner's approach is to consider the trace of the product of random matrices as the sum over the family of trajectories of $2s$ steps and then to compute the weights of these trajectories given by the mathematical expectation of the products of corresponding random variables.

The moments $M_{2s}^{(n,n)}$ of Wigner random matrices $A^{(n)}$ (1) in the limit $n \to \infty$ with infinitely increasing $s = s_n$ were studied in a long series of papers, where the eigenvalue distribution at the edge of the limiting spectra was studied in more and more details [2, 7, 8]. The crucial step has been performed in papers [19, 20], where the original Wigner's moment method has got a powerful and deep improvement. In these studies, the Tracy-Widom law for random matrices $A^{(n)}$ established in the case of normally distributed entries a_{ij} is shown to be true in the general case of arbitrary probability distribution of a_{ij} [21, 24]. This result is obtained by analysis of the high moments $M_{2s_n}^{(n,n)}$ in the limit $n, s_n \to \infty$ with $s_n = O(n^{2/3})$.

The high moments of large dilute random matrices $H^{(n,\rho_n)}$ (2) were studied in [10] in the asymptotic regime when $\rho = \rho_n = O(n^{\alpha})$ with $2/3 < \alpha < 1$. It was proved that the limiting expression of the moments $M_{2s_n}^{(n,\rho_n)}$ with $s_n = O(n^{2/3})$ coincides with that of the moments of the Wigner random matrices $M_{2s_n}^{(n,n)}$. This fact can be regarded as an evidence of the universal behavior of the local eigenvalue statistics for weakly dilute random matrices, i.e. when the dilution parameter ρ is sufficiently large. In the present paper we study the opposite asymptotic regime of strongly dilute random matrices, i.e. when the dilution parameter ρ_n tends to infinity as $n \to \infty$ but with much lower range than before, $\rho_n = O(n^{\alpha})$ with $\alpha < 1/5$. We show that in the limit

$$n, \rho_n \to \infty, \ \rho_n = o(n^{1/5}), \ s_n = \lfloor \chi \rho_n \rfloor, \quad \chi > 0, \tag{4}$$

where $\lfloor x \rfloor$ is the integer part of x, the limiting expressions of $M_{2s_n}^{(n,\rho_n)}$ are different from those obtained for the Wigner random matrix ensemble. This difference is due to the fact that the leading contribution to the moments $M_{2s}^{(n,\rho)}$ in the asymptotic regime (4) is given by the trajectories that generalize in certain sense the Catalan numbers that describe the moments of the Wigner ensemble. Up to our knowledge, these trajectories of the new type were not considered before. Their combinatorial properties are of their own interest and this fact has strongly motivated the work presented. In our studies, in particular, we obtain a number of explicit relations that were not known in the context of random matrices and plane rooted trees (see, for example, relation (9) below and formulas (106) and (108) of Sect. 5).

To make more compact the formulas we use, everywhere below we refer to the limiting transition (4) as $(n, s, \rho) \to \infty$. Our main result is given by the following statement.

Theorem 1 *Assume that* $V_2 = 1$ *and that for all* $1 \le i \le j$ *the random variables* a_{ij} *are bounded with probability 1,*

$$|a_{ij}| \le U. \tag{5}$$

There exists a constant $\chi_0 = \chi_0(U) > 0$ *such that for any given* $0 < \chi < \chi_0$ *the following upper bound holds in the limit* (4),

$$\limsup_{(n,s,\rho)\to\infty} \frac{1}{n t_s} M_{2s_n}^{(n,\rho_n)} \le 4e^{16\chi V_4}, \tag{6}$$

where

$$t_s = \frac{(2s)!}{s!\,(s+1)!}, \quad s = 0, 1, 2, \ldots \tag{7}$$

are the Catalan numbers. The moments $M_{2s_n}^{(n,\rho_n)}$ *are given by the following asymptotic relation,*

$$M_{2s_n}^{(n,\rho)} = n\hat{m}_{s_n}^{(\rho_n)}(1 + o(1)), \quad (n, s, \rho) \to \infty, \tag{8}$$

where the sequence $\{\hat{m}_s^{(\rho)}\}_{s\ge 0}$ *is such that its generating function* $F_\rho(z) = \sum_s z^s \hat{m}_s^{(\rho)}$ *verifies equation*

$$F_\rho(z) = 1 + z\left(F_\rho(z)\right)^2 + \frac{z^2 V_4}{\rho}\left(\frac{1}{1 - zF_\rho(z)}\right)^4 \tag{9}$$

with the initial condition $\hat{m}_0^{(\rho)} = 1$.

Remarks

1. We restrict the rate of ρ_n by $n^{1/5}$ (4) not to overload the technical part of the paper. In fact, it follows from the proof of Theorem 1 that relations (6) and (8) can be obtained with (4) replaced by the limit $\rho_n \to \infty$ such that $\rho_n = o(n^{1/2})$ (see formula (57) and the discussion below). Moreover, one can expect that Theorem 1 remains valid in the asymptotic regime when $\rho_n = n^\alpha$ with $0 < \alpha < 2/3$. This asymptotic regime is complementary to the one studied

in [10]. However, in the present paper we are aimed mostly at the lowest rates of ρ_n having a particular interest in the asymptotic regime when $\rho_n = O(\log n)$, $n \to \infty$.

2. In contrast to the technical restriction (4), it is not clear whether condition (5) can be essentially relaxed, especially in the case of the asymptotic regime when $\rho_n = O(\log n), n \to \infty$ and $s_n = \lfloor \chi \rho_n \rfloor$. However, a part of the estimates that concerns the tree-type walks can be proved under considerably less restricted conditions than (5) (see relation (66) of Sect. 4 below).
3. We will show that the numbers $\hat{m}_s^{(\rho)}$ are uniquely determined and verify the following upper bound [cf. (6)],

$$\frac{1}{\mathrm{t}_s} \hat{m}_s^{(\rho)} \leq 4e^{3V_4 s/\rho}. \tag{10}$$

Therefore the generating function $F_\rho(z)$ (9) exists and is bounded in absolute value for any given $\rho > 0$. Then it follows from (9) that the limiting function $f(z) = \lim_{\rho \to \infty} F_\rho(z)$ exists and verifies the following relation,

$$f(z) = 1 + z(f(z))^2. \tag{11}$$

This equation has a unique solution that determines the generating function of the Catalan numbers (7), $f(z) = \sum_{k \geq 0} \mathrm{t}_k z^k$.
4. Relation (8) can be rewritten in slightly more precise form. We will show that there exists a constant $C > 0$ such that the following relation holds

$$\limsup_{(n,s,\rho)\to\infty} \frac{\rho_n}{n \mathrm{t}_s} \left(M_{2s_n}^{(n,\rho_n)} - n\hat{m}_{s_n}^{(\rho_n)} \right) \leq C\chi e^{16\chi V_4} \tag{12}$$

in the limit $n, s_n, \rho_n \to \infty$ such that $s_n = \lfloor \chi \rho_n \rfloor$, $0 < \chi \leq \chi_0$ and $\rho_n = o(n^{1/6})$ (see Sect. 4.2 below). In fact, one can show that the left-hand side of relation (12) admits the asymptotic expansion in powers of ρ and that the first terms of this expansion are given by relation

$$\frac{1}{n\mathrm{t}_s} M_{2s}^{(n,\rho)} = \frac{1}{\mathrm{t}_s} \left(\hat{m}_s^{(\rho)} + \frac{1}{\rho} R_s^{(1)} + o(\rho^{-1}) \right), \tag{13}$$

where

$$R_s^{(1)} = \frac{4V_4^2}{\rho} \frac{(2s)!}{(s-4)!\,(s+4)!} + \frac{V_6}{\rho} \frac{(2s)!}{(s-3)!\,(s+3)!} + O(\rho^{-1})$$

and $s, \rho \to \infty$ are such that $s = \lfloor \chi \rho \rfloor$ with $\chi > 0$ (see Sect. 5.2).
5. As we will see, the numbers $\{\hat{m}_s^{(\rho)}\}_{k \geq 0}$ of (8) can be regarded as a generalization of the Catalan numbers $\mathrm{t}_k, k \geq 0$ in the following sense. The Catalan number t_k counts the half-plane rooted trees $\mathscr{T}_k$ of k edges. Regarding the chronological

run over $\mathcal{T}_k$, we get a closed walk of $2k$ steps such that in its graph each edge is passed two times, in there and back directions. Using this terminology, we can say that $\hat{m}_s^{(\rho)}$ represents the sum of weighs of all closed walks of $2s$ steps such that in their graphs each edge is passed either two or four times when counted in there and back directions. Also it is shown that in the corresponding graph the edges passed four times do not share a common vertex (see Sect. 4 for the rigorous definition of the tree-type $(2,4)^*$-walks). One of the consequences of this definition of numbers $\hat{m}_s^{(\rho)}$ is given by the following inequality (see also formula (105) of Sect. 5 below),

$$\hat{m}_s^{(\rho)} \geq \mathrm{t}_s + \frac{V_4}{\rho} \frac{(2s)!}{(s-2)!\,(s+2)!},$$

where the last fraction represents the number of closed tree-type walks of $2s$ steps such that their graphs contain exactly one edge passed four times. Then in the limiting transition (4) we get the following lower bound,

$$\liminf_{(n,s,\rho)\to\infty} \frac{1}{\mathrm{t}_s} \hat{m}_s^{(\rho)} \geq \liminf_{(n,s,\rho)\to\infty} \left(1 + V_4 \frac{s(s-1)}{\rho(s+2)}\right) = 1 + \chi V_4. \tag{14}$$

Let us discuss relations of the results of Theorem 1 with the spectral properties of large dilute random matrices $H^{(n,\rho)}$ (2). Regarding the ordered family of real eigenvalues of the matrix $H^{(n,\rho)}(\omega)$, $\lambda_1^{(n,\rho)}(\omega) \leq \cdots \leq \lambda_n^{(n,\rho)}(\omega)$, we denote its spectral norm by $\lambda_{\max}^{(n,\rho)}(\omega)$,

$$\|H^{(n,\rho)}(\omega)\| = \lambda_{\max}^{(n,\rho)}(\omega) = \max\{|\lambda_1^{(n,\rho)}(\omega)|, |\lambda_n^{(n,\rho)}(\omega)|\}.$$

The well-known analog of the Chebyshev inequality for the deviation probability

$$P\left(\lambda_{\max}^{(n,\rho)} \geq 2(1+\varepsilon)\right) \leq \frac{1}{(2(1+\varepsilon))^{2s}} E\left(\sum_{j=1}^{n} (\lambda_j^{(n,\rho)})^{2s}\right) = \frac{1}{(2(1+\varepsilon))^{2s}} M_{2s}^{(n,\rho)} \tag{15}$$

allows us to deduce from estimate (6) the following upper bound valid for all $n \geq n_0$ with some n_0,

$$P\left(\lambda_{\max}^{(n,\rho)} \geq 2(1+\varepsilon)\right) \leq 6e^{16V_4\chi} \frac{n\mathrm{t}_s}{(2(1+\varepsilon))^{2s}}.$$

Applying the Stirling formula to the Catalan numbers t_s (7), we get inequality

$$P\left(\lambda_{\max}^{(n,\rho)} \geq 2(1+\varepsilon)\right) \leq 4e^{16V_4\chi} \frac{n}{s^{3/2}(1+\varepsilon)^{2s}}. \tag{16}$$

Remembering that $s_n = \lfloor \chi \rho_n \rfloor$, we deduce from (16) that if the sequence $(\rho_n)_{n\geq 1}$ is such that $\rho_n / \log n \to \infty$ as $n \to \infty$, then for any given $\varepsilon > 0$

$$\sum_{n\geq n_0} P\left(\lambda_{\max}^{(n,\rho_n)} \geq 2(1+\varepsilon)\right) < \infty. \tag{17}$$

Relation (17) means that $P\left(\limsup_{n\to\infty} \lambda_{\max}^{(n,\rho_n)} \leq 2\right) = 1$ in this asymptotic regime.

Taking into account the fact that the semicircle law is valid for the eigenvalue distribution of $H^{(n,\rho_n)}$ in the limit $n, \rho_n \to \infty$ [14], it is not hard to conclude that with probability 1,

$$\lim_{n\to\infty} \|H^{(n,\rho_n)}\| = 2, \qquad \frac{\rho_n}{\log n} \to \infty,\ n \to \infty. \tag{18}$$

This statement slightly improves our earlier results [13], where the convergence of $\lambda_{\max}^{(n,\rho_n)}$ to 2 has been proved in the asymptotic regime when $\rho_n = (\log n)^{1+\delta}$ with given $\delta > 0$. Let us note however, that in the present article we prove Theorem 1 and (18) under condition (5) that is more restrictive than those imposed in paper [13].

Returning to inequalities (15) and (16), we can rewrite them in the following form

$$\Psi_n(x) = P\left(\lambda_{\max}^{(n,\rho)} \geq 2\left(1 + \frac{x}{h_n}\right)\right) \leq \frac{4n}{s^{3/2}} \exp\left\{2\chi\left(8V_4 - \frac{x\rho_n}{h_n}\right)\right\}. \tag{19}$$

The right-hand side of (19) shows that in the limit of infinite n, the probability to find eigenvalues of $H^{(n,\rho_n)}$ outside of the interval $(-2(1+x/h_n), 2(1+x/h_n)$ goes to zero provided h_n is much smaller than $\rho_n(\ln n)^{-1}$, i.e. the length of the corresponding interval is larger than $(\log n)/\rho_n$.

Neglecting the first factor in the right-hand side of (19), we can observe that to obtain a non-trivial and non-zero limit $\Psi_n(x) \to \Psi(x)$, it is natural to consider the scaling parameter h_n of the left-hand side of (19) to be of the order $O(\rho_n)$. This reasoning could be compared with the result widely known as the Tracy-Widom distribution for the maximal eigenvalue of Gaussian Unitary (and Orthogonal) Ensembles of random matrices of the form (1), where the limiting expression of $\Psi_n(x)$ is explicitly determined with the scaling factor $h_n = n^{-2/3}$ [24] (see also monograph [1]).

It should be stressed that in papers [21, 22], the existence of a non-trivial non-zero limit of the moments $\lim_{n\to\infty} M_{2s_n}^{(n,n)} = \mathscr{M}(\tau)$, $s_n = \lfloor \tau n^{2/3} \rfloor$ of the Wigner random matrices A_n (1) is shown to imply the Tracy-Widom distribution for the maximal eigenvalue $\lambda_{\max}$ in the case of arbitrarily distributed matrix entries a_{ij}. The limiting expression $\mathscr{M}(\tau)$ being independent from the particular values of the moments $V_{2l}, l \geq 2$ of a_{ij}, the result of [21] means a wide universality of the Tracy-Widom law for large random matrices. In paper [10] we have proved that the same limiting expression $\mathscr{M}(\tau) = \lim_{n\to\infty} M_{2s_n}^{(n,\rho_n)}$, $s_n = \lfloor \tau n^{2/3} \rfloor$ is valid for the ensemble of dilute random matrices (2) in the asymptotic regime of weak dilution, $\rho_n/n^{2/3} \to \infty$. In contrast, when the dilution becomes moderate, i.e. when $\rho_n = \zeta n^{2/3}$ with given $\zeta > 0$, the estimate from below takes the following form (see [10], Theorem 7.1),

$$\liminf_{n\to\infty} M_{2s_n}^{(n,\rho_n)} \geq \frac{4V_4}{\zeta\sqrt{\pi\tau}} e^{-e\tau^3}. \tag{20}$$

This inequality could be interpreted as an argument to support our earlier conjecture that the universality of the Tracy-Widom distribution discussed above cannot hold in the asymptotic regimes of moderate and strong dilution [10]. The fourth moment V_4 enters explicitly into the lower bound (20) while this is not so in the case of non-diluted or weakly diluted Wigner random matrices according to the statements of [21, 22].

Returning to the case of the strong dilution when $\rho_n = o(n^{1/5})$ (4), we observe that the difference with respect to the Wigner ensemble becomes even more striking the sequence $M_{2s_n}^{(n,\rho_n)}/(nt_s)$ diverges as the ratio s_n/ρ_n goes to infinity [see relation (13) and the lower bound (14)]. Moreover, we expect the exponential divergence of $M_{2s_n}^{(n,\rho_n)}/(nt_s)$ with respect to $s_n/\rho_n \approx \chi$ as χ tends to infinity (see relations (109) and (110) and their discussion at the end of Sect. 5.2). This behavior of $M_{2s_n}^{(n,\rho_n)}/(nt_s)$ can be regarded as one more argument to the conjecture that the scaling parameter h_n (19) at the edge of the spectrum of large strongly diluted random matrices has to be switched from $n^{-2/3}$ to another value related rather with ρ_n^{-1}. We postpone the study of this problem to subsequent publications.

2 Trajectories, Walks and Graphical Representations

In this section we describe the main components of the method we develop to study the high moments of dilute random matrices $H^{(n,\rho)}$ (2). In the pioneering works of E. Wigner (see e.g. [25]), it was proposed to consider the moments M_{2s} of random matrices as a weighted sum over paths of $2s$ steps. In the case of dilute random matrices, we can write that

$$M_{2s}^{(n,\rho)} = \sum_{i=1}^{n} E\left(H^{(n,\rho)}\right)_{ii}^{2s} = \sum_{\mathscr{I}_{2s}\in\mathbb{I}_{2s}(n)} \Pi_{a,b}(\mathscr{I}_{2s}) = \sum_{\mathscr{I}_{2s}\in\mathbb{I}_{2s}(n)} \Pi_a(\mathscr{I}_{2s})\,\Pi_b(\mathscr{I}_{2s}), \tag{21}$$

where the sequence $\mathscr{I}_{2s} = (i_0, i_1, \ldots, i_{2s-1}, i_0)$, $i_k \in \{1, 2, \ldots, n\}$ is regarded as a closed path of $2s$ steps (i_{t-1}, i_t) with the discrete time $t \in [0, 2s]$. We will also say that $\mathscr{I}_{2s}$ is a *trajectory* of $2s$ steps. The set of all possible trajectories of $2s$ steps over $\{1, \ldots, n\}$ is denoted by $\mathbb{I}_{2s}(n)$. The weights $\Pi_a(\mathscr{I}_{2s})$ and $\Pi_b(\mathscr{I}_{2s})$ are determined as the mathematical expectations of the products of corresponding random variables,

$$\Pi_a(\mathscr{I}_{2s}) = E\left(a_{i_0 i_1} \cdots a_{i_{2s-1} i_0}\right), \quad \Pi_b(\mathscr{I}_{2s}) = E\left(b_{i_0 i_1} \cdots b_{i_{2s-1} i_0}\right). \tag{22}$$

Here and below, we omit the superscripts in $b_{ij}^{(n,\rho)}$ when no confusion can arise.

In papers [19, 20] a deep and powerful generalization of the E. Wigner's approach was proposed by Ya. Sinai and A. Soshnikov to study the high moments of random matrices. Somehow different point of view has been developed to consider the ensembles of dilute random matrices [10, 11, 13]. The difference between these two approaches is related with the fact that the leading contribution to the moments $M_{2s}^{(n,\rho)}$ of the dilute random matrices $H^{(n,\rho)}$ (2) can be given by those trajectories $\mathscr{I}_{2s}$ that have a vanishing weight in the case of non-diluted Wigner random matrices A_n (1) studied in [19, 20]. In the first part of the present paper we use a combination of these two approaches to study the terms that determine the vanishing contribution to the moments of strongly diluted random matrices (see Sect. 3). On the second stage we develop a new method that allows us to prove that in the limit of infinitely increasing dimension, the non-zero contribution to the moments $M_{2s}^{(n,\rho)}$ is given by a new kind of tree-type walks such that their weight contains the factors V_2 and V_4 only (see Sect. 4). We refer to this kind of walks as to (2,4)-walks. To determine rigorously the corresponding classes of trajectories, we need to describe briefly the fundamental notions of the methods developed in papers mentioned above.

Regarding a trajectory $\mathscr{I}_{2s}$, one can determine a *walk*

$$\mathscr{W}_{2s} = \mathscr{W}_{2s}^{(\mathscr{I}_{2s})} = \{\mathscr{W}(t), t \in [0, 2s]\}, \quad \text{where} \quad [0, 2s] = \{0, 1, 2, \dots, 2s\},$$

that we define as a sequence of $2s + 1$ symbols (or equivalently, letters) from an ordered alphabet, say $\mathscr{A} = \{\alpha_1, \alpha_2, \dots\}$. The walk $\mathscr{W}_{2s}^{(\mathscr{I}_{2s})}$ is constructed with the help of the following rules of recurrence [11]. Given a trajectory $\mathscr{I}_{2s}$, we write that $\mathscr{I}_{2s}(t) = i_t$, $t \in [0, 2s]$ and consider a subset $\mathbb{U}(\mathscr{I}_{2s}; t) = \{\mathscr{I}_{2s}(t'), 0 \le t' \le t\} \subseteq \{1, 2, \dots, n\}$. We denote by $|\mathbb{U}(\mathscr{I}_{2s}; t)|$ its cardinality. Then

1. $\mathscr{W}_{2s}(0) = \alpha_1$;
2. if $\mathscr{I}_{2s}(t+1) \notin \mathbb{U}(\mathscr{I}_{2s}; t)$, then $\mathscr{W}_{2s}(t+1) = \alpha_{|\mathbb{U}(\mathscr{I}_{2s};t)|+1}$;

if there exists $t' \le t$ such that $\mathscr{I}_{2s}(t+1) = \mathscr{I}_{2s}(t')$, then $\mathscr{W}_{2s}(t+1) = \mathscr{W}_{2s}(t')$.

For example, $\mathscr{I}_{16} = (5, 2, 7, 9, 7, 1, 5, 2, 7, 9, 7, 2, 7, 2, 7, 1, 5)$ produces the walk

$$\mathscr{W}_{16} = (\alpha_1, \alpha_2, \alpha_3, \alpha_4, \alpha_3, \alpha_5, \alpha_1, \alpha_2, \alpha_3, \alpha_4, \alpha_3, \alpha_2, \alpha_3, \alpha_2, \alpha_3, \alpha_5, \alpha_1). \tag{23}$$

We say that the pair $(\mathscr{W}_{2s}(t-1), \mathscr{W}_{2s}(t))$ represents the tth step of the walk $\mathscr{W}_{2s}$ and that α_1 represents the *root* of the walk $\mathscr{W}_{2s}$.

Taking two trajectories $\mathscr{I}'_{2s}$ and $\mathscr{I}''_{2s}$ such that $\mathscr{W}_{2s}^{(\mathscr{I}'_{2s})} = \mathscr{W}_{2s}^{(\mathscr{I}''_{2s})} = \mathscr{W}_{2s}$, we say that they are equivalent, $\mathscr{I}'_{2s} \sim \mathscr{I}''_{2s}$. We denote by $\mathscr{C}_{\mathscr{W}_{2s}}$ the corresponding class of equivalence. It is clear that

$$|\mathscr{C}_{\mathscr{W}_{2s}}| = n(n-1)\cdots(n - |\mathbb{U}(\mathscr{I}_{2s}; 2s)| + 1). \tag{24}$$

Given $\mathscr{W}_{2s}$, one can introduce a *graphical representation* $g(\mathscr{W}_{2s}) = (\mathbb{V}_g, \mathbb{E}_g)$ that can be considered as a kind of multigraph with the set of vertices $\mathbb{V}_g = \{\alpha_1, \dots, \alpha_{|\mathbb{U}(\mathscr{I}_{2s};2s)|}\}$ and the set $\mathbb{E}_g$ of $2s$ oriented edges (or equivalently, arcs) labelled by $t \in \{1, \dots, 2s\}$. To describe the properties of $g(\mathscr{W}_{2s})$ in general situations, we will use the Greek letters $\alpha, \beta, \gamma, \dots$ instead of the symbols from $\mathscr{A}$. In this case, the root of the walk will be denoted by ϱ. In what follows, we refer to $g(\mathscr{W}_{2s})$ simply as to the *graph* of the walk $\mathscr{W}_{2s}$. If $\mathscr{W}_{2s}(t) = \gamma$, we say that γ is the *value of the walk* $\mathscr{W}_{2s}$ at the instant of time t.

Let us define the *current multiplicity* of the couple of vertices $\{\beta, \gamma\}$, $\beta, \gamma \in \mathbb{V}_g$ up to the instant t by the following variable

$$\mathfrak{m}_{\mathscr{W}}^{(\{\beta,\gamma\})}(t) = \#\{t' \in [1, t] :$$
$$(\mathscr{W}_{2s}(t'-1), \mathscr{W}_{2s}(t')) = (\beta, \gamma) \text{ or } (\mathscr{W}_{2s}(t'-1), \mathscr{W}_{2s}(t')) = (\gamma, \beta)\}$$

and say that $\mathfrak{m}_{\mathscr{W}}^{(\{\beta,\gamma\})}(2s)$ represents the total multiplicity of the couple $\{\beta, \gamma\}$.

The probability law of a_{ij} being symmetric, the weight of $\mathscr{I}_{2s}$ (22) is non-zero if and only if $\mathscr{I}_{2s}$ is such that in the corresponding graph of the walk $\mathscr{W}_{2s}^{(\mathscr{I}_{2s})}$ each couple $\{\alpha, \beta\}$ has an even multiplicity $\mathfrak{m}_{\mathscr{W}}^{(\{\alpha,\beta\})}(2s) = 0 (\mathrm{mod}\, 2)$. We refer to the walks of such trajectories as to the *even closed walks* [19] and denote by $\mathbb{W}_{2s}$ the set of all possible even closed walks of $2s$ steps. In what follows, we consider the even closed walks only and refer to them simply as to the walks.

It is natural to say that the pair $(\mathscr{W}_{2s}(t-1), \mathscr{W}_{2s}(t)) = \mathfrak{s}_t$ represents the *step of the walk* number t. Given $\mathscr{W}_{2s} \in \mathbb{W}_{2s}$, we say that the instant of time t is *marked* [19] if the couple $\{\alpha, \beta\} = \{\mathscr{W}_{2s}(t-1), \mathscr{W}_{2s}(t)\}$ has an odd current multiplicity $\mathfrak{m}_{\mathscr{W}}^{(\{\alpha,\beta\})}(t) = 1 (\mathrm{mod}\, 2)$. We also say that the corresponding step $\mathfrak{s}_t$ and the edge e_t of $g(\mathscr{W}_{2s})$ are marked. All other steps and edges are called the *non-marked* ones. Regarding a collection of the marked edges $\bar{\mathbb{E}}_g$ of $g(\mathscr{W}_{2s})$, we consider a multigraph $\bar{g}_s = (\bar{\mathbb{V}}_g, \bar{\mathbb{E}}_g)$. Clearly, $\bar{\mathbb{V}}_g = \mathbb{V}_g$ and $|\bar{\mathbb{E}}_g| = s$. It is useful to keep the time labels of the edges $\bar{\mathbb{E}}_g$ as they are in $\mathbb{E}_g$. Given two edges $e' = e_{t'}$ and $e'' = e_{t''}$ such that $t' < t''$, we write that $e' < e''$. Sometimes we denote $t' = t(e')$.

In general, $\bar{g}(\mathscr{W}_{2s})$ is a multigraph with multiple edges. Replacing the multiple edge by a simple one, we get a new graph that we refer to as the *skeleton* $S_{\bar{g}}$ of the graph $\bar{g}$.

Any even closed walk $\mathscr{W}_{2s} \in \mathbb{W}_{2s}$ generates a sequence θ_{2s} of s marked and s non-marked instants. Corresponding sequence of $2s$ signs $+$ and $-$ is known to encode a Dyck path of $2s$ steps [23]. We denote by $\theta_{2s} = \theta(\mathscr{W}_{2s})$ the Dyck path of $\mathscr{W}_{2s}$ and say that $\theta(\mathscr{W}_{2s})$ represents the *Dyck structure* of $\mathscr{W}_{2s}$.

Let us denote by Θ_{2s} the set of all Dyck paths of $2s$ steps. It is known that Θ_{2s} is in one-by-one correspondence with the set of all half-plane rooted trees $\mathcal{T}_s \in \mathbb{T}_s$ constructed with the help of s edges [23]. The correspondence between Θ_{2s} and $\mathbb{T}_s$ can be established with the help of the chronological run $\mathfrak{R}$ over the edges of $\mathcal{T}_s$. It is known that the cardinalities of $\mathbb{T}_s$, $s = 0, 1, 2, \ldots$ are given by the Catalan numbers, $|\mathbb{T}_s| = \mathrm{t}_s$ (7). We refer to the elements of $\mathbb{T}_s$ as to the *Catalan trees*. We consider the edges of the tree $\mathcal{T}_s$ as the oriented ones in the direction away from the root of $\mathcal{T}_s$.

Given a Catalan tree $\mathcal{T}_s \in \mathbb{T}_s$, one can label its vertices with the help of letters of $\mathcal{A}$ according to $\mathfrak{R}_{\mathcal{T}}$. The root vertex gets the label α_1 and each new vertex that has no label is labelled by the next in turn letter. We denote the walk obtained by $\mathring{\mathcal{W}}_{2s}[\mathcal{T}_s]$ and the corresponding Dyck path $\theta_{2s} = \theta(\mathring{\mathcal{W}}_{2s})$ will be denoted also as $\theta_{2s} = \theta(\mathcal{T}_s)$.

Any Dyck path θ_{2s} generates a sequence $(\xi_1, \xi_2, \ldots, \xi_s)$, $\xi_i \in \{1, 2, \ldots, 2s-1\}$ such that each step $\mathfrak{s}_{\xi_i}$, $1 \le i \le s$ of $\mathring{\mathcal{W}}_{2s}[\theta_{2s}]$ is marked. We denote this sequence by $\Xi_s = \Xi(\theta_{2s})$. Given Ξ_s and $\tau \in [1, s]$, one can uniquely reconstruct θ_{2s} and find corresponding instant of time $\xi_\tau \in \{1, \ldots, 2s-1\}$. We will say that the interval $[1, s]$ represents the *τ-marked instants* or *instants of marked time* that varies from 1 to s; sometimes we will simply say that $\tau \in [1, s]$ is the *marked instant* when no confusion can arise.

Given a walk $\mathcal{W}_{2s}$ and a letter β such that $\beta \in \mathbb{V}_g(\mathcal{W}_{2s})$, we say that the instant of time t' such that $\mathcal{W}_{2s}(t') = \beta$ represents an *arrival* $\mathfrak{a}$ at β. If t' is marked, we will say that the corresponding arrival $\mathfrak{a}(\beta)$ is the marked arrival at β. In $\mathcal{W}_{2s}$, there can be several marked arrival instants of time at β that we denote by $1 \le t_1^{(\beta)} < \cdots < t_N^{(\beta)}$. For any non-root vertex β, we have $N = N_\beta \ge 1$. The first arrival instant of time β is always the marked one. We can say that β is created at this instant of time. To unify the description, we assume that the root vertex ϱ is created at the zero instant of time $t_1^{(\rho)} = 0$ and add the corresponding zero marked instant to the list of the marked arrival instants at ϱ.

If $N_\beta \ge 2$, then we say that the N-plet $(t_1^{(\beta)}, \ldots, t_N^{(\beta)})$ of marked arrival instants of time represents the *self-intersection* of $\mathcal{W}_{2s}$, β is the *vertex of self-intersection*, and this self-intersection is of the *degree* N [19]. We say that the self-intersection degree $\varkappa(\beta)$ is equal to N and denote this by $\varkappa(\beta) = N_\beta$. If $N_\beta = 1$, then we will say that $\varkappa(\beta) = 1$.

Finally, let us consider a vertex β and a collection of the marked edges of the form (β, α_i). We say that this collection is the *exit cluster* of β and denote it by $\Delta(\beta)$,

$$\Delta(\beta) = \Delta_{\mathcal{W}}(\beta) = \{e \in \bar{\mathbb{E}}(\mathcal{W}_{2s}) : \; e = (\beta, \alpha_i)\}. \tag{25}$$

Sometimes we will say that $\Delta(\beta)$ is given by the collection of corresponding vertices α_i.

Given $\mathcal{W}_{2s}$, we can say that it is of the *tree-type structure* if the skeleton $S_{\bar{g}}$ of $g(\mathcal{W}_{2s})$ is a tree. Regarding a walk of the tree-type structure $\hat{\mathcal{W}}_{2s}$, we will say that it is a *(2,4)-walk* if the weight of the corresponding trajectory $\Pi(\mathcal{I}_{2s})$ (22) contains the factors V_2 and V_4 only. If the (2,4)-walk is such that in its graph the multiple edges passed four times have no vertices in common, we will say that this walk is the $(2,4)^*$-*walk*.

To complete this section, let us note that the number of the tree-type walks of $2s$ steps whose weight contains the factor V_2^s is given by the Catalan numbers t_s (7). It follows from (7) that these numbers verify the following recurrent relation

$$\mathrm{t}_k = \sum_{j=0}^{k-1} \mathrm{t}_{k-1-j}\,\mathrm{t}_j, \quad k \geq 1 \tag{26}$$

with the initial condition $\mathrm{t}_0 = 1$. As a by-product of the studies of the $(2,4)$-walks, we find a number of generalizations of relations (7) and (26).

3 Walks of Non-tree Type

Given a walk $\mathcal{W}_{2s}$ with $\bar{g}(\mathcal{W}_{2s}) = (\mathbb{V}_g, \bar{\mathbb{E}}_g)$, let us consider two arbitrary vertices $\alpha, \beta \in \mathbb{V}_g$. We denote by $\mathcal{E}_{\{\alpha,\beta\}}$ the collection of all edges $(\alpha, \beta) \in \bar{\mathbb{E}}(\mathcal{W}_{2s})$ and $(\beta, \alpha) \in \bar{\mathbb{E}}(\mathcal{W}_{2s})$ and determine the minimal edge $\tilde{e} = \min\{e : e \in \mathcal{E}_{\{\alpha,\beta\}}\}$. Let us assume that $\mathcal{E}_{\{\alpha,\beta\}}$ is non-empty and that $\tilde{e} = (\alpha, \beta)$. If the multi-edge $\mathcal{E}_{\{\alpha,\beta\}}$ contains the edge e_1 of the first arrival at β, $e_1 = e(\mathfrak{a}_1(\beta))$, then $\tilde{e} = e_1$ and we say that this edge $\tilde{e} = (\alpha, \beta)$ creates β and that $\tilde{e}$ is the *base edge of* β or simply the base edge.

3.1 *Classification of Vertices and Weights of Walks*

Let us consider the edge of the second arrival at β, $e_2 = e(\mathfrak{a}_2(\beta))$. If $e_2 = (\alpha, \beta)$, then we color it in green and say that β is the *green p-vertex*.

Let us consider the edge $e_2 = e(\mathfrak{a}_2) = (\gamma, \beta)$ of the second arrival at β such that $\gamma \neq \beta$. If e_2 is the minimal edge of the multi-edge $\mathcal{E}_{\{\beta,\gamma\}}$, then we say that β is the *blue r-vertex* and color e_2 in blue.

Let us consider the case when $\tilde{e} = \min\{e : e \in \mathcal{E}_{\{\gamma,\beta\}}\} = (\beta, \gamma)$. If $\tilde{e}$ is the edge of the first or the second arrival at γ, $\tilde{e} = \mathfrak{a}_i(\gamma)$ with $i = 1$ or $i = 2$, then we color $e_2 = (\gamma, \beta)$ in red and say that β is the *red q-vertex*. If $\tilde{e} = \mathfrak{a}_j(\gamma)$ with $j \geq 3$, then we color $e_2 = (\gamma, \beta)$ in blue and consider β as the blue r-vertex.

It is not hard to see that all edges of the second arrival to one or another vertex are colored and that their colors are uniquely determined. All remaining edges of $\bar{g}(\mathcal{W}_{2s})$ that are not the base or the color ones are referred as to the *grey u-edges*.

Lemma 1 *Let $\mathcal{I}_{2s}$ be such that the graph of its walk $\mathcal{W}_{2s} = \mathcal{W}(\mathcal{I}_{2s})$ contains r blue r-vertices, p green p-vertices, q red q-vertices. Also we assume that $\bar{g}(\mathcal{W}_{2s})$ has u grey u-edges. Then the weight of $\mathcal{I}_{2s}$* (22) *is bounded as follows*

$$\Pi_a(\mathcal{I}_{2s})\,\Pi_b(\mathcal{I}_{2s}) = \Pi_a(\mathcal{W}_{2s})\,\Pi_b(\mathcal{W}_{2s}) \le \left(\frac{V_2^2}{n^2}\right)^r \left(\frac{V_2U^2}{n\rho}\right)^{p+q}$$

$$\times \left(\frac{U^2}{\rho}\right)^u \left(\frac{V_2}{n}\right)^{s-u-2(r+p+q)}. \tag{27}$$

Remark 1 To make the statements of the present section and their proofs clearer, we keep the factors V_2 as they are remembering that $V_2 = 1$.

Proof The weight of the walk $\Pi_{a,b}(\mathcal{W}_{2s}) = \Pi_a(\mathcal{W}_{2s})\,\Pi_b(\mathcal{W}_{2s})$ is given by the product of weights of all existing multi-edges $\Pi_{a,b}(\mathcal{E}_{\{\delta,\epsilon\}})$. It is easy to see that the weight of the multi-edge can be estimated from above as follows,

$$\Pi_{a,b}(\mathcal{E}_{\{\delta,\epsilon\}}) \le \left(\frac{V_2}{n}\right)^{I_{b,r}} \left(\frac{U^2}{\rho}\right)^{I_{r,p}+I_q+u(\mathcal{E})},$$

where $I_{b,r} = 1$ if the minimal edge $\tilde{e}$ of $\mathcal{E}_{\{\delta,\epsilon\}}$ is either the base one or the blue one and zero otherwise, $I_{r,p}$ is equal to one if $\mathcal{E}_{\{\delta,\epsilon\}}$ contains a green edge and zero otherwise, I_q is equal to one if $\mathcal{E}_{\{\delta,\epsilon\}}$ contains a red edge and zero otherwise and $u(\mathcal{E}) = u(\mathcal{E}_{\{\delta,\epsilon\}})$ represents the number of the grey edges in $\mathcal{E}_{\{\delta,\epsilon\}}$. Due to this factorized upper bound, the weight of the walk can be estimated by the product of factors V_2/n and U^2/ρ that can be rearranged into the product with respect to all vertices of the graph $g(\mathcal{W}_{2s})$. This can be done by attributing the weights V_2/n to all base and blue edges of the graph $\bar{g}(\mathcal{W}_{2s})$ and the weights U^2/ρ to all green, red and grey edges of $\bar{g}(\mathcal{W}_{2s})$ and by attributing to each vertex β the product of weights of all edges that enter β. It is clear that any color vertex has exactly one edge of the second arrival that is of the same color as the vertex. This observation completes the proof of Lemma 1. □

3.2 *Tree-Type Walks and Walks of Non-tree Type*

Given a walk $\mathcal{W}_{2s}$, we can say that is it a *tree-type walk* if its graph $\bar{g}(\mathcal{W}_{2s})$ does not contain any blue r-vertex. We denote by $\hat{\mathbb{W}}_{2s}$ a collection of tree-type walks. If $\mathcal{W}_{2s}$ is such that its graph $\bar{g}(\mathcal{W}_{2s})$ contains at least one blue r-vertex, then we say that $\mathcal{W}_{2s}$ is of *non-tree type*. We denote a collection of all non-tree-type walks by $\tilde{\mathbb{W}}_{2s}$.

The following simple statement plays an important role in our studies.

Lemma 2 *If $\mathcal{W}_{2s}$ is such that its graphical representation $g(\mathcal{W}_{2s})$ has at least one red q-vertex, then $g(\mathcal{W}_{2s})$ contains at least one blue r-vertex and therefore is of non-tree type.*

We prove Lemma 2 in Sect. 5. Regarding the example walk $\mathscr{W}_{16}$ (23), we see that its graphical presentation contains two vertices of the self-intersection degree 2 (these are α_1 and α_4) and one vertex α_2 of the self-intersection degree 3. Among vertices of $\bar{g}(\mathscr{W}_{16})$, there is one p-vertex α_4 and one q-vertex α_2. The root vertex α_1 has one blue edge of the (mute) first arrival and one blue edge $e(6)$ of the second distinct arrival. So, the root vertex α_1 is the blue r-vertex and $\mathscr{W}_{16}$ is of non-tree type.

According to definitions of the tree-type and non-tree type walks, we can rewrite relation (21) in the form

$$M_{2s}^{(n,\rho)} = \tilde{\mathscr{Z}}_{2s}^{(n,\rho)} + \hat{\mathscr{Z}}_{2s}^{(n,\rho)}, \tag{28}$$

where

$$\tilde{\mathscr{Z}}_{2s}^{(n,\rho)} = \sum_{\mathscr{I}_{2s}:\ \mathscr{W}(\mathscr{I}_{2s}) \in \tilde{\mathbb{W}}_{2s}} \Pi_a(\mathscr{I}_{2s})\, \Pi_b(\mathscr{I}_{2s})$$

and

$$\hat{\mathscr{Z}}_{2s}(n,\rho) = \sum_{\mathscr{I}_{2s}:\ \mathscr{W}(\mathscr{I}_{2s}) \in \hat{\mathbb{W}}_{2s}} \Pi_a(\mathscr{I}_{2s})\, \Pi_b(\mathscr{I}_{2s}).$$

The following statements represents the main technical result of this section.

Theorem 2 *Under conditions of Theorem 1, the following relation holds*

$$\tilde{\mathscr{Z}}_{2s_n}(n,\rho) = O\left(nt_{s_n} V_2^{s_n} \frac{s_n^5}{n} \right), \quad (n,s,\rho) \to \infty. \tag{29}$$

Remark Observing that all terms of the right-hand side of the definition of $\hat{\mathscr{Z}}_{2s}^{(n,\rho)}$ are non-negative, we conclude that $\hat{\mathscr{Z}}_{2s}(n,\rho) \ge nt_s V_2^s = nt_s$. Therefore relation (29) implies the asymptotic estimate

$$\tilde{\mathscr{Z}}_{2s}(n,\rho) = o(\hat{\mathscr{Z}}_{2s}(n,\rho)), \quad (n,s,\rho) \to \infty. \tag{30}$$

We prove Theorem 2 in Sect. 3.5 below. In Sects. 3.3 and 3.4 we formulate necessary notions and auxiliary statements.

3.3 Diagrams $\mathscr{G}^{(c)}(\bar{\nu})$ and Their Realizations

Each walk $\mathscr{W}_{2s}$ generates a set of numerical data, $\bar{\nu} = (\nu_2, \nu_3, \ldots, \nu_s)$, where ν_k is the number of vertices β_i of $\bar{g}(\mathscr{W}_{2s})$ such that their self-intersection degree is equal to k, $\varkappa(\beta_i) = k$. To estimate the number of elements of the set $\tilde{\mathbb{W}}_{2s}$, we construct a kind of diagrams $\mathcal{G}^{(c)}(\bar{\nu})$.

To explain general principles of the estimates, let us start with the construction of *non-colored* diagram $\mathcal{G}(\bar{\nu})$. This diagram consists of $|\bar{\nu}| = \sum_{k=2}^{s} \nu_k$ vertices. We arrange these vertices in $s-1$ levels, the kth level contains ν_k vertices. Each vertex v of kth level is attributed by k *half-edges* that have heads attached to v but have no tails. Instead of the tail of each edge, we join a square box (or window) to it. Then any vertex v of this kth level has k *edge-boxes* (or edge-windows) attached.

Given $\mathcal{G}(\bar{\nu})$, one can attribute to its edge-windows the values from the set $\{1, 2, \dots, s\}$ such that there is no pair of windows with the same value. The diagram together with the corresponding values produces a realization of $\mathcal{G}(\bar{\nu})$ that we denote by $\langle \mathcal{G}(\bar{\nu}) \rangle_s$.

One of the principal components of the Sinai-Soshnikov method is given by the observation that an even closed walk $\mathcal{W}_{2s}$ can be completely determined by its values at the marked instant of time added by a family of rules that indicate the values of the walk at the non-marked instant of time. Given Dyck path θ_s and a realization $\langle \mathcal{G}(\bar{\nu}) \rangle_s$, the positions of the walk at the marked instants of time are completely determined.

The values at the non-marked instant of time are determined by a family of rules $\mathbb{Y}(\bar{\nu})$ that indicate the way to leave a vertex β of self-intersection with the help of the non-marked step out. It is shown in [19, 20] that if $\varkappa(\beta) = k$, then the number of the exit rules at this vertex is bounded as follows, $|\mathbb{Y}(\beta)| \le (2k)^k$. An additional proof of this upper bound was given in [9, 12]. No such rule as Υ is needed for the non-marked instants of time when the walk leaves a vertex of the self-intersection degree 1 because in this case the continuation of the run is uniquely determined. Then the total number of the rules can be estimated as follows,

$$|\mathbb{Y}(\bar{\nu})| \le \prod_{k=2}^{s} (2k)^{k\nu_k}. \tag{31}$$

The number of all possible realizations of $\mathcal{G}(\bar{\nu})$ is given by the following expression

$$\sum_{\langle \mathcal{G}(\bar{\nu}) \rangle_s} 1 = \frac{s!}{\nu_2!(2!)^{\nu_2}\, \nu_3!(3!)^{\nu_3} \cdots \, \nu_s!(s!)^{\nu_s}\, (s - \|\bar{\nu}\|)!},$$

where $\|\bar{\nu}\| = \sum_{k=2}^{s} k\nu_k$. It is easy to see that the following upper bound is true,

$$\sum_{\langle \mathcal{G}(\bar{\nu}) \rangle_s} 1 \le \prod_{k=2}^{s} \frac{1}{\nu_k!} \left(\frac{s^k}{k!} \right)^{\nu_k}.$$

Combining this inequality with (31), we conclude that the number of elements in $\mathbb{W}_{2s}(\bar\nu)$ can be estimated as follows,

$$|\mathbb{W}_{2s}| \le t_s \prod_{k=2}^{s} \frac{1}{\nu_k!}\left(\frac{(2k)^k s^k}{k!}\right)^{\nu_k} \le t_s \prod_{k=2}^{s} \frac{(C_1 s)^{k\nu_k}}{\nu_k!}, \quad \text{where} \quad C_1 = \sup_{k\ge 1} \frac{(2k+2)}{(k!)^{1/(k+1)}}. \tag{32}$$

We have introduced the constant C_1 in the form that simplifies further computations.

The upper bound (32) clearly explains the role of the diagrams $\mathscr{G}(\bar\nu)$ in the estimates of the number of walks. However, it is rather rough and does not give inequalities needed in the majority of cases of interest. In particular, the estimate (32) is hardly compatible with the upper bound of the weight of walks (27) in the case of dilute random matrices.

To improve the upper bound (32), we adapt the diagram technique to our model by introducing more informative diagrams based on $\mathscr{G}(\bar\nu)$. Also, we formulate a new filtering principle to estimate more accurately the number of walks. A kind of the filtration principle has been implicitly used already by Ya. Sinai and A. Soshnikov. The rigorous formulation of the filtration technique is given in [9]. In paper [10] it was adapted to the study of the moments of dilute random matrices.

Let us describe the construction of the *color diagram* $\mathscr{G}^{(c)}(\bar\nu,\bar p,\bar q)$ determined by parameters $\bar\nu = (\nu_2,\dots,\nu_s)$, $\bar p = (p_2,\dots,p_s)$ and $\bar q = (q_2,\dots,q_s)$. We start with the non-colored diagram $\mathscr{G}(\bar\nu)$ and consider ν_k vertices of the kth level of it. We fill the second edge-box attached to each vertex by using the set $\{1,\dots,s\}$. This can be done by

$$\frac{s!}{\nu_k!(s-\nu_k)!} \le \frac{s^{\nu_k}}{\nu_k!} = \frac{s^{r_k+p_k+q_k}}{\nu_k!} \tag{33}$$

ways. Then we color the ν_k vertices in blue, red and green colors by one of $\frac{\nu_k!}{r_k!p_k!q_k!}$ ways, where $r_k = \nu_k - p_k - q_k$. Then we color corresponding edge-boxes in grey, blue, red and green colors. The base edge-boxes of the first arrivals attached to blue vertices are colored in blue. Instead of boxes, the base edges of the red and green vertices get circles colored with respect to the color of the corresponding vertex.

Taking the empty $k-2$ edge-boxes attached to green or red vertex, we fill them with the values from the set $\{1,\dots,s\}$. This can be done by not more than $s^{k-2}/(k-2)!$ ways. Regarding the edges of the first arrivals at red and green vertices that remain empty, we replace corresponding boxes by circles colored according to the color of the vertex.

Let us consider $k-1$ empty edge-boxes attached to a blue vertex and fill them with the values from $\{1,\dots,s\}$. Ignoring the restriction of the edge-box of the second arrival, we estimate the number of ways to do this by expression $s^{k-1}/(k-1)!$.

This procedure being performed at each level independently, we get the following estimate from above of the number of different realizations of color diagrams,

$$\sum_{\langle \mathscr{G}^{(c)}(\bar{\nu},\bar{p},\bar{q})\rangle_s} 1 \le \prod_{k=2}^{s} \frac{1}{r_k!}\left(\frac{s^k}{(k-1)!}\right)^{r_k} \frac{1}{p_k!}\left(\frac{s^{k-1}}{(k-2)!}\right)^{p_k} \frac{1}{q_k!}\left(\frac{s^{k-1}}{(k-2)!}\right)^{q_k}. \tag{34}$$

The *filtration procedure* is follows: we consider a realization of the color diagram $\langle \mathscr{G}^{(c)}\rangle_s$ such that all grey, blue, red and green boxes of edge-windows of $\mathscr{G}^{(c)}$ are filled with different values of $\{1,\dots,s\}$ while the red and green circles of the first arrivals at the q-vertices and p-vertices remain empty.

Having a Dyck path θ_s and a rule $\Upsilon \in \mathbb{Y}(\bar{\nu})$ pointed out, we start the run of the walk $\mathscr{W}$ according θ_s, $\langle \mathscr{G}^{(c)}\rangle_s$ and Υ till the marked instant of the first p-edge or q-edge appear. Let us denote by v' the corresponding vertex of the diagram $\mathscr{G}^{(c)}$. Let us denote the marked instant mentioned above by τ' with $t' = \xi_{\tau'}$ and assume that the sub-walk $\mathscr{W}_{[0,t'-1]}$ get its end value $\beta = \mathscr{W}_{[0,t'-1]}(t'-1)$. Then at the instant of time t' the walk has to choose one of the admissible vertices from the set $\Gamma = \{\gamma_1,\dots,\gamma_L\}$ such that the edge (β,γ_j) possesses the properties of either p-edge or q-edge, respectively. Clearly, the set Γ depends on the color of the edge-box with τ'. Once the vertex γ_j is chosen, we take the marked instant of the first arrival at γ_j and record its value to the edge-box of the first arrival $\mathscr{O}_1(v')$. Clearly, the number of walk is bounded by $|\Gamma|$. This is why it is natural to say that we apply the filtering of all possible values to fill $\mathscr{O}_1(v')$.

Having chosen the value of $\mathscr{O}_1$, we continue the run of the walk, if it is possible, till the marked value of the second arrival at the next in turn red or green vertex v'' is seen. Then the filtering procedure is repeated. When all the walk is constructed, if it exists, we denote by $\langle\langle \mathscr{G}^{(c)}\rangle_s^{(b)}\rangle_{\mathscr{W}}$ the set of values in red and green circles obtained during this run of $\mathscr{W}$.

Lemma 3 *Given a realization of a color diagram $\langle \mathscr{G}^{(c)}(\bar{\nu},\bar{p},\bar{q})\rangle_s$, let us denote by $\mathbb{W}_{2s}(D,\langle \mathscr{G}^{(c)}(\bar{\nu},\bar{p},\bar{q})\rangle_s,\Upsilon)$ the set of walks $\mathscr{W}_{2s}$ that have this realization of $\mathscr{G}^{(c)}$ and follow the rule Υ and such that the maximal exit degree*

$$\mathscr{D}(\mathscr{W}_{2s}) = \max_{\beta\in\mathbb{V}(\mathscr{W}_{2s})} |\Delta(\beta)|$$

is equal to D, $\mathscr{D}(\mathscr{W}_{2s}) = D$. Then the number of possible realizations of the values of red and green circles of $\mathscr{G}^{(c)}$ admits the following upper bound,

$$|\langle\langle \mathscr{G}^{(c)}\rangle_s\rangle_{\mathscr{W}}| \le 2^{|\bar{q}|}\, D^{|\bar{p}|}, \tag{35}$$

where $|\bar{q}| = \sum_{k=2}^{s} q_k$ and $|\bar{p}| = \sum_{k=2}^{s} p_k$ and therefore

$$|\mathbb{W}_{2s}(D,\langle \mathscr{G}^{(c)}(\bar{\nu},\bar{p},\bar{q})\rangle_s,\Upsilon)| \le 2^{|\bar{q}|}\, D^{|\bar{p}|}\, \mathrm{t}_s. \tag{36}$$

Proof First let us prove (36) in the case when the color diagram $\mathcal{G}^{(c)}$ has one red vertex v and no the green ones. Following the filtration principle, we take a Dyck path θ_s and perform the run of the walk in accordance with the data given by the self-intersections of $\langle \mathcal{G}^{(c)}(\bar{\nu}, \bar{p}, \bar{q})\rangle_s^{(b)}$ and Υ till the value $\xi_{\tau'}$ appear, where τ' is attributed to the second arrival edge-box attached to v. By the definition, the edge $(\mathcal{W}(\xi_{\tau'} - 1), \mathcal{W}(\xi_{\tau'} - 1)) = (\beta, \gamma) = e$ is red only in the case when the edge $\tilde{e} = (\gamma, \beta)$ is the edge either of the first or the second arrival at γ and $\tilde{e} < e$. Therefore the sub-walk $\mathcal{W}_{[0,\xi_{\tau'}-1]}$ has not more that two vertices available to join at the instant $\xi_{\tau'}$. This explains the factor 2 in the left-hand side of (35).

In the case when v is the one green vertex and no the red ones, the sub-walk $\mathcal{W}_{[0,\xi_{\tau'}-1]}$ has the set $\Delta(\beta)$ completely determined, and the vertex to join at the instant $\xi_{\tau'}$ necessarily belongs to this set. This gives the factor D in the upper bound (35).

It is clear that the general case can be treated by the same reasoning and the upper bound (36) can be proved by recurrence. This observation completes the proof of Lemma 3. □

Now we can estimate the number of walks that have a color diagram $\mathcal{G}^{(c)}(\bar{\nu}, \bar{p}, \bar{q})$ and the maximal exit degree D,

$$|\mathbb{W}_{2s}(D, \mathcal{G}^{(c)}(\bar{\nu}, \bar{p}, \bar{q}))| \leq \mathrm{t}_s \prod_{k=2}^{s} \frac{(C_1 s)^{kr_k}}{r_k!} \frac{\left(D(C_1 s)^{k-1}\right)^{p_k}}{p_k!} \frac{\left(2(C_1 s)^{k-1}\right)^{q_k}}{q_k!}. \qquad (37)$$

This relation follows from inequalities (31), (34) and (36).

We will use Lemma 3 and a version of relation (37) in the proof of Theorem 2 below. However, to get the estimates we need, we have to show that the number of Catalan trees $\mathcal{T}_s$ generated by the elements of $\mathbb{W}_{2s}(D, \mathcal{G}^{(c)})$ is exponentially small with respect to the total number t_s of all $\mathcal{T}_s$ [20, 21]. To do this, we need to study the vertex of maximal exit degree of walks $\mathcal{W}_{2s}$ in more details.

3.4 Vertex of Maximal Exit Degree, Arrival Cells and BTS-Instants

Let us consider a walk $\mathcal{W}_{2s}$ and find the first letter that we denote by $\breve{\beta}$ such that

$$|\Delta(\breve{\beta})| = \mathcal{D}(\mathcal{W}_{2s}). \qquad (38)$$

We will refer to $\breve{\beta}$ as to the *vertex of maximal exit degree* and denote for simplicity $D = \mathcal{D}(\mathcal{W}_{2s})$. To classify the arrival instants at $\breve{\beta}$, we need to determine reduction procedures similar to those considered in [12] and further modified in [10]. Certain elements of the reduction procedure of [12] were independently introduced in paper [6].

3.4.1 Reduction Procedures and Reduced Sub-walks

Given $\mathscr{W}_{2s}$, let t' be the minimal instant of time such that

(i) the step $\mathfrak{s}_{t'}$ is the marked step of $\mathscr{W}_{2s}$;
(ii) the consecutive to $\mathfrak{s}_{t'}$ step $\mathfrak{s}_{t'+1}$ is non-marked;
(iii) $\mathscr{W}_{2s}(t'-1) = \mathscr{W}_{2s}(t'+1)$.

If such t' exists, we apply to the ensemble of steps $\mathfrak{S} = \{\mathfrak{s}_t, 1 \le t \le 2s, \mathfrak{s}_t \in \mathscr{W}_{2s}\}$ a reduction $\dot{\mathscr{R}}$ that removes from $\mathfrak{S}$ two consecutive elements $\mathfrak{s}_{t'}$ and $\mathfrak{s}_{t'+1}$; we denote $\dot{\mathscr{R}}(\mathfrak{S}) = \mathfrak{S}'$. The ordering time labels of elements of $\mathfrak{S}'$ are inherited from those of $\mathfrak{S}$.

The new sequence $\mathfrak{S}'$ can be regarded again as an even closed walk. We can apply to this new walk the reduction procedure $\dot{\mathscr{R}}$. Repeating this operation maximally possible number of times m, we get the walk

$$\dot{\mathscr{W}}_{2\dot{s}} = (\dot{\mathscr{R}})^m(\mathscr{W}_{2s}), \quad \dot{s} = s - m,$$

that we refer to as the *strongly reduced walk*. We denote $\dot{\mathfrak{S}} = (\dot{\mathscr{R}})^m(\mathfrak{S})$ and say that $\dot{\mathscr{R}}$ is the *strong reduction* procedure.

We introduce a *weak reduction* procedure $\breve{R}$ of $\mathfrak{S}$ that removes from $\mathfrak{S}_{2s}$ the pair $(\mathfrak{s}_{t'}, \mathfrak{s}_{t'+1})$ in the case when the conditions (i)–(iii) are verified and

(iv) $\mathscr{W}_{2s}(t') \ne \breve{\beta}$.

We denote by

$$\breve{\mathscr{W}}_{2\breve{s}} = (\breve{\mathscr{R}})^l(\mathscr{W}_{2s}), \quad \breve{s} = s - l \tag{39}$$

the result of the action of maximally possible number of consecutive weak reductions $\breve{\mathscr{R}}$ and denote $\breve{\mathfrak{S}} = (\breve{\mathscr{R}})^l(\mathfrak{S})$. In what follows, we sometimes omit the subscripts $2\dot{s}$ and $2\breve{s}$. Regarding the example walk $\mathscr{W}_{16}$ (23), we observe that $\breve{\beta} = \alpha_3$ and that the strongly and weakly reduced walks coincide and are as follows,

$$\dot{\mathscr{W}}_8 = \breve{\mathscr{W}}_8 = (\alpha_1, \alpha_2, \alpha_3, \alpha_5, \alpha_1, \alpha_2, \alpha_3, \alpha_5, \alpha_1).$$

Taking the difference $\breve{\mathfrak{S}} \setminus \dot{\mathfrak{S}} = \ddot{\mathfrak{S}}$, we see that it represents a collection of sub-walks, $\ddot{W} = \cup_j \ddot{\mathscr{W}}^{(j)}$. Each sub-walk $\ddot{\mathscr{W}}^{(j)}$ can be reduced by a sequence of the strong reduction procedures $\dot{\mathscr{R}}$ to an empty walk. We say that $\ddot{\mathscr{W}}^{(j)}$ is of the *Dyck-type* structure. It is easy to see that any $\ddot{\mathscr{W}}^{(j)}$ starts by a marked step and ends by a non-marked steps and there is no steps of $\dot{W}$ between these two steps of $\ddot{\mathscr{W}}^{(j)}$. We say that $\ddot{\mathscr{W}}^{(j)}$ is the *non-split* sub-walk.

It is not hard to see that the collection of steps $\breve{\mathfrak{S}} = \mathfrak{S} \setminus \check{\mathfrak{S}}$ is given by a collection of subsets $\breve{\mathfrak{S}} = \cup_k \breve{\mathfrak{S}}^{(k)}$, each of $\breve{\mathfrak{S}}^{(k)}$ represents a non-split Dyck-type sub-walk $\breve{\mathscr{W}}^{(k)}$,

$$\breve{W} = \cup_k \breve{\mathscr{W}}^{(k)}. \tag{40}$$

In this definition, we assume that each sub-walk $\breve{\mathscr{W}}^{(k)}$ is maximal by its length.

3.4.2 Arrival Instants and Dyck-Type Sub-walks Attached to $\breve{\beta}$

Given $\mathscr{W}_{2s}$, let us consider the instants of time $0 \le t_1 < t_2 < \dots t_R \le 2s$ such that for all $i = 1, \dots, R$ the walk arrives at $\breve{\beta}$ by the steps of $\check{\mathscr{W}}_{2\check{s}}$,

$$\mathscr{W}_{2s}(t_i) = \breve{\beta} \quad \text{and} \quad \mathfrak{s}_{t_i} \in \check{\mathscr{W}}_{2\check{s}}, \quad i = 1, 2, \dots, R. \tag{41}$$

We say that t_i are the $\check{t}$*-arrival* instants of time of $\mathscr{W}_{2s}$. Let us consider a sub-walk that corresponds to the subset $\mathfrak{S}_{[t_i+1, t_{i+1}]} = \{\mathfrak{s}_t, t_i + 1 \le t \le t_{i+1}\} \subseteq \mathfrak{S}$; we denote this sub-walk by $\mathscr{W}_{[t_i, t_{i+1}]}$. In general, we denote a sub-walk that is not necessary even and/or closed by $\mathscr{W}_{[t', t'']}$ also.

Let us consider the interval of time $[t_i + 1, t_{i+1} - 1]$ between two consecutive $\check{t}$-arrivals at $\breve{\beta}$. It can happen that $\mathscr{W}_{2s}$ arrives at $\breve{\beta}$ at some instants of time $t' \in [t_i + 1, t_{i+1} - 1]$, $\mathscr{W}_{2s}(t') = \breve{\beta}$. We denote by $\check{t}_{(i)}$ the maximal value of such t'.

Lemma 4 *The sub-walk $\mathscr{W}_{[t_i, \check{t}_{(i)}]}$ coincides with one of the maximal Dyck-type sub-walks $\breve{\mathscr{W}}^{(k')}$ of* (40).

Lemma 4 is proved in [10].

Let us consider a collection of all marked exit edges from $\breve{\beta}$ performed by the marked steps on the interval of time $[t_i, \check{t}_{(i)}]$ and denote this collection by $\check{\Delta}_i$. We say that $\check{\Delta}_j$ represents the *exit sub-clusters of Dyck type* attached to $\breve{\beta}$. Or simply that $\check{\Delta}_j$ are the *exit sub-clusters* of $\mathscr{W}_{2s}$. We denote their cardinalities by $d_j = |\check{\Delta}_j|$. The exit sub-clusters are ordered in natural way. To keep a unified description, we accept the existence of empty exit sub-clusters; then we get equality $D = \sum_{j=1}^{R} d_j, d_j \ge 0$. Clearly, any exit sub-cluster is attributed to a uniquely determined $\check{t}$-arrival instant at $\breve{\beta}$.

Regarding the reduced walk $\check{\mathscr{W}}_{2\check{s}}$ of $\mathscr{W}_{2s}$ (39), we can determine corresponding Dyck path $\check{\theta}_{2\check{s}} = \theta(\check{\mathscr{W}}_{2\check{s}})$ and the tree $\check{\mathscr{T}}_{\check{s}} = \mathscr{T}(\check{\theta})$. It is easy to show that $\check{\mathscr{T}}_{\check{s}}$ is a sub-tree of the original tree $\mathscr{T}_s = \mathscr{T}(\theta(\mathscr{W}_{2s}))$. One can introduce the difference $\breve{\mathscr{T}} = \mathscr{T}_s \setminus \check{\mathscr{T}}_{\check{s}}$ and say that it is represented by a collection of sub-trees $\breve{\mathscr{T}}^{(j)}$.

Returning to the Catalan tree $\mathscr{T}(\theta_{2s})$, let us consider the chronological run over it that we denote by $\mathfrak{R}_{\mathscr{T}}$. Then the $\check{t}$-arrival instant t_l (41) determines the step ϖ_l of $\mathfrak{R}_{\mathscr{T}}$. Also the corresponding vertex $\breve{v}_l$ of the tree $\mathscr{T}_s$ is determined. It is clear that $\breve{v}_l$ are not necessarily different for different l.

The sub-trees $\check{\mathscr{T}}^{(l)}$ are attached to $\check{\upsilon}_l$ and the chronological run over $\check{\mathscr{T}}^{(l)}$ starts immediately after the step ϖ_l is performed. We will say that these steps ϖ_l, $1 \le l \le R$ represent the *nest cells* from where the sub-trees $\check{\mathscr{T}}^{(l)}$, $1 \le l \le L$ grow. It is clear that the sub-tree $\mathscr{T}_l$ has $d_l \ge 0$ edges attached to its root ϱ_l and this root coincides with the vertex $\check{\upsilon}_l$. Returning to $\mathscr{W}_{2s}$, we will say that the arrival instants of time $\check{t}_l$ represent the *arrival cells* at $\check{\beta}$. In the next sub-section, we describe a classification of the arrival cells at $\check{\beta}$ that represents a natural improvement of the approach proposed in [12].

3.4.3 Classification of Arrival Cells at $\check{\beta}$

Let us consider a walk $\mathscr{W}_{2s}$ together with its reduces counterparts $\mathscr{W}_{2\dot{s}} = \dot{\mathscr{W}}$ and $\mathscr{W}_{2\check{s}} = \check{\mathscr{W}}$. Let t_i denote a $\check{t}$-arrival cell (30). If the step $\mathfrak{s}_{t_i}$ of $\mathscr{W}_{2s}$ is marked, then we say that t_i represents a *proper cell* at $\check{\beta}$. If the step $\mathfrak{s}_{t_i}$ is non-marked and $\mathfrak{s}_{t_i} \in \ddot{\mathscr{W}} = \check{\mathscr{W}} \setminus \dot{\mathscr{W}}$, then we say that t_i represents a *mirror cell* at $\check{\beta}$. If the step $\mathfrak{s}_{t_i} \in \dot{\mathscr{W}}$ is non-marked, then we say that t_i represents an *imported cell* at $\ddot{\beta}$.

Let us consider I proper cells $\ddot{t}_i$ such that $\mathfrak{s}_{\ddot{t}_i}$ belongs to $\ddot{\mathfrak{S}}$. We denote by x_i the corresponding marked instants, $x_i = \xi_{\ddot{t}_i}$, $1 \le i \le I$ and write that $\bar{x}_I = (x_1, \dots, x_I)$. It is easy to see that each proper cell x_i can be attributed by a number 1 or 0 in dependence of whether it produces a corresponding mirror cell at $\check{\beta}$ or not. We denote this number by $m_i \in \{0, 1\}$ and write that

$$M = \sum_{i=1}^{I} m_i$$

and $\bar{m}_I = (m_1, \dots, m_I)$. Clearly, $M \le I$.

Regarding the strongly reduced walk $\mathscr{W}_{2\dot{s}}$, we denote by $\dot{t}_k$ the proper cells such that the steps $\mathfrak{s}_{\dot{t}_k} \in \dot{\mathfrak{S}}$. Corresponding to $\dot{t}_k$ marked instants will be denoted by z_k, $1 \le k \le K$. Then $\bar{z}_K = (z_1, \dots, z_K)$ and the total self-intersection degree of $\check{\beta}$ is $\varkappa(\check{\beta}) = I + K$.

Given $\mathscr{W}_{2s}$ with non-empty set $\dot{\mathfrak{S}}$, there exists at least one pair of elements of $\dot{\mathfrak{S}}$ denoted by $(\mathfrak{s}', \mathfrak{s}'')$ such that $\mathfrak{s}'$ is a marked step of $\mathscr{W}_{2\dot{s}}$, $\mathfrak{s}''$ is the non-marked one and $\mathfrak{s}''$ follows immediately after $\mathfrak{s}'$ in $\dot{\mathfrak{S}}$. We refer to each pair of this kind as to the pair of *broken tree structure steps* of $\mathscr{W}_{2s}$ or in abbreviated form, the *BTS-pair* of $\mathscr{W}_{2s}$. If τ' is the marked instant that corresponds to $\mathfrak{s}'$, we will simply say that τ' is the *BTS-instant* of $\mathscr{W}_{2s}$ [12].

Regarding the strongly reduced walk $\dot{\mathscr{W}}$, let us consider a non-marked arrival step at $\check{\beta}$ that we denote by $\bar{\mathfrak{s}} = \mathfrak{s}_{\bar{t}}$. Then one can find a uniquely determined marked instant τ' such that all steps $\mathfrak{s}_t \in \dot{\mathfrak{S}}$ with $\xi_{\tau'} + 1 \le t \le \bar{t}$ are the non-marked ones. Let us denote by t'' the instant of time of the first non-marked step $\mathfrak{s}_{\bar{t}''} \in \hat{\mathfrak{S}}$ of this series of non-marked steps. Then $(\mathfrak{s}_{t'}, \mathfrak{s}_{t''})$ with $t' = \xi_{\tau'}$ is the BTS-pair of $\mathscr{W}_{2s}$ that corresponds to $\bar{t}$. We will say that $\bar{t}$ is attributed to the corresponding BTS-instant τ'.

It can happen that several arrival instants $\bar{t}_i$ are attributed to the same BTS-instant τ'. We will also say that the BTS-instant τ' *generates* the imported cells that are attributed to it.

Let us consider a BTS-instant τ such that $\mathscr{W}_{2s}(\xi_\tau) = \breve{\beta}$. As it is said above, there are K such marked instants denoted by z_k, $1 \le k \le K$. We refer to such BTS-instants as to the *local* ones. Assuming that a marked BTS-instant z_k generates $f'_k \ge 0$ imported cells, we denote by $\varphi_1^{(k)}, \dots, \varphi_{f'_k}^{(k)}$ the positive numbers such that

$$\mathscr{W}_{2s}(\xi_{z_k} + \sum_{j=1}^{l} \varphi_j^{(k)}) = \breve{\beta} \quad \text{for all} \quad 1 \le l \le f'_k. \tag{42}$$

If for some $\tilde{k}$ we have $f'_{\tilde{k}} = 0$, then we will say that $z_{\tilde{k}}$ does not generate any imported cell at $\breve{\beta}$. We denote $\bar{\varphi}^{(k)} = (\varphi_1^{(k)}, \dots, \varphi_{f'_k}^{(k)})$.

Let us consider a BTS-instant τ that generates imported cells at $\breve{\beta}$ and such that $\mathscr{W}_{2s}(\xi_\tau) \ne \breve{\beta}$. We denote these BTS-instants by y_j, $1 \le j \le J$ a say that y_j is a *remote* BTS-instant with respect to $\breve{\beta}$. Assuming that a marked BTS-instant y_j generates $f''_j + 1$ imported cells, $f''_j \ge 0$, we denote by $\Lambda_j, \psi_1^{(j)}, \dots, \psi_{f''_j}^{(j)}$ the positive numbers such that $\mathscr{W}_{2s}(\xi_{y_j} + \Lambda_j) = \breve{\beta}$ and

$$\mathscr{W}_{2s}\left(\xi_{y_j} + \Lambda_j + \sum_{l=1}^{k} \psi_l^{(j)}\right) = \breve{\beta} \quad \text{for all} \quad 1 \le k \le f''_j. \tag{43}$$

In this case we will say that the first arrival at $\breve{\beta}$ given by the instant of time $\xi_{y_j} + \Lambda_j$ represents the *principal* imported cell at $\breve{\beta}$. All subsequent arrivals at $\breve{\beta}$ given by (41) represent the *secondary* imported cells at $\breve{\beta}$. We will use denotations $\bar{y}_J = (y_1, \dots, y_J)$ and $\bar{\Lambda}_J = (\Lambda_1, \dots, \Lambda_J)$. We also denote $\bar{\psi}^{(j)} = (\psi_1^{(j)}, \dots, \psi_{f''_j}^{(j)})$.

We see that for a given walk $\mathscr{W}_{2s}$, the proper, mirror and imported cells at its vertex of maximal exit degree are characterized by the set of parameters, $(\bar{x}, \bar{m})_I$, $(\bar{z}, \Phi, \bar{f}')_K$, where $\Phi_K = (\bar{\varphi}^{(1)}, \dots \bar{\varphi}^{(K)}), \bar{f}'_K = (f'_1, \dots, f'_K)$ and $(\bar{y}, \bar{\Lambda}, \Psi, \bar{f}'')_J$, where $\Psi_J = (\bar{\psi}^{(1)}, \dots, \bar{\psi}^{(J)}), \bar{f}''_J = (f''_1, \dots, f''_J)$. We also denote

$$F' = \sum_{k=1}^{K} f'_k \quad \text{and} \quad F'' = \sum_{j=1}^{J} f''_j.$$

Summing up, we observe that the vertex $\breve{\beta}$ with the self-intersection degree $\varkappa(\breve{\beta}) = I + K$ has the total number of cells given by $R = I + M + K + J + F$, where I is the number of proper cells from the Dyck-type parts, M is the number of corresponding mirror cells, K is the number of local BTS-instants and J is the number of remote BTS-instants, F represents the number of imported cells at $\breve{\beta}$

generated by the local and remote BTS-instants, $F = F' + F''$. In what follows, we denote the family of the parameters described above by

$$\mathscr{P}_R = \left\{ (\bar{x}, \bar{m})_I, (\bar{y}, \bar{\Lambda}, \Psi, \bar{f}'')_J, (\bar{z}, \Phi, \bar{f}')_K \right\}. \tag{44}$$

3.5 *Proof of Theorem 2*

We are going to estimate the number of walks in the family of walks $\tilde{\mathbb{W}}_{2s}(D)$ that have a vertex of maximal exit degree D. We rewrite (28) in the following form

$$\tilde{\mathscr{Z}}_{2s}(n,\rho) = \sum_{D=1}^{s} \sum_{\mathscr{W}_{2s} \in \tilde{\mathbb{W}}_{2s}(D)} \Pi_a(\mathscr{W}_{2s})\, \Pi_b(\mathscr{W}_{2s})\, |\mathscr{C}_{\mathscr{W}_{2s}}|\,,$$

where $|\mathscr{C}_{\mathscr{W}_{2s}}|$ is given by (24). To estimate the number of elements in $\tilde{\mathbb{W}}_{2s}(D)$, we have to consider a kind of color diagrams that have a separate vertex $\breve{v}$ attributed by the parameters from the family $\mathscr{P}_R$, namely by $\bar{x}_I$ and $\bar{z}_K$. Also we have to incorporate into the diagram description the parameters $\bar{y}_J$. Thus we get a new type of color diagrams that we are going to determine.

3.5.1 Color Diagrams with a Vertex of Maximal Exit Degree

Let us consider a vertex $\breve{v}$ and attach to it $I + K$ edge-boxes. We denote by $\langle \breve{v}_{I,K} \rangle_s$ a realization of the values of marked instants that fill these boxes. Given $\bar{v}, \bar{p}$ and $\bar{q}$, we consider a realization of the corresponding color diagram $\langle \mathscr{G}^{(c)}(\bar{v}, \bar{p}, \bar{q}) \rangle_s$ and point out J edge-boxes that will provide the marked instants $\bar{y}$. Joining such a realization with chosen J edge-boxes $\langle \mathscr{G}^{(c)}_{\bar{y}}(\bar{v}, \bar{p}, \bar{q}) \rangle_s^{(b)}$ with $\langle \breve{v}_{I,K} \rangle_s$, we get a realizations of the diagram we need,

$$\langle \breve{\mathscr{G}}^{(c)}_{\bar{x},\bar{z},\bar{y}}(\bar{v}, \bar{p}, \bar{q}) \rangle_s^{(b)} = \langle \breve{v}_{I,K} \rangle_s \uplus \langle \mathscr{G}^{(c)}_J(\bar{v}, \bar{p}, \bar{q}) \rangle_s^{(b)} = \langle \breve{v} \uplus \mathscr{G}^{(c)} \rangle_s^{(b)}.$$

The last equality of the formula presented above introduces a denotation for a realization of the diagram we consider.

The number of different realizations of the color diagram $\mathscr{G}^{(c)}(\bar{v}, \bar{p}, \bar{q})$ is estimated by the right-hand side of (34). Regarding realizations $\langle \breve{v}_{I,K} \rangle_s$, we can write that

$$|\langle \breve{v}_{I,K} \rangle_s| \le \frac{s^{I+K}}{(I+K)!} 2^{I+K}, \tag{45}$$

where the last factor gives the upper bound for the choice of K elements among $I+K$ ones to be marked as the values of $\bar{z}_K$. The vertex $\breve{\beta}$ of the walk can be attributed by the weight

$$\Pi_a(\breve\beta)\,\Pi_b(\breve\beta) = \begin{cases} \dfrac{V_2}{n}, & \text{if } \varkappa(\breve\beta)=1,\\ \dfrac{1}{n^2\rho^{I+K-2}}\, V_2^2 U^{2(I+K)-4}, & \text{if } \breve\beta \text{ is an } r\text{-vertex},\\ \dfrac{1}{n\rho^{I+K-1}}\, V_2 U^{2(I+K)-2}, & \text{if } \breve\beta \text{ is a } p\text{-vertex or a } q\text{-vertex}. \end{cases} \tag{46}$$

In the first and in the third cases of (46), at least one blue r-vertex is necessarily present in $\mathscr{G}^{(c)}(\bar\nu,\bar p,\bar q)$.

Regarding $\langle \mathscr{G}^{(c)}(\bar\nu,\bar p,\bar q)\rangle_s$, one can choose J edge-boxes to be labeled as the values of the realization $\langle\bar y\rangle$ among $\sum_{k=2}^{s}(k-1)\nu_k = \|\bar\nu\|_1$ edges only. This is because the first arrival to a vertex cannot be the marked BTS-instant. The number of ways to choose J ordered places among $\|\bar\nu\|_1$ ordered edges can be estimated as follows,

$$\binom{\|\bar\nu\|_1}{J} \le \frac{\|\bar\nu\|_1^J}{J!} \le \frac{1}{h_0^J}\exp\{h_0\|\bar\nu\|_1\}, \tag{47}$$

where $h_0 > 1$ is a constant.

3.5.2 Exit Sub-clusters and Cells at $\breve\beta$

The maximal exit degree of a walk $\mathscr{W}_{2s} \in \mathbb{W}_{2s}(D)$ can be represented as follows, $D = \dot D + \ddot D + \breve D$, where $\dot D$ is the number of marked edges of the form $(\breve\beta,\gamma)$ that belong to the strongly reduced walk $\dot{\mathscr{W}}$ (39), $\ddot D$ represents the exit edges that belong to $\ddot{\mathscr{W}} = \mathscr{W}\setminus\dot{\mathscr{W}}$. It is known that $\dot D = F+J$ and that $F \le K$ [12] (see also Lemma 12 of [9]). Also we observe that $\ddot D = M$. Taking into account that $M \le I$, we can write that

$$\breve D = D - M - F - J \ge D - I - K - J. \tag{48}$$

The remaining $\breve D$ edges of $\bar{\mathbb{E}}(\mathscr{W}_{2s})$ belong to the exit sub-clusters of the Dyck-type sub-walks $\breve{\mathscr{W}}^{(k)}$ (40) attached to $\breve\beta$. They are distributed among R arrival cells at $\breve\beta$. We denote by $\bar d = (\breve d_1,\dots,\breve d_R)$ a particular distribution such that $\sum_{l=1}^{R}\breve d_l = \breve D$.

The number of cells R depends on $\langle\mathscr{G}^{(c)}\rangle_s^{(b)}$, θ_s and Υ. However, the inequalities used to get (48) show that

$$R = I+K+M+F+J \le 2I+2K+J = R^*. \tag{49}$$

Then the first relation of (48) implies that $\check{D}+R \leq D+R^*$ and we deduce from (49) that

$$\sum_{\bar{d}_R, |\bar{d}_R|=\check{D}} 1 = \binom{\check{D}+R-1}{R-1} \leq \binom{D+R^*-1}{R^*-1}.$$

Elementary analysis shows that if $D \geq 2$, then

$$\binom{D+R^*-1}{R^*-1} \leq h_0^{R^*} \sup_{R^*\geq 2} \frac{1}{h_0^{R^*-1}} \binom{D+R^*-1}{R^*-1} \leq h_0^{2I+2K+J} \exp\left\{\frac{eD}{h_0}\right\}, \ h_0 > e. \tag{50}$$

Indeed, using the standard estimates

$$\sqrt{2\pi n}\left(\frac{n}{e}\right)^n \leq n! \leq e\sqrt{n}\left(\frac{n}{e}\right)^n, \quad n \geq 1,$$

we can write that

$$\frac{1}{h_0^m}\frac{(D+m)!}{D!\,m!} \leq \frac{1}{h_0^m}\frac{e}{2\pi}\sqrt{\frac{D+m}{mD}}\left(1+\frac{m}{D}\right)^D\left(1+\frac{D}{m}\right)^m \leq \frac{e^m}{h_0^m}\left(1+\frac{D}{m}\right)^m,$$

where we take into account that $D+m \leq 2mD$. Then the last relation of (50) follows. Now we are ready to perform the estimates that prove Theorem 2.

3.5.3 Exponential Estimates and $\tilde{\mathscr{Z}}_{2s}$

In this subsection we estimate the contribution of the non-tree type walks $\tilde{\mathscr{Z}}_{2s}$ and prove relation (29) with the help of computations that are very similar to those used in the pioneering papers by Ya. Sinai and A. Soshnikov. The following statement can be regarded as the principal result of the method.

Lemma 5 *Given D, a realization of the color diagram* $\langle \check{v} \uplus \mathscr{G}^{(c)} \rangle_s^{(b)}$ *and a rule* Υ, *let us consider a family of walks* $\mathbb{W}_{2s}(D, \langle \check{v} \uplus \mathscr{G}^{(c)} \rangle_s, \Upsilon)$ *such that the vertex of the maximal exit degree given by* $\check{v}$ *has D exit edges of the form* $(\check{v}, \gamma_i)$, $i = 1, \ldots, D$. *Then*

$$|\mathbb{W}_{2s}(D, \langle \check{v} \uplus \mathscr{G}^{(c)} \rangle_s, \Upsilon)| \leq 2^{|\bar{q}|}\, D^{|\bar{p}|}\, \left(e^{\eta} h_0^2\right)^{I+K+J}\, e^{-\eta D + eD/h_0}\, \mathrm{t}_s, \tag{51}$$

where $\eta = \ln(4/3)$.

We prove Lemma 5 in Sect. 5. The walks we consider are of the non-tree type and therefore contain at least one blue r-vertex v_0. Let us divide the sum $\tilde{\mathscr{Z}}_{2s}$ in two parts in dependence whether $v_0 = \breve{v}$ or $v_0 \neq \breve{v}$,

$$\tilde{\mathscr{Z}}_{2s}(n,\rho) = \tilde{\mathscr{Z}}_{2s}^{(1)} + \tilde{\mathscr{Z}}_{2s}^{(2)}, \tag{52}$$

respectively. Then we can write that

$$\tilde{\mathscr{Z}}_{2s}^{(1)} = \sum_{D=1}^{s}\sum_{J=0}^{s}\left(\prod_{k=2}^{s}\sum_{r_k,p_k,q_k}\right)^{\star}\sum_{I,K:\,I+K\geq 1} |\mathscr{C}_{W_{2s}}|$$
$$\times \sum_{\breve{v}\uplus\mathscr{G}_s^{(c)}}\ \sum_{\langle\breve{v}\uplus\mathscr{G}^{(c)}\rangle_s}\ \sum_{\mathscr{W}_{2s}\in\mathbb{W}_{2s}(D,\langle\breve{v}\uplus\mathscr{G}^{(c)}\rangle_s)} \Pi_{a,b}(\mathscr{W}_{2s}), \tag{53}$$

where the star means that the values of r_k, p_k and q_k are such that $\sum(k-1)\nu_k \geq J$ and $\sum r_k \geq 1$. The first sum of the second line of (53) takes into account the choice of the J places in $\mathscr{G}^{(c)}$ to be marked as the edge-boxes of values y_j [see also (47)]; the second sum is performed over all possible realizations of the diagram $\breve{v} \uplus \mathscr{G}^{(c)}$ obtained with the help of the values from $\{1,\dots,s\}$ (see (34) for example).

Using relations (27), (37), (45) and (46), we deduce from (53) that

$$\tilde{\mathscr{Z}}_{2s}^{(1)} \leq \sum_{D=1}^{s}\sum_{J=0}^{s}\sum_{I+K\geq 1}\frac{(2s)^{I+K}}{(I+K)!}\,(2(I+K))^{I+K}\,h_0^{2(I+K)+J}\,e^{\eta(I+J+K)}$$
$$\times\left(\prod_{k=2}^{s}\sum_{r_k,p_k,q_k}\right)^{\star} e^{h_0\|\bar{\nu}\|_1}\,\frac{1}{r_k!}\left(\frac{(2k)^k s^k}{(k-1)!}\right)^{r_k}\frac{1}{p_k!}\left(\frac{(2k)^k D s^{k-1}}{(k-2)!}\right)^{p_k}\frac{1}{q_k!}\left(\frac{(2k)^k 2s^{k-1}}{(k-2)!}\right)^{q_k}$$
$$\times e^{-\eta D + eD/h_0}\,\mathrm{t}_s\cdot n^{s+1-(\|\bar{\nu}\|_1+(I+K-1))}$$
$$\times\frac{V_2U^{2(I+K-1)}}{n\,\rho^{I+K-1}}\left(\frac{V_2^2}{n^2}\right)^{r_k}\left(\frac{V_2U^2}{n\rho}\right)^{p_k+q_k}\left(\frac{U^2}{\rho}\right)^{(k-2)\nu_k}\left(\frac{V_2}{n}\right)^{s-\|\bar{\nu}\|-(I+K)}, \tag{54}$$

where we denoted $\|\bar{\nu}\| = \sum_{k=2}^{s} k\nu_k$.

Let us consider a constant [cf. (32)]

$$C_2 = \max\left\{\sup_{k\geq 2}\frac{2k}{((k-1)!)^{1/k}},\ \sup_{k\geq 2}\frac{(2k)^{k/(k-1)}}{((k-2)!)^{1/(k-1)}}\right\}$$

and denote

$$B = C_2h_0e^{h_0+\eta} = 4C_2h_0e^{h_0}/3,$$

where $h_0 > e$ will be determined below. Remembering that $s = \chi\rho$, we can deduce from (54) the following inequality,

$$\tilde{\mathscr{L}}_{2s}^{(1)} \le V_2^s \sum_{D=1}^{s} e^{-\eta D + eD/h_0} n\, \mathrm{t}_s \sum_{J=0}^{s} \sum_{I+K\ge 1} 2sB\left(2B\hat{U}^2\chi\right)^{I+K-1} \left(\prod_{k=2}^{s} \sum_{r_k,p_k,q_k}\right)^{\star} \frac{1}{r_k!}$$
$$\times \left(\frac{B\hat{U}^2 s^2}{n}(B\hat{U}^2\chi)^{k-2}\right)^{r_k} \frac{1}{p_k!}\left(D(B\hat{U}^2\chi)^{k-1}\right)^{p_k} \frac{1}{q_k!}\left(2(B\hat{U}^2\chi)^{k-1}\right)^{q_k}. \tag{55}$$

If χ is such that

$$2B\hat{U}^2\chi \le 1, \tag{56}$$

then (55) implies inequality

$$\tilde{\mathscr{L}}_{2s}^{(1)} \le 4Bs^3\left(\exp\left\{\frac{2Bs^2}{n}\right\} - 1\right) e^{4B\hat{U}^2\chi}\, n\mathrm{t}_s V_2^s \sum_{D=1}^{\infty} \exp\left\{-\eta + 2B\hat{U}^2\chi + e/h_0 D\right\}. \tag{57}$$

Remembering that $\eta = \ln(4/3) > 0.28$, we see that if

$$\frac{3C_2U^2h_0e^{h_0}}{V_2}\chi + \frac{e}{h_0} \le 0.28 \tag{58}$$

then

$$\tilde{\mathscr{L}}_{2s}^{(1)} = O(n\mathrm{t}_s V_2^s s^5/n) = o(n\mathrm{t}_s V_2^s) \tag{59}$$

in the limit $(n, s, \rho) \to \infty$ (4). Clearly, the choice of h_0 and χ such that

$$h_0 = 4e \quad \text{and} \quad \chi \le \frac{V_2}{400e^{4e+1}C_2U^2} \tag{60}$$

makes (56) and (58) valid. Let us note that more detailed analysis of the walks with maximal exit degree D show that the factor s^3 in the right-hand side of (57) could be eliminated. However, in the present paper we do not aim the maximal rate of ρ_n and therefore the upper bound (57) is sufficient for our purposes.

Let us consider the second term of (52). The sub-sum $\tilde{\mathscr{L}}_{2s}^{(2)}$ can be estimated from above by the expression given by the right-hand side of (54), where the sum over I, K is performed over the range $I+K \ge 2$ and the weight factor $V_2U^{2(I+K-1)}/(n\rho^{I+K-1})$ is replaced by $V_2^2U^{2(I+K-2)}/(n^2\rho^{I+K-2})$ [see relation (47)] and where the condition $\sum_k r_k \ge 1$ is omitted.

Then we can write that

$$\tilde{\mathscr{Z}}_{2s}^{(2)} \le nt_s V_2^s \sum_{D=1}^{s} e^{-\eta D + eD/h_0} \sum_{J=0}^{s} \sum_{I+K\ge 2} \frac{2s^2 B}{n} (2B\hat{u}^2\chi)^{I+K-2} \prod_{k=2}^{s} \sum_{r_k,p_k,q_k} \frac{1}{r_k!}$$
$$\times \left(\frac{B\hat{U}^2 s^2}{n} (B\hat{U}^2\chi)^{k-2} \right)^{r_k} \frac{1}{p_k!} \left(D(B\hat{U}^2\chi)^{k-1} \right)^{p_k} \frac{1}{q_k!} \left(2(B\hat{U}^2\chi)^{k-1} \right)^{q_k}.$$

If (60) is true, then we get the following upper bound

$$\tilde{\mathscr{Z}}_{2s}^{(2)} \le nt_s V_2^s \frac{4s^4 B}{n} e^{2B\hat{U}^2\chi} \sum_{D=1}^{\infty} \exp\left\{ -\eta + 2B\hat{U}^2\chi + e/h_0 D \right\}.$$

Then

$$\tilde{\mathscr{Z}}_{2s}^{(2)} = O(nt_s V_2^s s^4/n) = o(nt_s V_2^s) \tag{61}$$

under conditions of Theorem 1. Combining this estimate with the estimate of $\tilde{\mathscr{Z}}_{2s}^{(1)}$ (59), we get (29). Theorem 2 is proved.

4 Tree-Type Walks and $(2,4^\star)$-Walks

Let us consider the family $\hat{\mathbb{W}}_{2s}$ of tree-type walks and separate it into two non-intersecting subsets,

$$\hat{\mathbb{W}}_{2s} = \dot{\mathbb{W}}_{2s} \sqcup \ddot{\mathbb{W}}_{2s},$$

where $\dot{\mathbb{W}}_{2s}$ contains the walks $\mathscr{W}_{2s}$ such that their weights have the factors $V_2 = 1$ and V_4 only and the graph $\bar{g}(W_{2s})$ is such that the V_4-edges do not share a vertex in common. We also denote this set by $\mathbb{W}_{2s}^{(2,4^\star)} = \dot{\mathbb{W}}_{2s}$ and say that if $\mathscr{W}_{2s} \in \mathbb{W}_{2s}^{(2,4^\star)}$, then this $\mathscr{W}_{2s}$ is a tree-type $(2,4^\star)$-*walk*. We denote

$$\dot{\mathscr{Z}}_{2s}^{(n,\rho)} = \sum_{\mathscr{W}_{2s}\in \mathbb{W}_{2s}^{(2,4^\star)}} \Pi_{a,b}(\mathscr{W}_{2s})\, |\mathscr{C}_{\mathscr{W}_{2s}}|, \quad \ddot{\mathscr{Z}}_{2s}^{(n,\rho)} = \sum_{\mathscr{W}_{2s}\in \ddot{\mathbb{W}}_{2s}} \Pi_{a,b}(\mathscr{W}_{2s})\, |\mathscr{C}_{\mathscr{W}_{2s}}|$$

and $\hat{\mathscr{Z}}_{2s}^{(n,\rho)} = \dot{\mathscr{Z}}_{2s}^{(n,\rho)} + \ddot{\mathscr{Z}}_{2s}^{(n,\rho)}$. Let us point out that two following relations are true,

$$|\mathscr{C}_{\mathscr{W}_{2s}}| = n^{|\mathbb{V}(\mathscr{W}_{2s})|}(1+o(1)),\ n\to\infty \quad \text{and} \quad |\mathscr{C}_{\mathscr{W}_{2s}}| \le n^{|\mathbb{V}(\mathscr{W}_{2s})|}, \tag{62}$$

where $\mathbb{V}(\mathscr{W}_{2s})$ is the ensemble of vertices of the graph $\bar{g}(\mathscr{W}_{2s})$.

Theorem 3 *Under conditions of Theorem 1, the following upper bounds are true*

$$\limsup_{(n,s,\rho)\to\infty} \frac{1}{nt_s} \dot{\mathscr{Z}}_{2s}^{(n,\rho)} \leq 4\exp\{16V_4\chi\} \tag{63}$$

and

$$\limsup_{(n,s,\rho)\to\infty} \frac{\rho}{nt_s} \ddot{\mathscr{Z}}_{2s}^{(n,\rho)} \leq C\chi\exp\{16V_4\chi\}, \tag{64}$$

for all $0 < \chi \leq \chi_0 = \chi_0(U)$ *and* $C \geq C_0 = C_0(U)$, *where*

$$\chi_0(U) = \frac{1}{4^{11}U^2} \text{ and } C_0(U) = 3 \cdot 4^{16} U^6. \tag{65}$$

Remark Theorem 3 can be proved under conditions of Theorem 1 with (5) replaced by much less restrictive condition on the probability distribution of a_{ij} to be such that all its moments exist and are bounded as follows,

$$V_{2+2k} \leq k!\, V_2\, b_0^k, \quad k = 2, 3, \dots \tag{66}$$

with given $b_0 > 0$ (see also [13]). In this case the constants of (65) should be replaced by

$$\chi_0'(b_0) = \frac{1}{3 \cdot 2^{19} b_0} \quad \text{and} \quad C_0'(b_0) = 3 \cdot 4^{16} b_0^2, \tag{67}$$

respectively, where we assumed that (66) holds with $V_2 = 1$.

To describe the general structure of the tree-type walk, let us introduce several auxiliary notions. Regarding a sub-walk of $2a$ steps $\mathscr{W}_{2a}$ and its graph $\bar{g}(\mathscr{W}_{2a})$, let us denote by Γ_ϱ the ensemble of the multiple edges of $\bar{g}$ that make a connected component attached to the root ϱ. If the graph $\bar{g}(\mathscr{W}_{2a})$ has no other multiple edges than those of Γ_ϱ and the first step and the last step of $\mathscr{W}_{2a}$ are performed along the edges of Γ_ϱ, we say that $\mathscr{W}_{2a}$ is the element of the block of the first level $\mathbb{B}_a^{(1)}(\Gamma)$, $\Gamma = \Gamma_\varrho$,

$$\mathscr{W}_{2a} = \mathscr{B}_a \in \mathbb{B}_a^{(1)}(\Gamma).$$

We will say also that $\mathscr{W}_{2s}$ by itself is a block of the first level, when no confusion can arise.

We say that a walk $\mathscr{W}_{2b}$ is a block of the second level, $\mathscr{W}_{2s} = \mathscr{B}_b^{(2)}$, if it starts and ends with the steps along the root component of multiple edges Γ_ϱ and in $\mathscr{W}_{2b}$ there exists at least on sub-walk $\mathscr{W}_{2a}'$ that is the block of the first level.

By recurrence, we say that $\mathscr{W}_{2s}$ belongs to the block of the $(k+1)$th level, if it starts and end by the steps of Γ_ϱ and contains sub-walks $\mathscr{B}^{(l_1)}, \dots, \mathscr{B}^{(l_p)}$ such that $l_i \le k$ and there exists at least one l_j such that $l_j = k$.

In general, the tree-type walk $\mathscr{W}_{2s}$ is such that its graph $\bar{g}(\mathscr{W}_{2s})$ represents a tree $\mathscr{T}_h$ of $h \ge 0$ edges and along the chronological run over $\mathscr{T}_h$, the sub-walks $\mathscr{B}_{a_1}^{(l_1)}, \dots, \mathscr{B}_{a_p}^{(l_q)}$ of the levels $1 \le l_j \le s/3$ appear at the different moments $t_1, \dots t_q$, $t_j \in [0, 2h+1]$, $t_i \ne t_j$.

4.1 Proof of Theorem 3

Let us introduce the weight $\pi(\mathscr{W}_{2s}) = \prod_{e \in S_{\bar{g}}} V_{2m_e} \rho^{1-m_e}$, where $S_{\bar{g}}$ is the skeleton of the graph $\bar{g}(\mathscr{W}_{2s})$ and m_e is the multiplicity of the edge e.

Lemma 6 *Let us consider a family of walks* $\mathbb{W}^\diamond(m)$ *such that all edges of* $\bar{g}(\mathscr{W}_{2m})$ *are multiple and all of them are attached to the root* ϱ. *Then for any given* $0 < \chi \le \chi_0$ *and* $C \ge C_0$ (65) *the following estimate,*

$$P(m) = \sum_{\mathscr{W}_{2m} \in \mathbb{W}^\diamond(m)} \pi(\mathscr{W}_{2m}) \le \frac{C}{4^{4m}} \rho^{1 + \mathrm{I}_{\{m \ge 3\}}} \tag{68}$$

holds for all $1 \le m \le s = \lfloor \chi\rho \rfloor - 1$, *where* $\mathrm{I}_A = 1$ *if* A *is true and* $\mathrm{I}_A = 0$ *otherwise.*

Proof In the cases of $m = 2, 3, 4$ relation (68) can be deduced directly from relations

$$P(2) = \frac{V_4}{\rho}, \quad P(3) = \frac{V_6}{\rho^2}, \quad \text{and} \quad P(4) = \frac{3V_4^2}{\rho^2}.$$

In the general case of $m \ge 5$, we can write that

$$P(m) = \sum_{l=1}^{m-5} \binom{m-1}{l} \frac{V_{2+2l}}{\rho^l} P(m-1-l) + \binom{m-1}{3} \frac{V_{2m-6} V_6}{\rho^{m-2}} + \binom{m-1}{2} \frac{V_{2m-4} V_4}{\rho^{m-2}} + \frac{V_{2m}}{\rho^{m-1}}. \tag{69}$$

Using upper bound (5) and assuming that (68) holds for $P(m-1-l)$ of the right-hand side of (69), we can write that

$$P(m) \le \frac{C}{4^{4m} \rho^2} R,$$

where

$$R = \sum_{l=1}^{m-5} \frac{4^{4+4l}}{l!} \cdot \frac{V_2 U^{2l} s^l}{\rho^l} + \frac{(m-1)(m-2)(m-3)}{3!} \cdot \frac{4^{4m} V_2^2 U^{2m-4}}{C\rho^{m-4}}$$
$$+ \frac{(m-1)(m-2)}{2} \cdot \frac{4^{4m} V_2^2 U^{2m-4}}{C\rho^{m-4}} + \frac{4^{4m} V_2 U^{2m-2}}{C\rho^{m-3}}. \tag{70}$$

Denoting $\phi = 4^4 U^2 \chi$ and remembering that $V_2 = 1$, we deduce from (70) that

$$R \le 4^4\, \phi\, e^{\phi} + \frac{4^{16} U^4}{C} \phi \max_{m \ge 5} \frac{m(m-1)(m-2)}{6m^{m-4}} + \phi^2 \frac{4^{12} U^4}{Cm^{m-3}}.$$

It is easy to see that under conditions of Lemma 6, $R \le R_0 < 1$. Similar computations based on (69) show that Lemma 6 is true under conditions (66) and (67). □

Given a Catalan tree $\mathscr{T}_h$, let us consider the ensemble of vertices $\upsilon_1, \upsilon_2, \dots, \upsilon_q$ that have exit edges and denote by $\delta_1, \delta_2 \dots, \delta_q$ the number of such edges, $\delta_i \ge 1$. In this case we will say that the tree $\mathscr{T}_h$ has q inner vertices with exit clusters (or bushes) $\Delta_1, \dots, \Delta_q$ [cf. (25)]. Given $\bar{\mu} = (\mu_1, \dots, \mu_q)$ with $\mu_i \ge \delta_i$, let us consider the family of walks $\mathbb{W}^{\diamond}(\mathscr{T}_h, \bar{\mu})$ of $2h + 2\hat{m}$ steps, $\hat{m} = \mu_1 + \dots + \mu_q$ such that all edges of their graphs $\bar{g}(\mathscr{W}_{2h+2\hat{m}})$ are multiple, $S_{\bar{g}} = \mathscr{T}_h$ and the vertex υ_i has $\delta_i + \mu_i$ exit edges, $1 \le i \le q$.

Lemma 7 *Given $\bar{\mu} = (\mu_1, \dots, \mu_q)$ with $\mu_i \ge \delta_i$, consider the family of walks $\mathbb{W}^{\diamond}(\mathscr{T}_h, \bar{\mu}_q)$ of $2h + 2\hat{m}$ steps, $\hat{m} = |\bar{\mu}_q| = \mu_1 + \dots + \mu_q$ such that all edges of their graphs $\bar{g}(\mathscr{W}_{2h+2\hat{m}})$ are multiple, the skeleton of $\bar{g}$ is given by $S_{\bar{g}} = \mathscr{T}_h$ and the vertex υ_i has $\delta_i + \mu_i$ exit edges, $1 \le i \le q$. Then under conditions of Lemma 6,*

$$P(\mathscr{T}_h, \bar{\mu}_q) = \sum_{\mathscr{W}_{2h+2\hat{m}} \in \mathbb{W}^{\diamond}(\mathscr{T}_h, \bar{\mu}_q)} \pi(\mathscr{W}_{2h+2\hat{m}}) \le 4^{h+\hat{m}} \prod_{j=1}^{q} P(\delta_j + \mu_j). \tag{71}$$

Proof We prove Lemma 7 by recurrence. On the first step, let us consider the family of walks $\mathbb{W}^{\diamond}(p, \mu_0)$ such that each walk $\mathscr{W}_{2p+2\mu_0}$ has the graph $\bar{g}(\mathscr{W}_{2p+2\mu_0})$ with all edges attached to the root ρ, the skeleton $S_{\bar{g}}$ is a bush with p edges and each edge of $S_{\bar{g}}$ is multiple. Then

$$\sum_{\mathscr{W}_{2p+2\mu_0} \in \mathbb{W}^{\diamond}(p, \mu_0)} \pi(\mathscr{W}_{2p+2\mu_0}) \le P(p + \mu_0) \tag{72}$$

and (71) is true. Inequality (72) follows from the obvious observation that $\mathbb{W}^{\diamond}(p, \mu_0) \subset \mathbb{W}^{\diamond}(p + \mu_0)$.

On the next step of recurrence we consider the family of walks $\mathbb{W}^{\diamond}(p, \mu_0; \bar{\delta}_p, \bar{\mu}_p)$ such that their skeleton is given by a tree that has a bush with the root ϱ and p exit edges (ϱ, v_i), $1 \le i \le p$ and there are $\delta_i > 0$ main exit edges and μ_i additional exit edges at each v_i, $1 \le i \le p$. Given a sub-walk $\mathscr{W}_{2p+2\mu_0} \in \mathbb{W}^\diamond(p, \mu_0)$ of (72), we denote by $\kappa_1, \dots, \kappa_p$ the multiplicities of the p edges of its skeleton. It is not hard to see that

$$\sum_{\mathscr{W} \in \mathbb{W}^\diamond(p,\mu_0;\bar\delta_p,\bar\mu_p)} \pi(\mathscr{W}) = \sum_{\mathscr{W} \in \mathbb{W}^\diamond(p,\mu_0)} \pi(\mathscr{W}) \prod_{i=1}^{p} \left(\Xi^{(\kappa_i)}(\delta_i + \mu_i) \sum_{\mathscr{W} \in \mathbb{W}^\diamond(\delta_i,\mu_i)} \pi(\mathscr{W}) \right), \tag{73}$$

where $\Xi^{(\kappa_i)}(\delta_i + \mu_i)$ denotes the number of possibilities to perform $\delta_i + \mu_i$ exits from the vertex v_i after κ_i arrivals to v_i. In (73), we do not indicate the number of steps of corresponding walks that is obvious. Taking into account the upper estimate

$$\Xi^{(\kappa_i)}(\delta_i + \mu_i) = \binom{\delta_i + \mu_i + \kappa_i - 1}{\kappa_i - 1} \le 2^{\delta_i+\mu_i+\kappa_i}, \tag{74}$$

and relation (72), we deduce from (73) that

$$\begin{aligned} \sum_{\mathscr{W} \in \mathbb{W}^\diamond(p,\mu_0;\bar\delta_p,\bar\mu_p)} \pi(\mathscr{W}) &\le \sum_{\mathscr{W} \in \mathbb{W}^\diamond(p,\mu_0)} \pi(\mathscr{W}) \prod_{i=1}^{p} \left(2^{\delta_i+\mu_i+\kappa_i} \sum_{\mathscr{W} \in \mathbb{W}^\diamond(\delta_i,\mu_i)} \pi(\mathscr{W}) \right) \\ &\le 2^{p+\mu_0} P(p+\mu_0) \prod_{i=1}^{p} 2^{\delta_i+\mu_i} P(\delta_i + \mu_i). \end{aligned} \tag{75}$$

Remembering that $h = p + \delta_1 + \dots + \delta_p$ and $\hat m = \mu_0 + \mu_1 + \dots + \mu_p$, we see that (71) follows from (75).

Now it is clear how to proceed in the general case of the walks $\mathscr{W} \in \mathbb{W}^\diamond(\mathscr{T}; \bar\mu)$ whose skeleton is given by a tree $\mathscr{T}_h$. It is sufficient to consider the chronological run over $\mathscr{T}_h$ and to find the first inner vertex v_0 such that there are no inner vertices among its children $v_1, \dots, v_p$. In other words, the vertex v_0 is such that all of the exit edges (v_0, v_i) are leaves and u_i are the outer vertices. Then we can apply (75) to the corresponding sub-walks of $\mathscr{W}' \in \mathbb{W}^\diamond(p, \mu_0; \bar\delta_p, \bar\mu_p)$. After that we can consider the vertex v_0 as the outer one with respect to the reduced walk $\mathscr{W} \setminus \mathscr{W}'$ and repeat the reduction procedure by recurrence.

We see that in this process the vertex v_0 and the edges (v', v_0) are considered twice in the estimates of the form (74): first in the role of κ enters at v_i and then in the role of $\delta + \mu$ exits from v'. Therefore the base of the exponent 2 is replaced by 4 in the final estimate (71). We do not present the detailed computations here because they repeat those of (73) and (75). Lemma 7 is proved. □

As a corollary of Lemma 7, we deduce from (71) with the help of (68) that

$$P(\mathscr{T}_h, \bar{\mu}_q) \le \frac{C^q}{4^{3(h+\hat{m})}} \rho^{q+\sum_{i=1}^{q} \mathrm{I}_{\{\delta_i+\mu_i \ge 3\}}} \tag{76}$$

for any $C \ge C_0(U)$ determined by (65).

Lemma 8 *Let us denote by $B_s^{(1)}$ the sum of weights of all walks of 2s steps that represent the blocks of the first level. Then*

$$B_s^{(1)} = \sum_{\mathscr{W}_{2s} \in \mathbb{B}_s^{(1)}} \pi(\mathscr{W}_{2s}) = \dot{B}_s^{(1)} + \ddot{B}_s^{(1)},$$

where $\dot{B}_s^{(1)}$ is the sum over all walks that have only one multiple edge, this edge is the V_4-edge attached to the root,

$$\dot{B}_s^{(1)} = \frac{V_4}{\rho} T_{s-2}^{(3)}, \tag{77}$$

where

$$T_{s-2}^{(3)} = \sum_{\substack{a_1, a_2, a_3 \ge 0 \\ a_1+a_2+a_3=s-2}} \mathrm{t}_{a_1}\, \mathrm{t}_{a_2}\, \mathrm{t}_{a_3} = \mathrm{t}_s \frac{3s}{2(2s-1)}. \tag{78}$$

If $\rho \ge C$ (76), then

$$\ddot{B}_s^{(1)} \le \frac{C}{120\rho^2} \mathrm{t}_s. \tag{79}$$

Proof Taking into account the definition of the blocks of the first level and remembering that $V_2 = 1$, we can write that

$$B_s^{(1)} = \sum_{h=1}^{\lfloor s/2 \rfloor} \sum_{\mathscr{T}_h} \sum_{\hat{m} \ge 2h} \sum_{\substack{\mu_1,\dots,\mu_q \ge 1 \\ \mu_1+\dots+\mu_q=\hat{m}}} P(\mathscr{T}_h, \bar{\mu}_q)\, T_{s-h-\hat{m}}^{(2(h+\hat{m})-1)}, \tag{80}$$

where the sum is taken over all possible values of $\hat{m}$ and $\mu_i \ge 1$ and similarly to (78),

$$T_{s-h-\hat{m}}^{(2(h+\hat{m})-1)} = \sum_{\substack{a_i \ge 0 \\ a_1+\dots+a_{2(h+\hat{m})-1}=s-h-\hat{m}}} \prod_{i=1}^{2(h+\hat{m})-1} t_{a_i} \le 4^{2(h+\hat{m})} \mathrm{t}_s. \tag{81}$$

The upper bound (81) is proved in Sect. 5. Using (76), we deduce from (80) that

$$B_s^{(1)} \leq \mathrm{t}_s \sum_{h=1}^{s/2} \sum_{\mathscr{T}_h} \frac{C^q}{4^{2h}\rho^q} \prod_{i=1}^{q} \left(\sum_{\mu_i=1}^{\infty} \frac{1}{4^{2\mu_i}\rho^{\mathrm{I}_{\{\delta_i+\mu_i\geq 3\}}}} \right)$$

$$\leq \mathrm{t}_s \sum_{h=1}^{s/2} \sum_{\mathscr{T}_h} \frac{C^q}{4^{2h}\rho^q} \prod_{i=1}^{q} \left(\frac{1}{4^2\rho^{\mathrm{I}_{\{\delta_i\geq 2\}}}} + \sum_{\mu_i=2}^{\infty} \frac{1}{4^{2\mu_i}\rho^2} \right)$$

$$\leq \mathrm{t}_s \sum_{h=1}^{s/2} \sum_{\mathscr{T}_h} \frac{C^q}{4^{2h}(8\rho)^q} \prod_{i=1}^{q} \frac{1}{\rho^{\mathrm{I}_{\{\delta_i\geq 2\}}}}. \tag{82}$$

Regarding the last expression, we observe that if $h = 1$, then $q = 1$ and $\delta_1 = 1$ and we get the term (78) in this case, and therefore $\dot{B}_s^{(1)} \leq 3V_4\mathrm{t}_s/(4\rho)$. For the remaining terms of the sum over $h \geq 2$, we observe that if $q = 1$, then the tree $\mathscr{T}_h$ is determined uniquely with $\delta_1 \geq 2$. Therefore we can write that

$$\sum_{h\geq 2} \sum_{\mathscr{T}_h} \frac{C^q}{4^{2h}(8\rho)^q} \prod_{i=1}^{q} \frac{1}{\rho^{\mathrm{I}_{\{\delta_i\geq 2\}}}} \leq \sum_{h\geq 2} \frac{1}{4^{2h}} \left(\frac{C}{8\rho^2} + (\mathrm{t}_h - 1)\frac{C}{8\rho^2} \right) \leq \frac{C}{120\rho^2},$$

where we have used inequality $C/\rho \leq 1$. Relation (79) is proved. We prove relation (78) in Sect. 5. Lemma 8 is proved. □

It follows from (77), (78) and (79) that if $\rho \geq C$, then

$$B_s^{(1)} \leq \frac{V_4}{\rho}\,\mathrm{t}_s. \tag{83}$$

Lemma 9 *Let us denote by $B_s^{(k)}$ the sum of the weights of walks that represent the blocks of the kth level. If $\rho \geq C$ (65), then the following upper bound holds,*

$$B_s^{(k)} \leq \frac{\alpha \mathrm{t}_s}{\rho}\,\hat{\chi}^{k-1}, \tag{84}$$

where $\alpha = V_4$ and $\hat{\chi} = 128V_4\chi$ with χ determined by (65).

Proof We prove Lemma 9 by recurrence. The case of $k = 1$ is verified directly with the help of (83). In the general case, we can write that

$$B_s^{(k+1)} \leq \sum_{p\geq 1} \sum_{a+b_1+\cdots+b_p=s} B_a^{(1)} \binom{2a-1}{p} pB_{b_1}^{(k)} \prod_{i=2}^{p} \left(B_{b_i}^{(1)} + B_{b_i}^{(2)} + \cdots + B_{b_i}^{(k)} \right), \tag{85}$$

where $B_a^{(1)}$ counts the sub-walks attached to the root and $\binom{2a-1}{p}$ gives the upper bound of the possibilities to choose the instants to start the remaining sub-walks of $\mathscr{B}_{b_i}^{(l)}$; among them there is at least one block of the kth level, we denote the corresponding sum by $B_{b_1}^{(k)}$.

Using (83) and assuming that (84) can be applied to the right-hand side of (85), we obtain that

$$B_s^{(k+1)} \le \frac{\alpha}{\rho} \sum_{p\ge 1} \sum_{a+b_1+\dots+b_p=s} \mathrm{t}_a \mathrm{t}_{b_1} \cdots \mathrm{t}_{b_p} \frac{(2s\alpha)^p}{(p-1)!\rho^p} \hat{\chi}^{k-1} \left(\sum_{j=1}^{k} \hat{\chi}^{j-1} \right)^{p-1}$$

$$\le \frac{\alpha}{\rho} \mathrm{t}_s \frac{2\cdot 4^2 s\alpha}{\rho} \sum_{p\ge 1} \frac{1}{p!} \left(\frac{8\alpha\chi}{1-\hat{\chi}} \right)^{p-1} \le \frac{\alpha \mathrm{t}_s}{\rho} \hat{\chi}^k \frac{\exp\{8\alpha\chi/(1-\hat{\chi})\}}{4}. \quad (86)$$

It is easy to deduce from (65) that $\hat{\chi} < 1/2$ and $\exp\{16\alpha\chi\} < 4$. Then $B^{(k+1)} \le \alpha\hat{\chi}^k \mathrm{t}_s/\rho$ and Lemma 9 is proved. □

Lemma 10 *Denote by $\ddot{B}_s^{(k)}$ the sum of the weights of walks that represent the blocks of kth level and such that some of its sub-walks contains two or more V_4-edges that share a vertex or at least one V_{2l}-edge with $l \ge 3$. Then*

$$\ddot{B}_s^{(k)} \le \frac{\beta \mathrm{t}_s}{\rho^2} \hat{\chi}^{k-1}, \quad k \ge 1, \quad (87)$$

where $\beta = C/120$.

Proof Relation (87) with $k = 1$ is verified by (79). In the case of $k \ge 2$, we can use the denotations of (85) and write that

$$\ddot{B}^{(k+1)} \le \sum_{p\ge 1} \sum_{a+b_1+\dots+b_p=s} \binom{2a}{p} p \left(\ddot{B}_a^{(1)} B_{b_1}^{(k)} + B_a^{(1)} \ddot{B}_{b_1}^{(k)} \right) \prod_{i=2}^{p} \sum_{j_i=1}^{k} B_{b_i}^{(j_i)}$$

$$+ \sum_{p\ge 1} \sum_{a+b_1+\dots+b_p=s} \binom{2a}{p} p(p-1) B_a^{(1)} B_{b_1}^{(k)} \left(\ddot{B}_{b_2}^{(1)} + \dots + \ddot{B}_{b_2}^{(k)} \right) \prod_{i=3}^{p} \sum_{j_i=1}^{k} B_{b_i}^{(j_i)}. \quad (88)$$

Using (79), (83) and (84) and repeating computations of (86), we deduce from (88) that

$$\ddot{B}_s^{(k+1)} \le \mathrm{t}_s \frac{\beta\hat{\chi}^{k-1}}{\rho^2} \frac{4^3\alpha s}{\rho} \sum_{p\ge 1} \frac{1}{(p-1)!} \left(\frac{8\alpha\chi}{1-\hat{\chi}} \right)^{p-1}$$

$$+ \mathrm{t}_s \frac{\beta\hat{\chi}^{k-1}}{\rho^2} \left(\frac{8\alpha s}{\rho} \right)^2 \frac{4}{1-\hat{\chi}} \sum_{p\ge 2} \frac{1}{(p-2)!} \left(\frac{8\alpha\chi}{1-\hat{\chi}} \right)^{p-2}.$$

Then we can write that

$$\ddot{B}_s^{(k+1)} \le t_s \frac{\beta \hat{\chi}^k}{\rho^2} \left(\frac{1}{2} + \frac{2\alpha\chi}{1-\hat{\chi}} \right) \exp\left\{ \frac{8\alpha\chi}{1-\hat{\chi}} \right\}.$$

Taking into account the conditions (65), it is easy to see that $\ddot{B}_s^{(k+1)} \le t_s\beta\hat{\chi}^k/\rho^2$. Lemma 10 is proved. □

According to the general description of the tree-type walks given above and using the second relation of (62), we can write that

$$\hat{\mathscr{Z}}_{2s}^{(n,\rho)} \le n \sum_{h\ge 0} \sum_{\mathscr{T}_h} \sum_{p\ge 0} \binom{2h}{p} \sum_{l_1\ge 1,\dots,l_p\ge 1} \sum_{a_1+\dots+a_p=s-h} \prod_{j=1}^{p} B_{a_j}^{(l_j)},$$

where the sums over l_j and a_i are taken over all possible values such that $a_j \ge 2l_j$. Using the result of Lemma 9, we get for all $\rho \ge C$ the following estimate from above,

$$\hat{\mathscr{Z}}_{2s}^{(n,\rho)} \le n \sum_{h\ge 0} t_h \sum_{p\ge 0} \frac{1}{p!} \left(\frac{2s\alpha}{\rho(1-\hat{\chi})} \right)^p \sum_{a_1+\dots+a_p=s-h} t_{a_1}\cdots t_{a_p} \le 4nt_s \exp\left\{ \frac{8\alpha\chi}{1-\hat{\chi}} \right\}. \tag{89}$$

Then the upper bound (63) follows.

To prove relation (64), we observe that if a walk $\mathscr{W}_{2a}$ belongs to $\ddot{\mathbb{B}}_a^{(k)}$, then it belongs to $\mathbb{B}_a^{(k)}$. Therefore $\ddot{B}_a^{(k)} \le B_a^{(k)}$ and we can write that

$$\ddot{\mathscr{Z}}_{2s} \le n \sum_{h\ge 0} \sum_{p\ge 0} \sum_{\mathscr{T}_h} V_2^h \binom{2h}{p} \sum_{l_1\ge 1,\dots l_p\ge 1} \sum_{a_1+\dots+a_p=s-h} p\ddot{B}_{a_1}^{(l_1)} B_{a_2}^{(l_2)} \cdots B_{a_p}^{(l_p)}.$$

Then for sufficiently large $\rho \ge C$ we get the following upper bound,

$$\begin{aligned} \ddot{\mathscr{Z}}_{2s} &\le n \sum_{h\ge 0} \frac{2\beta s}{(1-\hat{\chi})\rho^2} t_h \sum_{p\ge 1} \frac{1}{(p-1)!} \left(\frac{2s\alpha}{\rho(1-\hat{\chi})} \right)^{p-1} \sum_{a_1+\dots+a_p=s-h} t_{a_1}\cdots t_{a_p} \\ &\le nt_s \frac{8\beta\chi}{\rho} \exp\left\{ \frac{8\alpha\chi}{1-\hat{\chi}} \right\} \end{aligned} \tag{90}$$

and relation (64) follows. Theorem 3 is proved.

4.2 Proof of Theorem 1

It follows from relations (28), (29) and Theorem 3 that

$$M_{2s_n}^{(n,\rho_n)} = \dot{\mathscr{Z}}_{2s_n}^{(n,\rho_n)} + \ddot{\mathscr{Z}}_{2s_n}^{(n,\rho_n)} + \tilde{\mathscr{Z}}_{2s_n}^{(n,\rho_n)} = \dot{\mathscr{Z}}_{2s_n}^{(n,\rho_n)}(1+o(1)).$$

Then the upper bound (6) follows from inequality (63). Also, it follows from Theorem 3 and the first relation of (62) that

$$\dot{\mathscr{Z}}_{2s_n}^{(n,\rho_n)} = n\hat{m}_{s_n}^{(\rho_n)}(1+o(1)), \tag{91}$$

where

$$\hat{m}_s^{(\rho)} = \sum_{\mathscr{W}_{2s}\in\mathbb{W}_{2s}^{(2,4^\star)}} \pi(\mathscr{W}_{2s}) \tag{92}$$

is the total weight of $(2,4^\star)$-walks of $2s$ steps.

Lemma 11 *The generating function $F_\rho(z) = \sum_{s\geq 0} z^s \hat{m}_s^{(\rho)}$ verifies Eq. (9).*

Proof Given a walk $\mathscr{W}_{2s} \in \mathbb{W}_{2s}^{(2,4^\star)}$, we will say that it is of $\mathscr{M}$-type. Let us consider the first edge $e_1 = (\varrho,\alpha)$ of the graph $\bar{g}(\mathscr{W}_{2s})$ of this walk. If e_1 is the V_2-edge, then $\mathscr{W}_{2s}$ splits into three parts, the sub-walk (ϱ,α,ϱ) and two $\mathscr{M}$-type sub-walks $\mathscr{W}_{2a}$ and $\mathscr{W}_{2b}$, $a+b=s-1$.

If the edge e_1 is the V_4-edge, then $\mathscr{W}_{2s}$ splits in five parts, the sub-walk $(\varrho,\alpha,\varrho,\alpha,\varrho)$ and four sub-walks of $\mathscr{S}$-type, $\mathscr{W}_{2a_i}$, $i=1,2,3,4$ and $a_1+a_2+a_3+a_4=s-4$. We say that the sub-walk $\mathscr{W}_{2a}$ is of $\mathscr{S}$-type if it is an $\mathscr{M}$-type sub-walk such that its graph $\bar{g}(\mathscr{W}_{2s})$ has the root ϱ' attached by V_2-edges only.

Let us denote by S_k the total weight of $\mathscr{S}$-type walks of $2k$ steps. It is clear that

$$S_k = \sum_{l=0}^{k} \sum_{a_1+\cdots+a_l=k-l} \hat{m}_{a_1}^{(\rho)}\cdots\hat{m}_{a_l}^{(\rho)}. \tag{93}$$

Elementary computation shows that (93) implies the following relation,

$$S(z) = \sum_{k=0}^{\infty} S_k z^k = \frac{1}{1-zF_\rho(z)}. \tag{94}$$

Taking into account the two possibilities described above, it is not hard to see that

$$F_\rho(z) = 1 + zV_2\left(F_\rho(z)\right)^2 + \frac{z^2V_4}{\rho}(S(z))^4.$$

This equality together with (94) implies (9). Lemma 11 is proved. □

It is easy to see that Lemma 11 combined with (91) completes the proof of Theorem 1. Let us discuss the corollaries of Theorem 1. Regarding the number $\hat{m}_s^{(\rho)}$, we consider its part $[\hat{m}_s^{(\rho)}]_p$ that contains the factor V_4^p. One can write that

$$[\hat{m}_s^{(\rho)}]_p = \left(\frac{V_4}{\rho}\right)^p \sum_{h \ge p} \binom{\mathscr{T}_h}{p}^{\star} T^{(2p)}_{s-h-p}, \tag{95}$$

where $\binom{\mathscr{T}_h}{p}^{\star}$ denotes the number of possibilities to choose p edges $\hat{e}_1, \dots, \hat{e}_p$ in the tree $\mathscr{T}_h$ such that they do not share a common vertex. The factor $T^{(2p)}_{s-h-p}$ counts the number of trees that can be attached to the extremities of the additional edges $\tilde{e}_i$ joined to $\hat{e}_i$, $i = 1, \dots, p$, respectively. Using corollary of Lemma 12 (see formula (103) below), we deduce from (95) that

$$[\hat{m}_s^{(\rho)}]_p \le \frac{1}{p!}\left(\frac{3sV_4}{\rho}\right)^p \sum_{h \ge p} \mathrm{t}_h\, \mathrm{t}_{s-h} \le \frac{1}{p!}\left(\frac{3sV_4}{\rho}\right)^p \mathrm{t}_{s+1}. \tag{96}$$

Then the upper bound (10) follows.

Regarding the right-hand sides of the upper bounds (57) and (61), it is easy to see that

$$\frac{1}{n\mathrm{t}_s} \tilde{\mathscr{Z}}_{2s}^{(n,\rho)} = O\left(\frac{s^5}{n}\right), \quad (n, s, r) \to \infty. \tag{97}$$

Regarding the total weight of the trajectories that belong to the classes of $\mathbb{W}_{2s}^{(2,4^*)}$ and using more precise version of (62), we conclude that

$$\dot{\mathscr{Z}}_{2s}^{(n,\rho)} = \sum_{p=0}^{s} [\hat{m}_s^{(\rho)}]_p \times \frac{n(n-1)\cdots(n-(s-p))}{n^{s-p}} \tag{98}$$

Using elementary inequality

$$\frac{(n-1)\cdots(n-(s-p))}{n^{s-p}} \le \exp\left\{-\frac{(s-p+1)^2}{2n} + \frac{(s-p)^3}{3n^2}\right\}, \quad \text{for all } s = o(\sqrt{n}),$$

we deduce from (96) and (98) that under conditions of Theorem 1 the following relation holds [cf. (91)],

$$\dot{\mathscr{Z}}_{2s}^{(n,\rho)} = n\hat{m}_s^{(\rho)} \left(1 + o(s^{-1})\right). \tag{99}$$

Assuming that $s = o(n^{1/6})$, we see that relations (29), (64) and (99) imply (12).

5 Auxiliary Statements

In this section we collect the auxiliary statements and prove lemmas needed for the proof of Theorems 2 and 3.

5.1 *Proof of Lemma 2*

Let us consider the q-vertex β such that the edge of the second arrival $e_2 = e(\mathfrak{a}_2) = (\beta, \alpha_2)$ is the minimal q-edge over the whole walk $\mathscr{W}_{2s}$. We denote by t_2 the instant of time such that $e_2 = e(t_2)$ and consider the sub-walk $\mathscr{W}_{[0,t_2-1]}$. The reasonings below concern $\mathscr{W}_{[0,t_2-1]} = \mathscr{W}^*$ only.

If the edge $[\beta, \alpha_2]$ represents the second distinct arrival at α_2 by $\mathscr{W}^*$, then α_2 is the blue r-vertex and we are done. Let us consider the case when $[\beta, \alpha_2] = E_1'$ is the first distinct arrival at α_2 by $\mathscr{W}^*$ and denote by $e'_{max} = \max\{e,\ e \in E_1'(\alpha_2)\}$. This edge e'_{max} is closed in $\mathscr{W}^*$ by a non-marked edge f. We consider two possible orientations of f separately.

Let us consider first the cases when $f = (\alpha_2, \beta)$. Then $\mathscr{W}^*$ has to go from β to α_2 after $t(f)$ to create the q-edge $e(t_2)$. It can arrive at α_2 only with by a non-marked step $h = (\gamma, \alpha_2), \gamma \neq \beta$ that closes the marked edge $(\alpha_2, \gamma) = \hat{e}$. Thus, the sub-walk $\mathscr{W}^*$ has to go from β to γ to perform h. If $\mathscr{W}^*$ arrives at γ by a marked edge (δ, γ), then γ is the blue r-vertex because $\delta \neq \alpha_2$. If $\mathscr{W}^*$ arrives at γ by a non-marked step (δ, γ), then this step closes a marked edge $\{\delta, \gamma\}$. If $\{d, \gamma\} = (\delta, \gamma)$, then γ is the blue r-vertex. If $\{\delta, \gamma\} = (\gamma, \delta)$, then we get the recurrence, where the couple a_2, γ is replaced by δ, γ. Since $\varkappa_{\mathscr{W}^*}(\beta) = 1$ byE_1, then this recurrence will be terminated before we come to β and the r-vertex will be specified.

Let us consider the case when $f = (\beta, \alpha_2)$. To perform this step, the sub-walk $\mathscr{W}^*$ has to go from α_2 to β before $t(f)$. Assume that it arrives at β by the step $h = (\gamma, \beta), \gamma \neq \alpha_2$ that has to be the non-marked one.

Let us consider first the case when $\gamma \neq \alpha_1$. The sub-walk has to go from α_2 to γ and arrive at γ by the step $g = (\delta, \gamma)$. If this step is marked, then γ is the blue r-vertex and we are done. If g is non-marked, then it closes the marked edge $\{\gamma, \delta\}$. If $\{\gamma, \delta\} = (\delta, \gamma)$, then γ is the blue r-vertex. If $\{\gamma, \delta\} = (\gamma, \delta)$, then we get a recurrence. Since $\varkappa_{\mathscr{W}^*}(\alpha_2) = 1$ by E_1, then this recurrence will be terminated by a blue r-vertex.

Finally, let us consider the case when $\gamma = \alpha_1$ and $h = (\alpha_1, \beta)$. Then the sub-walk has to go from α_2 to α_1 and arrive it by the step $g = (\gamma, \alpha_1)$. If this step is marked, then α_1 is the blue r-vertex. If g is non-marked, then either $\gamma = \epsilon$ or $\gamma \neq \epsilon$, where the edge $(\epsilon, \alpha_1) \in E_1(\alpha_1)$.

If $g = \epsilon$, then we get a recurrence with the couple α_1, β replaced by ϵ, α_1. Please note that the fact that (ϵ, α_1) generally is not the first arrival at α_1 does not alter this recurrence. Then we terminate with the blue r-edge.

If $\gamma \neq \epsilon$, then g closes a marked edge $\{\gamma, \alpha_1\}$. If $\{\gamma, \alpha_1\} = (\gamma, \alpha_1)$, then α_1 is the blue r-vertex. If $\{\gamma, \alpha_1\} = (\alpha_1, \gamma)$, then we get a recurrence that will terminate before α_2 and the blue r-vertex will be specified. Lemma 2 is proved.

5.2 *Catalan Trees and Exponential Estimates*

In papers [10, 13], the following statement is proved with the help of recurrent relation (26).

Lemma 12 *Consider the family of Catalan trees constructed with the help of s edges and such that the root vertex ϱ has d edges attached to it and denote by* $\mathrm{t}_s^{(d)}$ *its cardinality,*

$$\mathrm{t}_s^{(d)} = \sum_{u_1+\cdots+u_{d-1}+u_d=s-d} \mathrm{t}_{u_1}\,\mathrm{t}_{u_2}\cdots\,\mathrm{t}_{u_{d-1}}\,\mathrm{t}_{u_d},$$

where the sum runs over all possible $u_i \geq 0$. *Then the upper bound*

$$\mathrm{t}_s^{(d)} \leq e^{-\eta d}\,\mathrm{t}_s, \quad \eta = \ln(4/3) \tag{100}$$

is true for any given integers d and s such that $1 \leq d \leq s$.

Remark 2 We can say that $\mathrm{t}_s^{(d)}$ represents the number of Catalan trees such that their root vertex ϱ has the exit sub-cluster of cardinality d. It is not hard to deduce from (26) that the numbers $\{\mathrm{t}_s^{(d)}, 1 \leq d \leq s\}$ verify the following recurrent relation [13],

$$\mathrm{t}_s^{(d)} = \mathrm{t}_s^{(d-1)} - \mathrm{t}_{s-1}^{(d-2)}, \quad 3 \leq d \leq s \tag{101}$$

with the initial values $\mathrm{t}_s^{(1)} = \mathrm{t}_{s-1}$, $s \geq 1$ and $\mathrm{t}_s^{(2)} = \mathrm{t}_{s-1}$, $s \geq 2$.

Denoting $\mathrm{t}_s^{(d)} = T_{s-d}^{(d)}$, changing variables $k = s - d$, $p = d$ and using an elementary consequence of (7) $\mathrm{t}_{s+1} \leq 4\mathrm{t}_s$ that holds for any $s \geq 0$, we deduce from (100) the following inequalities,

$$T_k^{(p)} \leq e^{-\eta p}\,\mathrm{t}_{k+p} = \left(\frac{3}{4}\right)^p \mathrm{t}_{k+p} \leq 3^p\,\mathrm{t}_k. \tag{102}$$

Let us note that $T_k^{(p)}$ enumerates the family of Catalan trees constructed on p roots with the help of k edges. Another useful consequence of (100) is as follows,

$$T_{k-p}^{(2p)} \leq \left(\frac{3}{4}\right)^{2p} \mathrm{t}_{k+p} \leq 3^p \mathrm{t}_k. \tag{103}$$

These relations prove the upper bounds (81) and (96).

Regarding the numbers $R_s = T^{(3)}_{s-2}$ (78), it is easy to see that

$$\sum_{s=2}^{\infty} z^s R_s = z^2 f^3(z).$$

It follows from (11) that $zf^2(z) = f(z) - 1$. Using this equality several times, we obtain that

$$R_s = 3\frac{(2s-2)!}{(s-2)!(s+1)!}, \quad s \geq 2.$$

This proves the last relation of (78).

Let us denote by $\mathcal{N}_s^{(1,2)}$ the number of tree-type walks of $2s$ steps such that their graph contains one V_4-edge (α, β) and remaining $s-2$ edges are V_2-edges. Then it is not hard to see that [cf. (78)]

$$\mathcal{N}_s^{(1,2)} = \sum_{a+b_1+{}_2+b_3=s-2} (2a+1)\mathrm{t}_a\, \mathrm{t}_{b_1}\, \mathrm{t}_{b_2}\, \mathrm{t}_{b_3}, \quad s \geq 2,$$

where the factor $(2a+1)$ gives the number of choice of the root ρ in the sub-tree $\mathcal{T}_a$ attached to the vertex α while the remaining three sub-trees $\mathcal{T}_{b_i}$ are attached to the vertex β. Then the generating function

$$\Phi^{(1,2)}(z) = \sum_{s=2}^{\infty} z^s \mathcal{N}_s^{(1,2)}, \quad \mathcal{N}_2^{(1,2)} = 1$$

is given by expression

$$\Phi^{(1,2)}(z) = 2z^3 f'(z) f^3(z) + z^2 f^4(z). \tag{104}$$

Using (11), one can deduce from (104) that

$$\mathcal{N}_s^{(1,2)} = \frac{(2s)!}{(s-2)!\,(s+2)!} = \mathrm{st}_s\left(1 - \frac{3}{s+2}\right), \quad s \geq 2. \tag{105}$$

More generally, denoting by $N_s^{(1,m)}$ the family of even closed walks $\mathcal{W}_{2s}$ such that their graphs contain one edge of total multiplicity $2m$ and all other edges of multiplicity 2, we can write for its cardinality the following relation,

$$\mathcal{N}_s^{(1,m)} = |N_s^{(1,m)}| = \frac{(2s)!}{(s-m)!\,(s+m)!}, \quad s \geq m \geq 1. \tag{106}$$

To prove (106), let us introduce variables

$$D_k^{(m)} = \sum_{a+b_1+\cdots+b_{2m-1}=k-m} (2a+1)\mathrm{t}_a\mathrm{t}_{b_1}\cdots\mathrm{t}_{2m-1}, \quad k \geq m \geq 1,$$

such that $D_k^{(l)} = 0$ for all $0 \leq l < k$ and

$$E_k^{(m)} = \sum_{a+b_1+\cdots+b_{2m-1}=k-m} (2a+1)\mathrm{t}_a\mathrm{t}_{b_1}\cdots\mathrm{t}_{2m}, \quad k \geq m \geq 1,$$

such that $E_k^{(l)} = 0$ for all $0 \leq l < k$. It is not hard to see that $D_s^{(m)} = \mathscr{N}_s^{(1,m)}$. Indeed, regarding a vertex α, we attach to it m blue edges and get the sub-cluster Δ_m. Then we attach to α a green Catalan tree $\mathscr{T}_a$ and determine the root ϱ by choosing one of $2a+1$ instants of time of the chronological run over $\mathscr{T}_a$. Then we attach $2m-1$ red Catalan trees $\mathscr{T}_{b_i}$ at the remaining $2m-1$ instants of time of the chronological run over the sub-cluster Δ_m of blue edges. The chronological run over the tree obtained gives a walk that belongs to $N_s^{(1,m)}$.

According to these definitions, we get that

$$D_k^{(1)} = \sum_{a+b=k-1} (2a+1)\mathrm{t}_a\mathrm{t}_b, \quad k \geq 1$$

and $D_0^{(1)} = 0$. Then the generating function $\mathscr{D}^{(1)}(z) = \sum_{k\geq 0} z^k D_k^{(1)}$ is such that

$$\mathscr{D}^{(1)}(z) = 2z^2 f'(z)f(z) + z(f(z))^2 = zf'(z)$$

and therefore

$$D_k^{(1)} = \frac{(2k)!}{(k-1)!\,(k+1)!}, \quad k \geq 1.$$

In this computation, we have used the identity $2zf'(z)f(z) = f'(z) - (f(z))^2$ that follows from (11). We can also write that $D_k^{(1)} = k\mathrm{t}_k$. This relation is obvious because the number $D_k^{(1)}$ by its definition enumerates the ensemble of Catalan trees with one marked edge colored in blue.

It follows from the definition of $\{E_k^{(m)}\}_{k\geq 0}$ that the generating function $\mathscr{E}^{(1)}(z) = \sum_{k\geq 0} z^k E_k^{(1)}$ is given by the formulas

$$\mathscr{E}^{(1)}(z) = 2z^2 f'(z)(f(z))^2 + z(f(z))^3 = \frac{f'(z)}{2} - \frac{f(z)-1}{2z}$$

and therefore

$$E_k^{(1)} = \frac{k}{2}\mathrm{t}_{k+1}, \quad k \geq 0.$$

Using the fundamental recurrence (26), we can write that

$$\begin{aligned} D_k^{(m)} &= \sum_{a+b_1+\cdots+b_{2m-3}+c=k-m} (2a+1)\mathrm{t}_a\mathrm{t}_{b_1}\cdots\mathrm{t}_{2m-3} \sum_{b_{2m-2}+b_{2m-1}=c} \mathrm{t}_{2m-2}\,\mathrm{t}_{2m-1} \\ &= \sum_{a+b_1+\cdots+b_{2m-3}+c=k-m} (2a+1)\mathrm{t}_a\mathrm{t}_{b_1}\cdots\mathrm{t}_{2m-3}\mathrm{t}_{c+1}. \end{aligned}$$

Then we get the following equality,

$$D_k^{(m)} = E_k^{(m-1)} - D_k^{(m-1)}, \quad k \geq m \geq 2.$$

Similar computation shows that

$$E_k^{(m)} = D_{k+1}^{(m)} - E_k^{(m-1)}, \quad k \geq m \geq 2.$$

Using these recurrent relations together with the initial expressions given by $D_k^{(1)}$ and $E_k^{(1)}$, one can easily check that

$$D_k^{(m)} = \frac{(2k)!}{(k-m)!\,(k+m)!} \quad \text{and} \quad E_k^{(m)} = \frac{(2k+1)!}{(k+1-m)!\,(k+m)!}$$

for all $k \geq m \geq 1$. This proves relation (106).

It is interesting to note that with the help of the same reasoning as above, one can deduce from (101) the following relation,

$$\mathrm{t}_s^{(d)} = \begin{cases} (2m-1)\mathrm{t}_{s-m} \prod_{i=1}^{m-1} \dfrac{s+1-m-i}{s+1-i}, & \text{if } d = 2m-1,\ m \geq 1, \\ m\mathrm{t}_{s-m} \prod_{i=1}^{m-1} \dfrac{s-m-i}{s+1-i}, & \text{if } d = 2m,\ m \geq 0. \end{cases} \tag{107}$$

Let us denote by $\mathcal{N}_s^{(2,2)}$ the number of tree-type walks that contain two V_4-edges while the remaining ones are the V_2-edges. One can show that the generating function $\Phi_s^{(2,2)}$ is given by the following expression,

$$\Phi^{(2,2)}(z) = \frac{z^4}{2} f''(z) f^4(z) + 3z^4 f'(z) f^6(z).$$

Then

$$\mathcal{N}_s^{(2,2)} = \mathrm{t}_s \frac{k^4 + 8k^3 + 39k^2 + 12}{2(k+2)(k+3)} - 4^k + 3\frac{(2k)!}{(k-4)!\,(k+4)!}. \tag{108}$$

Although the right-hand side of (108) is not as compact as that of (106), we can easily deduce from them that

$$\lim_{s\to\infty} \frac{1}{s\mathrm{t}_s}\mathcal{N}_s^{(1,2)} = 1 \quad \text{and} \quad \lim_{s\to\infty} \frac{1}{s^2\mathrm{t}_s}\mathcal{N}_s^{(2,2)} = \frac{1}{2}.$$

These relations agree with the upper bounds (96) in the cases of $p = 1$ and $p = 2$. Moreover, one can put forward a conjecture that

$$\mathcal{N}_s^{(p,2)} = \frac{s^p}{p!}\,\mathrm{t}_s\,(1 + o(1)), \quad s \to \infty. \tag{109}$$

This allows one to expect that the following lower bound holds [cf. (14)],

$$\liminf_{(n,s,\rho)\to\infty} \frac{1}{\mathrm{t}_s}\,\hat{m}_s^{(\rho)} \geq e^{\chi V_4}. \tag{110}$$

Finally, let us note that a part of $\mathcal{N}_s^{(2,2)}$ given by

$$\ddot{\mathcal{N}}_s^{(2,2)} = 4\frac{(2s)!}{(s-4)!\,(s+4)!}$$

represents the number of $(2,4)$-walks of $2s$ steps such that have two V_4-edges with common vertex. It is easy to see that $\ddot{\mathcal{N}}_s^{(2,2)} = s\mathrm{t}_s(1 + o(1))$ and therefore these walks do not contribute to the limiting expression for $M_{2s}^{(\rho)}$ (8). This is in complete accordance with the definition of the numbers $\hat{m}_s^{(\rho)}$ as the total weight of the tree-type $(2,4^\star)$-walks. The terms $\ddot{\mathcal{N}}_s^{(2,2)}$ and $\mathcal{N}_s^{(1,3)}$ provide the leading contribution to the asymptotic expansion (13) given by $R_s^{(1)}$.

5.3 D-Lemma

In the present subsection we prove Lemma 5. Let us introduce an auxiliary collection of variables

$$\mathcal{H} = (\bar{m}_I, (\bar{\Lambda}, \Psi, \bar{f}'')_J, (\Phi, \bar{f}')_K)$$

that represents a part of parameters $\mathscr{P}_R(\bar{x}, \bar{y}, \bar{z}, \mathscr{H})$ (44) and consider its numerical realization $\langle \mathscr{H} \rangle$. Then relation (51) can be rewritten in the following form,

$$| \sqcup_{\langle \mathscr{H} \rangle} \mathbb{W}_{2s}(D, \langle \breve{v} \uplus \mathscr{G}^{(c)} \rangle_s, \langle \mathscr{H} \rangle, \Upsilon)| \le 2^{|\bar{q}|} D^{|\bar{p}|} \left(e^{\eta} h_0^2\right)^{I+J+K} e^{-\eta D + eD/h_0} \mathrm{t}_s, \tag{111}$$

where the disjoint union is taken over the set of all possible realizations $\mathbb{H} = \{\langle \mathscr{H} \rangle\}$. By the construction, the values of $\langle (\bar{x}, \bar{y}, \bar{z}) \rangle_s$ are determined by the realization of the color diagram $\langle \mathscr{G}^{(c)} \rangle_s$. As we will see below, the set $(\langle \bar{x}, \bar{y}, \bar{z} \rangle_s, \langle \mathscr{H} \rangle)$ uniquely determine the nest cells $\breve{v}_1, \dots, \breve{v}_R$ in the underlying trees $\mathscr{T}(\mathscr{W}_{2s})$ where the clusters $\check{\Delta}_1, \dots, \check{\Delta}_R$ are attached. Then we can apply inequalities of the form (106) to get the upper bound of the set of underlying trees that is exponential with respect to the sum $\sum_{i=1}^R d_i$, $d_i = |\check{\Delta}_i|$.

We prove (111) by recurrence with respect to R and $N = I + J + K$.

5.3.1 The Case of $R = 1$

If the total number of cells at $\breve{\beta}$ is equal to one, $R = 1$, then either $\mathscr{P}_1 = x_1$ or $\mathscr{P}_1 = z_1$ and the set of variables $\mathscr{H}$ is empty. For simplicity, we consider the former case such that $\langle x_1 \rangle_s = \tau_1 \ge 1$. Regarding a walk $\mathscr{W}_{2s}$ from the left-hand side of (111) and the corresponding tree $\mathscr{T}_s = \mathscr{T}(\mathscr{W}_{2s})$, we observe that its vertex $\breve{v}$ such that $\breve{v} = \mathfrak{R}(\xi_{\tau_1})$ is attached by a sub-cluster $\check{\Delta}_1$ of d_1 edges. It is easy to construct the corresponding family of trees $\mathbb{T}_s(\tau_1, d_1)$ with the help of the following procedure.

Let us take a root vertex $\mathfrak{b}_0$ and attach to it a linear branch $\mathscr{B}_l$ that consists of l edges and $l + 1$ vertices $\mathfrak{b}_0, \mathfrak{b}_1, \dots, \mathfrak{b}_l$. Regarding the set of vertices $\{\mathfrak{b}_0, \dots, \mathfrak{b}_{l-1}\}$, we attach to them the sub-trees $\mathscr{T}_{a_1}, \dots, \mathscr{T}_{a_l}$ with given $\bar{a} = (a_1, \dots, a_l)$ such that $|\bar{a}| = a_1 + \dots + a_l = \tau_1 - l$. We do this in the way that the sub-trees grow to the left of the branch $\mathfrak{B}_l$ with respect to the ascending chronological run over $\mathfrak{B}_{l-1}$ from $\mathfrak{b}_0$ to $\mathfrak{b}_{l-1}$.

We attach to the vertex $\breve{v} = \mathfrak{b}_l$ the sub-cluster $\check{\Delta}_1$ of $D = d$ edges. Using d vertices of $\check{\Delta}_1$, we attach to them d sub-trees $\mathscr{T}_{b_1}, \dots, \mathscr{T}_{b_d}$. Regarding the vertices $\mathfrak{b}_{l_1}, \dots, \mathfrak{b}_0$ as the descending part of the chronological run over $\mathscr{B}_{l-1}$, we construct on these vertices the sub-trees $\mathscr{T}_{c_1}, \dots, \mathscr{T}_{c_l}$, where b_i and c_j are such that $\sum_{i=1}^d b_i + \sum_{j=1}^l c_j = s - \tau_1 - d$.

Then we can write that

$$\mathrm{T}_s(\tau_1, d) = |\mathbb{T}_s(\tau_1, d)| = \sum_{l=1}^{\tau_1} \sum_{\substack{\bar{a}: \\ |\bar{a}| = \tau_1 - l}} \mathrm{t}_{a_1} \cdots \mathrm{t}_{a_l} \sum_{\substack{\bar{b}, \bar{c}: \\ |\bar{b}| + |\bar{c}| = s - \tau_1 - d}} \mathrm{t}_{b_1} \cdots \mathrm{t}_{b_d} \, \mathrm{t}_{c_1} \cdots \mathrm{t}_{c_l}. \tag{112}$$

It follows from Lemma 12 that

$$\sum_{\bar{b}:\, |\bar{b}|=m} \mathfrak{t}_{b_1} \cdots \mathfrak{t}_{b_d} \le e^{-\eta d}\, \mathfrak{t}_{m+d}. \tag{113}$$

Then we can deduce from (112) the following inequality,

$$\mathrm{T}_s(\tau_1, d) \le e^{-\eta d} \sum_{l=1}^{\tau_1} \sum_{\substack{\bar{a},b,\bar{c}:\ |\bar{a}|,b,|\bar{c}|\ge 0 \\ |\bar{a}|+b+|\bar{c}|=s-l}} \mathfrak{t}(\bar{a})\, \mathfrak{t}_b\, \mathfrak{t}(\bar{c}) = e^{-\eta d}\, \mathfrak{t}_s, \tag{114}$$

where we denoted $\mathfrak{t}(\bar{a}) = \mathfrak{t}_{a_1} \cdots \mathfrak{t}_{a_l}$. The last equality follows from the observation that the sum in the central part of (114) represents the cardinality of the set of Catalan trees $\mathscr{T}_s$ that have the vertex $\check{\upsilon}$ seen at the instant ξ_{τ_1} of the chronological run $\mathscr{R}\{\mathscr{T}\}$ colored in white, the others being the black ones. Clearly, given $\mathscr{T}_s$, there exists only one such vertex and therefore the cardinality of this family is equal to $\mathfrak{t}_s$.

Using inequality (114) with $d = D$ and applying the filtration estimate (36) to the number of realizations of the color diagram $\mathscr{G}^{(c)}(\bar{\upsilon}, \bar{p}, \bar{q})$, we get the following inequality

$$|\mathbb{W}_{2s}(D, \check{\upsilon}(x_1) \uplus \langle \mathscr{G}^{(c)}(\bar{\upsilon}, \bar{p}, \bar{q})\rangle_s^{(b)}, \Upsilon)| \le 2^{|\bar{q}|}\, D^{|\bar{p}|}\, e^{-\eta D}\, \mathfrak{t}_s \tag{115}$$

that implies the upper bound (111).

5.3.2 The Case of $R = 2, N = 2$

(a) Let us consider first the case when $R = 2, I = 2$. Then there is no mirror cells at $\breve{\beta}$ and $J = K = 0$. Therefore $\mathscr{P}_2$ (44) is such that $\bar{x} = (x_1, x_2)$ and $\langle \bar{x} \rangle_s = (\tau_1, \tau_2), 1 \le \tau_1 < \tau_2$ and the set $\mathscr{H}$ is empty. In this case the tree $\mathscr{T}_s = \mathscr{T}(\mathscr{W}_{2s})$ is such that the vertices $\check{\upsilon}_1$ and $\check{\upsilon}_2$ lie on two different branches $\mathscr{B}_1$ and $\mathscr{B}_2$, respectively. Let us describe the construction of the corresponding subset of trees and estimate its cardinality.

We start with the first branch $\mathscr{B}_1 = \mathscr{B}_{l_1}$ that contains l_1 edges and l_+1 vertices $\mathfrak{b}_0, \mathfrak{b}_1, \ldots, \mathfrak{b}_{l_1}$ and construct on its first l_1 vertices the sub-trees $\mathscr{T}_1, \ldots, \mathscr{T}_{a_{l_1}}$ such that $|\bar{a}| = \tau_1 - l_1$. Assuming that $D = d_1 + d_2$, we attach to $\mathfrak{b}_{l_1} = \check{\upsilon}_1$ the sub-cluster $\check{\Delta}_1$ of d_1 edges and join to each of d_1 vertices the sub-trees $\mathscr{T}_{b_1^{(1)}}, \ldots, \mathscr{T}_{b_{d_1}^{(1)}}$ such that $|\bar{b}^{(1)}| = m_1 \ge 0$.

Performing λ steps down along the descending part of $\mathscr{B}_1$, we stop at the vertex $\mathfrak{b}_{l_1-\lambda}, 1 \le \lambda \le l_1$ and attach to it the second branch $\mathscr{B}_2$ with vertices $\mathfrak{c}_0 = \mathfrak{b}_{l_1-\lambda}, \mathfrak{c}_1, \ldots, \mathfrak{c}_{l_2}$. Regarding vertices $\mathfrak{b}_{l_1-1}, \ldots, \mathfrak{b}_{l_1-\lambda}, \mathfrak{c}_1, \ldots \mathfrak{c}_{l_2-1}$, we construct attach to them the sub-trees $\mathscr{T}_{c_j^{(1)}}$ such that $|\bar{c}^{(1)}| = c_1^{(1)} + \ldots c_\lambda^{(1)} + c_{\lambda+1}^{(1)} + \cdots + c_{\lambda+l_2-1}^{(1)} = \tau_2 - \tau_1 - l_1 - l_2 - m_1$.

We construct the second sub-cluster $\check{\Delta}_{d_2}$ with $|\check{\Delta}_{d_2}| = d_2$ on the vertex $\mathfrak{c}_{l_2} = \check{\upsilon}_2$ and construct on the d_2 vertices obtained the sub-trees $\mathscr{T}_{b_1^{(2)}}, \dots, \mathscr{T}_{b_{d_2}^{(2)}}$. Finally, we join the sub-trees $\mathscr{T}_{c_1^{(2)}}, \dots, \mathscr{T}_{c_{l_1+l_2-\lambda}^{(2)}}$ to the vertices $\{\mathfrak{c}_{l_2-1}, \dots, \mathfrak{c}_1, \mathfrak{b}_{l_1-\lambda}, \dots, \mathfrak{b}_0\}$. Then we can write that

$$|\mathbb{T}_s(\tau_1, d_1; \tau_2, d_2)| = \sum_{l_1=1}^{\tau_1} \sum_{\lambda=1}^{l_1} \sum_{l_2=1}^{\tau_2-l_1-d_1} \sum_{\bar{a}:\,|\bar{a}|=\tau_1-l_1} \mathrm{t}(\bar{a}) \sum_{m_1=0}^{\tau_2-\tau_1-l_1} \sum_{\bar{b}^{(1)}:\,|\bar{b}^{(1)}|=m_1} \mathrm{t}(\bar{b}^{(1)})$$

$$\times \sum_{\bar{c}^{(1)}:\,|\bar{c}^{(1)}|=0}^{\tau_2-\tau_1-d_1-m_1-l_2} \mathrm{t}(\bar{c}^{(1)}) \sum_{m_2=0}^{s-\tau_2-d_2} \sum_{\bar{b}^{(2)}:\,|\bar{b}^{(2)}|=m_2} \mathrm{t}(\bar{b}^{(2)}) \sum_{\bar{c}^{(2)}:\,|\bar{c}^{(2)}|=s-\tau_2-d_2-m_2} \mathrm{t}(\bar{c}^{(2)}). \tag{116}$$

We apply (113) two times with respect to the sub-trees with the exit sub-clusters $\check{\mathrm{d}}_1$ and $\check{\mathrm{d}}_2$ and get the estimate

$$\sum_{\bar{b}^{(1)}:\,|\bar{b}^{(1)}|=m_1} \mathrm{t}(\bar{b}^{(1)}) \sum_{\bar{b}^{(2)}:\,|\bar{b}^{(2)}|=m_2} \mathrm{t}(\bar{b}^{(2)}) \le e^{-\eta(d_1+d_2)}\, \mathrm{t}_{m_1+d_1}\, \mathrm{t}_{m_2+d_2} \tag{117}$$

Substituting the right-hand side of (117) into (116), we get an expression similar to (114) with d replaced by $d_1 + d_2$. In this case, t_s is interpreted as the number of Catalan trees such that the vertices seen at the instants ξ_{τ_1} and ξ_{τ_2} are colored in white. The sum over all possible values of d_1 is estimated with the help of the right-hand side of (50). Using (35), it is easy to complete the proof of (115). Then (111) follows.

(b) Let us consider the case of two cells $\mathscr{P}_2 = (x_1, (y_1, \Lambda))$ such that $\langle (x_1, y_1) \rangle_s = (\tau_1, \tau_2)$ are given as well as $\langle \Lambda \rangle = \lambda'$. This means that $\check{\beta}$ is attributed by one proper cell and one imported cell. We first study the case when y_1 does not fill the edge-box attached to a red or to a green vertex of $\mathscr{G}^{(c)}$. There is no mirror cells at $\check{\beta}$ and therefore the vertices $\check{\upsilon}_1 = \mathfrak{R}(\xi_{\tau_1})$ and $\check{\upsilon}_2 = \mathfrak{R}(\xi_{\tau_2})$ are situated on different branches of $\mathscr{T}_s$. Let us briefly describe the construction of the corresponding tree $\mathscr{T}_s$ that is very similar to that we performed above.

Taking the root vertex $\mathfrak{b}_0$, we draw a branch $\mathscr{B}_1$ with the help of l_1 edges. Starting from the extreme vertex $\mathfrak{b}_{l_1}$, we descend by λ_1 steps till the vertex $\mathfrak{b}_{l_1-\lambda_1}$ and attach to it the second branch $\mathfrak{B}_2$ of l_2 edges. We attach to the vertex $\check{\upsilon}_1$ the sub-cluster d_1 of d_1 edges. We denote the skeleton obtained by $K(l_1, d_1, \lambda_1; l_2)$. Regarding the vertices of K, we construct on them the subtrees $\mathscr{T}(\bar{a})$, $\mathscr{T}(\bar{m}_1)$, $\mathscr{T}(\bar{b})$ and $\mathscr{T}(\bar{c}^{(1)})$ with properly chosen values. In this construction, we do not use the vertices of the descending part of K from the vertex $\check{\upsilon}_2$ to $\mathfrak{b}_0$. We denote the sub-tree obtained by $\grave{\mathscr{T}} = \mathscr{T}(\bar{a}, \bar{\beta}, \bar{c}^{(1)}, \bar{m}_1; K)$.

Now let us consider a sub-walk $\mathscr{W}_{[0,\check{\xi}_{\tau_2}-1]}(\grave{\mathscr{T}}) = \grave{\mathscr{W}}$ performed according to the rules of $\langle \mathscr{G}^{(c)} \rangle_s$ and Υ. The vertex $\check{\beta}$ is completely determined by the run of $\grave{\mathscr{W}}$ as well as the vertex $\mathscr{W}(\xi_{\tau_2}) = \gamma$. Therefore the path $\mathscr{L}$ from γ to $\check{\beta}$ by non-marked steps according to Υ, if it exists, is completely determined as well as its length $|\mathscr{L}| = \lambda_2$. This produces the indicator function $I_{\acute{\mathscr{W}}}(\lambda')$ that is equal to 1 if $\lambda' = \lambda_2$ and zero otherwise.

With λ_2 determined, we descend from $\check{\upsilon}_2$ to $\mathfrak{b}_0$ of K by λ_2 steps and attach to the vertex obtained the sub-cluster d_2 of d_2 edges. Then the family of sub-trees $\acute{\mathscr{T}}(d_2) = \mathscr{T}_{s-\tau_2}(\bar{c}^{(2)}, \bar{m}_2)(d_2)$ with the help of the remaining edges is constructed.

Regrading the sum over all values of $\langle \mathscr{H} \rangle = \lambda'$, we see that the cardinality of the set of trees obtained is given by the following expression [cf. (116)],

$$\sum_{\lambda'} \sum_{l_1,\lambda_1,l_2} \sum_{K(l_1,d_1,\lambda_1;l_2)} \sum_{\grave{\mathscr{T}}} I_{\acute{\mathscr{W}}}(\lambda') \sum_{\acute{\mathscr{T}}} 1. \tag{118}$$

Taking into account that

$$\sum_{\acute{\mathscr{T}}_{s-\tau_2}(\bar{c}^{(2)},\bar{m}_2)(d_2)} 1 \le e^{-\eta d_2} \sum_{\acute{\mathscr{T}}_{s-\tau_2}(\bar{c}^{(2)},\bar{m}_2)} 1$$

and that $\sum_{\lambda'} I_{\acute{\mathscr{W}}}(\lambda') = 1$, we conclude that the right-hand side of (118) is bounded by the sum

$$e^{-\eta d_2} \sum_{l_1,\lambda_1,l_2} \sum_{K(l_1,d_1,\lambda_1;l_2)} \sum_{\grave{\mathscr{T}}} \sum_{\acute{\mathscr{T}}} 1 \le e^{-\eta(d_1+d_2)}\, \mathfrak{t}_s. \tag{119}$$

To get the last inequality, we have used the same reasoning as that of (112), (113) and (114). Now it is clear that (35) and (50) together with (119) imply (111).

(c) Let us consider the case of $\langle \mathscr{P}_2 \rangle = (\tau_1, (\tau_2, \lambda'))$ such that the variable y_1 is attributed to the second arrival at the green vertex $\hat{\upsilon}$ of $\mathscr{G}^{(c)}$. The case when it is attributed to the red edge-box is similar and we do not discuss it here.

Let us denote by q' and p' the number of red and green vertices that lie to the left of $\hat{\upsilon}$ and by q'' and $p'' - 1$ the number of red and green vertices to the right of $\hat{\upsilon}$. We construct the skeleton K and the tree $\grave{\mathscr{T}}$ as it is described above. Then we perform the run of the sub-walk $\grave{\mathscr{W}} = \mathscr{W}^{(\grave{\mathscr{T}})}_{[0,t_2-1]}$, $t_2 = \xi_{\tau_2}$ following the prescriptions of $\langle \check{\upsilon} \uplus \mathscr{G}^{(c)} \rangle^{(b)}_s$ and the rule Υ. The vertex $\alpha = \grave{\mathscr{W}}(t_2 - 1)$ being determined, the exit cluster $\Delta(\alpha) = \{\gamma_1, \dots, \gamma_m\}$ is also uniquely determined.

At the instant of time ξ_{τ_2} the walk has to choose a vertex γ' from $\Delta(\alpha)$ such that γ' is situated on the distance of $\langle \Lambda \rangle = \lambda'$ non-marked steps from $\breve{\beta}$. Therefore the indicator function of (118) $I_{\acute{\mathscr{W}}}(\lambda')$ is replaced by $I^{(\gamma_1,\dots,\gamma_m)}_{\acute{\mathscr{W}}}(\lambda')$ that is non-zero only in the case when λ' takes one of the values that correspond to the length of one of the paths of non-marked edges from γ_i to $\breve{\beta}$, if such paths exist. Thus,

$$\sum_{\lambda'} I_{\grave{\mathscr{W}}}^{(\gamma_1,\dots,\gamma_m)}(\lambda') \le m = |\Delta(\alpha)| \le D. \tag{120}$$

Denoting $\mathbb{W}^*_{2s} = \mathbb{W}_{2s}(d_1, d_2; \langle\breve{v} \uplus \mathscr{G}^{(c)}\rangle_s, \langle\mathscr{H}\rangle, \Upsilon)$, we can write the following equality

$$\mathbb{W}^*_{2s} = \sqcup_{\grave{\mathscr{T}}} \{\mathscr{W}_{[0,\xi_{\tau_2}-1]}\} \otimes \{\langle\mathscr{W}(\xi_{\tau_2})\rangle_{\lambda'}\} \otimes \sqcup_{\acute{\mathscr{T}}} \left\{\mathscr{W}^{(\acute{\mathscr{T}})}_{[\xi_{\tau_2}+1,2s]}\right\},$$

where the curly brackets denote the families of realizations of corresponding sub-walks and $\langle\mathscr{W}(\xi_{\tau_2})\rangle_{\lambda'}$ indicates the set of possible values γ'. Then

$$\sum_{\lambda'} |\mathbb{W}^*_{2s}| \le \prod_{\grave{\mathscr{T}}} \#\{\mathscr{W}_{[0,\xi_{\tau_2}-1]}\} \times \sum_{\lambda'} \#\{\langle\mathscr{W}(\xi_{\tau_2})\rangle_{lambda'}\} \times e^{-\eta d_2}\, 2^{q''}\, D^{p''-1}\, \#\left\{\acute{\mathscr{T}}\right\},$$

where we have used inequality (119).

It follows from (120) that

$$\sum_{\lambda'} \#\{\langle\mathscr{W}(\xi_{\tau_2})\rangle_{lambda'}\} \le D.$$

Using this inequality and estimates

$$\#\left\{\mathscr{W}_{[0,\grave{\xi}_{\tau_2}-1]}\right\} \le 2^{q'}\, D^{p'}$$

and

$$\prod_{\acute{\mathscr{T}}} 1 \cdot \#\left\{\acute{\mathscr{T}}\right\} \le e^{-\eta d_1}\, \mathrm{t}_s, \tag{121}$$

we conclude that

$$\sum_{\lambda'} |\mathbb{W}^*_{2s}| \le 2^{|\bar{q}|}\, D^{|\bar{p}|}\, e^{-\eta(d_1+d_2)} \mathrm{t}_s.$$

Remembering (50), it is easy to show that (111) is true in the case under consideration.

5.3.3 The Cases of $N = 2, R \ge 3$

Let us consider $\mathscr{P}_3$ such that the first two cells at $\breve{\beta}$ are given by the instants $\langle(x_1, x_2)\rangle = (\tau_1, \tau_2)$ while the third one is represented by the mirror cell. The presence of this mirror cell with $m_2 = 1$ means that in the Dyck-type part of the walk, and in the corresponding tree the vertices $\check{\upsilon}_1 = \mathfrak{R}(\xi_{\tau_1})$ and $\check{\upsilon}_2 = \mathfrak{R}(\xi_{\tau_2})$ lie on the same branch of edges that starts at the root vertex $\mathfrak{b}_0$. Therefore the tree $\mathscr{T}$ is of the following structure: we choose a length l_1 and construct the branch $\mathscr{B}_1$ of l_1 edges that starts by $\mathfrak{b}_0$ and ends by $\check{\upsilon}_1 = \mathfrak{b}_{l_1}$. Then we attach to $\check{\upsilon}_1$ another linear branch $\mathscr{B}_2$ of l_2 edges that ends by $\check{\upsilon}_2$. We attach the exit sub-cluster $\check{\mathfrak{d}}_1$ to υ_1 at the instant l_1 of the chronological run $\mathfrak{R}(\mathscr{B}_1 \uplus \mathscr{B}_2)$ and the sub-cluster $\check{\mathfrak{d}}_3$ to the vertex υ_1 at the instant $l_1 + 2l_2 + 1$ of $\mathfrak{R}(\mathscr{B}_1 \uplus \mathscr{B}_2)$.

Then we attach the sub-cluster $\check{\mathfrak{d}}_2$ at the vertex $\check{\upsilon}_2$. Regarding $2(l_1 + l_2) + 1 + d_1 + d_2 + d_3$ vertices of the obtained skeleton, we attach to them the sub-trees of the total number of edges $s - (l_1 + l_2 + d_1 + d + 2 + d_3)$. Using three times inequalities of the from (113) we easily get exponential estimates of (112) in this case.

To complete the study of the initial step of the proof of Lemma 12, we consider the case of numerous imported cells of the form $\mathscr{P}_R = (z_1, (y_1, \Lambda, \psi_1, \dots, \psi_f))$, where $f = f_1''$ and $R = 3 + f$. We assume for simplicity that $\langle (z_1, y_1) \rangle = (\tau_1, \tau_2)$ and that $\tau_1 < \tau_2$. The reasoning presented below can be applied without any changes to the case of imported cells generated by the local BTS instants $\langle (z_1, z_2) \rangle = (\tau_1, \tau_2)$. Let us point out that in this situation either $f = 0$ or $f = 1$ [see inequality (49)]. However, we include into considerations the general case of greater values of f. Another remark is that we can ignore the presence of the proper cell $\langle z_1 \rangle = \tau_1$ with the exit sub-cluster $\check{\mathfrak{d}}_1$ at $\check{\beta}$ and consider the imported cells and corresponding exit sub-clusters only. We also assume for simplicity that y_2 is attributed to a blue r-vertex $\hat{\upsilon}$ of $\mathscr{G}^{(c)}$.

To get a realization of $\langle \mathscr{H} \rangle$, we take an integer f and then attribute numerical values to the variables $\Lambda, \psi_1, \dots, \psi_f$ given by $\lambda', \psi_1', \dots, \psi_f'$. Let us take a tree $\grave{\mathscr{T}} = \grave{\mathscr{T}}_{\tau_2}$ and consider a part of the chronological run over it $\mathfrak{R}_{[0,t'-1]}$ with $t' = \xi_{\tau_2}$. Following this run, we construct a sub-walk $\mathscr{W}_{[0,t'-1]}$ according to the rules prescribed by $\langle \check{\upsilon} \uplus \mathscr{G}^{(c)} \rangle_s^{(b)}$ and Υ. At the instant of time t', the walk has to join a vertex γ of $g(\mathscr{W}_{[0,t'-1]})$ prescribed by the values of marked instants of the edge-boxes attached to $\hat{\upsilon}$. This vertex γ is uniquely determined and therefore we are able to conclude whether the set of numerical data $f, (\lambda', \psi_1', \dots, \psi_f')$ is compatible with $\mathscr{W}_{[0,t'-1]}$ or not. We mean that it becomes clear whether there exists a path from γ to $\check{\beta}$ of λ' non-marked steps that the walk can perform according to the rules Υ or not. The same concern f consecutive returns to $\check{\beta}$ with the help of ψ_i' non-marked steps.

The $f + 1$ nest cells are uniquely determined in $\grave{\mathscr{T}}_{\tau_2}$ and the exit sub-clusters of the total cardinality $\check{D} = D - (f + 1)$ are to be distributed to these nest cells. Let us denote by $\bar{d}_{f+1}$ this distribution. We also denote by $\mathbb{T}_s(\grave{\mathscr{T}}_{\tau_2} \uplus \{\check{\mathfrak{d}}_1, \dots, \check{\mathfrak{d}}_{f+1}\})$ a collection of Catalan trees constructed over the base tree $\grave{\mathscr{T}}$ with the exit sub-clusters $\check{\mathfrak{d}}_j$ attached.

Using (114) and (117) several times, one can easily prove the exponential estimate for the number of trees

$$|\mathbb{T}_s(\grave{\mathscr{T}}_{\tau_2} \uplus \{\check{\mathfrak{d}}_1, \dots, \check{\mathfrak{d}}_{f+1}\})| \leq e^{-\eta \check{D}}\, |\mathbb{T}_s(\grave{\mathscr{T}}_{\tau_2})|.$$

By changing somehow the point of view, we can say that given $\langle \breve{v} \uplus \mathscr{G}^{(c)} \rangle_s^{(b)}$ and Υ, the set of all possible values of f and $\langle (\Lambda, \psi_1, \ldots, \psi_f) \rangle$ is filtered by the run of the walk $\mathscr{W}_{[0,t'-1]}$. The values f and $\check{D} = D - (f+1)$ depend on the realization of $\mathscr{W}_{[0,t'-1]}$. With the help of the filtration principle, we get the following inequality,

$$
\begin{aligned}
&| \sqcup_{\langle \mathscr{H} \rangle} \mathbb{W}_{2s}^{(\grave{\mathscr{T}})}(D, \langle \breve{v} \uplus \mathscr{G}^{(c)} \rangle_s^{(b)}, \langle \mathscr{H} \rangle, \Upsilon)| \\
&\quad \le 2^{|\bar{q}|}\, D^{|\bar{p}|} \sup_f \left\{ e^{\eta(f+1)} \binom{\check{D}+f}{f} \right\} e^{-\eta D} \sum_{\grave{\mathscr{T}}} |\mathbb{T}_s(\mathscr{T}_{\tau_2})|, \qquad (122)
\end{aligned}
$$

where the superscript $\grave{\mathscr{T}}$ means that the walks have this tree as the first part of the underlying trees. Taking into account the upper bound $f \le K = 1$ [see (49)], we can apply to the right-hand side of (122) relations (48) and (50) and write that

$$
\sup_f \left\{ e^{\eta(f+1)} \binom{\check{D}+f}{f} \right\} \le e^{2\eta}\, h_0^2\, e^{eD/h_0}. \qquad (123)
$$

Repeating the reasoning of (114), we get from (122) and (123) the upper bound (111).

5.3.4 General Step of Recurrence

The general step of the proof of (111) is to show that if this estimate is true for $N = I + J + K$, then it is true in the case of $N' = N + 1$, where $N' = I' + J' + K'$. Let us consider the case when $K' = K + 1$ and $I' = I$, $J' = J$. This means that if the set $(\bar{x}_I, \bar{y}_J, \bar{z}_k)$ is represented by N marked instants of time $\tau_1 < \tau_2 < \cdots < \tau_N$, then $\tau_{N+1} > \tau_n$ and $z_{K+1} = \tau_{N+1}$. Obviously, the numbers $f''_{K+1} = f$ and $\bar{\varphi}^{(K+1)} = (\varphi_1^{(K+1)}, \ldots, \varphi_f^{(K+1)})$ are also joined to the set of parameters $\langle \mathscr{P}_R \rangle$ (44).

Let us briefly describe the steps that we perform to get the estimate needed. Regarding the vertices and the edge-boxes of realization of the color diagram $\langle \mathscr{G}^{(c)}(\bar{p}, \bar{q}, \bar{\nu}) \rangle_s$, we separate the edge-boxes of each vertex into two groups in dependence of whether the values in the boxes are less than τ_{N+1} or greater than τ_{N+1}. Clearly, the vertex attached by the edge-box with τ_{N+1} plays a special role here. By this procedure, we obtain realizations of two sub-diagrams $\langle \grave{\mathscr{G}} \rangle$ and $\langle \acute{\mathscr{G}} \rangle$ determined in obvious way.

The underlying trees $\mathscr{T}_s = \mathscr{T}(\mathscr{W}_{2s})$ of the walks are of the following structure: there exists a branch $\mathscr{B}_{N+1}$ such that the descending path from the extreme vertex $\check{u}_{N+1}$ to the root $\mathfrak{b}_0$ is of the total length not less than $|\bar{\phi}^{(K+1)}| = \sum_{i=1}^f \varphi_i^{(K+1)}$. At the vertex $\check{v}_{N+1}$ and corresponding f vertices of the descending part of $\mathscr{B}_{N+1}$, the sub-clusters of the total number of D_{N+1} edges are attached. Then the remaining edges are used to construct sub-trees attached to $l_{N+1} + D_{N+1} - f$ vertices. We denote this part of $\mathscr{T}$ by $\acute{\mathscr{T}}$.

It is clear that the set of the walks under consideration can be represented in the form of the right-hand side of (118) with τ_2 replaced by τ_{N+1} and that the exponential estimate with the factor $e^{-\eta D_{N+1}}$ can be obtained for the family of trees $\{\acute{\mathscr{T}}\}$ (see also inequality (119), where q'' and p'' are determined with the help of sub-diagram $\acute{\mathscr{G}}$). Using (123), it is not hard to complete the proof of (111) in the case of $N' = N + 1$. We omit the detailed computations here because they repeat in major part those performed earlier in this sub-section (see also [10] for more discussion of the general step of recurrent estimates).

Acknowledgements The author is grateful to Dr. V. Vengerovsky for the careful reading of the earlier versions of the manuscript. Remarks, corrections and decent suggestions made by the anonymous referee as well as the general discussion are gratefully acknowledged.

References

1. G.W. Anderson, A. Guionnet, O. Zeitouni, *An Introduction to Random Matrices*. Cambridge Studies in Advanced Mathematics, vol. 118 (Cambridge University Press, Cambridge, 2010)
2. Z.D. Bai, Y.Q. Yin, Necessary and sufficient conditions for the almost sure convergence of the largest eigenvalue of Wigner matrices. Ann. Probab. **16**, 1729–1741 (1988)
3. B. Bollobás, *Random Graphs*. Cambridge Studies in Advances Mathematics, vol. 73 (Cambridge University Press, Cambridge, 2001)
4. L. Erdős, A. Knowles, H.-T. Yau, J. Yin, Spectral statistics of Erdős-Rényi Graphs II: eigenvalue spacing and the extreme eigenvalues. Commun. Math. Phys. **314**, 587–640 (2012)
5. L. Erdős, A. Knowles, H.-T. Yau, J. Yin, Spectral statistics of Erdős-Rényi graphs I: local semicircle law. Ann. Probab. **41**, 2279–2375 (2013)
6. O.N. Feldheim, S. Sodin, A universality result for the smallest eigenvalues of certain sample covariance matrices. Geom. Funct. Anal. **20**, 88–123 (2010)
7. Z. Füredi, J. Komlós, The eigenvalues of random symmetric matrices. Combinatorica **1**, 233–241 (1981)
8. S. Geman, A limit theorem for the norm of random matrices. Ann. Probab. **8**, 252–261(1980)
9. O. Khorunzhiy, High moments of large Wigner random matrices and asymptotic properties of the spectral norm. Random Oper. Stoch. Equ. **20**, 25–68 (2012)
10. O. Khorunzhiy, On high moments and the spectral norm of large dilute Wigner random matrices. J. Math. Phys. Anal. Geom. **10**, 64–125 (2014)
11. O. Khorunzhiy, M. Shcherbina, V. Vengerovsky, Eigenvalue distribution of large weighted random graphs. J. Math. Phys. **45**, 1648–1672 (2004)
12. O. Khorunzhiy, V. Vengerovsky, Even walks and estimates of high moments of large Wigner random matrices. Preprint (2008). arXiv:0806.0157
13. A. Khorunzhy, Sparse random matrices: spectral edge and statistics of rooted trees. Adv. Appl. Probab. **33**, 124–140 (2001)
14. A. Khorunzhy, B. Khoruzhenko, L. Pastur, M. Shcherbina, The large-n limit in statistical mechanics and the spectral theory of disordered systems, in *Phase Transitions and Critical Phenomena*, vol. 5 (Academic, London, 1992)
15. M. Krivelevich, B. Sudakov, The largest eigenvalue of sparse random graphs. Comb. Probab. Comput. **12**, 61–72 (2003)
16. M.L. Mehta, *Random Matrices* (Elsevier/Academic, Amsterdam, 2004)
17. A.D. Mirlin, Ya.V. Fyodorov, Universality of level correlation function of sparse random matrices. J. Phys. A **24**, 2273–2286 (1991)

18. G.J. Rodgers, A.J. Bray, Density of states of a sparse random matrix. Phys. Rev. B **37**, 3557–3562 (1988)
19. Ya. Sinai, A. Soshnikov, Central limit theorem for traces of large symmetric matrices with independent matrix elements. Bol. Soc. Brazil. Mat. **29**, 1–24 (1998)
20. Ya. Sinai, A. Soshnikov, A refinement of Wigner's semicircle law in a neighborhood of the spectrum edge for random symmetric matrices. Funct. Anal. Appl. **32**, 114–131 (1998)
21. A. Soshnikov, Universality at the edge of the spectrum in Wigner random matrices. Commun. Math. Phys. **207**, 697–733 (1999)
22. A. Soshnikov, A note on universality of the distribution of the largest eigenvalues in certain sample covariance matrices. J. Stat. Phys. **108**, 1033–1056 (2002)
23. R.P. Stanley, *Enumerative Combinatorics, Vol. 1*. Cambridge Studies in Advanced Mathematics, vol. 49 (Cambridge University Press, Cambridge, 2012)
24. C.A. Tracy, H. Widom, Level spacing distribution and the Airy kernel. Commun. Math. Phys. **159**, 151–174 (1994)
25. E. Wigner, Characteristic vectors of bordered matrices with infinite dimensions. Ann. Math. **62**, 548–564 (1955)

Dyson Processes on the Octonion Algebra

Songzi Li

Abstract We consider Brownian motion on symmetric matrices of octonions, and study the law of the spectrum. Due to the fact that the octonion algebra is nonassociative, the dimension of the matrices plays a special role. We provide two specific models on octonions, which give some indication of the relation between the multiplicity of eigenvalues and the exponent in the law of the spectrum.

1 Introduction

The study of the laws of the spectrum is one of the most important topics in random matrix theory. One may consider stochastic diffusion processes on specific set of matrices, for example symmetric or Hermitian matrices. Usually one considers the empirical measure of the spectrum, which is often again a stochastic diffusion process, called a Dyson process, see the works of Wigner [18], Mehta [14], Dyson [7], Anderson-Guionnet-Zeitouni [1], Erdös et al. [8–10], Forrester [11] and references therein.

Let us recall some classical results on this topic. Consider specific matrices with independent Gaussian elements: real symmetric ($\beta = 1$), Hermitian ($\beta = 2$) and real quaternionic ($\beta = 4$). Then the law of their eigenvalues $(\lambda_i)_{1\leq i\leq n}$, ordered as $\lambda_1 \leq \ldots \leq \lambda_n$, has a density with respect to the Lebesgue measure $d\lambda_1 \ldots d\lambda_n$ which is

$$C_{\beta,n}\exp(-\frac{\beta}{2}\sum_{j=1}^{n}\lambda_j^2)\prod_{1\leq j\leq k\leq n}|\lambda_k-\lambda_j|^{\beta}, \tag{1}$$

where $C_{\beta,n}$ are constants depending on β, n.

S. Li (✉)
Institut de Mathématiques de Toulouse, Université Paul Sabatier, 118 route de Narbonne, 31062 Toulouse, France

School of Mathematical Sciences, Fudan University, 220 Handan Road, 200433 Shanghai, China
e-mail: songzi.li@math.univ-toulouse.fr

C. Donati-Martin et al. (eds.), *Séminaire de Probabilités XLVIII*, Lecture Notes in Mathematics 2168, DOI 10.1007/978-3-319-44465-9_14

On the other hand, one can also consider random matrices with stochastic process as entries. In his paper [7], Dyson derived the stochastic equations of the eigenvalues of Hermitian matrices whose elements are independent complex Brownian motions (see also in this direction Anderson-Guionnet-Zeitouni [1], Mehta [14], Li-Li-Xie [13]). The stochastic process on the spectrum provides a dynamic way to study the law of the eigenvalues of the matrix with Gaussian entries. In fact there are two ways: one is through the law of the eigenvalues of matrices with Brownian motions as entries, considered at time $t = 1$; the other one is through the matrix whose elements are Ornstein-Uhlenbeck process, since when $t \to \infty$ the law of the matrix converges to a matrix with Gaussian entries, and the law of its spectrum is invariant through O-U process: in this case, the law of the spectrum may be seen as the invariant (in fact reversible, see Definition 4) law of the process. This is in general a much easier way to identify the law, since reversible measures are easy to identify through the knowledge of the generator.

Meanwhile, if we consider real symmetric matrix, Hermitian matrix and real quaternionics matrix as real ones of dimension respectively $n \times n$, $2n \times 2n$, $4n \times 4n$, the multiplicity of the eigenvalues is again 1, 2 and 4. This fact leads us to wonder whether this exponent factor in the density reflects the multiplicity of the eigenvalues. However, this is not true.

In a recent paper, Bakry and Zani [3], the authors considered real symmetric matrices whose elements are independent Brownian motions depending on some associative algebra structure of the Clifford type. Their computation of the law of the spectrum shows that, even though there is still the term $\prod_{1 \le j \le k \le n} |\lambda_k - \lambda_j|^\beta$ with $\beta = 1, 2, 4$, the factor β here reflects the structure of the algebra, known as Bott periodicity, rather than the dimension of the eigenspaces, which in this situation may be as large as we want.

The previous study on Dyson Brownian motion, including the work of Bakry and Zani [3] on Clifford algebra, mainly concentrated on the case where the underlying algebra is associative. It is therefore worth understanding how important this property is in the study of the related Dyson processes. The octonion algebra, which is nonassociative but only alternative, provides a good example for us to start with. Its structure differs from the Clifford one, although Clifford algebras with one or two generators coincide with complex numbers and quaternions, the Clifford algebra with three generators does not coincide with octonions, even if the algebras have the same real dimension 8. In his book [11, Sect. 1.3.5], Forrester mentions that the distribution (1) with $\beta = 8$ can be realized by 2×2 matrices on octonions, with Gaussian entries. It is therefore worth to look at the associated Dyson process, which could also provide this result through the study of its reversible measure.

There are only four normed division algebras: $\mathbb{R}$, $\mathbb{C}$, $\mathbb{H}$ and $\mathbb{O}$. We are familiar with $\mathbb{R}$, $\mathbb{C}$, and while the quaternion algebra $\mathbb{H}$ is noncommutative but associative,the octonion algebra $\mathbb{O}$ is nonassociative, but only alternative. Even though their properties are not so nice, octonions have some important connections to different fields of mathematics, such as geometry, topology and algebra. One interesting example is its role in the classification of simple Lie algebra. There are three infinite families of simple Lie algebras, coming from the isometry groups of the projective spaces $\mathbb{RP}^n$, $\mathbb{CP}^n$ and $\mathbb{HP}^n$. The remaining five simple Lie algebras were later discovered to be in connection with octonions: they come from the isometry groups of the projective planes over $\mathbb{O}$, $\mathbb{O}\otimes\mathbb{C}$, $\mathbb{O}\otimes\mathbb{H}$, $\mathbb{O}\otimes\mathbb{O}$ and the automorphism group of octonions. It is also worth to mention that, according to the independent work by Kervaire [12] and Bott-Milnor [5] in 1958, there are only four parallelizable spheres: $\mathbb{S}^0$, $\mathbb{S}^1$, $\mathbb{S}^3$ and $\mathbb{S}^7$, which correspond precisely to elements of unit norm in the normed division algebras of the real numbers, complex numbers, quaternions, and octonions. See more examples in the paper by Baez [2].

For the eigenvalue problem of matrices on octonions, Y.G. Tian proved in his paper [17] that 2×2 Hermitian matrix on octonions has two eigenvalues, each of them has multiplicity 8. For 3×3 Hermitian octonionic matrix, Dray-Manogue [6] and Okubo [16] showed that it has six eigenvalues with multiplicity 4. For 4×4 and 5×5 Hermitian octonionic matrices, there are only numerical results, indicating that the eigenvalues have multiplicity 2 [17]. It is still unknown for matrices in higher dimension. Following the analysis of Bakry and Zani [3], one may expect that the study of probabilistic models on matrices of octonions could give new insights in these directions.

In this paper, we consider Brownian motions on symmetric matrices of octonions. Due to the fact that octonions are nonassociative, and in contrast with the Clifford case, the dimension of the matrices plays a specific role. In fact, contrary to the real, complex and quaternionic cases, octonions do not give rise to infinite series of Lie groups but only specific ones, which are closely related to dimension 2. Thus the study of Dyson processes is mainly pertinent in this dimension, although we introduce another probabilistic model related to the octonion algebra, but with a special structure, see Sect. 4.2. To study the law of the spectrum of the matrices, we consider the processes on the characteristic polynomials $P(X)$, as introduced in the paper by Bakry and Zani [3]. Because of the specific structure of octonions, the traditional way to compute the law of the spectrum turns out to be quite hard, while computation on the process of $P(X)$ provides a simpler and more efficient method to see things clearly.

The paper is organized as follows. Section 2 gives an introduction to the basics of the octonion algebra; Sect. 3 explains briefly the language and tools of symmetric diffusion process; Sect. 4 states our main results, two specific models on octonions, and then we explain what is so special about dimension 2; Sect. 5 is devoted to the demonstration of the connection between the algebra structure and the Euclidean structure associated with the associated symmetric matrices, and the fact that the two exponents, multiplicity of eigenvalues and exponent in the law of eigenvalues, are not correlated.

2 The Octonion Algebra

In this section, we recall some facts about the octonion algebra, and we refer to [2] for more details. We start with a few definitions.

Definition 1 An algebra A is a division algebra if for any $a, b \in A$, with $ab = 0$, then either $a = 0$ or $b = 0$. A normed division algebra is a division algebra that is also a normed vector space with $\|ab\| = \|a\|\|b\|$.

Definition 2 An algebra A is alternative if the subalgebra generated by any two elements is associative. By a theorem of Artin [15], this is equivalent to the fact that for any $a, b \in A$, $(aa)b = a(ab)$, $\quad (ba)a = b(aa)$.

As mentioned earlier, there are only four normed division algebras, $\mathbb{R}$, $\mathbb{C}$, $\mathbb{H}$, $\mathbb{O}$. There is a nice way called "Cayley-Dickson construction" to produce this sequence of algebras: the complex number $a + ib$ can be seen as a pair of real numbers (a, b); the quaternions can be defined as a pair of complex number; and similarly the octonions is a pair of quaternions. As the construction proceeds, the property of the algebra becomes worse and worse: the quaternions are noncommutative but associative, while the octonions are only alternative but not associative.

Since octonions and Clifford algebra are both the algebra with dimension 2^n (in this case $n = 3$), which share some special property, we can use the presentation provided in Bakry and Zani [3] to describe the algebra structure on a basis of octonions, in order to simplify the computations. This presentation is not classical, and we shall therefore use the table below.

Define $E = \{1, 2, 3\}$, and let $\mathscr{P}(E)$ denote the set of the subsets of E. For every set $A \in \mathscr{P}(E)$,we associate a basis element ω_A in the octonion algebra, with $\omega_\emptyset =$ Id, the identity element. Then an element $x \in \mathbb{O}$ can be written in the form

$$x = \sum_A x_A \omega_A, \ x_A \in \mathbb{R},$$

and the product of two elements x and y is given by

$$xy = \sum_{A,B} x_A y_B \omega_A \omega_B.$$

It remains to define $\omega_A \omega_B$ for $A, B \in \mathscr{P}(E)$ through the following rule: denote by $A.B$ the symmetric difference $A \cup B \setminus (A \cap B)$, then $\omega_A \omega_B = (A|B)\omega_{A.B}$, where $(A|B)$ takes value in $\{-1, 1\}$. Then, the multiplication rule in the octonion algebra is defined by a sign table, which is as follows:

	$\emptyset$	$\{1\}$	$\{2\}$	$\{3\}$	$\{1,2\}$	$\{1,3\}$	$\{2,3\}$	$\{1,2,3\}$
$\emptyset$	1	1	1	1	1	1	1	1
$\{1\}$	1	-1	1	1	-1	-1	1	-1
$\{2\}$	1	-1	-1	1	1	-1	-1	1
$\{3\}$	1	-1	-1	-1	1	1	1	-1
$\{1,2\}$	1	1	-1	-1	-1	-1	1	1
$\{1,3\}$	1	1	1	-1	1	-1	-1	-1
$\{2,3\}$	1	-1	1	-1	-1	1	-1	1
$\{1,2,3\}$	1	1	-1	1	-1	1	-1	-1

In this table, the element (i,j) is the sign $(A_i|A_j)$, where A_i is the ith element in the first column, A_j is the jth element in the first row.

From the facts that for $A, B \neq \emptyset$, $\omega_A^2 = -1$ and $\omega_A\omega_B = -\omega_B\omega_A$, it is easy to get the following rules:

$$(A|A) = \begin{cases} -1, & A \neq \emptyset, \\ 1, & A = \emptyset, \end{cases}$$

$$(A|B) = -(B|A), \quad for \quad B \neq A, A, B \neq \emptyset\,.$$

It can be seen from the above table that $\mathbb{O}$ is an algebra, non-associative but alternative. Moreover $\mathbb{O}$ can be equipped with the Euclidean structure obtained by identifying $\mathbb{O}$ as a 8 dimensional (real) vector space via

$$x = \sum_A x_A\omega_A \mapsto (x_\emptyset, x_{\{1\}}, x_{\{2\}}, x_{\{3\}}, x_{\{1,2\}}, x_{\{1,3\}}, x_{\{2,3\}}, x_{\{1,2,3\}})\,,$$

so that the inner product and the norm are respectively:

$$\langle x, y\rangle = \sum_A x_A y_A\;, \quad \|x\| = (\sum_A x_A^2)^{1/2}\,,$$

so that $\{\omega_A,\ A \in \mathscr{P}(E)\}$ form a real orthonormal basis for the algebra $\mathbb{O}$.

Let us recall that to prove that $\mathbb{O}$ is a division algebra, it is usual to introduce the conjugate

$$x = \sum_A x_A\omega_A \mapsto x^* = \sum_A x_A\omega_A(A|A),$$

and observe that $(xy)^* = y^*x^*$, $xx^* = x^*x$ and $\|x\|^2 = xx^*$, so that $\|xy\|^2 = (xy)(xy)^* = (xy)(y^*x^*) = x(yy^*)x^* = \|x\|^2\|y\|^2$.

Although the previous table does not provide an associative algebra, the octonion algebra satisfies however some useful identities. In what follows, we shall make a strong use of Moufang identities, which are stated as follows: for elements x, y, z belongs to $\mathbb{O}$, we have

$$z(x(zy)) = (zxz)y,$$
$$((xz)y)z = x(zyz),$$
$$(zx)(yz) = (z(xy))z,$$
$$(zx)(yz) = z((xy)z)\,.$$

We shall mainly use this for the elements $\omega_A \in \mathbb{O}$, although Moufang identities provide more information than this.

According to the alternativity property, we can get some basic formulae about the sign table $\{(A|B)\}$ for octonions.

Lemma 1 *For $A, B, C, D \in \mathscr{P}(E)$, we have*

1. $(A.B|B) = (A|B)(B|B)$.
2. $(A.B|A)(A.B|B) = (A.B|A.B)$.
3. *If* $A.B \neq \emptyset$, $(A.C|A)(B.C|B) = -(A.C|B)(B.C|A)$.
4. *If* $A.B.C.D = \emptyset$,

$$(B.C|C)(C.D|D)(D.A|A)(A.B|B) = (B.D|B.D).$$

Proof The first one is just the result of alternativity:

$$(\omega_A\omega_B)\omega_B = (A|B)\omega_{A.B}\omega_B = (A.B|B)\omega_A,$$

while

$$(\omega_A\omega_B)\omega_B = \omega_A(\omega_B\omega_B) = (B|B)\omega_A,$$

Hence

$$(A.B|B) = (A|B)(B|B).$$

The second one can be easily proved by the first statement.

For the third statement, we first remark that, for any vector $x = \sum x_C\omega_C$, $x\omega_A$ is always orthogonal to $x\omega_B$ if $A \neq B$. Indeed, to see this, we may reduce to the case where $\|x\| = 1$, and then observe that the fact that the algebra is a division algebra shows that for any $y \in \mathbb{O}$, $y \mapsto xy$ is an orthogonal transformation. For $A, B \in \mathscr{P}(E)$,

$A.B \neq \emptyset$, and any $C \in \mathscr{P}(E)$, choose $D = A.B.C$ and set $x = \omega_C + \omega_D$. Then

$$\begin{aligned} 0 &= \langle x\omega_A, x\omega_B \rangle \\ &= \langle (C|A)\omega_{C.A} + (D|A)\omega_{D.A}, (C|B)\omega_{C.B} + (D|B)\omega_{D.B} \rangle \\ &= \langle (C|A)\omega_{C.A}, (D|B)\omega_{D.B} \rangle + \langle (D|A)\omega_{D.A}, (C|B)\omega_{C.B} \rangle \\ &= (C|A)(D|B) + (D|A)(C|B)\,. \end{aligned}$$

Since $D = A.B.C$, the above formula indicates that for $A, B, C \in \mathscr{P}(E)$, $A.B \neq \emptyset$,

$$(C|A)(A.B.C|B) + (A.B.C|A)(C|B) = 0.$$

By changing C into $A.C$, we get

$$(A.C|A)(B.C|B) + (B.C|A)(A.C|B) = 0,$$

then the third statement is proved.

For the last statement, denote

$$\Theta := (B.C|C)(C.D|D)(D.A|A)(A.B|B). \tag{2}$$

Then, from Moufang identities,

$$\begin{aligned} &((\omega_{B.C}\omega_C)(\omega_{D.A}\omega_A))((\omega_{C.D}\omega_D)(\omega_{A.B}\omega_B)) \\ &\quad = (B.C|C)(C.D|D)(D.A|A)(A.B|B)(\omega_B\omega_D)(\omega_C\omega_A) \\ &\quad = \Theta(B|D)(C|A)\omega_{B.D}\omega_{C.A} \\ &\quad = \Theta(B|D)(C|A)\omega_{B.D}\omega_{B.D} \\ &\quad = \Theta(B|D)(C|A)(B.D|B.D)\,. \end{aligned}$$

On the other hand,

$$\omega_A\omega_B = (A.B|A.B)(A|A)(B|B)\omega_B\omega_A\,,$$

and

$$\begin{aligned} &((\omega_{B.C}\omega_C)(\omega_{D.A}\omega_A))((\omega_{C.D}\omega_D)(\omega_{A.B}\omega_B)) \\ &\quad = (D|D)(B.C|B.C)(A|A)(A|A)(B|B)(A.B|A.B)((\omega_{B.C}\omega_C)(\omega_A\omega_{B.C}))((\omega_{A.B}\omega_D)(\omega_B\omega_{A.B})) \\ &\quad = (D|D)(B.C|B.C)(B|B)(A.B|A.B)(\omega_{B.C}(\omega_C\omega_A)\omega_{B.C})(\omega_{A.B}(\omega_D\omega_B)\omega_{A.B}) \\ &\quad = (D|D)(B.C|B.C)(B|B)(A.B|A.B)(C|A)(D|B)(\omega_{B.C}\omega_{C.A}\omega_{B.C})(\omega_{A.B}\omega_{D.B}\omega_{A.B}) \\ &\quad = (D|D)(B|B)(C|A)(D|B)\omega_{C.A}\omega_{D.B} \\ &\quad = (D|D)(B|B)(C|A)(D|B)(B.D|B.D)\,. \end{aligned}$$

Hence,

$$\Theta = (B|D)(C|A)(B.D|B.D)(D|D)(B|B)(C|A)(D|B)(B.D|B.D)$$
$$= (B|D)(D|D)(B|B)(D|B) = (B.D|B.D),$$

which ends the proof of the lemma.

For a $n \times n$ matrix on octonions, write it as $\mathscr{M} = \sum_A M^A \omega_A$, where $\{M^A\}$ are real $n \times n$ matrices. For an n dimensional vector $\sum_B X^B \omega_B$,

$$(\sum_A M^A \omega_A)(\sum X^B \omega_B) = \sum_{A,B} M^A X^B (A|B) \omega_{A.B} = \sum_{A,B} (A.B|B) M^{A.B} X^B \omega_A.$$

Therefore, $\mathscr{M}$ can be expressed by the real $8n \times 8n$ block matrix $\{M_{ij}^{A,B}\}$, where $M_{ij}^{A,B} = (A.B|B) M_{ij}^{A.B}$.

This leads to the following definition:

Definition 3 A $(2^3 \times n) \times (2^3 \times n)$ block matrix $M^{A,B}$ (where $A, B \in \mathscr{P}(E)$) is a real octonionic if $M^{A,B} = (A.B|B) M^{A.B}$, where $M^A = M^{A,\emptyset}$ is a family of 8 $n \times n$ square matrices. It is the real form of a matrix with octonionic entries. We shall denote it as $\mathscr{M} = \sum_A M^A \omega_A$.

Then, we shall say that an octonionic matrix is symmetric if its real form is symmetric. This corresponds to the fact that, for any $A \in \mathscr{P}(E)$, $(M^A)^t = (A|A) M^A$.

That is to say, $(M^{A,B})^t = (A.B|B)(M^{A.B})^t = M^{B,A} = (B.A|A) M^{A.B}$. Due to property 2 of Lemma 1, this leads to the fact that for any $A \in \mathscr{P}(E)$, $(M^A)^t = (A|A) M^A$, i.e. $M^\emptyset$ is symmetric while M^A is antisymmetric for any $A \neq \emptyset$.

It is worth to point out that since the octonion algebra is not associative, there is no matrix representation of the algebra structure for the octonions, and therefore the matrix multiplication of the real octonionic matrices does not corresponds to the octonionic multiplication of the associated matrices with octonion entries. Even the product of octonionic matrices is not octonionic in general.

The inverse of an octonionic matrix is in general not octonionic, and its exact structure is not easy to decipher; the octonionic property may not be preserved. The following lemma gives a condition for this last property to hold, and will play an important role in the rest of this paper.

Lemma 2 *Let* $M = \sum M^A \omega_A$ *be an octonionic matrix such that* $M^\emptyset$ *is invertible. Assume moreover that, for any* $A, B \in \mathscr{P}(E)$

$$M^A (M^\emptyset)^{-1} M^B = M^B (M^\emptyset)^{-1} M^A, \tag{3}$$

and that $\sum_C M^C (M^{\emptyset})^{-1} M^C$ *is invertible. Then, M is invertible and its inverse N is octonionic, satisfying* $N = \sum_A \omega_A N^A$, *with*

$$N^A = -N^{\emptyset} M^A (M^{\emptyset})^{-1}, \ \ for A \neq \emptyset, \tag{4}$$

$$N^{\emptyset} = (\sum_C M^C (M^{\emptyset})^{-1} M^C)^{-1}. \tag{5}$$

Proof In fact, assume the octonionic matrix $N^{A.B} = (A.B|B)N^{A.B} = (A.B|A)\tilde{N}^{A.B}$ is the inverse of M, where

$$N^{A.B} = (A.B|A.B)\tilde{N}^{A.B} = \begin{cases} -\tilde{N}^{A.B}, & A.B \neq \emptyset; \\ \tilde{N}^{\emptyset}, & A.B = \emptyset. \end{cases}$$

Then

$$\sum_C (A.C|A)(C.B|B)\tilde{N}^{A.C} M^{C.B} = 0, \ \ for A \neq B, \tag{6}$$

$$\sum_C (A.C|A)(C.A|A)\tilde{N}^{A.C} M^{C.A} = \text{Id}. \tag{7}$$

Changing C into $A.B.C$ in (6), we get $\sum_C (B.C|A)(A.C|B)\tilde{N}^{B.C} M^{C.A} = 0$, then it is enough to have

$$(A.C|A)(C.B|B)\tilde{N}^{A.C} M^{C.B} + (B.C|A)(A.C|B)\tilde{N}^{B.C} M^{C.A} = 0.$$

According to Lemma 1, it holds as soon as

$$\tilde{N}^{A.C} M^{C.B} = \tilde{N}^{B.C} M^{C.A}. \tag{8}$$

Choosing $C = B$ and then setting $D = A.C$, this leads to

$$\tilde{N}^D = N^{\emptyset} M^D (M^{\emptyset})^{-1}, \tag{9}$$

for every $D \in \mathscr{P}(E)$. Now choose $C = \emptyset$ in (8) and apply (9) to N^A and N^B, plugging into (8), this reduces to

$$N^{\emptyset} M^A (M^{\emptyset})^{-1} M^B = N^{\emptyset} M^B (M^{\emptyset})^{-1} M^A.$$

Now (7) is

$$\sum_D \tilde{N}^D M^D = \text{Id},$$

and using (9), this gives

$$N^{\emptyset}(\sum M^{C}(M^{\emptyset})^{-1}M^{C})^{-1} = \mathrm{Id},$$

which means that $N^{\emptyset}$ is invertible, and gives its inverse, such that (5) holds true. Then we can use (5) and (9) to get (4).

Remark 1 It is worth to observe for later use that if the matrix $\mathscr{M}$ on the octonions satisfies the assumptions of Lemma 2, then it is also the case of $\mathscr{M} - X\mathrm{Id}$.

3 Symmetric Diffusion Operators on Matrices

We introduce the basics on symmetric diffusion operators, in a simplified version adapted to our case. For further details see [4].

Let E be an open set in $\mathbb{R}^n$, endowed with a σ-finite measure μ and let $\mathscr{A}_0$ be the set of smooth compactly supported functions, or of polynomials functions on E. For any linear operator $\mathrm{L} : \mathscr{A}_0 \mapsto \mathscr{A}_0$, we define its carrè du champ operator as

$$\Gamma(f,g) = \frac{1}{2}\Big(\mathrm{L}(fg) - f\mathrm{L}(g) - g\mathrm{L}(f)\Big).$$

We have the following

Definition 4 A symmetric diffusion operator is a linear operator L: $\mathscr{A}_0 \oplus 1 \mapsto \mathscr{A}_0$, such that

1. $\mathrm{L}(1) = 0$,
2. $\forall f, g \in \mathscr{A}_0 \oplus 1,\ \int f\mathrm{L}(g)\, d\mu = \int g\mathrm{L}(f)\, d\mu$,
3. $\forall f \in \mathscr{A}_0, \Gamma(f,f) \geq 0$,
4. $\forall f = (f_1, \cdots, f_n)$, where $f_i \in \mathscr{A}_0$, Φ is a smooth function $\mathbb{R}^n \mapsto \mathbb{R}$ and $\Phi(0) = 0$,

$$\mathrm{L}(\Phi(f)) = \sum_i \partial_i\Phi(f)\mathrm{L}(f_i) + \sum_{i,j} \partial^2_{ij}\Phi(f)\Gamma(f_i,f_j). \tag{10}$$

Consider an open set $\Omega \subset E$, and a given system of coordinates (x^i), then we can write

$$\mathrm{L}(f) = \sum_{ij} g^{ij}(x)\partial^2_{ij}f + \sum_i b^i(x)\partial_i f,$$

where

$$g^{ij}(x) = \Gamma(x^i, x^j),\ b^i(x) = \mathrm{L}(x^i).$$

In this paper, we perform computations on the characteristic polynomial $P(X) = \det(M - XId)$ of a matrix M. Assume that we have some diffusion operator acting on the entries of a matrix M, described by the values of $\mathrm{L}(M_{ij})$ and $\Gamma(M_{ij}, M_{kl})$ for any (i,j,k,l). Then, we have,

$$\Gamma(\log P(X), \log P(Y)) = \sum_{i,j,k,l} \partial_{M_{ij}} \log(P(X)) \partial_{M_{kl}} \log(P(Y)) \Gamma(M_{ij}, M_{kl})$$

$$\mathrm{L}(\log P(X)) = \sum_{i,j} \partial_{M_{ij}} \log(P(X)) \mathrm{L}(M_{ij}) + \sum_{i,j,k,l} \partial_{M_{ij}} \partial_{M_{kl}} \log(P(X)) \Gamma(M_{ij}, M_{kl}) \,.$$

To compute $\partial_{M_{ij}} \log(P(X))$ and $\partial_{M_{ij}} \partial_{M_{kl}} \log(P(X))$ in the above formulae, we use [3, Lemma 6.1], which we quote here without proof.

Lemma 3 *Let $M = (M_{ij})$ be a matrix and M^{-1} be its inverse, on the set $\{\det M \neq 0\}$ we have*

$$\partial_{M_{ij}} \log \det M = M^{-1}_{ji},$$

$$\partial_{M_{ij}} \partial_{M_{kl}} \log \det M = -M^{-1}_{jk} M^{-1}_{li}.$$

Hence, with $M^{-1}(X) = (M - XId)^{-1}$,

$$\Gamma(\log P(X), \log P(Y)) = \sum_{i,j,k,l} M^{-1}(X)_{ji} M^{-1}(Y)_{lk} \Gamma(M_{ij}, M_{kl}), \tag{11}$$

$$\mathrm{L}(\log P(X)) = \sum_{i,j} M^{-1}_{ji}(X) \mathrm{L}(M_{ij}) - \sum_{i,j,k,l} M^{-1}_{jk}(X) M^{-1}_{li}(X) \Gamma(M_{ij}, M_{kl}). \tag{12}$$

According to Bakry and Zani [3], one can get from $\Gamma(P(X), P(Y))$ and $\mathrm{L}(P(X))$ informations about the multiplicities of the eigenvalues, and on the invariant measure of the operator L acting on $P(X)$:

If for some constants $\alpha_1, \alpha_2, \alpha_3$,

$$\mathrm{L}(P) = \alpha_1 P'' + \alpha_2 \frac{P'^2}{P}, \; \Gamma(\log P(X), \log P(Y)) = \frac{\alpha_3}{Y - X} \Big(\frac{P'(X)}{P(X)} - \frac{P'(Y)}{P(Y)} \Big). \tag{13}$$

and if there exists for some $a \in \mathbb{R}$, $a \neq 0$ which satisfies

$$a^2(\alpha_1 + \alpha_2) - a(\alpha_1 + \alpha_3) + \alpha_3 = 0, \tag{14}$$

then,

1. If a is a positive integer, it is the multiplicity of the eigenvalues of M;
2. Write $P(X) = \prod_{i=1}^{n}(X - x_i)^a$, the invariant measure for the operator L in the Weyl chamber $\{x_1 < \cdots < x_n\}$ is, up to a multiplicative constant,

$$d\mu = (\prod_{i<j}(x_i - x_j)^2)^{-\frac{a^2(\alpha_1+\alpha_2)}{\alpha_3}} d\mu_0,$$

where $d\mu_0$ is the Lebesgue measure.

4 Symmetric Matrices on Octonions

Our aim is to describe the law of the spectrum of the real form of symmetric matrices on octonions. The block matrix is $\mathscr{M} = ((A.B|B)M^{A.B})_{A,B\in\mathscr{P}(E)}$, satisfying $(M^A)^t = (A|A)M^A$ from the symmetry assumption.

We will focus on cases where the symmetry condition (3) of matrix $\mathscr{M} - X\mathrm{Id}$ is satisfied, i.e. where the matrix

$$U(X) := (\mathscr{M} - X\mathrm{Id})^{-1}$$

is octonionic (almost surely for the stochastic process under consideration).

Setting

$$P(X) := \det(\mathscr{M} - X\mathrm{Id}),$$

by Lemma 3 we have

$$\begin{aligned}\Gamma(\log P(X), \log P(Y)) &= \sum_{\substack{A,B,C,D\\ i,j,k,l}} U_{ji}^{B,A} U_{lk}^{D,C} \Gamma(M_{ij}^{A,B}, M_{kl}^{C,D}) \\ &= \sum_{\substack{A,B,C,D\\ i,j,k,l}} (A.B|A.B)(C.D|C.D) U_{ji}^{A.B} U_{lk}^{C.D} \Gamma(M_{ij}^{A.B}, M_{kl}^{C.D})\end{aligned} \tag{15}$$

where we used property 2 of Lemma 1 and

$$\mathrm{L}(\log P(X)) = \sum_{\substack{A,B\\ i,j}} U_{ji}^{A,B}(X)\mathrm{L}(M_{ij}^{A,B}) - \sum_{\substack{A,B,C,D\\ i,j,k,l}} U_{jk}^{B,C}(X) U_{li}^{D,A}(X) \Gamma(M_{ij}^{A,B}, M_{kl}^{C,D}). \tag{16}$$

For further use in both examples, we state (without proof) two preliminary lemmas. The first lemma collects some elementary facts, consequences of the definition, the symmetries and property 2 of Lemma 1.

Lemma 4 *For $F \in \mathscr{P}(E)$, we have*

$$\operatorname{tr} U(X)^F = \sum_i U(X)^F_{ii}\,, \tag{17}$$

$$\operatorname{tr} U(X) = 8 \operatorname{tr} U(X)^{\emptyset}\,, \tag{18}$$

$$\operatorname{tr}[U(X)^F U(Y)^F] = \sum_{ij} U(X)^F_{ji} U(Y)^F_{ij}\,, \tag{19}$$

$$(F|F) \operatorname{tr}[U(X)^F U(Y)^F] = \sum_{ij} U(X)^F_{ji} U(Y)^F_{ji}\,, \tag{20}$$

$$\operatorname{tr}[U(X)U(Y)] = 8 \sum_C (C|C) \operatorname{tr}[U(X)^C U(Y)^C]\,. \tag{21}$$

The second lemma gives expressions of the traces in terms of the characteristic polynomials. The first two identities are obtained by derivation from

$$\log P(X) = \operatorname{tr} \log(\mathscr{M} - X\mathrm{Id}) = -\operatorname{tr} \log U(X)\,,$$

and the third one is a consequence of the first one and the resolvent equation.

Lemma 5

$$\operatorname{tr} U(X) = \frac{P'(X)}{P(X)} \tag{22}$$

$$\operatorname{tr}(U(X)^2) = \frac{P'(X)^2}{P(X)^2} - \frac{P''(X)}{P(X)} \tag{23}$$

$$\operatorname{tr}[U(X)U(Y)] = \frac{1}{Y-X}\left(\frac{P'(X)}{P(X)} - \frac{P'(Y)}{P(Y)}\right). \tag{24}$$

4.1 The Dimension 2 Case

Consider $M = \sum M^A \omega_A$, where $\{M^A\}$ are matrices whose elements are independent Brownian motions. For $A \neq \emptyset$, due to symmetry of $\mathscr{M}$, $(M^A)^t = (A|A)M^A = -M^A$. Such matrices naturally satisfy the symmetry restriction 3 in dimension 2, since the 2×2 antisymmetric matrices are all of the form $\begin{pmatrix} 0 & -z \\ z & 0 \end{pmatrix}$, and they are therefore all proportional to each other. This property fails in higher dimensions. Set

$$\Gamma(M^A_{ij}, M^B_{kl}) = \frac{1}{2}\delta_{A,B}(\delta_{ik}\delta_{jl} + (A|A)\delta_{il}\delta_{jk}), \quad L(M^A_{ij}) = 0, \tag{25}$$

which reflects the symmetry of the matrices. Notice that the inverse matrix $U(X)$ is also symmetric with $(U^A)^t = (A|A)U^A$. We have the following result:

Proposition 1 *For the* 2×2 *symmetric matrix* $M = \sum M^A \omega_A$,

$$\Gamma(\log P(X), \log P(Y)) = \frac{8}{Y - X}\left(\frac{P'(X)}{P(X)} - \frac{P'(Y)}{P(Y)}\right), \tag{26}$$

$$\mathrm{L}(\log P(X)) = 3\left(\frac{P'(X)^2}{P(X)^2} - \frac{P''(X)}{P(X)}\right) - \frac{1}{2}\frac{P'(X)^2}{P(X)^2}. \tag{27}$$

Proof Proof of (26): From (15) and (25) we have

$$\begin{aligned}
&\Gamma(\log P(X), \log P(Y))\\
&= \sum_{\substack{A,B,C,D\\ AB=CD}} (A.B|A.B)(C.D|C.D)U(X)^{B.A}_{ji}U(Y)^{D.C}_{lk}\frac{1}{2}(\delta_{ik}\delta_{jl} + (A.B|A.B)\delta_{il}\delta_{jk})\\
&= 8\sum_F \mathrm{tr}(U(X)^F U(Y)^F)(F|F)\\
&= 8\,\mathrm{tr}(U(X)U(Y)) = \frac{8}{Y - X}\left(\frac{P'(X)}{P(X)} - \frac{P'(Y)}{P(Y)}\right),
\end{aligned}$$

where we applied (20), (19), (21) and (24). This ends the proof of (26).

Proof of (27): In the following, we write U is for $U(X)$. From (16) and (25), we have

$$\begin{aligned}
\mathrm{L}(\log P(X)) = &-\frac{1}{2}\sum_{B,C,D}(B.D|B.D)\sum_{ij}(U^{B.C}_{ji})^2\\
&-\frac{1}{2}\sum_{\substack{A,B,C,D\\ A.B.C.D=\emptyset}}(B.D|B.D)(A.B|A.B)(\sum_i U^{BC}_{ii})^2.
\end{aligned} \tag{28}$$

On the one hand, in view of (20) and (21)

$$\sum_{B,C,D}(B.D|B.D)\sum_{ij}\left(U^{B.C}_{ji}\right)^2 = -6\,\mathrm{tr}(U^2).$$

On the other hand, since for $F \neq \emptyset$, U^F is antisymmetric, so that $\operatorname{tr} U^F = 0$,

$$\sum_{\substack{A,B,C,D\\ A.B.C.D=\emptyset}} (B.D|B.D)(A.B|A.B)(\sum_i U_{ii}^{B.C})^2 = 8^2 \left(\operatorname{tr} U^{\emptyset}\right)^2.$$

Combined as in (28), these sums give:

$$\mathrm{L}(\log P(X)) = 3 \operatorname{tr} U^2 - 4 \cdot 8 \left(\operatorname{tr} U^{\emptyset}\right)^2 = 3 \operatorname{tr} U^2 - \frac{1}{2} \left(\operatorname{tr} U\right)^2, \tag{29}$$

which, in view of (22) and (23) ends the proof of (27) and then the proof of the Proposition.

Remark 2 By the results of above proposition and formula (10), it is easy to get

$$\mathrm{L}(P) = (11 - \frac{1}{2})\frac{P'(X)^2}{P(X)} - 11P''(X).$$

Now chose $\alpha_1 = -11, \alpha_2 = 11 - \frac{1}{2}$ and $\alpha_3 = 8$ in formula (13): the resulting value for a is $a = 8$. This shows that the multiplicity of the eigenvalues is 8. Assume ρ is the density of the invariant measure of L of the coordinates $\{x_i\}$ in the Weyl chamber, then according to our discussion in the previous section, we have

$$\rho = C \prod_{i<j} (x_i - x_j)^8.$$

4.2 Another Model in Any Dimension

We now provide another set of random octonionic matrices for which the symmetry condition (3) is automatically satisfied.

Let $M^{\emptyset}$ be a symmetric matrix with independent Brownian motions as its entries. For all $A, B \neq \emptyset$, let $M^A = M^B = \mathscr{A}$ be a random antisymmetric matrix with independent Brownian motion as its off diagonal entries. Then consider $M = M^{\emptyset}\omega_{\emptyset} + \mathscr{A}\sum_{C\neq\emptyset}\omega_C$. This model is similar to the Hermitian case considered in Bakry and Zani [3] (see Remark 4). Similarly to the Hermitian case, we set

$$\Gamma(M_{ij}^{\emptyset}, M_{kl}^{\emptyset}) = \frac{1}{2}(\delta_{ik}\delta_{jl} + \delta_{il}\delta_{jk}), \tag{30}$$

$$\Gamma(\mathscr{A}_{ij}, \mathscr{A}_{kl}) = \frac{1}{14}(\delta_{ik}\delta_{jl} - \delta_{il}\delta_{jk}), \tag{31}$$

$$\Gamma(M_{ij}^{\emptyset}, \mathscr{A}_{kl}) = 0, \ \mathrm{L}(M_{ij}^{A}) = 0. \tag{32}$$

Due to Lemma 1, for the inverse matrix $U(X) = (M - XI)^{-1}$, we have for every $C \neq \emptyset$

$$U^C = -U^{\emptyset} M^C (M^{\emptyset} - XI)^{-1} = -U^{\emptyset} \mathscr{A} (M^{\emptyset} - XI)^{-1},$$

which means for all $C \neq \emptyset$, U^C is the same, and we denote it by U_a.

Proposition 2 *For the matrix $M = M^{\emptyset}\omega_{\emptyset} + \sum_{C\neq\emptyset} \mathscr{A}\omega_C$ on the octonions,*

$$\Gamma(\log P(X), \log P(Y)) = \frac{8}{Y - X}\left(\frac{P'(X)}{P(X)} - \frac{P'(Y)}{P(Y)}\right),$$

$$\mathrm{L}(\log P) = -\frac{1}{8}\frac{P'(X)^2}{P(X)^2}.$$

Proof On the one hand, from (15) and (32)

$$\Gamma(\log P(X), \log P(Y)) = \frac{1}{2}S_1 + \frac{1}{14}S_2, \tag{33}$$

where

$$S_1 := \sum_{A=B,C=D}\sum_{i,j,k,l} U(X)^{\emptyset}_{ji}U(Y)^{\emptyset}_{lk}(\delta_{ik}\delta_{jl} + \delta_{il}\delta_{jk}),$$

$$S_2 := \sum_{A\neq B,C\neq D}\sum_{i,j,k,l} U(X)^{A.B}_{ji}U(Y)^{C.D}_{lk}(\delta_{ik}\delta_{jl} - \delta_{il}\delta_{jk}).$$

A careful computation, using the fact that $U(Y)^{\emptyset}$ is symmetric and $U(Y)_a$ is antisymmetric gives

$$S_1 = 2\cdot 8^2 \,\mathrm{tr}\left[U(X)^{\emptyset}U(Y)^{\emptyset}\right], \; S_2 = -2\cdot 7^2\cdot 8^2 \,\mathrm{tr}\left[U(X)_a U(Y)_a\right], \tag{34}$$

and then, using (33)

$$\Gamma(\log P(X), \log P(Y)) = 8^2 \,\mathrm{tr}\left[U(X)^{\emptyset}U(Y)^{\emptyset}\right] - 7\cdot 8^2 \,\mathrm{tr}\left[U(X)_a U(Y)_a\right].$$

Going back to (21) we see that

$$\mathrm{tr}\left[U(X)U(Y)\right] = 8\,\mathrm{tr}\left[U(X)^{\emptyset}U(Y)^{\emptyset}\right] - 7\cdot 8 \,\mathrm{tr}\left[U(X)_a U(Y)_a\right],$$

so that

$$\Gamma(\log P(X), \log P(Y)) = 8\,\mathrm{tr}\left[U(X)U(Y)\right] = \frac{8}{Y - X}\left(\frac{P'(X)}{P(X)} - \frac{P'(Y)}{P(Y)}\right).$$

On the other hand, with U for $U(X)$

$$\mathrm{L}(\log P) = -\frac{1}{2}\sum_{\substack{A=B,C=D\\ i,j,k,l}} (B.C|C)(D.A|A)(A.B|B)(C.D|D)U(X)^{B.C}_{jk}U(X)^{D.A}_{li}(\delta_{ik}\delta_{jl}+\delta_{il}\delta_{jk})$$

$$-\frac{1}{14}\sum_{\substack{A\neq B,C\neq D\\ i,j,k,l}} (B.C|C)(D.A|A)(A.B|B)(C.D|D)U(X)^{B.C}_{jk}U(X)^{D.A}_{li}(\delta_{ik}\delta_{jl}-\delta_{il}\delta_{jk})$$

$$=: -\frac{1}{2}S'_1 - \frac{1}{14}S'_2\,. \tag{35}$$

Let us first remark that

$$\sum_{ijkl} U^F_{jk}U^G_{li}(\delta_{ik}\delta_{jl}\pm\delta_{il}\delta_{jk}) = (F|F)\,\mathrm{tr}(U^F U^G) \pm (\mathrm{tr}\ U^F)(\mathrm{tr}\ U^G). \tag{36}$$

For the first part, we have

$$\begin{aligned} S'_1 &= \sum_{B,C}\mathrm{tr}(U^{B.C})^2 + \sum_{B,C}(B.C|B.C)[\mathrm{tr}\ U^{B.C}]^2 \\ &= 8\left[\mathrm{tr}(U^{\emptyset})^2 + 7\,\mathrm{tr}(U_a)^2\right] + 8[\mathrm{tr}\ U^{\emptyset}]^2. \end{aligned} \tag{37}$$

For the second part, going back to the notation (2) for Θ and applying (36) we have

$$S'_2 = \sum_{A\neq B,C\neq D} \Theta \times \left((B.C|B.C)\,\mathrm{tr}[U^{B.C}U^{A.D}] - \mathrm{tr}\ U^{B.C}\ \mathrm{tr}\ U^{A.D}\right). \tag{38}$$

Let us split the sum into four parts according to $B = C$ or not, and $A = D$ or not.

(i) When $A \neq D, B = C$, the sum vanishes. Indeed, in this case,

$$(B.C|B.C)\,\mathrm{tr}[U^{B.C}U^{A.D}] - (\mathrm{tr}\ U^{B.C}\ \mathrm{tr}\ U^{A.D}) = \mathrm{tr}[U^{\emptyset}U_a], \tag{39}$$

and $\Theta = (A.D|A)(A.B|B)(B.D|D)$ which is antisymmetric in A, B.

(ii) The same occurs when $A = D$ and $B \neq C$.

(iii) When $A = D$ and $B = C$,

$$(B.C|B.C)\,\mathrm{tr}[U^{B.C}U^{A.D}] - (\mathrm{tr}\ U^{B.C}\ \mathrm{tr}\ U^{A.D}) = \mathrm{tr}[U^{\emptyset}]^2 - (\mathrm{tr}\ U^{\emptyset})^2\,,$$

and $\Theta = (B.D|B.D)$ in view of property 4 of Lemma 1, so that the contribution is

$$\sum_{A\neq B, C\neq D, B=C, A=D} (B.D|B.D)\left((B.C|B.C)\operatorname{tr}[U^{B.C}U^{A.D}] - \operatorname{tr} U^{B.C}\operatorname{tr} U^{A.D}\right)$$

$$= -7\cdot 8\left(\operatorname{tr}[U^{\emptyset}]^2 - (\operatorname{tr} U^{\emptyset})^2\right).$$

(iv) When $A \neq D, B \neq C$,

$$(B.C|B.C)\operatorname{tr}[U^{B.C}U^{A.D}] - (\operatorname{tr} U^{B.C}\operatorname{tr} U^{A.D}) = -\operatorname{tr}(U_a)^2 .$$

With the help of some computer algebra, we get

$$\sum_{A\neq B, C\neq D, B\neq C, A\neq D} \Theta = 2^3\cdot 7^2 .$$

Finally all the contributions in (38) give

$$S_2' = -7\cdot 8\left(\operatorname{tr}[U^{\emptyset}]^2 - (\operatorname{tr} U^{\emptyset})^2\right) - 7^2\cdot 8\operatorname{tr}(U_a)^2 .$$

Going back to (35) and (37), we conclude, using again (32)

$$\mathrm{L}(\log P) = -8(\operatorname{tr} U^{\emptyset})^2 = -\frac{1}{8}(\operatorname{tr} U)^2 ,$$

which ends the proof of the proposition.

Remark 3 Similarly we have

$$\mathrm{L}(P(X)) = (8 - \frac{1}{8})\frac{P'(X)^2}{P} - 8P''(X) .$$

Chose $\alpha_1 = -8, \alpha_2 = 8 - \frac{1}{8}, \alpha_3 = 8$, we still obtain the multiplicity $a = 8$, while the density of the invariant measure of L is then

$$C\prod_{i<j}|x_i - x_j|^2.$$

Remark 4 Recall that in Bakry and Zani [3, Sect. 7.1], for a Hermitian matrix $H = M + iA$ with independent Brownian motions as its entries (where M is symmetric, A is anti-symmetric), we have

$$\Gamma(M_{ij}, M_{kl}) = \frac{1}{2}(\delta_{ik}\delta_{jl} + \delta_{il}\delta_{jk}),$$

$$\Gamma(A_{ij}, A_{kl}) = \frac{1}{2}(\delta_{ik}\delta_{jl} - \delta_{il}\delta_{jk}) .$$

In our model $M = M^{\emptyset}\omega_{\emptyset} + \mathscr{A}\sum_{C\neq\emptyset}\omega_C$, denote e the specific element in the octonion algebra, $e = \sum_{C\neq\emptyset}\omega_C$. Notice that

$$e^2 = -7,$$

which indicates that e works like i in the Hermitian matrices, just with a different variance. Therefore, this example is indeed similar to the case of Hermitian matrices.

Remark 5 In this remark we would like to discuss why the dimension 2 is so special. Consider a more general model: let

$$M = M^{\emptyset}\omega_0 + \sum_{C\neq\emptyset} M^C\omega_C,$$

with $M^C = x_C A_0$, where $M^{\emptyset}$ is a Brownian motion on symmetric matrices, $\{x_C\}$ a series of Brownian motions on $\mathbb{R}$, and $A_0 = \{a_{ij}\}$ a fixed anti-symmetric matrix. Obviously this model satisfies the symmetry condition (3). When M is a 2×2 matrix, it can be considered as a special example of the first case. Let $e = \sum_{C\neq\emptyset} x_C\omega_C$. Different from the previous model, in this case e can be considered as a Brownian motion on the basis of octonions satisfying $e^2 = -\sum_{C\neq\emptyset}|x_C|^2 = -|e|^2$. Therefore,

$$\begin{aligned}
\Gamma(M_{ij}^{\emptyset}, M_{kl}^{\emptyset}) &= \frac{1}{2}(\delta_{ik}\delta_{jl} + \delta_{il}\delta_{jk}),\\
\Gamma(M_{ij}^{A}, M_{kl}^{B}) &= \delta_{A=B}a_{ij}a_{kl}, A, B \neq \emptyset,\\
\Gamma(M_{ij}^{\emptyset}, M_{kl}^{A}) &= 0, A \neq \emptyset\,.
\end{aligned}$$

Similar computations yield

$$\begin{aligned}
\Gamma(\log P(X), \log P(Y)) = 8 &\sum_{A.B=\emptyset} \operatorname{tr}(U(X)^{\emptyset}U(Y)^{\emptyset})\\
&+ 8\sum_{A.B\neq\emptyset}(\operatorname{tr}(U(X)^{B.A}A_0)\operatorname{tr}(U(Y)^{B.A}A_0)),
\end{aligned}$$

$$\begin{aligned}
L(\log P) = &-\frac{1}{2}\sum_{A.C\neq\emptyset}\operatorname{tr}(U(X)^{A.C})^2 - 4\operatorname{tr}(U(X)^{\emptyset})^2 - 4(\operatorname{tr}U(X)^{\emptyset})^2\\
&+ 5\sum_{B.C\neq\emptyset}\operatorname{tr}(U(X)^{B.C}A_0U(X)^{B.C}A_0) + 56\operatorname{tr}(U(X)^{\emptyset}A_0U(X)^{\emptyset}A_0),
\end{aligned}$$

which are hard to describe in terms of P. When the matrix is 2×2, the following equalities hold: for $C \neq \emptyset$,

$$\begin{aligned} \operatorname{tr}(U(X)^C A_0)\operatorname{tr}(U(Y)^C A_0) &= -\operatorname{tr}(U(X)^C U(Y)^C), \\ \operatorname{tr}(U(X)^C A_0 U(X)^C A_0) &= -\operatorname{tr}(U(X)^C)^2, \\ \operatorname{tr}(U(X)^{\emptyset} A_0 U(X)^{\emptyset} A_0) &= \operatorname{tr}(U^{\emptyset})^2 - (\operatorname{tr} U^{\emptyset})^2 . \end{aligned}$$

which give rise to the results in the first model.

However, in higher dimensions, the above conditions are hard to satisfy. In fact when $n = 2$, it is enough to take $A_0 = \begin{pmatrix} 0 & -1 \\ 1 & 0 \end{pmatrix}$. Set $x_{\{1\}} = z$. By the formula

$$\begin{aligned} U^{\emptyset} &= (M^{\emptyset} + z^2 A_0 (M^{\emptyset}) - 1 A_0)^{-1}, \\ U^{\{1\}} &= -z U^{\emptyset} A_0 (M^{\emptyset}) - 1 . \end{aligned}$$

It is easily seen that $U^{\{1\}}$ is a 2×2 antisymmetric matrix and can be written as

$$U^{\{1\}} = \lambda A_0, \ \ \lambda = \frac{z}{z^2 - \det(M^{\emptyset})} .$$

Compare it with the expressions of $U^{\emptyset}$ and $U^{\{1\}}$, we have

$$A_0 = \frac{\lambda}{\lambda z^2 - z} M^{\emptyset} A_0 M^{\emptyset} .$$

Since $A_0^2 = -I$, this leads to $(M^{\emptyset} A_0)^2 = \frac{z - \lambda z^2}{\lambda} I = -\det(M^{\emptyset}) I$, which is impossible to hold for any symmetric matrix $M^{\emptyset}$ in higher dimensions. This restriction insures the first two conditions, and the third one is proved by this and the fact that $\operatorname{tr}(M^2) - (\operatorname{tr} M)^2 = -2\det(M)$ holds in dimension 2.

5 Some Remarks

Our two models provide examples where the multiplicity of eigenvalues and the exponent β in the law are not related, which is in accordance with the conclusion in Bakry and Zani [3], that the exponent reflects the structure of the algebra while the multiplicity of the eigenvalues is decided by the dimension of the eigenspaces.

As we have seen, the octonionic structure of the matrix plays an important role. For higher dimension, the problem may be studied by our method if we know the structure of the inverse matrix, which is not necessarily octonionic. The main obstacle is still the non-associativity, which prevents any matrix presentation for octonionic multiplication. Let us recall that the 3×3 matrices on octonions have been studied by Dray and Manogue [6] and Okubo [16] using algebraic method, showing that there are six eigenvalues with multiplicity 4. It it still an open problem to provide a probabilistic model in this case which would lead to this conclusion.

Acknowledgements I would like to thank my supervisor, Prof. Dominique Bakry, for his precious suggestions and helpful discussions. I am also indebted to my other supervisors, Prof. Jiangang Ying and Prof. Xiangdong Li, for their support and help. I would also like to express my gratitude to the referee for careful reading and suggestions.

This research was supported by China Scholarship Council.

References

1. G.W. Anderson, A. Guionnet, O. Zeitouni, An introduction to random matrices, in *Cambridge Studies in Advanced Mathematics*, vol. 118 (Cambridge University Press, Cambridge, 2010)
2. J. Baez, The octonions. Bull. Am. Math. Soc. **39**, 145–205 (2002)
3. D. Bakry, M. Zani, Dyson processes associated with associative algebras: the Clifford case, in *Geometric Aspects of Functional Analysis*. Lecture Notes in Mathematics, vol. 2116 (Springer, Cham, 2014), pp. 1–37
4. D. Bakry, I. Gentil, M. Ledoux, *Analysis and Geometry of Markov Diffusion Operators*. Grund. Math. Wiss., vol. 348 (Springer, Berlin, 2013)
5. R. Bott, J. Milnor, On the parallelizability of the spheres. Bull. Am. Math. Soc. **64**, 87–89 (1958)
6. T. Dray, C.A. Manogue, The octonionic eigenvalue problem. Adv. Appl. Clifford Algebras **8**(2), 341–364 (1998)
7. F.J. Dyson, A Brownian-motion model for the eigenvalues of a random matrix. J. Math. Phys. **3**, 1191–1198 (1962)
8. L. Erdős, S. Péché, J. Ramírez, B. Schlein, H.-T. Yau, Bulk universality for Wigner matrices. Commun. Pure Appl. Math. **63**, 895–925 (2010)
9. L. Erdős, J. Ramírez, B. Schlein, T. Tao, V. Vu, H.-T. Yau, Bulk universality for Wigner Hermitian matrices with subexponential decay. Math. Res. Lett. **17**, 667–674 (2010)
10. L. Erdős, A. Knowles, H.-T. Yau, J. Yin, The local semicircle law for a general class of random matrices. Electron. J. Probab. **18**(59), 58p. (2013)
11. P. Forrester, *Log-Gases and Random Matrices*. London Mathematical Society Monographs Series, vol. 34 (Princeton University Press, Princeton, 2010)
12. M. Kervaire, Non-parallelizability of the n-sphere for n > 7. Proc. Natl. Acad. Sci. U. S. A. **44**, 280–283 (1958)
13. S. Li, X. Li, Y. Xie, On the law of large numbers for the empirical measure process of Generalized Dyson Brownian motion (2014). arxiv preprint 1407.7234
14. M.L. Mehta. *Random Matrices*. Pure and Applied Mathematics (Amsterdam), vol. 142, 3rd edn. (Elsevier/Academic, Amsterdam, 2004)
15. R. Schafer, *An Introduction to Nonassociative Algebras* (Dover, New York, 1995)
16. O. Susumu, Eigenvalue problem for symmetric 3 dimension octonionic matrix. Adv. Appl. Clifford Algebras **9**(1), 131–176 (1999)
17. Y. Tian, Matrix representations of octonions and their applications. Adv. Appl. Clifford Algebras **10**(1), 61–90 (2000)
18. E.P. Wigner, On the distribution of the roots of certain symmetric matrices. Ann. Math. **67**, 325–327 (1958)

Necessary and Sufficient Conditions for the Existence of α-Determinantal Processes

Franck Maunoury

Abstract We give necessary and sufficient conditions for existence and infinite divisibility of α-determinantal processes. For that purpose we use results on negative binomial and ordinary binomial multivariate distributions.

AMS 2010 Subject Classifications 60G55, 60E07

1 Introduction

Several authors have already established necessary and sufficient conditions for existence of α-determinantal processes.

Macchi in [8] and Soshnikov in its survey paper [11] gave a necessary and sufficient condition for determinantal processes with self-adjoint kernels, which corresponds to the case $\alpha = -1$.

The same condition has also been established in a different way by Hough et al. in [7] in the case $\alpha = -1$. They have also given a sufficient condition of existence in the case $\alpha = 1$ and self-adjoint kernel.

In the special case when the configurations are on a finite space, the paper of Vere-Jones [13] provides necessary and sufficient conditions for any value of α.

Finally, Shirai and Takahashi have given sufficient conditions for the existence of an α-determinantal process for any values of α. However, in the case $\alpha > 0$, their sufficient condition (Condition B) in [10] does not work for the following example: the space is reduced to a single point space and the reference measure λ is a unit point mass. With their notations, the two kernels K and J_α are respectively reduced

F. Maunoury (✉)
Université Pierre et Marie Curie, LPMA, Case courrier 188, 4, place Jussieu, 75252 Paris Cedex 05, France

Telecom ParisTech, LTCI, 46, rue Barrault, 75634 Paris Cedex 13, France
e-mail: franck.maunoury@upmc.fr

C. Donati-Martin et al. (eds.), *Séminaire de Probabilités XLVIII*, Lecture Notes in Mathematics 2168, DOI 10.1007/978-3-319-44465-9_15

to two real numbers k and j_α, with

$$j_\alpha = \frac{k}{1+\alpha k}$$

We can choose $\alpha > 0$ and $k < 0$ such that $j_\alpha > 0$. Under these assumptions, Condition B is fulfilled but the obtained point process has a negative correlation function ($\rho_1(x) = k$), which has to be excluded, since a correlation function is an almost everywhere non-negative function.

We are going to strengthen Condition B of Shirai and Takahashi and obtain a necessary and sufficient condition in the case $\alpha > 0$. This is presented in Theorem 1.

Besides, in the case $\alpha < 0$, we extend the result of Shirai and Takahashi to the case of non self-adjoint kernels and show that the obtained condition is also necessary (Theorems 3 and 4). Moreover, we show that $-1/\alpha$ is necessarily an integer. This has been noticed by Vere-Jones in [12] in the case of configurations on a finite space.

We also give a necessary and sufficient condition for the infinite divisibility of an α-determinantal process for all values of α.

The main results are presented in Sect. 3. Section 2 introduces the needed notation. In Sect. 4, we write a multivariate version of a Shirai and Takahashi formulae on Fredholm determinant expansion. Sections 5 and 6 present the proofs of the results concerning respectively the cases $\alpha > 0$ and $\alpha < 0$. The proofs concerning infinite divisibility are presented in Sect. 7.

2 Preliminaries

Let E be a locally compact Polish space. A locally finite configuration on E is an integer-valued positive Radon measure on E. It can also be identified with a set $\{(M, \alpha_M) : M \in F\}$, where F is a countable subset of E with no accumulation points (i.e. a discrete subset of E) and, for each point in F, α_M is a non-null integer that corresponds to the multiplicity of the point M (M is a multiple point if $\alpha_M \geq 2$).

Let λ be a Radon measure on E. Let $\mathcal{X}$ be the space of the locally finite configurations of E. The space $\mathcal{X}$ is endowed with the vague topology of measures, i.e. the smallest topology such that, for every real continuous function f with compact support, defined on E, the mapping

$$\mathcal{X} \ni \xi \mapsto \langle f, \xi \rangle = \sum_{x \in \xi} f(x) = \int f d\xi$$

is continuous. Details on the topology of the configuration space can be found in [1].

We denote by $\mathcal{B}(\mathcal{X})$ the corresponding σ-algebra. A point process on E is a random variable with values in $\mathcal{X}$. We do not restrict ourselves to simple point processes, as the configurations in $\mathcal{X}$ can have multiple points.

For a $n \times n$ matrix $A = (a_{[ij})_{1 \le i,j \le n}$, set:

$$\det_\alpha A = \sum_{\sigma \in \Sigma_n} \alpha^{n-\nu(\sigma)} \prod_{i=1}^{n} a_{i\sigma(i)}$$

where Σ_n is the set of all permutations on $\{1, \dots, n\}$ and $\nu(\sigma)$ is the number of cycles of the permutation σ.

For a relatively compact set $\Lambda \subset E$, the Janossy densities of a point process ξ w.r.t. a Radon measure λ are functions (when they exist) $j_n^\Lambda : E^n \to [0, \infty)$ for $n \in N$, such that

$$j_n^\Lambda(x_1, \dots, x_n) = n!\, \mathbb{P}(\xi(\Lambda) = n)\, \pi_n^\Lambda(x_1, \dots, x_n)$$
$$j_0^\Lambda(\emptyset) = \mathbb{P}(\xi(\Lambda) = 0),$$

where π_n^Λ is the density with respect to $\lambda^{\otimes n}$ of the ordered set $(x_1, \dots, x_n)$, obtained by first sampling ξ, given that there are n points in Λ, then choosing uniformly an order between the points.

For $\Lambda_1, \dots, \Lambda_n$ disjoint subsets included in Λ, $\int_{\Lambda_1 \times \dots \times \Lambda_n} j_n^\Lambda(x_1, \dots, x_n)\lambda(dx_1) \dots \lambda(dx_n)$ is the probability that there is exactly one point in each subset Λ_i $(1 \le i \le n)$, and no other point elsewhere.

We recall that we have the following formula, for a non-negative measurable function f with support in a relatively compact set $\Lambda \subset E$:

$$\mathbb{E}(f(\xi)) = f(\emptyset) j_0^\Lambda(\emptyset) + \sum_{n=1}^{\infty} \frac{1}{n!} \int_{\Lambda^n} f(x_1, \dots, x_n) j_n^\Lambda(x_1, \dots, x_n) \lambda(dx_1) \dots \lambda(dx_n).$$

For $n \in \mathbb{N}$ and $a \in \mathbb{R}$, we denote $a^{(n)} = \prod_{i=0}^{n-1}(a - i)$.

The correlation functions (also called joint intensities) of a point process ξ w.r.t. a Radon measure λ are functions (when they exist) $\rho_n : E^n \to [0, \infty)$ for $n \ge 1$, such that for any family of mutually disjoint relatively compact subsets $\Lambda_1, \dots, \Lambda_d$ of E and for any non-null integers $n_1, \dots, n_d$ such that $n_1 + \dots + n_d = n$, we have

$$\mathbb{E}\left(\prod_{i=1}^{d} \xi(\Lambda_i)^{(n_i)} \right) = \int_{\Lambda_1^{n_1} \times \dots \times \Lambda_d^{n_d}} \rho_n(x_1, \dots, x_n) \lambda(dx_1), \dots, \lambda(dx_n).$$

Intuitively, for a simple point process, $\rho_n(x_1, \dots, x_n)\lambda(dx_1) \dots \lambda(dx_n)$ is the infinitesimal probability that there is at least one point in the vicinity of each x_i (each vicinity having an infinitesimal volume $\lambda(dx_i)$ around x_i), $1 \le i \le n$.

Let α be a real number and K a kernel from E^2 to $\mathbb{R}$ or $\mathbb{C}$. An α-determinantal point process, with kernel K with respect to λ (also called α-permanental point process) is defined, when it exists, as a point process with the following correlation functions $\rho_n, n \in \mathbb{N}$ with respect to λ:

$$\rho_n(x_1, \ldots, x_n) = \det_\alpha(K(x_i, x_j))_{1\le i,j\le n}.$$

We denote by $\mu_{\alpha,K,\lambda}$ the probability distribution of such a point process.

We exclude the case of a point process almost surely reduced to the empty configuration.

The case $\alpha = -1$ corresponds to a determinantal process and the case $\alpha = 1$ to a permanental process. The case $\alpha = 0$ corresponds to the Poisson point process. We suppose in the following that $\alpha \neq 0$.

We will always assume that the kernel K defines a locally trace class integral operator $\mathcal{K}$ on $L^2(E, \lambda)$. Under this assumption, one obtains an equivalent definition for the α-determinantal process, using the following Laplace functional formula:

$$\mathbb{E}_{\mu_{\alpha,K,\lambda}}\left[\exp\left(-\int_E f d\xi\right)\right] = \mathrm{Det}\left(\mathcal{I} + \alpha\mathcal{K}[1 - e^{-f}]\right)^{-1/\alpha} \tag{1}$$

where f is a compactly-supported non-negative function on E, $\mathcal{K}[1-e^{-f}]$ stands for $\sqrt{1-e^{-f}}\mathcal{K}\sqrt{1-e^{-f}}$, $\mathcal{I}$ is the identity operator on $L^2(E, \lambda)$ and Det is the Fredholm determinant. Details on the link between the correlation function and the Laplace functional of an α-determinantal process can be found in the Chap. 4 of [10]. Some explanations and useful formula on the Fredholm determinant are given in Chap. 2.1 of [10].

For a subset $\Lambda \subset E$, set: $\mathcal{K}_\Lambda = p_\Lambda \mathcal{K} p_\Lambda$, where p_Λ is the orthogonal projection operator from $L^2(E, \lambda)$ to the subspace $L^2(\Lambda, \lambda)$.

For two subsets $\Lambda, \Lambda' \subset E$, set: $\mathcal{K}_{\Lambda\Lambda'} = p_\Lambda \mathcal{K} p_{\Lambda'}$, and denote by $K_{\Lambda\Lambda'}$ its kernel. We have for any $x, y \in E$, $K_{\Lambda\Lambda'}(x, y) = \mathbb{1}_\Lambda(x)\, \mathbb{1}_{\Lambda'}(y)\, K(x, y)$.

When $\mathcal{I}+\alpha\mathcal{K}$ (resp. $\mathcal{I}+\alpha\mathcal{K}_\Lambda$) is invertible, $\mathcal{J}_\alpha$ (resp. $\mathcal{J}_\alpha^\Lambda$) is the integral operator defined by: $\mathcal{J}_\alpha = \mathcal{K}(\mathcal{I} + \alpha\mathcal{K})^{-1}$ (resp. $\mathcal{J}_\alpha^\Lambda = \mathcal{K}_\Lambda(\mathcal{I} + \alpha\mathcal{K}_\Lambda)^{-1}$) and we denote by J_α (resp. J_α^Λ) its kernel. Note that $\mathcal{J}_\alpha^\Lambda$ is not the orthogonal projection of $\mathcal{J}_\alpha$ on $L^2(\Lambda, \lambda)$.

3 Main Results

Theorem 1 *For $\alpha > 0$, there exists an α-permanental process with kernel K iff:*

- $\mathrm{Det}(\mathcal{I} + \alpha\mathcal{K}_\Lambda) \geq 1$, *for any compact set $\Lambda \subset E$*
- $\det_\alpha(J_\alpha^\Lambda(x_i, x_j))_{1\le i,j\le n} \geq 0$, *for any $n \in \mathbb{N}$, any compact set $\Lambda \subset E$ and any $\lambda^{\otimes n}$-a.e. $(x_1, \ldots, x_n) \in \Lambda^n$.*

Remark 1 Even when E is a finite set, note that the second condition of Theorem 1 consists in an infinite number of computations. Finding a simpler condition, that could be checked in a finite number of steps is still an open problem.

Theorem 2 *For $\alpha > 0$, if an α-permanental process with kernel K exists, then:*

$$\operatorname{Spec} \mathcal{K}_\Lambda \subset \{z \in \mathbb{C} : \operatorname{Re} z > -\frac{1}{2\alpha}\}, \text{ for any compact set } \Lambda \subset E.$$

We remark that this condition is equivalent to

$$\operatorname{Spec} \mathcal{J}_\alpha^\Lambda \subset \{z \in \mathbb{C} : |z| < \frac{1}{\alpha}\}, \text{ for any compact set } \Lambda \subset E$$

Theorem 3 *For $\alpha < 0$ and $\mathcal{K}$ an integral operator such that $\mathcal{I} + \alpha\mathcal{K}_\Lambda$ is invertible, for any compact set $\Lambda \subset E$, an α-determinantal process with kernel K exists iff the two following conditions are fulfilled:*

(i) $-1/\alpha \in \mathbb{N}$
(ii) $\det(J_\alpha^\Lambda(x_i, x_j))_{1 \le i,j \le n} \ge 0$, *for any $n \in \mathbb{N}$, any compact set $\Lambda \subset E$ and any $\lambda^{\otimes n}$-a.e. $(x_1, \ldots, x_n) \in \Lambda^n$.*

The arguments developed in the proof of Theorem 3 shows that actually $(ii) \implies (i)$. Consequently, Condition (ii) is itself a necessary and sufficient condition. It also implies that $\operatorname{Det}(\mathcal{I} + \beta\mathcal{K}_\Lambda) > 0$ for any $\beta \in [\alpha, 0]$ and any compact $\Lambda \subset E$.

Theorem 4 *For $\alpha < 0$ and $\mathcal{K}$ an integral operator such that for some compact set $\Lambda_0 \subset E$, $\mathcal{I} + \alpha\mathcal{K}_{\Lambda_0}$ is not invertible, an α-determinantal process with kernel K exists iff:*

(i') $-1/\alpha \in \mathbb{N}$
(ii') $\det(J_\beta^\Lambda(x_i, x_j))_{1 \le i,j \le n} \ge 0$, *for any $n \in \mathbb{N}$, any $\beta \in (\alpha, 0)$, any compact set $\Lambda \subset E$ and any $\lambda^{\otimes n}$-a.e. $(x_1, \ldots, x_n) \in \Lambda^n$.*

As in Theorem 3, we also have $(ii') \implies (i')$ and Condition (ii') is itself a necessary and sufficient condition.

Note that $\mathcal{I} + \alpha\mathcal{K}_{\Lambda_0}$ is not invertible if and only if there is almost surely at least one point in Λ_0.

Corollary 1 *For m a positive integer, the existence of a $(-1/m)$-determinantal process with kernel K is equivalent to the existence of a determinantal process with the kernel $\frac{K}{m}$.*

Corollary 2 *For $\alpha < 0$ and $\mathcal{K}$ a self-adjoint operator, an α-determinantal process with kernel K exists iff:*

- $-1/\alpha \in \mathbb{N}$
- $\operatorname{Spec} \mathcal{K} \subset [0, -1/\alpha]$

This result is well known in the case $\alpha = -1$ (see for example Hough et al. in [7]).

The sufficient part of this necessary and sufficient condition corresponds to condition A in [10] of Shirai and Takahashi.

Theorem 5 *For $\alpha < 0$, an α-determinantal process in never infinitely divisible.*

Theorem 6 *For $\alpha > 0$, an α-determinantal process is infinitely divisible iff*

- $\mathrm{Det}(\mathcal{I} + \alpha\mathcal{K}_\Lambda) \geq 1$, *for any compact set* $\Lambda \subset E$
- $\sum_{\sigma\in\Sigma_n:\nu(\sigma)=1}\prod_{i=1}^{n} J_\alpha^\Lambda(x_i, x_{\sigma(i)}) \geq 0$, *for any* $n \in \mathbb{N}$, *any compact set* $\Lambda \subset E$ *and* $\lambda^{\otimes n}$*-a.e.* $(x_1, \ldots, x_n) \in \Lambda^n$.

This theorem gives a more general condition for infinite-divisibility of an α-permanental process than the condition given by Shirai and Takahashi in [10].

Theorem 7 *For $\mathcal{K}$ a real symmetric locally trace class operator and $\alpha > 0$, an α-permanental process is infinitely divisible iff*

- $\mathrm{Det}(\mathcal{I} + \alpha\mathcal{K}_\Lambda) \geq 1$, *for any compact set* $\Lambda \subset E$
- $J_\alpha^\Lambda(x_1, x_2)\ldots J_\alpha^\Lambda(x_{n-1}, x_n)J_\alpha^\Lambda(x_n, x_1) \geq 0$, *for any* $n \in \mathbb{N}$, *any compact set* $\Lambda \subset E$ *and* $\lambda^{\otimes n}$*-a.e.* $(x_1, \ldots, x_n) \in \Lambda^n$.

Following Griffith and Milne's remark in [6], when an α-permanental process with kernel K exists and is infinitely divisible, we can replace J_Λ^α by $|J_\Lambda^\alpha|$ and obtain an α-permanental process with the same probability distribution.

Remark 2 In Theorems 1, 6 and 7 , the condition

$$\mathrm{Det}(\mathcal{I} + \alpha\mathcal{K}_\Lambda) \geq 1, \text{ for any compact set } \Lambda \subset E$$

can be replaced by

$$\mathrm{Det}(\mathcal{I} + \alpha\mathcal{K}_\Lambda) > 0, \text{ for any compact set } \Lambda \subset E.$$

4 Fredholm Determinant Expansion

In [10], Shirai and Takahashi have proved the following formula

$$\mathrm{Det}(\mathcal{I} - \alpha z\mathcal{K})^{-1/\alpha} = \sum_{n=0}^{\infty} \frac{z^n}{n!} \int_{E^n} \det_\alpha(K(x_i, x_j))_{1\leq i,j\leq n} \lambda(dx_1)\ldots\lambda(dx_n) \quad (2)$$

for a trace class integral operator $\mathcal{K}$ with kernel K and for $z \in \mathbb{C}$ such that $\|\alpha z\mathcal{K}\| < 1$. In the case where the space E is finite, this formula is also given by Shirai in [9].

As $z \mapsto \mathrm{Det}(\mathcal{I} - \alpha z\mathcal{K})$ is analytic on $\mathbb{C}$ and $z \mapsto z^{-1/\alpha}$ is analytic on $\mathbb{C}^*$, we obtain that $z \mapsto \mathrm{Det}(\mathcal{I} - \alpha z\mathcal{K}_{\Lambda,\alpha})^{-1/\alpha}$ is analytic on $\{z \in \mathbb{C} : \mathcal{I} - \alpha z\mathcal{K}_{\Lambda,\alpha} \text{ invertible}\}$.

Therefore, the formula can be extended to the open disc D, centered in 0 with radius $R = \sup\{r \in \mathbb{R}_+ : \forall z \in \mathbb{C}, |z| < r \Rightarrow \mathcal{I} - \alpha z\mathcal{K} \text{ is invertible}\}$.

D is the open disc of center 0 and radius $1/\|\alpha\mathcal{K}\|$, if the operator $\mathcal{K}$ is self-adjoint, but it can be larger if $\mathcal{K}$ is not self-adjoint.

As remarked by Shirai and Takahashi, the formula (2) is valid for any $z \in \mathbb{C}$ if $-1/\alpha \in \mathbb{N}$.

The following proposition extends (2) to a multivariate case.

Proposition 1 *Let $\Lambda \subset E$ be a relatively compact set, $\Lambda_1, \dots \Lambda_d$ mutually disjoint subsets of Λ and $\mathcal{K}$ a locally trace class integral operator with kernel K.*

We have the following formula

$$\mathrm{Det}\left(\mathcal{I} - \alpha \sum_{k=1}^{d} z_k \mathcal{K}_{\Lambda_k \Lambda}\right)^{-1/\alpha}$$
$$= \sum_{n_1,\dots,n_d=0}^{\infty} \left(\prod_{k=1}^{d} \frac{z_k^{n_k}}{n_k!}\right) \int_{\Lambda_1^{n_1}\times\cdots\times\Lambda_d^{n_d}} \det_\alpha (K(x_i,x_j))_{1\le i,j\le n}\, \lambda(dx_1)\dots\lambda(dx_n) \tag{3}$$

for any $z_1,\dots,z_d \in \mathbb{C}$, such that $\mathcal{I} - \alpha\gamma \sum_{k=1}^{d} z_k \mathcal{K}_{\Lambda_k \Lambda}$ is invertible for any complex number γ satisfying $|\gamma| < 1$ (n denotes $n_1 + \cdots + n_d$).

Proof We apply the formula (2) to the class trace operator $\sum_{k=1}^{d} z_k \mathcal{K}_{\Lambda_k \Lambda}$ and we use the multilinearity property of the α-determinant of a matrix with respect to its rows.

We obtain

$$\mathrm{Det}\left(\mathcal{I} - \alpha \sum_{k=1}^{d} z_k \mathcal{K}_{\Lambda_k \Lambda}\right)^{-1/\alpha}$$
$$= \sum_{n=0}^{\infty} \frac{1}{n!} \int_{E^n} \det_\alpha \left(\sum_{k=1}^{d} z_k K_{\Lambda_k \Lambda}(x_i,x_j)\right)_{1\le i,j\le n} \lambda(dx_1)\dots\lambda(dx_n)$$
$$= \sum_{n=0}^{\infty} \frac{1}{n!} \int_{E^n} \sum_{k_1,\dots k_n=1}^{d} \det_\alpha \left(z_{k_i} \mathbb{1}_{\Lambda_{k_i}}(x_i) \mathbb{1}_{\Lambda}(x_j) K(x_i,x_j)\right)_{1\le i,j\le n} \lambda(dx_1)\dots\lambda(dx_n)$$
$$= \sum_{n=0}^{\infty} \frac{1}{n!} \sum_{k_1,\dots k_n=1}^{d} \int_{\Lambda_{k_1}\times\cdots\times\Lambda_{k_n}} \det_\alpha \left(z_{k_i} K(x_i,x_j)\right)_{1\le i,j\le n} \lambda(dx_1)\dots\lambda(dx_n)$$
$$= \sum_{n=0}^{\infty} \frac{1}{n!} \sum_{k_1,\dots k_n=1}^{d} \left(\prod_{i=1}^{n} z_{k_i}\right) \int_{\Lambda_{k_1}\times\cdots\times\Lambda_{k_n}} \det_\alpha \left(K(x_i,x_j)\right)_{1\le i,j\le n} \lambda(dx_1)\dots\lambda(dx_n)$$

where we have used the fact that $K_{\Lambda_k \Lambda}(x_i, x_j) = \mathbb{1}_{\Lambda_k}(x_i)\, \mathbb{1}_{\Lambda}(x_j) K(x_i, x_j)$ for the equality between the first and the second line.

As the value of the α-determinant of a matrix is unchanged by simultaneous interchange of its rows and its columns, the product $z_1^{n_1} \ldots z_d^{n_d}$ where $n_1 + \ldots n_d = n$, will be repeated $\binom{n}{n_1 \ldots n_d}$ times. This gives the desired formula.

For a relatively compact set $\Lambda \subset E$ and $\Lambda_1, \ldots, \Lambda_d$ mutually disjoint subsets of Λ, the computation of the Laplace functional of an α-determinantal process for the function $f : (z_1, \ldots, z_d) \mapsto -\sum_{k=1}^{d} (\log z_k) \mathbb{1}_{\Lambda_k}$, with $z_1, \ldots, z_d \in (0, 1]$ gives thanks to (1):

$$\mathbb{E}_{\mu_{\alpha,K,\lambda}} \left[\prod_{k=1}^{d} z_k^{\xi(\Lambda_k)} \right] = \text{Det} \left(\mathcal{I} + \alpha \sum_{k=1}^{d} (1 - z_k)\, \mathcal{K}_{\Lambda_k \Lambda} \right)^{-1/\alpha} \tag{4}$$

which is the probability generating function (p.g.f.) of the finite-dimensional random vector $(\xi(\Lambda_1), \ldots, \xi(\Lambda_d))$.

For $\alpha < 0$, the formula (4) reminds the multivariate binomial distribution p.g.f. and for $\alpha > 0$, the multivariate negative binomial distribution p.g.f., given by Vere-Jones in [13], in the special case where the space E is finite.

5 α- Permanental Process ($\alpha > 0$)

Proof (Theorem 1)

We first prove that the conditions are necessary. We suppose that there exists an α-permanental process with $\alpha > 0$, kernel K defining the locally trace class integral operator $\mathcal{K}$.

By taking $d = 1$ in the formula (4), we have

$$\mathbb{E}_{\mu_{\alpha,K,\lambda}} \left(z^{\xi(\Lambda)} \right) = \text{Det} \left(\mathcal{I} + \alpha(1 - z)\, \mathcal{K}_{\Lambda} \right)^{-1/\alpha}$$

for any compact set $\Lambda \subset E$ and $z \in (0, 1]$.

Thus, $\text{Det}(\mathcal{I} + \alpha(1 - z)\mathcal{K}_\Lambda) \geq 1$ for $z \in (0, 1]$. By continuity (as $z \mapsto \text{Det}(\mathcal{I} + (1 - z)\mathcal{K}_\Lambda)$ is indeed analytic on $\mathbb{C}$), we obtain that $\text{Det}(\mathcal{I} + \alpha\mathcal{K}_\Lambda) \geq 1$, which is the first condition. This implies that for any compact set $\Lambda \subset E$, $\mathcal{I} + \alpha\mathcal{K}_\Lambda$ is invertible. Hence $\mathcal{J}_\alpha^\Lambda$ exists and we have, for any non-negative function f, with compact support included in Λ

$$\begin{aligned}
\mathbb{E}_{\mu_{\alpha,K,\lambda}} \left(\prod_{x \in \xi} e^{-f(x)} \right) &= \text{Det}(\mathcal{I} + \alpha\mathcal{K}[1 - e^{-f}])^{-1/\alpha} \\
&= \text{Det}(\mathcal{I} + \alpha\mathcal{K}_\Lambda(1 - e^{-f}))^{-1/\alpha} \\
&= \text{Det}(\mathcal{I} + \alpha\mathcal{K}_\Lambda)^{-1/\alpha}\, \text{Det}(\mathcal{I} - \alpha\mathcal{J}_\alpha^\Lambda e^{-f})^{-1/\alpha}
\end{aligned}$$

$$= \mathrm{Det}(\mathcal{I} + \alpha\mathcal{K}_\Lambda)^{-1/\alpha}$$
$$\times \sum_{n=0}^{\infty} \frac{1}{n!} \int_{\Lambda^n} \left(\prod_{i=1}^{n} e^{-f(x_i)} \right) \det{}_\alpha (J_\alpha^\Lambda(x_i, x_j))_{1 \le i,j \le n} \lambda(dx_1) \ldots \lambda(dx_n) \tag{5}$$

where we have used for the equality between the first and the second line the fact that $\mathrm{Det}(\mathcal{I} + \mathcal{AB}) = \mathrm{Det}(\mathcal{I} + \mathcal{BA})$, for any trace class operator $\mathcal{A}$, and any bounded operator $\mathcal{B}$.

As the Laplace functional defines a.e. uniquely the Janossy density of a point process, one obtains:

$$\det{}_\alpha (J_\alpha^\Lambda(x_i, x_j))_{1 \le i,j \le n} \ge 0 \; \lambda^{\otimes n}\text{-a.e. } (x_1, \ldots, x_n) \in E^n$$

$j_{\alpha,n}^\Lambda(x_1, \ldots, x_n) = \mathrm{Det}(\mathcal{I} + \alpha\mathcal{K}_\Lambda)^{-1/\alpha} \det_\alpha (J_\alpha^\Lambda(x_i, x_j))_{1 \le i,j \le n}$ is the Janossy density.

Conversely, if we assume $\mathrm{Det}(\mathcal{I} + \alpha\mathcal{K}_\Lambda)^{-1/\alpha} > 0$ and $\det_\alpha (J_\alpha^\Lambda(x_i, x_j))_{1 \le i,j \le n} \ge 0$ for any $n \in \mathbb{N}$, any compact set $\Lambda \subset E$ and any $\lambda^{\otimes n}$-a.e. $(x_1, \ldots, x_n) \in \Lambda^n$, the Janossy density will be correctly defined and, on any compact set Λ, we get the existence of a point process ξ_Λ with kernel K_Λ (see Proposition 5.3.II. in [2]—here the normalization condition is automatic by choosing $f = 0$ in (5)).

The restriction of a point process η, defined on $\Lambda' \subset E$, to a subspace $\Lambda \subset \Lambda'$ is the point process denoted $\eta|_\Lambda$, obtained by keeping the points in Λ and deleting the points in $\Lambda' \backslash \Lambda$.

For any compact sets $\Lambda, \Lambda' \subset E$, such that $\Lambda \subset \Lambda'$, ξ_Λ and $\xi_{\Lambda'}|_\Lambda$ have the same Laplace functional, because we have for any non-negative function f, with compact support included in Λ:

$$\begin{aligned} \mathbb{E}\left(\exp\left(-\int_\Lambda f d\xi_{\Lambda'}|_\Lambda \right) \right) &= \mathrm{Det}(\mathcal{I} + \alpha\mathcal{K}_{\Lambda'}[1 - e^{-f}])^{-1/\alpha} \\ &= \mathrm{Det}(\mathcal{I} + \alpha\mathcal{K}_\Lambda[1 - e^{-f}])^{-1/\alpha} \\ &= \mathbb{E}\left(\exp\left(-\int_\Lambda f d\xi_\Lambda \right) \right). \end{aligned}$$

Therefore, ξ_Λ and $\xi_{\Lambda'}|_\Lambda$ have the same probability distribution. We say that the family $(\mathcal{L}(\xi_\Lambda))$, Λ compact set included in E, is consistent.

Then, we can obtain a point process on the complete space E by the Kolmogorov existence theorem for point processes. See Theorem 9.2.X in [3] with $P_k(A_1, \ldots, A_k; n_1, \ldots, n_k) = \mathbb{P}\left(\xi_{\cup_{i=1}^k A_i}(A_1) = n_1, \ldots, \xi_{\cup_{i=1}^k A_i}(A_k) = n_k \right)$: as $\xi_{\cup_{i=1}^k A_i}$ is a point process, it follows that the properties (i), (iii), (iv) are fulfilled ; (ii) is fulfilled because the family $(\mathcal{L}(\xi_\Lambda))$, Λ compact set included in E, is consistent.

As we used in this second part of the proof, only the fact that $\mathrm{Det}(\mathcal{I} + \alpha\mathcal{K}_\Lambda)^{-1/\alpha} > 0$ (instead of $\mathrm{Det}(\mathcal{I} + \alpha\mathcal{K}_\Lambda)^{-1/\alpha} \geq 1$), the assertion in Remark 2 is also proved.

Proof (Theorem 2)
We suppose there exists an α-permanental process with $\alpha > 0$, kernel K defining the locally trace class integral operator $\mathcal{K}$.

Then, following the proof of the preceding theorem, we get that, for all $z \in [0, 1]$

$$\mathrm{Det}(\mathcal{I} + \alpha(1 - z)\mathcal{K}_\Lambda) = \mathrm{Det}(\mathcal{I} + \alpha\mathcal{K}_\Lambda)\,\mathrm{Det}(\mathcal{I} - \alpha z\mathcal{J}_\alpha^\Lambda) > 0.$$

As the power series of $\mathrm{Det}(\mathcal{I} - \alpha z\mathcal{J}_\alpha^\Lambda)^{-1/\alpha}$ has all its terms non-negative,

$$|(\mathrm{Det}(\mathcal{I} - \alpha z\mathcal{J}_\Lambda^\alpha)^{-1/\alpha}| \leq (\mathrm{Det}(\mathcal{I} - \alpha\,|z|\,\mathcal{J}_\Lambda^\alpha)^{-1/\alpha}.$$

If z_0 is a complex number with minimum modulus such that $(\mathrm{Det}(\mathcal{I} - \alpha z_0\mathcal{J}_\Lambda^\alpha) = 0$, by analyticity of $z \mapsto \mathrm{Det}(\mathcal{I} - \alpha z\mathcal{J}_\alpha^\Lambda)$ on $\mathbb{C}$ and $z \mapsto z^{-1}$ on $\mathbb{C}^*$, $\mathrm{Det}(\mathcal{I} - \alpha z\mathcal{J}_\alpha^\Lambda)^{-1/\alpha}$ converges for $|z| < |z_0|$ and diverges for $z = z_0$. Thus the series diverges in $z = |z_0|$ and $|z_0| > 1$. This means that the series converges for $|z| \leq 1$ thus, in this case, $\mathrm{Det}(\mathcal{I} - \alpha z\mathcal{J}_\alpha^\Lambda) > 0$.

This implies the necessary condition: $\mathrm{Spec}\,\mathcal{J}_\alpha^\Lambda \subset \{z \in \mathbb{C} : |z| < \frac{1}{\alpha}\}$.

As ν eigenvalue of $\mathcal{K}$ is equivalent to $\frac{\nu}{1 + \alpha\nu}$ eigenvalue of $\mathcal{J}$, and as, $\mathcal{K}$ and $\mathcal{J}$ being compact operators, their non-null spectral values are their eigenvalues, we get the other equivalent necessary condition:

$$\mathrm{Spec}\,\mathcal{K}_\Lambda \subset \{z \in \mathbb{C} : \mathrm{Re}\,z > -\frac{1}{2\alpha}\}.$$

6 α- Determinantal Process ($\alpha < 0$)

We recall the following remark, already made for example in [7].

Remark 3 If we define kernels only $\lambda^{\otimes 2}$-almost everywhere, there can be problems when we consider only the diagonal terms, as $\lambda^{\otimes 2}\{(x, x) : x \in \Lambda\} = 0$. For example, in the formula

$$\mathrm{tr}\,K_\Lambda = \int_\Lambda K(x, x)\lambda(dx),$$

$\operatorname{tr} K_\Lambda$ is not uniquely defined. To avoid this problem, we write the kernel K_Λ as follows:

$$K_\Lambda(x,y) = \sum_{k=0}^{\infty} a_k \varphi_k(x) \overline{\psi_k}(y)$$

where $(\varphi_k)_{k\in\mathbb{N}}$, $(\psi_k)_{k\in\mathbb{N}}$ are orthonormal basis in $L^2(\Lambda,\lambda)$ and $(a_k)_{k\in\mathbb{N}}$ is a sequence of non-negative real number, which are the singular values of the operator $\mathcal{K}_\Lambda$.

The functions φ_k and ψ_k, $k \in \mathbb{N}$, are defined λ-almost everywhere, but this gives then a unique value for the expression of type

$$\int_{\Lambda^n} F(K(x_i,x_j)_{1\le i,j\le n})\, G(x_1,\dots,x_n)\lambda(dx_1)\dots\lambda(dx_n)$$

where F is an arbitrary complex function from $\mathbb{C}^{n^2}$ and G is an arbitrary complex function from Λ^n.

With this remark, the quantities that appear with $F = \det_\alpha$ are well defined.

Lemma 1 *Let K be a kernel defined as in Remark 3 and defining a trace class integral operator $\mathcal{K}$ on $L^2(\Lambda,\lambda)$, where Λ is a non-λ-null compact set included in the locally compact Polish space E, λ be a Radon measure, n an integer and α a real number. Let F be a continuous function from $\mathbb{C}^{n^2}$ to $\mathbb{C}$. The three following assertions are equivalent*

(i) $F(K(x_i,x_j)_{1\le i,j\le n}) \ge 0\ \lambda^{\otimes n} - a.e.(x_1,\dots,x_n) \in \Lambda^n$

(ii) there exists a set $\Lambda' \subset \Lambda$ *such that* $\lambda(\Lambda\backslash\Lambda') = 0$ *and* $F((K(x_i,x_j))_{1\le i,j\le n}) \ge 0$

for any $(x_1,\dots,x_n) \in (\Lambda')^n$

(iii) there exists a version of K *such that* $F((K(x_i,x_j))_{1\le i,j\le n}) \ge 0$

for any $(x_1,\dots,x_n) \in \Lambda^n$

Proof (i) is clearly a consequence of (ii). We assume now that (i) is satisfied and we denote by N the $\lambda^{\otimes n}$-null set of n-tuples $(x_1,\dots,x_n) \in \Lambda^n$ such that $F((K(x_i,x_j))_{1\le i,j\le n}) < 0$. As in Remark 3, we write the kernel K as follows

$$K(x,y) = \sum_{k=0}^{\infty} a_k\varphi_k(x)\overline{\psi_k}(y) = \langle(\sqrt{a_k}\varphi_k)_{k\in\mathbb{N}}(x)|(\sqrt{a_k}\psi_k)_{k\in\mathbb{N}}(y)\rangle$$

where $(\varphi_k)_{k\in\mathbb{N}}$, $(\psi_k)_{k\in\mathbb{N}}$ are orthonormal basis in $L^2(\Lambda,\lambda)$, $(a_k)_{k\in\mathbb{N}}$ is a sequence of non-negative real number, which are the singular values of the operator $\mathcal{K}$ and $\langle .|.\rangle$ denote the inner product in the Hilbert space $l_2(\mathbb{C})$.

As $\mathcal{K}$ is trace class, we have $\sum_{k=0}^{\infty} a_k < \infty$. Hence:

$$\sum_{k=0}^{\infty} a_k|\varphi_k(x)|^2 < \infty \text{ and } \sum_{k=0}^{\infty} a_k|\psi_k(x)|^2 < \infty \ \lambda\text{-a.e. } x \in \Lambda$$

From Lusin's theorem, there exists an increasing sequence $(A_p)_{p\in\mathbb{N}}$ of compact sets included in Λ such that, for any $p \in \mathbb{N}$

$$(\sqrt{a_k}\varphi_k)_{k\in\mathbb{N}} \text{ and } (\sqrt{a_k}\psi_k)_{k\in\mathbb{N}} \text{ are continuous from } A_p \text{ to } l_2(\mathbb{C}) \text{ and } \lambda(\Lambda\backslash A_p) < \frac{1}{p}$$

Therefore the kernel $K : (x, y) \mapsto \left\langle (\sqrt{a_k}\varphi_k)_{k\in\mathbb{N}}(x) | (\sqrt{a_k}\psi_k)_{k\in\mathbb{N}}(y) \right\rangle$ is continuous on A_p^2.

As E is a Polish space, it can be endowed with a distance that we denote by d. We consider the sets

$$A'_p = \{x \in A_p : \forall r > 0, \lambda(B(x, r) \cap A_p) > 0\}$$
$$B_{p,n} = \{x \in A_p : \lambda(B(x, 1/n) \cap A_p) = 0\}$$

where $B(x, r)$ is the open ball in E of radius r centered at x and n is an integer.

Let $(x_k)_{k\in\mathbb{N}}$ be a sequence in $B_{p,n}$ converging to $x \in A_p$. Then we have, when $d(x, x_k) < 1/n$,

$$\lambda(B(x, 1/n - d(x, x_k) \cap A_p) \leq \lambda(B(x_k, 1/n) \cap A_p) = 0$$

Therefore $\lambda(B(x, 1/n) \cap A_p) = 0$ and $x \in B_{p,n}$: $B_{p,n}$ is closed, thus compact (as it is included in the compact set A_p).

The set of open balls $\{B(x, 1/n) : x \in B_{p,n}\}$ is a cover of $B_{p,n}$. Then, by compactness, $B_{p,n}$ can be covered by a finite numbers of such balls. As the intersections of A_p and any such a ball is a λ-null set, we get $\lambda(B_{p,n}) = 0$.

Hence we have: $\lambda(A'_p) = \lambda\left(A_p \backslash \cup_{n\in\mathbb{N}} B_{p,n}\right) = \lambda(A_p) > \lambda(\Lambda) - 1/p$.

Let $(x_1, \ldots, x_n) \in (A'_p)^n$. If $(x_1, \ldots, x_n) \notin N$, then $F((K(x_i, x_j))_{1\leq i,j\leq n}) \geq 0$.

Otherwise $(x_1, \ldots, x_n) \in N$. For any $i \in [\![1, n]\!]$ and any $r > 0$, we have

$$\lambda(A_p \cap B(x_i, r)) > 0, \text{ then } \lambda^{\otimes n}(A_p^n \cap B((x_1, \ldots, x_n), r)) = \lambda^{\otimes n}(\prod_{i=1}^{n}(A_p \cap B(x_i, r))) > 0.$$

where $B((x_1, \ldots, x_n), r)$ denotes the open ball of radius r centered at x, in E^n endowed with the distance $d((x_1, \ldots, x_n), (y_1, \ldots, y_n)) = \max_{1\leq i\leq n} d(x_i, y_i)$.

Then, as $\lambda^{\otimes n}(N) = 0$, for any $q \in \mathbb{N}$, there exists $(y_1^{(q)}, \ldots, y_n^{(q)}) \in A_p^n \cap B((x_1, \ldots, x_n), 1/q)\backslash N$ and thus $(y_1^{(q)}, \ldots, y_n^{(q)})$ converge to $(x_1, \ldots, x_n)$ when $q \to \infty$.

As $(y_1^{(q)}, \dots, y_n^{(q)}) \notin N$, $F((K(y_i^{(q)}, y_j^{(q)}))_{1 \le i,j \le n}) \ge 0$.

As K is continuous on A_p^2 and F is continuous on $\mathbb{C}^{n^2}$, we have that the function $(x_1, \dots, x_n) \mapsto F((K(x_i, x_j))_{1 \le i,j \le n})$ is continuous on A_p^n.

Hence we have: $F((K(x_i, x_j))_{1 \le i,j \le n}) \ge 0$.

Therefore, in all cases, if $(x_1, \dots, x_n) \in (A_p')^n$, $F((K(x_i, x_j))_{1 \le i,j \le n}) \ge 0$.

As $(A_p)_{p \in \mathbb{N}}$ is an increasing sequence, it is the same for $(A_p')_{p \in \mathbb{N}}$.

Hence we have: $\cup_{p \in \mathbb{N}} (A_p')^n = \left(\cup_{p \in \mathbb{N}} A_p'\right)^n$.

We obtain:

$$F((K(x_i, x_j))_{1 \le i,j \le n}) \ge 0 \text{ for any } (x_1, \dots x_n) \in \left(\cup_{p \in \mathbb{N}} A_p'\right)^n$$

As $\lambda(\Lambda \setminus \left(\cup_{p \in \mathbb{N}} A_p'\right)) = 0$, we finally obtain (ii) with $\Lambda' = \cup_{p \in \mathbb{N}} A_p'$.

We obtained that (i) and (ii) are equivalent conditions.

(i) is clearly a consequence of (iii). Assume now (ii). We will define a version K_1 of K satisfying the condition (iii).

As $\lambda(\Lambda) \ne 0$, $\Lambda' \ne \emptyset$. We set an arbitrary $x_0 \in \Lambda'$.

For $(x, x') \in \Lambda^2$, we define, $y = x$ if $x \in \Lambda'$, $y = x_0$ if $x \in \Lambda \setminus \Lambda'$, $y' = x'$ if $x' \in \Lambda'$, $y' = x_0$ if $x' \in \Lambda \setminus \Lambda'$ and $K_1(x, x') = K(y, y')$.

For $(x_1, \dots, x_n) \in \Lambda^n$, we define, for $1 \le i \le n$, $y_i = x_i$ if $x_i \in \Lambda'$ and $y_i = x_0$ if $x_i \in \Lambda \setminus \Lambda'$. Then we have, $F((K_1(x_i, x_j))_{1 \le i,j \le n}) = F((K(y_i, y_j))_{1 \le i,j \le n}) \ge 0$ and K_1 is a version of K satisfying the condition (iii).

Remark 4 Let $F_n, n \in \mathbb{N}$, be continuous functions from $\mathbb{C}^{n^2}$ to $\mathbb{C}$. For any non-$\lambda - null$ compact set Λ, the condition:

(i) $F_n((J_\alpha^\Lambda(x_i, x_j))_{1 \le i,j \le n}) \ge 0$, for any $n \in \mathbb{N}$ and $\lambda^{\otimes n}$-a.e. $(x_1, \dots, x_n) \in \Lambda^n$

can always be replaced by the equivalent conditions:

(ii) there exists a set $\Lambda' \subset \Lambda$ such that $\lambda(\Lambda \setminus \Lambda') = 0$ and $F_n((J_\alpha^\Lambda(x_i, x_j))_{1 \le i,j \le n}) \ge 0$, for any $n \in \mathbb{N}$ and $(x_1, \dots, x_n) \in (\Lambda')^n$.

or:

(iii) there exists a version of the kernel J such that $F_n((J_\alpha^\Lambda(x_i, x_j))_{1 \le i,j \le n}) \ge 0$, for any $n \in \mathbb{N}$ and $(x_1, \dots, x_n) \in \Lambda^n$.

Proof The proof of (ii) $\Longrightarrow$ (iii) is done in the same way as in Lemma 1. The other parts of the proof are a direct application of Lemma 1.

Proof (Proof that (i) is Necessary in Theorem 3)

This has been mentioned by Vere-Jones in [13] for the multivariate binomial probability distribution, which corresponds to a determinantal process with E being finite. To our knowledge, this has not been proved in other cases.

We consider the $n \times n$ matrix 1_n, whose elements are all equal to one.

We have: $\prod_{j=0}^{n-1}(1 + j\alpha) = 1 + \sum_{k=1}^{n-1} \sum_{1 \le j_1 < \dots < j_k \le n-1} j_1 \dots j_k \, \alpha^k$

We will show by induction on n that the number of permutations in Σ_n having $n - k$ cycles for $k \ne 0$ is $a_{nk} = \sum_{1 \le j_1 < \dots < j_k \le n-1} j_1 \dots j_k$: this is true for $n = 2$ and

$k = 1$. Assume it is true for a given $n \in \mathbb{N}^*$ and for any $k \in [\![1, n-1]\!]$. If we consider the permutations $\sigma \in \Sigma_{n+1}$ having $n+1-k$ cycles ($0 \le k \le n$), we have 2 cases:

- either $\sigma(n+1) = n+1$: there is exactly a_{nk} permutations corresponding to this case (with the convention $a_{nn} = 0$, for the case $k = n$),
- or $\sigma(n+1) \neq n+1$. Then, if we denote $\tau_{n+1\,\sigma(n+1)}$ the transposition in Σ_{n+1} that exchange $n+1$ and $\sigma(n+1)$, $\tau_{n+1\,\sigma(n+1)} \circ \sigma$ is a permutation having $n+1$ as fixed point and $n+1-k$ other cycles (with elements in $[\![1, n]\!]$): there is exactly $na_{n\,k-1}$ permutations corresponding to this case.

Then we have

$$\begin{aligned} a_{n+1\;n+1-k} &= a_{nk} + na_{n\,k-1} \\ &= \sum_{1 \le j_1 < \cdots < j_k \le n-1} j_1 \ldots j_k + \sum_{\substack{1 \le j_1 < \cdots < j_{k-1} \le n-1 \\ j_k = n}} j_1 \ldots j_k \\ &= \sum_{1 \le j_1 < \cdots < j_k \le n} j_1 \ldots j_k \end{aligned}$$

which is what we expected.

Thus: $\det_\alpha 1_n = \prod_{j=0}^{n-1}(1 + j\alpha)$.

If $\alpha < 0$ but $-1/\alpha \notin \mathbb{N}$, there exists therefore $n \in \mathbb{N}$ such that $\det_\alpha 1_n < 0$.

We suppose that there exists an α-determinantal process with $\alpha < 0$ but $-1/\alpha \notin \mathbb{N}$ and kernel K. Then we have $\det_\alpha(K(x_i, x_j))_{1 \le i,j \le n} \ge 0$ $\lambda^{\otimes n}$-a.e. $(x_1 \ldots, x_n) \in E^n$.

As we exclude the case of a point process having no point almost surely and there is a sequence of compact sets Λ_p such that $\cup_{p \in \mathbb{N}} \Lambda_p = E$, there exists a compact set $\Lambda \in E$ such that

$$\mathbb{E}(\xi(\Lambda)) = \int_\Lambda K(x, x)\lambda(dx) > 0.$$

Applying Lemma 1, we get that there exist a version K_1 of the kernel K such that $\det_\alpha(K_1(x_i, x_j))_{1 \le i,j \le n} \ge 0$ for any $(x_1 \ldots, x_n) \in \Lambda^n$. We also have:

$$\int_\Lambda K(x, x)\lambda(dx) = \int_\Lambda K_1(x, x)\lambda(dx) > 0.$$

Hence there exists $x_0 \in \Lambda$ such that $K_1(x_0, x_0) > 0$.

For $(x_1, \ldots, x_n) = (x_0, \ldots, x_0)$, we get:

$$\det_\alpha(K_1(x_i, x_j))_{1 \le i,j \le n} = K(x_0, x_0)^n \det_\alpha 1_n < 0$$

which is a contradiction. Therefore if $\alpha < 0$ and an α-determinantal process exists, then α must be in $\{-1/m : m \in \mathbb{N}\}$.

We consider a $d \times d$ square matrix A. If $n_1, \ldots, n_d$ are d non-negative integers, $A[n_1, \ldots, n_d]$ is the $(n_1 + \cdots + n_d) \times (n_1 + \cdots + n_d)$ square matrix composed of the block matrices A_{ij}:

$$A[n_1, \ldots, n_d] = \begin{pmatrix} A_{11} & A_{12} & \ldots & A_{1d} \\ A_{21} & A_{22} & \ldots & A_{2d} \\ \vdots & \vdots & \ddots & \vdots \\ A_{d1} & A_{d2} & \ldots & A_{dd} \end{pmatrix},$$

where A_{ij} is the $n_i \times n_j$ matrix whose elements are all equal to a_{ij} $(1 \leq i, j \leq d)$.

Lemma 2 *Given a $d \times d$ square matrix A, the following assertions are equivalent*

(i) $\det_{-1/m} A[n_1, \ldots, n_d] \geq 0, \ \forall n_1, \ldots, n_d \in \mathbb{N}$

(ii) $\det_{-1/m} A[n_1, \ldots, n_d] \geq 0, \ \forall n_1, \ldots, n_d \in \{0, \ldots, m\}$

(iii) $\det A[n_1, \ldots, n_d] \geq 0, \ \forall n_1, \ldots, n_d \in \mathbb{N}$

(iv) $\det A[n_1, \ldots, n_d] \geq 0, \ \forall n_1, \ldots, n_d \in \{0, 1\}$

Proof If there exists $k \in [\![1, d]\!]$ such that $n_k > 1$, the matrix $A[n_1, \ldots, n_d]$ has at least two identical rows and its determinant is null. So it is clear that (iii) and (iv) are equivalent.

We have:

$$\det(I + ZA)^m = \sum_{n_1,\ldots,n_d=0}^{\infty} m^{n_1+\ldots n_d} \left(\prod_{k=1}^{d} \frac{z_k^{n_k}}{n_k!} \right) \det_{-1/m} A[n_1, \ldots, n_d] \tag{6}$$

where $Z = \mathrm{diag}(z_1, \ldots, z_d)$ and $z_1, \ldots, z_d$ are d complex numbers. It is a special case of the formula (3) with $\alpha = -1/m$, finite space $E = [\![1, d]\!]$ and reference measure λ atomic, where each point of E has measure 1, $\Lambda_k = \{k\}$, for $k \in [\![1, d]\!]$, $\Lambda = E$. Indeed, $ZA = \sum_{k=1}^{d} z_k A_k$, where A_k is the $d \times d$ square matrix having the same k^{th} row as A and the other rows with all elements equal to 0. The matrix A corresponds to the operator $\mathcal{K}$, the matrix A_k corresponds to the operator $\mathcal{K}_{\Lambda_k \Lambda}$. Formula (6) also corresponds to the one given by Vere-Jones in [12].

We also have for $m = 1$:

$$\det(I + ZA) = \sum_{n_1,\ldots,n_d=0}^{1} \left(\prod_{k=1}^{d} \frac{z_k^{n_k}}{n_k!} \right) \det A[n_1, \ldots, n_d]. \tag{7}$$

as $\det A[n_1, \ldots, n_d] = 0$ if there exists $k \in [\![1, d]\!]$ such that $n_k > 1$.

(i) is equivalent to the fact that the multivariate power series (6) has all its coefficients non-negative.

(iii) is equivalent to the fact that the multivariate power series (7) has all its coefficients non-negative.

The power series (6) being the mth power of the power series (7), if there exists $k \in [\![1, d]\!]$ such that $n_k > m$, the coefficient of $\prod_{k=1}^{d} z^{n_k}$ is null. Therefore, (i) is equivalent to (ii).

For the same reason, we also have that (i) is a consequence of (iii).

Conversely, following Vere-Jones in [13], we can show by induction on the order of the matrix A, that the fact that the power series (6) has all its coefficients non-negative implies that the power series (7) has all its coefficient non negative.

This proves the equivalence between (i) and (iii).

Proposition 2 *Let $\alpha < 0$ and $\mathcal{K}$ be an integral operator such that $\mathcal{I} + \alpha\mathcal{K}_\Lambda$ is invertible, for any compact set $\Lambda \subset E$. An α-determinantal process with kernel K exists iff:*

$$\det_\alpha(J_\alpha^\Lambda(x_i, x_j))_{1\le i,j\le n} \ge 0, \textit{for any } n \in \mathbb{N}, \textit{ and any compact set } \Lambda$$
$$\lambda^{\otimes n}\textit{-a.e. } (x_1, \dots, x_n) \in \Lambda^n \tag{8}$$

Condition (8) *implies that* $-\frac{1}{\alpha} \in \mathbb{N}$ *and* $\mathrm{Det}(\mathcal{I} + \beta\mathcal{K}) > 0$ *for any* $\beta \in [\alpha, 0]$.

Proof We assume that there exists an α-determinantal process ξ with kernel K.

We already proved that it is necessary to have $-1/\alpha \in \mathbb{N}$.

By taking $d = 1$ in the formula (4), we have

$$\mathbb{E}\left(z^{\xi(\Lambda)}\right) = \mathrm{Det}\,(\mathcal{I} + \alpha(1-z)\,\mathcal{K}_\Lambda)^{-1/\alpha}$$

for any compact set $\Lambda \subset E$ and $z \in (0, 1]$. Then $\mathrm{Det}\,(\mathcal{I} + \alpha(1-z)\,\mathcal{K}_\Lambda) > 0$ for $z \in (0, 1]$, and by continuity, $\mathrm{Det}\,(\mathcal{I} + \alpha\,\mathcal{K}_\Lambda) \ge 0$. As we assumed that $\mathcal{I} + \alpha\mathcal{K}_\Lambda$ is invertible, we have necessarily $\mathrm{Det}\,(\mathcal{I} + \alpha\,\mathcal{K}_\Lambda) > 0$.

For any non-negative function f, with compact support included in Λ

$$\begin{aligned}
\mathbb{E}\left(\prod_{x\in\xi} e^{-f(x)}\right) &= \mathrm{Det}(\mathcal{I} + \alpha\mathcal{K}[1 - e^{-f}])^{-1/\alpha} \\
&= \mathrm{Det}(\mathcal{I} + \alpha\mathcal{K}_\Lambda)^{-1/\alpha}\,\mathrm{Det}(\mathcal{I} - \alpha\mathcal{J}_\alpha^\Lambda e^{-f})^{-1/\alpha} \\
&= \mathrm{Det}(\mathcal{I} + \alpha\mathcal{K}_\Lambda)^{-1/\alpha} \\
&\quad \times \sum_{n=0}^{\infty} \frac{1}{n!} \int_{\Lambda^n} \left(\prod_{i=1}^{n} e^{-f(x_i)}\right) \det_\alpha(J_\alpha^\Lambda(x_i, x_j))_{1\le i,j\le n} \lambda(dx_1)\dots\lambda(dx_n)
\end{aligned}$$

As the Laplace functional defines a.e. uniquely the Janossy density of a point process, one obtains:

$$\det_\alpha(J_\alpha^\Lambda(x_i,x_j))_{1\le i,j\le n} \ge 0 \ \lambda^{\otimes n}\text{-a.e. } (x_1,\dots,x_n) \in E^n$$

Conversely, we assume that the condition

$\det_\alpha(J_\alpha^\Lambda(x_i,x_j))_{1\le i,j\le n} \ge 0$, for any $n \in \mathbb{N}$, $\lambda^{\otimes n}$-a.e. $(x_1,\dots,x_n) \in \Lambda^n$ and any compact set Λ

is fulfilled. We have

$$\mathrm{Det}(\mathcal{I} - \alpha z \mathcal{J}_\alpha^\Lambda)^{-1/\alpha} = \sum_{n=0}^{\infty} \frac{z^n}{n!} \int_{\Lambda^n} \det_\alpha(J_\alpha^\Lambda(x_i,x_j))_{1\le i,j\le n} \lambda(dx_1)\dots\lambda(dx_n)$$

As $-1/\alpha \in \mathbb{N}$, this formula is valid for any $z \in \mathbb{C}$. Then we obtain for $z = 1$, $\mathrm{Det}(\mathcal{I} - \alpha\mathcal{J}_\alpha^\Lambda)^{-1/\alpha} \ge 0$.

We also have $(\mathcal{I} - \alpha\mathcal{J}_\alpha^\Lambda)(\mathcal{I} + \alpha\mathcal{K}_\Lambda) = (\mathcal{I} + \alpha\mathcal{K}_\Lambda)(\mathcal{I} - \alpha\mathcal{J}_\alpha^\Lambda) = \mathcal{I}$.

Then $\mathrm{Det}(\mathcal{I} - \alpha\mathcal{J}_\alpha^\Lambda) > 0$ and $\mathrm{Det}(\mathcal{I} + \alpha\mathcal{K}_\Lambda) > 0$.

Thus the Janossy density is correctly defined and, on any compact set Λ we get the existence of a point process with kernel K and reference measure λ.

Then it can be extended to the complete space E by the Kolmogorov existence theorem (see Theorem 9.2.X in [3]).

Proof (Theorem 3)

For any $m \in \mathbb{N}$, applying Lemma 2, we have for any compact set Λ

$$\det_{-1/m}(J_{-1/m}^\Lambda(x_i,x_j))_{1\le i,j\le n} \ge 0, \text{ for any } n \in \mathbb{N}, \text{ and any } (x_1,\dots,x_n) \in \Lambda^n$$

is equivalent to

$$\det(J_{-1/m}^\Lambda(x_i,x_j))_{1\le i,j\le n} \ge 0, \text{ for any } n \in \mathbb{N}, \text{ and any } (x_1,\dots,x_n) \in \Lambda^n$$

Now, assume we only have

$$\det_{-1/m}(J_{-1/m}^\Lambda(x_i,x_j))_{1\le i,j\le n} \ge 0, \text{ for any } n \in \mathbb{N},\ \lambda^{\otimes n}\text{-a.e. } (x_1,\dots,x_n) \in \Lambda^n.$$

By Lemma 1, for each $n \in \mathbb{N}$, there exists a set $\Lambda'_n \subset \Lambda$ such that $\lambda(\Lambda\backslash\Lambda'_n) = 0$ and $\det_{-1/m}(J_{-1/m}^\Lambda(x_i,x_j))_{1\le i,j\le n} \ge 0$ for any $(x_1\dots,x_n) \in (\Lambda'_n)^n$.

If $\Lambda' = \cap_{n\in\mathbb{N}}\Lambda'_n$, we have $\lambda(\Lambda\backslash\Lambda') = 0$ and $\det_{-1/m}(J_{-1/m}^\Lambda(x_i,x_j))_{1\le i,j\le n} \ge 0$ for any $n \in \mathbb{N}$ and $(x_1\dots,x_n) \in (\Lambda')^n$.

Then, by Lemma 2, we have: $\det(J_{-1/m}^\Lambda(x_i,x_j))_{1\le i,j\le n} \ge 0$, for any $n \in \mathbb{N}$ and $(x_1\dots,x_n) \in (\Lambda')^n$.

Therefore, we have

$$\det(J^{\Lambda}_{-1/m}(x_i,x_j))_{1\le i,j\le n} \ge 0, \text{ for any } n\in\mathbb{N}, \lambda^{\otimes n}\text{-a.e. } (x_1,\dots,x_n)\in\Lambda^n.$$

The converse is done through a similar proof, using Lemmas 1 and 2.
Thus, we obtain:

$$\det_\alpha(J^{\Lambda}_{\alpha}(x_i,x_j))_{1\le i,j\le n} \ge 0, \text{ for any } n\in\mathbb{N}, \lambda^{\otimes n}\text{-a.e. } (x_1,\dots,x_n)\in\Lambda^n$$

is equivalent to

$$\det(J^{\Lambda}_{\alpha}(x_i,x_j))_{1\le i,j\le n} \ge 0, \text{ for any } n\in\mathbb{N}, \lambda^{\otimes n}\text{-a.e. } (x_1,\dots,x_n)\in\Lambda^n$$

Theorem 3 is then a consequence of Proposition 2.

Proof (Theorem 4) We assume that there exists ξ an α-determinantal process with kernel K.

For $p\in(0,1)$, let ξ_p be the process obtained by first sampling ξ, then independently deleting each point of ξ with probability $1-p$.

Computing the correlation functions, we obtain that ξ_p is an α-determinantal process with kernel pK.

Thus we get from Theorem 3 that the conditions of the theorem must be fulfilled.

Conversely, we assume that these conditions are fulfilled. We obtain from Theorem 3 that an α-determinantal process ξ_p with kernel pK exists, for any $p\in(0,1)$.

We consider a sequence $(p_k)\in(0,1)^{\mathbb{N}}$ converging to 1 and a compact Λ.

$$\mathbb{E}(\exp(-t\xi_{p_k}(\Lambda)) = \mathrm{Det}(\mathcal{I}+\alpha p_k K_\Lambda(1-e^{-t}))^{-1/\alpha} \underset{k\to\infty}{\longrightarrow} \mathrm{Det}(\mathcal{I}+\alpha K_\Lambda(1-e^{-t}))^{-1/\alpha}$$

As $t\mapsto \mathrm{Det}(\mathcal{I}+\alpha K_\Lambda(1-e^{-t}))^{-1/\alpha}$ is continuous in 0, $(\mathcal{L}(\xi_{p_k}(\Lambda)))_{k\in\mathbb{N}}$ converge weakly. Thus $(\mathcal{L}(\xi_{p_k}(\Lambda)))_{k\in\mathbb{N}}$ is tight.

$\Gamma\subset\mathcal{X}$ is relatively compact if and only if, for any compact set $\Lambda\subset E$, $\{\xi(\Lambda): \xi\in\Gamma\}$ is bounded.

Let $(\Lambda_n)_{n\in\mathbb{N}}$ be an increasing sequence of compact sets such that $\cup_{n\in\mathbb{N}}\Lambda_n = E$.

As, for any $n\in\mathbb{N}$, $(\mathcal{L}(\xi_{p_k}(\Lambda_n)))_{k\in\mathbb{N}}$ is tight, we have that, for any $\epsilon>0$ and $n\in\mathbb{N}$, there exists $M_n>0$ such that for any $k\in\mathbb{N}, \mathbb{P}(\xi_{p_k}(\Lambda_n)>M_n)<\epsilon\, 2^{-n-1}$.

Let $\Gamma=\{\gamma\in\mathcal{X}: \forall n\in\mathbb{N}, \gamma(\Lambda_n)\le M_n\}$. It is a compact set and for any $k\in\mathbb{N}, \mathbb{P}(\xi_{p_k}\in\Gamma^c)<\epsilon$.

Therefore, $(\mathcal{L}(\xi_{p_k}))_{k\in\mathbb{N}}$ is tight. As E is Polish, $\mathcal{X}$ is also Polish (endowed with the Prokhorov metric). Thus there is a subsequence of $(\mathcal{L}(\xi_{p_k}))_{k\in\mathbb{N}}$ converging weakly to the probability distribution of a point process ξ. By unicity of the distribution of an α-determinantal process for given kernel and reference measure, ξ must be an α-determinantal process with kernel K, which gives the existence.

Lemma 3 *Let $\mathcal{J}$ be a trace class self-adjoint integral operator with kernel J. We have*

$$\det(J(x_i,x_j))_{1\le i,j\le n} \ge 0 \textit{ for any } n \in \mathbb{N}, \lambda^{\otimes n}\textit{-a.e.}(x_1,\dots,x_n) \in \Lambda^n$$

if and only if

$$\operatorname{Spec}\mathcal{J} \subset [0,\infty)$$

Proof If we assume that the operator $\mathcal{J}$ is positive, the kernel can be written as follows:

$$J(x,y) = \sum_{k=0}^{\infty} a_k\varphi_k(x)\overline{\varphi_k}(y)$$

where $a_k \ge 0$ for $k \in \mathbb{N}$.

Hence:

$$\det(J(x_i,x_j))_{1\le i,j\le n} \ge 0 \text{ for any } n \in \mathbb{N}, \text{ and any } (x_1,\dots,x_n) \in \Lambda^n$$

Conversely, assume that

$$\det(J(x_i,x_j))_{1\le i,j\le n} \ge 0 \text{ for any } n \in \mathbb{N}, \lambda^{\otimes n}\text{-a.e.}(x_1,\dots,x_n) \in \Lambda^n.$$

From formula (2) with $\alpha = -1$, we have then for any $z \in \mathbb{C}$

$$\operatorname{Det}(\mathcal{I} + z\mathcal{J}) = \sum_{n=0}^{\infty} \frac{z^n}{n!}\int_{E^n} \det(J(x_i,x_j))_{1\le i,j\le n}\lambda(dx_1)\dots\lambda(dx_n). \tag{9}$$

As $\mathcal{J}$ is assumed to be self-adjoint, its spectrum is included in $\mathbb{R}$. Thanks to (9), it is impossible to have an eigenvalue in $\mathbb{R}_-^*$, as the power series has all its coefficients real non-negative and the first coefficient ($n = 0$) is real positive. Hence $\operatorname{Spec}\mathcal{J} \subset [0,\infty)$.

Proof (Corollary 2)

We assume: $-1/\alpha \in \mathbb{N}$ and $\operatorname{Spec}\mathcal{K} \subset [0,-1/\alpha]$. Then we have, as $\mathcal{K}$ is self-adjoint, that for any compact set Λ, $\operatorname{Spec}\mathcal{K}_\Lambda \subset [0,-1/\alpha]$. Then $\operatorname{Det}(\mathcal{I}+\beta\mathcal{K}_\Lambda) > 0$ for any $\beta \in (\alpha,0]$.

If $\mathcal{I} + \alpha K_\Lambda$ is invertible for any compact set $\Lambda \subset E$, we have $\operatorname{Spec} J_\alpha^\Lambda \subset [0,\infty)$ and J_α^Λ is a trace class self adjoint operator for any compact set Λ.

Then, applying Lemma 3, we get that

$\det(J(x_i, x_j))_{1\leq i,j\leq n} \geq 0$ for any $n \in \mathbb{N}$, compact set Λ and $\lambda^{\otimes n}$-a.e.$(x_1, \ldots, x_n) \in \Lambda^n$

Using Theorem 3, we get the existence of an α-determinantal process with kernel K.

When there exists a compact set Λ_0 such that $\mathcal{I} + \alpha K_{\Lambda_0}$ is not invertible, by the same line of proof, we obtain the announced result, using Theorem 4.

Conversely, we assume that there exists an α-determinantal process with kernel K.

Then, from Theorem 3 or 4, we get that $-1/\alpha \in \mathbb{N}$.

If $\mathcal{I} + \alpha K_\Lambda$ is invertible for any compact set $\Lambda \subset E$, we have $\operatorname{Spec} J_\alpha^\Lambda \subset [0, \infty)$, using Theorem 3 and Lemma 3. Then $\operatorname{Spec} K_\Lambda \subset [0, -1/\alpha) \subset [0, -1/\alpha]$, for any compact set Λ.

If there exists a compact set Λ_0 such that $\mathcal{I} + \alpha K_{\Lambda_0}$ is not invertible, we have $\operatorname{Spec} J_\beta^\Lambda \subset [0, \infty)$ for any compact set Λ and any $\beta \in (\alpha, 0)$, using Theorem 4 and Lemma 3. Then $\operatorname{Spec} K_\Lambda \subset [0, -1/\beta)$ for any $\beta \in (\alpha, 0)$. Therefore $\operatorname{Spec} K_\Lambda \subset [0, -1/\alpha]$ for any compact set Λ.

As K is self-adjoint, this implies in both cases that $\operatorname{Spec} K \subset [0, -1/\alpha]$.

Remark 5 Using the known result in the case $\alpha = -1$ (see for example Hough et al. in [7]) and Corollary 1, one obtains a direct proof of Corollary 2.

7 Infinite Divisibility

Proof (Theorem 5) For $\alpha < 0$, we have proved that it is necessary to have $-1/\alpha \in \mathbb{N}$. If an α-determinantal process was infinitely divisible, with $\alpha < 0$, it would be the sum of N i.i.d αN-determinantal processes for any $N \in \mathbb{N}^*$, as it can be seen for the Laplace functional formula (1). This would imply that $-1/(N\alpha) \in \mathbb{N}$, for any $N \in \mathbb{N}^*$, which is not possible. Therefore, an α-determinantal process with $\alpha < 0$ is never infinitely divisible.

Some characterization on infinite divisibility have also been given in [4] in the case $\alpha > 0$.

Proof (Theorem 6) For $\alpha > 0$, assume that $\operatorname{Det}(\mathcal{I} + \alpha K_\Lambda) \geq 1$ and

$$\sum_{\sigma\in\Sigma_n:\nu(\sigma)=1} \prod_{i=1}^{n} J_\alpha^\Lambda(x_i, x_{\sigma(i)}) \geq 0,$$

for any compact set $\Lambda \subset E$, $n \in \mathbb{N}$ and $\lambda^{\otimes n}$-a.e. $(x_1, \ldots, x_n) \in \Lambda^n$.

Then we have:

$$\sum_{\sigma\in\Sigma_n:\nu(\sigma)=k}\prod_{i=1}^{n}J_\alpha^\Lambda(x_i,x_{\sigma(i)}) = \sum_{\substack{\{I_1,\dots,I_k\}\\ \text{partition of } [\![1,n]\!]}}\ \sum_{\substack{\sigma_1\in\Sigma(I_1),\dots,\sigma_k\in\Sigma(I_k):\\ \nu(\sigma_1)=\dots=\nu(\sigma_k)=1}}\ \prod_{q=1}^{k}\prod_{i\in I_q}J_\alpha^\Lambda(x_i,x_{\sigma_q(i)})$$

$$= \sum_{\substack{\{I_1,\dots,I_k\}\\ \text{partition of } [\![1,n]\!]}}\ \prod_{q=1}^{k}\left(\sum_{\substack{\sigma\in\Sigma(I_q):\\ \nu(\sigma)=1}}\ \prod_{i\in I_q}J_\alpha^\Lambda(x_i,x_{\sigma(i)})\right)\geq 0,$$

for any compact set $\Lambda\subset E$, $n\in\mathbb{N}$, $k\in[\![1,n]\!]$ and $\lambda^{\otimes n}$-a.e. $(x_1,\dots,x_n)\in\Lambda^n$, where, for a finite set I, $\Sigma(I)$ denotes the set of all permutations on I.

Then, for any $N\in\mathbb{N}^*$ and any compact set $\Lambda\in E$, $\det_{N\alpha}(J_\alpha^\Lambda(x_i,x_j)/N)_{1\leq i,j\leq n}\geq 0$. From Theorem 1, we get that there exists a $(N\alpha)$-permanental process with kernel K/N. This means that an α-permanental process with kernel K is infinitely divisible.

Conversely, if we assume an α-permanental process with kernel K is infinitely divisible, we get the existence of a $N\alpha$-permanental process with kernel K/N, for any $N\in\mathbb{N}^*$.

From Theorem 1, we have that $\mathrm{Det}(\mathcal{I}+\alpha K_\Lambda)\geq 1$ for any compact set $\Lambda\in E$.

We also have

$$\frac{1}{(N\alpha)^{n-1}}\det_{N\alpha}(J_\alpha^\Lambda(x_i,x_j))_{1\leq i,j\leq n}\geq 0,$$

for any $N\in\mathbb{N}^*$, any $n\in\mathbb{N}$, any compact set $\Lambda\in E$ and $\lambda^{\otimes n}$-a.e. $(x_1,\dots,x_n)\in\Lambda^n$.

When N tends to ∞, we obtain:

$$\sum_{\sigma\in\Sigma_n:\nu(\sigma)=1}\prod_{i=1}^{n}J_\alpha^\Lambda(x_i,x_{\sigma(i)})\geq 0,$$

which is the desired result.

Proof (Theorem 7) We use the argument of Griffiths in [5] and Griffiths and Milne in [6]. Assume

$$\sum_{\sigma\in\Sigma_n:\nu(\sigma)=1}\prod_{i=1}^{n}J_\alpha^\Lambda(x_i,x_{\sigma(i)})\geq 0,$$

for any $n\in\mathbb{N}$ and any $(x_1,\dots,x_n)\in\Lambda^n$.

The condition $J_\alpha^\Lambda(x_1,x_2)\dots J_\alpha^\Lambda(x_{n-1},x_n)J_\alpha^\Lambda(x_n,x_1)\geq 0$ is satisfied for the elementary cycles, i.e. cycles such that $J_\alpha^\Lambda(x_i,x_j)=0$ if $i<j+1$ and ($i\neq 1$ or $j\neq n$). Then it can be extended to any cycle by induction, using $J_\alpha^\Lambda(x_i,x_j)=J_\alpha^\Lambda(x_j,x_i)$.

With Lemma 1, we can then extend the proof to the case when

$$\sum_{\sigma\in\Sigma_n:\nu(\sigma)=1}\prod_{i=1}^{n}J_\alpha^\Lambda(x_i,x_{\sigma(i)})\geq 0,$$

for any $n\in\mathbb{N}$ and $\lambda^{\otimes n}$-a.e. $(x_1,\ldots,x_n)\in\Lambda^n$.

Remark 6 Note that the argument from Griffiths and Milne in [5] and [6] is only valid for real symmetric matrices.

Acknowledgements I express sincere thanks to Nathalie Eisenbaum for her support in the writing of this paper. I also thank the referee for his valuable comments that helped in improving the exposition.

References

1. S. Albeverio, Y.G. Kondratiev, M. Röckner, Analysis and geometry on configuration spaces. J. Funct. Anal. **154**(2), 444–500 (1998)
2. D.J. Daley, D. Vere-Jones, *An Introduction to the Theory of Point Processes - Volume I: Elementary Theory and Methods*, 2nd edn. (Springer, New York, 2003)
3. D.J. Daley, D. Vere-Jones, *An Introduction to the Theory of Point Processes - Volume II: General Theory and Structures*, 2nd edn. (Springer, New York, 2008)
4. N. Eisenbaum, H. Kaspi, On permanental processes. Stochastic Process. Appl. **119**(5), 1401–1415 (2009)
5. R.C. Griffiths, Multivariate gamma distributions. J. Multivar. Anal. **15**, 13–20 (1984)
6. R.C. Griffiths, R.K. Milne, A class of infinitely divisible multivariate negative binomial distributions. J. Multivar. Anal. **22**, 13–23 (1987)
7. J.B. Hough, M. Krishnapur, Y. Peres, B. Virág, Determinantal processes and independence. Probab. Surv. **3**, 206–229 (2006)
8. O. Macchi, The coincidence approach to stochastic point processes. Adv. Appl. Probab. **7**, 83–122 (1975)
9. T. Shirai, Remarks on the positivity of α-determinants. Kyushu J. Math. **61**, 169–189 (2007)
10. T. Shirai, Y. Takahashi, Random point fields associated with certain Fredholm determinants. I: fermion, Poisson and boson point processes. J. Funct. Anal. **205**(2), 414–463 (2003)
11. A. Soshnikov, Determinantal random point fields. Russ. Math. Surveys **55**, 923–975 (2000)
12. D. Vere-Jones, A generalization of permanents and determinants. Linear Algebra Appl. **111**, 119–124 (1988)
13. D. Vere-Jones, Alpha-permanents and their applications to multivariate gamma, negative binomial and ordinary binomial distributions. New Zealand J. Math. **26**, 125–149 (1997)

Filtrations of the Erased-Word Processes

Stéphane Laurent

Abstract We define a class of erased-word processes and prove that the polyadic filtration generated by such a process is standard. This is shown by firstly constructing a generating process of innovations in the case of a finite alphabet equipped with the uniform probability measure, and then by deriving the general case with the help of the tools of Vershik's theory of filtrations in discrete negative time.

1 The Filtration of the Erased-Word Process

An erased-word process is depicted on Fig. 1. It is a stochastic process indexed by the set of negative integers $-\mathbb{N}$, and consists in picking at random a word W_n with $|n|$ letters at time n and then to obtain the next word W_{n+1} by deleting at random one letter of W_n (thus the final word W_0 is the empty word). More precisely, given a Lebesgue probability space (A, μ), and calling A the *alphabet*, the erased-word process on (A, μ) is the Markov process $(W_n, \eta_n)_{n \leqslant 0}$ whose law is defined as follows: for every $n \leqslant -1$,

- W_n is a random word on A made of $|n|$ letters i.i.d. according to μ;
- η_{n+1} is a random variable uniformly distributed on $\{1, 2, \ldots, |n|\}$ and independent of the past σ-field $\sigma(W_m, \eta_m; m \leqslant n)$;
- W_{n+1} is obtained by deleting the η_{n+1}-th letter of W_n.

The filtration $\mathcal{F}$ generated by the erased-word process $(W_n, \eta_n)_{n \leqslant 0}$ is defined by $\mathcal{F}_n = \sigma(W_m, \eta_m; m \leqslant n)$. We will sometimes term the η_n as the *erasers*. According to definition given below, the sequence $(\eta_n)_{n \leqslant 0}$ made of the erasers is a *process of innovations* of the filtration $\mathcal{F}$.

Definition 1 Let $\mathcal{F}$ be a filtration. A random variable η_n that is independent of $\mathcal{F}_{n-1}$ and such that $\mathcal{F}_n = \mathcal{F}_{n-1} \vee \sigma(\eta_n)$ is called an *innovation* of $\mathcal{F}$ (more precisely, we should say an innovation at time n, but thanks to the subscript in η_n this is not a point worth quibbling about). A sequence $(\eta_n)_{n \leqslant 0}$ of independent random variables

S. Laurent (✉)
e-mail: laurent_step@yahoo.fr

C. Donati-Martin et al. (eds.), *Séminaire de Probabilités XLVIII*, Lecture Notes in Mathematics 2168, DOI 10.1007/978-3-319-44465-9_16

$$\cdots \quad W_{-3} = bac \xrightarrow{\eta_{-2}=3} W_{-2} = ba \xrightarrow{\eta_{-1}=1} W_{-1} = a \xrightarrow{\eta_0=1} W_0 = \varnothing$$

Fig. 1 A trajectory of the erased-word process

such that each η_n is an innovation of $\mathcal{F}$ at time n, is called a *process of innovations* of $\mathcal{F}$.

When such a process of innovations exist, it defines the local structure of the filtration: for any other process of innovations $(\eta'_n)_{n\leqslant 0}$, the two random variables η_n and η'_n possibly generate two different σ-fields $\sigma(\eta_n)$ and $\sigma(\eta'_n)$, but there is a Boolean isomorphism between the measure algebras $\big(\sigma(\eta_n), \mathbb{P}\big)$ and $\big(\sigma(\eta'_n), \mathbb{P}\big)$ for every $n \leqslant 0$. Details about this point can be found in [1] and [5]. Thus, any possible innovation η'_n of the filtration of the erased-word process is uniformly distributed on $|n| + 1$ values, similarly to the eraser η_n. For this reason, the filtration $\mathcal{F}$ of the erased-word process is said to be $\big(|n| + 1\big)$*-adic*, and it belongs to the class of *poly-adic* filtrations, according to definition below.

Definition 2 A filtration $\mathcal{F}$ is *poly-adic* if there exists a process of innovations $(\eta_n)_{n\leqslant 0}$ of $\mathcal{F}$ such that each η_n is uniformly distributed on a finite set.

The poly-adicity will play an important role in the proof of theorem below, which is the main result of this article.

Theorem 1 *For any Lebesgue alphabet* (A, μ)*, the filtration of the erased-word process is of product type, that is to say, it is generated by a process of innovations.*

Let us comment this theorem. Consider the filtration $\mathcal{E}$ generated by the process of innovations $(\eta_n)_{n\leqslant 0}$. Obviously $\mathcal{E} \subset \mathcal{F}$, but $\mathcal{E} \neq \mathcal{F}$ because $\mathcal{E}_n$ is independent of W_n for every $n \leqslant 0$. But that does not mean that $\mathcal{E}$ and $\mathcal{F}$ are not isomorphic. Theorem 1 asserts that there exists another process of innovations $(\tilde{\eta}_n)_{n\leqslant 0}$ which generates $\mathcal{F}$ (then called a *generating process of innovations*), and this says that $\mathcal{E}$ and $\mathcal{F}$ are isomorphic. Thus $\mathcal{F}$, which is bigger than $\mathcal{E}$, is no more than $\mathcal{E}$ up to isomorphism.

This theorem together with Kolmogorov's zero-one law imply that the tail σ-field $\mathcal{F}_{-\infty} := \cap \mathcal{F}_n$ is degenerate. But it is not difficult to directly prove the degeneracy of $\mathcal{F}_{-\infty}$ with the help of the reverse martingale convergence theorem (this proof would be similar to the one given in [2] for the dyadic split-word process), whereas the proof of Theorem 1, even in the simpler case when A is finite and μ is uniform (see below our three demonstration steps), is not easy. The motivation of Theorem 1 is precisely the surprising fact that it is not a trivial result once we know that $\mathcal{F}_{-\infty}$ is degenerate: it is known that for any type of poly-adicity (such as the $\big(|n| + 1\big)$-adicity), there exist some filtrations whose tail σ-fields are degenerate but for which there does not exist any generating process of innovations. Thus, such a filtration is locally isomorphic to $\mathcal{F}$ and, similarly to $\mathcal{F}$, has a degenerate σ-field, but is not isomorphic to $\mathcal{F}$. This surprising fact has been discovered by Vershik [8–11], who developed a theory to characterize the existence of a generating process of innovations for poly-adic filtrations.

To give a better idea of the subtlety of Theorem 1, we mention that the opposite conclusion holds for a process seemingly close to the erased-word process. Namely, this process is similar to the erased-words process except that at each time, the letter deleted at random is either the first one or the last one with equal probabilities of $1/2$. It generates a dyadic (2-adic) filtration which is not of product type although its tail σ-field is degenerate. This result is not explicitly written in the literature but Heicklen and Hoffman's proof of their main result in [3] implicitly relies on it.

We will use the tools of the theory of filtrations developed by Vershik to derive the general case in Theorem 1 from the particular case when μ is uniform on a finite alphabet A. More precisely, our theorem will be proved in three steps:

1. we will prove Theorem 1 in the case when μ is the uniform probability measure on a finite alphabet A using a 'bare-hands' approach, that is, we will construct a generating process of innovations in this case;
2. using some tools of Vershik's theory, we will prove Theorem 1 in the case when μ is the Lebesgue measure on $A = [0, 1]$;
3. again using some tools of Vershik's theory, we will derive the general case of Theorem 1 from the case when μ is the Lebesgue measure on $A = [0, 1]$.

The main theorem of Vershik's theory that will be used is the equivalence between the existence of a generating process of innovations and *standardness* in the case of poly-adic filtrations:

Theorem 2 *A poly-adic filtration is of product type if and only if it is standard.*

Different definitions of standardness are used in the literature. Probabilists usually consider that a standard filtration is by definition a filtration which can be *immersed* in a filtration of product type [2, 5], and this definition directly yields that filtrations of product type are standard. The deep assertion of Theorem 2 is the reciprocal fact. Under the usual assumption that the final σ-field $\mathcal{F}_0$ of the filtration $\mathcal{F}$ is essentially separable, standardness is known to be characterized by a criterion discovered by Vershik, called *Vershik's standardness criterion* or *the Vershik property* [2, 6, 7]. A filtration satisfying the Vershik property is also said to be *Vershikian*. We will write an easy proposition (Proposition 1) about the Vershik property to derive step 2 from step 1 in the proof of Theorem 1. Then step 3 will be derived from step 2 and from the heritability property of standardness under immersion, with the help of Theorem 2.

The standardness property will be more precisely explained in Sect. 3, at time we will resort to it.

In Sect. 4 we derive standardness of the multidimensional Pascal filtration from standardness of the erased-word filtration. This filtration arises when one observes the evolution of the numbers of occurrences of each letter in the erased-word process.

It is worth mentioning that the filtration of the erased-word process can be viewed as the filtration induced by an ergodic central measure on a Bratteli graph, because there is a recent interest in the study of standardness of such filtrations [4, 12, 13]. This Bratteli graph arises by taking as probability space of the

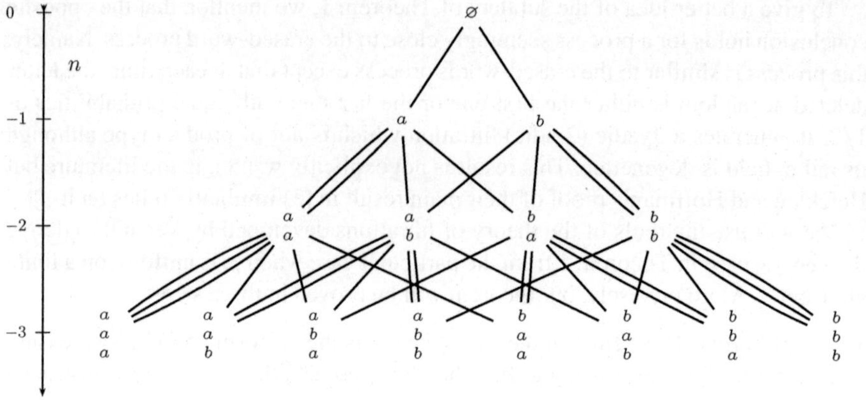

Fig. 2 The Bratteli graph of the erased-word filtration

erased-word process a so-called canonical space in the theory of stochastic process, that is, a space representing the trajectories of the erased-word process, equipped with a probability measure ν such that picking a trajectory according to ν defines the law of the erased-word process. The Bratteli graph is shown on Fig. 2 for a two-letters alphabet $A = \{a, b\}$.

The graph is graded: the vertices at each level $n \leqslant 0$ correspond to the possible states of the random word W_n, in particular it has a unique vertex $\varnothing$ at level 0. The edges connecting a vertex at level $n-1$ to a vertex at level n correspond to the possible values of the eraser η_n. In this way, a trajectory of the erased-word process corresponds to an infinite path in the graph, starting from the root vertex $\varnothing$, and the law of the erased-word process defines a probability measure ν on the set Γ of such paths. This graph is termed as Bratteli because in addition to be graded, each vertex at level $n \leqslant 0$ is connected to at least one vertex at level $n-1$, and each vertex at level $n \leqslant -1$ is connected to at least one vertex at level n. Thus, a trajectory of the erased-words process can be viewed as an infinite path in Γ taken at random according to ν, and the σ-field $\mathcal{F}_n$ is the one generated by the path observed up to level n. We can similarly define the filtration $\mathcal{F}$ for any Bratteli graph and a given probability measure ν on the space Γ of its infinite paths. In our case where ν is the law of the erased-word process, it is an *ergodic central measure* with the terminology of Vershik [12, 13]. The measure ν is said to be *central* when for every given path observed up to level n, the remaining finite piece of the path from the vertex picked at level n to the root vertex $\varnothing$ at level 0 is taken uniformly on the set of all such finite pieces of path (see [4] for more details). This obviously holds in our case because of the poly-adicity of $\mathcal{F}$. The measure ν is said to be *ergodic* when the tail σ-field $\mathcal{F}_{-\infty}$ is degenerate, and as already said before, this property holds in our case as a consequence of standardness but it is not difficult to prove it directly.

2 Discrete Uniform Case

Throughout this section, we assume that A is finite and μ is the uniform probability measure on A. We will prove Theorem 1 in this case by a more or less explicit construction of a generating process of innovations $(\tilde{\eta}_n)_{n\leqslant 0}$. We also set $\kappa = \#A$ and we fix a total order on A. Then we denote $A = \{a_1, \dots, a_\kappa\}$ where a_i is the i-th letter of A.

2.1 *Ingredients of the Construction*

The main ingredient of the construction is the *canonical coupling*. It is very easy to roughly explain what is the canonical coupling with the help of the picture shown on Fig. 3, but it is a bit tedious to write its rigorous definition. Below, we split the description of the canonical coupling into three paragraphs: we firstly define the *canonical word* (the periodic word at right on Fig. 3), then we introduce the notation $N_i^-(w)$ for the number of occurrences of the i-th letter of w to the left of position i, and finally we define the canonical coupling of a word (the permutation shown on Fig. 3, induced by the word at left).

The *canonical word* of length ℓ on A is the word $\tilde{w} \in A^\ell$ in which the letters of A appear in the order and repeat periodically: the i-th letter $\tilde{w}(i)$ of the canonical word $\tilde{w}$ is the r-th letter a_r of A if $i \equiv r \pmod{\kappa}$. For example, the canonical word of length 8 on $A = \{a_1, a_2, a_3\}$, shown at right on Fig. 3, is the word $a_1a_2a_3a_1a_2a_3a_1a_2$.

Given a word w and a position in w, that is to say an index i of one letter of w, we denote by $N_i^-(w) = \sum_{k=1}^{i-1} \mathbb{1}_{\{w(k)=w(i)\}}$ the number of occurrences of the i-th letter of w to the left of position i. If $\tilde{w}$ is the canonical word, then $N_i^-(\tilde{w})$ is the quotient in the Euclidean division of $i-1$ by $\kappa = \#A$.

The *canonical coupling* ϕ_w of a word w on the finite ordered alphabet $A = \{a_1, \dots, a_\kappa\}$ is the permutation illustrated on Fig. 3 and rigorously defined as follows. Let $\tilde{w}$ be the canonical word on A having the same length ℓ as w. The canonical coupling ϕ_w is a permutation of the set $\{1, \dots, \ell\}$ of positions in w. Its construction is made in two steps:

Fig. 3 A canonical coupling

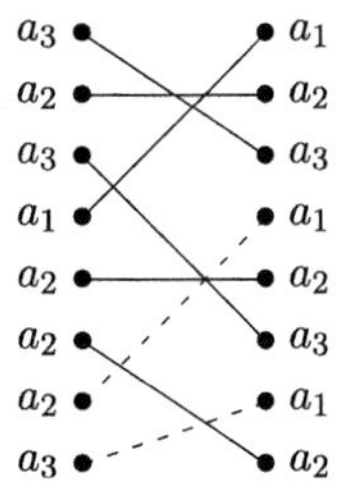

- *First step.* Take a position $i \in \{1, \dots, \ell\}$ in w. If $w(i) = a_r$ then we set $\phi_w(i) = r + \kappa N_i^-(w)$ provided $r + \kappa N_i^-(w) \leqslant \ell$, i.e. when $N_i^-(w)$ is strictly less than the number of occurrences of a_r in the canonical word $\tilde{w}$. Then $w(i) = \tilde{w}(\phi_w(i))$ for all such i. This step is illustrated on Fig. 3 by the solid lines.
- *Second step.* After performing the first step for every possible i, we assign the remaining positions in w to the remaining positions in $\tilde{w}$ in the increasing way. This step is illustrated on Fig. 3 by the dashed lines.

The last ingredient of the construction are the *cascaded permutations*. Consider the canonical coupling $\phi_{W_{n_0}}$ of W_{n_0} for an arbitrary small n_0, providing a correspondence between W_{n_0} and the canonical word of length $|n_0|$ denoted by $\tilde{w}_{n_0}$. Figure 3 is helpful to keep in mind that $\phi_{W_{n_0}}$ represents one-to-one connections between the letters of W_{n_0} and the letters of $\tilde{w}_{n_0}$. In parallel to $(W_{n_0}, W_{n_0+1}, \dots, W_0)$, we construct a sequence of erased words $(W'_{n_0}, W'_{n_0+1}, \dots, W'_0)$, starting from $W'_{n_0} = \tilde{w}_{n_0}$ and erasing one letter at each step as follows. At time $n = n_0+1$, the word W_{n_0+1} is obtained by deleting the η_{n_0+1}-th letter of W_{n_0}, and we delete the corresponding $\phi_{W_{n_0}}(\eta_{n_0+1})$-th letter of the canonical word $\tilde{w}_{n_0} = W'_{n_0}$, thereby obtaining a subword W'_{n_0+1} of W'_{n_0} having the same length as W_{n_0+1}. Thus $\eta'_{n_0+1} := \phi_{W_{n_0}}(\eta_{n_0+1})$ is the first eraser in the parallel erased-word sequence $(\tilde{w}_{n_0}, W'_{n_0+1}, \dots, W'_0)$, and its realization fully determines the realization of the random word W'_{n_0+1}. Moreover, by deleting the connection between η_{n_0+1} and η'_{n_0+1} in the canonical coupling $\phi_{W_{n_0}}$, we obtain a new permutation $\phi_{W_{n_0}, \eta_{n_0+1}}$ representing one-to-one connections between the letters of W_{n_0+1} and the letters of W'_{n_0+1}. Then we continue so on (this is illustrated on Fig. 4):

- At each time $n \in \{n_0 + 1, \dots, -1\}$ we have a word W'_n of length $|n|$ and a permutation $\phi_{W_{n_0}, \eta_{n_0+1}, \dots, \eta_n}$ representing one-to-one connections between the letters of W'_n and the letters of W_n.
- At time $n + 1$ we have a word W'_{n+1} obtained by deleting the η'_{n+1}-th letter of W'_n, where $\eta'_{n+1} = \phi_{W_{n_0}, \eta_{n_0+1}, \dots, \eta_n}(\eta_{n+1})$, and this provides a new permutation $\phi_{W_{n_0}, \eta_{n_0+1}, \dots, \eta_{n+1}}$ connecting the letters of W'_{n+1} to the letters of W_{n+1}.

Figure 4 illustrates the "cascaded" permutations $\phi_{W_{n_0}, \eta_{n_0+1}, \dots, \eta_n}$ initiated at time $n_0 = -5$ by the canonical coupling $\phi_{W_{n_0}}$ and sequentially obtained from the erasers η_{n_0+1} and η_{n_0+2}. By this way the random word W'_n is measurable with respect to $\sigma(\eta'_{n_0} + 1, \dots, \eta'_n)$. Lemma 1 in the next section shows that the η'_n are innovations

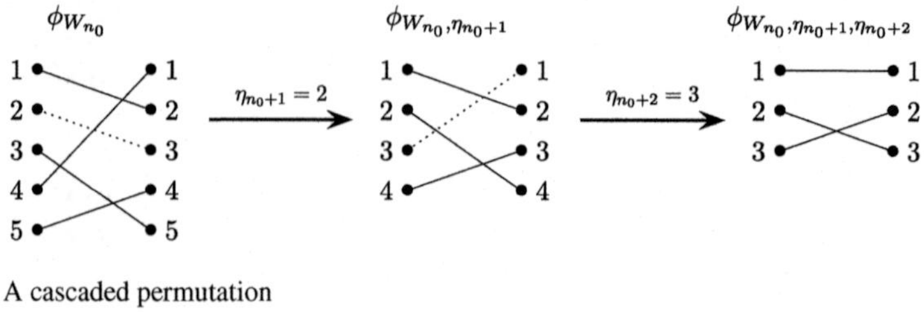

Fig. 4 A cascaded permutation

of $\mathcal{F}$ and $W_n = W'_n$ with probability as high as desired when the construction starts from an arbitrary small n_0, and this will allow us to construct a generating process of innovations.

2.2 Key Lemma

A generating process of innovations of $\mathcal{F}$ will be derived from Lemma 1 below. The following construction, already sketched in the previous section, is used in the statement of this lemma. Let f_n be the function from $A^{|n|+1} \times \{1, \ldots, |n| + 1\}$ to $A^{|n|}$ defined by setting $f_n(w, e)$ to the word obtained by deleting the e-th letter of w. This function represents the update from W_{n-1} to W_n because of the equality $W_n = f_n(W_{n-1}, \eta_n)$. Now, consider a random vector $(\eta'_{n_0+1}, \ldots, \eta'_0)$ having the same law as $(\eta_{n_0+1}, \ldots, \eta_0)$. Then define a Markov process $\big(Y_n(n_0)\big)_{n_0 \leqslant n \leqslant 0}$ by the initial condition $Y_{n_0}(n_0) = \tilde{w}_{n_0}$ (the canonical word of length $|n_0|$), and by the inductive relation

$$Y_{n+1}(n_0) := f_{n+1}\big(Y_n(n_0), \eta'_{n+1}\big).$$

Setting $W'_n = Y_n(n_0)$, the process $\big((W'_{n_0+1}, \eta'_{n_0+1}), \ldots, (W'_0, \eta'_0)\big)$ has the same distribution as the process $\big((W_{n_0+1}, \eta_{n_0+1}), \ldots, (W_0, \eta_0)\big)$ conditionally to $W_{n_0} = \tilde{w}_{n_0}$. Lemma 1 below shows that $W_n = W'_n$ with probability as high as desired when the construction starts from an arbitrary small n_0 and when we use the innovations η'_n encountered in the previous section when we defined the cascaded permutations.

Lemma 1 *Let $n_0 < 0$ be an integer and $\phi_{W_{n_0}}$ be the canonical coupling of W_{n_0}. For every integer $m \in [n_0 + 1, -1]$ let $\phi_{W_{n_0}, \eta_{n_0+1}, \ldots, \eta_m}$ be, as explained in Sect. 2.1, the cascaded permutation initiated by $\phi_{W_{n_0}}$ and sequentially obtained from $\eta_{n_0}, \ldots, \eta_m$, and define $\eta'_{m+1} = \phi_{W_{n_0}, \eta_{n_0+1}, \ldots, \eta_m}(\eta_{m+1})$ for every $m \in [n_0, -1]$. Then $(\eta'_n)_{n_0+1 \leqslant n \leqslant 0}$ has the same law as $(\eta_n)_{n_0+1 \leqslant n \leqslant 0}$ and each η'_n is, just as η_n, an innovation of $\mathcal{F}$, that is, η'_n is independent of $\mathcal{F}_{n-1}$ and $\mathcal{F}_n = \mathcal{F}_{n-1} \vee \sigma(\eta'_n)$. Moreover, with the notations above,*

$$\mathbb{P}\big(W_n \neq Y_n(n_0)\big) \to 0 \quad \text{as } n_0 \to -\infty$$

for every $n \leqslant 0$, where $\tilde{w}_{n_0}$ is the canonical word of length $|n_0|$.

Proof It is easy to check that η'_n is an innovation as any other $\mathcal{F}_{n-1}$-measurable random permutation of η_n. The word W_n is a subword of W_{n_0}, and we denote by $Q_{n_0,n} \subset \{1, \ldots, |n_0|\}$ the set of positions in the word W_{n_0} forming its subword W_n. Moreover, by construction of the cascaded permutations, $W'_n = Y_n(n_0)$ is a subword of $\tilde{w}_{n_0}$ and the set of positions in $\tilde{w}_{n_0}$ forming W'_n is the image of $Q_{n_0,n}$ by the canonical coupling $\phi_{W_{n_0}}$. We can check the intuitively clear fact that $Q_{n_0,n}$ is independent of W_{n_0} and is uniform on the subsets of $\{1, \ldots, |n_0|\}$ having size

$|n|$. Indeed, there is a bijective correspondence between the erasers $(\eta_{n_0+1}, \ldots, \eta_n)$ and the $(n - n_0)$-tuple listing the successive positions in the word W_{n_0} of the letters deleted at times $n_0 + 1, \ldots, n$. The set of all theses positions is exactly the complement of $Q_{n_0,n}$ in $\{1, \ldots, |n_0|\}$. Therefore there is a correspondence between $(\eta_{n_0+1}, \ldots, \eta_n)$ and $Q_{n_0,n}$, and consequently there is independence between $Q_{n_0,n}$ and W_{n_0}. Moreover, this correspondence between $(\eta_{n_0+1}, \ldots, \eta_n)$ and $Q_{n_0,n}$ is surjective and $(n-n_0)!$ to one, wherefrom follows the uniformity of the law of $Q_{n_0,n}$.

Now, to abbreviate notations, set $p = |n_0|$, $q = |n|$, $Q = Q_{n_0,n}$ and $W = W_{n_0}$. Thus we have seen that Q is a random variable independent of W and uniformly distributed on the subsets of $\{1, \ldots, p\}$ having size q. With these abbreviated notations, the main statement of the lemma is rephrased by

$$\pi(p, q) := \mathbb{P}\big(W_{|Q} = \tilde{w}_{|\phi_W(Q)}\big) \longrightarrow 1 \quad \text{as } p \to +\infty,$$

where ϕ_w is the canonical coupling of a word w and $\tilde{w}$ is the canonical word of length p, and we use the notation $w_{|J}$ to denote the subword of a word w obtained by keeping only those of its letters whose indices belong to the subset J.

Recall the notation $\kappa = \#A$. To show that $\pi(p, q) \to 1$ when $p \to \infty$, we introduce the three events

$$\begin{aligned}
E_1 &= \big\{\max(Q) \leqslant p - p^{3/4} - \kappa\big\}, \\
E_2 &= \big\{\forall (i, j) \in Q^2, i = j \text{ or } |i - j| \geqslant 3p^{3/4}\big\}, \\
E_3 &= \big\{\forall i \in Q, i - 1 - p^{3/4} \leqslant \kappa N_i^-(W) \leqslant i - 1 + p^{3/4}\big\},
\end{aligned}$$

and we are going to show that

$$E_1 \cap E_2 \cap E_3 \subset \big\{W_{|Q} = \tilde{w}_{|\phi_W(Q)}\big\}$$

if p is sufficiently large, and

$$\mathbb{P}(E_1^c \cup E_2^c \cup E_3^c) \xrightarrow[p\to\infty]{} 0.$$

We firstly show the inclusion. As a first step, we show that $W(i) = \tilde{w}\big(\phi_W(i)\big)$ for every $i \in Q$ on the event $E_1 \cap E_3$. Consider $i \in Q$ and assume that $W(i) = a_r$. On the event E_3,

$$r + \kappa N_i^-(W) \leqslant \kappa + i - 1 + p^{3/4},$$

and $\kappa + i - 1 + p^{3/4} \leqslant p$ on the event E_1. Thus, by definition of the canonical coupling, $\phi_W(i) = r + \kappa N_i^-(W)$ on the event $E_1 \cap E_3$ and $W(i) = \tilde{w}\big(\phi_W(i)\big)$ for every $i \in Q$. Moreover,

$$1 + (i - 1 - p^{3/4}) \leqslant r + \kappa N_i^-(W) \leqslant \kappa + (i - 1 + p^{3/4}),$$

on the event E_3, therefore ϕ_W satisfies the following property on $E_1 \cap E_3$:

$$\forall i \in Q, \quad \left|i - \phi_W(i)\right| \leqslant \kappa - 1 + p^{3/4}.$$

Consequently, if we are on $E_1 \cap E_2 \cap E_3$ and if p is sufficiently large so that $3p^{3/4} \geqslant 2\big(\kappa + p^{3/4}\big)$, then the restriction of ϕ_W to Q is increasing, and finally $W_{|Q} = \tilde{w}_{|\phi_W(Q)}$, as desired.

It remains to show that $\mathbb{P}(E_1^c \cup E_2^c \cup E_3^c) \xrightarrow[p\to\infty]{} 0$. The following upper bound of $\mathbb{P}(E_1^c)$ is easily obtained:

$$\mathbb{P}(E_1^c) \leqslant q\frac{p^{3/4} + \kappa + 1}{p} \xrightarrow[p\to\infty]{} 0.$$

The following upper bound of $\mathbb{P}(E_2^c)$ is obtained by sampling the elements of Q without replacement:

$$\begin{aligned}\mathbb{P}(E_2^c) &\leqslant \frac{6p^{3/4}+2}{p-1} + \frac{2(6p^{3/4}+2)}{p-2} + \cdots + \frac{(q-1)(6p^{3/4}+2)}{p-q+1}\\ &\leqslant \frac{(q-1)^2(6p^{3/4}+2)}{p-q+1} \xrightarrow[p\to\infty]{} 0.\end{aligned}$$

To find an upper bound of $\mathbb{P}(E_3^c)$, we call I_k the k-th element of Q for every $k \in \{1,\ldots,q\}$. Conditionally to $I_k = i$, the number of occurrences $N_{I_k}^-(W)$ has the binomial distribution with size $i-1$ and probability of success $1/\kappa$, because of the independence between W and Q. Therefore, using Bienaymé-Chebyshev's inequality,

$$\begin{aligned}\mathbb{P}\left(\left|N_{I_k}^-(W) - \frac{I_k - 1}{\kappa}\right| > \frac{p^{3/4}}{\kappa} \;\middle|\; I_k = i\right) &\leqslant \left(\frac{\kappa}{p^{3/4}}\right)^2 (i-1)\frac{1}{\kappa}\left(1 - \frac{1}{\kappa}\right)\\ &< \frac{p\kappa}{p^{3/2}} = \frac{\kappa}{\sqrt{p}},\end{aligned}$$

and this being true for every $i \in \{1,\ldots,p\}$, one also has

$$\mathbb{P}\left(\left|N_{I_k}^-(W) - \frac{I_k - 1}{\kappa}\right| > \frac{p^{3/4}}{\kappa}\right) < \frac{\kappa}{\sqrt{p}}.$$

By summing this equality over all $k \in \{1,\ldots,q\}$,

$$\mathbb{P}(E_3^c) < \frac{q\kappa}{\sqrt{p}} \xrightarrow[p\to\infty]{} 0,$$

and the proof is over. □

2.3 *Proof of Standardness*

We finish to prove that the erased-word process $(W_n, \eta_n)_{n\leqslant 0}$ generates a filtration of product type in the discrete uniform case. This can be quickly proved from Lemma 1 with the help of *Vershik's first level criterion* and Proposition 2.22 in [5]. Lemma 1 says that each random variable W_n satisfies Vershik's first level criterion. Since $(\eta_n)_{n\leqslant 0}$ is a process of innovations of $\mathcal{F}$, and since the σ-fields $\sigma(W_n, \eta_{n+1}, \ldots, \eta_0)$ increase to $\mathcal{F}_0$ as $n \to -\infty$, Proposition 2.22 in [5] ensures that $\mathcal{F}$ satisfies Vershik's first level criterion, and then $\mathcal{F}$ is of product type by Vershik's theorem (Theorem 2.25 in [5])

But we have not stated Vershik's first level criterion in the present paper, and we can give a self-contained proof that $\mathcal{F}$ is of product type by constructing a generating process of innovations. First recall that $W'_n = Y_n(n_0)$ in Lemma 1 is measurable with respect to $\sigma(\eta'_{n_0+1}, \ldots, \eta'_n)$. Then, given a sequence $(\delta_k)_{k\leqslant 0}$ of real numbers $\delta_k > 0$ satisfying $\delta_k \to 0$ as $k \to -\infty$, recursively applying Lemma 1 provides a strictly increasing sequence $(n_k)_{k\leqslant 0}$ of integers with $n_0 = -1$ and an innovation process $(\tilde{\eta}_n)_{n\leqslant 0}$ such that:

(i) $(\tilde{\eta}_{n_k+1}, \ldots, \tilde{\eta}_n) = \tau_{W_{n_k},n}(\eta_{n_k+1}, \ldots, \eta_n)$ for every $k < 0$ and every integer $n \in [n_k + 1, 0]$, where each $\tau_{w,n}$ is a permutation of $\{1, \ldots, |n_k|\} \times \cdots \times \{1, \ldots, |n| + 1\}$;
(ii) for every $k \leqslant 0$ there is a random word $\tilde{W}_{n_k}$ measurable with respect to $\sigma(\tilde{\eta}_{n_{k-1}+1}, \ldots, \tilde{\eta}_{n_k})$ and satisfying $\mathbb{P}(W_{n_k} \neq \tilde{W}_{n_k}) < \delta_k$.

Now we check that $(\tilde{\eta}_n)_{n\leqslant 0}$ generates $\mathcal{F}$. It suffices to construct, for each $n \leqslant 0$ and every $\delta > 0$, a pair of random variables $(\hat{W}_n, \hat{\eta}_n)$ that is measurable with respect to $\sigma(\ldots, \tilde{\eta}_{n-1}, \tilde{\eta}_n)$ and that satisfies $\mathbb{P}\big((W_n, \eta_n) \neq (\hat{W}_n, \hat{\eta}_n)\big) < \delta$. To do so, let k be sufficiently small in order that $\delta_k < \delta$ and $n_k < n$. Then define

$$(\hat{\eta}_{n_k+1}, \ldots, \hat{\eta}_n) = \tau^{-1}_{\tilde{W}_{n_k},n}(\tilde{\eta}_{n_k+1}, \ldots, \tilde{\eta}_n)$$

and define $\hat{W}_m$ for $m \in [n_k, n]$ by initially setting $\hat{W}_{n_k} = \tilde{W}_{n_k}$ and recursively setting $\hat{W}_{m+1} = f_{m+1}(\hat{W}_m, \hat{\eta}_{m+1})$ (the functions f_m were introduced before Lemma 1). Now, $(\hat{\eta}_{n_k+1}, \ldots, \hat{\eta}_n) = (\eta_{n_k+1}, \ldots, \eta_n)$ on the event $\{W_{n_k} = \tilde{W}_{n_k}\}$, hence $\hat{W}_n = W_n$ on this event too.

3 Vershikian Tools and the General Case

Here we finish the proof of Theorem 1 by following steps 2 and 3 announced in the introduction.

Consider the erased-word process $(W_n, \eta_n)_{n\leqslant 0}$ in the case when $A = [0, 1]$ and μ is the Lebesgue measure on A, and denote by $\mathcal{G}$ the filtration it generates. In order to prove that $\mathcal{G}$ is of product type (step 2), the idea consists in approximating this process by an erased-word process on a finite alphabet with equiprobable letters, known to generate a filtration of product type by the previous section (step 1). Then the tools of Vershik's theory of filtrations will allow to conclude.

For every integer $k \geqslant 1$, let $f_k: A \to A$ be the function defined by $f_k(x) = 2^{-k}\lfloor 2^k x \rfloor$. Then f_k sends the Lebesgue measure μ to the uniform probability measure on the finite alphabet $A_k := \left\{0, \frac{1}{2^k}, \ldots, \frac{2^k-1}{2^k}\right\}$. Applying f_k to each letter of a word w on A gives a word on A_k denoted by $f_k(w)$. Then the process $\left(f_k(W_n), \eta_n\right)_{n\leqslant 0}$ is an erased-word process generating a filtration of product type by Sect. 2. Moreover, denoting by $\mathcal{G}^k$ this filtration, the sequence of σ-fields $(\mathcal{G}_0^k)_{k\geqslant 1}$ is increasing and $\bigvee_{k=1}^{+\infty}\mathcal{G}_0^k = \mathcal{G}_0$

We give two ways to prove that $\mathcal{G}$ is of product type from the fact that each $\mathcal{G}^k$ is of product type. The first one uses *Vershik's first level criterion*, as the proof of the similar result 2.45 in [5] about the split-word processes. Vershik's first level criterion is known to be equivalent to productness (see [5]), hence we know it is satisfied by each filtration $\mathcal{G}^k$. Moreover, every innovation η'_n of $\mathcal{G}^k$ at time n is also an innovation of $\mathcal{G}$, because by Lemma 2.4 in [5] it can be written $\eta'_n = \Phi(\eta_n)$ where Φ is a $\mathcal{G}^k_{n-1}$-measurable random bijection from $\{1, \ldots, |n| + 1\}$ to some finite set of size $|n| + 1$. Thus, every random variable in $\cup_k L^1(\mathcal{G}_0^k)$ belongs to the set of random variables in $L^1(\mathcal{G}_0)$ satisfying Vershik's first level criterion with respect to $\mathcal{G}$. But this set is closed in $L^1(\mathcal{G}_0)$ by Proposition 2.7 in [5], consequently $\mathcal{G}$ satisfies Vershik's first level criterion.

The second proof we give relies on a more general result stated in our original Proposition 1 below. As we have seen, the key point in the previous proof is the fact that every innovation of $\mathcal{G}^k$ is also an innovation of $\mathcal{G}$ and it is very specific to our situation. This fact implies that each $\mathcal{G}^k$ is immersed in $\mathcal{G}$ (see Lemma 1.6 in [5]; that means here that the process $\left(f_k(W_n), \eta_n\right)_{n\leqslant 0}$ is Markovian with respect to $\mathcal{G}$), and this is the key point of the second proof.

Proposition 1 *Let $\mathcal{F}$ be a filtration. If there exists a sequence of Vershikian filtrations $(\mathcal{F}^k)_{k\geqslant 1}$ such that the sequence of σ-fields $(\mathcal{F}_0^k)_{k\geqslant 1}$ is increasing and satisfies $\bigvee_{k=1}^{+\infty}\mathcal{F}_0^k = \mathcal{F}_0$, and if each $\mathcal{F}^k$ is immersed in $\mathcal{F}$, then $\mathcal{F}$ is Vershikian.*

Proof Saying that $\mathcal{F}^k$ is Vershikian means by definition that the final σ-field $\mathcal{F}_0^k$ is Vershikian with respect to the filtration $\mathcal{F}^k$, but thanks to Lemma 4.1 in [6], this tantamounts to say that $\mathcal{F}_0^k$ is Vershikian with respect to the filtration $\mathcal{F}$ because of the immersion of $\mathcal{F}^k$ in $\mathcal{F}$. Now, because of $\bigvee_{k=1}^{+\infty}\mathcal{F}_0^k = \mathcal{F}_0$, Lemma 4.2 in [6] (closedness of the set of Vershikian random variables) shows that $\mathcal{F}_0$ is Vershikian with respect to $\mathcal{F}$, that is to say $\mathcal{F}$ is Vershikian. □

Thus, we know that $\mathcal{G}$ is standard by Proposition 1 and by the equivalence between standardness and the Vershik property. By Theorem 2, we conclude that $\mathcal{G}$ is of product type.

Step 2 of the proof of Theorem 1 is achieved. Step 3 (the general case) is easily achieved with the help of Theorem 2. Consider an arbitrary Lebesgue alphabet (A, μ) and take a measurable function f: $[0, 1] \to A$ sending the Lebesgue measure to μ. Then the process $\big(f(W_n), \eta_n\big)_{n\leqslant 0}$ is the erased-word process on (A, μ), and the filtration it generates is immersed in $\mathcal{G}$. We conclude that this filtration is of product type by using Theorem 2 and the heritability of standardness under immersion, an immediate consequence of the definition of standardness (see [2] or [5]). Now step 3 is achieved and the proof of Theorem 1 is over.

4 Standardness of the Multidimensional Pascal Filtration

The d-dimensional Pascal filtration is introduced in [4]. It is the filtration generated by the Markov chain $(V_n)_{n\leqslant 0}$ whose distribution depends on a given probability vector $(\theta_1, \ldots, \theta_d)$, where $d \geqslant 2$ is a finite integer or $d = \infty$, and is defined as follows:

- *(instantaneous distributions)* the random variable V_n has the multinomial distribution on

$$\mathbb{V}_n^d = \big\{v \in \mathbb{N}^d \mid v(1) + \cdots + v(d) = |n|\big\}$$

  with success probability vector $(\theta_1, \ldots, \theta_d)$;
- *(Markov transitions)* the transition laws from n to $n+1$ are

$$\mathcal{L}(V_{n+1} \mid V_n = v) = \sum_{i=1}^{d} \frac{v(i)}{|n|}\delta_{v-e_i}, \tag{1}$$

  where e_i is the vector in $\mathbb{R}^d$ whose i-th term is 1 and all the other ones are 0. In other words, given $V_n = \big(v(1), \ldots, v(d)\big)$, coordinate i is picked at random with probability $\frac{v(i)}{|n|}$ and V_{n+1} is obtained by subtracting 1 to this coordinate.

The case when $d = 2$ is illustrated on Figs. 5 and 6 (with p playing the role of θ_1), and the case when $d = 3$ is illustrated on Fig. 7. We refer to [4] for more detailed explanations.

It has been shown in [4] that the filtration generated by the d-dimensional Pascal random walk is standard for any d and any $(\theta_1, \ldots, \theta_d)$. This result is straightforwardly derived from our Theorem 1 and from the heritability property of standardness under immersion, already mentioned in the introduction and in Sect. 3. Indeed, taking the erased-word process $(W_n, \eta_n)_{n\leqslant 0}$ on an alphabet A with d letters and equipped with the probability μ whose masses are given by the probability vector $(\theta_1, \ldots, \theta_d)$, and defining the function $f_n\colon A^{|n|} \to \mathbb{V}_n$ as the one returning the list of the numbers of occurrences of each letter of A in a given word of length $|n|$, then the process $\big(f_n(W_n), \eta_n\big)_{n\leqslant 0}$ is the d-dimensional Pascal random walk defined by the probability vector $(\theta_1, \ldots, \theta_d)$, and the filtration $\mathcal{F}$ it generates is immersed in the filtration $\mathcal{G}$ generated by the erased-word process $(W_n, \eta_n)_{n\leqslant 0}$ because $\big(f_n(W_n), \eta_n\big)_{n\leqslant 0}$ is Markovian with respect to $\mathcal{G}$. Then standardness of the

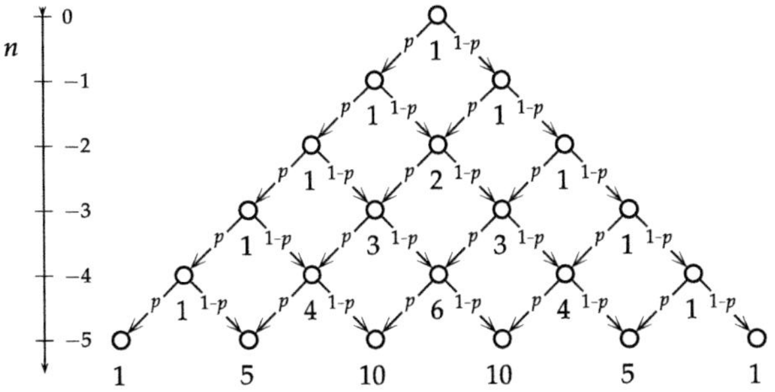

Fig. 5 2-dimensional Pascal random walk, directed from $n = 0$ to $n = -\infty$

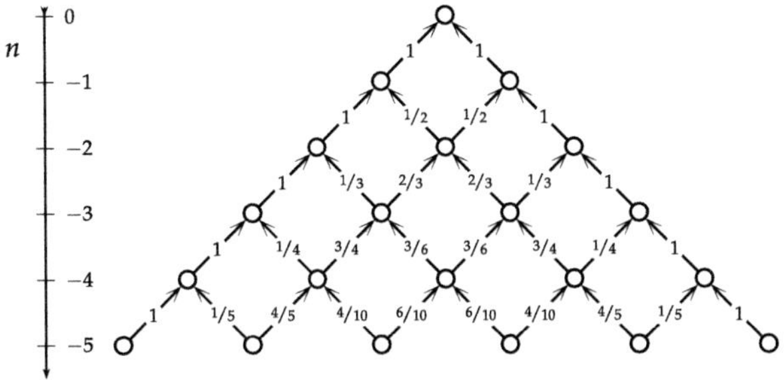

Fig. 6 2-dimensional Pascal random walk, directed from $n = -\infty$ to $n = 0$

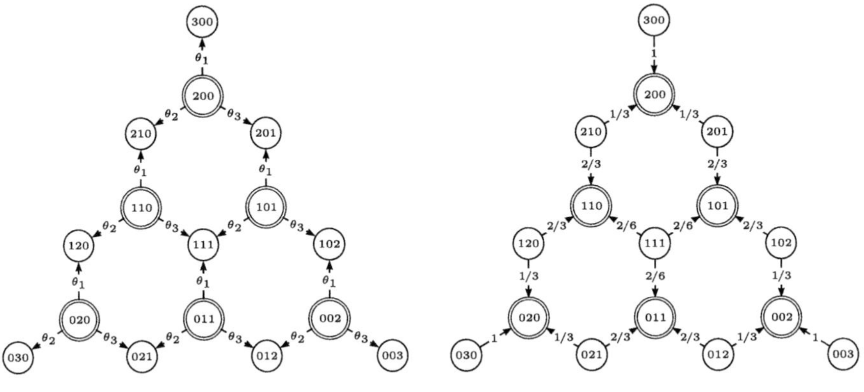

Fig. 7 Step in the 3-dimensional Pascal random walk: from $n = -2$ to $n = -3$ (*left*), and from $n = -3$ to $n = -2$ (*right*)

d-dimensional Pascal filtration $\mathcal{F}$ results from Theorem 1, from the obvious fact that standardness holds for filtrations of product type, and from the heritability property of standardness under immersion.

Acknowledgements I thank Christophe Leuridan for his helpful comments on the first version of this paper, and the referee for his/her meticulous review which led to substantial improvements of the article.

References

1. M. Barlow, M. Émery, F. Knight, S. Song, M. Yor, Autour d'un théorème de Tsirelson sur des filtrations browniennes et non-browniennes, in *Séminaire de probabilités XXXII*. Springer Lectures Notes in Mathematics, vol. 1686 (Springer, Berlin, 1998), pp. 264–305
2. M. Émery, W. Schachermayer, On Vershik's standardness criterion and Tsirelson's notion of cosiness, in *Séminaire de Probabilités XXXV*. Springer Lectures Notes in Mathematics, vol. 1755 (Springer, Berlin, 2001), pp. 265–305
3. D. Heicklen, C. Hoffman, $[T, T^{-1}]$ is not standard. Ergodic Theory Dyn. Syst. **18**, 875–878 (1998)
4. É. Janvresse, S. Laurent, T. de la Rue, Standardness of the filtration of a monotonic Markov process. arXiv:1501.02166 (2015). Markov Processes and Related Fields, to appear
5. S. Laurent, On standardness and I-cosiness, in *Séminaire de Probabilités XLIII*. Springer Lecture Notes in Mathematics, vol. 2006 (Springer, Berlin, 2010), pp. 127–186
6. S. Laurent, On Vershikian and I-cosy random variables and filtrations. Teoriya Veroyatnostei i ee Primeneniya **55**, 104–132 (2010). Also published in: Theory Probab. Appl. **55**, 54–76 (2011)
7. S. Laurent, Vershik's intermediate level standardness criterion and the scale of an automorphism, in *Séminaire de Probabilités XLV*. Springer Lecture Notes in Mathematics, vol. 2078 (Springer, Berlin, 2013), pp. 123–139
8. A.M. Vershik, Theorem on lacunary isomorphisms of monotonic sequences of partitions. Funktsional'nyi Analiz i Ego Prilozheniya **2**(3), 17–21 (1968). English translation: Functional Analysis and Its Applications **2**(3), 200–203 (1968)
9. A.M. Vershik, Decreasing sequences of measurable partitions, and their applications. Dokl. Akad. Nauk SSSR **193**, 748–751 (1970). English translation: Soviet Math. Dokl. **11**, 1007–1011 (1970)
10. A.M. Vershik, Approximation in measure theory (in Russian). Ph.D. Dissertation, Leningrad University (1973). [expanded and updated version: [11]]
11. A.M. Vershik, The theory of decreasing sequences of measurable partitions (in Russian). Algebra i Analiz **6**(4), 1–68 (1994). English translation: St. Petersburg Mathematical Journal **6**(4), 705–761 (1995)
12. A.M. Vershik, The problem of describing central measures on the path spaces of graded graphs. arXiv:1408.3291 (2013)
13. A.M. Vershik, Intrinsic metric on graded graphs, standardness, and invariant measures. Zapiski Nauchn. Semin. POMI **421**, 58–67 (2014)

Projections, Pseudo-Stopping Times and the Immersion Property

Anna Aksamit and Libo Li

Abstract Given two filtrations $\mathbb{F} \subset \mathbb{G}$, we study under which conditions the $\mathbb{F}$-optional projection and the $\mathbb{F}$-dual optional projection coincide for the class of $\mathbb{G}$-optional processes with integrable variation. It turns out that this property is equivalent to the immersion property for $\mathbb{F}$ and $\mathbb{G}$, that is every $\mathbb{F}$-local martingale is a $\mathbb{G}$-local martingale, which, equivalently, may be characterised using the class of $\mathbb{F}$-pseudo-stopping times. We also show that every $\mathbb{G}$-stopping time can be decomposed into the minimum of two barrier hitting times.

1 Introduction

The study of pseudo-stopping times started in the paper by Williams [11]. The author describes there an example of a non-stopping time τ which has the optional stopping property, namely, for every uniformly integrable martingale M, $\mathbb{E}(M_\tau) = \mathbb{E}(M_0)$. Let us recall this example here. Let B be a Brownian motion and define:

$$T_1 := \inf\{t : B_t = 1\} \quad \text{and} \quad \sigma := \sup\{t \leq T_1 : B_t = 0\}.$$

Therefore σ is the last zero of the process B before it reaches one. Let τ be the time of the maximum of B over $[0, \sigma]$, that is

$$\tau := \sup\{t < \sigma : B_t = B_t^*\} \quad \text{with} \quad B_t^* := \sup_{s \leq t} B_s.$$

Then, as shown in [11], τ has the optional stopping property. Such random times were then called pseudo-stopping times and further studied by Nikeghbali and Yor

A. Aksamit (✉)
Mathematical Institute and Oxford Man Institute, University of Oxford, Oxford, UK
e-mail: anna.aksamit@maths.ox.ac.uk

L. Li
Department of Mathematics and Statistics, University of New South Wales, Sydney, NSW, Australia
e-mail: libo.li@unsw.edu.au

C. Donati-Martin et al. (eds.), *Séminaire de Probabilités XLVIII*, Lecture Notes in Mathematics 2168, DOI 10.1007/978-3-319-44465-9_17

in [10]. In particular, it was shown in [10] that a finite random time τ is a pseudo-stopping time if and only if the optional projection of the process $\mathbb{1}_{[\tau,\infty[}$ coincides with its dual optional projection. We want to study the conditions under which the later property holds not only for $\mathbb{1}_{[\tau,\infty[}$ but for a larger class of processes. In other words, the main motivation of this work is to better understand the property that the optional projection is equal to the dual optional projection for processes of integrable variation, which is not true in general.

We work on a filtered probability space $(\Omega, \mathscr{A}, \mathbb{F}, \mathbb{P})$, where $\mathbb{F} := (\mathscr{F}_t)_{t\geq 0}$ denotes a filtration satisfying the usual conditions and we set $\mathscr{F}_\infty := \bigvee_{t\geq 0} \mathscr{F}_t \subset \mathscr{A}$. A process that is not necessarily adapted to the filtration $\mathbb{F}$ is said to be *raw*. As convention, for any martingale, we work always with its càdlàg modification, while for any random process $(X_t)_{t\geq 0}$, we set $X_{0-} = 0$ and $X_\infty = \lim_{t\to\infty} X_t$ a.s, if it exists.

Let $\mathbb{G} := (\mathscr{G}_t)_{t\geq 0}$ be another filtration such that $\mathbb{F} \subset \mathbb{G}$, that is for each $t \geq 0$, $\mathscr{F}_t \subset \mathscr{G}_t$.

The aim is to study under which conditions the $\mathbb{F}$-optional projection and the $\mathbb{F}$-dual optional projection coincide for the class of $\mathbb{G}$-optional processes with integrable variation. It turns out that this is closely related to the *immersion property* from the theory of enlargement of filtrations. A filtration $\mathbb{F}$ is said to be *immersed* in $\mathbb{G}$ and we write $\mathbb{F} \hookrightarrow \mathbb{G}$ if every $\mathbb{F}$-local martingale is a $\mathbb{G}$-local martingale. Often the immersion property is called the *hypothesis* $(\mathscr{H})$ in the literature. We refer the reader to Brémaud and Yor [4] for further discussions and other conditions equivalent to the immersion property.

The results of this paper are also motivated by the study of the converse implications to the following known observations in the literature. Let us define a filtration $\mathbb{F}^\tau := (\mathscr{F}_t^\tau)_{t\geq 0}$ as the progressive enlargement of $\mathbb{F}$ with τ, i.e. the smallest right-continuous filtration containing $\mathbb{F}$ such that τ is a stopping time, that is

$$\mathscr{F}_t^\tau := \bigcap_{s>t} (\mathscr{F}_s \vee \sigma(\tau \wedge s)).$$

In the reduced form approach to credit risk modelling (see Bielecki et al. [3]), given a filtration $\mathbb{F}$, a popular way to model the default time τ is to use a barrier hitting time of an $\mathbb{F}$-adapted increasing process and an independent barrier. It is known that a random time constructed in this fashion has the property that the filtration $\mathbb{F}$ is immersed in $\mathbb{F}^\tau$. It is also known that the property that $\mathbb{F} \hookrightarrow \mathbb{F}^\tau$ implies that every $\mathbb{F}^\tau$-stopping time is an $\mathbb{F}$-pseudo stopping time. In fact, the authors of [10] have observed that, given two filtrations $\mathbb{F}$ and $\mathbb{G}$ such that $\mathbb{F}$ is immersed in $\mathbb{G}$, every $\mathbb{G}$-stopping time is an $\mathbb{F}$-pseudo-stopping time.

The main contributions of this work are condition (v) of Theorems 1 and 2. In Theorem 2 we show that the converse of the observation made by the authors of [10] is true: if every $\mathbb{G}$-stopping time is $\mathbb{F}$-pseudo-stopping time then $\mathbb{F}$ is immersed in $\mathbb{G}$. Hence, it provides an alternative characterization of the immersion property based on pseudo-stopping times. Furthermore in Theorem 2 we show that the immersion property for $\mathbb{F}$ and $\mathbb{G}$ is equivalent to the property that the $\mathbb{F}$-optional projection and $\mathbb{F}$-dual optional projection coincide for the class of $\mathbb{G}$-optional processes with integrable variation.

As an application of Theorem 1 (v), which gives another equivalent characterisation for pseudo-stopping times, and Theorem 2, we provide in Proposition 1 an alternative proof to a result regarding the immersion property and the progressive enlargement with honest times. The advantage of our method is that we do not use specific structures of the progressive enlargement and the characterization of predictable sets as done in Jeulin [9].

As another application, assuming that $\mathbb{F}$ is immersed in $\mathbb{G}$, we show in Theorem 3 that every $\mathbb{G}$-stopping time can be written as the minimum of two $\mathbb{G}$-stopping times, one of which is a barrier hitting time of an $\mathbb{F}$-adapted increasing process, where the barrier is *'almost'* independent, and the other is an $\mathbb{F}$-pseudo-stopping time whose graph is contained in the union of the graphs of a family of $\mathbb{F}$-stopping times.

2 Characterisation of Pseudo-Stopping Times

The main object of interest in this section is the class of pseudo-stopping times. We start with recalling the definition of pseudo-stopping times from [10], with a slight modification, that is the random time is allowed to take the value infinity.

Definition 1 A random time τ is an $\mathbb{F}$-*pseudo-stopping time* if for every uniformly integrable $\mathbb{F}$-martingale M, we have $\mathbb{E}(M_\tau) = \mathbb{E}(M_0)$.

The main tools used in this study are the (dual) optional projections onto the filtration $\mathbb{F}$. We record here some known results from the general theory of stochastic processes. For more details on the theory the reader is referred to He et al. [7] or Jacod and Shiryaev [8] and for specific results from the theory of enlargement of filtrations to Jeulin [9].

For any locally integrable variation process V, we denote the $\mathbb{F}$-optional projection of V by ${}^oV$ and the $\mathbb{F}$-dual optional projection of V by V^o. It is known that the process $N^V := {}^oV - V^o$ is a uniformly integrable $\mathbb{F}$-martingale with $N_0^V = 0$ and ${}^o(\Delta V) = \Delta V^o$.

We specialize the above notions to the study of random times. For an arbitrary random time τ, we set $A := \mathbb{1}_{[\![\tau,\infty[\![}$ and define

- the supermartingale Z associated with τ, $Z := {}^o(\mathbb{1}_{[\![0,\tau[\![}) = 1 - {}^oA$,
- the supermartingale $\widetilde{Z}$ associated with τ, $\widetilde{Z} := {}^o(\mathbb{1}_{[\![0,\tau]\!]}) = 1 - {}^o(A_-)$,
- the martingale $m := 1 - ({}^oA - A^o)$.

These processes are linked through the following relationships:

$$Z = m - A^o \quad \text{and} \quad \widetilde{Z} = m - A^o_-.$$

We present in Theorem 1 an extension of Theorem 1 from Nikeghbali and Yor [10]. We extend their result in two directions. Firstly, we allow for non-finite pseudo-stopping times. Secondly, Theorem 1 from [10] states that if either all $\mathbb{F}$-martingales are continuous or the random time τ avoids all finite $\mathbb{F}$-stopping times, i.e. $\Delta A^o = 0$, then the random time τ is an $\mathbb{F}$-pseudo-stopping time if and only if the process Z is a decreasing $\mathbb{F}$-predictable process. We will remove these additional assumptions and present another equivalent characterization based on the process $\widetilde{Z}$ instead of Z in condition (v) of Theorem 1. We point out that the equivalence of condition (v) of Theorem 1 is one of the key results of this paper.

Before presenting Theorem 1 let us give a motivating example of a random time which is not a pseudo-stopping time and $Z = \widetilde{Z}$ is decreasing but not predictable. It illustrates the importance of the *càglàd* property in condition (v) of Theorem 1.

Example 1 Let N be a Poisson process with intensity λ and jump times $(T_n)_n$. Consider the random time $\tau = \frac{1}{2}(T_1 + T_2)$. Then we obtain

$$\mathbb{E}(N_{\tau\wedge 1} - \lambda(\tau \wedge 1)) < \mathbb{E}(N_{T_1\wedge 1} - \lambda(T_1 \wedge 1)) = 0$$

which implies that τ is not a pseudo-stopping time. Furthermore we compute

$$\widetilde{Z}_t = Z_t = \mathbb{1}_{\{T_1>t\}} + \mathbb{1}_{\{T_1\leq t\}}\mathbb{1}_{\{T_2>t\}}e^{-\lambda(t-T_1)}$$

hence $\widetilde{Z} = Z$ is a decreasing and càdlàg process. This example is further studied in Proposition 5.3 in [2].

Theorem 1 *The following are equivalent:*

(i) τ *is an* $\mathbb{F}$*-pseudo-stopping time;*
(ii) $A^o_\infty = \mathbb{P}(\tau < \infty \,|\, \mathscr{F}_\infty)$;
(iii) $m = 1$ *or equivalently* ${}^oA = A^o$;
(iv) for every $\mathbb{F}$*-local martingale* M*, the process* M^τ *is an* $\mathbb{F}^\tau$*-local martingale;*
(v) the process $\widetilde{Z}$ *is a càglàd decreasing process.*

Before proceeding to the proof of Theorem 1, we give an auxiliary lemma which characterizes the main property of our interest, that is, given a process of finite variation, when is its optional projection equal to the dual optional projection.

Lemma 1 *Given a raw locally integrable increasing process* V*, the following are equivalent:*

(i) ${}^o(V_-)$ *is a càglàd increasing process;*
(ii) ${}^o(V_-) = V^o_-$;
(iii) ${}^o(V_-) = {}^oV_-$;
(iv) ${}^oV = V^o$.

Proof For any raw locally integrable increasing process V, from classic theory we know that the process $N^V := {}^oV - V^o$ is a uniformly integrable martingale with $N^V_0 = 0$ and ${}^o(\Delta V) = \Delta V^o$. As a consequence we have

$$N^V = {}^o(V_-) - V^o_- \quad \text{and} \quad N^V_- = {}^oV_- - V^o_-. \tag{1}$$

If ${}^o(V_-)$ is a càglàd increasing process, then from (1), we see that N^V is a predictable martingale of finite variation, therefore is constant and equal to zero, since predictable martingales are continuous which shows $(i) \implies (ii)$ and $(i) \implies (iv)$. To prove $(iv) \implies (ii)$, it is enough to use the definition of N^V. Since N^V is càdlàg, we know that $N^V \equiv 0$ if and only if $N^V_- \equiv 0$. This fact combined with (1) gives the equivalence between *(ii)* and *(iii)* and the equivalence between *(ii)* and *(iv)*. □

Proof of Theorem 1 To see that *(i)* and *(ii)* are equivalent, suppose τ is an $\mathbb{F}$-pseudo-stopping time. Then, by properties of optional and dual optional projection, for any uniformly integrable $\mathbb{F}$-martingale M we have

$$\mathbb{E}(M_\tau \mathbb{1}_{\{\tau<\infty\}}) = \mathbb{E}(\int_{[0,\infty)} M_s dA^o_s) = \mathbb{E}(M_\infty A^o_\infty).$$

Therefore, the equality, $\mathbb{E}(M_\tau) = \mathbb{E}(M_\infty)$ holds true for every uniformly integrable $\mathbb{F}$-martingale M if and only if $A^o_\infty = \mathbb{P}(\tau < \infty \mid \mathcal{F}_\infty)$, since $\mathbb{E}(M_\tau) = \mathbb{E}(M_\infty(A^o_\infty + \mathbb{P}(\tau = \infty|\mathcal{F}_\infty)))$.

On the other hand, we have ${}^oA_\infty = \lim_{s\to\infty} \mathbb{P}(\tau \le s \mid \mathcal{F}_s) = \mathbb{P}(\tau < \infty \mid \mathcal{F}_\infty)$ a.s., and from the definition of m, we note that *(ii)* holds if and only if *(iii)* holds, that is $m = 1$ or equivalently ${}^oA = A^o$. The equivalence of *(iii)* and (v) follows directly from Lemma 1.

To see that $(i) \implies (iv)$, let M be a uniformly integrable $\mathbb{F}$-martingale. For any $\mathbb{F}^\tau$-stopping time ν, from Dellacherie et al. [5, p. 186], we know there exists an $\mathbb{F}$-stopping time σ such that $\tau \wedge \nu = \tau \wedge \sigma$. Therefore, from the definition of pseudo-stopping time,

$$\mathbb{E}(M_{\tau\wedge\nu}) = \mathbb{E}(M_{\tau\wedge\sigma}) = \mathbb{E}(M_0),$$

which shows that M^τ is a uniformly integrable $\mathbb{F}^\tau$-martingale by Theorem 1.42 [8]. The implication $(iv) \implies (i)$ is straightforward. □

Remark 1 The importance of the càglàd property in condition (v) of Theorem 1 is illustrated in Example 1. From this example we also see that a decreasing supermartingale Z is not sufficient to ensure that the time is a pseudo-stopping time. We would also like to point out that condition (v) in Theorem 1 is crucial when working with non-continuous filtrations and it is used later in the proof of Proposition 1.

3 Main Results and Applications

In this section, we formulate in Theorem 2 our main result which provides the necessary and sufficient conditions for the property that the $\mathbb{F}$-dual optional projection and $\mathbb{F}$-optional projection of any $\mathbb{G}$-optional process of integrable variation coincide.

As a part of this result we derive a new characterization of the immersion property in terms of pseudo-stopping times.

Theorem 2 *Given filtrations* $\mathbb{F}$ *and* $\mathbb{G}$ *such that* $\mathbb{F} \subset \mathbb{G}$*, the following are equivalent,*

(i) the $\mathbb{F}$*-dual optional projection of any* $\mathbb{G}$*-optional process of integrable variation is equal to its* $\mathbb{F}$*-optional projection;*
(ii) every $\mathbb{G}$*-stopping time is an* $\mathbb{F}$*-pseudo-stopping time;*
(iii) the filtration $\mathbb{F}$ *is immersed in* $\mathbb{G}$*.*

Proof The implication $(i) \implies (ii)$ follows directly from Theorem 1 (iii).

Let us now show the implication $(iii) \implies (i)$. Under the immersion property, the $\mathbb{F}$-optional projection of any bounded $\mathbb{G}$-optional process is equal to its optional projection on to the constant filtration $\mathscr{F}_\infty$ (see Bremaud and Yor [4]). More explicitly, for any given locally integrable increasing $\mathbb{G}$-adapted process V, we have ${}^o(V_-)_\sigma = \mathbb{E}(V_{\sigma-}|\mathscr{F}_\infty)$ for any $\mathbb{F}$-stopping time σ. From this we see that the process ${}^o(V_-)$ is increasing càglàd and (i) follows from Lemma 1.

To show $(ii) \implies (iii)$, suppose that M is a uniformly integrable $\mathbb{F}$-martingale and ν is any $\mathbb{G}$-stopping time. Since every $\mathbb{G}$-stopping time is an $\mathbb{F}$-pseudo-stopping time, we have $\mathbb{E}(M_\nu) = \mathbb{E}(M_0)$ for every $\mathbb{G}$-stopping time ν, which by Theorem 1.42 in [8], implies that M is a uniformly integrable $\mathbb{G}$-martingale.

The theorem is now proved, however, for the sake of completeness, let us directly show that $(iii) \implies (ii)$. To this end, let M be any uniformly integrable $\mathbb{F}$-martingale and ν a $\mathbb{G}$-stopping time. Then, from the immersion property, M is a uniformly integrable $\mathbb{G}$-martingale and $\mathbb{E}(M_\nu) = \mathbb{E}(M_0)$, which implies ν is an $\mathbb{F}$-pseudo-stopping time. □

We now give two applications of our main results in Theorems 1 and 2. An important class of random times is the class of honest times. A random time τ is an $\mathbb{F}$-honest time if for every $t > 0$ there exists an $\mathscr{F}_t$-measurable random variable τ_t such that $\tau = \tau_t$ on $\{\tau < t\}$. In Proposition 1 we relate pseudo-stopping times with honest times and, as an application of Theorem 1 (v) combined with the equivalence between (ii) and (iii) in Theorem 2, we recover a new proof of a result regarding honest times and the immersion property found in Jeulin [9]. Therein the result is obtained by computing explicitly the $\mathbb{G}$-semimartingale decompositions of $\mathbb{F}$-martingales. The equivalence $(i) \iff (ii)$ in Proposition 1 was already presented in Proposition 6 in [10] under the simplifying assumption that all $\mathbb{F}$-martingales are continuous and the proof therein uses distributional arguments. Here, we show that a similar result can be obtained in full generality by using sample path properties based on Theorem 1 (v).

Proposition 1 *Let* τ *be a random time. The following conditions are equivalent,*

(i) τ *is equal to an* $\mathbb{F}$*-stopping time on* $\{\tau < \infty\}$*,*
(ii) τ *is an* $\mathbb{F}$*-pseudo-stopping time and an* $\mathbb{F}$*-honest time.*

In particular if τ *is an* $\mathbb{F}$*-honest time which is not equal to an* $\mathbb{F}$*-stopping time on* $\{\tau < \infty\}$ *and a* $\mathbb{G}$*-stopping time for some filtration* $\mathbb{G} \supset \mathbb{F}$ *then* $\mathbb{F}$ *is not immersed in* $\mathbb{G}$.

Proof The implication $(i) \implies (ii)$ is obvious so we show only $(ii) \implies (i)$. Given that τ is a honest time, by Proposition 5.2. in [9], we have that $\tau = \sup\{t : \widetilde{Z}_t = 1\}$ on $\{\tau < \infty\}$. On the other hand, by Theorem 1 (v), the pseudo-stopping time property of τ implies that $\widetilde{Z} = 1 - A^o_-$. Therefore, on $\{\tau < \infty\}$,

$$\tau = \sup\{t : \widetilde{Z}_t = 1\} = \sup\{t : A^o_{t-} = 0\} = \inf\{t : A^o_t > 0\},$$

so, τ is equal to an $\mathbb{F}$-stopping time on $\{\tau < \infty\}$.
Therefore if τ is an $\mathbb{F}$-honest time which is not equal to an $\mathbb{F}$-stopping time on $\{\tau < \infty\}$ and a $\mathbb{G}$-stopping time for some filtration $\mathbb{G} \supset \mathbb{F}$ then, by Theorem 2, $\mathbb{F}$ is not immersed in $\mathbb{G}$. □

In the remaining, given that $\mathbb{F} \hookrightarrow \mathbb{G}$, we show that every $\mathbb{G}$-stopping time can be written as the minimum of two barrier hitting times for which the $\mathscr{F}_\infty$-conditional distribution of the barriers can be computed. The proof of our final result given in Theorem 3 relies on the equivalence $(i) \iff (iii)$ in Theorem 2.

Theorem 3 *Assume that* $\mathbb{F} \hookrightarrow \mathbb{G}$ *and let* τ *be a* $\mathbb{G}$*-stopping time. Then* τ *can be written as* $\tau_c \wedge \tau_d$*, where:*

(i) The random time τ_c *is a* $\mathbb{G}$*-stopping time which avoids all finite* $\mathbb{F}$*-stopping times. Denote by* $A^{c,o}$ *the* $\mathbb{F}$*-dual optional projection of the process* $\mathbb{1}_{[\![\tau_c,\infty[\![}$*. Then the* $\mathscr{F}_\infty$*-conditional distribution of* $A^{c,o}_{\tau_c}$ *is uniform on the interval* $[0, A^{c,o}_\infty)$*, with an atom of size* $1 - A^{c,o}_\infty$ *at* $A^{c,o}_\infty$*, that is*

$$\mathbb{P}(A^{c,o}_{\tau_c} \le u | \mathscr{F}_\infty) = u\mathbb{1}_{\{u < A^{c,o}_\infty\}} + \mathbb{1}_{\{u \ge A^{c,o}_\infty\}}.$$

(ii) The random time τ_d *is a* $\mathbb{G}$*-stopping time whose graph is contained in the disjoint union of the graphs of the jump times of the process* A^o *given by* $(\sigma_k)_{k\in\mathbb{N}}$*. Denote by* $A^{d,o}$ *the* $\mathbb{F}$*-dual optional projection of the process* $\mathbb{1}_{[\![\tau_d,\infty[\![}$*. Then*

$$\mathbb{P}(A^{d,o}_{\nu_d} = u | \mathscr{F}_\infty) = \sum_k \mathbb{1}_{\{A^{d,o}_{\sigma_k} = u\}} \Delta A^{d,o}_{\sigma_k}.$$

Before proceeding to the proof of Theorem 3, we show that in fact any random time τ can be written as a barrier hitting time of an $\mathbb{F}$-adapted increasing process given the appropriate barrier. We refer the reader to Remark 3.2 in Gapeev [6] where the author considers the situation where the process A^o is strictly increasing. We will demonstrate this result with no assumptions on A^o.

Lemma 2 *A random time* τ *can be written as the barrier hitting time of the process* A^o *with the barrier* A^o_τ*, that is* $\tau = \inf\{t > 0 : A^o_t \ge A^o_\tau\}$.

Proof We first define another random time τ^* by setting

$$\tau^* := \inf\{t > 0 : A^o_t \geq A^o_\tau\}.$$

To see that $\tau^* = \tau$ (it is obvious that $\tau^* \leq \tau$), we use Lemma 4.2 of [9] which states that the left-support of the measure dA, i.e.,

$$\{(\omega, t) : \forall \varepsilon > 0 \quad A_t(\omega) > A_{t-\varepsilon}(\omega)\} = [\![\tau]\!]$$

belongs to the left-support of dA^o, i.e., to the set $\{(\omega, t) : \forall \varepsilon > 0 \quad A^o_t(\omega) > A^o_{t-\varepsilon}(\omega)\}$. □

Proof of Theorem 3 For any $\mathbb{G}$-stopping time τ and the set $D := \{\Delta A^o_\tau > 0\} \in \mathcal{G}_\tau$, we see that τ can be written as $\tau_c \wedge \tau_d$, where $\tau_c := \tau \mathbb{1}_{D^c} + \infty \mathbb{1}_D$ and $\tau_d := \tau \mathbb{1}_D + \infty \mathbb{1}_{D^c}$. The random times τ_c and τ_d are therefore $\mathbb{G}$-stopping times, where τ_c avoids finite $\mathbb{F}$-stopping times and the graph of τ_d is contained in the graphs of the jump times of A^o. For more details on this decomposition of a random time see [1].

Given τ is a $\mathbb{G}$-stopping time that avoids all finite $\mathbb{F}$-stopping times. The $\mathcal{F}_\infty$-conditional distribution of A^o_τ is given by

$$\mathbb{E}(\mathbb{1}_{\{A^o_\tau \leq u\}} | \mathcal{F}_\infty) = \mathbb{E}(\mathbb{1}_{\{A^o_\tau \leq u\}} | \mathcal{F}_\infty) \mathbb{1}_{\{u < A^o_\infty\}} + \mathbb{1}_{\{u \geq A^o_\infty\}}.$$

Let us set C to be the right inverse of A^o, then the first term in the right hand side above is

$$\begin{aligned} \mathbb{E}(\mathbb{1}_{\{A^o_\tau \leq u\}} \mathbb{1}_{\{C_u < \infty\}} | \mathcal{F}_\infty) &= \mathbb{E}(\mathbb{1}_{\{\tau \leq C_u\}} \mathbb{1}_{\{C_u < \infty\}} | \mathcal{F}_{C_u}) \\ &= {}^o A_{C_u} \mathbb{1}_{\{C_u < \infty\}} \\ &= A^o_{C_u} \mathbb{1}_{\{C_u < \infty\}} \\ &= u \mathbb{1}_{\{u < A^o_\infty\}} \end{aligned}$$

where we apply Theorem 2 in the third equality, while the last equality follows from the fact that $A^o_{C_u} = u$, since A^o is continuous except, perhaps, at infinity. This implies that the $\mathcal{F}_\infty$-conditional distribution of A^o_τ is uniform on $[0, A^o_\infty)$.

On the other hand, given τ is a $\mathbb{G}$-stopping time whose graph is contained in the graphs of the jump times of A^o given by $(\sigma_k)_{k \in \mathbb{N}}$. Then

$$\begin{aligned} \mathbb{P}(A^o_\tau = u | \mathcal{F}_\infty) &= \sum_k \mathbb{P}(\{\tau = \sigma_k\} \cap \{A^o_{\sigma_k} = u\} | \mathcal{F}_\infty) \\ &= \sum_k \mathbb{1}_{\{A^o_{\sigma_k} = u\}} \mathbb{P}(\tau = \sigma_k | \mathcal{F}_\infty) \\ &= \sum_k \mathbb{1}_{\{A^o_{\sigma_k} = u\}} \Delta A^o_{\sigma_k} \end{aligned}$$

where the last equality follows from the fact that $\mathbb{F} \hookrightarrow \mathbb{G}$. □

Remark 2 As a special case of Theorem 3, if τ is a finite $\mathbb{G}$-stopping time that avoids finite $\mathbb{F}$-stopping times, then A^o_τ is independent of $\mathscr{F}_\infty$ and uniformly distributed on the interval $[0, 1]$. In this case, the $\mathbb{G}$-stopping time τ is a barrier hitting time of an $\mathbb{F}$-adapted increasing process, with the barrier being independent from $\mathscr{F}_\infty$. This is a class of random times widely used in credit risk modelling to model default times.

Acknowledgements The authors wish to thank Monique Jeanblanc, Jan Obłój, Marek Rutkowski and the anonymous referee for their careful readings and valuable advices on the writing of this paper and acknowledge the generous financial supports of *'Chaire Marchés en Mutation'*, Fédération Bancaire Française. The second author also wishes to thank Arturo Kohatsu-Higa for his hospitality in Japan and acknowledge the grants of the Japanese government. The research of the first author has been partially supported by the European Research Council under the European Union's Seventh Framework Programme (FP7/2007–2013)/ERC grant agreement no. 335421.

References

1. A. Aksamit, T. Choulli, M. Jeanblanc, Classification of random times and applications. Preprint (2016), available at http://arxiv.org/abs/1605.03905
2. A. Aksamit, T. Choulli, J. Deng, M. Jeanblanc, Arbitrages in a progressive enlargement setting, in *Arbitrage, Credit and Informational Risks*, ed. by C. Hillairet, M. Jeanblanc, Y. Jiao (World Scientific, Singapore, 2014), pp. 53–86
3. T. Bielecki, M. Jeanblanc, M. Rutkowski, *Credit Risk Modeling.* Osaka University CSFI Lecture Notes Series, vol. 2 (Osaka University Press, Osaka, 2009)
4. P. Brémaud, M. Yor, Changes of filtrations and of probability measures. Probab. Theory Relat. Fields **45**, 269–295 (1978)
5. C. Dellacherie, B. Maisonneuve, P.A. Meyer, *Probabilites et Potentiel*, vol. 5 (Hermann, Paris, 1992)
6. P.V. Gapeev, Some extensions of Norros' lemma in models with several defaults. in *Inspired by Finance, the Musiela Festschrift*, ed. by Yu.M. Kabanov, M. Rutkowski, Th. Zariphopoulou (Springer, Cham, 2014), pp. 273–281
7. S.W. He, J.G. Wang, J.A. Yan, *Semimartingale Theory and Stochastic Calculus* (Science Press, New York; CRC Press, Boca Raton, 1992)
8. J. Jacod, A.N. Shiryaev, *Limit Theorems for Stochastic Processes* (Springer, Berlin/Heidelberg, 2003)
9. T. Jeulin, *Semi-martingales et grossissement d'une filtration.* Lecture Notes in Mathematics, vol. 833 (Springer, Berlin/Heidelberg, 1980)
10. A. Nikeghbali, M. Yor, A definition and some characteristic properties of pseudo-stopping times. Ann. Prob. **33**, 1804–1824. (2005)
11. D. Williams, A 'non-stopping' time with the optional stopping property. Bull. Lond. Math. Soc. **34**, 610–612 (2002)

Stationary Random Fields on the Unitary Dual of a Compact Group

David Applebaum

Abstract We generalise the notion of wide-sense stationarity from sequences of complex-valued random variables indexed by the integers, to fields of random variables that are labelled by elements of the unitary dual of a compact group. The covariance is positive definite, and so it is the Fourier transform of a finite central measure (the spectral measure of the field) on the group. Analogues of the Cramer and Kolmogorov theorems are extended to this framework. White noise makes sense in this context and so, for some classes of group, we can construct time series and investigate their stationarity. Finally we indicate how these ideas fit into the general theory of stationary random fields on hypergroups.

1 Introduction

There are many important classes of stochastic process that have been systematically developed, both because of their mathematical vitality, and their importance for applications. These include, for example, Markov chains, branching processes, and diffusion processes. The emphasis in this paper is on discrete-time (wide-sense) stationary, complex-valued processes $(X_n, n \in \mathbb{Z})$, so that $\mathbb{E}(|X_n|^2) < \infty$ and

$$\mathbb{E}(X_m\overline{X_n}) = \mathbb{E}(X_{m-n}\overline{X_0}), \tag{1}$$

for all $m, n \in \mathbb{Z}$. These may be used to model fluctuations from some fixed background signal. Stationarity is a vital ingredient in the theory of time series (see e.g. [6]) which has a wide range of applications, including economics and climate science.

A stationary process is characterised by its covariance function $C(n) = \mathbb{E}(X_n\overline{X_0})$, which is positive-definite, and so by the Herglotz theorem, there is a finite measure

D. Applebaum (✉)
School of Mathematics and Statistics, University of Sheffield, Hicks Building, Hounsfield Road, Sheffield S3 7RH, England
e-mail: D.Applebaum@sheffield.ac.uk

C. Donati-Martin et al. (eds.), *Séminaire de Probabilités XLVIII*, Lecture Notes in Mathematics 2168, DOI 10.1007/978-3-319-44465-9_18

μ on the torus $\mathbb{T}$, known as the spectral measure of the process, for which

$$C(n) = \int_{\mathbb{T}} e^{-2\pi in\theta}\mu(d\theta),$$

for all $n \in \mathbb{N}$.

If we are interested in describing the interaction of chance with symmetry, then it is natural to consider stationary random fields on a group G, i.e. mappings $X : G \to L^2(\Omega, \mathbb{C})$ for which[1]

$$\mathbb{E}(X(hg_1)\overline{X(hg_2)}) = \mathbb{E}(X(g_1)\overline{X(g_2)}),$$

for all $g_1, g_2, h \in G$. The study of these, and related objects on homogeneous spaces, seems to have begun with work by A.M. Yaglom in the late 1950s (see e.g. [15]); recently there have been monograph treatments and new applications to e.g. earthquake modelling and the study of the cosmic background radiation left over from the Big Bang [13, 14].

In this paper, we suggest that, although replacing $\mathbb{Z}$ as the index of a stationary field by a group G is mathematically highly productive, it may not be the most natural generalisation. As was discussed above, the spectral measure of a stationary process is defined on the torus $\mathbb{T}$; this is the simplest compact group, and its dual group is $\mathbb{Z}$. We propose that $\mathbb{T}$ should be replaced by a general (and so, not necessarily abelian) compact group, so that the role of $\mathbb{Z}$ is now played by the unitary dual $\widehat{G}$ of G. Note that $\widehat{G}$ is not itself a group if G fails to be abelian.

In Sect. 2 of this paper we generalise the definition (1) to random fields over $\widehat{G}$. Indeed we say that a field $(Y_\pi, \pi \in \widehat{G})$ is stationary if

$$\mathbb{E}(Y_{\pi_1}\overline{Y_{\pi_2}}) = \mathbb{E}(Y_{\pi_1 \otimes \pi_2^*}\overline{Y_\epsilon}),$$

for all $\pi_1, \pi_2 \in \widehat{G}$, where π^* is the irreducible representation that is conjugate to π, and ϵ is the trivial representation. Some justification as to why this is a sensible generalisation of (1) will be provided. We also define the covariance function and show that it is the Fourier transform (in the group-theoretic sense) of a finite central measure on G, which we call the spectral measure of the field. We establish a Cramer-type representation of stationary fields as stochastic integrals with respect to orthogonally scattered random measures on G, and we prove a theorem of Kolmogorov-type to the effect that every positive-definite function on $\widehat{G}$ is the covariance of a stationary random field on $\widehat{G}$.

[1] We only write down the left-invariant case here, but of course right-invariance is equally valid.

We have already pointed out that $\widehat{G}$ is not in general a group, but it is a hypergroup [4] and we discuss this in Sect. 4. There is an existing literature on stationary random fields on hypergroups [9, 11, 12] which this current work complements. We make some observations:

1. The definition of stationarity for general hypergroups is quite non-intuitive. But in our case, the parallel with the classical case is very direct.
2. The duality between the hypergroup $\widehat{G}$ and the group G is manifest in the relationship between the stationary field and its spectral measure. There is a rich structure here that merits further investigation, and which could lead to new examples of the important class of central measures on compact groups.
3. The key process of "white noise" may not exist in general hypergroups. But it always does in our case. This means that, at least for some classes of compact groups, we may develop a theory of time series on their unitary duals, and investigate stationarity. Some examples for the case of the dual of $SU(2)$ are considered in Sect. 3 of this paper.

Notation If A is a complex-valued matrix, then $\text{tr}(A)$ is its trace (i.e. the sum across the leading diagonal). If U is a topological space, then $\mathscr{B}(U)$ is the Borel σ-algebra of U (i.e. the smallest σ-algebra containing all open sets). Haar integrals of suitable functions f on a compact group G are written $\int_G f(\sigma)d\sigma$.

2 Definition and Main Results

Let G be a compact (second countable, Hausdorff) topological group, $\widehat{G}$ be its unitary dual, comprising equivalence classes of irreducible unitary representations of G, and $\widehat{G_F}$ be the set of equivalence classes of finite-dimensional unitary representations of G (each with respect to unitary conjugation). Since G is compact, $\widehat{G}$ is countable and $\widehat{G} \subset \widehat{G_F}$.[2] We denote the trivial representation of G by $\epsilon \in \widehat{G}$. The *character* χ_π of $\pi \in \widehat{G_F}$ is defined by

$$\chi_\pi(g) = \text{tr}(\pi(g))$$

for each $g \in G$, and it is consequence of the celebrated Peter-Weyl theorem that $\{\chi_\pi, \pi \in \widehat{G}\}$ is a complete orthonormal basis in the complex Hilbert space $L^2_c(G)$ of all central (i.e. conjugate invariant) square-integrable (with respect to normalised Haar measure) functions on G. It follows that we may decompose each $\pi \in \widehat{G_F}$ as

$$\pi = \bigoplus_{\pi' \in \widehat{G}} M(\pi, \pi')\pi', \tag{2}$$

[2]We refer to a standard text, such as [5], for all facts about compact groups quoted herein. See also the account for probabilists in [2].

where

$$M(\pi, \pi') = \int_G \chi_\pi(g^{-1})\chi_{\pi'}(g)dg, \quad (3)$$

is the *multiplicity* of π' is π. Of course $M(\pi, \pi') \in \mathbb{Z}_+$ vanishes for all but finitely many $\pi' \in \widehat{G}$. The conjugate representation associated to $\pi \in \widehat{G_F}$ is denoted π^* and the tensor product of the representations π_1 and π_2 is $\pi_1 \otimes \pi_2$. Note that for all $g \in G$,

$$\chi_{\pi^*}(g) = \overline{\chi_\pi(g)} \quad , \quad \chi_{\pi_1 \otimes \pi_2}(g) = \chi_{\pi_1}(g)\chi_{\pi_2}(g). \quad (4)$$

Proposition 1 *For all* $\pi_1, \pi_2 \in \widehat{G}$,

$$M(\epsilon, \pi_1 \otimes \pi_2^*) = \delta_{\pi_1, \pi_2}.$$

Proof Using (3), (4), and orthonormality of characters, we have

$$\begin{aligned} M(\epsilon, \pi_1 \otimes \pi_2^*) &= \int_G \chi_{\pi_1}(g)\overline{\chi_{\pi_2}}(g)dg \\ &= \delta_{\pi_1, \pi_2}. \end{aligned}$$

Let $(\Omega, \mathscr{F}, P)$ be a probability space. A mapping $Y : \widehat{G_F} \to L^2(\Omega, \mathscr{F}, P; \mathbb{C})$ is said to be a *decomposable random field* on $\widehat{G_F}$ if it satisfies

$$Y_\pi = \sum_{\pi' \in \widehat{G}} M(\pi, \pi')Y_{\pi'},$$

with respect to the decomposition (2). Clearly such a field is uniquely determined by its values on $\widehat{G}$. We say that such a field is *(wide-sense) stationary* if

$$\mathbb{E}(Y_{\pi_1}\overline{Y_{\pi_2}}) = \mathbb{E}(Y_{\pi_1 \otimes \pi_2^*}\overline{Y_\epsilon}), \quad (5)$$

for all $\pi_1, \pi_2 \in \widehat{G}$. The motivation for the definition (5) comes from the well-known case $G = \mathbb{T} = [0, 2\pi), \widehat{G} = \mathbb{Z}$. In that case the irreducible representation corresponding to $\pi_1 = n$ is uniquely determined by the character $\theta \to e^{in\theta}$, and the character associated to $\pi_1 \otimes \pi_2^*$, where $\pi_2 = m$, is precisely $\theta \to e^{i(n-m)\theta}$.

Remark Clearly if the random field is stationary, then $\mathbb{E}(|Y_\pi|) < \infty$, for all $\pi \in \widehat{G}$. It may seem strange to some readers that we do not impose some additional stationarity condition on the means, i.e. that the quantity $\mathbb{E}(Y_\pi)$ does not depend on $\pi \in \widehat{G}$, or even that the field is *centred*, in that $\mathbb{E}(Y_\pi) = 0$, for all $\pi \in \widehat{G}$. Here we follow Doob [7, p. 95], who, in the classical case $G = \mathbb{T}, \widehat{G} = \mathbb{Z}$ wrote, "Usually the added condition that $\mathbb{E}(X_s)$ does not depend on s is imposed. This condition is unnatural mathematically, and has nothing to do with the essential properties of interest in these processes, and we shall therefore not impose it."

If Y is a stationary random field on $\widehat{G}$, we define its *covariance function* $C_Y : \widehat{G_F} \to \mathbb{C}$ by

$$C_Y(\pi) = \mathbb{E}(Y_\pi \overline{Y_\epsilon}),$$

for all $\pi \in \widehat{G_F}$, and we note that it is a decomposable mapping on $\widehat{G_F}$ in that

$$C_Y(\pi) = \sum_{\pi' \in \widehat{G}} M(\pi, \pi') C_Y(\pi'),$$

with respect to (2).

We recall from [8] that $\Phi : \widehat{G} \to \mathbb{C}$ is *positive definite* if for all $N \in \mathbb{N}, \pi_1, \ldots, \pi_N \in \widehat{G}$ and $c_1, \ldots, c_N \in \mathbb{C}$,

$$\sum_{m,n=1}^{N} c_m \overline{c_n} \sum_{\pi \in \widehat{G}} M(\pi, \pi_m \otimes \pi_n^*) \Phi(\pi) \geq 0.$$

If Φ extends to a mapping $\widehat{G_F} \to \mathbb{C}$ that is decomposable, then we have the equivalent condition

$$\sum_{m,n=1}^{N} c_m \overline{c_n} \Phi(\pi_m \otimes \pi_n^*) \geq 0. \tag{6}$$

Proposition 2 *If Y is a stationary random field on $\widehat{G}$, then its covariance function C_Y is positive definite.*

Proof Using (6) and (5), we find that

$$\begin{aligned}
\sum_{m,n=1}^{N} c_m \overline{c_n} C_Y(\pi_m \otimes \pi_n^*) &= \sum_{m,n=1}^{N} c_m \overline{c_n} \mathbb{E}(Y_{\pi_m \otimes \pi_n^*} \overline{Y_\epsilon}) \\
&= \sum_{m,n=1}^{N} c_m \overline{c_n} \mathbb{E}(Y_{\pi_m} \overline{Y_{\pi_n}}) \\
&= \mathbb{E}\left(\left| \sum_{n=1}^{N} c_n Y_{\pi_n} \right|^2 \right) \geq 0.
\end{aligned}$$

It follows from Proposition 2 and the Bochner theorem (Theorem 5.5 in [8]) that there exists a unique finite Radon central measure μ_Y defined on the Borel σ-algebra of G for which

$$C_Y(\pi) = \int_G \chi_\pi(g)\mu_Y(dg) \tag{7}$$

for all $\pi \in \widehat{G}$. We call μ_Y the *spectral measure* of the random field Y.

As an example, consider the *white noise* $Z : \widehat{G_F} \to L^2(\Omega, \mathcal{F}, P)$ which is defined to be a decomposable random field which is *uncorrelated* in that

$$\mathbb{E}(Z_\pi \overline{Z_{\pi'}}) = \delta_{\pi,\pi'},$$

for all $\pi, \pi' \in \widehat{G}$. It follows from Proposition 1 that Z is stationary and the spectral measure is easily seen to be (normalised) Haar measure on G.

The next result gives a *Cramer representation* for the field.

Theorem 1 *If $(Y_\pi, \pi \in \widehat{G})$ is a stationary random field, then there exists an orthogonally scattered random measure Γ_Y on G so that for all $\pi \in \widehat{G}$,*

$$Y_\pi = \int_G \chi_\pi(g)\Gamma_Y(dg) \text{ a.s.}. \tag{8}$$

Furthermore, $\mathbb{E}(|\Gamma_Y(A)|^2) = \mu_Y(A)$ for all $A \in \mathcal{B}(G)$, and Γ_Y is a.s. central in that for each $g \in G$,

$$P(\Gamma_Y(gAg^{-1}) = \Gamma_Y(A)) = 1.$$

Proof This is along standard lines. We sketch the details following the argument given in [10, pp. 46–47] for the classical case.

Let $\mathcal{M}$ be the closed subspace of $L^2(\Omega, \mathcal{F}, P; \mathbb{C})$ generated by $\{Y_\pi, \pi \in \widehat{G}\}$. Consider the linear mapping V from the complex linear span of $\{Y_\pi, \pi \in \widehat{G}\}$ into $L^2_c(G, \mu_Y) := L^2_c(G, \mathcal{B}(G), \mu_Y : \mathbb{C})$ (where the subscript c, indicates the restriction to central functions) given by

$$V\left(\sum_{j=1}^n \alpha_j Y_{\pi_j}\right) = \sum_{j=1}^n \alpha_j \chi_{\pi_j}.$$

It is straightforward to check that V is isometric. Since the set of all finite linear combinations of characters is dense in $L^2_c(G, \mu_Y)$, it follows that V extends to a unitary isomorphism between $\mathcal{M}$ and $L^2_c(G, \mu_Y)$. For each $A \in \mathcal{B}(G)$, define

$$\Gamma_Y(A) = V^* \mathbf{1}_A.$$

Then it is straightforward to check that Γ_Y has the desired properties. Moreover, for all $f \in L^2(G, \mu_Y)$,

$$V\left(\int_G f(g)\Gamma_Y(dg)\right) = f.$$

Then for all $\pi \in \widehat{G}$,

$$V\left(\int_G \chi_\pi(g)\Gamma_Y(dg)\right) = \chi_\pi(g) = VY_\pi,$$

and we thus obtain (8).

We also have a reconstruction theorem of Kolmogorov type:

Theorem 2 *Given a positive definite function $\Phi : \widehat{G} \to \mathbb{C}$ with $\Phi(\epsilon) = 1$, there exists a stationary random field $(Y_\pi, \pi \in \widehat{G_F})$ having covariance Φ.*

Proof By the Bochner theorem of [8], there exists a unique finite Radon central measure μ on $(G, \mathscr{B}(G))$ so that for all $\pi \in \widehat{G}$,

$$\Phi(\pi) = \int_G \chi_\pi(g)\mu(dg),$$

and the normalisation $\Phi(\epsilon) = 1$ ensures that μ is a probability measure. Now define $(Y_\pi, \pi \in \widehat{G})$ on the probability space $(G, \mathscr{B}(G), \mu)$ by the prescription $Y_\pi = \chi_\pi$, for each $\pi \in \widehat{G_F}$. Then the field is automatically decomposable and is stationary since

$$\begin{aligned}
\mathbb{E}(Y_{\pi_1}\overline{Y_{\pi_2}}) &= \int_G \chi_{\pi_1}(g)\overline{\chi_{\pi_2}}(g)\mu(dg) \\
&= \int_G \chi_{\pi_1 \otimes \pi_2^*}(g)\mu(dg) \\
&= \mathbb{E}(Y_{\pi_1 \otimes \pi_2^*}\overline{Y_\epsilon}),
\end{aligned}$$

and $Y_\epsilon = \chi_\epsilon = 1$.

Let ρ be a central probability measure on G. Then its central Fourier transform $\widehat{\rho} : \widehat{G} \to \mathbb{C}$ is positive definite, and so is the covariance of a stationary random field by Theorem 2. So Theorem 2 tells us that there are a rich variety of stationary random fields on $\widehat{G}$. One important example is white noise, as discussed in Sect. 2. For further examples, suppose that G is a compact, connected Lie group and that ρ is Gaussian, so that $\rho(d\sigma) = k_1(\sigma)d\sigma$, where $(k_t, t \geq 0)$ is the heat kernel on G.

Then for each $\pi \in \widehat{G}$,

$$\widehat{\rho}(\pi) := \int_G \chi_\pi(g)\rho(dg) = d_\pi e^{-\kappa_\pi},$$

where d_π is the dimension of the complex linear space in which π acts, and $\{\kappa_\pi, \pi \in \widehat{G}\}$ is the Casimir spectrum. A large class of non-Gaussian infinitely divisible central measures on G may be obtained by subordination of the heat kernel (see e.g. [1] or Chap. 4 of [2]).

3 Examples: Time Series

It is interesting to seek examples in the case where G is a rank-one, connected, compact Lie group. Then the lattice of weights is a subset of the real line, and so inherits an ordering that can be used to develop a theory of time series, by analogy with the familiar one on the group of integers. As an example, let us consider the group $G = SU(2)$. In this case $\widehat{G}$ is in one-to-one correspondence with the set $\mathbb{Z}_+$ (with 0 corresponding to ϵ) and we may consider the AR(1) process defined for each $n \in \mathbb{Z}_+, \lambda \in \mathbb{C}$ by

$$Y_n = \lambda Y_{n-1} + Z_n, \tag{9}$$

where $Y_{-1} := 0$.

Here we may take the index $n \in \mathbb{Z}_+$ as labelling the unique equivalence class of irreducible representations having representation space with dimension $n + 1$.

We show that this process cannot be stationary.

Define the backwards shift operator B on the linear space generated by $\{Y_n, n \in \mathbb{Z}_+\}$ by

$$BY_n = Y_{n-1}.$$

Then

$$\begin{aligned} Y_n &= (I - \lambda B)^{-1} Z_n \\ &= \sum_{k=0}^{n} \lambda^k Z_{n-k}. \end{aligned} \tag{10}$$

Note that in contrast to the familiar case of $G = \mathbb{T}$, no condition is needed on λ to obtain the moving average representation (10) as this series is finite. It follows easily from (10) that $(Y_n, n \in \mathbb{Z}_+)$ has covariance

$$\mathbb{E}(Y_{n+h}\overline{Y_n}) = \sum_{k=0}^{n+h}\sum_{l=0}^{n} \lambda^k \overline{\lambda}^l \mathrm{e}(Z_{n+h-k}\overline{Z_{n-l}})$$

$$= \overline{\lambda^{-h}} \sum_{k=0}^{n} |\lambda|^{2(k+h)}$$

$$= \begin{cases} \lambda^h \left(\dfrac{1-|\lambda|^{2n+2}}{1-|\lambda|^2}\right) & \text{if } |\lambda| \neq 1 \\ (n+1)\overline{\lambda^{-h}} & \text{if } |\lambda| = 1. \end{cases}$$

On the other hand, consider the MA(q) process on $\widehat{SU(2)}$ given by

$$Y_n = \sum_{k=0}^{q} \beta_k Z_{n-k},$$

where $\beta_k \in \mathbb{C}$ for all $k \in \mathbb{Z}_+$. Then by standard arguments, $(Y_n, n \in \mathbb{Z}_+)$ is easily seen to be a stationary random field with covariance function

$$\mathbb{E}(Y_{n+h}\overline{Y_n}) = \begin{cases} \sum_{k=0}^{q-h} \beta_{k+h}\overline{\beta_k} & \text{if } 0 \leq h \leq q \\ 0 & \text{if } h > q. \end{cases}$$

4 The Hypergroup Connection

Let K be a non-empty locally compact Hausdorff space which is equipped with an involution $x \to x'$. Let $M^b(K)$ be the complex linear space of all bounded, complex Radon measures on K. We say that $(K, *)$ is a *hypergroup* if there is a binary operation $*$ defined on $M^b(K)$ with respect to which $M^b(K)$ is a algebra, and if certain axioms hold. We state only one of these here; that there must exist a "neutral element" $e \in K$ so that for all $x \in K$,

$$\delta_x * \delta_e = \delta_e * \delta_x = \delta_x,$$

where δ_x is the Dirac mass at x. The others may be found on p. 9 of [4]; they will play no direct role in the sequel. Examples are locally compact groups (where $*$ is the usual convolution of measures), double coset spaces and the unitary dual of a compact group (see below). The hypergroup is said to be *discrete* if K is equipped with the discrete topology, and *commutative* if $\mu_1 * \mu_2 = \mu_2 * \mu_1$ for all $\mu_1, \mu_2 \in M^b(K)$.

Now let $\widehat{G}$ be the unitary dual of a compact group G. It becomes a discrete, commutative hypergroup, with neutral element ϵ and involution $\pi \to \pi^*$, under the convolution:

$$\delta_{\pi_1} * \delta_{\pi_2} = \sum_{\pi \in \widehat{G}} M(\pi_1 \otimes \pi_2, \pi)\delta_\pi, \tag{11}$$

for each $\pi_1, \pi_2 \in \widehat{G}$, relative to the decomposition

$$\pi_1 \otimes \pi_2 = \bigoplus_{\pi \in \widehat{G}} M(\pi_1 \otimes \pi_2, \pi)\pi.$$

The convolution (11) is extended to general measures in $M^b(\widehat{G})$ by taking weak limits of linear combinations. Note that this is not the same convolution as that given in [4, p. 13], where the following is found:

$$\delta_{\pi_1} *' \delta_{\pi_2} = \sum_{\pi \in \widehat{G}} \frac{d_\pi}{d_{\pi_1} d_{\pi_2}} M(\pi_1 \otimes \pi_2, \pi)\delta_\pi, \tag{12}$$

with d_π being the dimension of the representation space corresponding to $\pi \in \widehat{G}$.

Following Sect. 8.2 in [4], the survey article [9] and the original source [11] we define a *stationary random field* over a commutative hypergroup K to be a mapping $X : K \to L^2(\Omega, \mathcal{F}, P; \mathbb{C})$ which has covariance

$$C(a, b) = \mathbb{E}(X_a \overline{X_b}),$$

that satisfies the stationarity condition:

$$C(a, b) = \int_K C(x, e)(\delta_a * \delta_{b'})(dx), \tag{13}$$

for each $a, b \in K$.

Now suppose that $(Y_\pi, \pi \in \widehat{G})$ is a stationary random field on $\widehat{G}$ in the sense of (5). We will show that it is also stationary in the hypergroup sense, by using the convolution (11) to define the hypergroup structure. Note that if we used (12) then this assertion would be false. It is enough to show that (13) is satisfied. Indeed for all $\pi_1, \pi_2 \in \widehat{G}$,

$$\begin{aligned} C(\pi_1, \pi_2) &= \mathbb{E}(Y_{\pi_1} \overline{Y_{\pi_2}}) \\ &= \mathbb{E}(Y_{\pi_1 \otimes \pi_2^*} \overline{Y_\epsilon}) \end{aligned}$$

$$= C(\pi_1 \otimes \pi_2^*, \epsilon)$$
$$= \sum_{\pi \in \widehat{G}} M(\pi_1 \otimes \pi_2^*, \pi) C(\pi, \epsilon)$$
$$= \int_{\widehat{G}} C(\pi, \epsilon)(\delta_{\pi_1} * \delta_{\pi_2^*})(d\pi),$$

as required.

Let $\mathcal{M}_Y(\widehat{G})$ be the closure in $L^2(\Omega, \mathcal{F}, P; \mathbb{C})$ of $\{Y_\pi, \pi \in \widehat{G}\}$. Following [9, 11, 12], for fixed $\pi' \in \widehat{G}$, we define the *translation operator* $\tau_{\pi'}$ *associated to* Y to be the linear contraction in $\mathcal{M}_Y(\widehat{G})$ obtained by continuous linear extension of the prescription

$$\tau_{\pi'}(Y_\pi) = Y_{\pi \otimes \pi'}, \tag{14}$$

for each $\pi \in \widehat{G}$. Note that because of the rather concrete context in which we work, the definition (14) is much more transparent than that in the general hypergroup case. A useful list of properties of such operators is collected in Theorem 2 of Leitner [12].

Now let $\mathcal{A}$ be a family of subsets of $\widehat{G}$. For each $A \in \mathcal{A}$, let $\mathcal{M}_Y(A)$ be the closure of the linear span of $\{Y_\pi, \pi \in A\}$, and $\mathcal{M}_Y := \bigcap_{A \in \mathcal{A}} \mathcal{M}_Y(A)$. We say that the stationary field Y is $\mathcal{A}$*-singular* if $\mathcal{M}_Y = \mathcal{M}_Y(\widehat{G})$, $\mathcal{A}$*-regular* if $\mathcal{M}_Y = \{0\}$, and $\mathcal{A}$*-adapted* if $\tau_\pi(\mathcal{M}_Y) \subseteq \mathcal{M}_Y$ for all $\pi \in \widehat{G}$. The following abstract version of the *Wold decomposition* is proved for general commutative hypergroups in Theorem 2.2.5.2 of [9]. We will be content to state the result.

Theorem 3 (The Wold Decomposition) *If Y is an $\mathcal{A}$-adapted stationary random field, then there is a unique orthogonal decomposition*

$$Y_\pi = Y_\pi^{(1)} + Y_\pi^{(2)}, \tag{15}$$

for all $\pi \in \widehat{G}$, where $Y^{(1)}$ is $\mathcal{A}$-regular, and $Y^{(2)}$ is $\mathcal{A}$-singular.

If G is a rank one, connected, compact Lie group then it is natural to choose $\mathcal{A}$ in accordance with the lattice structure, e.g. for $G = SU(2)$, $\mathcal{A} = \{A_n, n \in \mathbb{Z}_+\}$, where $A_n := \{0, 1, \ldots, n\}$.

We conjecture that there is a generalisation to this context of the classical result that can be found e.g. in Chap. 4 of [10], whereby the absolutely continuous measure μ_1 and the singular measure μ_2 which arise in the Lebesgue decomposition of the spectral measure of Y are themselves the spectral measures of the processes $Y^{(1)}$ and $Y^{(2)}$ (respectively) of (15).

Acknowledgements The author would like to thank N.H. Bingham for stimulating discussions about prediction, in relation to the ideas in [3], that inspired the work of this note. He is also indebted to H. Heyer for very helpful comments.

References

1. D. Applebaum, Infinitely divisible central probability measures on compact Lie groups - regularity, semigroups and transition kernels. Ann. Prob. **39**, 2474–96 (2011)
2. D. Applebaum, Probability on compact lie groups, in *Probability and Stochastic Modelling*, vol. 70 (Springer International Publishing Switzerland, Berlin, 2014)
3. N.H. Bingham, Szegö's theorem and its probabilistic descendants. Probab. Surv. **9**, 287–324 (2012)
4. W.R. Bloom, H. Heyer, *Harmonic Analysis of Probability Measures on Hypergroups* (de Gruyter, Berlin, 1995)
5. T. Bröcker, T. tom Dieck, *Representations of Compact Lie Groups* (Springer, New York, 1985)
6. P.J. Brockwell, R.A. Davis, *Time Series: Theory and Methods*, 2nd edn. (Springer, New York, 1991) [1st edn. 1987]
7. J.L. Doob, *Stochastic Processes* (Wiley, New York, 1953)
8. H. Heyer, A Bochner-type representation of positive definite mappings on the dual of a compact group. Commun. Stoch. Anal. **7**, 459–479 (2013)
9. H. Heyer, Random fields and hypergroups, in *Real and Stochastic Analysis, Current Trends* ed. by M.M. Rao (World Scientific, Singapore, 2013), pp. 85–182
10. J. Lamperti, *Stochastic Processes*. Applied Mathematical Sciences, vol. 23 (Springer, New York, 1977)
11. R. Lasser, M. Leitner, Stochastic processes indexed by hypergroups I. J. Theor. Probab. **2**, 301–311 (1989)
12. M. Leitner, Stochastic processes indexed by hypergroups II. J. Theor. Probab. **4**, 321–332 (1991)
13. A. Malyarenko, *Invariant Random Fields on Spaces with a Group Action* (Springer, Berlin, 2013)
14. D. Marinucci, G. Peccati, *Random Fields on the Sphere - Representation, Limit Theorems and Cosmological Applications*. London Mathematical Society Lecture Note Series, vol. 389 (Cambridge University Press, Cambridge, 2011)
15. A.M. Yaglom, Second-order homogeneous random fields, in *Proceedings of 4th Berkeley Symposium Mathematical Statistics and Probability*, vol. II (University of California Press, Berkeley, 1961), pp. 593–622

On the Spatial Markov Property of Soups of Unoriented and Oriented Loops

Wendelin Werner

Abstract We describe simple properties of some soups of *unoriented* Markov loops and of some soups of *oriented* Markov loops that can be interpreted as a spatial Markov property of these loop-soups. This property of the latter soup is related to well-known features of the uniform spanning trees (such as Wilson's algorithm) while the Markov property of the former soup is related to the Gaussian Free Field and to identities used in the foundational papers of Symanzik, Nelson, and of Brydges, Fröhlich and Spencer or Dynkin, or more recently by Le Jan.

1 Introduction

Symanzik and then Nelson have pioneered the study of Euclidean field theory more than 40 years ago [13, 17]. In their approach, measures on random paths and loops play an important role and led to further important developments such as in the work of Brydges, Fröhlich and Spencer [1] (see also Dynkin [4, 5]). In all these papers, a gas of closed loops is used to represent partition functions and correlation structures of random fields.

The present note will be in the same spirit, but the focus will be on this random gas of loops itself as the main object of interest, rather than viewing it as a combinatorial diagrammatic tool to evaluate quantities related to fields. We will in particular focus on the role of orientation of loops and describe a particular simple property of such random configurations of unoriented loops as well as for random configurations of oriented loops. These properties are very directly related to the combinatorial features used in the aforementioned papers as well as to some features in the more recent study by Le Jan (in particular in Sects. 7 and 9 of [9]), who focuses more on properties of the occupation times of these soups.

These gases of loops, or loop-soups (as they have been called in [8]) are a random Poissonian (i.e. non-interacting) collection of random unrooted loops in a domain, that can be associated naturally to a Markov process or a discrete-time Markov chain (see [9] and the references therein). When one discovers the configurations of the

W. Werner (✉)
Department of Mathematics, ETH Zürich, Rämistr 101, 8092 Zürich, Switzerland
e-mail: wendelin.werner@math.ethz.ch

C. Donati-Martin et al. (eds.), *Séminaire de Probabilités XLVIII*, Lecture Notes in Mathematics 2168, DOI 10.1007/978-3-319-44465-9_19

loop-soup within a given sub-domain U of the entire domain in which the soup is defined, one observes on the one hand loops that are entirely contained in U (which form a loop-soup in U), and on the other hand, excursions in U that are parts of loops that do not entirely stay in U. Note that different such excursions can belong to the same loop or not, depending on the configuration outside of U. The Markovian property that we shall discuss basically describes how to randomly complete the missing pieces into the loops i.e. it describes the conditional distribution of the loop-soup outside of U when conditioning on these excursions of the loop-soup in U. As we shall see, this takes a nice "Markovian form" in two special cases:

- When one considers the loops to be oriented, and the intensity of the loop-soup to be the one that relates it to the partition function of uniform spanning trees i.e. to the number of spanning trees (and to Wilson's algorithm [20] to generate them uniformly at random, see e.g. [6, 7, 19, 20]).
- In the case where the chain is reversible, if one considers the loops to be unoriented, and chooses the intensity to be the one that relates the loop-soup to the Gaussian Free Field (for instance via their partition functions—and in fact the occupation time of a continuous-time version of the loop-soup then corresponds exactly to the square of the GFF, see [9]).

In those two cases, the only relevant information in order to complete the excursions in U into loops is the family of all endpoints of the excursions on ∂U, and not how these endpoints are connected by the excursions within U (nor which excursion endpoint is connected to which other by an excursion). In other words, the trace of the discrete loop-soup inside U and outside of U are conditionally independent given their trace on ∂U (more precisely, given their trace on the edges between U and the complement of U).

Let us illustrate another instance of the spatial Markov property in an impressionistic and heuristic way via the following figures (Figs. 1 and 2): We consider a

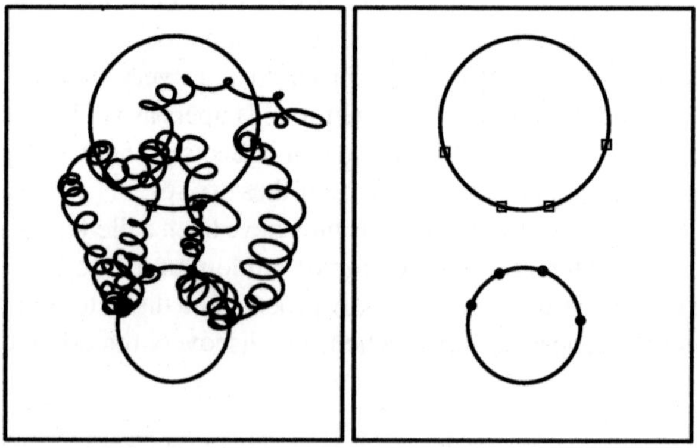

Fig. 1 The unoriented loop(s) in the soup that touch both circles, and the endpoints of their (four in this case) crossings between the two circles

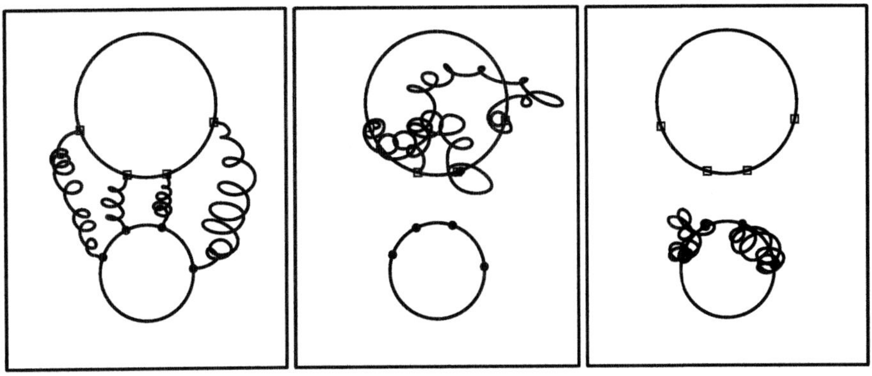

Fig. 2 Conditionally on the set of endpoints of crossings on each of the two circles, these three pictures, corresponding to different parts (excursions and bridges) of the loops that touch both circles, are independent conditionally on the endpoints on the circles

loop-soup of unoriented loops in the inside of the rectangle, of well-chosen intensity (related to the partition function of the GFF). In this loop-soup, only finitely many loops do touch the two circles, and in each such loop, there are an even number of "crossings" from one circle to the other. The statement in the caption of Fig. 2 is the type of result that we will derive.

To conclude this introduction, let us briefly mention that of the motivations for the present work is to explore the relation between the natural "Markovian" structures emerging from the loop-soups with the theory of local sets for the discrete and continuous GFF, as defined by Schramm and Sheffield in [15].

2 Background and Definitions

In this section, we recall standard facts about Markov loops and loop-soups, make some elementary comments about the orientation/non-orientation of loops, and we define the natural measures on Markov bridges that we will need.

2.1 The Measure on Unrooted Oriented Loops

Let us consider a discrete oriented graph Γ, where each vertex x has a finite number $d(x)$ of outgoing edges, so that it is possible to define simple random walk on Γ ($d(x)$ is however not necessarily the same for all x). Note that there could be "several" parallel edges from a vertex x to a vertex y. Also, as opposed to the unoriented case, the could be an edge from x to y but no edge from y to x.

We say that $l = (l_0, e_1, l_1, l_2, \dots, l_{n-1}, e_{n-1}, l_n)$ for $n \geq 1$ is a rooted loop with $n = |l|$ steps in Γ if l_0, l_1, l_{n-1} are sites of the graph, if $l_0 = l_n$ and if for all $i \in \{1, \dots, n\}$, e_i denotes an edge from l_{i-1} to l_i in the graph. Let us notice that in the case of parallel edges in the graph, the information about which oriented edges were used are part of the information contained in the loop.

We can note that the probability $p(l)$ that a random walk starting from l_0 follows exactly this loop during its first n steps is exactly $1/\prod_{i=0}^{n-1} d(l_i)$. We define the measure ρ on rooted loops by $\rho(l) = p(l)/n$. Note that this is not a probability measure (a loop l might for instance contain another loop as its first steps if it visits l_0 several times before time n; furthermore, we sum over all possible starting points l_0 in the graph).

The quantity $\rho(l)$ remains unchanged if one changes the root of the loop (if one considers the loop $(l_i, e_{i+1}, l_{i+1}, \dots, l_n, e_1, l_1, \dots, l_i)$ instead of l), which leads naturally to the definition of *an unrooted loop* L as an equivalence class of rooted loops, where two loops are equivalent as soon as they are obtained from one another by rerooting. The measure μ on unrooted oriented loops is then the image of the measure ρ under the mapping that maps each rooted oriented loop to its equivalence class of unrooted loops. This is the loop-measure that has been used and studied extensively in recent years, in connection with loop-erased random walks, Gaussian Free Fields, Dynkin's isomorphism theorems and in the continuous two-dimensional (Brownian) setting, with conformal loop ensembles and SLE curves, see e.g. [9, 19] and the references therein.

In many cases, the number of different rooted loops in the same equivalence class of unrooted loops is the length $n(l) = |l|$ of the loop (one possible root per step on the loop). However, when a loop l consist exactly of the concatenation of $J \geq 1$ copies of exactly the same loop, ie, $n = Jn_1$ and l is exactly the concatenation of J copies of $(l_0, \dots, l_{n_1})$ (and $J = J(l)$ is the maximal such number—note that this number is also invariant under rerooting of l so that we can view it as a function of L), then the number of rooted loops that give rise to the same unrooted loop as l is $n/J(L)$. Hence, the general formula for μ is $\mu(L) = p(l)/J(l)$, when l is any loop in the equivalence class L.

In the sequel, we will refer to loops $l = (l_0, e_1, \dots, l_n)$ (or their equivalence class) such that $J(l) = 1$ as single loops, and we say that the loop l^k defined as the concatenation of k copies of l i.e. as $(l_0, e_1, \dots, l_{n-1}, l_0, e_1, \dots, l_{n-1}, \dots, l_{n-1}, e_n, l_0)$ with $J(l^k) = k$ is its k-fold multiple.

2.2 The Measure on Unrooted Unoriented Loops

In the previous subsection, the graph was oriented, and all our loops (rooted and unrooted) were oriented. Let us now consider an unoriented graph, where each vertex x has a finite number $d(x)$ of outgoing edges (here a single edge from x to x would be counted twice, and we also allow parallel edges between two sites x and y). Then the previous quantity $p(l)$ remains unchanged when one changes the orientation of the loop; indeed, if one defines the time-reversal $\hat{l} := (l_n, e_n, l_{n-1}, \ldots, l_1, e_1, l_0)$, then $p(l) = 1/\prod_{i=1}^{n} d(l_i) = p(\hat{l})$.

We now define an unrooted unoriented loop as the equivalence class of oriented rooted loops, where two such loops are said to be equivalent as soon as they are obtained from one another by rerooting and possibly by time-reversal. Or alternatively, we say that an unrooted unoriented loop is the equivalence class of unrooted oriented loops, modulo time-reversal.

We then define the measure ν on unrooted unoriented loops to be the image of $\rho/2$ under the mapping that maps each rooted oriented loop onto to its equivalence class of unrooted and unoriented loops. The measure ν is of course just the unoriented projection of $\mu/2$.

When the time reversal $\hat{l}$ of a rooted oriented loop l is not in the same unrooted oriented class of loops as l, then there will be twice more rooted oriented loops in the same class $\tilde{L}$ of unoriented unrooted loops of l than in its class L of oriented unrooted loops, so that $\nu(\tilde{L}) = \mu(L)$. It however can happen that l and $\hat{l}$ define the same oriented unrooted loop L (for instance when the loop l is the concatenation of a loop with its time-reversal). In that case, $\nu(\tilde{L}) = \mu(L)/2$. We define $\tilde{J}(\tilde{L}) = J(L)$ or $2J(L)$ depending on whether $L \neq \hat{L}$ or not, so that $\nu(\tilde{L}) = p(l)/\tilde{J}(\tilde{L})$ for all $\tilde{L}$.

All the previous definitions have also straightforward counterparts and generalizations for general Markov processes (not necessarily random walks)—the processes would need to be reversible for the unoriented loops—, and in continuous time and/or in continuous space. Note that as soon as one deals with continuous time, the multiplicity issues (raised by the fact that J is not constant) do not exist. One fundamental example is of course the Brownian loop measure that gives rise to the loop-soup, as introduced in [8]. Other examples include the Brownian loops on cables systems associated to discrete graphs, as studied in [10].

Since our purpose here is to give an elementary presentation of the resampling property of loop-soups, we have opted in the present paper to state and explain things in the most transparent settings (random walk loops on regular graphs, where all points in Γ have the same number g of outgoing edges—which we will from now on assume—, and Brownian loops). The generalization of the proofs to continuous-time and discrete space Markov processes do not require any new idea.

2.3 Loop-Soups

For a given graph, one can define simple natural random objects out of the measures on loops. For each $\alpha > 0$, one can define a Poisson point process of loops, with intensity given by α times the measure μ on loops. This is the loop-soup, as introduced in the Brownian setting in [8] and studied more recently in the discrete setting in [9]. It is also the gas of loops that was already used in [1, 17].

Of course, when one samples a soup of (unrooted) oriented loops according to the loop measure $\alpha\mu$, and one forgets about the orientation of the loops, one gets a soup of unrooted unoriented loops with intensity $2\alpha\nu$, and conversely, one can recover the former by choosing at random the orientation of each loop. In order to avoid confusions, we will use the letters α to denote the intensity of soups of oriented loops (i.e. with intensity measure $\alpha\mu$) and c to denote the intensity of soups of unoriented loops (i.e. with intensity measure $c\nu$). The natural relation between c and α is then $c = 2\alpha$.

We will not recall all the properties of these loop-soups, but we would like to stress the following points:

- The soup of oriented loops with intensity $\alpha = 1$ is very closely related to uniform spanning trees. In particular, the loops in such a loop-soups correspond exactly to the family of loops that have been erased when performing Wilson's algorithm to sample a uniform spanning tree in Γ. And in this context, it is somewhat more natural to consider oriented loops.
- The soup of unoriented loops with intensity $c = 1$ is very closely related to the Gaussian Free Field in Γ and its square. In this context, because one looks only at the cumulated occupation times of the loops, it is in fact somewhat more natural to consider unoriented loops (as the orientation is not needed to define the occupation time measure).

With this notation, the UST is related to $c = 2$ and the GFF to $c = 1$, and more generally, in two dimensions, in the conformal field theory language, the value of c corresponds to the absolute value of the central charge of the corresponding models.

Suppose now that $L_1, \ldots, L_k$ are k different oriented unrooted loops, and let $\mathcal{U}_1, \ldots, \mathcal{U}_k$ denote the respective number of occurrences of these loops in an unrooted loop-soup with intensity $\alpha\mu$. These are k independent Poisson random variables with respective means $\alpha\mu(L_1), \ldots, \alpha\mu(L_k)$, so that

$$P(\mathcal{U}_1 = u_1, \ldots, \mathcal{U}_k = u_k) = \prod_{j=1}^{k} ((\alpha\mu(L_j))^{u_j} e^{-\alpha\mu(L_j)}/u_j!).$$

In the special case where $\alpha = 1$, the α^{u_j} terms disappear, and we get

$$\frac{P(\mathcal{U}_1 = u_1, \ldots, \mathcal{U}_k = u_k)}{P(\mathcal{U}_1 = \ldots = \mathcal{U}_k = 0)} = \prod_{j=1}^{k} \frac{(p(L_j)/J(L_j))^{u_j}}{u_j!}.$$

Similarly, if we are considering instead a loop-soup of unoriented loops with intensity ν (i.e. for $c = 1$), the very same formula holds, i.e. if $\tilde{L}_1, \ldots, \tilde{L}_k$ are k different unoriented loops, and if $\tilde{\mathcal{U}}_1, \ldots, \tilde{\mathcal{U}}_k$ denote the respective number of occurrences of these loops in a soup of unrooted loops with intensity ν, then

$$\frac{P(\tilde{\mathcal{U}}_1 = u_1, \ldots, \tilde{\mathcal{U}}_k = u_k)}{P(\tilde{\mathcal{U}}_1 = \ldots = \tilde{\mathcal{U}}_k = 0)} = \prod_{j=1}^{k} \frac{(p(L_j)/\tilde{J}(\tilde{L}_j))^{u_j}}{u_j!}.$$

2.4 Random Bridges

Recall that in order to slightly simplify notations and some of our considerations, we are from now going to assume that (both in the oriented and in the unoriented cases), the graph Γ will be such that each site has the same number g of outgoing edges. Note that this is not really a restriction, because it is for instance always possible starting from an unoriented graph Γ where each site x has $d(x)$ outgoing edges, with $\sup_x d(x) \leq g$, to add $(g - d(x))$ stationary edges from x to x to the graph, without changing the behavior of the random walks (and this leads to the natural way to extend the results to the case of graphs with non-constant degree).

Let us first suppose that Γ is an oriented graph. Consider now a subgraph $D \subset \Gamma$ and two points x and y in D. We say that a bridge b from x to y in D is a finite nearest-neighbour path (keeping track of the oriented edges used) in D starting at x and finishing at y. We call $n(b)$ the length (number of jumps) of b. A bridge from x to x is allowed to have a zero length.

Suppose now that the Green's function $G_D(x, y)$ is positive and finite. Recall that this is the mean number of visits at y before exiting D, by a random walk starting at x. In other words, it is the sum over all bridges from x to y in D of $g^{-n(b)}$. We can therefore define a probability measure on bridges from x to y in D, that assigns a probability $g^{-n(b)}/G_D(x, y)$ to each bridge b.

Suppose now that we are given N points $x_1, \ldots, x_N$ and N points $y_1, \ldots, y_N$ in D. We say that a the family of paths $b^1, \ldots, b^N$ is an ordered bridge in D from $X = (x_1, \ldots, x_N)$ onto $Y = (y_1, \ldots, y_N)$ if each b^j is a bridge from x_j to y_j in D. We also define $G_D(X, Y) = \prod_{j=1}^N G_D(x_j, y_j)$ and when this quantity is not equal to zero nor infinite, we define the probability measure on ordered bridges from X to Y in D to be obtained by taking N independent bridges from x_j to y_j respectively.

An unordered bridge from X to Y is defined to be the knowledge of a permutation s from $\{1, \ldots N\}$ and of an ordered bridge from X to $Y^s = (y_{s(1)}, \ldots, y_{s(N)})$. We now define a probability measure $B^D_{X,Y}$ on unordered bridges from X to Y in D as follows:

1. First sample a permutation σ so that the probability of $\sigma = s$ is proportional to $G_D(X, Y^s)$.
2. Then, conditionally on σ, sample the ordered bridge from X to Y^σ according to the probability measure on ordered bridges in D described above.

For this to make sense, we need that for at least one s, $G_D(X, Y^s) > 0$. This procedure basically samples an unordered bridge from X to Y in such a way that the probability of a given unordered bridge is proportional to g^{-K}, where K denote the sum of the length of the N bridges that form the generalized bridge. Mind that in the present setting, when $y_2 = y_3$ say, we do count the same collection of N bridges (corresponding to interchanging y_2 and y_3) twice in our partition function, because they correspond to different permutations.

Let us now suppose that the graph Γ is not oriented. In the previous definition, each bridge has an implicit orientation (from x to y). On the other hand, the image under time-reversal (i.e. consider $\hat{b}_j = b_{n-j}$) of the bridge probability from x to y in D is exactly the bridge probability from y to x in D (note that we use here the fact that x and y have the same number of outgoing edges g). One can therefore define the probability measure on unoriented bridges in D joining x and y to be the law obtained by considering $B^D_{x,y}$ and then forgetting about the time-orientation.

Suppose now that $Z = (z_1, \ldots, z_{2N})$ are $2N$ points in D. An unoriented Z-bridge is the knowledge of a pairing t of $\{1, \ldots, 2N\}$ (this is a permutation that contains only cycles of length exactly 2—and we say that i and $t(i)$ are paired—we will denote the N pairs of t by $(t^1_1, t^2_1), \ldots, (t^1_N, t^2_N)$ using some lexicographic rule), and of N unoriented bridges joining the N pairs $(z_{t^1_k}, z_{t^2_k})$ for $k \le N$.

For each Z, we then define the measure B^D_Z on unoriented unordered Z-bridges as follows:

1. We first sample a pairing τ in such a way that the probability of a given pairing t is proportional to $\prod_{k=1}^N G_D(z_{t^1_k}, z_{t^2_k})$.
2. When $\tau = t$, we then sample an N independent (unoriented) bridges in D joining the two points of each of the N pairs $(z_{t^1_k}, z_{t^2_k})$.

Again, this only makes sense if for at least one pairing t, $\prod_k G_D(z_{t^1_k}, z_{t^2_k})$ is positive. Then, the definition just means that we sample a Z-bridge in such a way that the probability of a given Z-bridge is just proportional to g^{-K} where K denote the sum of the length of the N bridges that form this Z-bridge.

These definitions of bridges can be trivially extended to the Brownian settings (both in d-dimensional space as well as on cable systems), provided that no two z_j's coincide (in the unoriented bridges) and that no x_i is equal to an y_j (for the oriented bridges) so that the Green's functions involved are all finite. The only difference is that the distribution of an individual bridge from x to y is done in two steps:

1. First, sample the time-length T of the Brownian bridge according to the probability measure $p_{D,t}(x,y)dt/G_D(x,y)$, where $p_{D,t}(x,y)$ is the density at y of the law of a Brownian motion at time t, starting from x and killed upon exiting D.
2. Then, conditionally on T, sample a usual Brownian bridge from x to y and time-length T, conditioned to stay in D.

3 Partial Resampling of Soups, and Spatial Markov Properties

We now describe various instances of the partial resampling properties of loop-soups, and discuss some consequences.

3.1 Partial Resampling of Soups of Oriented Loops at $\alpha = 1$

Let us suppose that Γ is an oriented graph of degree g as before, and that $D \subset \Gamma$ is a subgraph of Γ where the Green's function is finite. We are going to describe the resampling property of the soup of oriented loops with intensity $\alpha = 1$. Suppose that F_1 and F_2 are two disjoint finite set of vertices in our graph. When one considers a loop-soup in D, then the number of loops in the loop-soup that do intersect both F_1 and F_2 is a Poisson random variable $\mathscr{M} = \mathscr{M}(F_1, F_2)$ with finite mean equal to the μ-mass of the set of loops that intersect both F_1 and F_2. We denote the family of $\mathscr{M}$ loops that intersect both F_1 and F_2 by $\overline{\mathscr{L}}$ (the information in $\overline{\mathscr{L}}$ includes how many occurrences of any given oriented unrooted loop that intersects F_1 and F_2 there are). We will write $\overline{\mathscr{L}} = (\mathscr{L}_1, \dots, \mathscr{L}_{\mathscr{M}})$, where the chosen order of the loops in the family follows some lexicographic (deterministic) rule, so that the information provided by $\overline{\mathscr{L}}$ and $(\mathscr{L}_1, \dots, \mathscr{L}_{\mathscr{M}})$ are identical.

When L is an unrooted loop that intersects F_1 and F_2, we can consider the finitely many portions of L that are of the type $(a_0, e_1, a_1, a_1, \dots, a_k)$ where the points a_0, a_k are in F_2, where $\{a_1, \dots, a_{k-1}\} \cap F_2 = \emptyset$ and at least one of the a_i is in F_1. In other words, these are the excursions of L away from F_2 that do reach F_1. We allow $a_0 = a_k$, or the excursion to be the entire loop (which happens if L visits F_2 only once) and it can also happen that the same excursion occurs several times in the same loop.

When we sample $\overline{\mathscr{L}}$, we call η the collection of all excursions of its loops. We can again decide to order them in some lexicographic predetermined deterministic way, so that we can write $\eta = (\eta_1, \dots, \eta_{\mathscr{N}})$ (again, it is important that if a given piece appears several times in the loop-soup, then it appears several times in this list as well). Note that $\mathscr{N} \geq \mathscr{M}$ because each loop that intersects F_1 and F_2 contains at least one such excursion. The pieces $\eta_1, \dots, \eta_{\mathscr{N}}$ might be part of $\mathscr{N}$ different loops (in which case $\mathscr{N} = \mathscr{M}$), but they could also be all parts of the same loop (in which case $\mathscr{M} = 1$). Of course, the probability that $\mathscr{N} = \mathscr{M} = 0$ is also positive.

Observe that one intuitive way to discover all these excursions is in fact to explore all the loops "starting" from their intersection points with F_1, in both the positive time-direction and the negative time-direction, until reaching F_2 in both directions.

Each of the pieces η_j are naturally oriented as parts of oriented loops, and we can define their respective starting points $\mathscr{Y}_j$ and endpoints $\mathscr{X}_j$ (note that all these points are on F_2). The missing parts of the loops that the η's are part of will therefore be bridges in the complement of F_1, that join each of the $\mathscr{X}_j$'s to a $\mathscr{Y}_{\sigma(j)}$ for a permutation σ i.e. the missing part will be an unordered bridge β from the vector $\mathscr{X} = (\mathscr{X}_1, \dots, \mathscr{X}_{\mathscr{N}})$ to the vector $\mathscr{Y} = (\mathscr{Y}_1, \dots, \mathscr{Y}_{\mathscr{N}})$ in $D \setminus F_1$. Now, the resampling result in this case goes as follows:

Proposition 1 *The conditional distribution of β given η is exactly the unordered bridge measure $B^{D \setminus F_1}_{\mathscr{X}, \mathscr{Y}}$.*

Note that this conditional distribution is fully described by the vectors $\mathscr{X}$ and $\mathscr{Y}$ (i.e. it depends on η just as a function of $\mathscr{X}$ and $\mathscr{Y}$), which is one of the main

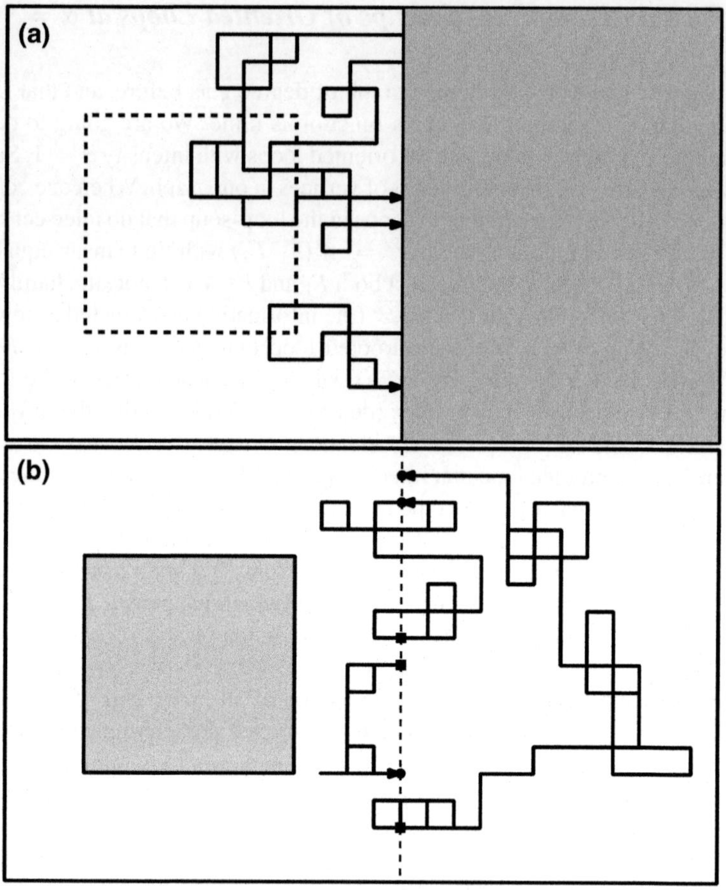

Fig. 3 Illustration for Proposition 1: discovering (**a**) the oriented excursions away from the right part that reach the small square, (**b**) sampling the three oriented bridges in the complement of the small square

features of this result. In other words, conditionally on $\mathscr{X}$ and $\mathscr{Y}$, η and β are independent. In particular, the number of actual loops that are being created by β when one concatenates it with η does not intervene in the conditional distribution, which is a specific feature of this $\alpha = 1$ case.

Let us comment on the case where $F_2 = D \setminus F_1$: If one then conditions on the number of jumps of the loop-soup on each edges from a point in F_1 to a point of F_2 (one then gets a collection $(\mathscr{X}_j', \mathscr{X}_j)_{j\le\mathscr{N}}$ of jumps from $\mathscr{X}_j' \in F_1$ to $\mathscr{X}_j \in F_2$), and on the number of jumps of the loop-soup on each edge from a point of F_2 to a point of F_1 (one then gets a collection $(\mathscr{Y}_j, \mathscr{Y}_j')_{j\le\mathscr{N}}$ of jumps from $\mathscr{Y}_j \in F_2$ to $\mathscr{Y}_j' \in F_1$), then the conditional distribution of the missing pieces in F_2 and in F_1 are independent, and there are respectively the unordered bridge measure in F_2 from $\mathscr{X}$ to $\mathscr{Y}$ (this corresponds to β), and the unordered bridge measure from $\mathscr{Y}'$ to $\mathscr{X}'$ in

F_1 (this corresponds to η without the first and last jumps of each excursion). This can be interpreted as a spatial Markov property of the occupation field on oriented edges (the random function that assigns to each oriented edge the total number of jumps of the soup along this edge) of the $\alpha = 1$ soup of oriented loops. We will discuss this again at the end of this section.

In the same spirit, we can in fact "symmetrize" also Proposition 1 also when F_2 is a subset of the complement of F_1. Let us then define the collection of crossings $\eta_{1\to2}$ to be the parts of the loops in the loop-soup of the type $a_0, e_1, \ldots, a_n$ with $a_0 \in F_1$, $a_n \in F_2$ and $a_1, \ldots, a_{n-1} \in D \setminus (F_1 \cup F_2)$. We also define $\eta_{2\to1}$ similarly, and note that there are as many crossings from F_1 to F_2 as there are crossings from F_2 to F_1. Let $\mathscr{X}$ (resp. $\mathscr{X}'$) denote the vector of endpoints of $\eta_{1\to2}$ (resp. $\eta_{2\to1}$) and $\mathscr{Y}$ (resp. $\mathscr{Y}'$) the vector of starting points of $\eta_{2\to1}$ (resp. $\eta_{1\to2}$). Then, we can note that $\mathscr{X}$ and $\mathscr{Y}$ are exactly the same as the ones defined in Proposition 1, while $\mathscr{X}'$ and $\mathscr{Y}'$ correspond to those that one obtains when interchanging F_1 and F_2. Furthermore, $\eta_{1\to2}$ and $\eta_{2\to1}$ are fully determined by η (or alternatively by the symmetric family η' of excursions outside of F_1 that do reach F_2). It follows readily from Proposition 1 that:

Proposition 2 *Conditionally on $\eta_{1\to2}$ and on $\eta_{2\to1}$, the missing parts of the loops that they are part of (these are the loops of the $\alpha = 1$ soup of oriented loops that intersect both F_1 and F_2) are described by two independent unordered bridges with conditional distributions $B^{D\setminus F_1}_{\mathscr{X},\mathscr{Y}}$ and $B^{D\setminus F_2}_{\mathscr{X}',\mathscr{Y}'}$.*

Note that the other loops in the loop-soup (i.e. the loops that either do not intersect at least one of the two sets F_1 or F_2) are just described by a loop-soup in the complement of F_1 and a loop-soup in the complement of F_2, that are coupled to share exactly the same loops that stay in $D \setminus (F_1 \cup F_2)$.

Let us now prove Proposition 1.

Proof Let us consider a family E of N excursions, such that $P(\eta = E) > 0$ and such that the N excursions $E_1, \ldots, E_N$ of E are all different. Then if $\eta = E$ and $\tilde{\mathscr{L}} = \overline{L}$, all the loops in $\overline{L}$ are simple, and they do occur necessarily exactly once (and not more). Hence, for such an $\overline{L}$, the probability that $\overline{\mathscr{L}} = \overline{L}$ is proportional to $g^{-\overline{n}(\overline{L})}$ where $\overline{n}$ is the sum of the lengths of the loops in $\overline{L}$ (and the proportionality constant does not depend on $\overline{L}$).

On the other hand, if X and Y are the vector of end-points of E, the $B^{D\setminus F_1}_{X,Y}$-probability to sample a unordered bridge that gives rise exactly to $\overline{L}$ when concatenating it to E is proportional to g^{-K} (where $K = \overline{n}(\overline{L}) - \overline{n}(E)$ is the total length of the generalized bridge), because there is just one permutation per bridge that works. It therefore follows immediately that conditionally on $\eta = E$, the distribution of the missing bridges is indeed $B_{\mathscr{X},\mathscr{Y}}$ in $D \setminus F_1$.

Instead of treating directly the case of multiple occurrences of the same excursions in η, we will use the following trick (a similar idea can be used to show the fact that the loops erased during Wilson algorithm do correspond exactly to an oriented loop-soup, see for instance [19]). We choose a very large integer W (that is going to tend to infinity), and we decide to replace the graph Γ by the graph

Γ^W, which is obtained by keeping the same set of vertices as Γ, but where each edge of Γ is replaced by W copies of itself. In this way, each site has now gW outgoing edges instead of g. There is of course a straightforward relation between random walks, loops and bridges on Γ^W and on Γ. For instance, a loop-soup (resp. bridge, resp. excursion) on Γ^W is directly projected on a loop-soup (resp. bridge, resp. excursion) on Γ.

Let us couple loop-soups with intensity $\alpha = 1$ in all of the Γ^W's on the same probability space, in such a way that the projections of the loop-soups in Γ^W onto Γ (in the sense described above) are the same for all W's. We fix also F_1, F_2, and define (with obvious notation), $\mathscr{L}^W$, η^W, $\overline{L}^W$ etc. Note that the vectors of extremal points $\mathscr{X}$ and $\mathscr{Y}$ are then the same for all η^W's.

We can also note that the probability that some edge is used more than once in the loop-soup does tend to 0 as $W \to \infty$. The probability that all excursions in η^W are different therefore tends to 1 as $W \to \infty$.

But conditionally on the fact that all excursions in η^W are different (applying our previous result to Γ^W), we know that the conditional distribution of $\overline{\mathscr{L}}^W \setminus \eta^W$ given η^W is the bridge probability measure from $\mathscr{X}$ to $\mathscr{Y}$ in $D^W \setminus F_1$. Projecting this onto Γ, we get that the conditional distribution of β given η^W (on the event that in η^W, no two excursions are the same) is the unordered bridge measure $B_{\mathscr{X},\mathscr{Y}}$ in $D \setminus F_1$.

If $U(W)$ is the event that no two excursions of η^W appear twice, we therefore get that, conditionally on $\eta = E$ and $U(W)$, the conditional distribution of β is the unordered bridge measure $B_{\mathscr{X},\mathscr{Y}}$ in $D \setminus F_1$. We now just let $W \to \infty$, which concludes the proof of the proposition.

3.2 *Partial Resampling of Soups of Unoriented Loops at c = 1*

Let us now come back to the setting where the graph Γ is unoriented. When one considers a soup of unoriented loops with intensity ν (recall that this corresponds to $c = 1$ or $\alpha = 1/2$ i.e. to a soup of oriented loops with intensity $\mu/2$ where we forget the orientation of each loop). We denote the collection of unoriented loops that intersect both F_1 and F_2 by $\overline{\mathscr{L}} = (\tilde{\mathscr{L}}_1, \ldots, \tilde{\mathscr{L}}_{\mathscr{M}})$, the corresponding collection of (unoriented) excursions by $\eta = (\tilde{\eta}_1, \ldots, \tilde{\eta}_{\mathscr{N}})$ and the endpoints of these $\mathscr{N}$ excursions by $\mathscr{Z} = (\mathscr{Z}_1, \ldots, \mathscr{Z}_{2\mathscr{N}})$. The missing parts of the (unoriented) loops are unoriented paths that join each $\mathscr{Z}_i$ to exactly one other $\mathscr{Z}_j$, so that β is an unordered $\mathscr{Z}$-bridge in $D \setminus F_1$.

Note again that it is intuitively possible to explore the excursions $\tilde{\eta}_j$ "starting" from their intersections with F_1 in both directions, until hitting F_2 (and in this way, one did yet discover the missing parts β).

Proposition 3 *The conditional distribution of β given η is exactly the unordered unoriented bridge measure $B_{\mathscr{Z}}$ in $D \setminus F_1$.*

Just as in the oriented case, we stress that an important feature in this statement is that this conditional distribution is a measurable function of the vector $\mathscr{Z}$ (the other information on the excursions are not needed). We will further comment on this in the next subsection.

Proof We will follow the same idea as in the proof of the oriented case. As in the unoriented case, when the N pieces $\tilde{E}_1, \ldots, \tilde{E}_N$ of E are all different, the statement is almost immediate (for each good ordered bridge, only one pairing works in order to complete E into $\overline{L}$, and the probability to complete these N pieces into $\overline{L}$ is therefore proportional to g^{-K} where K is the difference between the total number of jumps in the loop-configuration and in E).

We then use the same trick with copying each edge a large number of times. The very same argument the works, almost word for word.

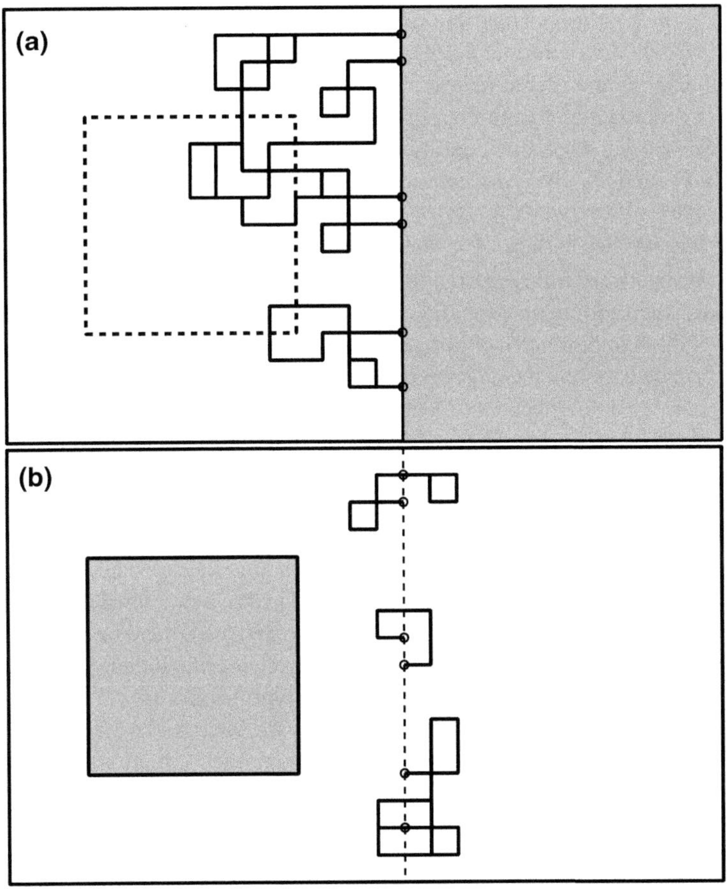

Fig. 4 Illustration of Proposition 3: discovering (**a**) the unoriented excursions away from the right part that reach the small square, (**b**) sampling the three unoriented bridges in the complement of the small square

3.3 Spatial Markov Properties

The particular case where F_2 is the complement of F_1 is also of interest for the soup of unoriented loops. Let us for instance describe how things work for the occupation times of loop-soups (which is the main focus of the papers of Le Jan [9]). If one then conditions on the numbers of jumps of the loop-soup on all edges between a point in F_1 and a point of F_2 (in either direction—the loops being unoriented there is anyway no direction), then the conditional distribution of the parts β in F_2 of the loops that intersect both F_1 and F_2 is described by Proposition 3 and it is a unordered unoriented bridge in F_2 (and it is in fact fully described by the knowledge of the number of jumps along the edges between F_1 and F_2, i.e. this conditional distribution is a function of these number jumps of the edges between F_1 and F_2). But, the situation is symmetric and we can interchange the roles of F_1 and F_2; we therefore conclude that given β and the numbers of jumps along the edges between F_1 and F_2, the conditional distribution of η' defined to be the collection η where one has removed the two extremal jumps of each η_j (these are the jumps between F_1 and F_2), is that of an unordered unoriented bridge in F_1 (and the law of this bridge is also fully described by the number of jumps between F_1 and F_2).

In other words, when one conditions on these number of jumps along the edges between F_1 and F_2, we can enumerate these jumps (using some deterministic lexicographic rule) by $(\mathscr{Z}_j', \mathscr{Z}_j)_{j \le 2\mathscr{N}}$ where $\mathscr{Z}_j' \in F_1$ and $\mathscr{Z}_j \in F_2$. Then, the conditional distribution of η' and β are conditionally independent unordered bridges, respectively following the unordered bridge measures $B_{\mathscr{Z}'}^{D\setminus F_2}$ and $B_{\mathscr{Z}}^{D\setminus F_1}$. In particular, when adding on top of this the loop-soups in F_1 and the loop-soups in F_2, it follows that conditionally on the occupation times (i.e. on the number of jumps N_e across each edge) on the edges between F_1 and F_2, the occupation times on sites and edges in F_1 is independent of the occupation times on sites and edges in F_2. We can rephrase this property in the following sentence: The occupation time field on edges of the soup of unoriented loops for $c = 1$ does satisfy the spatial Markov property.

We can note that if U is a non-negative function of the occupation time field on the edges of the form $U((N_e)) = \prod_e u_e(N_e)$, such that the expectation of U (for the $c = 1$ loop-soup) is equal to one, then if we define the new probability measure Q on occupation times on edges by $dQ/dP((N_e)) = U((N_e))$, then the spatial Markov property also holds for Q. This can be used to represent a modification of the Markov chain (i.e. different walks with non-uniform jump probabilities).

If we consider an unoriented graph, but that we interpret as an oriented graph (each unoriented edge defines an oriented edge in each direction), on which we define an $\alpha = 1$ soup of oriented loops, then we can also reformulate the results of Sect. 3.1 in a similar way. More precisely, for each edge, we can define the total number of jumps $N_1(e)$ by the soup in one direction of e, and $N_2(e)$, the number of jumps in the opposite direction. Then, if we define $N_e := ((N_1(e), N_2(e))$, this two-component occupation time field on edges of the $\alpha = 1$ soup of oriented loops satisfies the spatial Markov property in the same sense as above.

Let us now come back to the study of the loops themselves, and not just of the cumulated occupation time of the soup. As in the oriented case, we can also (when F_2 is a subset of F_1) rephrase Proposition 3 in a more symmetric way, involving the crossings between F_1 and F_2. We define $\eta_{1\leftrightarrow 2}$ the set of (unoriented) parts of loops in the $c = 1$ loop-soup that join a point of F_1 to a point of F_2 and otherwise stay in the complement of $F_1 \cup F_2$, and we denote by $\mathscr{X}$ the vector of endpoints of these crossings in F_2, and by $\mathscr{X}'$ the set of endpoints in F_1. Then:

Proposition 4 *Conditionally on $\eta_{1\leftrightarrow 2}$, the missing parts of the unoriented loops that these crossings are part of (these are the loops in the loop-soup that intersect both F_1 and F_2) are described by two independent unordered unoriented bridges with respective conditional distributions $B_{\mathscr{X}}^{D\setminus F_1}$ and $B_{\mathscr{X}'}^{D\setminus F_2}$.*

Figure 5 that illustrates the corresponding result in the Brownian case, can also be used to illustrate this result.

It is also easy to generalize Proposition 4 and Proposition 2 to more than two sets F_1 and F_2 (and have instead n disjoint sets $F_1, \dots, F_n$). For instance, in the unoriented case, one then conditions on the set $\eta_{\leftrightarrow}$ of all crossings from any F_i to any other F_j that also stay in the complement of all the other F_k's. These crossings define n vectors $\mathscr{X}^1, \dots, \mathscr{X}^n$ (where $\mathscr{X}^j$ is a list of the even number of endpoints on F_j of the aforementioned crossings). Conditionally on $\eta_{\leftrightarrow}$, the missing parts of the loops (that are the loops in the loop-soup that touch at least two different F_j's) are described by n conditionally independent unordered unoriented bridges with respective distributions $B_{\mathscr{X}_j}^{D' \cup F_j}$ (where $D = D \setminus \cup_i F_i$) for $j \le n$.

Such decompositions of the loops in the soup that intersect disjoint compact sets into crossings + conditionally independent unordered bridges, can be immediately transcribed to the case of Brownian loops on the cable system associated to this graph as studied in [10]; we leave this as a simple exercise to the reader. This is all of course closely related to the Markov property of the Gaussian Free Field, as well as to Dynkin's isomorphism theorem [4] via the relation between the square of the GFF and the loop-soup (see e.g., [9] and the references therein for background).

With such Markovian-type properties in hand, a natural next step is to define random sets that play the role of stopping times for one-dimensional Markov processes. In the setting of the discrete GFF, these are the local sets as defined in [15], and that turned out to be very useful concepts. Just as for one-dimensional stopping times, there are several possible ways to define them, depending on what precise filtration on considers. In the present case (we do here describe the definitions in the unoriented loop-soup for $c = 1$, but the oriented case would be almost identical), one can for instance say that:

- A random set of points $\mathscr{F}$ is a stopping set for the occupation time field filtration, if for any F_1, the event $\{\mathscr{F} = F_1\}$ is measurable with respect to the occupation time field on all edges adjacent to F_1.
- A random set of points $\mathscr{F}$ is a stopping set for the loop-soup filtration, if for any F_1, the event $\{\mathscr{F} = F_1\}$ is measurable with respect to the trace of the loop-soup

on all edges adjacent to F_1 (i.e. it is measurable with respect to the set of loops that are fully contained in F_1 and the set of excursions η defined above, when F_2 is the complement of F_1.

- A random set of points $\mathscr{F}$ is a stopping set for the loop-soup, if for any F_1 such that $P(\mathscr{F} = F_1) > 0$, conditionally on the event $\{\mathscr{F} = F_1\}$, the distribution of the loop-soup outside of F_1 consists of the union of an independent loop-soup in the complement F_2 of F_1 and of a set of bridges in F_2, with law described as above via the end-points of the excursions η in F_1.

Clearly, the first definition implies the second one, which implies the third one by Proposition 3 (the third property for the first two definitions can be viewed as a "strong Markov property" of these fields), but the converse is not true (the last definition allows the use of "external randomness" in the definition of $\mathscr{F}$ (while the second does not), and the second one allows features of individual loop (while the first does not).

3.4 Brownian Loop-Soup Decompositions

The previous results have almost identical counterparts in the setting of oriented Brownian loop-soups with intensity $\alpha = 1$ and unoriented Brownian loop-soups with intensity $c = 1$.

Suppose that D is an open set in d-dimensional space, such that the (Dirichlet) Green's function in D is finite (away from the diagonal). Suppose that F_1 and F_2 are two disjoint compact sets in D, that are both non-polar for Brownian motion (i.e. Brownian motion started away from these sets has a non-zero probability to hit them). Then, we can again define:

1. The law of unordered oriented Brownian bridges in $D \setminus F_1$ from a finite family $X = (x_1, \ldots, x_n)$ of points to another such family $Y = (y_1, \ldots, y_n)$, and the law of unordered unoriented Z-Brownian bridges in $D \setminus F_1$ from a finite family of points $Z = (z_1, \ldots, z_{2n})$ to itself (in the latter case, points of Z are paired, like in the random walk case). This works as long as all Green's functions involved are finite (which is the case as soon as all $x_i \neq y_j$ for all i, j, and that $z_i \neq z_j$ for all $i \neq j$).
2. The set η of $\mathscr{N}$ oriented (resp. unoriented) excursions of the loops in an oriented (resp. unoriented) loop-soup with intensity $\alpha = 1$ (resp. $c = 1$) away from F_2, that reach F_1. In the ordered case, we call their endpoints vector $\mathscr{X} = (\mathscr{X}_1, \ldots, \mathscr{X}_{\mathscr{N}})$ and their starting point vector $\mathscr{Y}$, and in the unoriented case, we call $\mathscr{Z} = (\mathscr{Z}_1, \ldots, \mathscr{Z}_{2\mathscr{N}})$ the extremity vector.

Then, the Brownian counterparts of Proposition 1 and of Proposition 3 go as follows:

Proposition 5

- *For the soup of oriented Brownian loops with* $\alpha = 1$*: Conditionally on* η*, the missing pieces of the loops (that the pieces* η *are part of) are distributed like an unordered Brownian bridge from* $\mathscr{X}$ *to* $\mathscr{Y}$ *in* $D \setminus F_1$.
- *For the soup of unoriented Brownian loops with* $c = 1$*: Conditionally on* η*, the missing pieces of the loops are distributed like an unordered unoriented* $\mathscr{Z}$*-Brownian bridge in* $D \setminus F_1$.

And as before, one can derive the more symmetric results: For instance, if F_1 and F_2 are two disjoint compact subsets of D, we can define the crossings from F_1 to F_2 and vice-versa in the oriented case, and the crossings between F_1 and F_2 in the unoriented case. When one conditions on these crossings, one can then complete the picture with two conditionally independent unordered oriented bridges (in the oriented case) or by two conditionally independent unordered unoriented bridges (in the unoriented case). We illustrate this result in Figs. 5 and 6 (here we consider the oriented case, D is the rectangle, F_1 is the small circle and F_2 the large circle). Conditionally on the points (and their status—square or circle depending on the orientation of the loops) on the two circles, the three pictures in Fig. 6 are independent (this is the oriented version of Fig. 2).

In the context of two-dimensional continuous systems, clusters of loops in a loop-soup are interesting to study, as pointed out in [18]; it has been proved in [16] that boundaries of such clusters for $c \leq 1$ form Conformal Loop Ensembles with parameter $\kappa = \kappa(c)$, where $\kappa(1) = 4$. The CLE_4 (and the SLE_4 curves) is also known (see [3, 15]) to be related quite directly to the Gaussian Free Field. The role of the $c = 1$-clusters of loops in the framework of cable-systems and in relation to

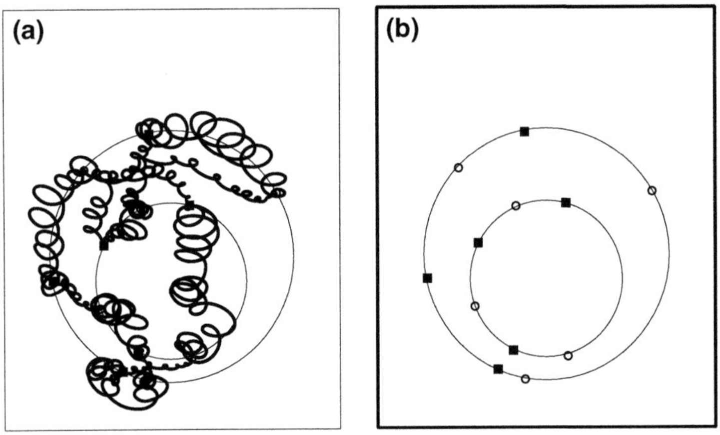

Fig. 5 Sketch of the oriented Brownian case: (**a**) the two oriented loops that touch the two circles, (**b**) keeping only the endpoints of these crossing on each circle, with trace of the orientation

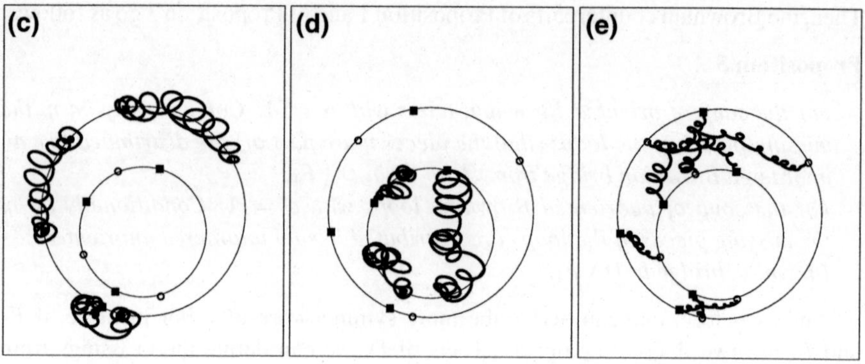

Fig. 6 (**c**) The outer bridges joining each circle point to a square point, (**d**) sampling the inner bridges joining each circle point to a square point, (**e**) the six crossings, joining a circle point to a square point. The final loops are oriented so that the crossings from small to large circle go from a circle point to a square point

the Gaussian Free Field has been pointed out by Lupu [10] (the clusters provides a direct link between the loop-soups and the Gaussian Free Field itself, rather than just to its square). The present result sheds some light on the recently derived [14] decomposition of critical $2d$ loop-soup clusters (for $c = 1$) in terms of Poisson point processes of Brownian excursions (we refer to [14] for comments and questions).

4 Resampling for Continuous-Time Loop-Soups, the GFF and Random Currents

We devote now a short separate section on the case of discrete continuous-time loop-soups, that have been studied by Le Jan [9]. As we shall see, in that setting, it is natural to consider the conditioned distribution of the loop-soup (unoriented for $c = 1$ i.e. $\alpha = 1/2$, or oriented for $\alpha = 1$) given the value of their local times on a given family of sites. Some of the results are very closely related to Dynkin's isomorphism theorem (i.e. it will be a pathwise version of a generalization of it). Just as previously, we will describe the case of simple random walk on the graph where each point x has the same number g of outgoing edges, but the results can easily be generalized to the case of general Markov chains. Some of following considerations will be reminiscent of the arguments in [9] (Sects. 7 and 9 in particular). In the first subsections, we will focus on the case of unoriented loop-soups, and we will briefly indicate the similar type of results that one gets in the oriented case.

4.1 *Slight Reformulation of the Resampling Property of the Discrete Loop-Soup*

We can start with the same setting as before, with the graphs Γ and $D \subset \Gamma$, the random walk on this graph killed upon hitting $\Gamma \setminus D$, and its Green's function $G_D(\cdot, \cdot)$. In the previous sections, we chose for expository reasons (as this was for instance the natural preparation for the Brownian case) to study loops in the loop-soup that visit two different sets of sites F_1 and F_2. But in fact, the following setting is a little more natural and more general: Consider now a family $e_1, \ldots, e_n$ of edges of D, and the graph D' obtained by removing these n edges from D. We can now sample an unoriented loop-soup (for $c = 1$), and observe the numbers $N_1, \ldots, N_n$ of jumps along those n unoriented edges. We now want to know the conditional distribution of the entire loop-soup given this information. In particular, we would like to know how these $N_1 + \cdots + N_n$ jumps are hooked together into loops (clearly, the loop-soup in D' consisting of the loops that use none of these n edges is independent of $\overline{N} := (N_1, \ldots, N_n)$).

We can associate to $\overline{N}$ the vector $\mathscr{X}$ consisting of the $2N_1 + \cdots + 2N_n$ endpoints of these jumps. Once we label them, we can as before the collection β of pairing and bridges that join them in the loop-soup. Note that the bridge is allowed to contain no jump when one pairs two identical end-points. We can also define the unordered bridge measures in D' (corresponding to paths that use no edge of $D \setminus D'$) as before. Then, exactly as before, one can prove the following version of the resampling:

Proposition 6 *The conditional distribution of β given $N_1, \ldots, N_n$ is exactly the unordered unoriented bridge measure $B_{\mathscr{X}}$ in D'.*

Note that for some choices of family of edges $e_1, \ldots, e_N$, it can happen that an even number of endpoints of the discovered jumps are at a certain vertex where no neighboring edge is in D'. In that case, the bridge measure pairs these jumps at random and the corresponding bridge is anyway the empty bridge from x to x. A trivial example is of course the case where $e_1, \ldots, e_n$ are all the edges of D. Then, the proposition just says that the conditional distribution of the loops given the occupation time measure is obtained by just pairing at random the incoming edges at each site. "Loops can exchange their hats uniformly at random at each site".

This reformulation makes it clear that in the discrete time setting, the Markov property of the occupation time field is really a Markov property on the edges (which is not surprising, given that the field is actually naturally defined on the edges).

4.2 *Continuous-Time Loops*

Following Le Jan's approach [9], we now introduce the associated continuous-time Markov chain, for each site x, the chain stays an exponential waiting time of mean $1/g$ before jumping along one of the g outgoing edges chosen at random (for expository reasons, we describe this in the case where each edge has the same number of outgoing edges). Note that we allowed stationary edges in the graph, so that the continuous-time Markov chain can also "jump" along those (and we can keep track of these jumps, even if they do not affect the occupation time at sites). As pointed out by Le Jan, the loop-soup of such continuous-time loops for $\alpha = 1/2$ is particularly interesting, as its cumulated occupation time (on sites) is exactly the square of a Gaussian Free Field on this graph (here one may introduce one or more killing point, so that the loop-soup occupation-time is finite, and the free field with boundary value 0 at this point is well-defined). In this setting, the loops of the discrete Markov chain do correspond exactly to loops of the continuous-time chain, but the latter also contains some additional stationary loops, that just stay at one single point without jumping during their entire life-time.

When one considers a continuous-time loop and a finite set of vertices in the graph that it does visit, one can cut-out from the loops the time that it does spend at these points and obtain a finite sequence of excursions away from this set. This corresponds to the usual excursion theory of continuous-time Markov processes (an excursion from x to y will be a path that jumps out of x at time 0 and jumps into y at the endpoint of the excursion). One can the introduce the natural excursion measure $\mu_{x,y}^A$, which is the natural measure on set of unoriented excursions that go from x to y while avoiding all the points in A (it corresponds to the discrete excursion measure that puts a mass g^{-n} to such an excursion with n jumps, and one then adds $n - 1$ independent exponential waiting times at the $(n - 1)$ points inside the excursions.

One can view the continuous-time Markov chain as the limit when $M \to \infty$ of the discrete-time Markov chain on a graph D^M, where one has added to each site x, M stationary edges from x to itself (when one renormalizes time by $1/M$, the geometric number of successive jumps along these added stationary edges from x to x before jumping on another edge, does converges to the exponential random variables)—this approach is for instance used in [19] in order to derive the properties of the continuous-time chains and loop-soups from the properties of the discrete-time loop-soups. Let us now consider a finite set of points $x_1, \ldots, x_n$ in the graph, and for a given M, we condition on the $N_1, \ldots, N_n$ of jumps by the loop-soup along the stationary unoriented edges $e_1, \ldots, e_n$. More precisely N_1 will denote the total number of jumps in the loop-soup along the M added stationary edges from x to x. Note that because both end-points of a stationary edge are the same, these N_1 jumps correspond to $2N_1$ jump-endpoints, that are all at x_1. We can now apply Proposition 6 to this case; this describes the distribution of how to complete and hook up these $N_1 + \cdots + N_n$ jumps into unoriented loops in order to recover the loops in the loop-soup that they correspond to. One has to pair all these $2N_1 + \cdots + 2N_n$ endpoints.

Mind that as M gets large, the mass of the trivial excursion from x_1 to x_1 with zero life-time is always 1, while the mass of (unoriented) excursions with at least one jump along the "non-added" nM stationary edges neighboring these points from x_1 to some x_j that stays away from $\{x_1, \ldots, x_n\}$ during the entire positive lifetime (if it is positive) will be of order $1/M$ (unless all neighbors of x_1 are in $\{x_1, \ldots, x_n\}$ in which case this quantity is zero) and that the set of excursions from x_1 to x_j that visit at least one of the points of $\{x_1, \ldots, x_n\}$ during its positive life-time is of the order of $O(1/M)^2$. It is a simple exercise that we safely leave to the reader to check that in the $M \to \infty$ limit, the discrete Markovian description becomes the following:

Proposition 7 *If we consider the continuous-time Markov chain loop-soup and condition on the total occupation time $l(x_1), \ldots, l(x_n)$ at the n points $x_1, \ldots, x_n$, then the unoriented excursions away from this set of points by the loop-soup will be distributed exactly like a Poisson point process of excursions with intensity $\mu_l = (1/2) \times \sum_{i \le j} l(x_i) l(x_j) \mu_{x_i, x_j}^{x_1, \ldots, x_n}$ conditioned on the event that the number of excursions starting or ending at each of the n points $x_1, \ldots, x_n$ is even.*

The particular case where the set of points $\{x_1, \ldots, x_n\}$ is the whole vertex set is again of some interest: The conditional distribution of the number of unoriented jumps on the edges given the occupation time field on the vertices is a collection of independent Poisson random variables with respective means $l(x_i)l(x_j)$, conditioned by the event that for all site x, the total number of jumps on the incoming edges at x is even. This is exactly the random current distribution associated with the Ising model. For some further comments on this relation with random currents, the GFF and Ising, we refer to [11].

4.3 Relation with Dynkin's Isomorphism

It should be of course noted that this decomposition is closely related Dynkin's isomorphism (see [4, 5, 12] and the references therein), except that one here conditions here on the value of the square of the GFF instead of the value of the GFF itself. The previous result implies (when one only looks at occupation times and not at the loop-soup itself) that conditionally on the value of the square of the GFF at the set of points $\{x_1, \ldots, x_n\}$, the square of the value of the GFF at the other points is the sum of the occupation times of the conditioned Poisson point process of excursions with an independent squared GFF in the remaining (smaller) domain.

If one however conditions the GFF at the n sites to be all equal to the same value t, then one can consider instead a graph where all these points are identified as a single point and note that when the GFF on the new graph conditioned to have value t at that point is distributed as the GFF on the initial graph, conditioned to have value t at each of the n points. One can apply the previous statement to that new graph and note that the conditioning on the event that the number of excursions-extremities at each boundary site is even then disappears, because when there is just one such site, this number is anyway even (each excursion from this point to itself has two endpoints). Here it is however essential that the signs of all these values are the same (because if one identifies them into a single point, then they will anyway correspond to the same value of the GFF, not just to the same value of its square.

In summary, conditioning by the value of the square of the GFF gives rise to the parity conditioning, but it is also possible to condition on the actual value of the GFF and the parity conditioning becomes irrelevant when one looks at the occupation times only. Note that Dynkin's isomorphism then follows, because in the latter case, the conditional distribution of the square of the GFF at the other points (which is therefore the square of the GFF in this smaller domain with boundary conditions given by these conditioned boundary values) will be the sum of the contribution of the loops that only visit those points (which is a squared GFF in the remaining domain) with the occupation time of the Poisson point process of excursions, while the conditioned GFF is a GFF with some prescribed boundary conditions, that can be viewed as the sum of a GFF in the complement of the set of marked points with the deterministic harmonic extension of these boundary values.

4.4 The Oriented Case

One can follow almost word for word the same strategy to study the conditional distribution of oriented continuous-time loop-soups at $\alpha = 1$ given their cumulated local time at sites. In that case, the excursions will be oriented, and the conditional distribution of the excursions away from these points will be a Poisson point process conditioned on the event that for each site, the number of incoming excursions is equal to the number of outgoing ones.

The particular case where the set of points is the whole vertex set is again interesting. The conditional distribution of the set of jumps will be independent Poisson on each oriented edge, but conditioned on the fact that the number of incoming jumps at each site is going to its number of outgoing jumps. We leave all the details and further results to the interested reader.

Note We found out that the recently posted preprint [2] by Camia and Lis describes some ideas that are similar to the present paper (which was prepared totally independently of [2]).

Acknowledgements The support of SNF grant SNF-155922, of the Clay foundation, and the hospitality of the Isaac Newton Institute in Cambridge (where the present work has been carried out) are gratefully acknowledged. The author is also part of the NCCR-Swissmap.

I also thank Yves Le Jan for comments on the first version of this preprint.

References

1. D. Brydges, J. Fröhlich, T. Spencer, The random walk representation of classical spin systems and correlation inequalities. Commun. Math. Phys. **83**, 123–150 (1982)
2. F. Camia, M. Lis, Non-backtracking loop soups and statistical mechanics on spin networks (2015). arXiv:1507.05065
3. J. Dubédat, SLE and the free field: partition functions and couplings. J. Am. Math. Soc. **22**, 995–1054 (2009)
4. E.B. Dynkin, Markov processes as a tool in field theory. J. Funct. Anal. **50**, 167–187 (1983)

5. E.B. Dynkin, Gaussian and non-Gaussian random fields associated with Markov processes. J. Funct. Anal. **55**, 344–376 (1984)
6. G.F. Lawler, Loop-erased random walk, in *Perplexing Problems in Probability*. Festschrift in Honor of Harry Kesten, Progress in Probability, vol. 44 (Birkhäuser, Boston, 1999), pp. 197–217
7. G.F. Lawler, V. Limic, *Random Walks: a Modern Introduction* (Cambridge University Press, Cambridge, 2010)
8. G.F. Lawler, W. Werner, The Brownian loop soup. Probab. Theory Relat. Fields **128**, 565–588 (2004)
9. Y. Le Jan, *Markov Paths, Loops and Fields*. Lecture Notes in Mathematics, vol. 2026 (Springer, Berlin, 2011)
10. T. Lupu, From loop clusters and random interlacement to the free field. Ann. Probab. **44**(3), 2117–2146 (2016)
11. T. Lupu, W. Werner, A note on Ising random currents, Ising-FK, loop-soups and the Gaussian free field. Electron. Commun. Probab. **21** (2016)
12. M.B. Marcus, J. Rosen, *Markov Processes, Gaussian Processes, and Local Times* (Cambridge University Press, Cambridge, 2006)
13. E. Nelson, The free Markoff field. J. Funct. Anal. **12**, 211–227 (1973)
14. W. Qian, W. Werner, Decomposition of two-dimensional loop-soup clusters (2015). arXiv:1509.01180
15. O. Schramm, S. Sheffield, A contour line of the continuous Gaussian free field. Probab. Theory Relat. Fields **157**, 47–80 (2013)
16. S. Sheffield, W. Werner, Conformal loop ensembles: the Markovian characterization and the loop-soup construction. Ann. Math. **176**, 1827–1917 (2012)
17. K. Symanzik, Euclidean quantum field theory, in *Local Quantum Theory*, ed. by R. Jost (Academic, New York, 1969)
18. W. Werner, SLEs as boundaries of clusters of Brownian loops. C.R. Math. Acad. Sci. Paris **337**, 481–486 (2003)
19. W. Werner, *Topics on the Gaussian Free Field*. Lecture Notes (Springer, Berlin, 2014)
20. D.B. Wilson, Generating random spanning trees more quickly than the cover time, in *Proceedings of the Twenty-Eighth Annual ACM Symposium on Theory of Computing* (Association for Computing Machinery, New York, 1996), pp. 296–303

LECTURE NOTES IN MATHEMATICS

Springer

Editors in Chief: J.-M. Morel, B. Teissier;

Editorial Policy

1. Lecture Notes aim to report new developments in all areas of mathematics and their applications – quickly, informally and at a high level. Mathematical texts analysing new developments in modelling and numerical simulation are welcome.

   Manuscripts should be reasonably self-contained and rounded off. Thus they may, and often will, present not only results of the author but also related work by other people. They may be based on specialised lecture courses. Furthermore, the manuscripts should provide sufficient motivation, examples and applications. This clearly distinguishes Lecture Notes from journal articles or technical reports which normally are very concise. Articles intended for a journal but too long to be accepted by most journals, usually do not have this "lecture notes" character. For similar reasons it is unusual for doctoral theses to be accepted for the Lecture Notes series, though habilitation theses may be appropriate.

2. Besides monographs, multi-author manuscripts resulting from SUMMER SCHOOLS or similar INTENSIVE COURSES are welcome, provided their objective was held to present an active mathematical topic to an audience at the beginning or intermediate graduate level (a list of participants should be provided).

   The resulting manuscript should not be just a collection of course notes, but should require advance planning and coordination among the main lecturers. The subject matter should dictate the structure of the book. This structure should be motivated and explained in a scientific introduction, and the notation, references, index and formulation of results should be, if possible, unified by the editors. Each contribution should have an abstract and an introduction referring to the other contributions. In other words, more preparatory work must go into a multi-authored volume than simply assembling a disparate collection of papers, communicated at the event.

3. Manuscripts should be submitted either online at www.editorialmanager.com/lnm to Springer's mathematics editorial in Heidelberg, or electronically to one of the series editors. Authors should be aware that incomplete or insufficiently close-to-final manuscripts almost always result in longer refereeing times and nevertheless unclear referees' recommendations, making further refereeing of a final draft necessary. The strict minimum amount of material that will be considered should include a detailed outline describing the planned contents of each chapter, a bibliography and several sample chapters. Parallel submission of a manuscript to another publisher while under consideration for LNM is not acceptable and can lead to rejection.

4. In general, **monographs** will be sent out to at least 2 external referees for evaluation.

   A final decision to publish can be made only on the basis of the complete manuscript, however a refereeing process leading to a preliminary decision can be based on a pre-final or incomplete manuscript.

   Volume Editors of **multi-author works** are expected to arrange for the refereeing, to the usual scientific standards, of the individual contributions. If the resulting reports can be forwarded to the LNM Editorial Board, this is very helpful. If no reports are forwarded or if other questions remain unclear in respect of homogeneity etc, the series editors may wish to consult external referees for an overall evaluation of the volume.

5. Manuscripts should in general be submitted in English. Final manuscripts should contain at least 100 pages of mathematical text and should always include

   – a table of contents;
   – an informative introduction, with adequate motivation and perhaps some historical remarks: it should be accessible to a reader not intimately familiar with the topic treated;
   – a subject index: as a rule this is genuinely helpful for the reader.
   – For evaluation purposes, manuscripts should be submitted as pdf files.

6. Careful preparation of the manuscripts will help keep production time short besides ensuring satisfactory appearance of the finished book in print and online. After acceptance of the manuscript authors will be asked to prepare the final LaTeX source files (see LaTeX templates online: https://www.springer.com/gb/authors-editors/book-authors-editors/manuscriptpreparation/5636) plus the corresponding pdf- or zipped ps-file. The LaTeX source files are essential for producing the full-text online version of the book, see http://link.springer.com/bookseries/304 for the existing online volumes of LNM). The technical production of a Lecture Notes volume takes approximately 12 weeks. Additional instructions, if necessary, are available on request from lnm@springer.com.

7. Authors receive a total of 30 free copies of their volume and free access to their book on SpringerLink, but no royalties. They are entitled to a discount of 33.3 % on the price of Springer books purchased for their personal use, if ordering directly from Springer.

8. Commitment to publish is made by a *Publishing Agreement*; contributing authors of multiauthor books are requested to sign a *Consent to Publish form*. Springer-Verlag registers the copyright for each volume. Authors are free to reuse material contained in their LNM volumes in later publications: a brief written (or e-mail) request for formal permission is sufficient.

Addresses:
Professor Jean-Michel Morel, CMLA, École Normale Supérieure de Cachan, France
E-mail: moreljeanmichel@gmail.com

Professor Bernard Teissier, Equipe Géométrie et Dynamique,
Institut de Mathématiques de Jussieu – Paris Rive Gauche, Paris, France
E-mail: bernard.teissier@imj-prg.fr

Springer: Ute McCrory, Mathematics, Heidelberg, Germany,
E-mail: lnm@springer.com

Printed by Printforce, the Netherlands